MW00984489

Organophosphorus Chemistry

Volume 26

A Specialist Periodical Report

Organophosphorus Chemistry

Volume 26

A Review of the Recent Literature Published between July 1993 and June 1994

Senior Reporters

D. W. Allen, *Sheffield Hallam University*
B. J. Walker, *The Queen's University of Belfast*

Reporters

C. W. Allen, *University of Vermont, U.S.A.*
O. Dahl, *University of Copenhagen, Denmark*
R. S. Edmundson, *formerly of University of Bradford*
J. A. Grasby, *University of Sheffield*
C. D. Hall, *King's College, London*
D. M. Williams, *MRC Laboratory of Molecular Biology, Cambridge*

THE ROYAL SOCIETY OF CHEMISTRY

ISBN 0-85404-304-7
ISSN 0306-0713

Published by the Royal Society of Chemistry,
Thomas Graham House, Science Park, Cambridge CB4 4WF

Introduction

We are delighted to welcome Dr Jane Grasby and Dr David Williams as co-authors of the chapter on Nucleotides and Nucleic Acids which returns after its absence from volume 25. We hope that Jane and David will contribute to future volumes.

Publications which cover wide areas of organophosphorus chemistry have appeared. These include a whole edition of *Chemical Reviews* (1994, **94**, issue 5) devoted to Organophosphorus Chemistry and the Proceedings of the XIIth International Conference on Phosphorus Chemistry (ICPC XII - Toulouse, 1992) which have been published in *Phosphorus Sulfur and Silicon* (1993, volumes 75, 76, and 77). Both of these contain much of interest and illustrate the current importance of the subject. ICPC XIII takes place this year (1995) in Jerusalem and we will report on this in volume 27. A new edition of Johnson's excellent book "Ylides and Imines of Phosphorus" has been published almost thirty years after the original and is indispensable to anyone who studies or uses ylides, phosphorus-stabilised carbanions or imines.

A significant increase in the volume of published work in the area of phosphine and related chemistry has been noted. Worthy of special mention are the first example of chiral resolution of a secondary phosphine which is chiral at phosphorus and, for followers of chemical fashion, inevitably the first phosphonium derivative of Buckminsterfullerene (C_{60}) has been described.

Despite the overall decline in output, the year has produced further interesting developments in the field of hypervalent phosphorus chemistry. These include phosphoranes containing acetylenic links, the synthesis and enantiomeric separation of a bicyclic phosphorane with chirality only at phosphorus, and a wide range of phosphoranes incorporated within macrocyclic ring systems (Houlla *et al.*). In addition Lattman has reported on phosphoranes bound to transition metals or enclosed within calixarene skeletons and Holmes and his co-workers have made further substantial contributions on the X-ray crystal structures of phosphoranes containing eight-membered rings.

In the chemistry of tervalent phosphorus acid derivatives the development most worthy of comment is the remarkable spontaneous formation of two-co-ordinate amino-iminophosphines from phosphorus triamides.

Apart from the continued and expanding interest in the synthesis of functionalised phosphonic acids, which is largely driven by their increasingly important biological properties, activity in the pentavalent phosphorus acid area is somewhat subdued. In addition to reports of a large number of compounds with wide ranging biological activity, highlights include the formation of phosphate triesters from elemental phosphorus or other simple inorganic sources, the preparation of callixarene derivatives and, *à la mode*, the synthesis of fullerene phosphonic acids. Also worthy of note are the developments in asymmetric synthesis with the Abramov preparation of (α-hydroxyalkyl)phosphonic acids and the preparation of an even more stable dithioxophosphorane.

The nucleoside phosphate area continues to be dominated by the synthesis of potential anti-viral agents. Several novel and highly efficient pro-drug forms of these compounds have been developed to facilitate their delivery and to allow administration of nucleoside analogues known to be active only as their 5'-phosphate derivatives. Several novel nucleoside phosphate, 3',5'-cyclic phosphate, and triphosphate derivatives have been prepared. The last of these have provided some excellent HIV-RT inhibitors. In addition, methods for the synthesis of sugar-modified fluorescently-labelled 2',3'- dideoxynucleoside-5'-phosphates have been described.

In the area of oligonucleotide chemistry, the award of the Nobel Prize for Chemistry to Michael Smith and Kary B. Mullis, for the discovery of site-directed mutagenesis, and the polymerase chain reaction, respectively, is noteworthy. Both techniques rely heavily on the ability to synthesise DNA. This general area is dominated by the potential of oligonucleotides and their modified analogues as new chemotherapeutic agents and this application has been the driving force for the synthesis of phosphate-, sugar-, and base-modified nucleic acids. Insight into the sequence-specific recognition of DNA by drugs has come from the work of groups led by Dervan and Wemmer.

The various phosphorus-based olefination methods continue to be used as widely as ever and further subtleties of stereo-control have been developed. Interest in the Wittig mechanism continues and evidence has been reported supporting the earlier suggestion that there is a contribution from single electron-transfer in certain cases. However, surprisingly few detailed investigations on the mechanisms of other phosphorus-based methods of alkene synthesis have been carried out. Interestingly, β-hydroxyalkylphosphines are reported to undergo *anti*-elimination on treatment with triethylamine and phosphorus trichloride to provide alkenes, probably *via* an *epi*-phosphonium salt intermediate.

Activity in phosphazene chemistry has increased significantly over recent years with an additional fifty citations being noted for this time period. The focus is evenly divided between acyclic, cyclic and polymeric materials with a decline in phosphathiazenes and other mixed element ring systems. Areas of particular attention in acyclic materials include the aza-Wittig reaction and metallophosphazenes, while in the high polymer area biologically-related and ionic, conducting materials remain the most important application areas. The single most notable study involves the mapping of the electron density in the cyclophosphazene from high quality X-ray work. This investigation is as close to an unambiguous verification of the so-called "island model" as one is likely to obtain.

D W Allen and B J Walker

Contents

CHAPTER 1 **Phosphines and Phosphonium Salts**
By D.W. Allen

1	Phosphines	1
1.1	Preparation	1
1.1.1	From Halogenophosphines and Organometallic Reagents	1
1.1.2	Preparation of Phosphines from Metallated Phosphines	2
1.1.3	Preparation of Phosphines by Addition of P-H to Unsaturated Compounds	8
1.1.4	Preparation of Phosphines by Reduction	10
1.1.5	Miscellaneous Methods of Preparing Phosphines	10
1.2	Reactions of Phosphines	14
1.2.1	Nucleophilic Attack at Carbon	14
1.2.2	Nucleophilic Attack at Halogen	16
1.2.3	Nucleophilic Attack at Other Atoms	16
1.2.4	Miscellaneous Reactions of Phosphines	19
2	Halogenophosphines	22
2.1	Preparation	22
2.2	Reactions	22
3	Phosphine Oxides and Related Chalcogenides	24
3.1	Preparation	24
3.2	Reactions	26
3.3	Structural and Physical Aspects	28
3.4	Phosphine Chalcogenides as Ligands	28
4	Phosphonium Salts	29
4.1	Preparation	29
4.2	Reactions	31
5	p_π-Bonded Phosphorus Compounds	33

6 Phosphirenes, Phospholes, and Phosphinines 40

References 47

CHAPTER 2 **Pentaco-ordinated and Hexaco-ordinated Compounds**
By C.D. Hall

1 Introduction 63

2 Acyclic Phosphoranes 64

3 Cyclic Phosphoranes 66

4 Hexaco-ordinated Phosphorus Compounds 75

References 79

CHAPTER 3 **Tervalent Phosphorus Acid Derivatives**
By O. Dahl

1 Introduction 81

2 Nucleophilic Reactions 81
2.1 Attack on Saturated Carbon 81
2.2 Attack on Unsaturated Carbon 82
2.3 Attack on Nitrogen, Chalcogen, or Halogen 82

3 Electrophilic Reactions 86
3.1 Preparation 86
3.2 Mechanistic Studies 88
3.3 Use for Nucleotide Synthesis 92
3.4 Use for Sugar Phosphate, Phospholipid, or Phosphopeptide Synthesis 98
3.5 Miscellaneous 100

4 Reactions involving Two-co-ordinate Phosphorus 100

5 Miscellaneous Reactions 104

References 108

CHAPTER 4 **Quinquevalent Phosphorus Acids**
By R.S. Edmundson

1 Phosphoric Acids and their Derivatives 112
1.1 Synthesis of Phosphoric Acids and their Derivatives 112
1.2 Reactions of Phosphoric Acids and their Derivatives 120

2 Phosphonic and Phosphinic Acids and their Derivatives 130
2.1 Synthesis of Phosphonic and Phosphinic Acids and their Derivatives 130
2.1.1 Phosphonic and Phosphinic Halides and Related Compounds 130
2.1.2 Alkyl, Aralkyl, and Cycloalkyl Acids 132
2.1.3 Alkenyl, Alkynyl, Aromatic, Heterocyclic, and Related Acids 135
2.1.4 Halogenoalkyl and Related Acids 136
2.1.5 Hydroxyalkyl and Epoxyalkyl Acids 138
2.1.6 Oxoalkyl Acids 148
2.1.7 Nitroalkyl and Aminoalkyl Acids 153
2.1.8 Sulphur and Selenium Containing Compounds 161
2.1.9 Phosphorus-Nitrogen Bonded Compounds 163
2.1.10 Compounds of Potential Pharmacological Interest 163
2.2 The Reactions of Phosphonic and Phosphinic Acids 170

References 185

CHAPTER 5 **Nucleotides and Nucleic Acids**
By J.A. Grasby and D.M. Williams

Introduction 196

1 Mononucleotides 196
1.1 Nucleoside Acyclic Phosphates 196
1.2 Nucleoside Cyclic Phosphates 212

2 Nucleoside Polyphosphates 214

3 Oligo- and Poly-nucleotides 220
3.1 DNA Synthesis 220
3.2 RNA Synthesis 223
3.3 The Synthesis of Modified Oligodeoxynucleotides and Modified Oligoribonucleotides 226
3.3.1 Oligonucleotides Containing Modified Phosphodiester Linkages 226
3.3.2 Oligonucleotides Containing Modified Sugars 232
3.3.3 Oligonucleotides Containing Modified Bases 235

4 Linkers 246

5 Nucleic Acid Triple Helices and Other Unusual DNA and RNA Structures 248

6 Interactions and Reactions of Nucleic Acids with Small Molecules 251

7 Determination of Nucleic Acid Structures 255

References 256

CHAPTER 6 Ylides and Related Compounds
By B.J. Walker

1 Introduction 265

2 Methylenephosphoranes 265
2.1 Preparation and Structure 265
2.2 Reactions of Methylenephosphoranes 267
2.2.1 Aldehydes 267
2.2.2 Ketones 270
2.2.3 Miscellaneous Reactions 274

3 The Structure and Reactions of Phosphonate Oxide Anions 279

4 The Structure and Reactions of Phosphonate Anions 279

5 Selected Applications in Synthesis 286
5.1 Carbohydrates 286
5.2 Carotenoids, Retenoids, Pheromones and Polyenes 286
5.3 Leukotrienes, Prostaglandins and Related Compounds 290
5.4 Macrolides and Related Compounds 290
5.5 Nitrogen Heterocycles 294
5.6 Steroids and Related Compounds 299
5.7 Terpenes 299
5.8 Vitamin D Analogues and Related Structures 299
5.9 Miscellaneous Reactions 299

References 306

CHAPTER 7 **Phosphazenes**
By C.W. Allen

1 Introduction 312

2 Acyclic Phosphazenes 312

3 Cyclophosphazenes 323

4 Cyclophospha(thia)zenes and Related Systems 333

5 Miscellaneous Phosphazene Ring Systems Including Metallocycles 335

6 Poly(phosphazenes) 338

7 Crystal Structures of Phosphazenes 349

References 358

Author Index 372

Abbreviations

AIBN	bisazoisobutyronitrile
CIDNP	Chemically Induced Dynamic Nuclear Polarization
CNDO	Complete Neglect of Differential Overlap
Cp	cyclopentadienyl
DAD	diethyl azodicarboxylate
DBN	1,5-diazabicyclo[4.3.0]non-5-ene
DBU	1,5-diazabicyclo[5.4.0]undec-5-ene
DCC	dicyclohexylcarbodi-imide
DIOP	[(2,2-dimethyl-1,3-dioxolan-4,5-diyl)bis-(methylene)]bis(diphenylphosphine)
DMF	dimethylformamide
DMSO	dimethyl sulphoxide
DMTr	4,4'-dimethoxytrityl
EDTA	ethylenediaminetetra-acetic acid
E.H.T.	Extended Huckle Treatment
ENU	N-ethyl-N-nitrosourea
FID	Free Induction Decay
g.l.c.-m.s.	gas-liquid chromatography-mass spectrometry
HMPT	hexamethylphosphortriamide
h.p.l.c.	high-performance liquid chromatography
i.r.	infrared
L.F.E.R.	Linear Free-Energy Relationship
MINDO	Modified Intermediate Neglect of Differential Overlap
MMTr	4-monomethoxytrityl
MO	Molecular Orbital
MS-Cl	mesitylenesulphonyl chloride
MS-nt	mesitylenesulphonyl 3-nitro-1,2,4-triazole
MS-tet	mesitylenesulphonyltetrazole
NBS	N-bromosuccinimide
n.q.r.	nuclear quadrupole resonance
p.e.	photoelectron
PPA	polyphosphoric acid
SCF	Self-Consistent Field
TBDMS	t-butyldimethylsilyl
TDAP	tris(diethylamino)phosphine
TFAA	trifluoroacetic acid
TfO	trifluoromethanesulphonic anhydride
THF	Tetrahydrofuran
Thf	2-tetrahydrofuranyl
Thp	2-tetrahydropyranyl
TIPS	tetraisopropyldisiloxanyl
t.l.c.	thin-layer chromatography
TPS-Cl	tri-isopropylbenzenesulphonyl chloride
TPS-nt	tri-isopropylbenzenesulphonyl-3-nitro-1,2,4-triazole
TPS-tet	tri-isopropylbenzenesulphonyltetrazole
TsOH	toluene-p-sulphonic acid
u.v.	ultraviolet

1
Phosphines and Phosphonium Salts

BY D. W. ALLEN

1 Phosphines

1.1 Preparation

1.1.1 From Halogenophosphines and Organometallic Reagents.- Organolithium reagents continue to dominate this route to tertiary phosphines. The reactions of 1-dimethylamino-8-lithionaphthalene and *o*-lithio-N,N-dimethylbenzylamine, respectively, with phosphorus tribromide have given the phosphines (1) and (2). Structural studies of both components reveal weak N:→P coordinative interactions. Treatment of (1) with iodomethane results in a slow quaternization at phosphorus, whereas the related reaction of (2) results in quaternization at nitrogen.[1] The diphosphine (3) has been prepared by the reaction of 1,8-dilithionaphthalene with chlorodiphenylphosphine.[2] The reactions of dilithium reagents with dichlorophenylphosphine have given the benzophosphepines (4) and (5).[3] Metallation at the benzylic carbons of 2,2'-dimethyl-1,1'-binaphthalene using *t*-butyllithium and potassium *t*-butoxide in the presence of TMEDA, followed by addition of a dichlorophosphine, provides a route to the chiral phosphepin system (6), which has been resolved *via* a chiral palladium complex.[4] Two reports have appeared of the synthesis of phosphinocarbamates (7) from the reactions of chlorodiphenylphosphine with the lithium enolates of N,N-disubstituted acetamides.[5,6] The reaction of chlorodiphenylphosphine, coordinated to a metal carbonyl acceptor, with the lithium reagent derived from 2,2,6-trimethyl-4H-1,3-dioxin-4-one has given the coordinated functionalised phosphine (8), capable of transformation to other functionalised phosphines. Attempts to carry out the above synthesis using free, uncoordinated, chlorodiphenylphosphine resulted only in transfer of chlorine to the carbanion.[7] New types of chiral phosphinoamine ligands, e.g., (9), have been obtained by the reactions of doubly metallated chiral diamines with dichlorophenylphosphine, and isolated as the diastereoisomeric borane complexes which can subsequently be separated by chromatography, prior to treatment with, e.g., morpholine, to free the phosphine centre.[8] A range of bis(diarylphosphino)alkanes (10), in which the aryl groups bear chiral amino or alkoxy substituents, has been obtained from the reactions of the appropriate bis(dichlorophosphino)alkane with aryllithium reagents.[9,10] The new chiral iminophosphine (11) has been isolated from the reaction of a lithiated fenchone-pinacolone mixed azine with chlorodiphenylphosphine.[11] Interest has continued in the synthesis of chiral phosphinoamines which involve a metallocene subunit, e.g., (12)[12] and (13)[13], the standard approach being the reaction of a lithiated aminoalkylmetallocene with chlorodiphenylphosphine. A mixture of isomeric tris(diphenylphosphino)ferrocenes (14) has been isolated from the reaction of lithiated diphenyl-(ferrocenyl)phosphine with chlorodiphenylphosphine. Individual isomers have been obtained by repeated chromatographic techniques.[14] Metallation at carbon of trimethylphosphine followed by treatment with chlorodi-*t*-butylphosphine has given the unsymmetrically substituted diphosphinomethane (15). The latter undergoes metallation at the bridging carbon, and subsequent treatment with halogenophosphine has provided the triphosphine (16).[15]

Grignard reagents have also been employed in the above approach to phosphines. Treatment of mesityldichlorophosphine with 1 mole of (-)menthylmagnesium chloride at -78°C gives an equimolar mixture of two diastereoisomers of the monochlorophosphine (17), reduction of which with lithium aluminium hydride affords the related diastereoisomeric secondary phosphine. Recystallisation of the latter from acetonitrile containing a proton scavenger resulted in the preferential crystallisation of one of the diastereoisomers, thus providing the first chiral resolution of a secondary phosphine which is chiral at phosphorus.[16] A Grignard reagent derived from diacetylene has been used in the synthesis of the phosphinoalkyne (18).[17] Grignard routes have also been employed in the synthesis of the diphosphines (19)[18] and the amino-functionalised phosphine (20). The latter is easily transformed into the related N-methacryl derivative, (21), of interest as a potentially polymerisable phosphine.[19]

Mixed lithium-sodium organometallic reagents derived from carbaborane systems have been used to prepare 7-vertex *nido*-phosphacarbaboranes.[20] The tetrazolylphosphines (22) have been obtained from the reactions of potassium tetrazolates with diphenylchlorophosphine.[21]

1.1.2 *Preparation of Phosphines from Metallated Phosphines.*- Alkali metal-free phosphide anions have been shown to be formed under synthetically useful conditions in the equilibria between primary or secondary phosphines with the Schwesinger bases (23).[22] Techniques have been developed for the preparation of alkali metal diphenylphosphide reagents in high purity, as evidenced by ^{31}P nmr studies. The same paper reports a study of the course of the reactions between potassium diphenylphosphide and a series of aryl-, *n*-alkyl- and neopentyl-halides. The results provide the first evidence of the involvement of single electron transfer (SET) processes in the reactions of alkyl halides. This pathway is dominant in the case of neopentyl-type iodides, but plays only a minor role in the related reactions of neopentyl-type bromides and chlorides. No evidence was adduced of the involvement of SET processes in the reactions of unhindered

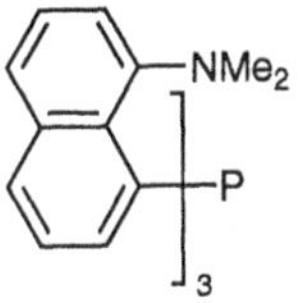

NMe_2 P 3 (1)

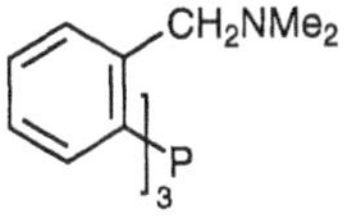

CH_2NMe_2 P 3 (2)

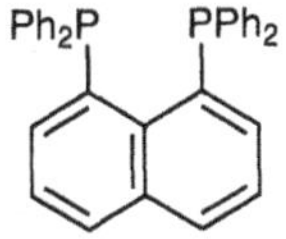

Ph_2P PPh_2 (3)

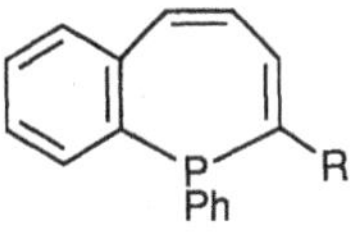

P Ph R (4) R = H or $SiMe_3$

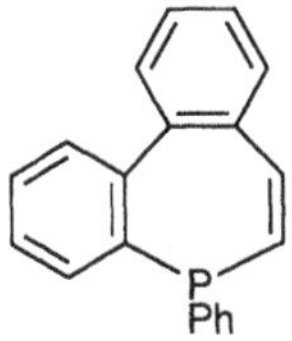

P Ph (5)

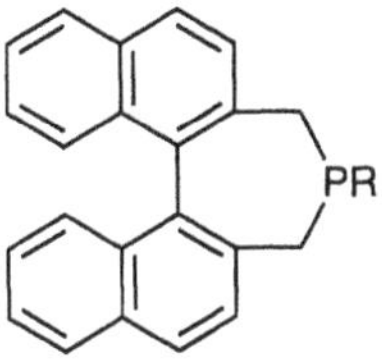

PR (6) R = Me or Ph

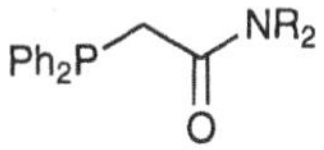

Ph_2P NR_2 O (7) R = Me or Ph

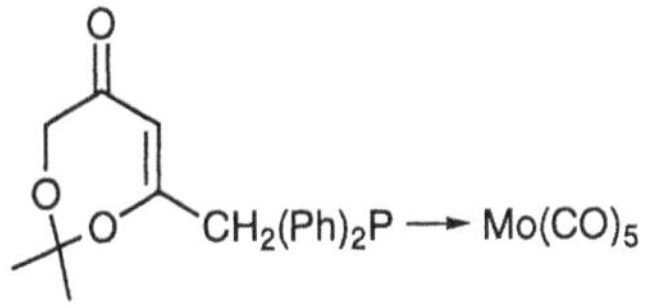

O O O $CH_2(Ph)_2P \rightarrow Mo(CO)_5$ (8)

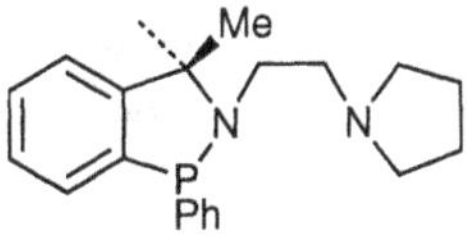

Me N N P Ph (9)

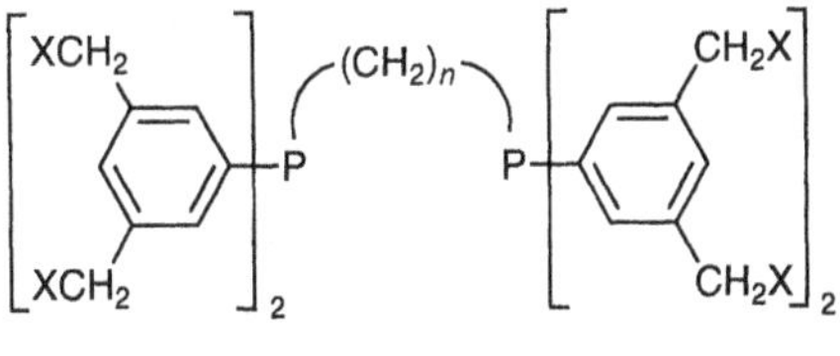

XCH_2 XCH_2 P $(CH_2)_n$ P CH_2X CH_2X 2 2

(10) n = 1,2, or 4
X = NR_2^* or OR^*

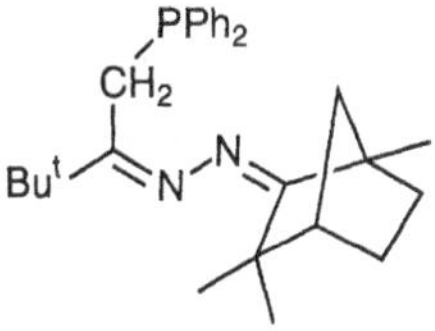

PPh_2 CH_2 Bu^t N N (11)

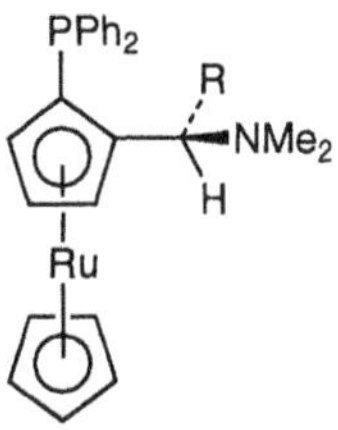

PPh_2 R NMe_2 H Ru (12)

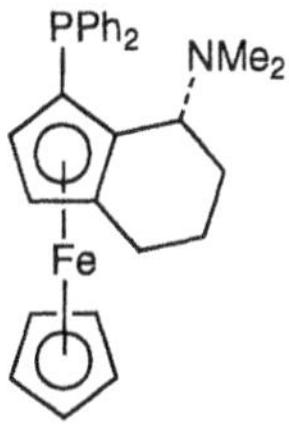

PPh_2 NMe_2 Fe (13)

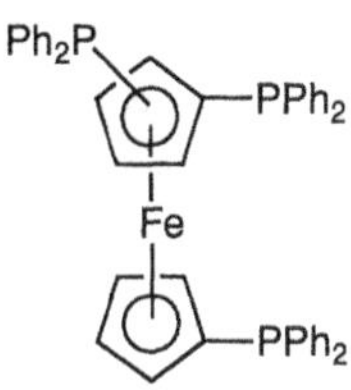

Ph_2P PPh_2 Fe PPh_2 (14)

alkyl halides.[23] The same group has also reported evidence of the involvement of a SET pathway in the reactions of 1-halonorbornanes with the diphenylphosphide ion in ether solvents, in the absence of light, and at room temperature.[24] Electron-transfer processes have also been identified in the reactions of cycloalkyl halides with diphenylphosphide ions in liquid ammonia.[25] Ultrasound has been shown to promote the reactions of *p*-iodoanisole, and 1-halonaphthalenes, with the diphenylphosphide ion in liquid ammonia, which proceed *via* the $S_{RN}1$ mechanism, to give triarylphosphines in good yield.[26] Structural studies of metallo-organophosphide reagents have continued to appear.[27-30]

The reactions of metallo-phosphide reagents with alkyl halides and related sulphonate esters have been applied extensively in the synthesis of tertiary phosphines. Full details of the synthesis of C_2-symmetric bis(phospholane) ligands, e.g., (24), have now appeared.[31] The phosphide-mesylate route has been used in the preparation of the enantiomerically pure *cis*- and *trans*-3-diphenylphosphino-4-hydroxytetrahydrofuran systems (25) from L-ascorbic acid and D-isoascorbic acid, respectively.[32] The reaction of lithium diphenylphosphide with a ditosylate is the key step in the synthesis of the diphosphine (26).[33] Both halide and tosylate variants have been employed in the preparation of a range of tripod-like triphosphine ligand systems (27).[34,35] The phosphinoalkylisocyanurate (28) has been obtained from the reaction of lithium diphenylphosphide with the related 3-bromopropylisocyanurate.[36] Treatment of 1-chloro-2-methylsulphinylethane with lithium diphenylphosphide has given the phosphine (29), which is chiral at sulphur, and which has been resolved *via* a chiral amine-palladium(II) complex. Above 110°C, (29) eliminates the sulphur centre to form diphenylvinylphosphine. The related arsine behaves in the same way.[37] The lithium phosphinoformates (30) have been obtained from the reactions of lithiophosphide reagents and carbon dioxide. In protic media, rapid decarboxylation occurs with the formation of the related primary or secondary phosphine. With dimethylsulphate, the esters (31) are formed.[38] Phosphide reagents have been employed in the synthesis of a range of ω-phosphinoalkylcyclopentadienyl systems, e.g., (32), of interest as novel ligands.[39] The reaction of the lithiophosphide reagent (33) with chiral styrene oxide results in the formation of the chiral phosphirane system (34). The free phosphirane is accessible on decomplexation with diphos.[40] Dilithiophenylphosphide (35) has been used in the synthesis of three and four-membered phosphorus heterocycles. With 1,2-dichloroethane, the simple phosphirane (36) is formed. This compound shows no tendency to polymerise on standing at 0°C over a period of a month. Treatment of (35) with 1,3-dichloropropane provides the previously unknown phosphetan (37) in 13% yield. Unlike the phosphirane, the latter undergoes a rapid ring-opening polymerisation at room temperature, to form the poly(phenylphosphino)propane system (38).[41] The dilithio-diphosphide reagent (39) has been employed in the synthesis of a range of chiral,

$Me_2PCH_2PBu^t_2$

(15)

$[Bu^t_2PCH_2]_2PMe$

(16)

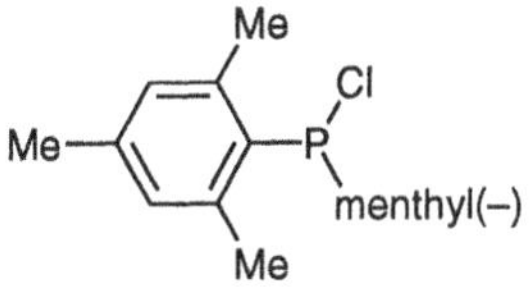

(17)

$Ph_2P-C\equiv C-C\equiv CH$

(18)

$(MeOCH_2CH_2CH_2)_2P$ $P(CH_2CH_2CH_2OMe)_2$

(19)

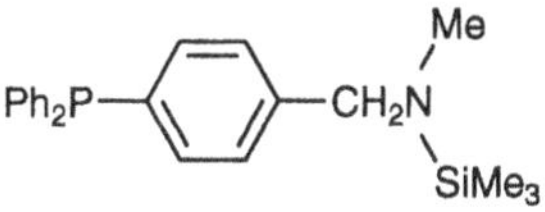

(20)

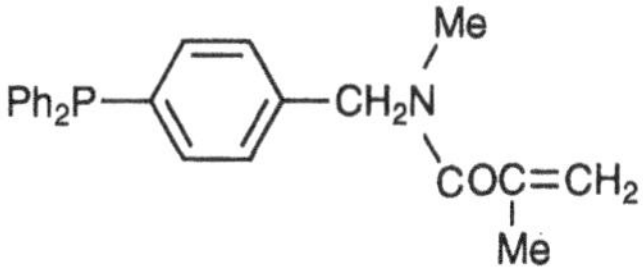

(21)

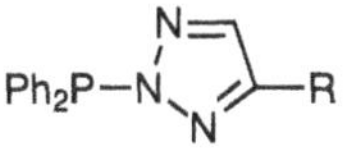

(22) R = Me or Ph

$(Me_2N)_3P=N[P=N]_3P=NBu^t$

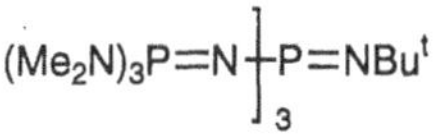

(23)

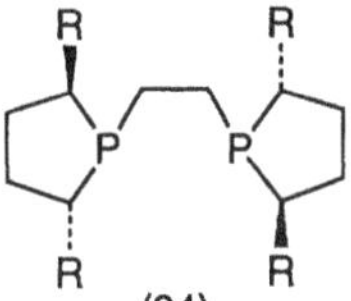

(24)

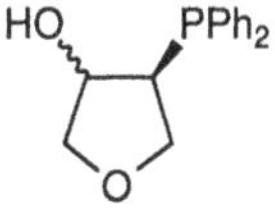

(25)

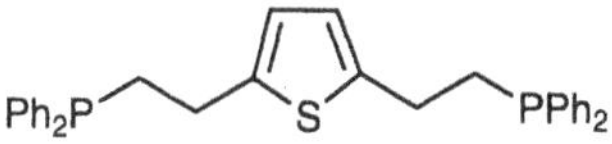

(26)

$RC[CH_2PPh_2]_3$

(27) R = H, pentyl, or 8-nonenyl

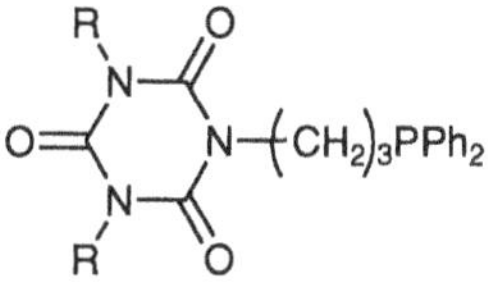

(28) R = allyl

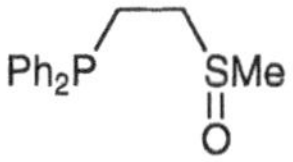

(29)

$R^1R^2PCOOLi$

(30) R^1 = Ph, Cy, Pr^i, or Et; R^2 = H or R^1

$R^1R^2PCOOMe$

(31)

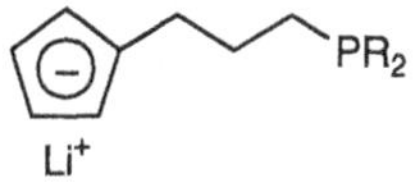

(32) R = Ph, Pr^i, or Bu^t

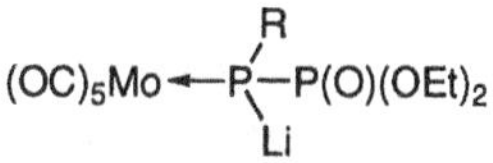

$(OC)_5Mo\leftarrow P(R)(Li)-P(O)(OEt)_2$

(33) R = menthyl

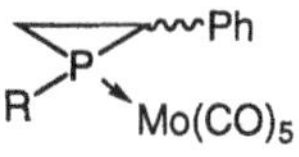

(34)

$PhPLi_2$

(35)

macrocyclic diphosphines, e.g., (40).[42,43] Improved conditions for the lithium-promoted cleavage of phenyl groups from $\alpha\omega$-bis(diphenylphosphino)alkanes have been developed, leading to the bis(secondary) phosphines (41) with minimal contamination from products arising from P-CH_2 cleavage.[44] The diphosphine (42), isolated as a mixture of diastereoisomers, has been obtained from the reaction of *trans*-1,2-dichloroethylene with lithium methyl(phenyl)phosphide. The latter reagent has also been used in the synthesis of the unsymmetrical diphosphine (43) from *o*-chlorophenyl(diphenyl)phosphine.[45] The diphosphine (43) and other related systems (44) have also been obtained from the reactions of *o*-chlorophenyl(diphenyl)phosphine with sodiophosphide reagents.[45,46] Both lithio- and sodio-phosphide reagents have been used in the synthesis of a range of new chiral quadridentate ligands, e.g., (45), from *o*-aminophenyl(diphenyl)phosphine.[47] An improved route to the potentially binucleating tridentate ligand, 6-diphenylphosphino-2,2'-bipyridyl, (46), has been developed, which involves the reaction of 6-chloro-2,2'-bipyridyl with lithium diphenylphosphide.[48] A detailed investigation of the products of the reaction of the phosphide reagent (47) with 1,2-dibromoethane has been reported. At low temperatures, the ylide (48) is the main product, together with the bromophosphine (49). Only with an excess of the phosphide reagent is the polyphosphine (50) formed. On raising the temperature, the ylide (48) decomposes to form a mixture of cyclopolyphosphines.[49] A convenient route to γ-ketophosphines (51) is provided by the reaction of lithium diphenylphosphide with $\alpha\beta$-enones, followed by hydrolysis.[50]

Cyclophosphines have been isolated from the reactions of 1-adamantyldihalophosphines with sodium.[51] The reaction of 1,2-difluoro-3-hydroxymethylbenzene with sodium diphenyl-phosphide has given the functionalised diphosphine (52), subsequently linked to a chloromethylated polystyrene support *via* the hydroxymethyl group.[52] A route to the 2,7-dihydro-1H-phosphepine system (53) is provided by the reaction of a (Z,Z)-1,6-dibromohexa-2,4-diene with disodiumphenylphosphide.[53] The new chiral phosphine (54) has been prepared by the metallophosphide-tosylate route using a mixed sodium-potassium diphenylphosphide reagent obtained from cleavage of triphenylphosphine.[54].

Two groups have reported syntheses of the chiral *o*-diphenylphosphino-oxazolines (55) which involve the displacement of fluorine from the related *o*-fluorophenyloxazolines using potassium diphenylphosphide.[55,56] High yields of *o*-substituted arylphosphines, e.g., (56), have been obtained in a similar manner from the reactions of *o*-substituted aryl fluorides with potassium diphenylphosphide.[57] Displacement of chlorine from a chloroarene by potassium diphenylphosphide in liquid ammonia is a key step in the synthesis of the phosphinobiaryl (57).[58] The reaction of potassium diphenylphosphide with 1,2-bis(diphenylphosphino)ethyne generates the salt (58), treatment of which with chlorodiphenylphosphine gives the tetrakis(diphenyl-

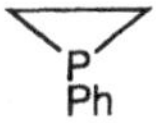

$[P(Ph)CH_2CH_2CH_2]_n$

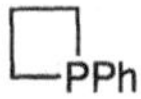

(36) (37) (38)

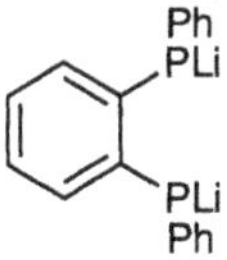

$PhP(H)(CH_2)_nPHPh$

(39) 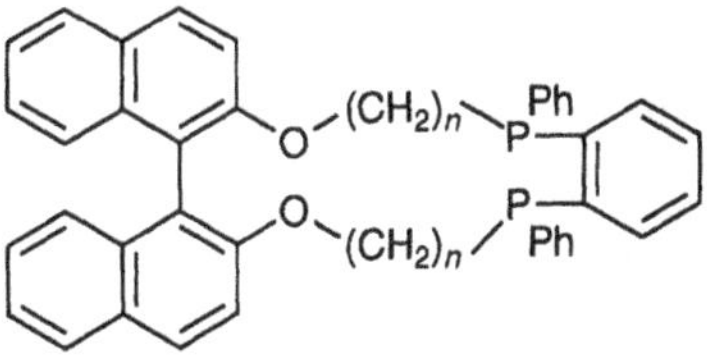 (40) n = 2 or 5 (41) n = 2, 3, 4, or 6

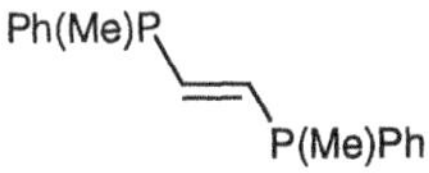

(42) 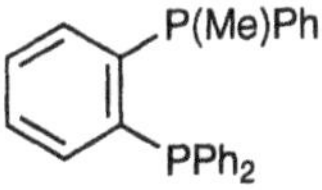 (43) 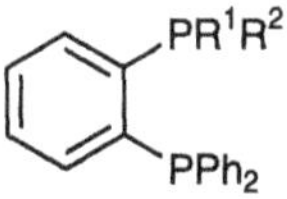 (44) R^1 = Me or Bu; R^2 = Me or Ph

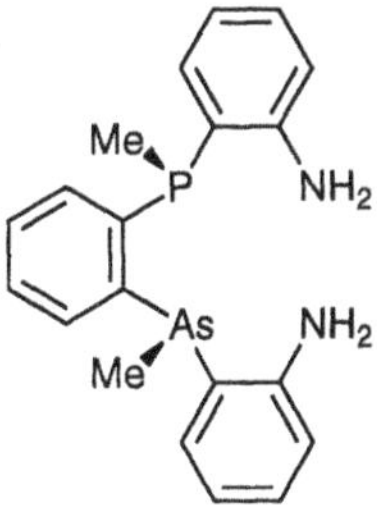

$(Bu^t_2P)_2Li$

(45) (46) (47)

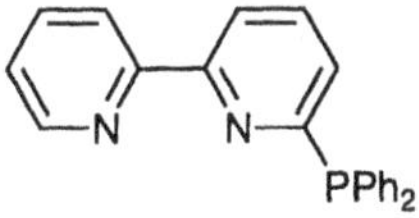

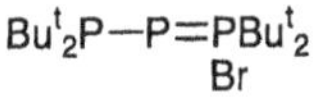

$Bu^t_2P-P{=}PBu^t_2$ Br

$(Bu^t_2P)_2PBr$

$(Bu^t_2P)_2P-P(PBu^t_2)_2$

(48) (49) (50)

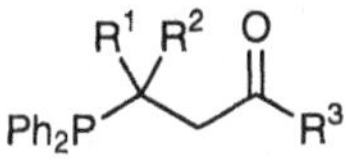

(51) R^1 = Me or Ph; R^2 = H or Me; R^3 = Me, Ph, or Bu^t

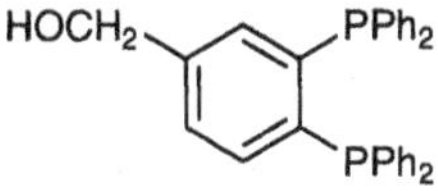

(52) 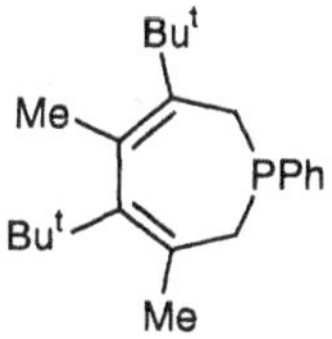 (53)

phosphino)butatriene (59).[59] Potassiophosphide reagents generated from diarylphosphines in superbasic media (potassium hydroxide in DMSO) have been employed in the synthesis of a range of tripod-like polyphosphine ligands, e.g., (60).[60] Treatment of 2-(tris chloromethyl)ethanol with diphenylphosphine in the presence of potassium *t*-butoxide in THF results in the formation of the diphosphine (61), which, with lithium diphenylphosphide, undergoes ring-opening to form the functionalised triphosphine (62).[61] A new route to the water-soluble triarylphosphine (63) is offered by the reaction of potassium *p*-fluorobenzenesulphonate with phosphine in the KOH-DMSO superbasic medium.[62] A further report has appeared of the preparation of tristyrylphosphine (together with its oxide, and as a mixture of stereoisomers) from the reaction of phenylacetylene with red phosphorus in superbasic media.[63]

Strong interest continues to be shown in the synthesis and structural studies of organophosphino derivatives of the *p*-block elements, notably boron,[64-70] aluminium,[71] gallium, indium, and thallium,[72-80] and metallophosphide reagents have been used extensively in this work. Organophosphido derivatives of zinc,[81] thorium,[82] and zirconium[83] have also been described.

1.1.3 Preparation of Phosphines by Addition of P-H to Unsaturated Compounds.- The reaction of red phosphorus with concentrated aqueous potassium hydroxide at 50°C provides a source of phosphine which can then be added to alkynes to form alkyl- and alkenyl-phosphines. Usually a mixture of products results, although one component often predominates. In the presence of DMSO as a cosolvent, a one-pot procedure is possible, even in the presence of air! The overall process is assisted by ultrasonic irradiation.[84] A pressurised *in-situ* ^{31}P-nmr technique has been used to study the radical-promoted addition of phosphine to linear alkenes.[85] A range of phosphines, some of them water-soluble, has been prepared by the radical-promoted additions of phenylphosphine, diphenylphosphine, and 1,2-bis(phenylphosphino)ethane to alkynols, alkyne ethers, and unsaturated carboxylic acids, esters and β-lactones.[86] Photochemically-initiated addition of dimethylphosphine to vinyl- and allyl-phosphines has provided routes to a series of tetraphosphines, e.g., (64).[87] The bidentate mixed donor ligand (65) is formed in a related addition of dimethylphosphine to N,N-dimethylallylamine.[88] Base-promoted addition of diphenylphosphine to the unsaturated diphosphine (66) in the coordination sphere of a metal has given complexes of the triphosphine (67).[89] Addition of the borane adduct of diphenylphosphine to unsaturated systems derived from tartaric acid is the key step in the stereoselective synthesis of a range of new chiral diphosphines, e.g., (68). A similar addition to an unsaturated dilactone features in the synthesis of the chiral system (69). In each case, the final stage is decomplexation of the initially-isolated borane adducts.[90] Addition of phenylphosphine, coordinated to a chromium carbonyl acceptor, to dialkynyl-phosphines and

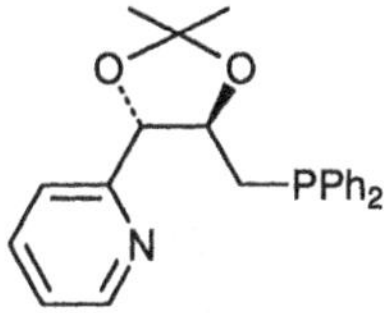

O O N PPh_2

(54)

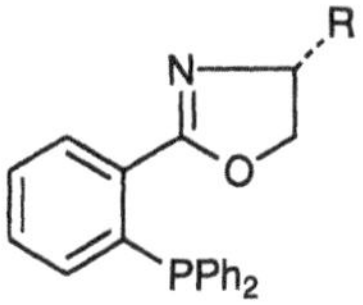

R N O PPh_2

(55) R = Me, Pr^i, CH_2Ph, or Ph

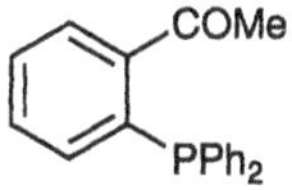

COMe PPh_2

(56)

Ph_2P OH

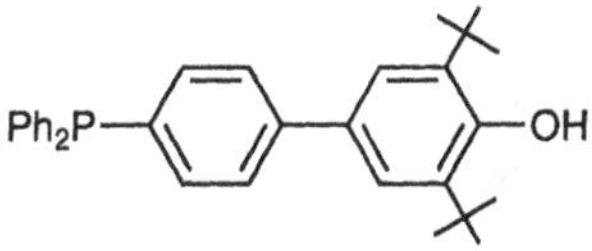

(57)

$K[Ph_2PC{\equiv}C{-}C(PPh_2){=}CPPh_2]$

(58)

$(Ph_2P)_2C{=}C{=}C{=}C(PPh_2)_2$

(59)

$MeC(CH_2PAr_2)_3$

(60)

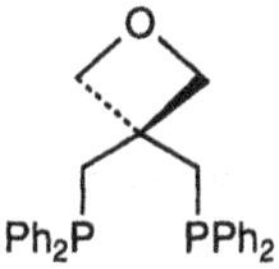

O Ph_2P PPh_2

(61)

$HOCH_2C(CH_2PPh_2)_3$

(62)

P SO_3K 3

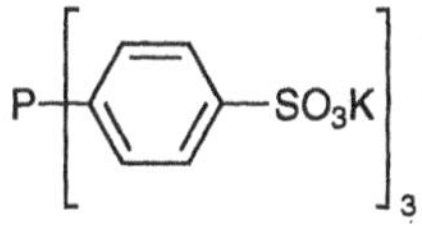

(63)

$P(CH_2CH_2PMe_2)_3$

(64)

$Me_2PCH_2CH_2CH_2NMe_2$

(65)

$(Ph_2PCH_2)_2C{=}CH_2$

(66)

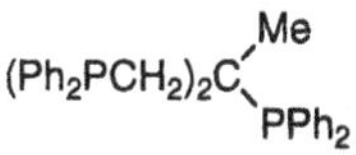

$(Ph_2PCH_2)_2C(Me)PPh_2$

(67)

PPh_2 R O O R PPh_2

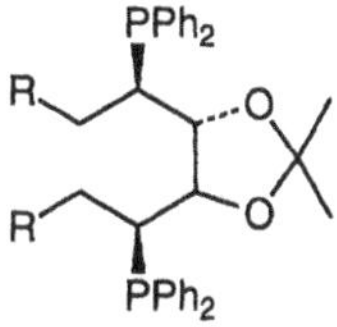

(68) R = CN, CO_2Me, CH_2OH, or CO_2H

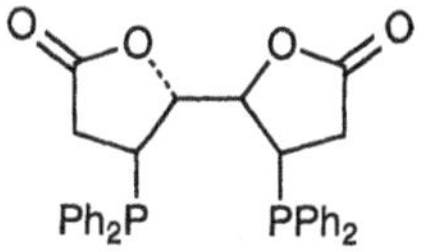

O O O O Ph_2P PPh_2

(69)

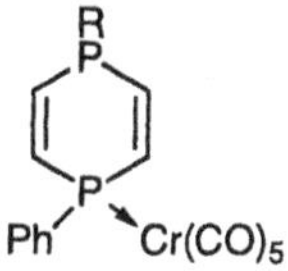

R P P Ph $Cr(CO)_5$

(70)

Bu Bu Sn P Ph $Cr(CO)_5$

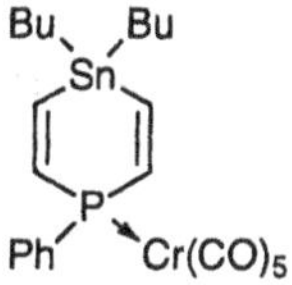

(71)

-stannanes, has given the heterocyclic systems (70) and (71).[91]

1.1.4 *Preparation of Phosphines by Reduction.*- Trichlorosilane, usually used in conjunction with triethylamine, remains the reagent of choice for the reduction of tertiary phosphine oxides, which is commonly the final step in a synthetic sequence leading to phosphines. Among new phosphines prepared in this way are the P-menthylphosphetan (72),[92] the mixed donor phosphines (73),[93] the chiral functionalised phosphines (74),[94] (75),[95] and (76),[96] and the diphosphines (77),[97] (78),[98] and (79).[99] Titanium(IV) isopropoxide has been shown to catalyse the reduction of phosphine oxides by triethoxysilane or, more conveniently, in view of the toxicity of the latter, by polymethylhydrosiloxane. Both systems promote reduction with retention of configuration at phosphorus.[100] A patent has described the use of silicon (or silicon alloys) for the reduction of dichlorophosphoranes or phosphine oxides in *o*-dichlorobenzene at 165-170°C, the reduction of the dichlorophosphoranes being far more efficient than that of the phosphine oxides.[101] Both trichlorosilane and lithium aluminium hydride have been used in the reduction of bis(1-adamantyl)phosphinyl chloride to the related secondary phosphine.[102] The dinaphthophospholium salts (80) are reduced by lithium aluminium hydride in THF to form the phosphines (81).[103] A combination of lithium aluminium hydride with cerium trichloride has been used in the synthesis of the phosphines (82) from the corresponding phosphine oxides. Treatment of the stereoisomeric phosphino-alcohols (82) with phosphorus trichloride results in an *anti*-elimination process to give alkenes, in contrast to the *syn*-elimination observed in the Horner-Wittig reactions of the corresponding phosphine oxides. The *anti*-elimination process is believed to proceed *via* phosphiranium salt intermediates.[104] The reduction of the bis(phosphino)maleic anhydride (83) with lithium aluminium hydride yields the new chiral diphosphinolactone (84).[105] The diphosphine (85) has been obtained in chiral forms by the desulphurisation of the individual diastereoisomers of the corresponding disulphide using tributylphosphine. An alternative route to (85) involves the use of borane as a protecting group at phosphorus, the borane being easily removed on treatment with diethylamine.[106] The diastereoisomeric phosphinous ester-boranes (86) have been reduced to the chiral borane adduct (87) of chiral secondary phosphines, using lithium naphthalenide.[107] Further examples have been described of the synthesis of unsymmetrical DIOP-type ligands, e.g., (88), in which cyclohexyl substituents at phosphorus are introduced by catalytic reduction of phenyl groups.[108]

1.1.5 *Miscellaneous Methods of Preparing Phosphines.*- The preparation, and uses in catalysis, of the various types of water-soluble phosphines, have been reviewed.[109] The direct sulphonation of diphos gives a complex mixture of sulphonated products, from which the water-

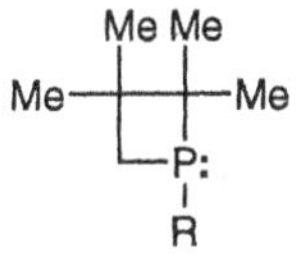

Me Me
Me Me
P:
R

(72) R = L-menthyl

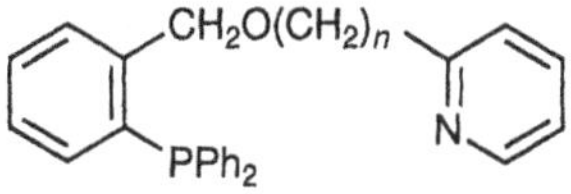

$CH_2O(CH_2)_n$
N
PPh_2

(73) *n* = 1–3

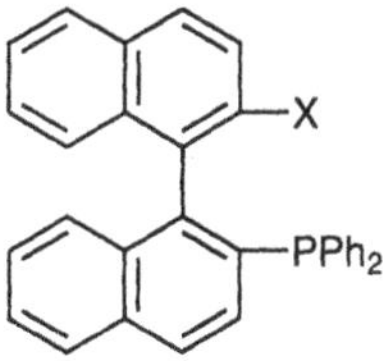

X
PPh_2

(74) X = H, CH_2OH, or CO_2Me

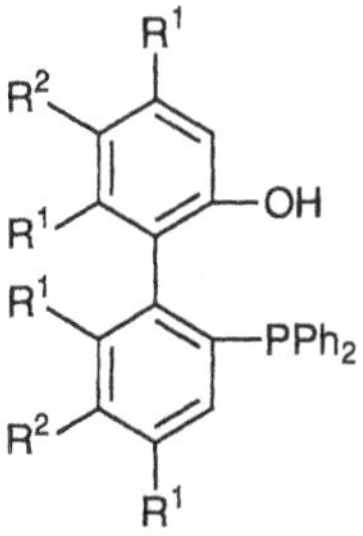

R^1 R^2 R^1 OH R^1 PPh_2 R^2 R^1

(75) R^1 = H or Me
R^2 = H or Cl

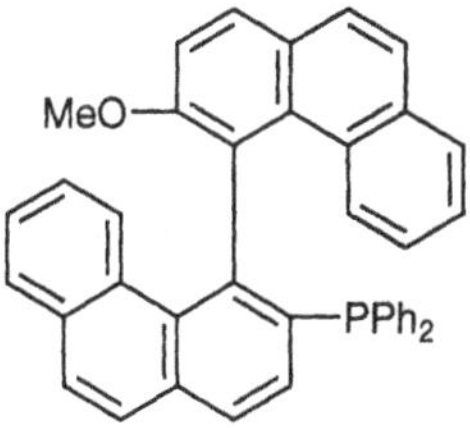

MeO
PPh_2

(76)

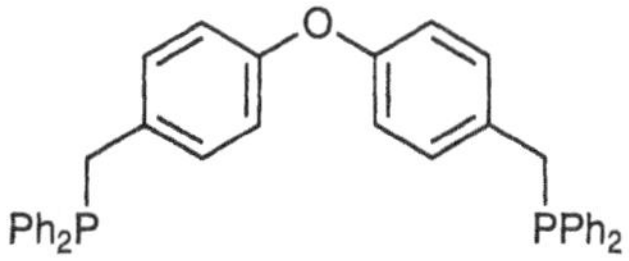

O
Ph_2P PPh_2

(77)

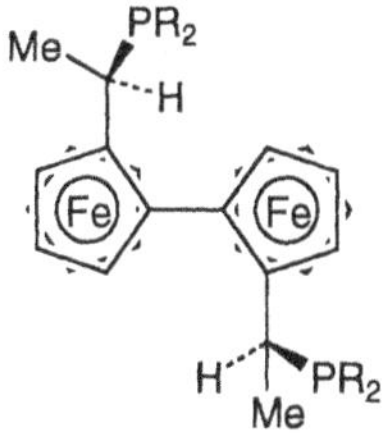

Me PR_2 H
Fe Fe
H PR_2 Me

(78)

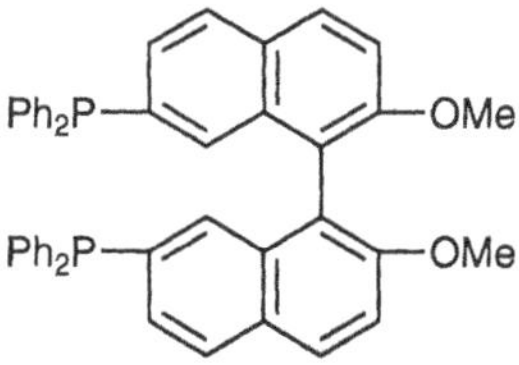

Ph_2P OMe
Ph_2P OMe

(79)

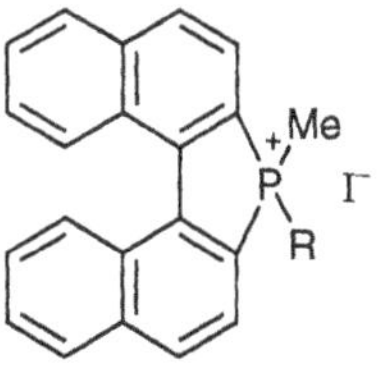

Me P R I^-

(80) R = Me, Et, or Ph

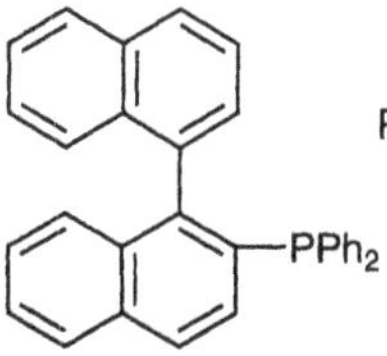

PPh_2

(81)

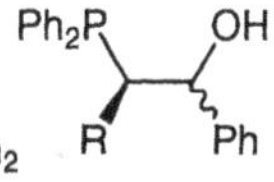

Ph_2P OH
R Ph

(82)

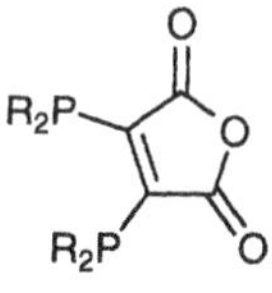

O
R_2P O
R_2P O

(83) R = menthyl

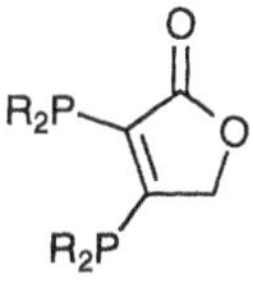

O
R_2P O
R_2P

(84)

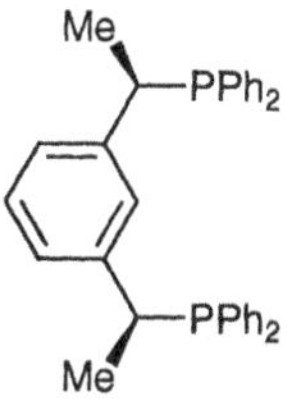

Me PPh_2
Me PPh_2

(85)

soluble chelating diphosphine (89) has been isolated in gram-quantities by fractional precipitation, in an overall yield of 30%.[110] Oxidation at phosphorus has been found to aid the introduction of bridgehead substituents in a series of new chiral diphosphines (90), the phosphine functionality being restored at the end of the synthetic sequence by a reduction step.[111] Esterification of the hydroxyl groups of the chiral diphosphine (91), using aryl dichlorophosphites or arylboronic acids, has given the new DIOP-analogues, (92)[112] and (93).[113,114] Further developments in the synthesis of phosphatriptycene systems have occurred, a range of thieno-fused systems (94) having been prepared, carbon-phosphorus bonds being formed in the reactions of thienyl-lithium reagents with triphenylphosphite.[115,116] Treatment of N,N-dialkyl(trichlorovinyl)amines with butyl-lithium followed by chlorodiorganophosphines has given the aminoalkynylphosphines (95).[117] The chemistry of phosphinomethanide anions has received further study, notably by Karsch's group.[118-121] The issue of C- versus P-nucleophilicity of such ambidentate anions has been addressed in a study of their reactivity towards organoelement chlorides, which result in the formation of *either* tetraheteroelement-substituted methanes *or* isomeric phosphorus ylides which involve an element-phosphorus bond.[119]. Selective carbon-carbon, carbon-phosphorus, and phosphorus-phosphorus bond formation has been observed in the reactions of phosphino-methanides with dicyclopentadienyltitanium dichloride and bismuth trichloride,[120] and zirconaphospha heterocycles, e.g., (96), have been obtained in related reactions with dicyclopentadienylzirconium dichloride.[121] Complexes of phosphinomethanide ligands with chromium and samarium have also been reported.[122] A phosphinomethanide generated by metallation at carbon of the dimethylaminomethylborane adduct of trimethylphosphine has been used in the synthesis of new heterocyclic systems, e.g., (97)[123] and (98).[124] Both P-phenyl 2- and 3-phospholenes undergo regiospecific metallation at the 2-position on treatment with dicyclopentadienylzirconiumhydridochloride, to form the hydrozirconation product (99). Treatment of the latter with chlorophosphines has given the 2-phosphinophospholanes (100), each as a single diastereoisomer.[125]

A series of diazoalkylphosphines (101) has been obtained from the reactions of diazomethyl compounds, $RCHN_2$, with chlorodiorganophosphines in the presence of lithium diisopropylamide.[126] The unsaturated triphosphine (102) is converted into the cyclopropane system (103) on treatment with dimethylsulphonium methylide.[127] Treatment of the Mannich base (104) with phosphine in methanol containing catalytic quantities of nickel(II) chloride and tris(hydroxymethyl)phosphine has given the functionalised benzylic phosphine (105).[128] Acylphosphines have been obtained from the reactions of primary or secondary adamantylphosphines with acid chlorides in the presence of a base.[129] Easy access to the chiral, unsymmetrical diphosphine (106) is given by the reaction of dicyclohexylphosphine with a

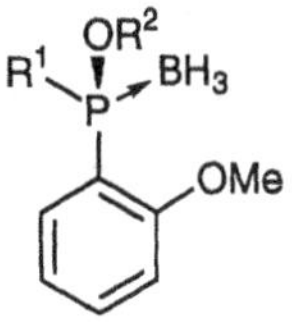

(86) R^1 = Bu^t or Cy
R^2 = (–)-menthyl

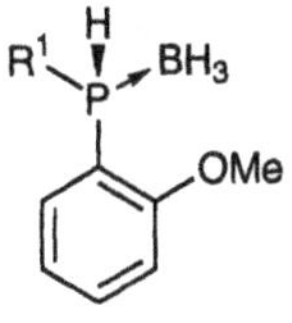

(87)

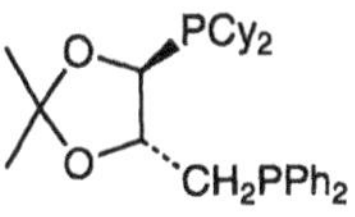

(88)

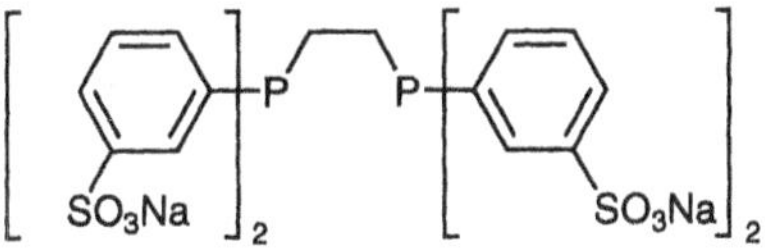

(89)

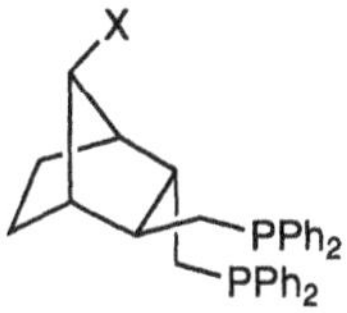

(90) X = OH, OCH_2CH_2OMe, OCH_2COOH, $OCH_2CO_2Bu^t$, or OCH_2CH_2OH

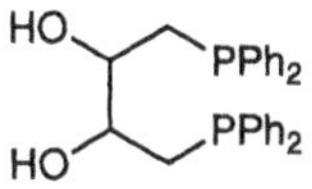

(91)

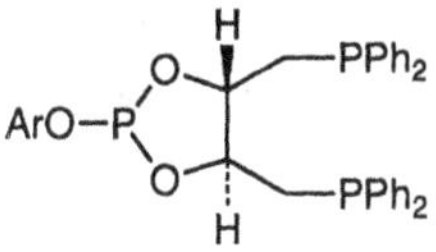

(92)

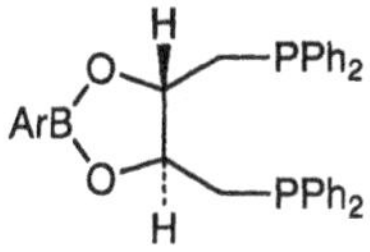

(93)

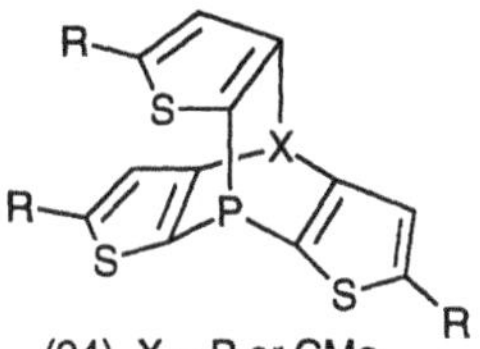

(94) X = P or CMe
R = H or Me

$R^1_2P-C\equiv C-NR^2_2$

(95) R^1 = Pr^i or Bu^t
R^2 = Me or Et

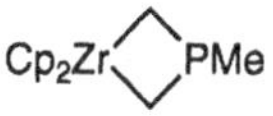

(96)

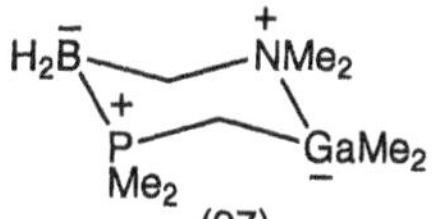

(97)

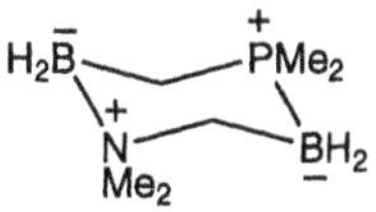

(98)

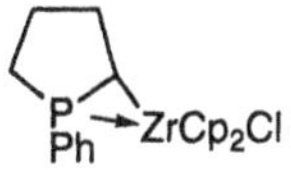

(99)

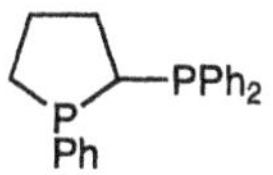

(100)

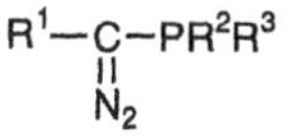

(101) R^1 = Bu^t or Me_3Si
R^2, R^3 = alkyl or R_2N

$(Ph_2P)_2C{=}CHPPh_2$

(102)

(103)

ferrocenylphosphine bearing a chiral dimethylaminoalkyl substituent.[130] A range of tertiary phosphines bearing 2-(1-methylimidazolyl) groups e.g., (107), has been obtained from the reactions of N-methylimidazole with chlorophosphines in the presence of diethylamine.[131] In a similar manner, the N,N-dimethylhydrazone of thiophen-2-carboxylaldehyde has given the phosphines (108) on treatment with phosphorus tribromide in the presence of diethylamine.[132] The pyridylphosphine (109) has been isolated from the reaction of a mixed pyridine-amine solvent with phosphine in the presence of copper(II) acetate or chloride in the presence of air.[133] A range of bis(phosphinomethyl)amines (110) has been obtained by treatment of bis(hydroxymethyl)phosphonium salts with primary amines.[134] The cage-like aminophosphonium salt (111) undergoes reductive cleavage on treatment with sodium in liquid ammonia, to form the new bicyclic ligand (112).[135] Procedures for the synthesis of phosphine-modified cyclodextrins have been developed.[136,137] Further examples of iminophosphines, e.g., (113), have been obtained from the Schiffs-base reactions of primary amines with *o*-diphenylphosphinobenzaldehyde.[138] Polysiloxane-bound ether-phosphine ligands have been prepared by the copolymerisation of monomeric ether-phosphines, e.g., (114), with tetraethoxysilane, under sol-gel conditions.[139] A minor modification of the synthesis of the very basic phosphine (115) has been reported, together with further studies of its coordination chemistry and a structural study of the phosphine and its oxide, the latter hydrogen-bonded to two molecules of water in the solid state.[140] A new and highly enantioselective synthetic route to P-chiral phosphines and diphosphines has been developed from the chiral hydroxythiol (116). The latter is converted to the chiral thiophosphonthioate (117), which then undergoes ring-opening on treatment with *o*-lithioanisole with retention of configuration at phosphorus to form the thiophosphinate (118). Treatment of the latter with methyl-lithium proceeds with inversion at phosphorus, leading to the isolation of the chiral phosphine sulphide (119), readily converted to the related chiral phosphine. Alternatively, (119) may be oxidatively coupled to form the chiral diphosphine sulphide (120), again easily converted to the parent diphosphine.[141]

1.2 Reactions of Phosphines

1.2.1 Nucleophilic Attack at Carbon.- Whereas the diphenylphosphinoimidazole (121) undergoes normal alkylation at phosphorus on treatment with iodomethane, the related bis(heteroaryl)phosphine (122) undergoes quaternization at the pyridine-like nitrogen of the ring system.[142] Triphenylphosphine acts as a nucleophilic catalyst in "umpolung"-style carbon-carbon bond-forming reactions involving nucleophilic γ-addition to alkynes bearing electron-withdrawing groups, presumably *via* initial nucleophilic attack of the phosphine at the triple bond.[143] The

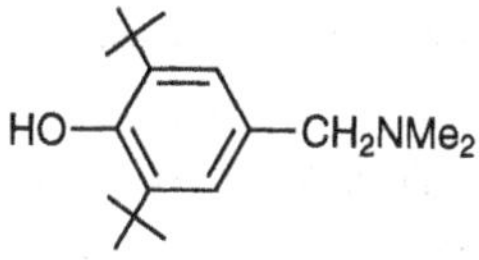

HO— —CH_2NMe_2

(104)

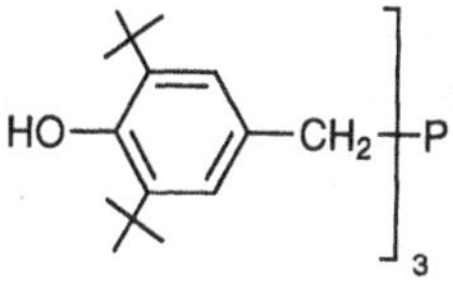

HO— —CH_2— P]$_3$

(105)

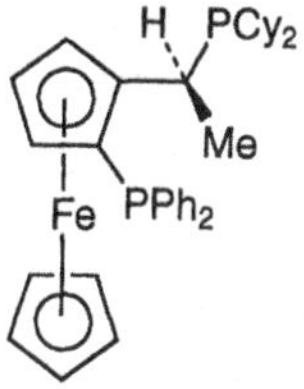

H PCy_2 Me Fe PPh_2

(106)

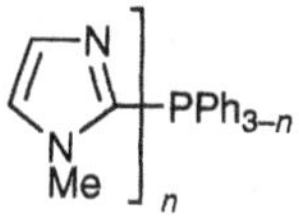

N N Me —PPh_{3-n}]$_n$

(107) n = 1 or 2

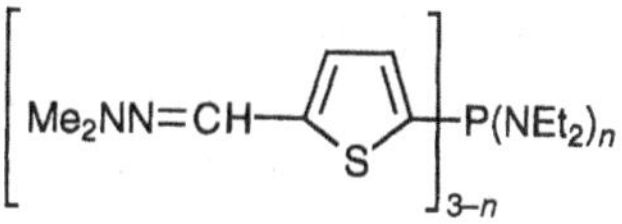

[$Me_2NN{=}CH$— S —$P(NEt_2)_n$]$_{3-n}$

(108) n = 1 or 2

N —P]$_3$

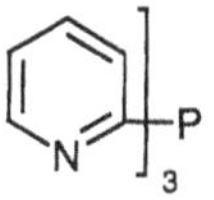

(109)

$[R^1_2PCH_2]_2NR^2$

(110) R^1 = Ph or Cy
R^2 = CH(Me)Ph,
CH(Me)CO_2Et,
CH_2CH_2OH, or allyl

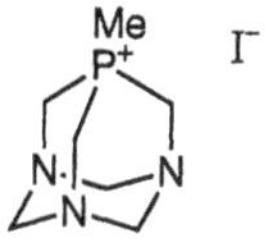

Me P^+ N N N I^-

(111)

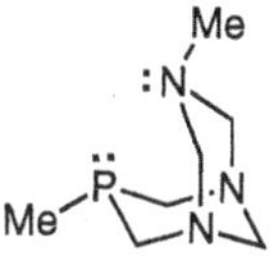

Me :N Me P N N

(112)

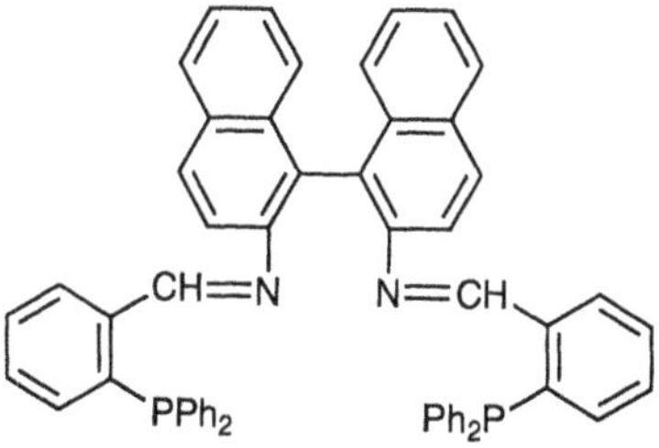

CH=N N=CH PPh_2 Ph_2P

(113)

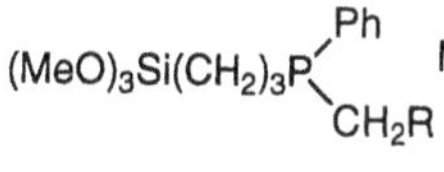

$(MeO)_3Si(CH_2)_3P$ Ph CH_2R

(114) R = CH_2COMe

O ,

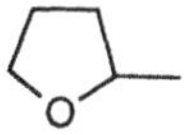

or

O O

MeO— OMe OMe —P]$_3$

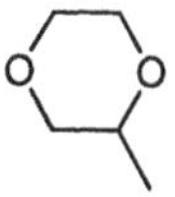

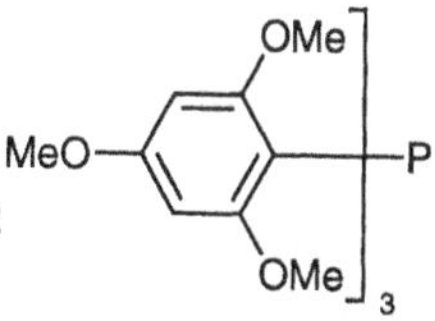

(115)

well-established reactions of triarylphosphines with alkynes bearing electron-withdrawing groups in protic solvents, which proceed *via* the formation of vinylphosphonium salts and their subsequent hydrolysis, have been applied in the synthesis of di-deuterated alkenes under aqueous (D_2O) conditions using water-soluble, sulphonated triarylphosphines. In the presence of a slight excess of phosphine, the initially formed *cis*-alkenes are converted into the *trans*-isomers.[144] Triphenylphosphine has also been shown to catalyse the acid-promoted isomerisation of enynes to (E,E,E)-trienes.[145] The phosphonio-sulphonate betaine (123) has been isolated from the reaction of a water-soluble phosphine with benzaldehyde in aqueous hydrochloric acid.[146] Phosphonio-sulphate betaines (124) formed initially in the reactions of phosphines with cyclic sulphate esters, undergo subsequent decomposition to form an alkene, sulphur trioxide, and phosphine oxide. The stereochemistry of the alkene is related to that of the initial cyclic sulphate.[147] A key step in the addition reactions of carbon dioxide, carbon disulphide, and carbodimides with the phosphinoylide (125) is attack by trivalent phosphorus at the carbon atom of the addend to form the heterocyclic systems (126).[148]

1.2.2 Nucleophilic Attack at Halogen.- Bromotriphenylphosphonium tribromide (127) has been prepared by the reaction of triphenylphosphine with bromine in dichloromethane. In the solid state, this exists in the form of discrete ions.[149] Adducts of tertiary phosphines with iodine monobromide have also been explored. In solution, the ionic iodophosphonium bromide structure (128) predominates, whereas in the solid state, the increasingly familiar four-coordinate "linear" arrangement, predominantly (129), is preferred.[150] Triphosgene has been used as a reagent for halogenation of phosphines, the reaction with one third of a mole of triphosgene resulting in the formation of the related dichlorophosphorane.[151] A mixture of triphosgene and triphenylphosphine in dichloromethane at room temperature converts primary and secondary alcohols into the related alkyl chlorides.[152] The reaction of the aminomethylphosphine (130) with carbon tetrachloride in ether at *ca* -20° to 0°C affords the salt (131), in which the methyleneiminium form is the principal contributor to the ground state structure.[153] Chlorophosphonium ylides, e.g., (132), have been isolated from the reactions of tertiary (α-silylalkyl)phosphines (133) with carbon tetrachloride.[154] A mixture of the phosphonium salt (134) and chlorodiethylphosphine is formed in the reaction of diethylphosphine with carbon tetrachloride.[155] A few examples of new synthetic applications of tertiary phosphine-positive halogen reagents have also appeared.[156-158]

1.2.3 Nucleophilic Attack at Other Atoms.- Routes for the preparation of borane adducts of polyfunctional phosphines have been developed.[159] The bicyclic base DABCO is recommended

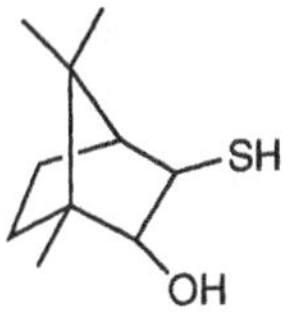

(116)

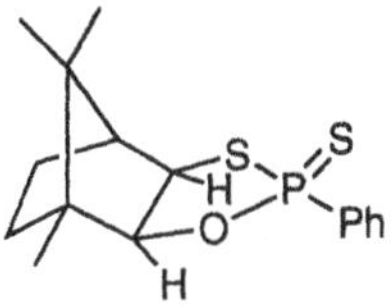

(117)

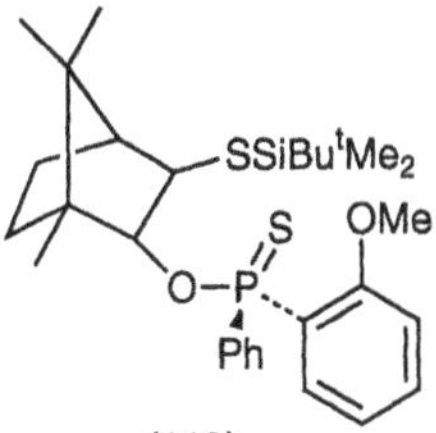

(118)

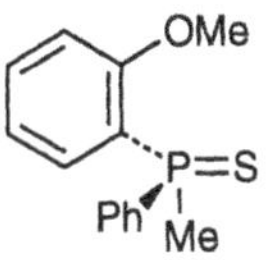

(119)

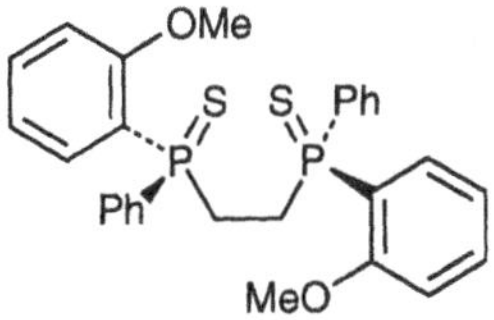

(120)

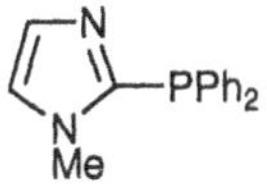

(121)

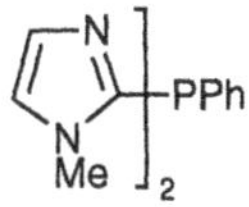

(122)

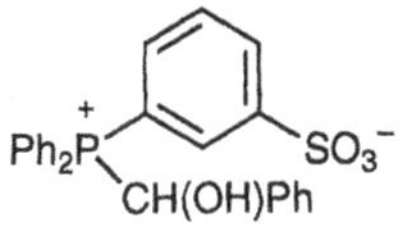

(123)

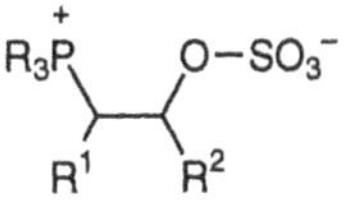

(124)

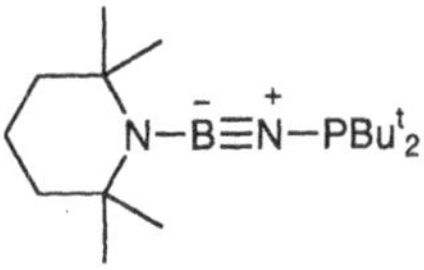

(125)

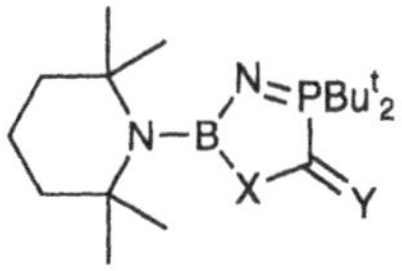

(126) X = Y, O, S, or NR

$[Ph_3\overset{+}{P}Br]$ Br_3^-

(127)

$[R_3\overset{+}{P}I]$ Br^-

(128) R = alkyl or aryl

$[R_3P—I—Br]$

(129)

$Bu^t_2PCH_2NMe_2$

(130)

$[Bu^t_2P{\cdots}CH{\cdots}NMe_2]^+$ Cl^-

(131)

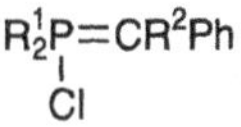

(132) $R^1 = Bu^t$ or NEt_2

R^2 = Ph or $SiMe_3$

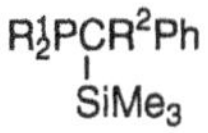

(133)

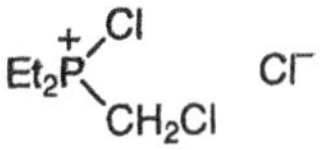

(134)

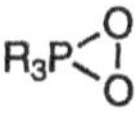

(135)

$Ph_3\overset{+}{P}—S—S_n\text{-}R$ RS^-

(136)

for the removal of borane protecting groups from phosphine-borane adducts.[160] The solid state structure of the trimethylsilylborane adduct of tricyclohexylphosphine has been reported.[161]

Phosphadioxirane intermediates (135) have been implicated in the reactions of tertiary phosphines with singlet oxygen, which result predominantly in the formation of the phosphine oxide, together with other minor products.[162] A kinetic study of the reaction of triphenylphosphine with dioxolanes (cyclic peroxides) is consistent with direct insertion of the phosphine into the peroxide bond to form a phosphorane intermediate, which then suffers hydrolysis to form the related phosphine oxide and a β-hydroxyketone.[163] In the presence of a palladium(II) catalyst, triphenylphosphine can be oxidised to the phosphine oxide with molecular oxygen at room temperature and pressure.[164] The reactions of primary and secondary phosphines, bearing 1-adamantyl substituents, with hydrogen peroxide, sulphur and selenium, have been studied, the related phosphine chalcogenides being isolated.[102,165] A mass-spectrometric study of the reaction between triphenylphosphine (and triethylphosphite) with sulphur is consistent with the accepted mechanism of sulphurization.[166] Two routes participate in the reactions of triphenylphosphine with di-, tri- and tetrasulphides, and also with elemental sulphur, in hydroxylic solvents. On the one hand, nucleophilic attack of RS^- occurs at the sulphur β to phosphorus in the initially formed salt (136), to form the phosphine sulphide and a dialkylsulphide having one less sulphur atom than the original. The alternative is attack by the solvent at the phosphorus atom of (136) to form triphenylphosphine oxide and monoalkylpolysulphur cleavage products. No phosphine oxide was isolated from the reaction of triphenylphosphine with elemental sulphur under aqueous conditions, and the bulk of the evidence suggests the operation of an intramolecular concerted process.[167] Tri-isopropylphosphine has been shown to cleave disulphide links in albumin,[168] and tributylphosphine has been used to desulphurize a ferrocenylpolysulphide.[169] The reaction between the cyclic triphosphine (137) and selenium in benzene solution gives initially the insertion product (138), which, on heating in the presence of selenium for a longer period is converted into the selenadiphosphirane (139).[170]

A study of the Staudinger reaction between triphenylphosphine and *o*-substituted aryl azides has shown that the nature of the products depends very strongly on the *o*-substituent. Thus, e.g., whereas with *o*-azidoacetophenone, the related iminophosphorane (140) is formed, the corresponding reaction of the azide (141) results in the formation of the heterocyclic system (142).[171] In related work, it has been shown that 2-azido-3-vinyl-1,4-naphthoquinones also behave abnormally in their reactions with triphenylphosphine, undergoing reduction of the azido group and oxidation of the unsaturated carbon-carbon side chain.[172] Phosphazides (143), having a linear structure, rather than iminophosphoranes, have been isolated from the reactions of tri-isopropylphosphine with *t*-alkylazides.[173] Iminophosphoranes have also been isolated from the reactions of tertiary phosphines with isoxazolequinone systems.[174] The reaction of tertiary phosphines with 1-bromo-1-phenyldiazirines proceeds *via* initial ring-opening to form the intermediate imidoylnitrene (144), which, in the presence of additional phosphine forms the salt (145), or, in the presence of a stannylphosphine, undergoes cyclisation to form the heterocyclic system (146), with loss of bromine as a bromostannane.[175]

The adducts of tertiary phosphines with dialkyl azodicarboxylates ("Mitsunobu reagents") continue to attract attention. Combination of the adduct of triphenylphosphine and diethyl azodicarboxylate with a 1,2,5-thiadiazolidine-1,1-dioxide in THF at room temperature has led to the isolation of the zwitterion (147), fully characterised by X-ray techniques. This species acts as a source of the triphenylphosphonio moiety in Mitsunobu-like processes, promoting coupling between alcohols and carboxylic acids in the usual way. The zwitterion does not form in the absence of the azodiester in the initial system.[176] Applications of Mitsunobu reagents in synthesis have continued to appear, having been used to prepare polyamines,[177] imides,[178,179] aryl ethers,[180] chiral aminoether ligands,[181] guanidines,[182] theophyllines,[183] α-arylglycosides,[184] and for the N-alkylation of indoles[185] and imidazoles.[186] An usual S_N2'-process operates in the synthesis of 2-alkylidene-1,3-propanediols *via* Mitsunobu chemistry.[187] Assignment of configuration in the 13α-methylgonane series has been aided by an understanding of the stereochemistry of Mitsunobu processes.[188] Mitsunobu-promoted coupling of a dipeptide to a hydroxymethylated polystyrene support has been shown to be facilitated by the presence of a tertiary amine.[189] A modified Mitsunobu system, involving the combination of triphenylphosphine and di-isopropyl azodicarboxylate, has been used in the synthesis of phosphinate esters.[190]

1.2.4 Miscellaneous Reactions of Phosphines.- Theoretical treatments have appeared of a number of strained heterocyclic phosphine systems, notably phospha[3]radialene (148),[191] triphospha[1,1,1]propellane (149),[192] and the tetraphosphacubane system (150).[193] The reactivity of tetrakis-*t*-butyltetraphosphacubane has been studied in superacid media, with respect to protonation, alkylation, and alkynylation. Stable monophosphonium ions have been characterised.[194]

The chemistry of pyridylphosphines has been reviewed.[195] Further studies have appeared of the reactions of phosphines bearing 2-pyridyl or 2-benzothiazolyl substituents with butyl-lithium, which result in the formation of biaryl coupling products *via* an initial nucleophilic attack at phosphorus.[196] A study of the de-diazoniation of arenediazonium salts using triphenylphosphine (and trialkylphosphites) indicates that the reactions proceed *via* a radical-chain mechanism, initiated by single electron transfer from the phosphine to the diazonium salt, to give a phosphonium radical cation, and an aryl radical. The phosphonium radical cation then reacts with the solvent with the eventual formation of the related phosphine oxide.[197] Electron-impact-induced *ortho*-effects have been observed in the mass spectrometric fragmentation of *o*-carboxyphenyldiphenylphosphine.[198] The geometrical isomers of the phosphines (151) have been separated and characterised.[199] The chemistry of lithiated diazomethylphosphines has been discussed in a review concerned with nitrile imines.[200] Cycloaddition of tetrahalo-*o*-quinones with the diazomethylphosphines (152) has given a range of new diazoalkylphosphoranes (153).[201] With trimethyl-aluminium, -gallium, and -indium, the phosphinylcarbene (154) gives rise to the novel ylides (155).[202] A series of α-phosphinoenolates (156) has been obtained from the reactions of acylphosphines with butyl-lithium and converted into the related dicyclopentadienyl-zirconium derivatives.[203] The Mannich condensation reactions of tris(hydroxymethyl)phosphine with an aminopolyether, followed by oxidation at phosphorus, provide flexible side-chain systems for enzyme immobilisation without loss of catalytic activity.[204] A series of monoaryliminobis-(diphenylphosphino)methanes (157) has been obtained *via* the reactions of the related monotrimethylsilylimino systems with nitrofluoroarenes and arylfluorononitriles.[205] Electrolysis of phosphines in the presence of an alcohol in a single compartment cell results in the deoxygenation of the alcohol to form the parent hydrocarbon. The reactions proceed *via* initial anodic oxidation of the phosphine to form the alkoxyphosphonium salt, which then suffers cathodic reduction to form the hydrocarbon and phosphine oxide.[206] Combinations of tributylphosphine with acid anhydrides have again been used for the acylation of alcohols, the phosphine having some advantages over alternative catalysts such as *p*-dimethylaminopyridine.[207] Trialkylphosphines have also found application as catalysts in addition and ring-cleavage reactions of cyclobutenones,[208] and in the decarboxylation of formate esters.[209] Whereas the silicon-carbon bond in the phosphinomethylsilane (158) is very readily cleaved by protic reagents, that in the related phosphinoethane (159) is much more stable, and considerable exploration of the chemistry of the latter has taken place, with particular reference to the design of the ligands suitable for attachment to the surfaces of solid supports.[210] Trimethylgermyl- and -stannyl-phosphines (160) have been shown to suffer cleavage of the phosphorus-metalloid bond on treatment with hexachlorodisilane, giving the trichlorosilylphosphines (161), together with Si-Ge, and Si-Sn systems.[211] Further examples of cleavage of phosphorus-carbon bonds in phosphines coordinated to transition metal catalyst systems have been described.[212,213] Full details of carbon-carbon coupling reactions undergone by coordinated acetylenic phosphines have now appeared.[214] The coordination chemistry of *o*-diphenylphosphinocarbaborane systems has started to develop.[215] A thallium(I) derivative of diphenylphosphinotetramethylcyclopentadiene (162) has been obtained by treatment of the phosphine with thallium(I) ethoxide.[216]

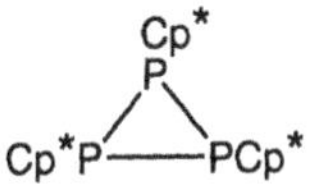

(137) $Cp^* = Me_5C_5$

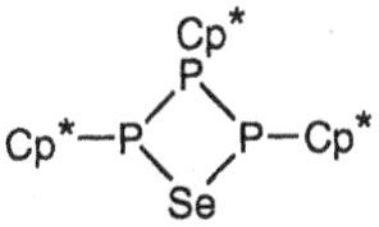

(138)

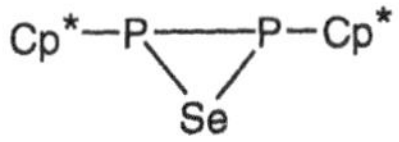

(139)

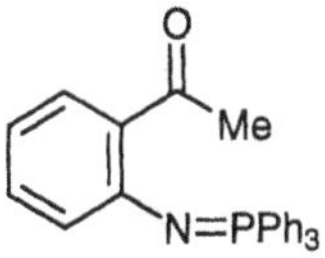

(140)

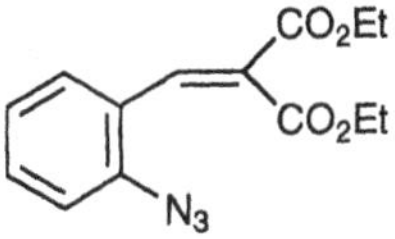

(141)

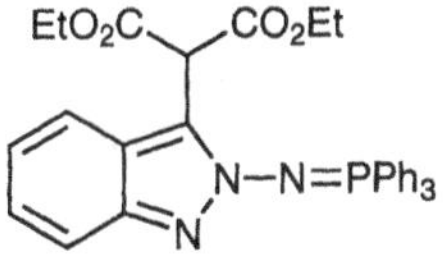

(142)

(143) $R^1 = Pr^i$, $R^2 = CR_3$

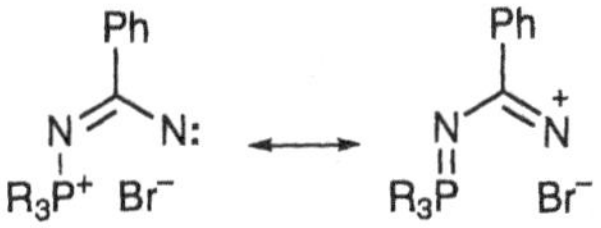

(144)

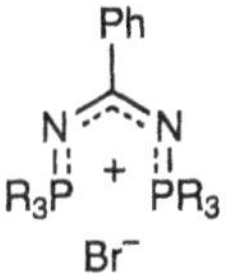

(145)

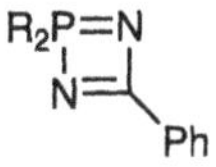

(146)

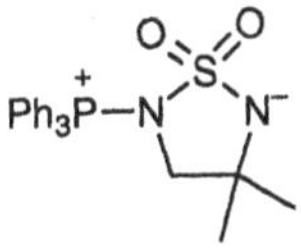

(147)

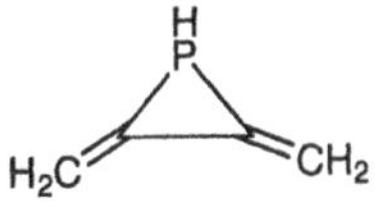

(148)

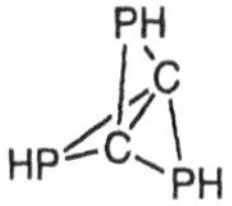

(149)

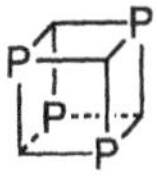

(150)

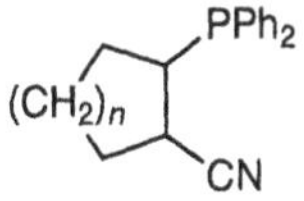

(151) n = 1 or 2

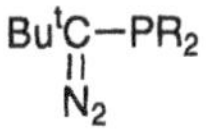

(152) R = Me, Pr^i, Bu^t, or NEt_2

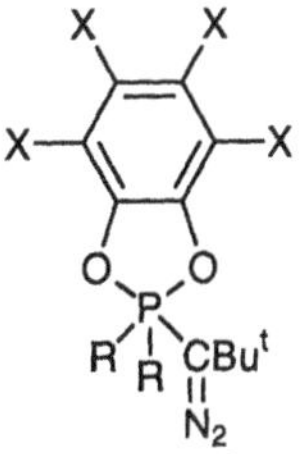

(153) X = Cl or Br

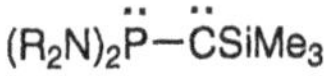

(154)

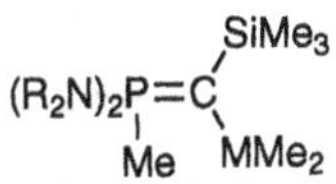

(155) M = Al, Ga, or In

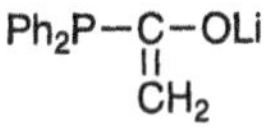

(156)

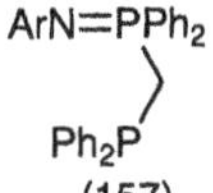

(157)

2 Halogenophosphines

2.1 Preparation.- A range of monofluorophosphines, e.g., (163) and (164), has been obtained from the reactions of organolithium reagents with dichlorofluorophosphine.[217] A new route to methyldifluorophosphine is provided by the reaction of methyldichlorophosphine with anhydrous hydrogen fluoride at low temperatures.[218] Treatment of N-phosphinoimidazoles (165) with acyl fluorides provides a general route to mono- and di-fluorophosphines.[219] Halogenophosphines bearing the 1-chloroadamant-3-yl substituent, e.g., (166), have been obtained from the reaction of 1,3-dehydroadamantane with organodichlorophosphines.[220] (1-Adamantyl)-dichlorophosphine has been shown to undergo halogen exchange on treatment with either bromo- or iodo-trimethylsilane, thereby providing the related dibromo- and di-iodo-phosphines in good yield.[221] Further reports have appeared of the synthesis of halogenophosphino derivatives of indolizine by the reactions of 2-substituted indolizines with phosphorus halogenides in the presence of a base.[222] In a similar manner, treatment of 2-propylindazole with phosphorus tribromide or phenyldibromophosphine in the presence of pyridine results in phosphorylation at the 3-position, to give, e.g., (167).[223]

2.2 Reactions.- The influence of the solvent on the course of disproportionation of organodifluorophosphines has been explored, some compounds following a completely different pathway on heating in solution in chloroform compared to that observed on heating in the absence of solvent.[224] Evidence for the reversible formation of organotetraiodophosphoranes from iodine and organodiiodophosphines has been adduced from solution and solid state NMR data, and structural studies.[225] In the presence of an electrogenerated zerovalent nickel-phosphine complex, chlorodiphenylphosphine undergoes a cross-coupling reaction with 2-bromopyridine, leading to the isolation of the phosphine oxide (168).[226] Polyphosphazenes have been obtained from the reactions of fluorophosphines with silylazides.[227] The reaction of bis(dichlorophosphino)methane with tetrachloro-*o*-benzoquinone gives the phosphorane (169) which undergoes further transformation on treatment with base.[228] Further examples of the formation of phosphorus-oxygen and phosphorus-nitrogen bonds from the reactions of halogenophosphines with alcohols, and amino compounds, respectively, have appeared, leading to the isolation of a range of new systems, e.g., the chiral ligands (170),[229] phosphorylated 2-aminopyridines,[230] N-phosphinoguanidines (171),[231] N-phosphinothioureas,[232] and N-phosphinobenzimidazoles.[233] A study of the reactions of chlorodiphenylphosphine with benzoylhydrazine has also appeared.[234] The reaction of the salt (172) with chlorodiphenylphosphine has given the heterocyclic system (173).[235] The propargylchlorophosphine (174) undergoes rearrangement on standing at 25°C in

$Ph_2PCH_2SiMe_2H$

(158)

$Ph_2PCH_2CH_2SiMe_2H$

(159)

R_2PMMe_3

(160) R = Pr^i or Bu^t
M = Ge or Sn

R_2PSiCl_3

(161)

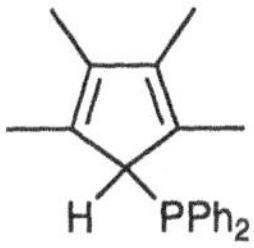

H PPh$_2$

(162)

$(PhC{\equiv}C)_2PF$

(163)

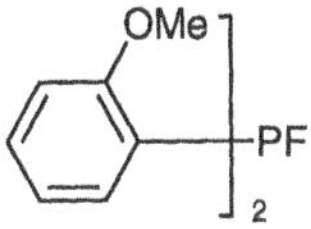

OMe
PF
2

(164)

R_nP N N $3-n$

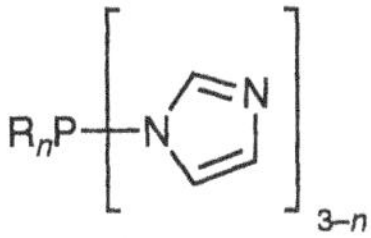

(165) n = 1 or 2
R = alkyl or aryl

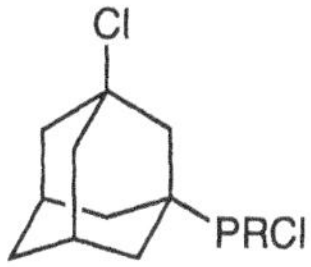

Cl
PRCl

(166) R = Me or Ph

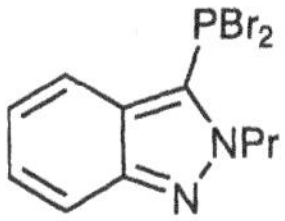

PBr_2
NPr
N

(167)

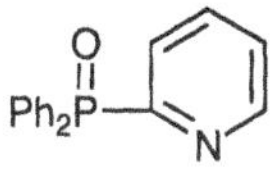

O
Ph_2P
N

(168)

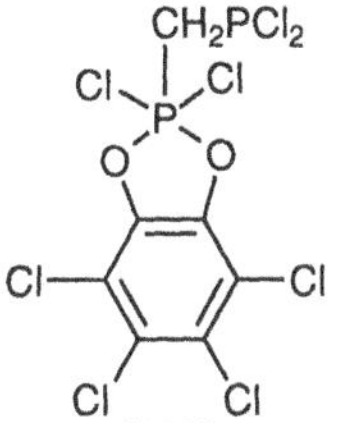

CH_2PCl_2
Cl Cl
P
O O
Cl Cl
Cl Cl

(169)

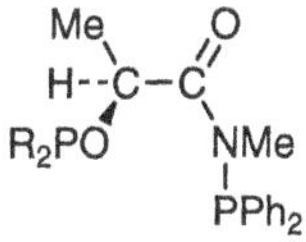

Me O
H--C–C
R_2PO NMe
PPh_2

(170) R = cyclopentyl

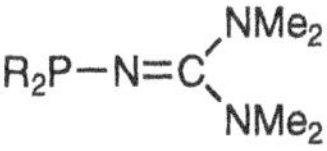

$R_2P–N{=}C(NMe_2)_2$

(171) R = Pr^i

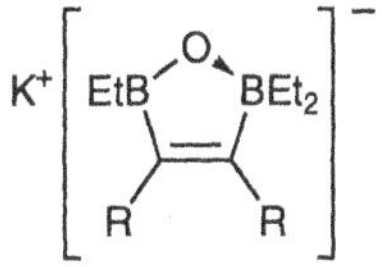

K^+ [EtB O BEt$_2$ R R]$^-$

(172) R = Et or Ph

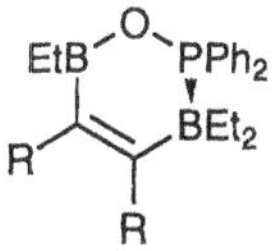

EtB O PPh$_2$
BEt$_2$
R R

(173)

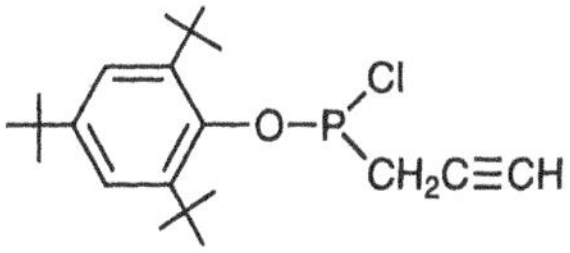

O–P Cl
$CH_2C{\equiv}CH$

(174)

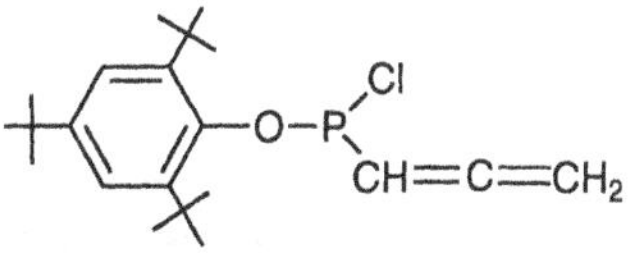

O–P Cl
$CH{=}C{=}CH_2$

(175)

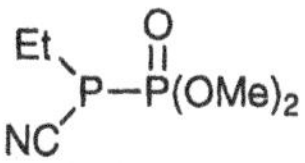

Et O
P–P(OMe)$_2$
NC

(176)

X
R–PH$_2$

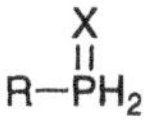

(177) R = alkyl or aryl
X = S or Se

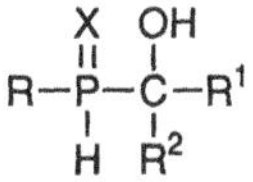

X OH
R–P–C–R^1
H R^2

(178)

the presence of triethylamine in THF to form the allenyl system (175).[236] Formation of P-P bonds occurs in the reactions of organodicyanophosphines with dialkylphosphites or their sodium salts. Thus, e.g., the reaction of ethyldicyanophosphine with dimethylphosphite leads to the formation of (176), together with hydrogen cyanide.[237]

3 Phosphine Oxides and Related Chalcogenides

3.1 Preparation.- Primary phosphine sulphides and selenides (177) have been obtained from the reactions of the primary phosphine with sulphur or selenium. If the reaction is carried out in the presence of a carbonyl compound, the secondary phosphine chalcogenides (178), are isolated.[238] Isomeric mono- and di-chalcogenides have been isolated from the reactions of the diphosphines (179) with sulphur or selenium[239] A selective procedure for the synthesis of the mono-sulphide and -selenide (180) has been developed, which involves the direct reaction of the related diphosphinoamine with the respective chalcogen in toluene or hexane.[240] The metallophosphines (181) react normally with hydrogen peroxide or elemental sulphur to give the related metallophosphine-oxides and -sulphides.[241] The bicyclic polyphosphine dioxide (182) has been isolated from the reaction of tetra-t-butylhexaphosphine with cumene hydroperoxide.[242] Several new polyphosphorus sulphide cage systems have been prepared.[243,244] Enantiomerically pure styryl- and dienyl-phosphine oxides (183) have been obtained from palladium catalysed Heck coupling reactions of the related chiral vinylphosphine oxides.[245] Treatment of the selenophosphinate (184) with phenyllithium, followed by alkyl halide, leads to the chiral phosphine oxides (185), with retention of configuration at phosphorus.[246] Multigram quantities of enantiomerically pure tertiary phosphine oxides are accessible in a two step process which involves Raney nickel reduction of the enantiomerically pure phosphinothioic acid (186) to give the chiral secondary phosphine oxide (187), treatment of which with lithium di-isopropylamide followed by an alkyl halide results in the formation of chiral tertiary phosphine oxides, e.g., (188).[247] 2-(Pyridyl)ethylphosphine oxides, e.g., (189), have been isolated from the reactions of red phosphorus in superbasic media, such as KOH-DMSO, with vinylpyridines.[248] Alkyl bis-(hydroxymethyl)phosphine oxides (190) are among the products of the reaction of alcohols with phosphine in the presence of platinum salt.[249] A synthetic route to the sulphide (191) of diphenylphosphinoserine, a useful intermediate for the synthesis of phosphinopeptides, has been developed.[250] Reduction of the phosphinate (192) with sodium dihydrobis(2-methoxyethoxy)-aluminate, followed by acidification, has given the first heterocyclic phosphorus analogue of a ketose, (193).[251] A range of tertiary phosphine oxides possessing juvenile hormone activity (194) has been prepared.[252] The dianionic systems (195) have received further application in the

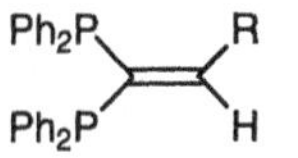

(179) R = Me or Ph

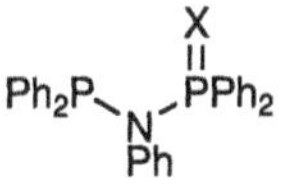

(180) X = S or Se

$Cp_2M(CO)PR_2$

(181) M = Nb or Ta
R = Me or Ph

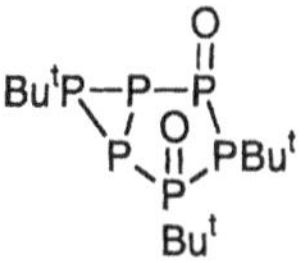

(182)

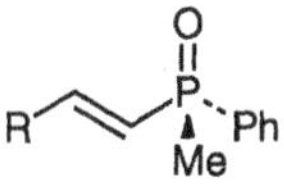

(183) R = aryl or vinyl

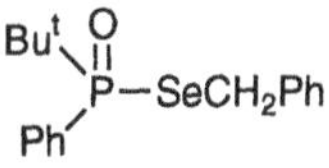

(184)

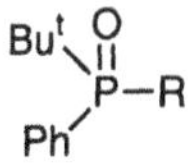

(185)

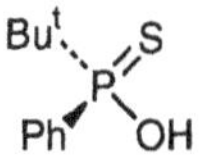

(186)

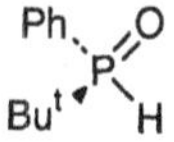

(187)

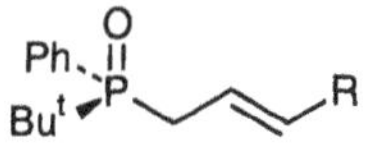

(188) R = H or Me

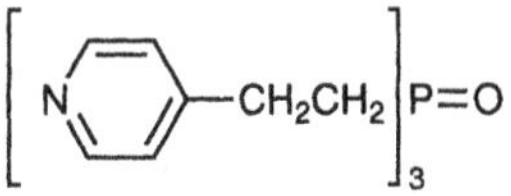

(189)

$RCH_2P(O)(CHOHR)_2$

(190)

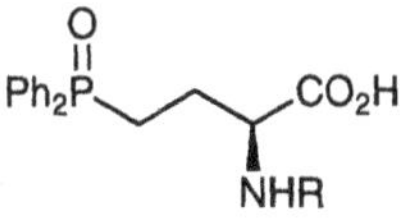

(191)

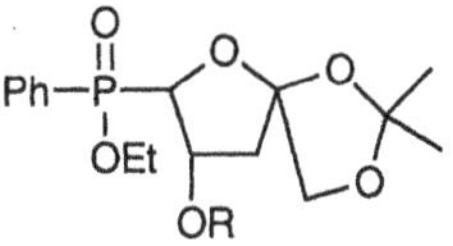

(192)

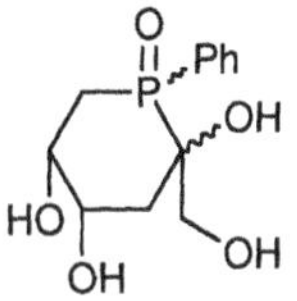

(193)

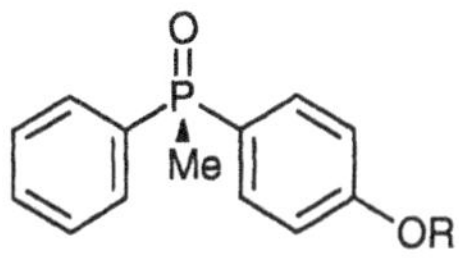

(194) R =

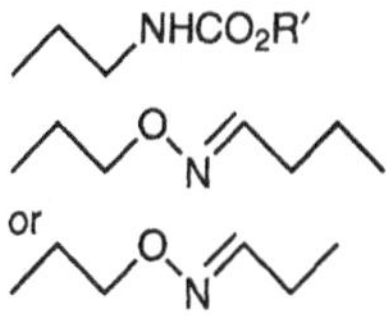

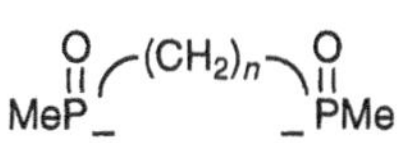

(195) n = 2 or 3

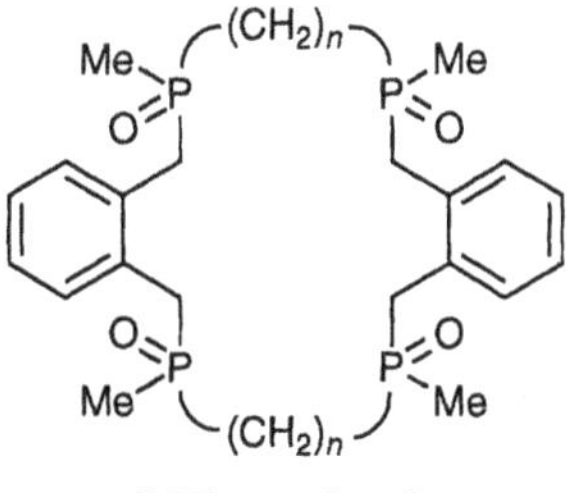

(196) n = 2 or 3

$(H_2NCH_2)_n P(O)Me_{3-n}$

(197)

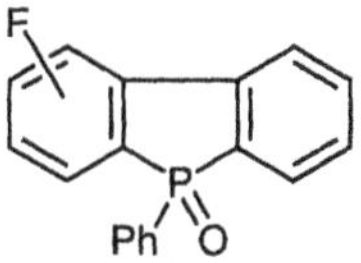

(198)

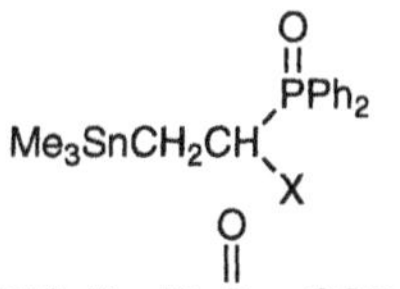

(199) X = $Ph_2P(O)$ or COR

synthesis of new macrocyclic phosphine oxides, e.g., (196), and the related phosphines.[253,254] A series of aminoalkylphosphine oxides (197) has been prepared by the Gabriel synthesis using the related chloromethyl compounds.[255] Oxidation of *cis*-$[Pt(PPh_3)_2Cl_2]$ with chlorine or sodium persulphate in neutral solution in the presence of hydrogen fluoride and a silver ion catalyst is reported to result in the formation of platinum(II) complexes of triphenylphosphine oxide, and, surprisingly, the monofluorodibenzophosphole oxides (198).[256] Interest has continued in the synthesis of phosphine oxides bearing organotin substituents at carbon. The phosphine oxides (199) are accessible from the reactions of iodomethyltrimethylstannane with the appropriate phosphinylmethanide anion.[257]

3.2 Reactions.- The diphenylphosphinoyl group activates the adjacent *o*-methoxy substituent in the naphthalene system (200) towards nucleophilic replacement by carbon, nitrogen, and oxygen nucleophiles. This approach has been utilised in the synthesis of the chiral system (201), capable of reduction to the related chiral chelating diphosphine.[258] Sequential treatment of the phosphine oxide (202) with methyllithium and *t*-butyllithium, followed by acidification, results in an anionic cyclisation to form the benzazaphospholine system (203).[259] Treatment of the phosphinate (204) with lithium bis(trimethylsilyl)amide results in a 1,4-phosphinyl migration with the formation of the bis(phosphine oxide) (205). The latter, on heating in benzene in the presence of a trace of pyridine, undergoes a reverse rearrangement to form a mixture of the phosphinates (204) and (206).[260] The fluoroalkylphosphine oxide (207) (accessible *via* the addition of diphenylphosphine oxide to the N-acetylimine of hexafluoro-acetone) undergoes a very facile cleavage of the phosphorus-alkyl carbon bond on treatment with *either* water *or* methanol.[261] Both acylphosphine oxides (208) and (209) undergo cleavage of the phosphorus-acyl bond on irradiation with ultraviolet light during laser flash photolysis.[262] The bis(phosphine oxide) (210) (obtained *via* Diels-Alder addition of $Ph_2P(O)C{\equiv}CP(O)Ph_2$ to anthracene) has been shown to act as a novel host system in forming crystalline 1:1 inclusion complexes with a wide variety of aliphatic solvent molecules. The bis(oxide) (210), (and its solvent complexes), undergoes a di-π-methane photorearrangement to form the dibenzosemibull-valene system (211) as the sole product. Irradiation of the ethanol complex of (210), in crystalline form, leads to (211) in >90% enantiomeric excess, an example of an "absolute asymmetric synthesis" in which an achiral reagent is converted into an optically active product without the intervention of an external chiral agent. Surprisingly, in contrast, irradiation of the related ethyl acetate complex of (210) yields racemic (211).[263] The iodide ion-iodine redox couple plays a crucial catalytic single electron transfer role in the photo-redox reaction between diphenylphosphine oxide and 10-methylacridinium iodide, which occurs with visible light in

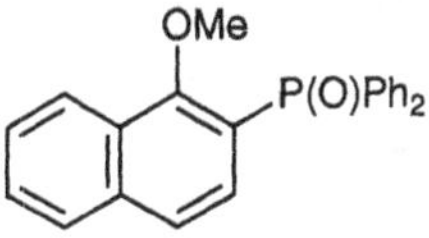

(200)

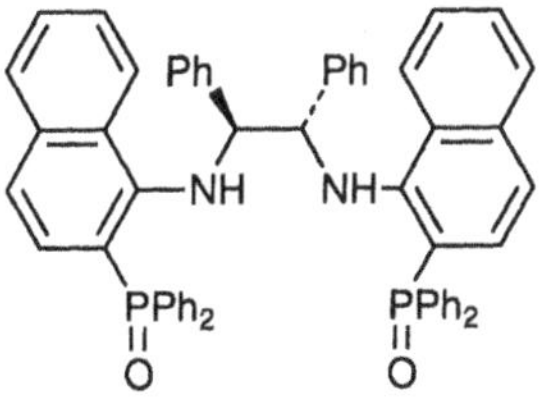

(201)

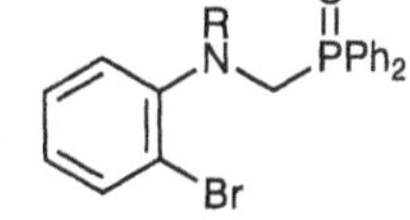

(202) R = Me, Pr^i, or CH_2Ar

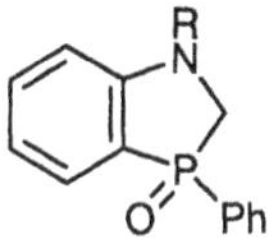

(203)

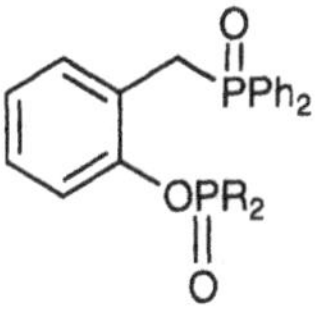

(204) R = Et, Pr, Bu, Ph, or CH_2Cl

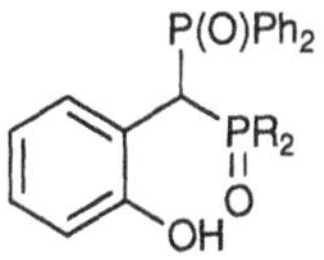

(205)

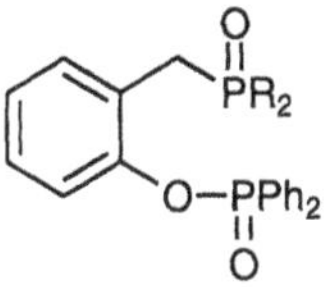

(206)

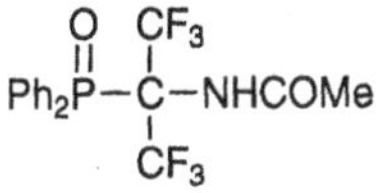

(207)

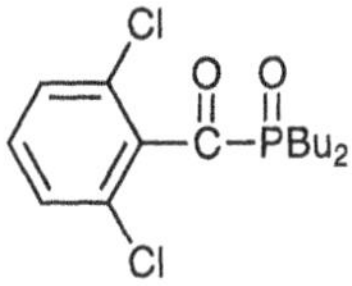

(208)

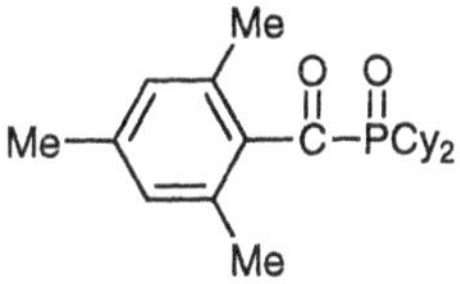

(209)

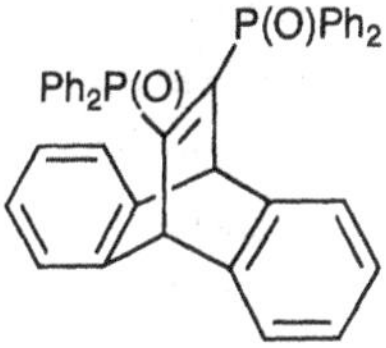

(210)

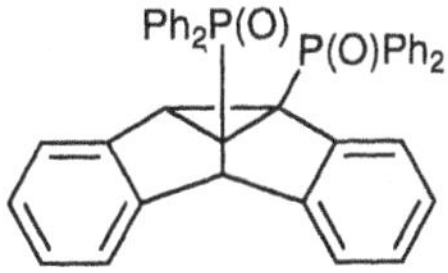

(211)

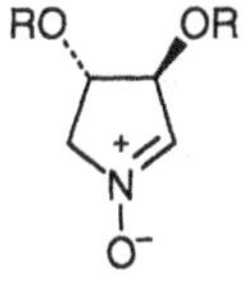

(212) R = Me, CH_2OMe, Bu^t, or $SiPh_2Bu^t$

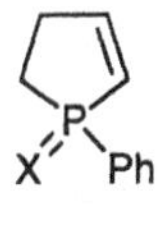

(213) X = O or S

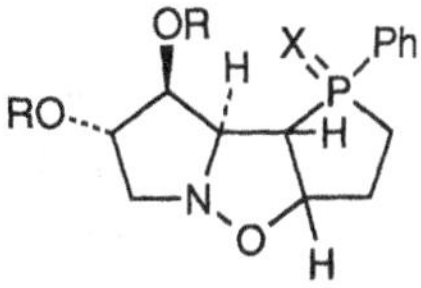

(214)

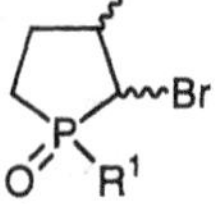

(215) R^1 = Ph or MeO R^2 = H or Me

(216) $R_3\overset{+}{P}{-}SMe$ Tf^- R = Ph or NMe_2

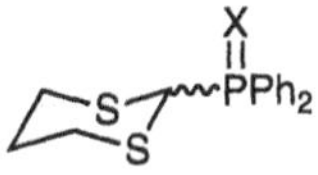

(217)

aqueous acetonitrile, to form diphenylphosphinic acid and 10-methylacridan as the final products.[264] The 1,3-dipolar cycloaddition of chiral nitrones (212) to the 2-phospholene oxide or sulphide (213) proceeds with a remarkable kinetic resolution with the preferred formation of one diastereoisomer of the adduct (214), and the unreacted R-enriched enantiomer of the starting phospholene.[265] Addition of bromine to 2-phospholene oxides in aqueous solvents results in the formation of the 2-bromo-systems (215).[266] A new stereoselective interconversion of thiophosphoryl and selenophosphoryl compounds has been developed, involving the intermediacy of related methylthio- and methylseleno-phosphonium salts. Thus, e.g., treatment of a phosphine sulphide with methyl triflate yields the methylthiophosphonium salt (216), which on treatment with sodium hydrogen selenide in ether affords the related phosphine selenide. Work with chiral systems has established that the reactions proceed with predominant retention of configuration at phosphorus.[267] There has also been considerable interest in the use of phosphine oxides as synthetic reagents, with particular reference to Horner-Wittig routes, and, to this end, quite a number of papers describing the elaboration of the alkyl side chain of alkyldiphenylphosphine oxides have appeared.[268-273] These will be discussed elsewhere in this volume.

3.3 Structural and Physical Aspects.- Interest has continued in studies of the conformational preferences of 1,3-dithianes (217) bearing phosphoryl, thiophosphoryl, and selenophosphoryl substituents at the 2-position.[274] This area has also been reviewed.[275] The study of hydrogen-bonded adducts of phosphine oxides has also continued to develop. Structural and spectroscopic studies of the adducts of triphenylphosphine oxide with nitroarylcarboxylic acids[276] and di(organosulphonyl)amines[277] have been reported. A theoretical study of the proton affinity of phosphoryl compounds has also been described.[278] Macrocyclic systems, e.g., (218), bearing phosphine oxide and sulphoxide groups, have been synthesised and their ability to recognise hydrogen-bonding substrates investigated.[279-282] The X-ray structure of one enantiomer of the chiral phosphine oxide (219) has been determined, enabling an assignment of absolute configuration in menthyl phosphinylacetates.[283] An electron impact mass spectral study of phosphine sulphides and selenides has established structural effects on the course of fragmentation.[284]

3.4 Phosphine Chalcogenides as Ligands.- Spectroscopic evidence has been obtained for the existence of a reversible coordinative interaction between a phosphine oxide (in both the free and metal-coordinated state) and sulphur dioxide.[285] Intramolecular coordinative interactions between the phosphoryl group and both fluoroboron and fluorophosphorane acceptors, e.g., (220), have been established by spectroscopic techniques.[286,287] The coordination chemistry of the diphosphine monochalcogenides (221) towards transition metal acceptors has been explored,[288] and the crystal structure of a molybdenum(VI) complex of the diphosphine dioxide (222) has been reported.[289] A structural study of tris(N,N-dimethylaminomethyl)phosphine oxide (223) in both the free and metal-coordinated state has shown that coordination of the ligand (which binds in a bidentate manner, involving the oxygen and one of the nitrogen atoms) brings about only very small changes in bond lengths and bond angles.[290] Solution ^{119}Sn, ^{31}P, and ^{1}H nmr studies in non-polar solvents have shown that whereas the sulphide and selenide of the heterocyclic system (224) involve intramolecular coordinative interactions between tin and sulphur or selenium, the related phosphine oxide prefers to bind intermolecularly. In the solid state, a similar situation applies.[291] Structural studies of adducts of triphenylphosphine oxide with monoorganotrichlorostannanes have also been reported.[292] Various types of phosphine oxides are able to complex alkali and alkaline earth metal cations, and to facilitate their transfer across aqueous/organic interfaces.[293] Organogallium complexes of phosphine oxide ligands have been isolated from the reactions of organogallium peroxo derivatives with phosphines.[294]

4 Phosphonium Salts

4.1 Preparation.- Conventional quaternization reactions have been employed in the synthesis of the macrocyclic phosphonium salt (225),[295] the heteroarylmethylphosphonium salt (226),[296] and new examples of phosphonium cascade systems having phosphorus in a variety of oxidation states at the core.[297] In the reactions of bromo- and iodo-trifluoromethanes with tributylphosphine in acetonitrile, the trifluoromethylphosphonium salt (227) is formed. However, the corresponding reaction between iodotrifluoromethane and tributylphosphine in the gas phase yields dibutyl(trifluoromethyl)phosphine.[298] The difluoroallylphosphonium salt (228) is formed *via* a S_N2' pathway in the reaction of 3,3-difluoroallyl bromide with triphenylphosphine.[299] Stable thiophthalylphosphonium salts (229) have been obtained from the reactions of triphenylphosphine with the related 1H-thiophthalylium carbonium ions.[300] An improved route to triphenylphosphonium bromide (230) is provided by heating triphenylphosphine under reflux in *t*-butylbromide containing catalytic quantities of tetrabutylammonium bromide. Heating the hydrobromide (230) at 150°C in xylene results in thermal decomposition, providing a useful laboratory source of hydrogen bromide.[301] The salt (231) is formed cleanly in the reaction of the precursor iminophosphine with hydrogen iodide.[302] Direct current electrolysis of a solution of buckminsterfullerene, C_{60}, in *o*-dichlorobenzene containing tetraphenylphosphonium chloride, has yielded the first phosphonium fulleride (232), isolated as black, air-stable crystals.[303] A zwitterionic phosphonium dicarbaborane system has also been characterised.[304] The nickel(II)-

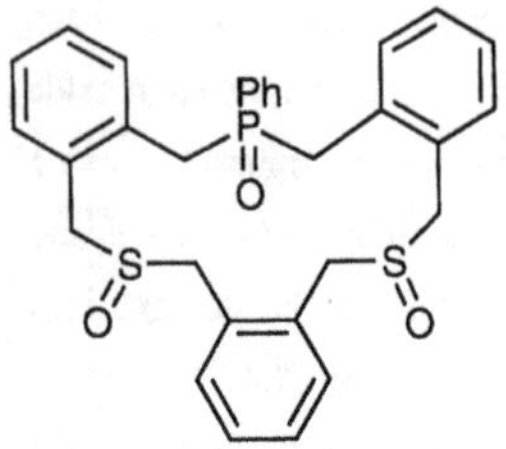

(218)

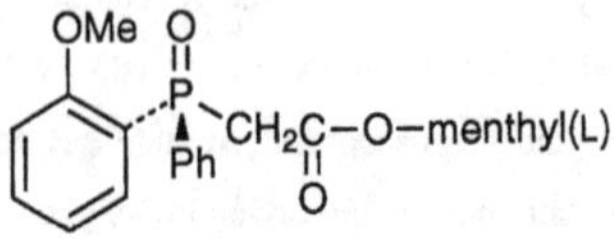

(219)

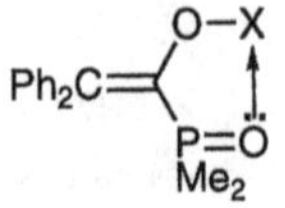

(220) X = BF_2 or RPF_3

$Ph_2(CH_2)_nP(X)Ph_2$

(221) n = 1 or 2
X = O or S

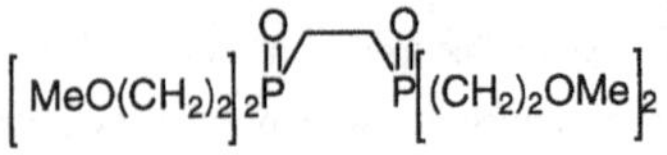

(222)

$[Me_2NCH_2]_3P{=}O$

(223)

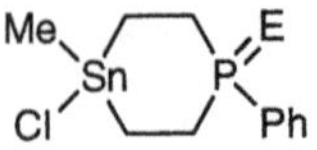

(224) E = O, S, or Se

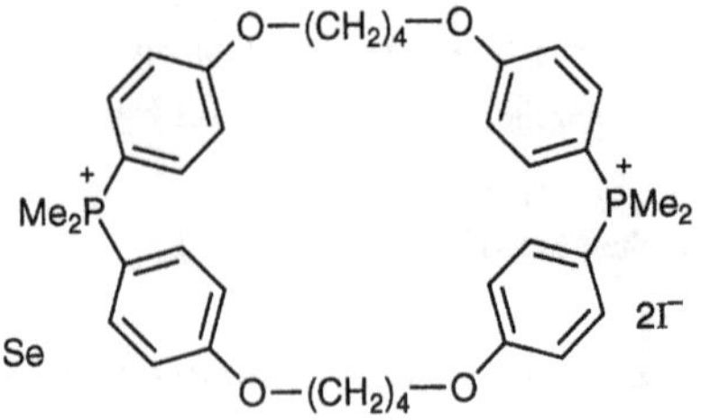

(225)

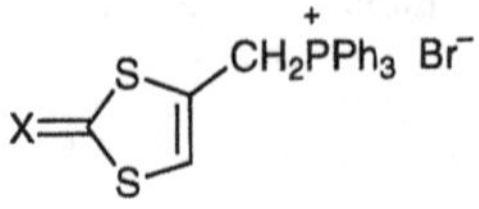

(226) X = O or S

$Bu_3\overset{+}{P}CF_3\ X^-$

(227) X = Br or I

$CF_2{=}CHCH_2\overset{+}{P}Ph_3\ Br^-$

(228)

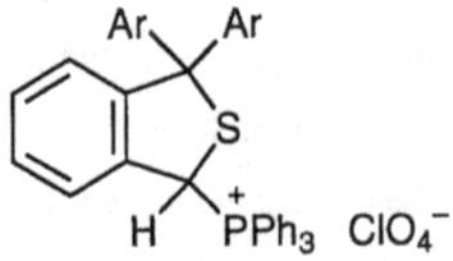

(229)

$Ph_3\overset{+}{P}H\ Br^-$

(230)

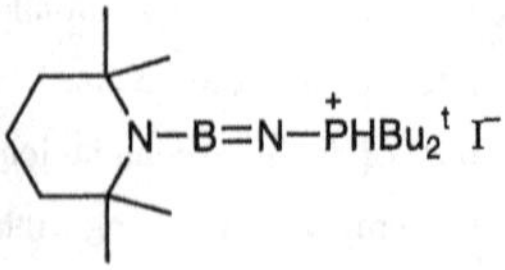

(231)

$Ph_4\overset{+}{P}\ C_{60}^{\bullet -}\cdot Ph_4\overset{+}{P}\ Cl^-$

(232)

$Ph_n\overset{+}{P}(NEt_2)_{3-n}Ar\ Y^-$

(233)

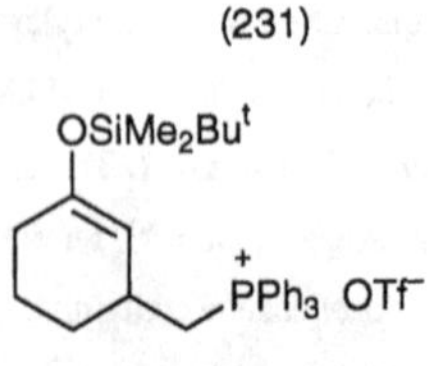

(234)

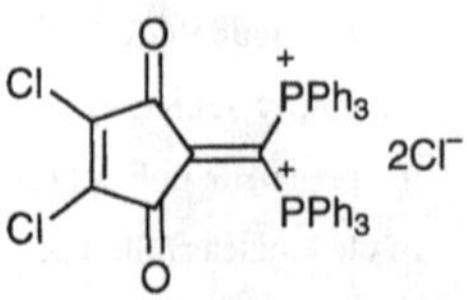

(235)

catalysed Horner route to arylphosphonium salts has been applied in the synthesis of the dialkylaminophosphonium salts (233).[305] The phosphonium salt (234) has been isolated from the reaction of bicyclo[4,1,0]-2-heptanone with triphenylphosphine in the presence of *t*-butyl-dimethylsilyl triflate.[306] The diphosphonium salt (235) is formed in the reaction of perchloro-2-methylene-4-cyclopentene-1,3-dione with triphenylphosphine in acetone.[307] Alkylation of a stabilised ylide using α,ω-dibromoalkanes has been employed in the synthesis of the cycloalkyl-triphenylphosphonium salts (236).[308] Diprotonation of the phosphinoylides (237) provides the diphosphoniomethanes (238).[309] The reaction of a water-soluble sulphonated triphenylphosphine with activated alkenes, e.g., maleic or fumaric acid, leads to the phosphonium zwitterion (239).[310] Ring-opening of functionalised oxiranes with triphenylphosphine in phenol as solvent provides a route to a range of new functionalised hydroxyalkylphosphonium salts, e.g., (240).[311] Procedures for the optical resolution of 2- and 3-hydroxyalkylphosphonium salts have been established.[312] A one-pot procedure for the synthesis of triphenylvinylphosphonium salts involves the reaction of oxiranes with triphenylphosphonium tetrafluoroborate, followed by addition of acetyl or oxalyl chloride.[313] Further reports have appeared of the synthesis of heteroaryl-phosphonium salts, e.g., (241)[314] and (242),[315] from the reactions of appropriately functionalised vinyltriphenylphosphonium salts.[314-319] Interest has also continued in the synthesis of phosphonium salts in which the phosphonium centre is directly attached to a carbon-based framework coordinated to a metal, e.g., (243)[320] and (244).[321] A range of phosphonium salts has been prepared in which single or double alkyl chains of various lengths (C_{10} to C_{18}) are attached to phosphorus. Many of these compounds have been found to show high levels of microbial activity, and are superior in this respect to the related ammonium salts.[322] Triphenyl-phosphonium trichlorogermanate (245) has been obtained in quantitative yield from the reactions of the adduct of triphenylphosphine and germanium tetrachloride with tributyltin hydride.[323]

4.2 Reactions.- Further studies of solvent effects in the alkaline hydrolysis of tetraphenyl-phosphonium chloride have appeared.[324] A variety of products has been isolated from the reactions of halomethylphosphonium halides with phosphines.[325] Acylation of fluorinated phosphoranium salts (246) with perfluorinated acyl chlorides has provided the (Z)-perfluoro-betaines (247), in high yield.[326] Equilibrium acidities have been measured for a range of alkyl-triphenylphosphonium salts, together with a study of the oxidation potentials of the related ylides.[327] The propargylphosphonium salt (248) has been polymerised to give the conjugated polyelectrolyte (249) in quantitative yield.[328] Treating the salt (250) with phenylacetylene at 100°C results in the formation of the conjugated ene-yneylphosphonium salt (251).[329] The tetraphenyl-phosphonium ion has been shown to bind to β-cyclodextrin systems.[330] The diphosphonium

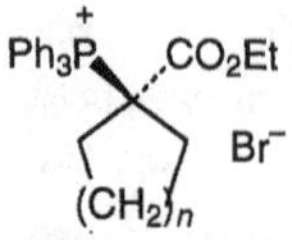

$Ph_3\overset{+}{P}$ CO_2Et $(CH_2)_n$ Br^-

(236) n = 2 or 3

$R_2P-C(Y)=PPh_3$

(237) Y = CO_2Et or tosyl
R = Ph or Bu

$R_2\overset{+}{P}H-CHY-\overset{+}{P}Ph_3$ $2X^-$

(238)

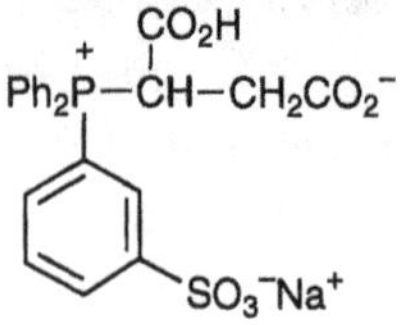

$Ph_2\overset{+}{P}-CH(CO_2H)-CH_2CO_2^-$ $SO_3^-Na^+$

(239)

$Ph_3\overset{+}{P}CH_2CHOHCH_2OH$ Br^-

(240)

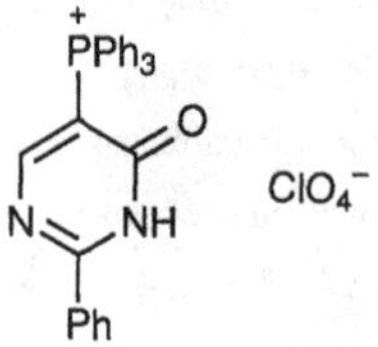

$\overset{+}{P}Ph_3$ O N NH Ph ClO_4^-

(241)

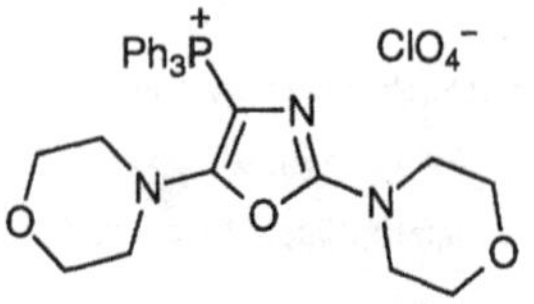

$Ph_3\overset{+}{P}$ N O N N O O ClO_4^-

(242)

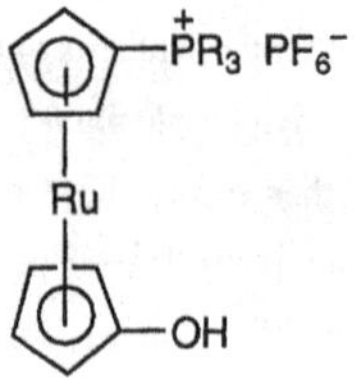

$\overset{+}{P}R_3$ PF_6^- Ru OH

(243)

$RCH-\overset{+}{P}Ph_3$ PF_6^- $Mn(CO)_3$

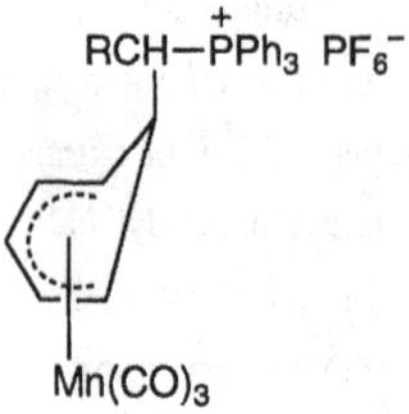

(244)

$Ph_3\overset{+}{P}H$ $GeCl_3^-$

(245)

$Bu_3\overset{+}{P}-\overset{-}{C}(F)-\overset{+}{P}Bu_3$ X^-

(246)

$Bu_3\overset{+}{P}-C(F)=C(R_F)-O^-$

(247)

$HC{\equiv}CCH_2\overset{+}{P}Ph_3$ Br^-

(248)

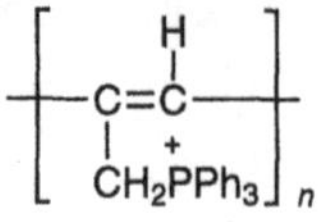

$[-C(CH_2\overset{+}{P}Ph_3)=C(H)-]_n$ Br^-

(249)

$Ph_3\overset{+}{P}CH_2C{\equiv}CPh$ Br^-

(250)

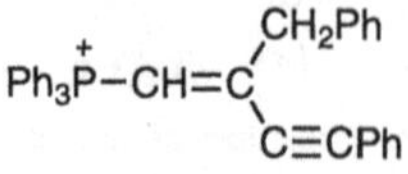

$Ph_3\overset{+}{P}-CH=C(CH_2Ph)(C{\equiv}CPh)$ Br^-

(251)

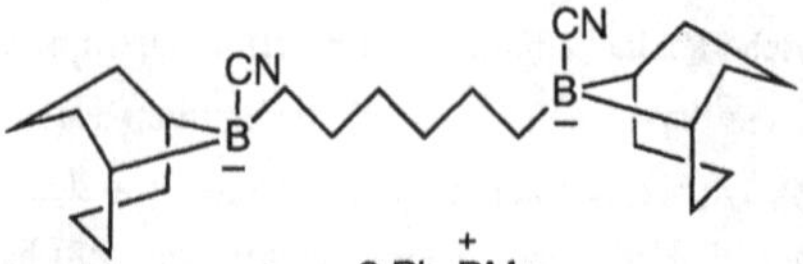

CN B CN B $2\ Ph_3\overset{+}{P}Me$

(252)

$Bu_4\overset{+}{P}$ $H_2F_3^-$

(253)

cyanoborate system (252) acts as a host towards furan, forming either a solid or a liquid clathrate, depending on water content. An X-ray study of the furan adduct has shown that the phosphonium group plays a key structural role in the design of the clathrate, and lends itself to future design modifications.[331] In the presence of N-halosuccinimides and the salt (253), alkenes are converted to the corresponding halofluorides.[332] The phosphonium anhydride reagent (254) aids the direct borohydride reduction of alcohols to the related hydrocarbon by facilitating the formation of an intermediate alkoxyphosphonium salt.[333] Photolysis of arylmethyltriphenylphosphonium salts results in the formation of a variety of products derived from both arylmethyl-radical and -cation intermediates, formed *via* different pathways, depending on the structure of the phosphonium ion, the nature of the counterion, and the solvent.[334] Both radical and ionic pathways have also been implicated in the course of photolysis of the arylmethylsulphanylphosphonium salts (255), which proceeds with predominant carbon-sulphur bond cleavage, and only minor sulphur-phosphorus cleavage.[335] The ferrocenylmethylphosphonium salt (256) undergoes photolysis in the presence of water, alcohols, or secondary amines to form triphenylphosphine oxide and a range of products derived from the ferrocenylmethyl carbocation.[336] The same group has also reported a solid state structural study of a ferrocenylphosphonium salt, the results of which indicate that the ferrocenyl group acts as a weakly electron-donating substituent.[337] An X-ray study of the structure of iodomethyltriphenylphosphonium iodide has also been reported, showing the presence of the monomeric system $Ph_3P^+CH_2I\ I^-$.[338] The betaine (257) forms an adduct with tribenzyltin chloride in which the oxygen bearing the negative charge becomes coordinated to the metal.[339] Interest has been shown in the extent of ion-pairing in phosphonium salts in both solution and solid states,[340] and, exceptionally, in the case of trimethylphosphonium bromide, in the vapour phase.[341] Fast atom bombardment and tandem mass spectrometric techniques have been developed for the structural characterisation of phosphonium salts.[342] More information on anomeric effects between sulphur and phosphorus has been obtained from conformational studies of the dithiane-phosphonium salts (258).[343] Two adducts of the phosphonium cation (259) with TCNQ have been characterised.[344]

5 p_π-Bonded Phosphorus Compounds

Progress over the past decade in the chemistry of p_π-p_π systems involving phosphorus, and the heavier group 14 and 15 elements, has been reviewed.[345] The new "push-pull" diphosphene (260), bearing the N-(9-fluorenyl)-N-mesitylamino group as an electron-rich stabilising group, has been prepared,[346] and structurally characterised.[347] Another variant on the usual diphosphene-stabilising groups is displayed in (261), in which the 2,4-di-*t*-butyl-6-(methoxymethyl)phenyl group may provide both steric and coordinative interactions.[348] Similar arguments may apply to the 2,4-di-*t*-butyl-6-methoxyphenyl group present in (262), which has been shown to combine with sulphur to form a mixture of the related thiadiphosphirane (263) and the diphosphene monosulphide (264).[349] The selenadiphosphirane (265) is formed in the reaction of the related diphosphene with elemental selenium.[350] The reaction of the aminoaryldiphosphene (266) with phosphorus pentachloride leads to the 1,2-dihalo(aryl)diphosphine (267), which undergoes a base-induced dehydrohalogenation to give the P-chloroazadiphosphiridine (268).[351] The diphosphene (269) is formed as a transient intermediate in the reaction of tris-*t*-butylcyclotriphosphine with di-*p*-tolylditelluride, which results in the formation of (270). When the reaction is conducted in the presence of 2,3-dimethylbutadiene as trapping agent, the adduct (271) is formed.[352] The chemistry of diphosphenes bearing complexed metallo substituents at phosphorus,[353-355] and the coordination chemistry of both stable and reactive diphosphenes,[356-358] has continued to develop.

A range of phospha-alkenes (272) stabilised by the presence of the fluoromesityl substituent has been prepared.[359] The tris-*t*-butylcyclopropenyl group has been shown to be effective as a stabilising substituent in a range of phospha-alkenes and 1-phospha-allenes, e.g., (273).[360,361] In a similar vein, a heavily substituted cyclopropyl group has been introduced as a stabilising substituent at carbon in the phospha-alkene system (274).[362] The diphospha[4]-radialene system (275) has been prepared and shown to have a planar and highly symmetrical structure. The ultraviolet spectrum of (275) indicates the presence of an extended π-system.[363] The first "free" 1-phospha-allyl anion has been prepared and structurally characterised in the form of its lithium salt (276).[364] Interest has continued in the generation and characterisation of short-lived phospha-alkenes and related systems. Both phospha-alkenes and phospha-alkynes have been characterised by photo-electron spectroscopy in the gas phase elimination of hydrogen chloride from alkyldichlorophosphines.[365] The base-induced rearrangement of vinylphosphines under high dilution conditions has given transient phospha-alkenes which have been trapped with dienes, thereby providing a useful route to cyclic phosphines, e.g., (277).[366] Base-induced dehydrochlorination of chloromethyl(vinyl)phosphine and 2-chloromethylvinylphosphine has enabled the characterisation of the phosphabutadienes (278) and (279), respectively.[367] The unstabilised 1-phospha-allene system (280) is formed in a similar way by dehydrochlorination of (1-chlorovinyl)methylphosphines.[368] The phospha-alkene (281) has been shown to behave as an enophile in its reactions with alkenes, which lead to the unsaturated phosphines (282).[369] The 1,2-dihydro-1,3-diphosphete system (283) is formed in a cycloaddition process between various phospha-alkenes and phospha-alkynes.[370] Both phospha-alkenes and phospha-alkynes have been shown to act as trapping partners for benzyne and its precursor, phenylium 2-carboxylate. Thus, e.g., in the presence of benzyne, the phospha-alkyne (284) forms the new phospha-alkene system

$(Ph_3\overset{+}{P})_2O$ 2 OTf^-

(254)

$ArCH_2S\overset{+}{P}Ph_3 X^-$

(255)

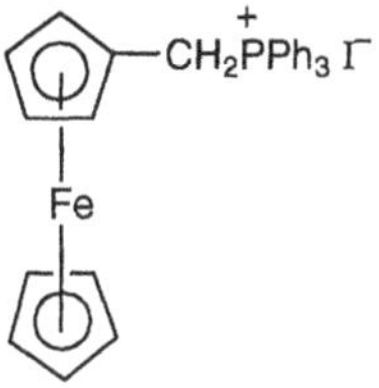

$CH_2\overset{+}{P}Ph_3 I^-$ Fe

(256)

$Ph_3\overset{+}{P}CH_2CH_2COO^-$

(257)

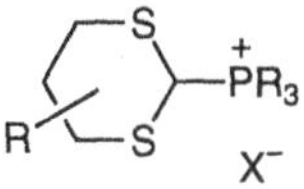

S S R $\overset{+}{P}R_3$ X^-

(258)

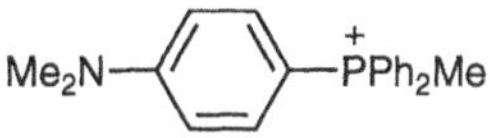

Me_2N $\overset{+}{P}Ph_2Me$

(259)

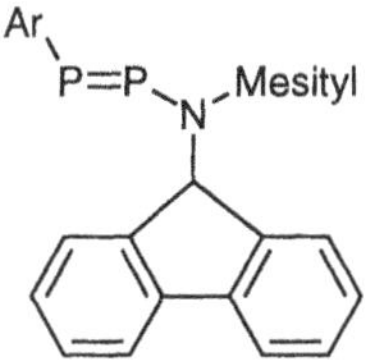

Ar P=P N Mesityl

(260)
Ar = 2,4,6-$(CF_3)_3C_6H_2$

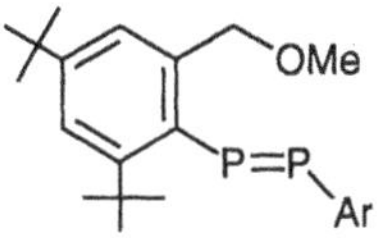

OMe P=P Ar

(261)
Ar = 2,4,6-$Bu^t_3C_6H_2$

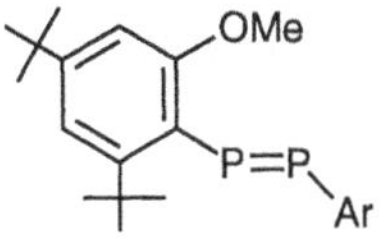

OMe P=P Ar

(262)
Ar = 2,4,6-$Bu^t_3C_6H_2$

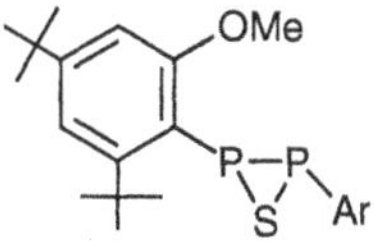

OMe P–P S Ar

(263)
Ar = 2,4,6-$Bu^t_3C_6H_2$

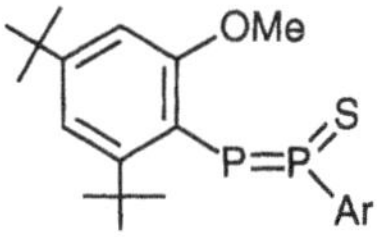

OMe P=P S Ar

(264)
Ar = 2,4,6-$Bu^t_3C_6H_2$

Se P P Ar Ar

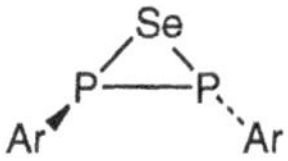

(265)
Ar = 2,4,6-$(CF_3)_3C_6H_2$

ArNHP=PAr

(266)
Ar = 2,4,6-$Bu^t_3C_6H_2$

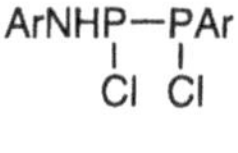

ArNHP–PAr
Cl Cl

(267)

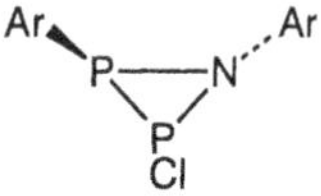

Ar P N Ar P Cl

(268)

$Bu^tP{=}PBu^t$

(269)

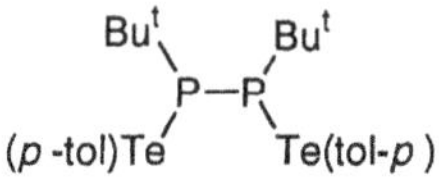

Bu^t Bu^t P–P (*p*-tol)Te Te(tol-*p*)

(270)

P Bu^t P Bu^t

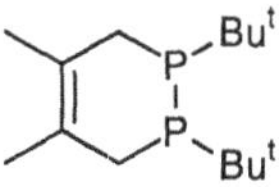

(271)

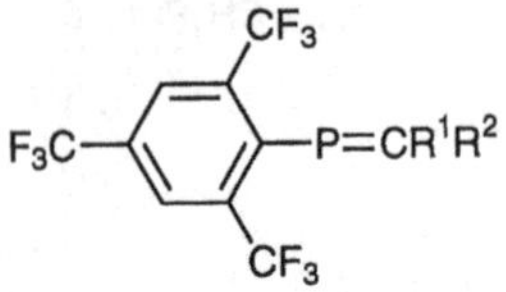

CF_3 F_3C P=CR^1R^2 CF_3

(272) R^1 = Cl, Ph, or $SiMe_3$
R^2 = H or Cl

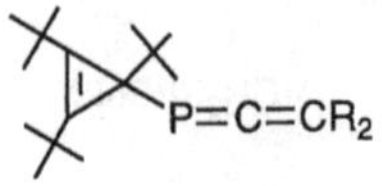

P=C=CR_2

(273) R = Ph or Mes

R^2 R^1 R^3 C=$PSiMe_3$ Me_3SiO

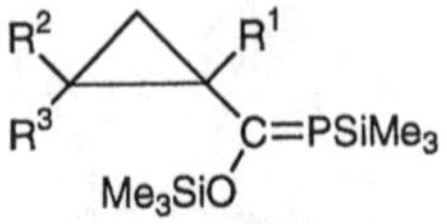

(274) R^1 = H or Me
R^2 = H, Me, or Cl
R^3 = H or Cl

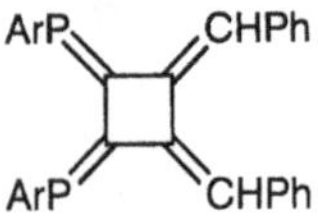

ArP CHPh ArP CHPh

(275)
Ar = 2,4,6-$Bu^t{}_3C_6H_2$

ArP Ph Li(dme)$^+$

(276)
Ar = 2,4,6-$Bu^t{}_3C_6H_2$

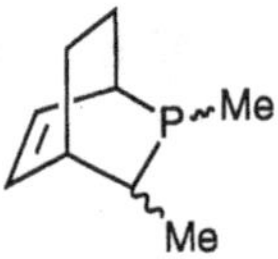

P Me Me

(277)

CH_2=P–CH=CH_2

(278)

CH_2=CH-CH=Ph

(279)

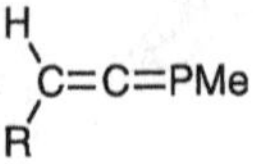

H C=C=PMe R

(280) R = H or Me

PhP=$C(SiMe_3)_2$

(281)

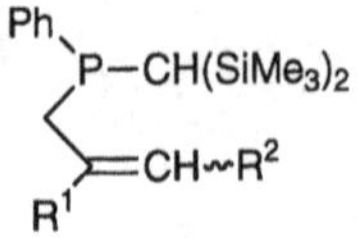

Ph P–$CH(SiMe_3)_2$ =CHR^2 R^1

(282) R^1 = CN, CO_2Me, Ph
alkyl, or OMe

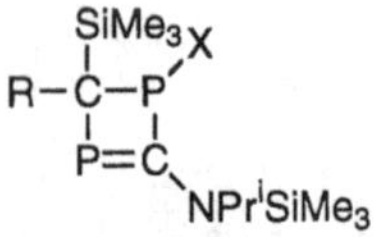

$SiMe_3$ X R–C–P P=C NPr^iSiMe_3

(283) R = $SiMe_3$ or Ph
X = Cl or Br

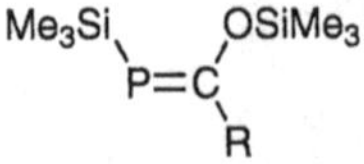

Me_3Si $OSiMe_3$ P=C R

(284)

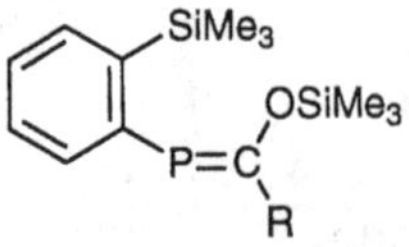

$SiMe_3$ $OSiMe_3$ P=C R

(285)

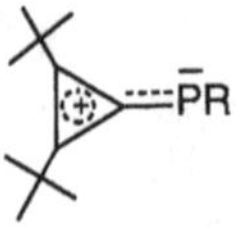

PR

(286)

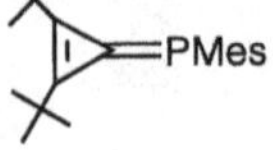

=PMes

(287)

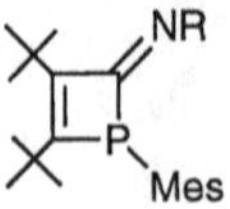

NR P Mes

(288)

R R– –P=CI_2 R

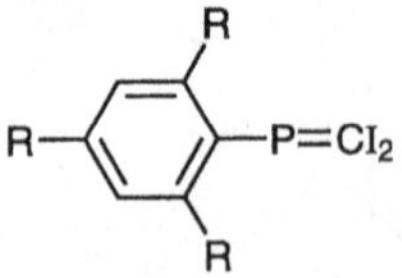

(289) R = Me, Et, Pr^i, or Bu^t

$Bu^t{}_2PHCF_2PHCF_3$

(290)

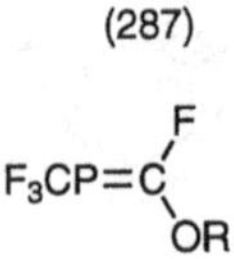

F F_3CP=C OR

(291)

–P=CHPh

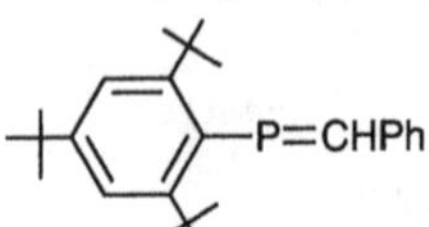

(292)

ArP=C=PAr

(293) Ar = 2,4,6-$Bu^t{}_3C_6H_2$

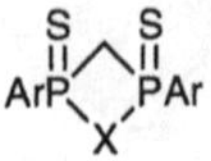

S S ArP PAr X

(294) X = O or S

(285).[371] Full details have appeared of studies of the inversion of the usual polarity of the P=C bond when the alkene carbon is incorporated into the cyclopropenium system, as in the phosphatriafulvenes (286). Reactions with nucleophilic reagents take place at the cyclopropenium centre.[372] Ring expansion occurs in the reaction of the mesitylphosphatriafulvene (287) with azides, with the formation of the 1H-2-iminophosphete system (288).[373] The stability, and reactivity towards bromine, of a range of C-iodophospha-alkenes (289) have been shown to depend on the size of the aryl substituents.[374] The diphosphine (290) is formed as a mixture of diastereo-isomers in the addition of *t*-butylphosphine to perfluoro-2-phosphapropene, $F_3CP=CF_2$.[375] The addition reactions of alcohols to the phospha-alkenes (291) have given rise to a variety of products, including novel trifluoromethylphosphines and phospha-alkenes.[376] Irradiation of single crystals of the phospha-alkene (292) with X-rays have been shown to generate phosphorus-centred radicals.[377] Treatment of the diphospha-allene (293) with sulphur under wet conditions has given rise to a mixture of isomers of the heterocyclic systems (294).[378] Electrochemical oxidation of phosphabutatrienes bearing *p*-dimethylaminoaryl substituents at the terminal carbon results in formation of the P-P coupling products (295).[379] A route to the methylenephosphonium salt (296) has been developed from the reaction of a phosphinocarbene with triethylboron.[380] The methylenephosphonium salt (297) has been shown to exist in equilibrium with the fused four-membered ring system (298). In the presence of pyridine, the latter rearranges to the benzo-phosphetanium salt (299).[381] The methylenediphosphenium cation (300) has been stabilised in the form of a complex (301) with triphenylphosphine.[382] C-phosphinophospha-alkenes (302) are formed in the thermal ring-opening of diphosphiranes, a theoretical study suggesting the intermediacy of diradical species.[383] Theoretical treatments of a wide range of phospha-alkene and related systems have also appeared.[384-391] Weber's group has continued to explore the reactivity of phospha-alkenes bearing a complexed metallo-substituent at phosphorus.[392-394] Interest has also continued in the coordination chemistry of phospha-alkenes and related systems.[395-400] The reactivity of coordinated phospha-alkenes has also received attention.[401] The mechanism of the transition metal-promoted conversion of phospha-alkenes to phospha-alkynes has received attention.[402,403]

The novel phospha-alkyne system (303) is formed in the reaction of lithium bis(trimethylsilyl)phosphide with O,O'-diethyldithiocarbamate.[404] C-chlorophosphaethyne, ClC≡P, has been generated by the pyrolysis of trichloromethyldichlorophosphine over granulated zinc at 550°C, and characterised by infrared spectroscopy.[405] A bimolecular proton transfer mechanism has been suggested for the base-promoted isomerism of alkynyl- and alkenyl-phosphines to the phospha-alkynes (304).[406] The potential of phospha-alkynes as novel building blocks in heterocyclic chemistry has been reviewed.[407] New examples of phospha-alkyne-derived heterocyclic systems include the 14π-annulene (305)[408] and the diphosphatricyclobenzoheptane (306).[409] The bicyclic system (307) has been isolated from the reactions of phospha-alkynes with a polyfluorinated cyclohexadienone.[410] A product in which two B_{10} units are linked by a Bu^{t}C-PH bridge has been obtained from the reaction of decaborane with Bu^{t}C$\equiv$P.[411] Detailed structural studies of phospha-alkynes bearing bulky dialkylamino substituents at carbon have been reported.[412,413] The coordination chemistry of phospha-alkynes continues to develop with the discovery of further modes of coordination.[414-417] Dimerisation and even trimerisation of the phospha-alkyne Bu^{t}C$\equiv$P in the coordination sphere of a metal occurs to form 1,3-diphospha-cyclobutadiene and related complexes.[418-420] Metal complexes involving bridging cyaphide (C$\equiv$P$^-$) and arylisocyaphide (ArP$\equiv$C) ligands have also been characterised.[421,422] A theoretical treatment of bonding in the phospha-ethyne cation HCP^+ has appeared.[423]

The chemistry of iminophosphenes also continues to attract attention. An unexpected thermodynamic preference for a two-coordinate R-P=N-R system over a three coordinate aminophosphine is revealed in the spontaneous elimination of a secondary amine from (308) to form the iminophosphene system (309).[424] A range of new, sterically crowded iminophosphenes (310) has been prepared by established methods. Depending on the steric demand of the substituents, the latter undergo a reversible [2+1] cyclodimerisation to form the azadiphosphiridine imines (311).[425] Both phosphirenimines (312) and diphosphetenes (313) have been isolated from the cycloaddition reactions of iminophosphenes with alkynes.[426,427] A series of cycloadducts of fluorinated ketones and a silylated iminophosphene has also been characterised.[428] An iminophosphene bearing a diazaphospholyl substituent at phosphorus (314) has been prepared.[429] Substitution reactions of a P-chloro iminophosphene with alkoxide reagents have received attention.[430] The conformation of iminophosphenes in solution has been explored by polarizability studies.[431] The chemistry of the aryliminophosphenium cation (315) has also continued to develop; a series of cationic P,N-heterocyclic systems has been prepared therefrom by [n+2] cycloaddition processes,[432] and curious π-complexes with benzene and toluene characterised.[433]

A theoretical comparison of the B=N and B=P π-systems supports earlier findings that the latter bond is stronger, and a genuine p_π-p_π interaction.[434] Isocyanates have been shown to insert into the P=Si bond of phospha-silenes to form the phospha-alkene systems (316). A macrocyclic system has been isolated from the related reaction with 1,6-di-isocyanohexane.[435] A range of resonance-stabilised monomeric thioxophosphenes (317) has been prepared.[436] Methods for the generation of the phosphenite (318) have also been explored.[437]

Interest continues in the development of the chemistry of phosphinidenes and phosphenium salts, and also that of their metal complexes. A molecular orbital study of the

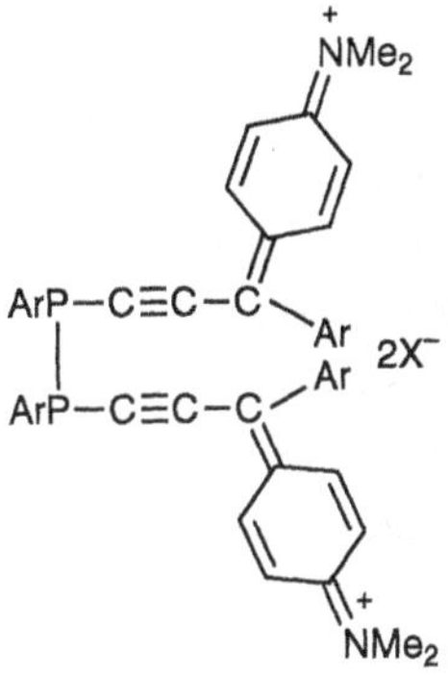

(295)

$(R_2N)_2\overset{+}{P}{=}C(SiMe_3)_2$ $CF_3SO_3^-$

(296)

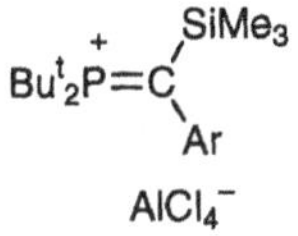

$Bu^t_2\overset{+}{P}{=}C(SiMe_3)Ar$ $AlCl_4^-$

(297)

$AlCl_4^-$

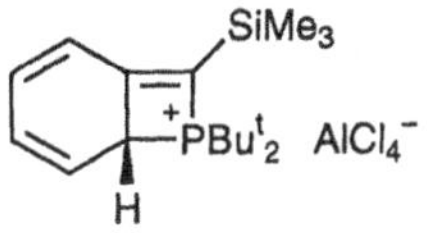

(298)

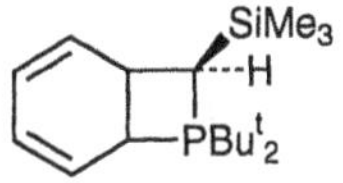

(299)

$[P{=}C(SiMe_3)_2]^+$

(300)

$[Ph_3P{\rightarrow}P{=}C(SiMe_3)_2]^+$ $AlCl_4^-$

(301)

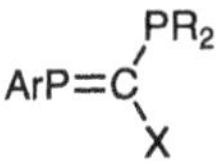

$ArP{=}C(PR_2)X$

(302) Ar = 2,4,6-$Bu^t_3C_6H_2$
or $C(SiMe_3)_2$
X = Cl, Me, or Ph

$Li^+[P{\equiv}C{-}S]^-$

(303)

$RCH_2C{\equiv}P$

(304)

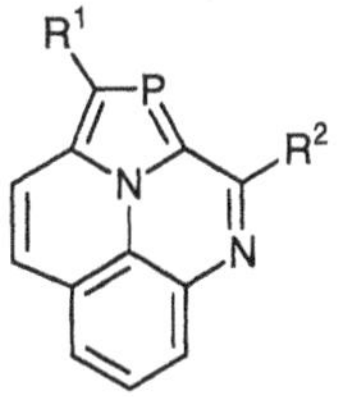

(305) R^1 = Bu^t, adamantyl,
or CMe_2Et
R^2 = Me, Ph, 2-naphthyl, or Bu^t

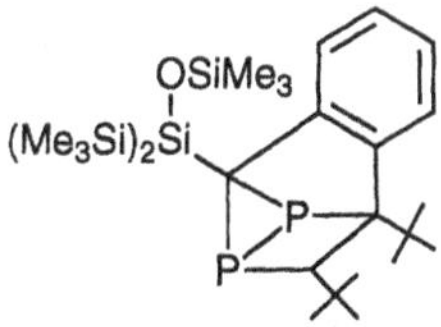

(306)

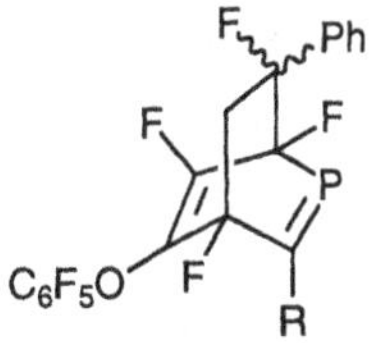

(307) R = Pr^i_2N or Bu^t

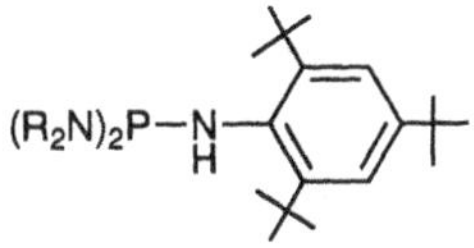

(308) R = Me, Et, Pr^i
or R_2N = piperidyl

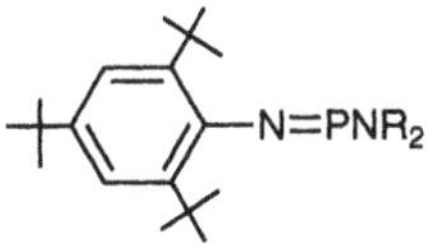

(309) R = Me, Et, Pr^i
or R_2N = piperidyl

vinylphosphinidene to phosphapropyne rearrangement has appeared.[438] Metal complexes of phosphinidenes undergo cycloaddition reactions with both phospha-alkenes and alkenes, to form uncomplexed diphosphiranes, e.g., (319)[439] and complexed phosphiranes (320),[440] respectively. The intermediacy of metal phosphinidene complexes is suspected in the reactions of zirconium and hafnium organophosphido systems.[441,442] The potential of tantalum(V) phosphinidene complexes as phospha-Wittig reagents has been reviewed.[443] Further evidence of the dependence of the stability of phosphenium salts on the nature of the anion has appeared.[444] Transfer of substituents from zirconium to phosphorus occurs in the reactions of zirconocene derivatives with the phosphenium salt (321).[445] The reactivity of metal-complexed phosphenium ions continues to be a fruitful area of study.[446-450]

Considerable progress has been made in the stabilisation of σ^3-λ^5-p_π-bonded systems by intramolecular coordination. Treatment of the dichlorophosphine (322) with lithium aluminium hydride followed by elemental sulphur gives the disulphur system (323), which is stable for several days in non-polar solvents, but dimerises in the solid state. Such species have long been postulated as intermediates in sulphurization reactions using Lawesson's reagents, and, indeed, it has now been demonstrated that (323) will convert a carbonyl compound into its thiocarbonyl equivalent.[451] Other coordinatively stabilised systems discovered by Yoshifuji's group are (324),[452] (325),[453] and (326). Treatment of the latter with sulphur in the presence of triethylamine gives the mixed system (327), which on subsequent treatment with tris(dimethyl-amino)phosphine gives the thermally stable λ^2-system (328).[454] New nitrogen-containing systems include (329)[455] and (330).[456] Quin's group has reported further studies of the generation of metaphosphates and related species, (331), by fragmentation reactions of bicyclic systems.[457-459]

6 Phosphirenes, Phospholes, and Phosphinines

Cycloaddition of an aryl(amino)carbene-molybdenum complex with the phospha-alkene $ClP{=}C(SiMe_3)_2$ provides a route to the complexed lH-l-aza-2-phosphirene system (332).[460] On treatment with lithium di-isopropylamide, the diphosphirenium salt (333) undergoes ring-opening to form the phospha-alkene (334). With *t*-butylisocyanide, (333) affords the four-membered ring-system (335).[461] Ring-expansion only occurs in the reactions of the phosphirenes (336) with phenyldichloroarsine which affords the arsaphosphetene system (337). Treatment of the latter with tributylphosphine gives the neutral system (338). This suffers ring-opening in the presence of lithium in THF to form the reagent (339), which, on sequential treatment with carbon disulphide, iodomethane, and further lithium metal affords the 1,3-arsaphospholide system (340), from which a diarsadiphosphaferrocene has been prepared.[462] The phosphirene complex (341)

$R^1—P=N—R^2$

(310) R^1 = CMe_2Et, $CMeEt_2$, or CEt_3
R^2 = 2,4,6-$Bu^t_3C_6H_2$, Et_3C, or PhCH=CH—

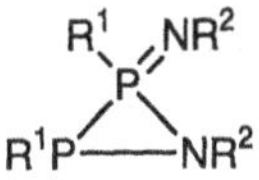

(311)

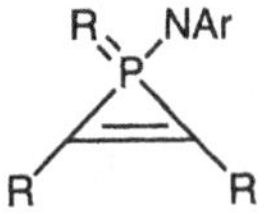

(312)

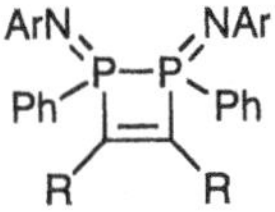

(313)

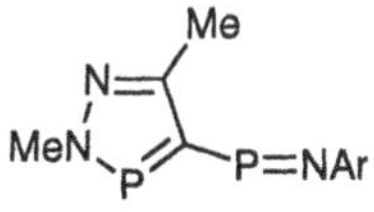

(314) Ar = 2,4,6-$Bu^t_3C_6H_2$

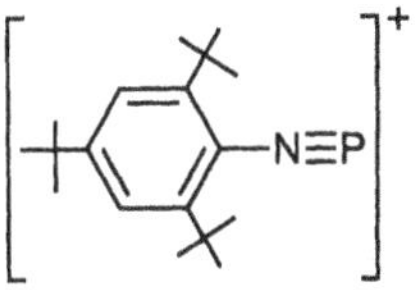

(315)

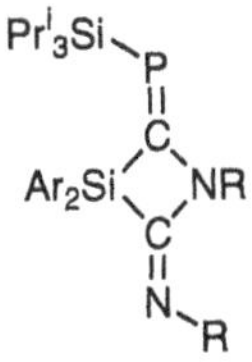

(316) Ar = 2,4,6-$Pr^i_3C_6H_2$
R = Cy or Mes

(317) R = Me, Et, or Ar

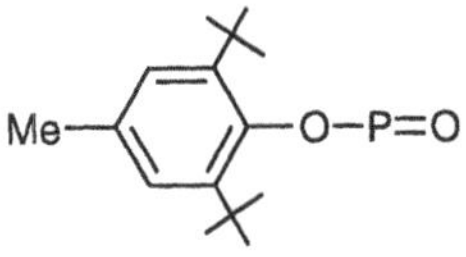

(318)

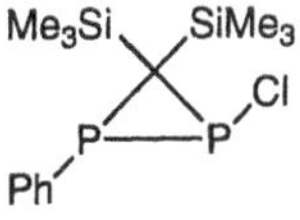

(319)

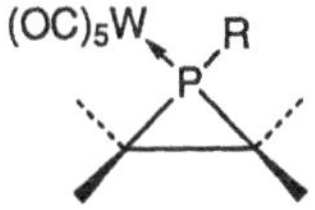

(320)

(321) $(Pr^i_2N)_2\overset{+}{P}$ $CF_3SO_3^-$

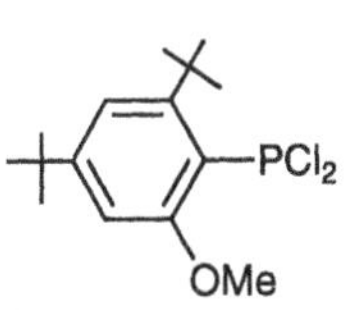

(322)

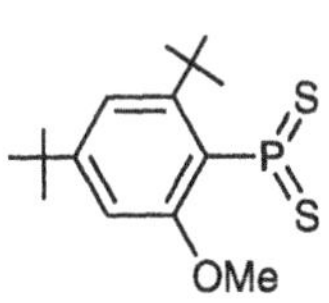

(323)

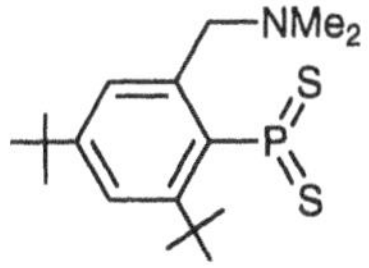

(324)

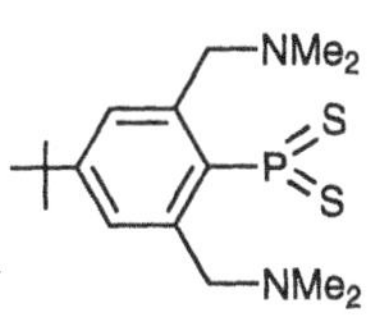

(325)

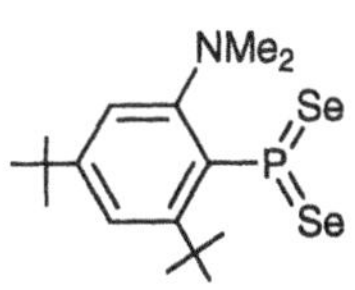

(326)

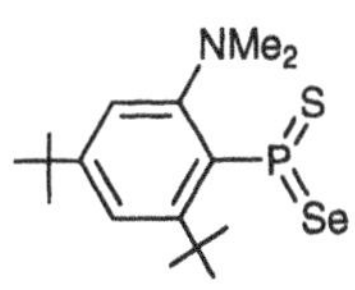

(327)

also undergoes ring-expansion when treated with ethyl propiolate in the presence of a palladium(0) complex to form the functionalised phosphole (342), which can be decomplexed on treatment with "diphos". The phosphorus substituent is cleaved in the presence of lithium di-isopropylamide to form the related lithiophospholide (343), which is capable of further elaboration.[463] *Sec*-butyllithium selectively deprotonates one of the methyl groups of the phosphole-borane adduct (344) to give an allylic anion which reacts with electrophiles at *either* the α-endocyclic *or* γ-exocyclic position, thereby providing a route to functionalised phosphole systems.[464] The dibromophosphole (345) is a key intermediate for the synthesis of α-linked quaterphospholes and precursor biphospholes.[465] Improved conditions for the generation of the phospholide reagent (346) have been developed as part of a route to new chiral diphospholes, e.g. (347).[466,467] Previously reported phosphole tetramers involving P,P-bonds are cleaved on treatment with sodium naphthalenide to give phospholide reagents, e.g. (348), used subsequently in reactions with α,ω-dihaloalkanes to form macrocyclic systems.[468] The synthesis of the benzophosphole system has attracted considerable attention in the past year. Treatment of the readily accessible zirconium reagent (349) with phosphorus trichloride yields the P-halofunctional reagent (350), readily converted to the related metallobenzophospholide reagent which has been used in the synthesis of a range of new systems.[469] A simple route to the lithiobenzophospholides (351) is afforded by the reaction of diphenylacetylene with monolithioaryl phosphides.[470] Two closely related routes to the benzophosphole (352) have been developed.[471,472] Treatment of 2,2'-dilithio-1,1'-binaphthyl with organodichlorophosphines provides a route to a series of dinaphthophospholes (353), which are only atropisomerically stable below room temperature.[473]

A theoretical study of [1,5]-sigmatropic H-shifts in cyclopentadiene, pyrrole, and phosphole has revealed a low energy barrier for the 1H-phosphole → 2H-phosphole rearrangement.[474] The same group has also compared the relative energies of isomeric methyl-(and vinyl-)phospholes, and showed that substitution is preferred at phosphorus rather than ring carbon, thereby minimising steric interactions.[475] The position of equilibrium between 1H- and 2H-phospholes has been studied using diphenylacetylene as a selective trapping agent for the 2H-species. This study has shown that 1-phenylphosphole equilibrates with 5-phenyl-2H-phosphole at 150°C.[476] The [4+2]-dimer (354) of 3,4-dimethyl-2H-phosphole has been shown to react with maleic anhydride to give the adduct (355) as a single isomer. Subsequent alkaline hydrolysis of the latter affords the new highly-water soluble phosphine (356).[477] A review of ring-expansion reactions leading to six- and seven-membered ring heterocyclic phosphorus systems is much concerned with the chemistry of phospholes.[478] The adducts of 1,2,5-triphenylphosphole with dimethyl acetylenedicarboxylate have been reinvestigated by X-ray cystallographic techniques; an earlier spectroscopic structural assignment by Tebby's group was confirmed, but others

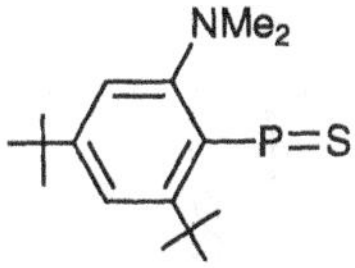

NMe_2 P=S

(328)

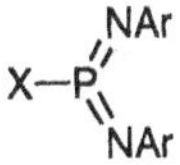

NAr X—P NAr

(329) X = halogen
Ar = 2,4,6-$Bu^t{}_3C_6H_2$

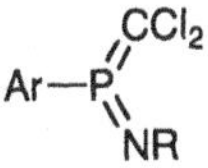

CCl_2 Ar—P NR

(330)

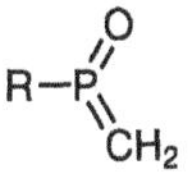

O R—P CH_2

(331) R = Ph, Me, or OR

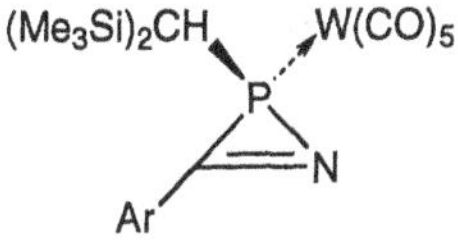

$(Me_3Si)_2CH$ $W(CO)_5$ P N Ar

(332)

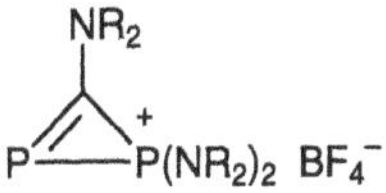

NR_2 P $\overset{+}{P}(NR_2)_2$ BF_4^-

(333) R = Pr^i

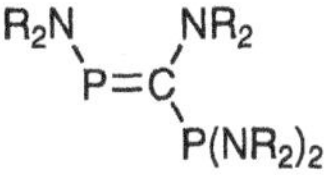

R_2N NR_2 P=C $P(NR_2)_2$

(334) R = Pr^i

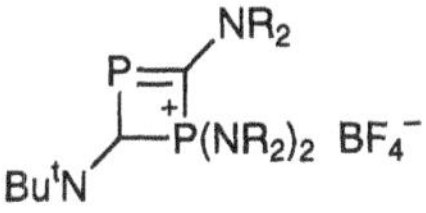

NR_2 P $\overset{+}{P}(NR_2)_2$ BF_4^- Bu^tN

(335)

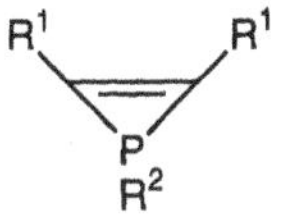

R^1 R^1 P R^2

(336) $R^1 = R^2 = Ph$
$R^1 = Ph$; $R^2 = Me$
$R^1 = Et$; $R^2 = Ph$

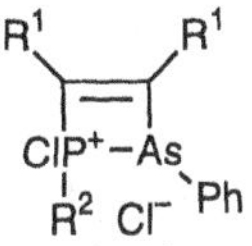

R^1 R^1 ClP^+—As R^2 Cl^- Ph

(337)

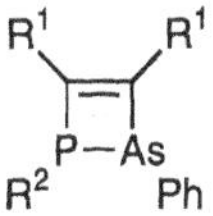

R^1 R^1 P—As R^2 Ph

(338)

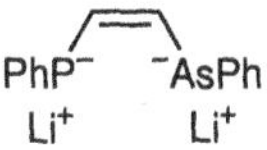

PhP^- ^-AsPh Li^+ Li^+

(339)

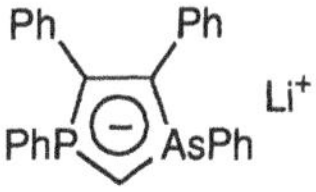

Ph Ph PhP AsPh Li^+

(340)

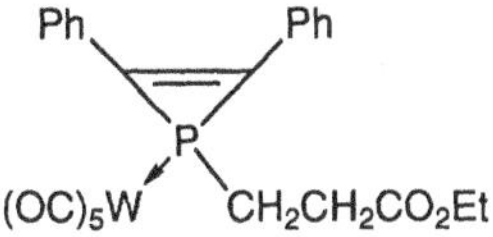

Ph Ph P $(OC)_5W$ $CH_2CH_2CO_2Et$

(341)

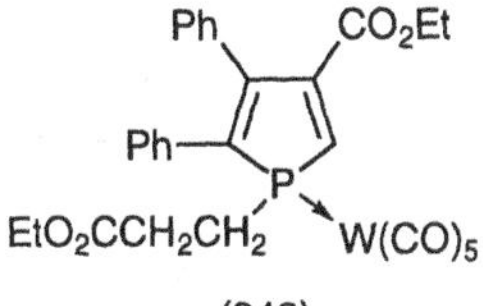

Ph CO_2Et Ph P $EtO_2CCH_2CH_2$ $W(CO)_5$

(342)

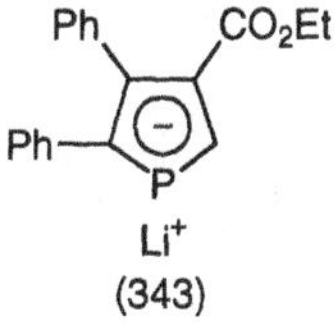

Ph CO_2Et Ph P Li^+

(343)

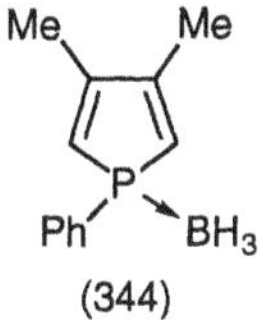

Me Me P Ph BH_3

(344)

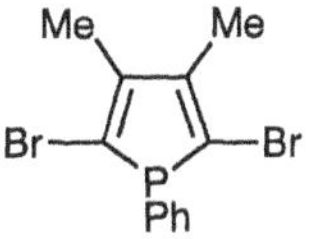

Me Me Br Br P Ph

(345)

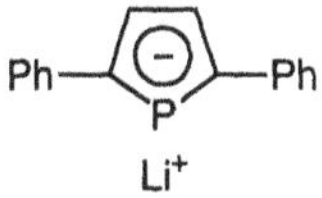

Ph P Ph Li^+

(346)

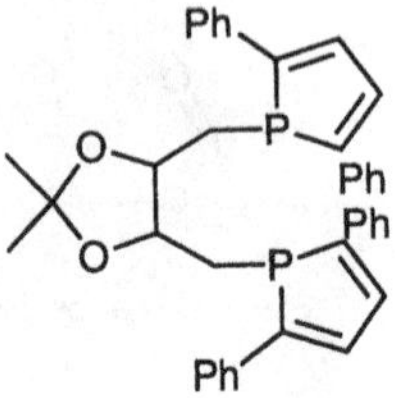

(347)

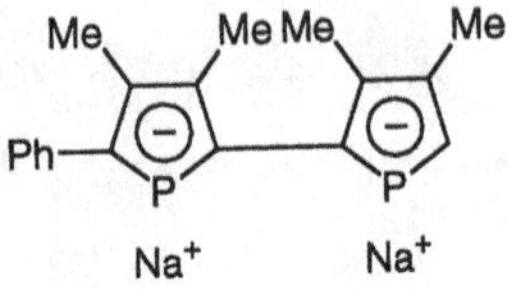

(348)

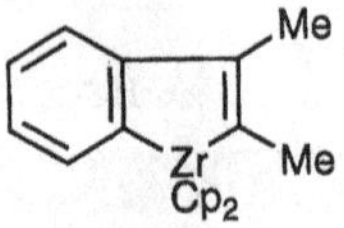

(349)

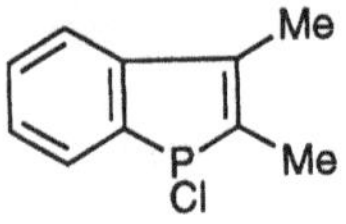

(350)

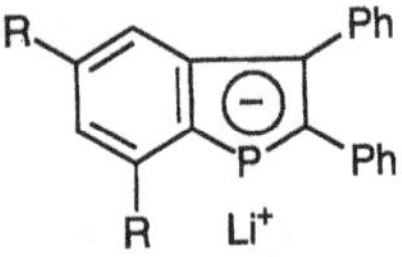

(351) R = Bu^t or H

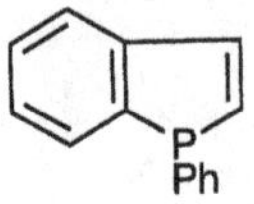

(352)

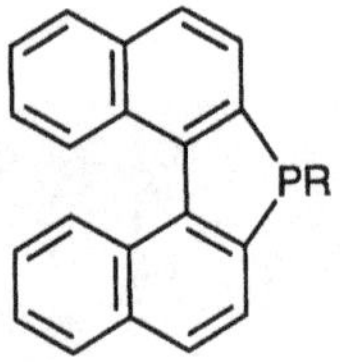

(353) R = H, Me, Ph, Et, CH_2Ph, $CH_2CH(Me)Et$, or neomenthyl

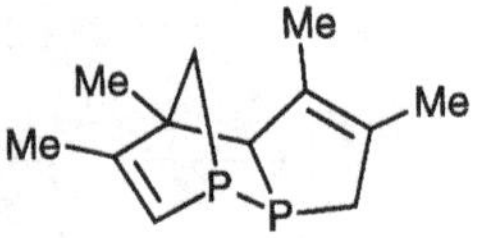

(354)

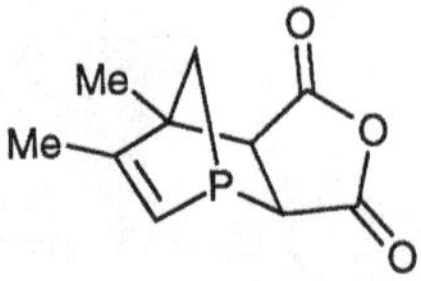

(355)

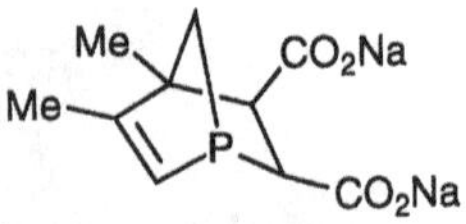

(356)

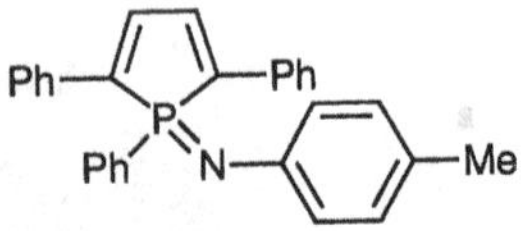

(357)

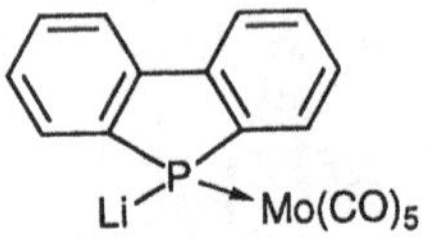

(358)

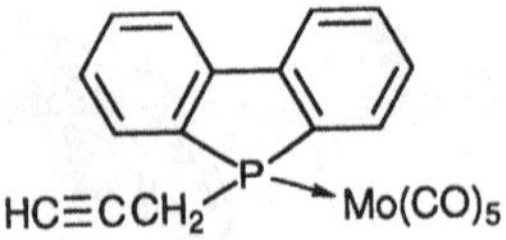

(359)

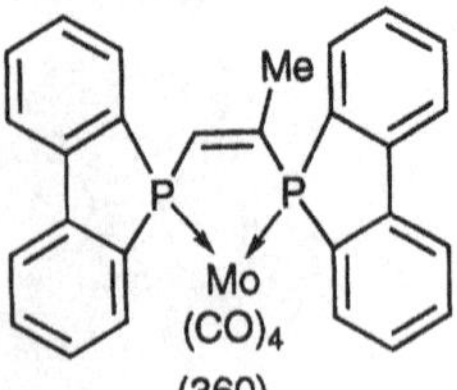

(360)

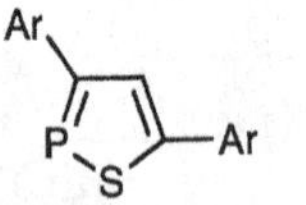

(361)

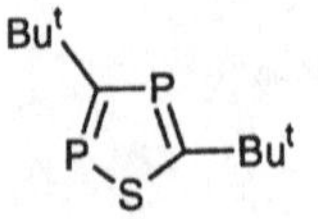

(362)

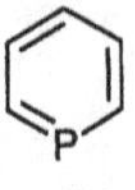

(363)

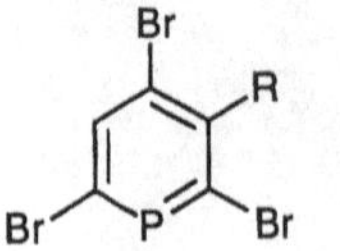

(364) R = H or Me

needed revision, and a new product was also characterised.[479] The aryliminophosphole (357) has been obtained from the reaction of 1,2,5-triphenylphosphole with *p*-tolyl azide.[480] The reactions of the complexed lithiodibenzophosphole (358) with various electrophilic reagents have been studied. With propargyl bromide, the related complexed propargyldibenzophosphole (359) is formed. On heating in pyridine, this is transformed into the complexed bis(dibenzophosphole) system (360).[481]

Interest in heterophospholes has continued. Routes to the 1,2-thiaphosphole (361),[482] and the 1-thia-2,4-diphosphole (362)[483,484] have been reported. New routes to polyazaphospholes have also been explored, and their reactivity studied.[485-489] The coordination chemistry of polyphospholide anions has also continued to receive attention.[490-493] Interest has been maintained in the coordination chemistry of simple phospholes and related phospholide anions.[494-496] Complexes of 1-substituted-3,4-dimethylphospholes with platinum(II) acceptors undergo thermal dimerisation of the phosphole ligands *either* in solution at 145°C *or* in the solid state at 140°C, and three product-types have been characterised by X-ray crystallographic techniques.[497]

The parent phosphinine (363) is formed in 40% yield by the flash vacuum thermolysis of vinyldiallylphosphine at 700°C/10^{-3} mm. Under the same conditions, triallylphosphine gives rise to the parent phospha-alkyne HC≡P.[498] A theoretical treatment of (363) has predicted bond lengths at variance with those deduced from a combined electron diffraction and microwave study.[499] Several papers have appeared describing studies of the reactivity of halogeno-phosphinines. Treatment of the tribromophosphinines (364) with heteroaryltrimethylstannanes in the presence of palladium(0) results in the introduction of heteroaryl substituents in the 2,6-positions, as in (365). Related reactions have also been employed for the synthesis of phosphinines bearing phosphino- and alkynyl-substituents.[500] The insertion of zirconocene into the carbon-halogen bond of 2-halogenophosphinines, followed by treatment with electrophiles, also provides a route to 2-functionalised phosphinines.[501] A new route to 2,2'-biphosphinines (366) is provided by the reactions of 2-halogenophosphinines with lithium tetramethyl-piperidide.[502] A ring expansion route from 1H-phosphole oxides has been developed for the synthesis of 1-alkoxy-di- and tetra-hydrophosphinine-1-oxides, e.g. (367).[503] Interest in the coordination chemistry of phosphinines has continued, and evidence that the phosphinine system is a powerful π-acceptor is mounting.[504-506] In the λ^5-phosphorin area, the new chelating diphosphine derivative (368) has been prepared.[507] In air, it undergoes oxidation to form the related dioxide (369).[508]

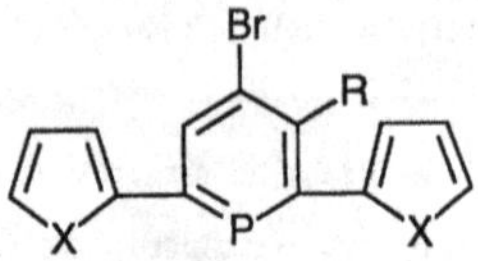

Br
R
X P X

(365) X = O, S, or NMe

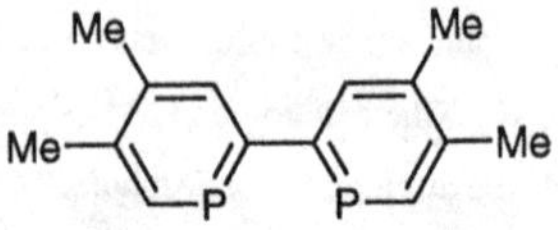

Me Me
Me Me
P P

(366)

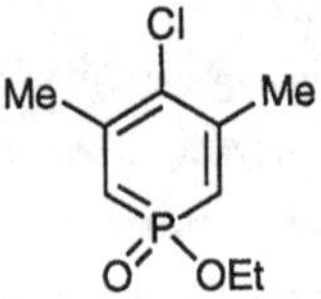

Cl
Me Me
P
O OEt

(367)

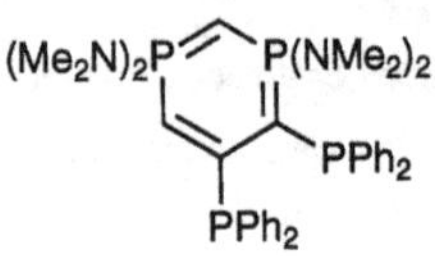

$(Me_2N)_2P$ $P(NMe_2)_2$
PPh_2
PPh_2

(368)

$(Me_2N)_2P$ $P(NMe_2)_2$
$P(O)Ph_2$
$P(O)Ph_2$

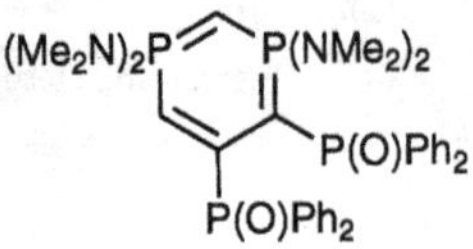

(369)

References

1 C. Chuit, R. J. P. Corriu, P. Monforte, C. Reyé, J-P. Declercq, and A. Dubourg, *Angew. Chem., Int. Ed. Engl.*, 1993, **32**, 1430.
2 R. D. Jackson, S. James, A. G. Orpen, and P. G. Pringle, *J. Organomet. Chem.*, 1993, **458**, C3.
3 S. Yasuike, H. Ohta, S. Shiratori, J. Kurita, and T. Tsuchiya, *J. Chem. Soc., Chem. Commun.*, 1993, 1817.
4 S. Gladiali, A. Dore, D. Fabbri, O. De Lucchi, and M. Manassero, *Tetrahedron: Asymmetry*, 1994, **5**, 511.
5 D. Matt, N. Sutter-Beydoun, J-P. Brunette, F. Balegroune, and D. Grandjean, *Inorg. Chem.*, 1993, **32**, 3488.
6 J. Andrieu, P. Braunstein, and A. D. Burrows, *J. Chem. Res.* (S), 1993, 380.
7 K. A. Richie, D. K. Bowsher, J. Popinski, and G. M. Gray, *Polyhedron*, 1994, **13**, 227.
8 G. Brenchley, E. Merifield, M. Wills, and M. Fedouloff, *Tetrahedron Lett.*, 1994, **35**, 2791.
9 H. Brunner, J. Fürst, and J. Ziegler, *J. Organomet. Chem.*, 1993, **454**, 87.
10 H. Brunner and J. Fürst, *Tetrahedron*, 1994, **50**, 4303.
11 S. D. Perera, B. L. Shaw, and M. Thornton-Pett, *J. Chem. Soc., Dalton Trans.*, 1994, 713.
12 T. Hayashi, A. Ohno, S. Lu, Y. Matsumoto, E. Fukuyo, and K. Yanagi, *J. Am. Chem. Soc.*, 1994, **116**, 4221.
13 B. Jedlicka, C. Kratky, W. Weissensteiner, and M. Widhalm, *J. Chem. Soc., Chem. Commun.*, 1993, 1329.
14 I. R. Butler, L. J. Hobson, S. M. E. Macan, and D. J. Williams, *Polyhedron*, 1993, **12**, 1901.
15 J. Krill, I. V. Shevchenko, A. Fischer, P. G. Jones, and R. Schmutzler, *Chem. Ber.*, 1993, **126**, 2379.
16 A. Bader, M. Pabel and S. B. Wild, *J. Chem. Soc., Chem. Commun.*, 1994, 1405.
17 H. J. Bestmann, D. Hadawi, H. Behl, M. Bremer, and F. Hampel, *Angew. Chem., Int. Ed. Engl.*, 1993, **32**, 1205.
18 A. Herbowski and E. A. Deutsch, *J. Organometallic Chem.*, 1993, **460**, 19.
19 A. Nikitidis and C. Andersson, *Phosphorus, Sulfur, Silicon, Related Elem.*, 1993, **78**, 141.
20 W. Keller, B. A. Barnum, J. W. Bausch, and L. G. Sneddon, *Inorg. Chem.*, 1993, **32**, 5058.
21 A. Schmidpeter and G. Jochem, *Z. Naturforsch., B: Chem. Sci.*, 1993, **48**, 93.
22 F. Uhlig, W. Uhlig, and M. Dargatz, *Phosphorus, Sulfur, Silicon, Related Elem.*, 1993, **84**, 181; F. Uhlig, B. Puschner, E. Herrmann, B. Zobel, H. Bernhardt, and W. Uhlig, *Phosphorus, Sulfur, Silicon, Related Elem.*, 1993, **81**, 155.
23 E. C. Ashby, R. Gurumurthy, and R. W. Ridlehuber, *J. Org. Chem.*, 1993, **58**, 5832.
24 E. C. Ashby, X. Sun, and J-L. Duff, *J. Org. Chem.*, 1994, **59**, 1270.
25 M. A. Nazareno, S. M. Palacios, and R. A. Rossi, *J. Phys. Org. Chem.*, 1993, **6**, 421.
26 P. G. Manzo, S. M. Palacios, and R. A. Alonso, *Tetrahedron Lett.*, 1994, **35**, 677.
27 G. Becker, B. Eschbach, D. Käshammer, and O. Mundt, *Z. Anorg. Allg. Chem.*, 1994, **620**, 29.
28 M. Driess, U. Winkler, W. Imhof, L. Zsolnai, and G. Huttner, *Chem. Ber.*, 1994, **127**, 1031.
29 M. Westerhausen and W. Schwarz, *Z. Anorg. Allg. Chem.*, 1994, **620**, 304.

30 G. Stieglitz, B. Neumueller, and K. Dehnicke, *Z. Naturforsch., B: Chem. Sci.*, 1993, **48**, 27.
31 M. J. Burk, J. E. Feaster, W. A. Nugent, and R. L. Harlow, *J. Am. Chem. Soc.*, 1993, **115**, 10125.
32 A. Börner, J. Holz, J. Ward, and H. B. Kagan, *J. Org. Chem.*, 1993, **58**, 6814.
33 M. Alvarez, N. Lugan, and R. Mathieu, *Inorg. Chem.*, 1993, **32**, 5652.
34 B. C. Janssen, A. Asam. G. Huttner, V. Sernau, and L. Zsolnai, *Chem. Ber.*, 1994, **127**, 501.
35 A. Muth, A. Asam, G. Huttner, A. Barth, and L. Zsolnai, *Chem. Ber.*, 1994, **127**, 305.
36 I. P. Romanova, S. G. Fattakhov, and V. S. Reznik, *Zh. Obshch. Khim.*, 1993, **63**, 1567 (*Chem. Abstr.*, 1994, **120**, 191 835).
37 S. Y. M. Chooi, S-Y. Siah, P-H. Leung, and K. F. Mok, *Inorg. Chem.*, 1993, **32**, 4812.
38 K. Diemert, T. Hahn, W. Kuchen, and P. Tommes, *Phosphorus, Sulfur, Silicon, Related Elem.*, 1993, **83**, 65.
39 R. T. Kettenbach, W. Bonrath, and H. Butenschön, *Chem. Ber.*, 1993, **126**, 1657.
40 A. Marinetti, F. Mathey, and L. Ricard, *Organometallics*, 1993, **12**, 1207.
41 Y. B. Kang, M. Papel, A. C. Willis, and S. B. Wild, *J. Chem. Soc., Chem. Commun.*, 1994, 475.
42 M. Widhalm, and G. Klintschar, *Tetrahedron: Asymmetry*, 1994, **5**, 189.
43 M. Widhalm, H. Kalchhauser, and H. Kählig, *Helv. Chim. Acta*, 1994, **77**, 409.
44 B. R. Kimpton, W. McFarlane, A. S. Muir, and P. G. Patel, *Polyhedron*, 1993, **12**, 2525.
45 G. A. Bowmaker and J. P. Williams, *Aust. J. Chem.*, 1994, **47**, 451.
46 N. Gabbitas, G. Salem, M. Sterns, and A. C. Willis, *J. Chem. Soc., Dalton Trans.*, 1993, 3271.
47 R. J. Doyle, G. Salem, and A. C. Willis, *J. Chem. Soc., Chem. Commun.*, 1994, 1587.
48 J. S. Field, R. J. Haines, C. J. Parry, and S. H. Sookraj, *S. Afr. J. Chem.*, 1993, **46**, 70; *Polyhedron*, 1993, **12**, 2425.
49 I. Kovacs and G. Fritz, *Z. Anorg. Allg. Chem.*, 1994, **620**, 1.
50 B. Demerseman, R. Le Lagadec, B. Guilbert, C. Renouard, P. Crochet, and P. H. Dixneuf, *Organometallics*, 1994, **13**, 2269.
51 J. R. Goerlich and R. Schmutzler, *Z. Anorg. Allg. Chem.*, 1994, **620**, 173.
52 N. R. Champness, W. Levason, R. D. Oldroyd, and D. J. Gulliver, *J. Organomet. Chem.*, 1994, **465**, 275.
53 J. G. Walsh, P. J. Furlong, and D. G. Gilheany, *J. Chem. Soc., Chem. Commun.*, 1993, 67.
54 G. Chelucci, M. A. Cabras, C. Botteghi, and M. Marchetti, *Tetrahedron: Asymmetry*, 1994, **5**, 299.
55 G. J. Dawson, C. G. Frost, J. M. J. Williams, and S. J. Coote, *Tetrahedron Lett.*, 1993, **34**, 3149.
56 J. V. Allen, G. J. Dawson, C. G. Frost, J. M. J. Williams, and S. J. Coote, *Tetrahedron*, 1994, **50**, 799.
57 S. J. Coote, G. J. Dawson, C. G. Frost and J. M. J. Williams, *Synlett*, 1993, 509.
58 C. Combellas, H. Marzouk, C. Suba, and A. Thiébault, *Synthesis*, 1993, 788.
59 H. Schmidbaur, S. Manhart, and A. Schier, *Chem. Ber.*, 1993, **126**, 2389.
60 A. Muth, O. Walter, G. Huttner, A. Asam, L. Zsolnai, and C. Emmerich, *J. Organomet. Chem.*, 1994, **468**, 149.
61 Th. Seitz, A. Muth, G. Huttner, Th. Klein, O. Walter, M. Fritz, and L. Zsolnai, *J. Organomet. Chem.*, 1994, **469**, 155.

62 O. Herd, K. P. Langhans, O. Stelzer, N. Weferling, and W. S. Sheldrick, *Angew. Chem., Int. Ed. Engl.*, 1993, **32**, 1058.

63 N. K. Gusarova, S. N. Arbuzova, S. I. Shaikhudinova, S. F. Malysheva, T. N. Rakhmatulina, S. V. Zinchenko, V. I. Dmitriev, and B. A. Trofimov, *Zh. Obshch. Khim.*, 1993, **63**, 1753 (*Chem. Abstr.*, 1994, **120**, 298 729).

64 B. Riegel, A. Pfitzner, G. Heckmann, H. Binder, and E. Fluck, *Z. Anorg. Allg. Chem.*, 1994, **620**, 8.

65 H. Noeth, S. Staude, M. Thomann, and R. T. Paine, *Chem. Ber.*, 1993, **126**, 611.

66 D. Dou, B. Kaufmann, E. N. Duesler, T. Chen, R. T. Paine, and H. Noth, *Inorg. Chem.*, 1993, **32**, 3056.

67 D. Dou, M. Fan, E. N. Duesler, H. Noth, and R. T. Paine, *Inorg. Chem.*, 1994, **33**, 2151.

68 P. Frankhauser, M. Driess, H. Pritzkow, and W. Sierbert, *Chem. Ber.*, 1994, **127**, 329.

69 T. J. Groshens, K. T. Higa, R. Nissan, R. J. Butcher, and A. J. Freyer, *Organometallics*, 1993, **12**, 2904.

70 G. Müller and J. Lachmann, *Z. Naturforsch., B: Chem. Sci.*, 1993, **48**, 1248.

71 R. L. Wells, A. T. McPhail, and M. F. Self, *Organometallics*, 1993, **12**, 3333.

72 D. A. Atwood, V. O. Atwood, A. H. Cowley, H. R. Gobran, R. A. Jones, T. M. Smeal, and C. J. Carrano, *Organometallics*, 1993, **12**, 3517.

73 K. Niediek and B. Neumuller, *Chem. Ber.*, 1994, **127**, 67.

74 R. L. Wells, M. F. Self, A. T. McPhail, S. R. Aubuchon, R. C. Woudenberg, and J. P. Jasinski, *Organometallics*, 1993, **12**, 2832.

75 M. D. B. Dillingham, J. A. Burns, J. Byers-Hill, K. D. Gripper, W. T. Pennington, and G. H. Robinson, *Inorg. Chim. Acta*, 1994, **216**, 267.

76 A. A. Naiini, Y. Han, M. Akinc, and J. G. Verkade, *Inorg. Chem.*, 1993, **32**, 5394.

77 R. L. Wells, A. T. McPhail, L. J. Jones, and M. F. Self, *Polyhedron*, 1993, **12**, 141.

78 J. A. Burns, M. D. B. Dillingham, J. B. Hill, K. D. Gripper, W. T. Pennington, and G. H. Robinson, *Organometallics*, 1994, **13**, 1514.

79 O. T. Beachley, S-H. L. Chao, M. R. Churchill, and C. H. Lake, *Organometallics*, 1993, **12**, 3992.

80 G. Müller and J. Lachmann, *Z. Naturforsch., B: Chem. Sci.*, 1993, **48**, 1544.

81 A. J. Edwards, M. A. Paver, P. R. Raithby, C. A. Russell, and D. S. Wright, *Organometallics*, 1993, **12**, 4687.

82 P. G. Edwards, M. B. Hursthouse, K. M. A. Malik, and J. S. Parry, *J. Chem. Soc., Chem. Commun.*, 1994, 1249.

83 J. Ho, Z. Hou, R. J. Drake, and D. W. Stephan, *Organometallics*, 1993, **12**, 3145; J. Ho, R. Rousseau, and D. W. Stephan, *Organometallics*, 1994, **13**, 1918.

84 D. Semenzin, G. Etemad-Moghadam, D. Albouy, and M. Koenig, *Tetrahedron Lett.*, 1994, **35**, 3297.

85 H. Song. C. Xia, and D. Li, *Huaxue Xuebao*, 1993, **51**, 815 (*Chem. Abstr.*, 1993, **119**, 250 052).

86 K. Heesche-Wagner and T. N. Mitchell, *J. Organomet. Chem.*, 1994, **468**, 99.

87 N. Bampos, L. D. Field, B. A. Messerle, and R. J. Smernik, *Inorg. Chem.*, 1993, **32**, 4084.

88 L. D. Field and I. J. Luck, *Tetrahedron Lett.*, 1994, **35**, 1109.

89 J. L. Bookham, W. Clegg, W. McFarlane, and E. S. Raper, *J. Chem. Soc., Dalton Trans.*, 1993, 3567.

90 Y. Gourdel, P. Pellon, L. Toupet, and M. Le Corre, *Tetrahedron Lett.*, 1994, **35**, 1197.

91 H. Lang, U. Lay, and W. Imhof, *J. Organomet. Chem.*, 1993, **444**, 53.

92 A. Marinetti and L. Ricard, *Tetrahedron*, 1993, **49**, 10291.

93 K. Tani, M. Yabuta, S. Nakamura, and T. Yamagata, *J. Chem. Soc., Dalton Trans.*, 1993, 2781.
94 Y. Uozumi, N. Suzuki, A. Ogiwara, and T. Hayashi, *Tetrahedron*, 1994, **50**, 4293.
95 T. Higashizima, N. Sakai, K. Nozaki, and H. Takaya, *Tetrahedron Lett.*, 1994, **35**, 2023.
96 T. Hayashi, H. Iwamura, Y. Uozoumi, Y. Matsumoto, and F. Ozawa, *Synthesis*, 1994, 526.
97 M. Fujita, J. Yazaki, T. Kuramochi, and K. Ogura, *Bull. Chem. Soc. Jpn.*, 1993, **66**, 1837.
98 M. Sawamura, R. Kuwano, and Y. Ito, *Angew. Chem., Int. Ed. Engl.*, 1994, **33**, 111.
99 T. Horiuchi, T. Ohta, M. Stephan, and H. Takaya, *Tetrahedron: Asymmetry*, 1994, **5**, 325.
100 T. Coumbe, N. J. Lawrence, and F. Muhammad, *Tetrahedron Lett.*, 1994, **35**, 625.
101 T. Wettling, Eur. Pat. Appl., EP 548 682 (*Chem. Abstr.*, 1993, **119**, 139 547).
102 J. R. Goerlich and R. Schmutzler, *Phosphorus, Sulfur, Silicon, Related Elem.*, 1993, **81**, 141.
103 D. Fabbri, S. Gladiali, and O. De Lucchi, *Synth. Commun.*, 1994, **24**, 1271.
104 N. J. Lawrence and F. Muhammad, *J. Chem. Soc., Chem. Commun.*, 1993, 1187.
105 A. Tillack, M. Michalik, D. Fenske, and H. Goesmann, *J. Organomet. Chem.*, 1993, **454**, 95.
106 F. Gorla, L. M. Venanzi, and A. Albinati, *Organometallics*, 1994, **13**, 43.
107 T. Imamoto, M. Matsuo, T. Nonomura, K. Kishikawa, and M. Yanagawa, *Heteroatom Chem.*, 1993, **4**, 475.
108 T. Morimoto, M. Chiba, and K. Achiwa, *Chem. Pharm. Bull.*, 1993, **41**, 1149.
109 W. A. Herrmann and C. W. Kohlpainter, *Angew. Chem., Int. Ed. Engl.*, 1993, **32**, 1524.
110 T. Bartik, B. B. Dunn, B. Bartik, and B. E. Hanson, *Inorg. Chem.*, 1994, **33**, 164.
111 A. Yamazaki, T. Morimoto, and K. Achiwa, *Tetrahedron: Asymmetry*, 1993, **4**, 2287.
112 A. Börner, R. Kadyrov, M. Michalik, and D. Heller, *J. Organomet. Chem.*, 1994, **470**, 237.
113 A. Borner, J. Ward, K. Kortus, and H. B. Kagan, *Tetrahedron: Asymmetry*, 1993, **4**, 2219.
114 L. B. Fields and E. N. Jacobsen, *Tetrahedron: Asymmetry*, 1993, **4**, 2229.
115 A. Ishii, I. Takaki, J. Nakayama, and M. Hoshino, *Tetrahedron Lett.*, 1993, **34**, 8255.
116 A. Ishii, R. Yoshioka, J. Nakayama, and M. Hoshino, *Tetrahedron Lett.*, 1993, **34**, 8259.
117 N. V. Lukashev, P. E. Zhichkin, E. A. Tarasenko, Yu. N. Luzikov, and M. A. Kazankova, *Zh. Obshch. Khim.*, 1993, **63**, 1767 (*Chem. Abstr.*, 1994, **120**, 245 323).
118 H. H. Karsch, G. Grauvogl, P. Mikulcik, P. Bissinger, and G. Müller, *J. Organomet. Chem.*, 1994, **465**, 65.
119 H. H. Karsch, R. Richter, and A. Schier, *Z. Naturforsch., B: Chem. Sci.*, 1993, **48**, 1533.
120 H. H. Karsch, B. Deubelly, G. Grauvogl, and G. Müller, *J. Organomet. Chem.*, 1993, **459**, 95.
121 H. H. Karsch, G. Grauvogl, M. Kawecki, and P. Bissinger, *Organometallics*, 1993, **12**, 2757.
122 S. Hao, J-I. Song, H. Aghabozorg, and S. Gambarotta, *J. Chem. Soc., Chem. Commun.*, 1994, 157.
123 K. M. Webb and N. E. Miller, *J. Organomet. Chem.*, 1993, **460**, 1.
124 S. W. Dougherty and N. E. Miller, *Inorg. Chem.*, 1993, **32**, 5889.

125 M. Zablocka, F. Boutonnet, A. Igau, F. Dahan, J. P. Majoral, and K. M. Pietrusiewicz, *Angew. Chem., Int. Ed. Engl.*, 1993, **32**, 1735.
126 H. J. Ness, H. Keller, T. Facklam, A. Herrmann, J. Welsch, U. Bergstrasser, H. Heydt, and M Regitz, *J. Prakt. Chem./Chem. Ztg.*, 1993, **335**, 589.
127 H. Schmidbaur, S. Manhart, and A. Schier, *Chem. Ber.*, 1993, **126**, 2259.
128 N. A. Mukmeneva, V. Kh. Kadyrova, E. N. Cherezova, and A. S. Sharifullin, *Zh. Obshch. Khim.*, 1993, **63**, 1909 (*Chem. Abstr.*, 1994, **120**, 191 855).
129 J. R. Goerlich, C. Muller, and R. Schmutzler, *Phosphorus, Sulfur, Silicon, Related Elem.*, 1993, **85**, 193.
130 A. Togni, C. Breutel, A. Schnyder, F. Spindler, H. Landert, and A. Tijani, *J. Am. Chem. Soc.*, 1994, **116**, 4062.
131 A. A. Tolmachev, A. A. Yurchenko, M. G. Semenova, and N. G. Feshchenko, *Zh. Obshch. Khim.*, 1993, **63**, 714 (*Chem. Abstr.*, 1993, **119**, 180 912).
132 A. A. Tolmachev, S. P. Ivomin, A. V. Kharchenko, and E. S. Kozlov, *Zh. Obshch. Khim.*, 1993, **63**, 222 (*Chem. Abstr.*, 1993, **119**, 95642).
133 Ya. A. Dorfman, L. V. Levina, and A. K. Borangazieva, *Kinet. Katal.*, 1993, **34**, 470 (*Chem. Abstr.*, 1994, **120**, 191 842).
134 J. Fawcett, P. A. T. Hoye, R. D. W. Kemmitt, D. J. Law, and D. R. Russell, *J. Chem. Soc., Dalton Trans.*, 1993, 2563.
135 B. Assmann, K. Angermaier, and H. Schmidbaur, *J. Chem. Soc., Chem. Commun.*, 1994, 941.
136 M. T. Reetz and J. Rudolph, *Tetrahedron: Asymmetry*, 1993, **4**, 2405.
137 M. Sawamura, K. Kitayama, and Y. Ito, *Tetrahedron: Asymmetry*, 1993, **4**, 1829.
138 W-K. Wong, J-X. Gao, W-T. Wong, W. C. Cheng, and C-M. Che, *J. Organomet. Chem.*, 1994, **471**, 277; H. Brunner and J. Fürst, *Inorg. Chim. Acta*, 1994, **220**, 63.
139 E. Lindner, M. Kemmler, H. A. Mayer, and P. Wegner, *J. Am. Chem. Soc.*, 1994, **116**, 348.
140 K. R. Dunbar and S. C. Haefner, *Polyhedron*, 1994, **13**, 727.
141 E. J. Corey, Z. Chen, and G. J. Tanoury, *J. Am. Chem. Soc.*, 1993, **115**, 11,000.
142 A. A. Tolmachev, A. A. Yurchenko, A. B. Rozhenko, and M. G. Semenova, *Zh. Obshch. Khim.*, 1993, **63**, 1911 (*Chem. Abstr.*, 1994, **120**, 245 328).
143 B. M. Trost and C-J. Li, *J. Am. Chem. Soc.*, 1994, **116**, 3167.
144 C. Larpent and G. Meignan, *Tetrahedron Lett.*, 1993, **34**, 4331.
145 S. D. Rychnovsky and J. Kim, *J. Org. Chem.*, 1994, **59**, 2659.
146 D. J. Darensbourg, F. Joo, A. Kathó, J. N. W. Stafford, A. Bényei, and J. H. Reibenspies, *Inorg. Chem.*, 1994, **33**, 175.
147 K. S. Kim, Y. H. Joo, I. W. Kim, K. R. Lee, D. Y. Cho, M. Kim, and I. H. Cho, *Synth. Commun.*, 1994, **24**, 1157.
148 J. Kroner, H. Nöth, K. Polborn, H. Stolpmann, M. Tacke, and M. Thomann, *Chem. Ber.*, 1993, **126**, 1995.
149 H. Vogt, S. I. Trojanov, and V. B. Rybakov, *Z. Naturforsch., B: Chem. Sci.*, 1993, **48**, 258.
150 N. Bricklebank, S. M. Godfrey, C. A. McAuliffe, and R. G. Pritchard, *J. Chem. Soc., Dalton Trans.*, 1993, 2261.
151 A. Wells, *Synth. Commun.*, 1994, **24**, 1715.
152 I. A. Rivero, R. Somanathan, and L. H. Hellberg, *Synth. Commun.*, 1993, **23**, 711.
153 O. I. Kolodyazhnyi and A. N. Chernega, *Zh. Obshch. Khim.*, 1992, **62**, 2670 (*Chem. Abstr.*, 1993, **119**, 117 379).
154 O. I. Kolodyazhnyi and S. N. Ustenko, *Zh. Obshch. Khim.*, 1993, **63**, 711 (*Chem. Abstr.*, 1993, **119**, 180 910).
155 P. Majewski, *Phosphorus, Sulfur, Silicon, Related Elem.*, 1993, **85**, 41.
156 Y. Kikugawa and K. Mitsui, *Chem. Lett.*, 1993, 1369.

157 G. Hodosi, B. Podanyi, and J. Kuszmann, *Carbohydrate Res.*, 1992, **230**, 327.
158 Y. Arredondo, R. Pleixats, and M. Moreno-Manas, *Synth. Commun.*, 1993, **23**, 1245.
159 H. Schmidbaur, A. Stützer, P. Bissinger, and A. Schier, *Z. Anorg. Allg. Chem.*, 1993, **619**, 1519.
160 H. Brisset, Y. Gourdel, P. Pellon, and M. Le Corre, *Tetrahedron Lett.*, 1993, **34**, 4523.
161 A. Blumenthal, P. Bissinger, and H. Schmidbaur, *J. Organomet. Chem.*, 1993, **462**, 107.
162 S. Tsuji, M. Kondo, K. Ishiguro, and Y. Sawaki, *J. Org. Chem.*, 1993, **58**, 5055.
163 A. L. Baumstark, P. C. Vasquez, and Y. Chen, *Heteroat. Chem.*, 1993, **4**, 175.
164 C. W. Lee, J. S. Lee, N. S. Cho, K. D. Kim, S. M. Lee, and J. S. Oh, *J. Mol. Catal.*, 1993, **80**, 31.
165 J. R. Goerlich and R. Schmutzler, *Z. Anorg. Allg. Chem.*, 1994, **620**, 898.
166 J. Pang, L. Huang, G. Wu, D. Dai, and G. Zhu, *Youji Huaxue*, 1993, **13**, 377 (*Chem. Abstr.*, 1993, **119**, 250 056).
167 M. C. Demarcq, *J. Chem. Res.* (S), 1993, 450; (M) 3052.
168 G. F. Shaw, A. A. Isab, J. D. Hoeschele, M. Starich, J. Locke, P. Schulteis, and J. Xiao, *J. Am. Chem. Soc.*, 1994, **116**, 2254.
169 C. P. Galloway and T. B. Rauchfuss, *Angew. Chem., Int. Ed. Engl.*, 1993, **32**, 1319.
170 P. Jutzi, N. Brusdeilins, H-G. Stammler, and B. Neumann, *Chem. Ber.*, 1994, **127**, 997.
171 P. Molina, C. Conesa, A. Alias, A. Arques, M. D. Velasco, A. L. Llamas-Saiz, and C. Foces-Foces, *Tetrahedron*, 1993, **49**, 7599.
172 P. Molina, A. Tarraga, M. J. del Bano, and A. Espinosa, *Leibigs Ann. Chem.*, 1994. 223.
173 J. R. Goerlich, M. Farkens, A. Fischer, P. G. Jones, and R. Schmutzler, *Z. Anorg. Allg. Chem.*, 1994, **620**, 707.
174 S. Rodriguez-Morgade, T. Torres, and P. Vazquez, *Synthesis*, 1993, 1235.
175 G. Alcaraz, A. Baceiredo, M. Nieger, and G. Bertrand, *J. Am. Chem. Soc.*, 1994, **116**, 2159.
176 J. L. Castro, V. G. Matassa, and R. G. Ball, *J. Org. Chem.*, 1994, **59**, 2289.
177 M. L. Edwards, D. M. Stemerick, and J. R. McCarthy, *Tetrahedron*, 1994, **50**, 5579.
178 M. A. Walker, *Tetrahedron Lett.*, 1994, **35**, 665.
179 C. Simon, S. Hosztafi, and S. Makleit, *Tetrahedron Lett.*, 1993, **34**, 6475.
180 K. C. Santhosh and K. K. Balasubramanian, *Synth. Commun.*, 1994, **24**, 1049.
181 M. Okuda and K. Tomioka, *Tetrahedron Lett.*, 1994, **35**, 4585.
182 D. S. Dodd and A. P. Kozikowski, *Tetrahedron Lett.*, 1994, **35**, 977.
183 J. Reisch, A. R. R. Rao, and C. O. Usifoh, *Monatshefte für Chemie*, 1994, **125**, 79.
184 A. Sobti and G. A. Sulikowski, *Tetrahedron Lett.*, 1994, **35**, 3661.
185 S. S. Bhagwat and C. Gude, *Tetrahedron Lett.*, 1994, **35**, 1847.
186 M. Botta, V. Summa, G. Trapassi, E. Monteagudo, and F. Corelli, *Tetrahedron: Asymmetry*, 1994, **5**, 181.
187 A. B. Charette and B. Côté, *Tetrahedron Lett.*, 1993, **34**, 6833.
188 G. Neef, A. Seeger, and H. Vierhufe, *Synth. Commun.*, 1993, **23**, 931.
189 L. S. Richter and T. R. Gadek, *Tetrahedron Lett.*, 1994, **35**, 4705.
190 D. A. Campbell and J. C. Bermak, *J. Org. Chem.*, 1994, **59**, 658.
191 S. M. Bachrach, *J. Phys. Chem.*, 1993, **97**, 4996.
192 M. T. Nguyen, E. Van Praet, and L. G. Vanquickenborne, *Inorg. Chem.*, 1994, **33**, 1153.
193 S. M. Bachrach and L. M. Perriott, *Tetrahedron Lett.*, 1993, **34**, 6365.
194 K. K. Laali, M. Regitz, M. Birkel, P. J. Stang, and C. M. Crittell, *J. Org. Chem.*, 1993, **58**, 4105.

195 G. R. Newkome, *Chem. Rev.*, 1993, **93**, 2067.
196 Y. Uchida, M. Kawai, H. Masauji, and S. Oae, *Heteroat. Chem.*, 1993, **4**, 421.
197 S. Yasui, M. Fujii, C. Kawano, Y. Nishimura, K. Shioji, and A. Ohno, *J. Chem. Soc., Perkin Trans.*, **2**, 1994, 177.
198 R. Srinivas, G. K. Rao, and R. V. Viswanadha, *Org. Mass. Spectrom.*, 1993, **28**, 267.
199 E. J. Fredericks, M. J. Gindling, L. C. Kroll, and B. N. Storhoff, *J. Organomet. Chem.*, 1994, **465**, 289.
200 G. Bertrand and C. Wentrup, *Angew. Chem., Int. Ed. Engl.*, 1994, **33**, 527.
201 H-J. Nees, U. Bergstrasser, H. Heydt, and M. Regitz, *Heteroat. Chem.*, 1993, **4**, 525.
202 A. H. Cowley, F. P. Gabbai, C. J. Carrano, L. M. Mokry, M. R. Bond, and G. Bertrand, *Angew. Chem., Int. Ed. Engl.*, 1994, **33**, 578.
203 P. Veya, C. Floriani, A. Chiesi-Villa, and C. Guastini, *Organometallics*, 1994, **13**, 208.
204 W. Henderson, H. H. Petach, and K. Sarfo, *J. Chem. Soc., Chem. Commun.*, 1994, 245.
205 K. V. Katti, B. D. Santarsiero, A. A. Pinkerton, and R. G. Cavell, *Inorg. Chem.*, 1993, **32**, 5919.
206 H. Maeda, T. Maki, K. Eguchi, T. Koide, and H. Ohmori, *Tetrahedron Lett.*, 1994, **35**, 4129.
207 E. Vedejs, N. S. Bennett, L. M. Conn, S. T. Diver, M. Gingras, S. Lin, P. A. Oliver, and M. J. Peterson, *J. Org. Chem.*, 1993, **58**, 7286.
208 A. Cammers-Goodwin, *J. Org. Chem.*, 1993, **58**, 7619.
209 F. R. Vega, J-C. Clement, and H. des Abbayes, *Tetrahedron Lett.*, 1993, **34**, 8117.
210 R. D. Brost, G. C. Bruce, S. L. Grundy, and S. R. Stobart, *Inorg. Chem.*, 1993, **32**, 5195.
211 R. Martens and W. W. du Mont, *Chem. Ber.*, 1993, **126**, 1115.
212 A. J. Deeming and M. B. Smith, *J. Chem. Soc., Dalton Trans.*, 1993, 2041.
213 K. Yang, J. M. Smith, S. G. Bott, and M. G. Richmond, *Organometallics*, 1993, **12**, 4779.
214 D. K. Johnson, T. Rukachaisirikul, Y. Sun, N. J. Taylor, A. J. Canty, and A. J. Carty, *Inorg. Chem.*, 1993, **32**, 5544.
215 F. Teixidor, C. Vinas, M. M. Abad, M. Lopez, and J. Casabo, *Organometallics*, 1993, **12**, 3766.
216 H. Schumann, T. Ghodsi, L. Esser, and E. Hahn, *Chem. Ber.*, 1993, **126**, 591.
217 T. G. Meyer, P. G. Jones, and R. Schmutzler, *Z. Naturforsch., B: Chem. Sci.*, 1993, **48**, 875.
218 R. Minkwitz, A. Kornath, D. Lennhoff, and H. Preut, *Z. Anorg. Allg. Chem.*, 1994, **620**, 509.
219 W. Dabkowski, J. Michalski, J. Wasiak, and F. Cramer, *J. Chem. Soc., Dalton Trans.*, 1994, 817.
220 B. I. No, Yu. L. Zotov, R. M. Petruneva, Z. M. Sabirov, V. N. Urazbaev, and Yu. B. Monakov, *Dokl. Akad. Nauk*, 1993, **328**, 199 (*Chem. Abstr.*, 1993, **119**, 95 630).
221 V. V. Miroshnichenko, R. I. Yurchenko, and N. G. Feshchenko, *Zh. Obshch. Khim.*, 1993, **63**, 231 (*Chem. Abstr.*, 1993, **119**, 72 712).
222 A. A. Tolmachev, A. A. Yurchenko, E. S. Kozlov, V. A. Shulezhko, and A. M. Pinchuk, *Heteroat. Chem.*, 1993, **4**, 343.
223 A. A. Tolmachev, M. G. Semenova, A. I. Sviridon, A. A. Yurchenko, and N. G. Feshchenko, *Zh. Obshch. Khim.*, 1993, **63**, 1344 (*Chem. Abstr.*, 1994, **120**, 134 643).

224 R. Schmutzler, L. Heuer, and D. Schomburg, *Phosphorus, Sulfur, Silicon, Related Elem.*, 1993, **83**, 149.
225 W-W. du Mont, V. Stenzel, J. Jeske, P. G. Jones, A. Sebald, S. Pohl, W. Saak, and M. Bätcher, *Inorg. Chem.*, 1994, **33**, 1502.
226 Yu. G. Budnikova, Yu. M. Kargin, and T. R. Novoselova, *Zh. Obshch. Khim.*, 1993, **63**, 1308 (*Chem. Abstr.*, 1994, **120**, 134 642).
227 L. Riesel, R. Friebe, and D. Sturm, *Z. Anorg. Allg. Chem.*, 1993, **619**, 1685.
228 A. Fischer, P. G. Jones, I. Neda, R. Schmutzler, and I. V. Shevchenko, *Z. Anorg. Allg. Chem.*, 1994, **620**, 908.
229 A. Roucoux, F. Agbossou, A. Mortreux, and F. Petit, *Tetrahedron: Asymmetry*, 1993, **4**, 2279.
230 M. K. Grachev, G. I. Kurochkina, A. R. Bekker, L. K. Vasyanina, and E. E. Nifant'ev, *Zh. Obshch. Khim.*, 1993, **63**, 338 (*Chem. Abstr.*, 1993, **119**, 160 385).
231 M. K. Grachev, G. I. Kurochkina, A. R. Bekker, L. K. Vasyanina, and E. E. Nifant'ev, *Zh. Obshch. Khim.*, 1993, **63**, 948 (*Chem. Abstr.*, 1993, **119**, 250 040).
232 M. Gruber, P. G. Jones, and R. Schmutzler, *Phosphorus, Sulfur, Silicon, Related Elem.*, 1993, **80**, 195.
233 V. Yu Iorish, M. K. Grachev, A. R. Bekker, and E. E. Nifant'ev, *Zh. Obshch. Khim.*, 1993, **63**, 783 (*Chem. Abstr.*, 1994, **120**, 107 132).
234 D. A. Haristos, *Chem. Chron.*, 1993, **22**, 23 (*Chem. Abstr.*, 1993, **119**, 203 502).
235 R. Koster, G. Seidel, and G. Müller, *Chem. Ber.*, 1993, **126**, 2211.
236 H. Lang, U. Lay, M. Leise, and L. Zsolnai, *Z. Naturforsch., B: Chem. Sci.*, 1993, **48**, 27.
237 R. M. K. Deng and K. B. Dillon, *Polyhedron*, 1993, **12**, 1767.
238 F. Uhlig, E. Herrmann, D. Schädler, G. Ohms, G. Grossmann, S. Besser, and R. Herbst-Irmes, *Z. Anorg. Allg. Chem.*, 1993, **619**, 1962.
239 J. L. Bookham, F. Conti, H. C. E. McFarlane, W. McFarlane, and M. Thornton-Pett, *J. Chem. Soc., Dalton Trans.*, 1994, 1791.
240 M. S. Balakrishna, R. Klein, S. Uhlenbrock, A. A. Pinkerton, and R. G. Cavell, *Inorg. Chem.*, 1993, **32**, 5676.
241 S. Challet, J. C. Leblanc, and C. Moise, *New J. Chem.*, 1993, **17**, 251.
242 M. Baudler, P. Koch, and C. Wiaterek, *Z. Anorg. Allg. Chem.*, 1993, **619**, 1973.
243 B. W. Tattershall and N. I. Kendall, *Polyhedron*, 1994, **13**, 1507.
244 R. Blachnik and P. Lönnecke, *Z. Anorg. Allg. Chem.*, 1994, **620**, 167.
245 K. M. Pietrusiewicz, M. Kuznikowski, and M. Koprowski, *Tetrahedron: Asymmetry*, 1993, **4**, 2143.
246 T. Kawashima, H. Iwanaga, and R. Okazaki, *Chem. Lett.*, 1993, 1531.
247 R. K. Haynes, R. N. Freeman, C. R. Mitchell, and S. C. Vonwiller, *J. Org. Chem.*, 1994, **59**, 2919.
248 N. K. Gusarova, B. A. Trofimov, S. F. Malysheva, S. N. Arbuzova, S. I. Shaikhudinova, V. I. Dmitriev, A. V. Polubentsev, and A. I. Albanov, *Zh. Obshch. Khim.*, 1993, **63**, 53 (*Chem. Abstr.*, 1993, **119**, 139 327).
249 Y. A. Dorfman, L. V. Levina, and E. Zh. Aibasov, *Zh. Obshch. Khim.*, 1993, **63**, 1552 (*Chem. Abstr.*, 1994, **120**, 217 868).
250 S. R. Gilbertson, G. Chen, and M. McLoughlin, *J. Am. Chem. Soc.*, 1994, **116**, 4481.
251 T. Hanaya, R. Okamoto, Y. V. Prikhod'ko, M. A. Armour, A. M. Hogg, and H. Yamamoto, *J. Chem. Soc., Perkin Trans. 1*, 1993, 1663.
252 O. I. Kolodiazhnyi, V. N. Zemlianoy, L. I. Baranova, and G. V. Shurubura, *Phosphorus, Sulfur, Silicon, Related Elem.*, 1993, **82**, 137.

253 C. Toulhoat, M. Vidal, and M. Vincens, *Phosphorus, Sulfur, Silicon, Related Elem.*, 1993, **78**, 119.
254 C. Toulhoat, M. Vincens, and M. Vidal, *Bull. Soc. Chim. Fr.*, 1993, **130**, 647.
255 G. Hagele, S. Varbanov, J. Ollig, and H. W. Kropp, *Z. Anorg. Allg. Chem.*, 1994, **620**, 914.
256 N. V. Vorob'ev-Desyatovskii, V. M. Adamov, A. D. Misharev, Yu. A. Teterin, and M. L. Sosul'nikov, *Zh. Obshch. Khim.*, 1993, **63**, 304 (*Chem. Abstr.*, 1993, **119**, 203 571).
257 F. Richter and H. Weichmann, *J. Organomet. Chem.*, 1994, **466**, 77.
258 T. Hattori, J. Sakamoto, N. Hayashizaka, and S. Miyano, *Synthesis*, 1994, 199.
259 A. Couture, E. Deniau, and P. Grandclaudon, *J. Chem. Soc., Chem. Commun.*, 1994, 1329.
260 A. N. Bovin, A. N. Yarkevich, A. V. Kharitonov, and E. N. Tsvetkov, *J. Chem. Soc., Chem. Commun.*, 1994, 973.
261 A. Yu. Aksinenko, A. N. Pushin, and V. B. Sokolov, *Phosphorus, Sulfur, Silicon, Related Elem.*, 1993, **84**, 249.
262 T. Majima and W. Schnabel, *Reza Kagaku Kenkyu*, 1992, **14**, 88 (*Chem. Abstr.*, 1993, **119**, 8874).
263 T. Y. Fu, Z. Liu, J. R. Scheffer, and J. Trotter, *J. Am. Chem. Soc.*, 1993, **115**, 12202.
264 S. Yasui, K. Shioji, A. Ohno, and M. Yoshihara, *Chem. Lett.*, 1993, 1393.
265 A. Brandi, S. Cicchi, A. Goti, M. Koprowski, and K. M. Pietrusiewicz, *J. Org. Chem.*, 1994, **59**, 1315.
266 M. Yamashita, A. Iida, H. Mizuno, Y. Miyamoto, T. Morishita, N. Sata, K. Kiguchi, A. Yabui, and T. Oshikawa, *Heteroat. Chem.*, 1993, **4**, 553.
267 J. Omelanczuk, *Tetrahedron*, 1993, **49**, 8887.
268 T. Fujimoto, R. Nakao, Y. Hotei, K. Ohta, and I. Yamamoto, *J. Chem. Res.*, (S), 1993, 486.
269 S. K. Armstrong, E. W. Collington, J. G. Knight, A. Naylor, and S. Warren, *J. Chem. Soc., Perkin Trans. 1*, 1993, 1433.
270 J. Clayden, E. W. Collington, J. Elliott, S. J. Martin, A. B. McElroy, S. Warren, and D. Waterson, *J. Chem. Soc., Perkin Trans. 1*, 1993, 1849.
271 J. Clayden and S. Warren, *J. Chem. Soc., Perkin Trans. 1*, 1993, 2913.
272 S. K. Armstrong, E. W. Collington, and S. Warren, *J. Chem. Soc., Perkin Trans. 1*, 1994, 515.
273 J. Clayden and S. Warren, *J. Chem. Soc., Perkin Trans. 1*, 1994, 1529.
274 M. Mikolajczyk, P. P. Graczyk, and M. W. Wieczorek, *J. Org. Chem.*, 1994, **59**, 1672.
275 M. Mikolajczyk, *Phosphorus, Sulfur, Silicon, Related Elem.*, 1993, **74**, 311.
276 D. E. Lynch, G. Smith, N. J. Calas, C. H. L. Kennard, A. K. Wittaker, K. S. Jack, and A. C. Willis, *Aust. J. Chem.*, 1993, **46**, 1535.
277 A. Blaschette, T. Hamann, D. Henschel, and P. G. Jones, *Z. Anorg. Allg. Chem.*, 1993, **619**, 1945.
278 S. N. Slabzhennikov, L. N. Alexieko, and E. I. Matrosov, *Izv. Akad. Nauk, Ser. Khim.*, 1992, 2579 (*Chem. Abstr.*, 1993, **119**, 250 042).
279 P. B. Savage, S. K. Holmgren, J. M. Desper, and S. H. Gellman, *Pure Appl. Chem.*, 1993, **65**, 461.
280 P. B. Savage, S. K. Holmgren, and S. H. Gellman, *J. Am. Chem. Soc.*, 1993, **115**, 7900.
281 P. B. Savage and S. H. Gellman, *J. Am. Chem. Soc.*, 1993, **115**, 10448.
282 P. B. Savage, S. K. Holmgren, and S. H. Gellman, *J. Am. Chem. Soc.*, 1994, **116**, 4069.

283 K. M. Pietrusiewicz and W. Wieczorek, *Phosphorus, Sulfur, Silicon, Related Elem.*, 1993, **82**, 99.
284 R. Miranda, I. Salas, C. Salceda, L. Velasco, A. Cabrera, and M. Salmon, *Org. Mass Spectrom.*, 1993, **28**, 593.
285 S. M. Godfrey, C. A. McAuliffe, G. C. Ranger, and D. G. Kelly, *J. Chem. Soc., Daltron Trans.*, 1993, 2809.
286 M. Well, A. Fischer, P. G. Jones, and R. Schmutzler, *Chem. Ber.*, 1993, **126**, 1765.
287 M. Well, A. Fischer, P. G. Jones, and R. Schmutzler, *Phosphorus, Sulfur, Silicon, Related Elem.*, 1993, **80**, 157.
288 T. C. Blagborough, R. Davis, and P. Ivison, *J. Organomet. Chem.*, 1994, **467**, 85.
289 A. Herbowski, T. Lis, and E. A. Deutsch, *J. Organomet. Chem.*, 1993, **460**, 25.
290 P. Vojtisek, J. Podlahova, K. Maly, and J. Hasek, *Coll. Czech. Chem. Commun.*, 1993, **58**, 1354.
291 H. Weichmann and J. Meunier-Piret, *Organometallics*, 1993, **12**, 4097.
292 M. Gielen, H. Pan, and E. R. T. Tiekink, *Bull. Soc. Chim. Belg.*, 1993, **102**, 447.
293 K. Ogura, S. Kihara, S. Umetani, and M. Matsui, *Bull. Chem. Soc. Jpn.*, 1993, **66**, 1971.
294 M. B. Power, J. W. Ziller, and A. R. Barron, *Organometallics*, 1993, **12**, 4908.
295 A. W. Schwabacher, S. Zhang, and W. Davy, *J. Am. Chem. Soc.*, 1993, **115**, 6995.
296 T. Nozdryn, J. Cosseau, A. Gorgues, M. Jubault, J. Orduna, S. Uriel, and J. Garin, *J. Chem. Soc., Perkin Trans. 1*, 1993, 1711.
297 R. Engel, K. Rengen, and C. S. Chan, *Heteroat. Chem.*, 1993, **4**, 181.
298 L. Riesel, K. Lauritsen, and H. Vogt, *Z. Anorg. Allg. Chem.*, 1994, **620**, 1099.
299 W. Qiu and D. J. Burton, *J. Fluorine Chem.*, 1993, **65**, 143.
300 D. A. Oparin and B. O. Khodorkovskii, *Zh. Org. Khim.*, 1993, **29**, 168 (*Chem. Abstr.*, 1994, **120**, 30 633).
301 X-H. Wang and M. Schlosser, *Synthesis*, 1994, 479.
302 H. Nöth, H. Stolpmann, and M. Thomann, *Chem. Ber.*, 1994, **127**, 81.
303 U. Bilow and M. Jansen, *J. Chem. Soc., Chem. Commun.*, 1994, 403.
304 R. Kivekas, R. Sillanpaa, F. Teixidor, C. Vinas, and J. A. Ayllon, *Acta Chem. Scand.*, 1994, **48**, 117.
305 E. A. Krasil'nikova, V. V. Sentemov, and E. L. Gavrilova, *Zh. Obshch. Khim.*, 1993, **63**, 848 (*Chem. Abstr.*, 1993, **119**, 250 037).
306 P. H. Lee, J. S. Kim. I-S. Han, *Bull. Korean Chem. Soc.*, 1993, **14**, 424.
307 V. P. Talzi and G. V. Mongustova, *Zh. Obshch. Khim.*, 1992, **62**, 2015 (*Chem. Abstr.*, 1993, **119**, 49 466).
308 R. Araya-Maturana and F. Castaneda, *Phosphorus, Sulfur, Silicon, Related Elem.*, 1993, **81**, 165.
309 I. V. Leont'eva, I. M. Aladzheva, O. V. Bykhovskaya, P. V. Petrovskii, M. Yu. Antipin, Yu. T. Struchkov, T. A. Mastryukova, and M. I. Kabachknik, *Zh. Obshch. Khim.*, 1993, **63**, 621 (*Chem. Abstr.*, 1993, **119**, 226 053).
310 A. Benyei, J. N. W. Stafford, A. Katho, D. J. Darensbourg, and F. Joo, *J. Mol. Catal.*, 1993, **84**, 157.
311 F. Plenat, D. Grelet, V. Ozon, and H. J. Cristau, *Synlett*, 1994, 269.
312 K. Okuma, Y. Tanaka, H. Ohta, and H. Matsuyama, *Bull. Chem. Soc. Jpn.*, 1993, **66**, 2623.
313 K. Okuma, M. Ono, and H. Ohta, *Bull. Chem. Soc. Jpn.*, 1993, **66**, 1308.
314 O. B. Smoli, S. Ya. Panchishin, and B. S. Drach, *Zh. Obshch. Khim.*, 1993, **63**, 1184 (*Chem. Abstr.*, 1993, **119**, 271 262).
315 V. S. Brovarets, R. N. Vydzhak, and B. S. Drach, *Zh. Obshch. Khim.*, 1993, **63**, 80 (*Chem. Abstr.*, 1993, **119**, 117 363).

316 V. S. Brovarets, R. N. Vydzhak, T. K. Vinogradova, and B. S. Drach, *Zh. Obshch. Khim.*, 1993, **63**, 87 (*Chem. Abstr.*, 1993, **119**, 117 395).
317 V. S. Brovarets, R. N. Vydzhak, K. V. Zyuz, and B. S. Drach, *Zh. Obshch. Khim.*, 1993, **63**, 1266 (*Chem. Abstr.*, 1994, **120**, 191 654).
318 V. S. Brovarets, R. N. Vydzhak, and B. S. Drach, *Zh. Obshch. Khim.*, 1993, **63**, 1053 (*Chem. Abstr.*, 1994, **120**, 134 380).
319 L. Van Meervelt, R. N. Vydzhak, V. S. Brovarets, N. I. Mishchenko, and B. S. Drach, *Tetrahedron*, 1994, **50**, 1889.
320 K. Kirchner, K. Mereiter, K. Mauthner, and R. Schmid, *Inorg. Chim. Acta*, 1994, **217**, 203.
321 S. G. Lee, K. Y. Chung, T. S. Yoon, and W. Shin, *Organometallics*, 1993, **12**, 2873.
322 A. Kanazawa, T. Ikeda, and T. Endo, *Antimicrob. Agents Chemother.*, 1994, **38**, 945 (*Chem. Abstr.*, 1994, **121**, 83 482).
323 Q. Zhu, K. L. Ford, and E. J. Roskamp, *Heteroat. Chem.*, 1993, **3**, 647.
324 M. T. Hanna, F. Y. Khalil, and S. M. Beder, *Phys. Chem. (Peshawar, Pak.)*, 1992, **11**, 115 (*Chem. Abstr.*, 1994, **120**, 245 341).
325 K. Lauritsen, H. Vogt, and L. Riesel, *Z. Anorg. Allg. Chem.*, 1994, **620**, 1103.
326 D. J. Burton and I. N. Jeong, *J. Fluorine Chem.*, 1993, **62**, 259.
327 X-M. Zhang and F. G. Bordwell, *J. Am. Chem. Soc.*, 1994, **116**, 968.
328 Y-S. Gal, *J. Chem. Soc., Chem. Commun.*, 1994, 327.
329 R. A. Khachatryan, G. A. Mkrtchyan, A. M. Torgomyan, and M. G. Indzhikyan, *Zh. Obshch. Khim.*, 1993, **63**, 2151, (*Chem. Abstr.*, 1994, **120**, 298 754).
330 M. D. Johnson, C. A. McIntosh, and V. C. Reinsborough, *Aust. J. Chem.*, 1994, **47**, 187.
331 R. Hunter, R. H. Haueisen, and A. Irving, *Angew. Chem., Int. Ed. Engl.*, 1994, **33**, 566.
332 Y. Uchibori, M. Umeno, H. Seto, and H. Yoshioka, *Chem. Lett.*, 1993, 673.
333 J. B. Hendrickson, M. Singer, and M. S. Hussoin, *J. Org. Chem.*, 1993, **58**, 6913.
334 C. Imrie, T. A. Modro, E. R. Rohwer, and C. C. P. Wagener, *J. Org. Chem.*, 1993, **58**, 5643.
335 C. Imrie, T. A. Modro, and C. C. P. Wagener, *J. Chem. Res.*, (*S*), 1994, 222.
336 C. Imrie, T. A. Modro, and C. C. P. Wagener, *J. Chem. Soc., Perkin Trans. 2*, 1994, 1379.
337 C. Imrie, T. A. Modro, and P. H. Van Rooyen, *Polyhedron*, 1994, **13**, 1677.
338 H. Vogt, K. Lauritsen, L. Riesel, M. von Loewis, and G. Reck, *Z. Naturforsch., B: Chem. Sci.*, 1993, **48**, 1760.
339 S. W. Ng and V. G. Kumar Das, *Main Group Met. Chem.*, 1993, **16**, 81.
340 G. V. Ratovskii, S. L. Belaya, and S. V. Zinchenko, *Zh. Obshch. Khim.*, 1992, **62**, 2147 (*Chem. Abstr.*, 1993, **119**, 28 235).
341 A. C. Legon and J. C. Thorn, *J. Chem. Soc., Faraday Trans.*, 1993, **89**, 3319.
342 J. Claereboudt, W. Baeten, H. Geise, and M. Claeys, *Org. Mass Spectrom.*, 1993, **28**, 71.
343 P. P. Graczyk and M. Mikolajczyk, *Phosphorus, Sulfur, Silicon, Related Elem.*, 1993, **78**, 313.
344 C. Garrigou-Lagrange, M. Lequan, R. M. Lequan, and V. M. Yartsev, *J. Chim. Phys.*, 1993, **90**, 1749.
345 N. C. Norman, *Polyhedron*, 1993, **12**, 2431.
346 M. Abe, K. Toyota, and M. Yoshifuji, *Heteroat. Chem.*, 1993, **4**, 427.
347 M. Yoshifuji, M. Abe, K. Toyota, I. Miyahara, and K. Hirotsu, *Bull. Chem. Soc. Jpn.*, 1993, **66**, 3831.
348 M. Yoshifuji, K. Kamijo, and K. Toyota, *Bull. Chem. Soc. Jpn.*, 1993, **66**, 3440.
349 M. Yoshifuji, D. L. An, K. Toyota, and M. Yasunami, *Chem. Lett.*, 1993, 2069.

350 H. Voelker, U. Pieper, H. W. Roesky, and G. M. Sheldrik, *Z. Naturforsch., B: Chem. Sci.*, 1994, **49**, 255.
351 E. Niecke, B. Kramer, M. Nieger, and H. Severin, *Tetrahedron Lett.*, 1993, **34**, 4627.
352 H. U. Meyer, T. Severengiz, and W. W. du Mont, *Bull. Soc. Chim. Fr.*, 1993, **130**, 691.
353 L. Weber, H. Misiak, S. Buchwald, H-G. Stammler, and B. Neumann, *Organometallics*, 1994, **13**, 2139.
354 L. Weber, I. Schumann, H-G. Stammler, and B. Neumann, *J. Organomet. Chem.*, 1993, **443**, 175.
355 L. Weber and R. Kirchhoff, *Organometallics*, 1994, **13**, 1030.
356 A. L. Rizopoulos and M. P. Sigalas, *New J. Chem.*, 1994, **18**, 197.
357 B. Hansert and H. Vahrenkamp, *J. Organomet. Chem.*, 1993, **460**, C19.
358 S. Kurz and E. Hey-Hawkins, *J. Organomet. Chem.*, 1993, **462**, 203.
359 K. B. Dillon and H. P. Goodwin, *J. Organomet. Chem.*, 1994, **469**, 125.
360 T. Wegmann, M. Hafner, and M. Regitz, *Chem. Ber.*, 1993, **126**, 2525.
361 M. Hafner, T. Wegmann, and M. Regitz, *Synthesis*, 1993, 1247.
362 A. B. Kostitsyn, O. M. Nefedov, H. Heydt, and M. Regitz, *Synthesis*, 1994, 161.
363 K. Toyota, K. Tashiro, and M. Yoshifuji, *Angew. Chem., Int. Ed. Engl.*, 1993, **32**, 1163.
364 E. Niecke, M. Nieger, and P. Wenderoth, *J. Am. Chem. Soc.*, 1993, **115**, 6989.
365 H. Bock and M. Bankman, *Z. Anorg. Allg. Chem.*, 1994, **620**, 418.
366 A-C. Gaumont, J-C. Guillemin, and J-M. Denis, *J. Chem. Soc., Chem. Commun.*, 1994, 945.
367 J-C. Guillemin, J-L. Cabioch, X. Morise, J-M. Denis, S. Lacombe, D. Gonbeau, G. Pfister-Guillouzo, P. Guenot, and P. Savignac, *Inorg. Chem.*, 1993, **32**, 5021.
368 J-C. Guillemin, T. Janati, J-M. Denis, P. Guenot, and P. Savignac, *Tetrahedron Lett.*, 1994, **35**, 245.
369 M. Suzuki, K. Sho, K. Otani, T. Haruyama, and T. Saegusa, *J. Chem. Soc., Chem. Commun.*, 1994, 1191.
370 R. Streubel, M. Frost, M. Nieger, and E. Niecke, *Bull. Soc. Chim. Fr.*, 1993, **130**, 642.
371 B. Breit, and M. Regitz, *Chem. Ber.*, 1993, **126**, 1945.
372 E. Fuchs, F. Krebs, H. Heydt, and M. Regitz, *Tetrahedron*, 1994, **50**, 759.
373 W. Eisfeld, M. Slany, U. Bergsträsser, and M. Regitz, *Tetrahedron Lett.*, 1994, **35**, 1527.
374 S. J. Goede, M. A. Dam, and F. Bickelhaupt, *Rec. Trav. Chim. Pays-Bas*, 1994, **113**, 278.
375 J. Grobe, T. Grosspietsch, D. Le Van, B. Krebs, and M. Läge, *Z. Naturforsch., B: Chem. Sci.*, 1993, **48**, 1203.
376 J. Grobe, D. Le Van, and G. Lange, *Z. Naturforsch., B: Chem. Sci.*, 1993, **48**, 58.
377 S. N. Bhat, T. Berclaz, A. Jouati, and M. Geoffroy, *Helv. Chim. Acta*, 1994, **77**, 372.
378 K. Toyota, Y. Ishikawa, K. Shirabe, M. Yoshifuji, K. Okada, and K. Hirotsu, *Heteroat. Chem.*, 1993, **4**, 279.
379 G. Märkl, P. Kreitmeier, and R. Daffner, *Tetrahedron Lett.*, 1993, **34**, 7045.
380 G. Alcaraz, R. Reed, A. Baceiredo, and G. Bertrand, *J. Chem. Soc., Chem. Commun.*, 1993, 1354.
381 U. Heim, H. Pritzkow, U. Fleischer, and H. Grützmacher, *Angew. Chem., Int. Ed. Engl.*, 1993, **32**, 1359.
382 G. David, E. Niecke, M. Nieger, and J. Radseck, *J. Am. Chem. Soc.*, 1994, **116**, 2191.

383 M. J. Hervé, G. Etemad-Moghadam, M. Gouygou, D. Gonbeau, M. Koenig, and G. Pfister-Guillouzo, *Inorg. Chem.*, 1994, **33**, 596.
384 W. W. Schoeller, W. Haug, and J. Strutwolf, *Bull. Soc. Chim. Fr.*, 1993, **130**, 636.
385 M. T. Nguyen, L. Landuyt, and L. G. Vanquickenborne, *J. Chem. Soc., Faraday Trans.*, 1994, 1771.
386 M. T. Nguyen, L. Landuyt, and L. G. Vanquickenborne, *Chem. Phys. Lett.*, 1993, **212**, 543.
387 O. Treutler, R. Ahlrichs, and M. Soleilhavoup, *J. Am. Chem. Soc.*, 1993, **115**, 8788.
388 N. J. Fitzpatrick, D. F. Brougham, P. J. Groarke, and M. T. Nguyen, *Chem. Ber.*, 1994, **127**, 969.
389 M. Yu. Antipin, A. N. Chernega, and Yu. T. Struchkov, *Phosphorus, Sulfur, Silicon, Related Elem.*, 1993, **78**, 289.
390 A. N. Chernega, V. V. Pen'kovskii, and V. D. Romanenko, *Zh. Obshch. Khim.*, 1993, **63**, 60 (*Chem. Abstr.*, 1993, **119**, 139 328).
391 L. N. Alekseiko, V. V. Penkovsky, and V. I. Kharchenko, *Teor. Eksp. Khim.*, 1992, **28**, 329 (*Chem. Abstr.*, 1993, **119**, 250 051).
392 L. Weber, R. Kirchoff, and R. Boese, *Chem. Ber.*, 1993, **126**, 1963.
393 L. Weber, A. Rühlicke, H-G. Stammler, and B. Neumann, *Organometallics*, 1993, **12**, 4653.
394 L. Weber and A. Rühlicke, *J. Organomet. Chem.*, 1994, **470**, C1.
395 P. B. Hitchcock, R. M. Matos, M. F. Meidine, J. F. Nixon, B. F. Trigo Passos, D. Barion, and E. Niecke, *J. Organomet. Chem.*, 1993, **461**, 61.
396 A. Jouaiti, M. Geoffroy, and G. Bernardinelli, *J. Chem. Soc., Dalton Trans.*, 1994, 1685.
397 D. J. Brauer, A. Ciccu, J. Fischer, G. Hessler, and O. Stelzer, *J. Organomet. Chem.*, 1993, **462**, 111.
398 L. Weber and R. Kirchhoff, *Organometallics*, 1994, **13**, 1030.
399 M. Gouygou, J-C. Daran, B. Heim, and Y. Jeannin, *J. Organomet. Chem.*, 1993, **460**, 219.
400 M. Yoshifuji, K. Toyota, T. Uesugi, I. Miyahara, and K. Hirotsu, *J. Organomet. Chem.*, 1993, **461**, 81.
401 A. Marinetti, L. Ricard, F. Mathey, M. Siany, and M. Regitz, *Tetrahedron*, 1994, **49**, 10279.
402 H. Jun and R. J. Angelici, *Organometallics*, 1993, **12**, 4265.
403 H. Jun, V. G. Young, and R. J Angelici, *Organometallics*, 1994, **13**, 2444.
404 G. Becker and K. Hubler, *Z. Anorg. Allg. Chem.*, 1994, **620**, 405.
405 D. McNaughton and D. N. Bruget, *J. Mol. Spectrosc.*, 1993, **161**, 336.
406 W. Dong, S. Lacombe, D. Gonbeau, and G. Pfister-Guillouzo, *New J. Chem.*, 1994, **18**, 629.
407 M. Regitz, *J. Heterocyclic Chem.*, 1994. **31**, 663.
408 U. Bergsträsser, J. Stannek, and M. Regitz, *J. Chem. Soc., Chem. Commun.*, 1994, 1121.
409 B. Breit, R. Boese, and M. Regitz, *J. Organomet. Chem.*, 1994, **464**, 41.
410 J. Grobe, D. Le Van, B. Broschk, and L. S. Kobrina, *Tetrahedron Lett.*, 1993, **34**, 4619.
411 F. Meyer, P. Paetzold, and U. Englert, *Chem. Ber.*, 1994, **127**, 93.
412 A. N. Chernega, G. N. Koidan, A. P. Marchenko, and A. A. Korkin, *Heteroat. Chem.*, 1993, **4**, 365.
413 G. Becker, M. Bohringer, R. Gleiter, K-H. Pfeifer, J. Grobe, D. Le Van, and M. Hegemann, *Chem. Ber.*, 1994, **127**, 1041.
414 D. Carmichael, S. I. Al-Resayes, and J. F. Nixon, *J. Organomet. Chem.*, 1993, **453**, 207.

415 S. I. Al-Resayes, C. Jones, M. J. Maah and J. F. Nixon, *J. Organomet. Chem.*, 1994, **468**, 107.

416 S. I. Al-Resayes and J. F. Nixon, *Inorg. Chim. Acta*, 1993, **212**, 265.

417 P. Binger, F. Sandmeyer, C. Krüger, J. Kuhnigk, R. Goddard, and G. Erker, *Angew. Chem., Int. Ed. Engl.*, 1994, **33**, 197.

418 P. B. Hitchcock, M. J. Maah, J. F. Nixon, and M. Green, *J. Organomet. Chem.*, 1994, **466**, 153.

419 F. G. N. Cloke, K. R. Flower, P. B. Hitchcock, and J. F. Nixon, *J. Chem. Soc., Chem. Commun.*, 1994, 489.

420 P. B. Hitchcock, C. Jones, and J. F. Nixon, *Angew. Chem., Int. Ed. Engl.*, 1994, **33**, 463.

421 H. Jun and R. J. Angelici, *Organometallics*, 1994, **13**, 2454.

422 L. Weber, I. Schumann, T. Schmidt, H-G. Stammler, and B. Neumann, *Z. Anorg. Allg. Chem.*, 1993, **619**, 1759.

423 F. T. Chau, Y. W. Tang, and X. Song, *THEOCHEM*, 1993, **99**, 233.

424 N. Burford, J. A. C. Clyburne, S. Mason, and J-F. Richardson, *Inorg. Chem.*, 1993, **32**, 4988.

425 D. Barion, C. Gärner-Winkhaus, M. Link, M. Nieger, and E. Niecke, *Chem. Ber.*, 1993, **126**, 2187.

426 A. D. Averin, N. V. Lukashev, M. A. Kazankova, and I. P. Beletskaya, *Mendeleev. Commun.*, 1993, 68.

427 M. Link, E. Niecke, and M. Nieger, *Chem. Ber.*, 1994, **127**, 313.

428 F. U. Seifert and G. V. Röschenthaler, *Z. Naturforsch., B: Chem. Sci.*, 1993, **48**, 1089.

429 V. D. Romanenko, V. L. Rudzevich, A. O. Gudima, M. Sanchez, A. B. Rozhenko, A. N. Chernega, and M. R. Mazières, *Bull. Soc. Chim. Fr.*, 1993, **130**, 726.

430 V. D. Romanenko, G. V. Reitel, L. S. Kachkovskaya, M. Mikolajczyk, J. Omieyanczuk, and W. Pierlikowska, *Zh. Obshch. Khim.*, 1993, **63**, 1182 (*Chem. Abstr.*, 1994, **120**, 8670).

431 I. I. Patsanovskii, Yu. Z. Stepanova, E. A. Ishmaeva, V. D. Romanenko, and L. N. Markowskii, *Zh. Obshch. Khim.*, 1993, **63**, 561 (*Chem. Abstr.*, 1993, **119**, 203 489).

432 G. David, E. Niecke, M. Nieger, V. von der Gönna, and W. W. Schoeller, *Chem. Ber.*, 1993 **126**, 1513.

433 N. Burford, J. A. C. Clyburne, P. K. Bakshi, and T. S. Cameron, *J. Am. Chem. Soc.*, 1993, **115**, 8829.

434 T. L. Allen and W. H. Fink, *Inorg. Chem.*, 1993, **32**, 4230.

435 M. Driess and H. Pritzkow, *J. Chem. Soc., Chem. Commun.*, 1993, 1585.

436 G. Jochem, H. Nöth, and A. Schmidpeter, *Angew. Chem., Int. Ed. Engl.*, 1993, **32**, 1089.

437 L. D. Quin, S. Jankowski, J. Rudzinski, A. G. Sommese, and X-P. Wu, *J. Org. Chem.*, 1993, **58**, 6212.

438 D. J. Berger, P. P. Gaspar, R. S. Grev, and F. Mathey, *Organometallics*, 1994, **13**, 640.

439 R. Strubel, N. H. T. Huy, and F. Mathey, *Synthesis*, 1993, 763.

440 J-T. Hung, S-W. Yang, G. M. Gray, and K. Lammertsma, *J. Org. Chem.*, 1993, **58**, 6786.

441 J. Ho, T. L. Breen, A. Ozarowski, and D. W. Stephan, *Inorg. Chem.*, 1994, **33**, 865.

442 Z. Hou, T. L. Breen, and D. W. Stephan, *Organometallics*, 1993, **12**, 3158.

443 P. P. Power, *Angew. Chem., Int. Ed. Engl.*, 1993, **105**, 850.

444 N. Burford, P. Losier, C. Macdonald, V. Kyrimis, P. K. Bakshi, and T. S. Cameron, *Inorg. Chem.*, 1994, **33**, 1434.

445 F. Boutonnet, M. Zablocka, A. Igau, J-P. Majoral, J. Jaud, and K. M. Pietrusiewicz, *J. Chem. Soc., Chem. Commun.*, 1993, 1487.
446 H. Lang, M. Winter, M. Leise, O. Walter, and L. Zsolnai, *J. Chem. Soc., Chem. Commun.*, 1994, 595.
447 H. Nakazawa, Y. Yamaguchi, and K. Miyoshi, *J. Organomet. Chem.*, 1994, **465**, 193.
448 M. Leise, L. Zsolnai, and H. Lang, *Polyhedron*, 1993, **12**, 1257.
449 H. Lang and M. Leise, *J. Organomet. Chem.*, 1993, **456**, C4.
450 M. Leise, H. Lang, W. Imohof, and L. Zsolnai, *Chem. Ber.*, 1993, **126**, 1077.
451 M. Yoshifuji, D-L. An, K. Tokota, and M. Yasunami, *Tetrahedron Lett.*, 1994, **35**, 4379.
452 M. Yoshifuji, K. Kamijo, and K. Toyota, *Tetrahedron Lett.*, 1994, **35**, 3971.
453 M. Yoshifuji, A. Otoguro, and K. Toyota, *Bull. Chem. Soc. Jpn.*, 1994, **67**, 1503.
454 M. Yoshifuji, S. Sangu, M. Hirano, and K. Toyota, *Chem. Lett.*, 1993, 1715.
455 A. Ruban, M. Nieger, and E. Niecke, *Angew. Chem., Int. Ed. Engl.*, 1993, **32**, 1419.
456 W. Schilbach, V. von der Gönna, D. Gudat, M. Nieger, and E. Niecke, *Angew. Chem., Int. Ed. Engl.*, 1994, **33**, 982.
457 L. D. Quin, Z-P. Wu, N. D. Sadanani, I. Lukes, A. S. Ionkin, and R. O. Day, *J. Org. Chem.*, 1994, **59**, 120.
458 L. D. Quin, J. S. Tang, G. S. Quin, and G. Keglevich, *Heteroat. Chem.*, 1993, **4**, 189.
459 G. Keglevich, K. Ujszaszy, L. D. Quin, and G. S. Quin, *Heteroat. Chem.*, 1993, **4**, 559.
460 R. Streubel, J. Jeske, P. G. Jones, and R. Herbst-Irmer, *Angew. Chem., Int. Ed. Engl.*, 1994, **33**, 80.
461 M. Soleilhavoup, A. Baceiredo, F. Dahan, and G. Bertrand, *J. Chem. Soc., Chem. Commun.*, 1994, 337.
462 M. L. Sierra, C. Charrier, L. Ricard, and F. Mathey, *Bull. Soc. Chim. Fr.*, 1993, **130**, 521.
463 A. E. Ferao, B. Deschamps, and F. Mathey, *Bull. Soc. Chim. Fr.*, 1993, **130**, 695.
464 N. H. T. Huy and F. Mathey, *Organometallics*, 1994, **13**, 925.
465 E. Deschamps, L. Ricard, and F. Mathey, *Angew. Chem., Int. Ed. Engl.*, 1994, **33**, 1158.
466 J-J. Brunet, M. Gomez, H. Hajouji, and D. Neibecker, *J. Organomet. Chem.*, 1993, **463**, 205.
467 J. J. Brunet, M. Gomez, H. Hajouji, and D. Neibecker, *Phosphorus, Sulfur, Silicon, Related Elem.*, 1993, **85**, 207.
468 F. Laporte, F. Mercier, L. Ricard, and F. Mathey, *J. Am. Chem. Soc.*, 1994, **116**, 3306.
469 F. Nief and L. Ricard, *J. Organomet. Chem.*, 1994, **464**, 149.
470 M. Niecke, M. Nieger, and P. Wenderoth, *Angew. Chem., Int. Ed. Engl.*, 1994, **33**, 353.
471 A. J. Ashe and P. M. Savla, *J. Organomet. Chem.*, 1993, **461**, 1.
472 J. Kurita, M. Ishii, S. Yasuike, and T. Tsuchiya, *J. Chem. Soc., Chem. Commun.*, 1993, 1309.
473 A. Dore, D. Fabbri, S. Gladiali, and O. De Lucchi, *J. Chem. Soc., Chem. Commun.*, 1993, 1124.
474 S. M. Bachrach, *J. Org. Chem.*, 1993, **58**, 5414.
475 S. M. Bachrach and L. Perriott, *J. Org. Chem.*, 1994, **59**, 3394.
476 F. Laporte, F. Mercier, L. Ricard, and F. Mathey, *Bull. Soc. Chim. Fr.*, 1994, **130**, 843.
477 F. Mercier and F. Mathey, *J. Organomet. Chem.*, 1993, **462**, 103.

478 G. Keglevich, *Synthesis*, 1993, 931.
479 M. B. Hocking and F. W. van der Voort Maarschalk, *Can. J. Chem.*, 1993, **71**, 1873.
480 G. Keglevich, *Magy. Kem. Lapja*, 1992, **47**, 456 (*Chem. Abstr.*, 1993, **119**, 117 337).
481 W. L. Wilson, N. W. Alcock, E. C. Alyea, S. Song, and J. H. Nelson, *Bull. Soc. Chim. Fr.*, 1993, **130**, 673.
482 S. Motoki, T. Sakai, I. Shinoda, T. Saito, and T. Uchida, *Chem. Lett.*, 1993, 1563.
483 E. Lindner, C. Haase, H. A. Mayer, M. Kemmler, R. Fawzi, and M. Steimann, *Angew. Chem., Int. Ed. Engl.*, 1993, **32**, 1424.
484 E. Lindner, T. Schlenker, and C. Haase, *J. Organomet. Chem.*, 1994, **464**, C31.
485 A. Baceiredo, M. Nieger, E. Niecke, and G. Bertrand, *Bull. Soc. Chim. Fr.*, 1993, **130**, 757.
486 G. Stieglitz, B. Neumüller, and K. Dehnicke, *Z. Naturforsch., B: Chem. Sci.*, 1993, **48**, 730.
487 A. Schmidpeter, F. Steinmueller, and E. Ya. Zabotina, *J. Prakt. Chem./Chem. Ztg.*, 1993, **335**, 458.
488 Y. K. Rodi, L. Lopez, C. Malavaud, M-T. Boisdon, and J-P. Fayet, *Can. J. Chem.*, 1993, **71**, 1200.
489 Y. K. Rodi, L. Lopez, J. Bellan, J. Barrans, and E. M. Essassi, *Phosphorus, Sulfur, Silicon, Related Elem.*, 1993, **85**, 224.
490 P. B. Hitchcock, R. M. Matos, and J. F. Nixon, *J. Organomet. Chem.*, 1993, **462**, 319.
491 J. F. Nixon and G. J. D. Sillett, *J. Organomet. Chem.*, 1993, **461**, 237.
492 C. Müller, R. Bartsch, A. Fischer, and P. G. Jones, *J. Organomet. Chem.*, 1993, **453**, C16.
493 P. Gradoz, D. Baudry, M. Ephritikhine, M. Lance, M. Nierlich, and J. Vigner, *J. Organomet. Chem.*, 1994, **466**, 107.
494 D. Carmichael, L. Ricard, and F. Matthey, *J. Chem. Soc., Chem. Commun.*, 1994, 1167.
495 L. Brunet, F. Mercier, L. Ricard, and F. Mathey, *Angew. Chem., Int. Ed. Engl.*, 1994, **33**, 742.
496 A. Houlton, R. M. G. Roberts, J. Silver, and J. Zakrzewski, *J. Organomet. Chem.*, 1993, **456**, 107.
497 W. L. Wilson, J. A. Rahn, N. W. Alcock, J. Fischer, J. H. Frederick, and J. H. Nelson, *Inorg. Chem.*, 1994, **33**, 109.
498 P. Le Floch and F. Mathey, *J. Chem. Soc., Chem. Commun.*, 1993, 1295.
499 V. Jones and G. Frenking, *Chem. Phys. Lett.*, 1993, **210**, 211.
500 P. Le Floch, D. Carmichael, L. Ricard, and F. Mathey, *J. Am. Chem. Soc.*, 1993, **115**, 10665.
501 P. Le Floch, L. Ricard, and F. Mathey, *J. Chem. Soc., Chem. Commun.*, 1993, 789.
502 P. Le Floch, L. Ricard, and F. Mathey, *Bull. Soc. Chim. Fr.*, 1994, **131**, 330.
503 G. Keglevich, L. Toke, A. Kovacs, G. Toth, and K. Ujszaszy, *Heteroat. Chem.*, 1993, **4**, 61.
504 M. Shiotsuka and Y. Matsuda, *Chem. Lett.*, 1994, 351.
505 C. Elschenbroich, M. Nowotny, J. Kroker, A. Behrendt, W. Massa, and S. Wocadlo, *J. Organomet. Chem.*, 1993, **459**, 157.
506 C. Elschenbroich, F. Baer, E. Bilger, D. Mahrwald, M. Nowotny, and B. Metz, *Organometallics*, 1993, **12**, 3373.
507 E. Fluck, K. Bieger, G. Heckmann, and B. Neumüller, *Z. Anorg. Allg. Chem.*, 1994, **620**, 483.
508 E. Fluck, K. Bieger, G. Heckmann, and B. Neumüller, *J. Organomet. Chem.*, 1993, **459**, 73.

2
Pentaco-ordinated and Hexaco-ordinated Compounds

BY C. D. HALL

Introduction - Despite the overall decline in output, the year has produced further interesting developments in the field of hypervalent phosphorus chemistry. The Symposium on Main Group Chemistry (Texas, 1993) - dedicated to Professor Alan H.Cowley, was reported in Volumne 87 of Phosphorus, Sulfur and Silicon and included a number of lectures on pentaco-ordinate phosphorus chemistry. A new intramolecular cyclisation of (1) to form phosphoranes (2) containing the P-H bond was reported and properties of these molecules were discussed.[1] The anions derived by treatment with base, coordinated metal ions - particularly nickel, to form highly active catalysts for the oligomerisation of alkenes. Reactions of phosphoranes containing the P-H bond with borane were reported by Contreras and comparisons were made with the corresponding boron heterocycles.[2] In a presentation dealing with main group chemistry within macrocyclic rings and baskets, Lattman *et al.* discussed the pentacoordinate trigonal bipyramidal (*tbp*) geometry about phosphorus coordinated within cyclen as in (3) which was maintained when bound to a transition metal as in (4) or (5).[3] They also reported that a calix-[4]-arene appears to stabilise hexacoordinate phosphorus under certain circumstances (*vide infra*). Formation of hypervalent compounds in silicon and phosphorus chemistry has been reviewed by Holmes *et al.* by comparing the reactions of phosphites and chlorosilanes with diols.[4] Crystal structures and solution nmr behaviour have also been compared in an excellent review by Holmes leading to a catalogue of ^{31}P nmr data for mono- and bicyclic phosphoranes containing six- to eight-membered rings.[5] The effects caused by ring size, the number of rings, the nature of the ring heteroatoms and the electronegativity of the attached exocyclic substituents were correlated with structural and bonding features, particularly the changes in P-O π-bonding. The formation of hexacoordinated phosphorus through incorporation of sulphur as a ring heteratom was proposed as a potentially important factor in enzyme-phosphate reactions at sites where sulphur may be present.

The reactions of phosphorus chlorides with imines have been reviewed[6] and lead in some instances (e.g.with 6) to the formation of pentacordinate phosphorus compounds (e.g. 8). On the whole, however, this chemistry involves tri- and tetracoordinate phosphorus. Finally in the introductory section, it should be mentioned that the stereoelectronic effect in oxyphosphorane species has been reconsidered in a recent theoretical paper.[7] The properties of various oxyphosphoranes (e.g. 9a-c) were examined by ab-initio M.O. calculations, the energies of the stationary points being evaluated at the MP2 level with the 6-31 + G* basis. The analysis indicated that the orientation of the equatorial methoxyl group in (9c) determined the mode of formation/cleavage of the axial PO^2/PO^5 bond. The dependence of the reactivity on the conformation of the equatorial PO^3 bond is in accord with predictions based on the stereoelectronic effect but the new calculations show that the "long-postulated n-σ^* interaction is not predominant in stereoelectronic control of the oxyphosphorane system."

2. Acyclic Phosphoranes - The tetrafluorophosphate anion (PF_4^-) was prepared as its tetramethylammonium salt from Me_4NF and PF_3 with acetonitrile, trifluoromethane or excess PF_3 as solvent.[8] The structure of the PF_4^- anion was studied theoretically and by variable temperature ^{31}P and ^{19}F nmr, infrared and Raman spectroscopy and by single crystal X-ray diffraction. The molecule possesses a pseudo *tbp* configuration with the axial bonds longer than the equatorial bonds and with a sterically active free valence electron pair. In solution it undergoes an intramolecular exchange process by the Berry mechanism and the vibrational frequencies for PF_4^- in solid Me_4NPF_4 are in excellent agreement with those calculated for the free gaseous anion.

Tertiary phosphine adducts of mixed halogens, R_3PIBr with R=alkyl or aryl, have been synthesised and their structures determined in the solid state and in solution.[9] The crystal structure data show that the molecule with R=Ph is a further example of the four coordinate molecular "spoke" structure, Ph_3P-X-X, previously established for Ph_3PX_2 (X=Br or I) and Ph_3AsI_2. Furthermore, X-ray powder diffraction studies indicate that the molecule is predominantly Ph_3P-I-Br rather than Ph_3P-Br-I. Raman spectra confirm the dominance of the P-I bond with minor peaks attributable to the P-Br structure. In $CDCl_3$ solution, $^{31}P\{^1H\}$ nmr studies indicate complete ionisation of the compounds to $R_3P^+I\ Br^-$ with no evidence for the corresponding $R_3P^+Br\ I^-$ and no sign of pentacoordinate species. The molecular "spoke" structure, however, is disputed by Gates *et al.* in a recent paper[10a] dealing with the Raman spectra of Ph_nPBr_{5-n} with n = 1-3. Two ionic modifications of $PhPBr_4$ were identified by Raman spectroscopy whereas Ph_2PBr_3 was found to exist in only one ionic form, $Ph_2P^+Br_2\ Br^-$. Furthermore, although agreeing that there is no evidence for pentacoordinate structures, Gates *et al.* claim that "evidence for the ionic formulation of $Ph_3{}^+PBr\ Br^-$ is overwhelming in contrast to the interpretation........ by McAuliffe[10b] and co-workers." It seems that the debate surrounding this issue has still a little way to run.

Phosphorylation of hexafluoropropanol (10) with PCl_5 or *tris*-(hexafluoroisopropoxy)-dichlorophosphorane (11) in the absence of base gave *tetrakis*-(hexafluoroisopropoxy)-chlorophosphorane (12) for the first time.[11] Subsequent reaction with benzaldehyde (13) or chloral (14) gave pentaalkoxyphosphoranes (15) and (16) respectively.

A number of acyclic σ-dialkoxyphosphoranes (18) have been prepared in moderate to high yields by the reaction of bromotriphenylphosphonium bromide (17) with the relevant sodium alkoxide in alcohol and ethereal solvents.[12] The method offers obvious advantages over the peroxide route and in some cases ($R=Pr^i$) higher yields than the Mitsunobu reaction but at the

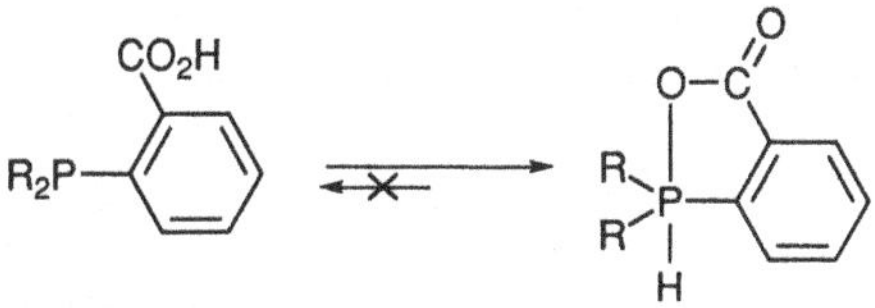

(1) R = c-C_6H_{11} (2)

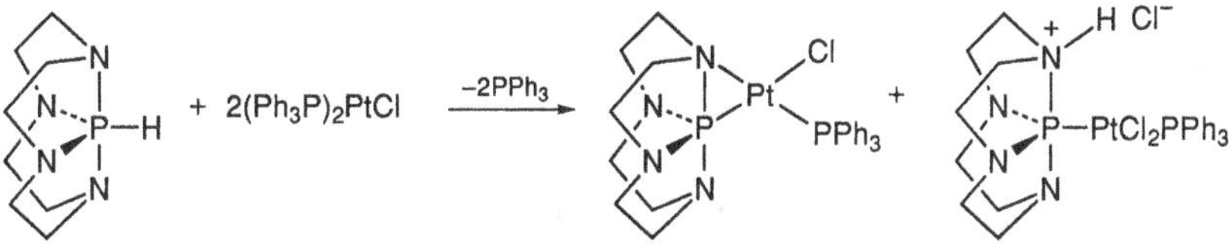

$+ 2(Ph_3P)_2PtCl \xrightarrow{-2PPh_3}$

(3) (4) (5)

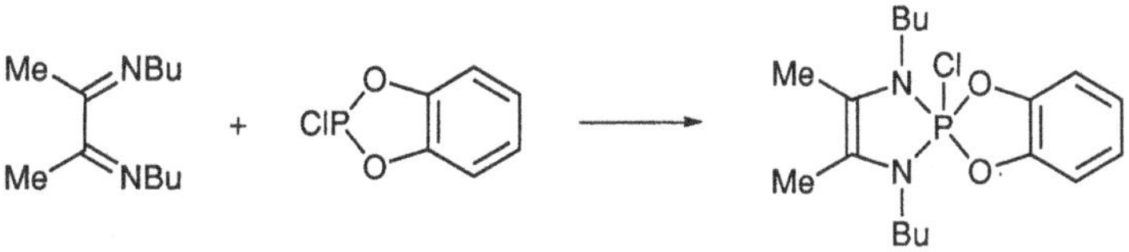

(6) (7) (8)

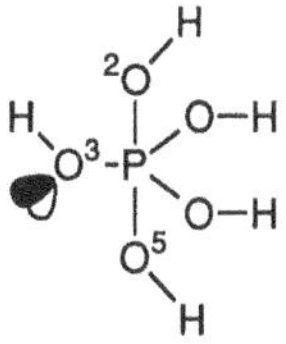

(9a) 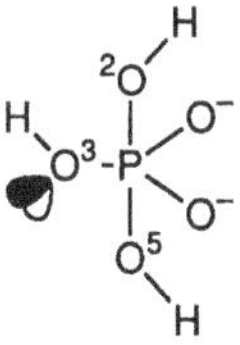 (9b) 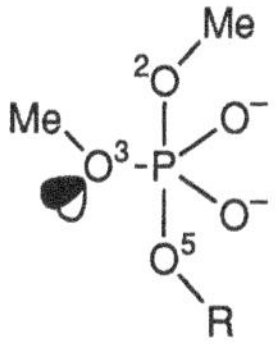 (9c)

$$4(CF_3)_2CHOH + PCl_5 \longrightarrow [(CF_3)_2CHO]_4PCl + 4HCl$$

(10) (12)

$$(10) + [(CF_3)_2CHO]_3PCl_2 \longrightarrow (12) + 2HCl$$

(11)

$$PhCHO + (12) \underset{60\ ^\circ C}{\overset{20\ ^\circ C}{\rightleftharpoons}} [(CF_3)_2CHO]_4POCHClPh$$

(13) (15)

$$Cl_3CCHO + (12) \xrightarrow{60\text{–}80\ ^\circ C} [(CF_3)_2CHO]_4P\text{–}OCHClCCl_3$$

(14) (16)

moment is less versatile than the sulphenate ester route pioneered by Denney[13] since the latter may be used with a whole range of tricoordinate phosphorus compounds.

Hydroxyphosphoranes (20) may be prepared by the oxidative addition of diols (19ab) to triphenylphosphine.[14] An analogous reaction with *o*-aminophenol (21) gave a 90% yield of the hydroxyaminophosphorane (22) which was stable to hydrolysis. The hydroxyphosphoranes (20 ab) were also obtained by the reaction of triphenylphosphine oxide with the diols using benzene as solvent. Unfortunately the paper lacks nmr data.

A remarkable *bis*-phosphorane (25) containing two hypervalent centres each with five P-C bonds has been synthesised by Furukawa et.al. by the reaction of dilithioacetylene (24) with two moles of the phosphonium salt (23).[15] Phosphorane (27) may be prepared by an analogous method from (26) and both pentacoordinate compounds react readily with water to form phenylacetylene.

3. Cyclic Phosphoranes - The reaction of dioxetanes (28a-c) with ylides (29a-d) gave phosphonium alkoxides (30a-f) in equilibrium with the pentacoordinate dioxa-2,5-phosphorinanes (31a-f).[16] Hydrolysis *via* the hydroxyphosphoranes (32a-f) gave the phosphine oxides (33a-f).

The reaction of the benzophosphole (34) with chlorine gave the pentacoordinated adduct (35) whereas the analogous diazabenzophosphole (36) gave the tetracoordinate adduct (37).[17] Both react with antimony pentachloride to give the phosphonium hexachloroantimonates (38) and (39).

In a sequel to earlier work on (40) and (41) which were shown to have *tbp* and octahedral geometry respectively, Holmes *et al.*synthesised (42) and showed by X-ray crystallography that the combination of Me and Bu^t groups led to an octahedral configuration with the sulphur atom of the eight-membered ring entering the coordination sphere.[18] The ^{1}H and ^{19}F solution nmr spectra were unable to distinguish the *tbp* and octahedral structures unequivocally but a ^{31}P nmr signal at -82.2ppm was consistent with octahedral structure {^{31}P for (41) = -82.4ppm} whereas the *tbp* structure of (40) gave $\delta^{31}P$ = -77.3ppm.

The reaction of *tris*-(dimethylamino)phosphine (43) with the *bis*-phenol (44) in the presence of N-chlorodiisopropylamine (45) gave the pentacoordinate compound (46) which was shown by X-ray crystallography to have a *tbp* structure displaced 29% towards a rectangular pyramid. The pentacoordinate structure was maintained in solution as evidenced by a ^{31}P nmr signal at -76.5ppm close to that observed for (40). The lack of P-S coordination was attributed to strong P-N π-bonding. Compound (49) was formed by an analogous reaction between (47) and (48) in the presence of (45) but attempts to obtain a structure by X-ray crystallography were unsuccessful. In solution, however, the ^{31}P nmr spectrum pointed to an octahedral structure with a $\delta^{31}P$ value of -82.0ppm indicating S-P coordination.[19]

A milestone paper has appeared on the characterisation of an optically active pentacoordinate phosphorane with asymmetry only at phosphorus.[20] This was achieved by separating the diastereomeric phosphoranes [50-(R)p + (S)p] into prisms and needles by repeated recrystallisation from MeOH-H_2O and then reducing the separate diastereomers with

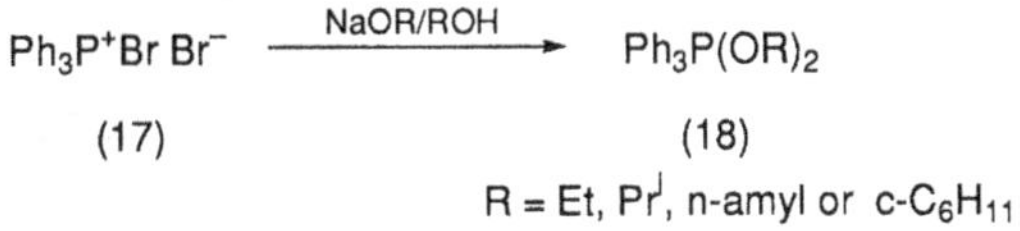

$$Ph_3P^+Br\ Br^- \xrightarrow{NaOR/ROH} Ph_3P(OR)_2$$

(17) (18)

R = Et, Pr^i, n-amyl or c-C_6H_{11}

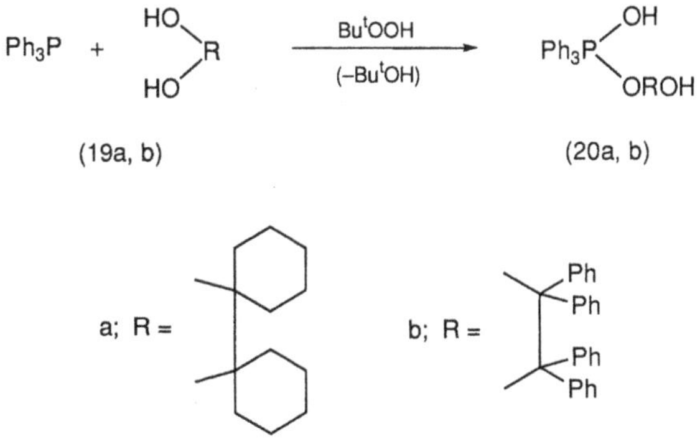

$$Ph_3P + HO{-}R{-}OH \xrightarrow[(-Bu^tOH)]{Bu^tOOH} Ph_3P(OH)(OROH)$$

(19a, b) (20a, b)

a; R =

b; R =

Ph Ph Ph Ph

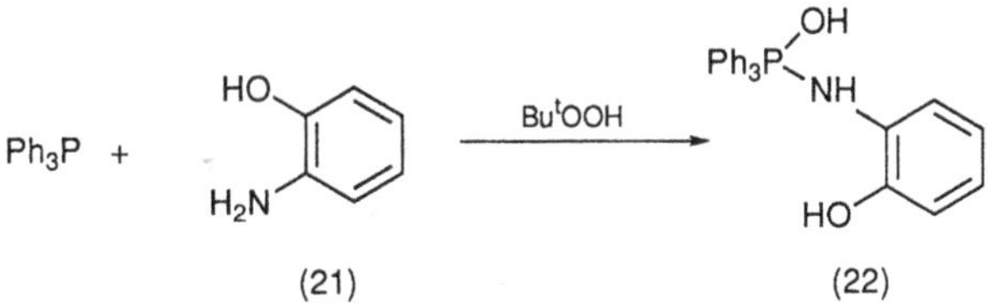

$$Ph_3P + \text{HO}, \text{H}_2\text{N} \xrightarrow{Bu^tOOH} Ph_3P(OH)(NH{-}C_6H_4{-}OH)$$

(21) (22)

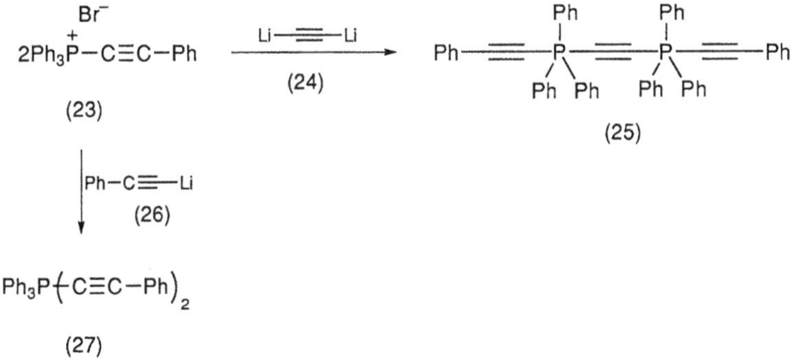

$$2Ph_3\overset{+}{P}{-}C{\equiv}C{-}Ph\ Br^- \xrightarrow[(24)]{Li{-}C{\equiv}C{-}Li} Ph{-}C{\equiv}C{-}PPh_3{-}C{\equiv}C{-}PPh_3{-}C{\equiv}C{-}Ph$$

(23) (25)

$$\downarrow Ph{-}C{\equiv}C{-}Li\ (26)$$

$$Ph_3P(C{\equiv}C{-}Ph)_2$$

(27)

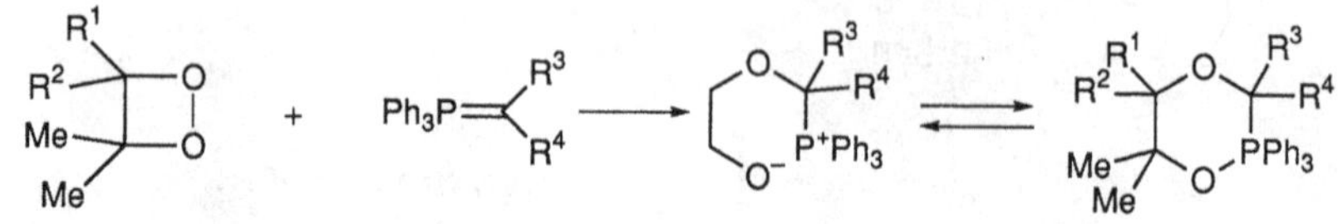

R^1 R^2 O O Me Me + Ph_3P R^3 R^4 → O R^3 R^4 P^+Ph_3 O^- ⇌ R^1 R^2 O R^3 R^4 Me Me O PPh_3

(28a–c) (29a–d) (30a–f) (31a–f)

a: $R^1 = R^2 = H$
b; $R^1 = Me, R^2 = H$
c; $R^1 = R^2 = Me$

a: $R^3 = R^4 = H$
b; $R^3 = Me, R^4 = H$
c; $R^3 = R^4 = Me$
d; $R^3 = Ph, R^4 = H$

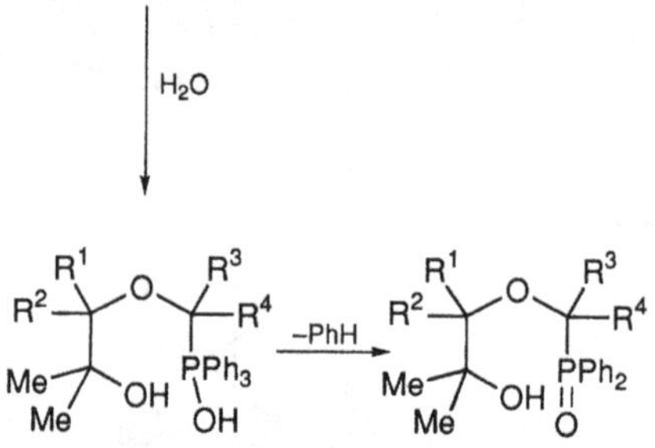

H_2O

R^1 R^2 O R^3 R^4 Me Me OH PPh_3 OH —–PhH→ R^1 R^2 O R^3 R^4 Me Me OH PPh_2 O

(32a–f) (33a–f)

a; $R^1, R^2, R^3, R^4 = H$
b; $R^1 = Me, R^2, R^3, R^4 = H$
c; $R^1, R^2 = Me, R^3, R^4 = H$
d; R^1 $R^2 = H, R^3 = Me, R^4 = H$
e; $R^1, R^2 = H, R^3, R^4 = Me$
f; $R^1, R^2, R^4 = H, R^3 = Ph$

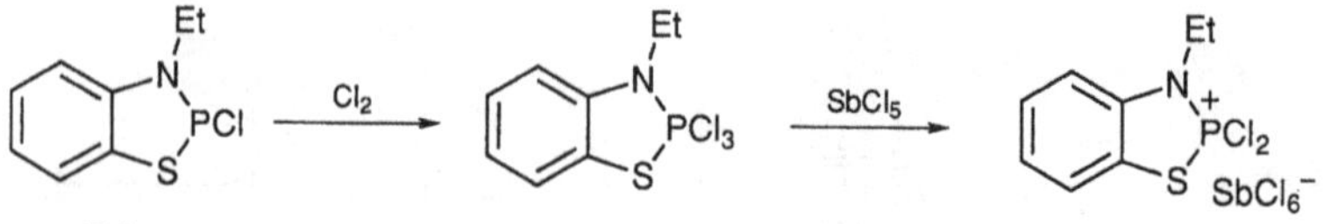

Et N PCl S —Cl_2→ Et N PCl_3 S —$SbCl_5$→ Et N^+ PCl_2 S $SbCl_6^-$

(34) (35) (38)

Et N PCl N Et —Cl_2→ Et N^+ PCl_2 N Et Cl^- —$SbCl_5$→ Et N^+ PCl_2 N Et $SbCl_6^-$

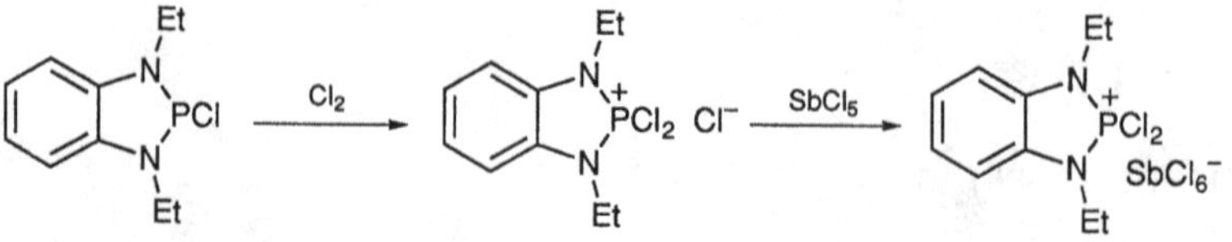

(36) (37) (39)

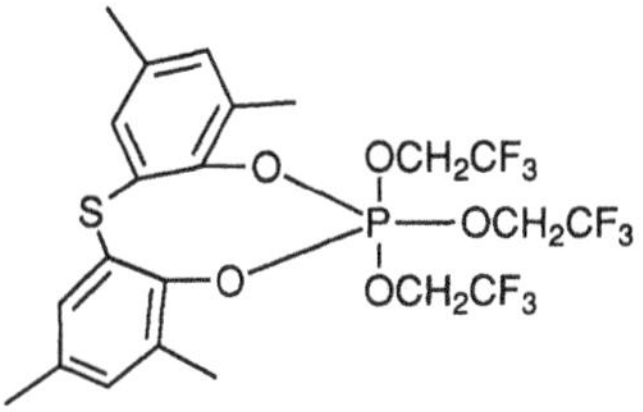

(40) 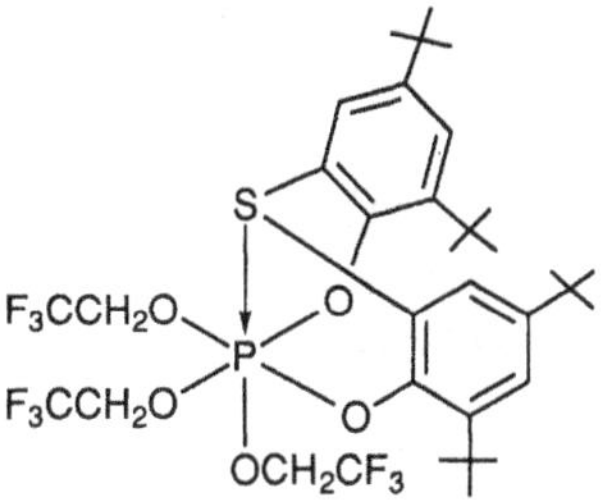 (41)

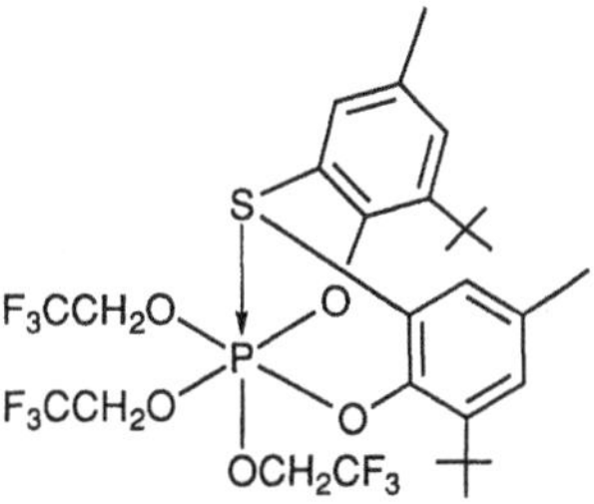

(42)

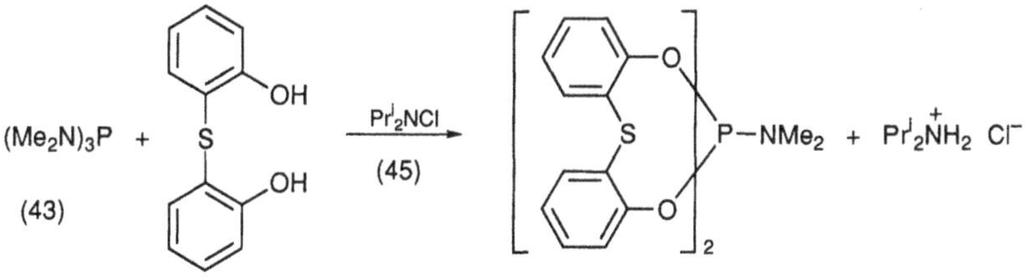

$(Me_2N)_3P$ + (44) $\xrightarrow[(45)]{Pr^i_2NCl}$ (46) + $Pr^i_2\overset{+}{N}H_2\ Cl^-$

(43)

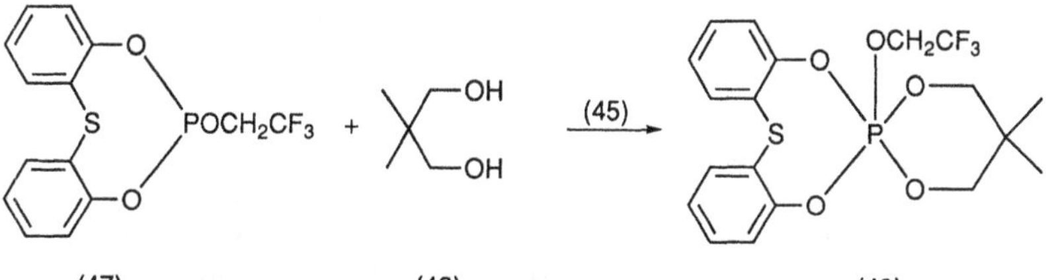

(47) + (48) $\xrightarrow{(45)}$ (49)

$LiAlH_4$ to afford [51-(S)p] and [51-(R)p]. The absolute stereochemistry about phosphorus as determined by X-ray crystallography *of the prisms* was found to be (R)p and the specific rotations of [51-(R)p] and [51-(S)p]) were $[\alpha]_{436}$ +108 (c 1.02, $CHCl_3$, 21°C) and $[\alpha]_{436}$ -107 (c 0.83, $CHCl_3$, 21°C) respectively. The rate of epimerisation of the compounds was determined in toluene and pyridine and the $\Delta G^{\ddagger}$ values at 373K were 33.8 kcal mol^{-1} and 33.5 kcal mol^{-1} in the respective solvents.

The first example of a phosphorane containing two oxaphosphetane rings (52) has been synthesised by Kawashima and X-ray crystallography supported by multinuclear nmr indicate a distorted *tbp* configuration with the oxygen atoms apical as expected. Thermolysis of the phosphorane (in d_8 toluene) at 190°C in a sealed tube was complete after 24h and gave quantitative yields of the alkene (53) and the vinylphosphinic acid (55) *via* (54).[21]

There follows a series of four papers by Houalla *et al.* on further extensions of the Atherton-Todd reaction to the synthesis of pentacoordinated phosphorus compounds with the phosphorus as a heteratom within a wide variety of macrocyclic structures. The first paper[22] describes the reaction of 1-hydridobicyclophosphoranes containing a secondary amine group (e.g. 56) with (57) which leads to a cyclic *bis*(bicyclophosphorane)-(58), as well as the symmetrical bicyclophosphorane (59). The opportunity to exploit the reaction in the synthesis of novel macrocyclic structures was too good to miss and compounds such as (62) were prepared from the corresponding *bis*-phosphorane (60) and the *bis*-phenol (61).[23]

The reaction was then extended to a series of novel *bis*-phosphoranes (63a-e), two macrocyclic *tris*-phosphoranes (64ab) and three *tetrakis*-phosphoranes (65ab) and (66)[24] all of which were characterised by multinuclear nmr and sometimes by elemental analysis or mass spectrometry. In the fourth paper, *bis*-bicyclic phosphoranes such as (69a-c) were obtained through Michael addition reactions between the *bis*-hydridobicyclophosphoranes (67a-c) with diacrylic diesters (e.g. 68).[25] Again characterisation was largely by multinuclear nmr.

The reaction of (70) with benzenesulphenate esters (71a-h) at 14°C in benzene gave a mixture of the phosphoranes (72a-h) and (73) which on heating between 40 and 50°C for several hours gave high yields of (72a-h) through an exchange reaction between (73) and the corresponding alcohol. The stable oxazaphosphoranes (72a-h) were isolated by distillation and characterised by elemental analysis, mass spectrometry and multinuclear nmr.[26]

An attempt to prepare the tricyclic phosphorane (78) by reaction of (74) with the disulphenate ester (75) in fact gave the dimer (77) *via* (76).[27] An X-ray crystal structure of (77) revealed a highly symmetrical molecule with *tbp* geometry at both phosphorus atoms and with the 12-membered ring attached to phosphorus diequatorially. The ring is strongly puckered about phosphorus (O-P-O∠=110°) and highly mobile at ambient on the nmr time scale. Variable temperature nmr gave $\Delta G^{\ddagger}$ (231K) = 10.1 kcal mol^{-1} for the coalescence of the OCH_2 groups.

Reactions of diastereomeric mixtures of imino-oxaphospholenes (79a-c) with hexafluoroacetone (80) gave in each case only one diastereomer of each of the bicyclicphosphoranes (81a-c). The structures of (81a) and (81b) were shown by X-ray crystallography to have distorted *tbp* geometries with the oxygen atoms apical and the O-P-O angles ca. 169°. The compounds were also fully characterised by analysis, mass spectrometry

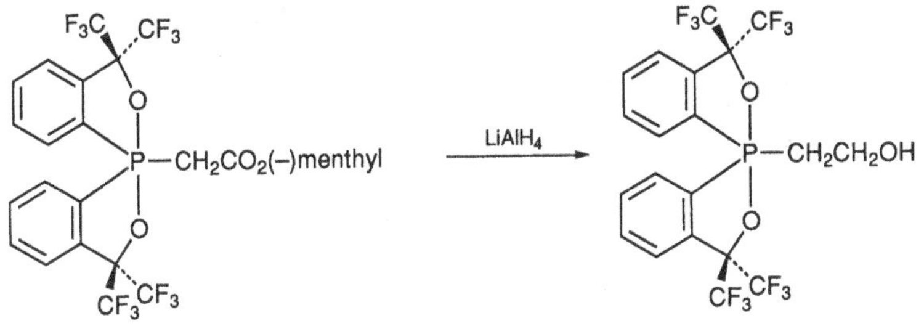

LiAlH4

(50)-$(R)_P$; (50)-$(S)_P$

(51)-$(S)_P$; (51)-$(R)_P$

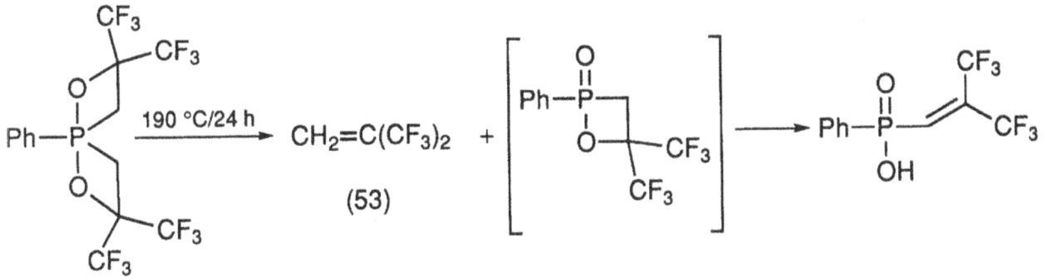

190 °C/24 h

$CH_2{=}C(CF_3)_2$

(53)

(52)

(54)

(55)

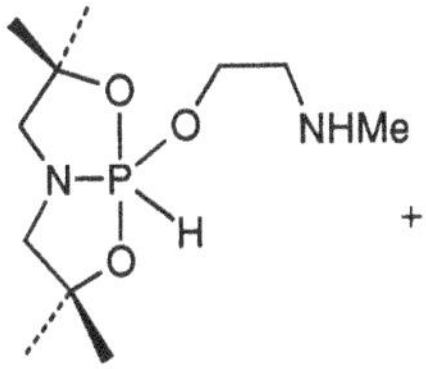

(56)

$\delta^{31}P = -35.7$ ($^1J_P = 796.8$)

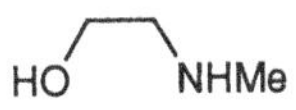

(57)

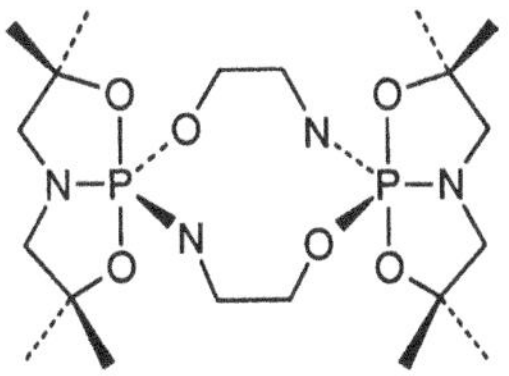

(58)

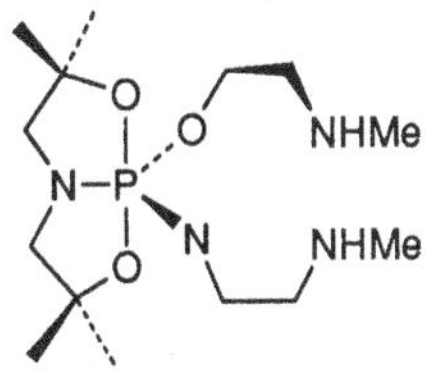

(59)

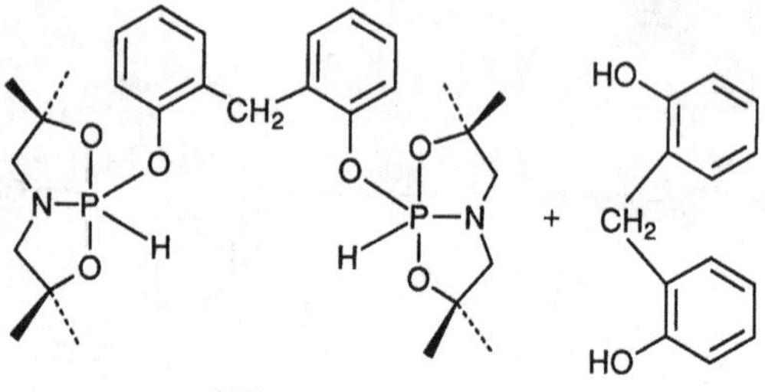

(60) (61)

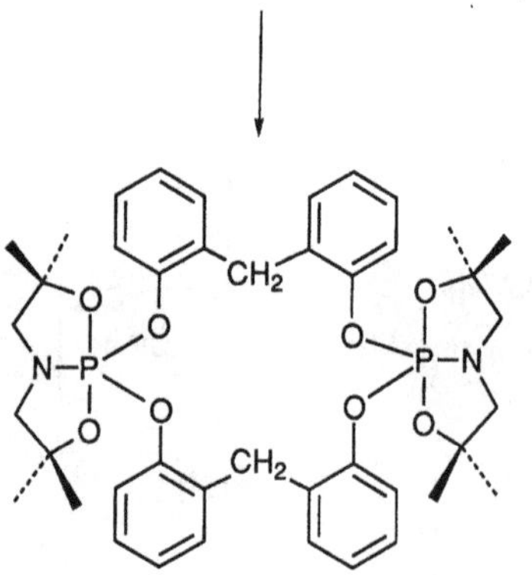

(62)

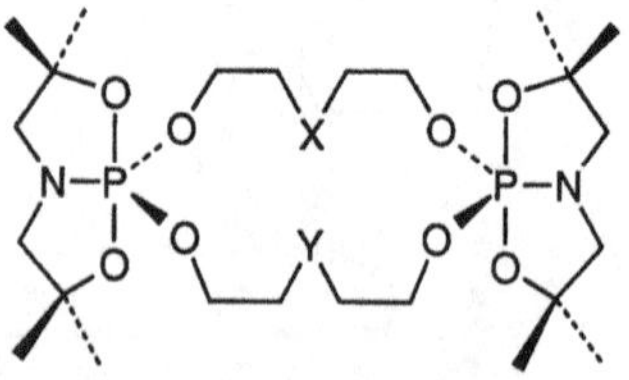

(63) a; X = S, Y = NBu^t
b; X = O, Y = S
c; X = NPh, Y = NBu^t
d; X = NMe, Y = O
e; X = NMe, Y = NPh

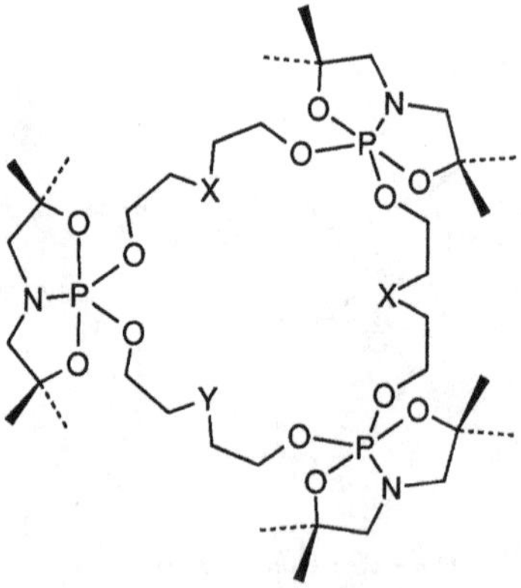

(64) a; X = S, Y = NBu^t
b; X = O, Y = S

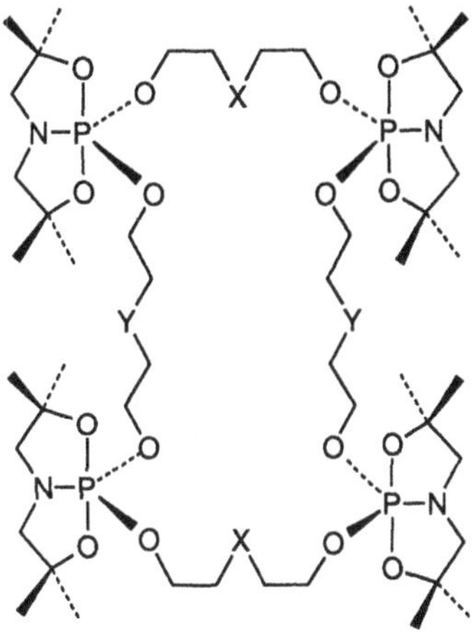

(65) a; X = S, Y = NBu^t
b; X = NPh, Y = NBu^t

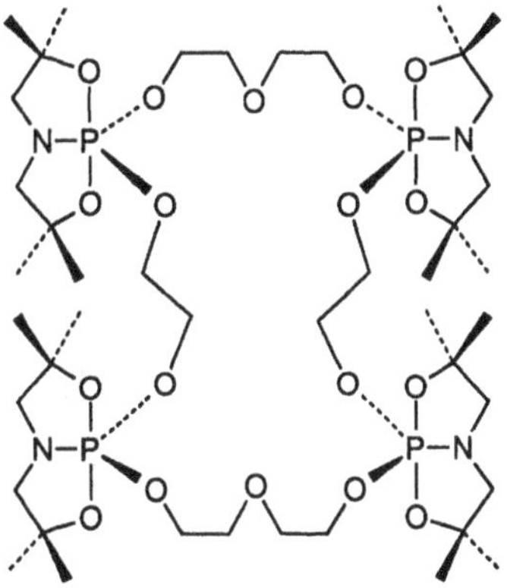

(66)

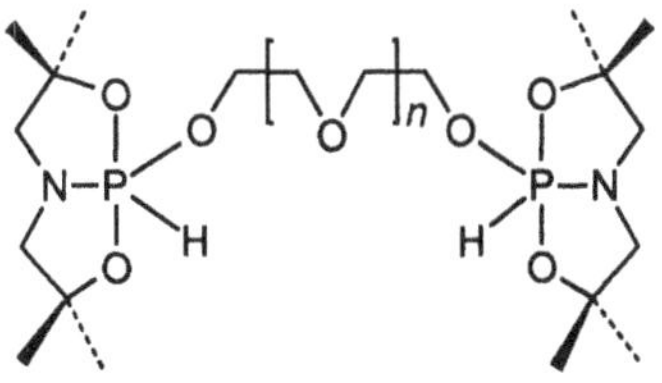

(67) a; $n = 1$
b; $n = 2$
c; $n = 3$

+

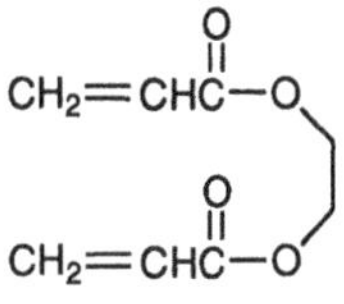

$CH_2{=}CHC(O){-}O$

$CH_2{=}CHC(O){-}O$

(68)

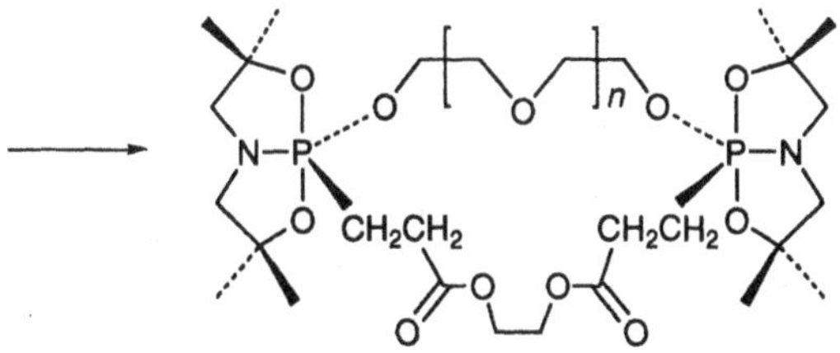

(69a–c)

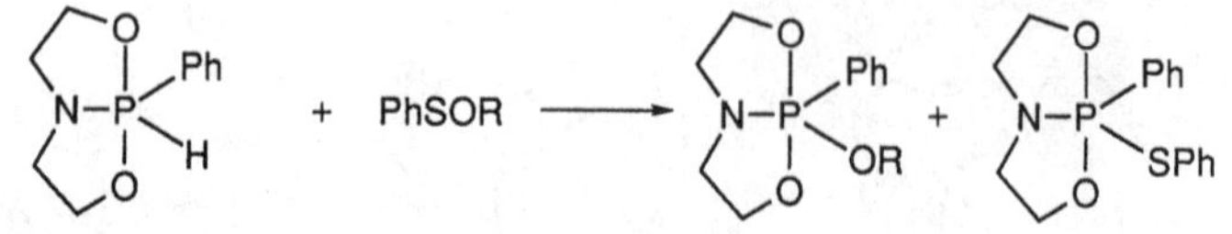

(70) + PhSOR → (72a–h) + (73)

(71a–h)

a; R = Me
b; R = Et
c; R = Pr^n
d; R = Pr^i
e; R = Bu^n
f; R = Bu^i
g; R = Pen^n
h; R = Hex^n

(73) + ROH —40–50 °C→ (72a–h) + PhSH

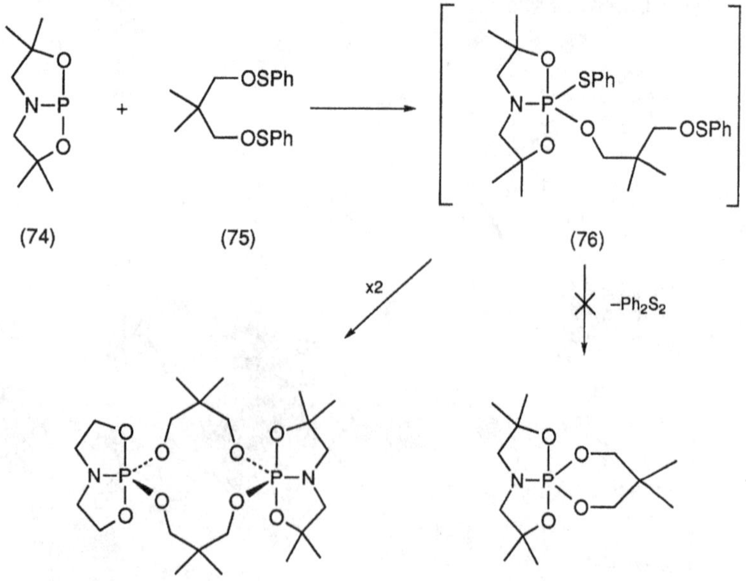

(74) + (75) → [(76)]

OSPh, OSPh, SPh, OSPh

(76) —x2→ (77)

(76) —✗ $-Ph_2S_2$→ (78)

and multinuclear nmr.[28]

During a study of possible donor-acceptor interactions in N,N',N'-trimethylethylenediamine substituted compounds it was found that the λ^5P-spirophosphorane (82) showed no evidence for intramolecular interaction by ^{1}H nmr.[29]

The reaction of the oxazaphospholine (83) with alcohols or phenols occurred with cleavage of the exocyclic P-N bond to form (84) followed by N-P migration of the diethylamino group to form (85) which dimerised to (86).[30]

As a result of a search for anti-viral (particularly anti-HIV) agents, Katalenic *et al.* discovered a novel type of nucleoside analogue (90) containing pentacoordinate phosphorus by the reaction of (87) with triphenyl phosphite (88) with (89) as the crucial intermediate.[31]

A useful article has appeared on the published crystal structures of metallated phosphoranes.[32] Of the fifteen structures reviewed the majority when compared to the non-metallated phosphoranes exhibit a marked preference for *tbp* geometry and the review suggests that this may be due to the π-donating abilities of the metal substituents. In a few cases, however, the metallated phosphoranes exhibit distortion along the turnstile coordinate in contrast to the Berry mechanism followed by the non-metallated phosphoranes. It is suggested that the considerable steric demands of the metal substituents in the equatorial plane of the phosphoranes may be responsible for the exceptional behaviour.

Hexacoordinate Phosphorus Compounds - Details of the chemistry of phosphorus containing calixarenes have appeared.[33] The conversion of the hexacoordinate phosphorus calixarene (91a) to the tricoordinate structure (93a) by heating or by treatment with CF_3CO_2H, occurs through the spectroscopically characterised intermediate (92a). Treatment of (91a) with butyllithium followed by methyl trifluoromethanesulphate led to (94a) the trimethylammonium analogue of (91a). When the t-butyl groups in the starting calixarene (R= Bu^t) were replaced by hydrogen (R=H) an analogous series of compounds (93b and 94b) could be prepared. Significantly higher temperatures were required to convert (91) to (93) with R= Bu^t and both the tricoordinate species were converted to the potassium salts (95ab) by treatment with potassium t-butoxide.When (93b) was reacted with n-butyllithium however, a product analogous to (95b) was formed together with another hexacoordinate structure (96) in which the bottom of the basket had reclosed by attack of the butyl anion on phosphorus followed by transfer of the hydrogen from the OH group to phosphorus. An X-ray structure of (91b) showed the calixarene in the cone conformation with phosphorus in an octahedral geometry lying inside the basket about 14pm above the plane of the four calixarene oxygens.

New ^{19}F and ^{31}P nmr evidence has appeared[34] on the partial hydrolysis in CH_2Cl_2 of PF_6^- to $PO_2F_2^-$ with the necessary water arising from the reagent ($AgPF_6$) and from traces in the solvent. In conclusion, some significant crystallographic evidence has appeared for seven coordinate phosphorus.[35] The reaction of excess 8-dimethylamino-1-naphthyllithium (97) with PBr_3 gave (98) which was shown by ^{1}H nmr to contain two sets of diastereotopic methyl groups due to coordination of the three nitrogen atoms with phosphorus. This was confirmed by X-ray crystallography which showed (98) to have a helicoidal shape with a pseudo three-fold

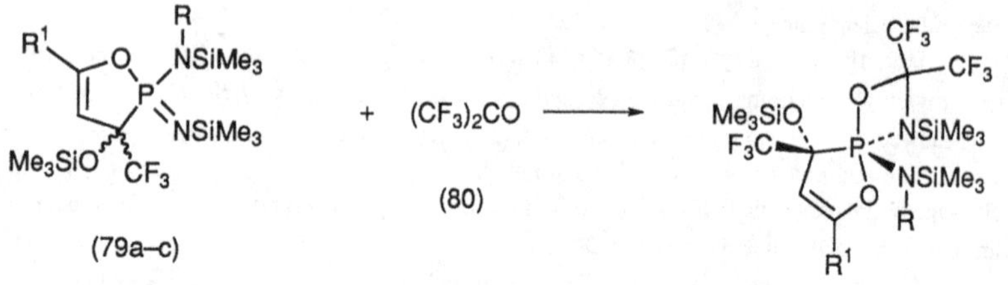

$(CF_3)_2CO$

(79a–c) + (80) → (81a–c)

a; $R^1 = CF_3$, R = H
b; R^1 = Me, R = H
c; $R^1 = CF_3$, $R = SiMe_3$

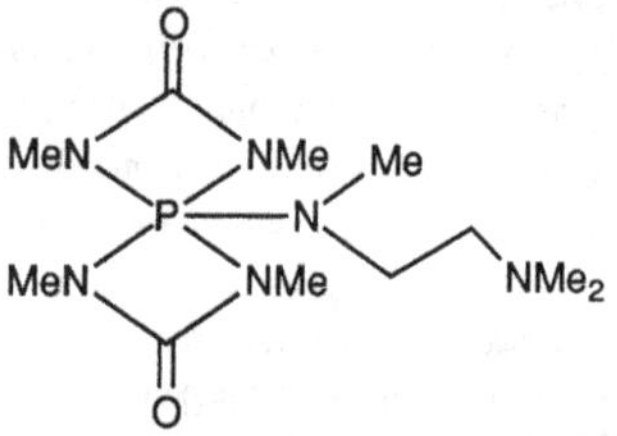

(82)

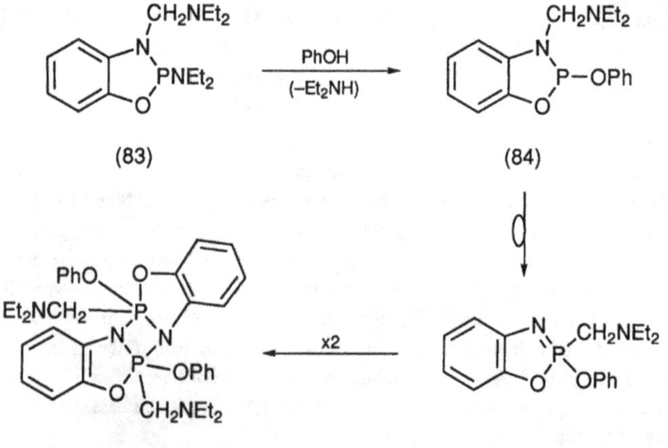

(83) —PhOH ($-Et_2NH$)→ (84) → (85) —x2→ (86)

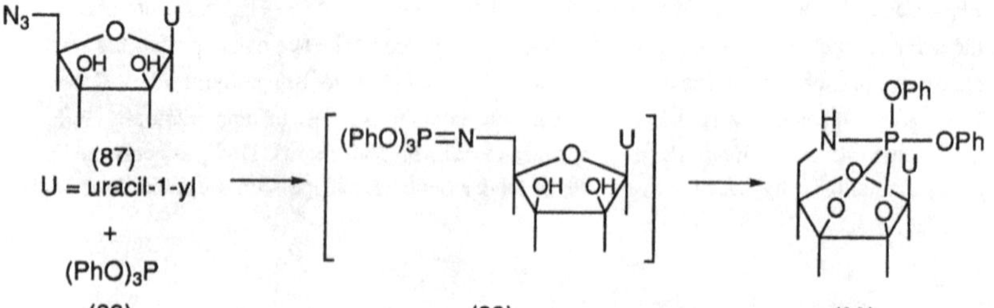

(87) U = uracil-1-yl + $(PhO)_3P$ (88) → [(89)] → (90)

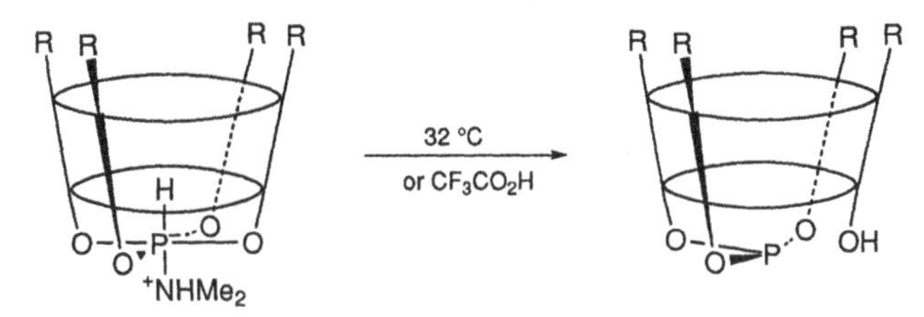

(91ab) $\delta^{31}P = -120$ (R = Bu^t)

a; R = Bu^t

b; R = H

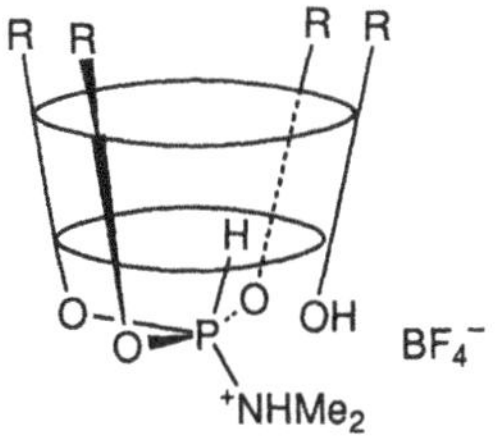

(93ab) $\delta^{31}P = +113$ (R = Bu^t)

R R R R

H

O O P O OH BF_4^-

$^{+}NHMe_2$

(92a)

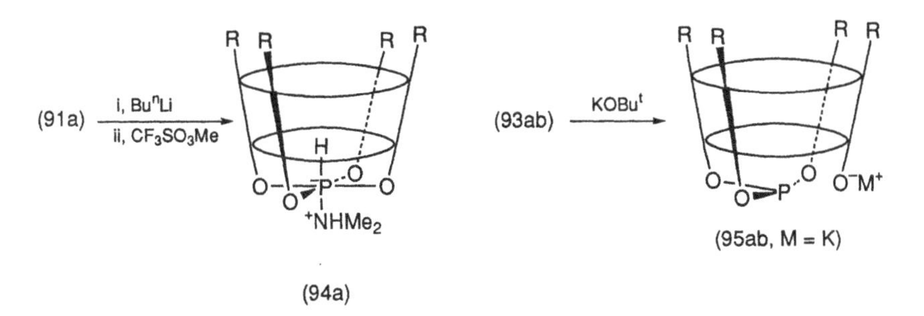

(94a)

(95ab, M = K)

(93b) $\xrightarrow{Bu^nLi}$ R R R R O O P O O^-M^+ + R R R R H O O P O O Li^+ Bu

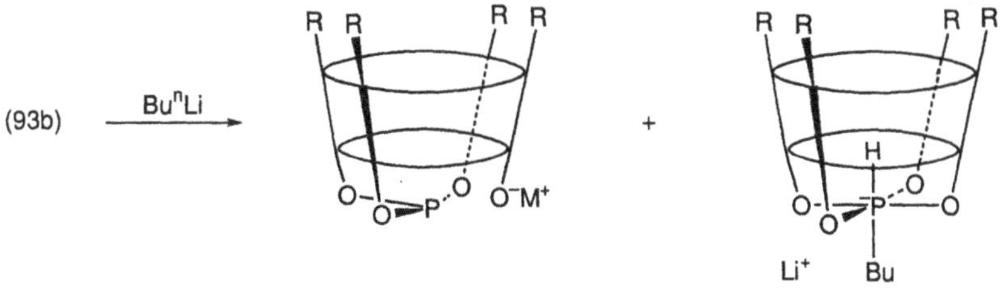

(95b, M = Li)

(96)

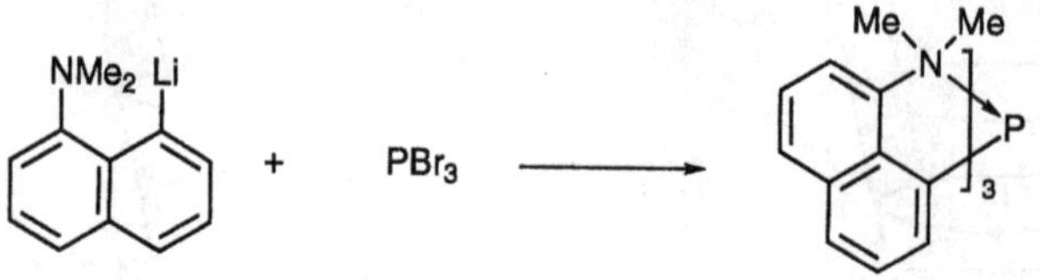

NMe$_2$ Li

+ PBr_3 →

Me Me N P 3

(97) (98)

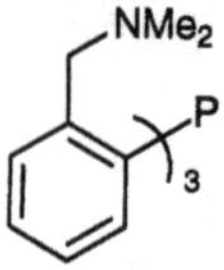

NMe$_2$ P 3

(99)

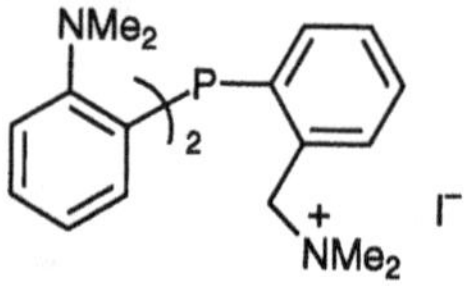

NMe$_2$ P 2 $\overset{+}{N}Me_2$ I^-

(100)

Me Me N $\overset{+}{P}Me$ I^- 3

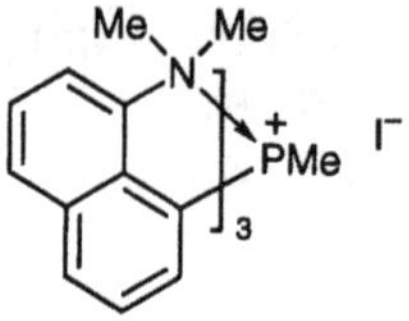

(101)

rotation axis. The pyramidal configuration about phosphorus is retained and the P-N bond distances (280.5, 284.4, and 285.3pm) are significantly shorter than the sum of the van der Waals radii (340pm). Thus the phosphorus is formally seven-coordinate with three carbon atoms, three nitrogen atoms, and a lone pair. An analogous compound (99) revealed similar X-ray crystallographic features showing that both phosphanes adopted the same geometry even when there was no steric constraint (as found in 98) to force the NMe_2 donor groups to interact with phosphorus. The flexibility of the 2-dimethylamino group in (99) however, explains why the P-N bond distances are slightly longer (299.9-307.1pm) than in (98) and also why the ^1H nmr of the NMe_2 groups in (98) appear as a sharp singlet whereas two sets of signals are seen in (97) even at 90^oC. Both compounds are quaternised by methyl iodide but whereas (99) gives the ammonium salt (100), phosphane (98) quaternises at phosphorus to give (101).

REFERENCES

1. R. A. Kemp, *Phosphorus, Sulfur & Silicon*, 1994, **87**, 83.
2. R. Contreras, *Phosphorus, Sulfur & Silicon*, 1994, **87**, 49.
3. D. V. Khasnis, J. M. Burton, J. D. McNeil, H. Zhang and M. Lattman, *Phosphorus, Sulfur & Silicon*, 1994, **87**, 93.
4. R. R. Holmes, T. K. Prakasha and R. O. Day, *Phosphorus, Sulfur & Silicon*, 1994, **87**, 59.
5. R. R. Holmes and T. K. Prakasha, *Phosphorus, Sulfur & Silicon*, 1993, **80**, 1.
6. A. M. Kibardin and A. N. Pudovik, *Russian J. Gen. Chem. Engl. trans.*, 1993, **63**, (11) 1687.
7. T. Uchimaru, S. Tsuzuki, J. W. Storer, K. Tanabe and K. Taira, *J. Org. Chem.*, 1994, **59**, 1835.
8. K. O. Christe, D. A. Dixon, H. P. A. Mercier, J. C. P. Sanders, G. J. Schrobilgen and W. W. Wilson, *J. Am. Chem. Soc.*, 1994, **116**, 2850.
9. N. Bricklebank, S. M. Godfrey, C. A. McAuliffe and R. G. Pritchard, *J. Chem. Soc., Dalton Trans.*, 1993, 2261.

10a. M. A. H. A. Al-Juboori, P. N. Gates and A. S. Muir, *J. Chem. Soc., Dalton Trans.*, 1994, 1441.

10b. N. Bricklebank, S. M. Godfrey, A. G. Mackie, C. A. McAuliffe and R. G. Pritchard, *J. Chem. Soc., Chem Commun.*, 1992, 355.

11. V. F. Mironov and I. V. Konovalova, *Russian J. Gen. Chem. Engl. trans.*, 1993, **63**, (11), 1760.
12. I. Mathieu-Pelta and S. A. Evans, Jr., *J. Org. Chem.*, 1994, **59**, 2234.
13. D. B. Denney, D. Z. Denney, P. J. Hammond, L. Z. Liu, and Y. P. Wang, *J. Org. Chem.*, 1983, **48**, 2159, and papers quoted therein.
14. V. A. Dodonov, S. N. Zaburdyaeva, and D. G. Fominskii, *Russian Chem. Bull.*, 1993, **42**, (10), 1703.
15. S. Ogawa, Y. Tajiri and N. Furukawa, *Tet. Lett.*, 1993, **34** , (5), 839.
16. W. Adam, H. M. Harrer and A. Treiber, *J. Am. Chem. Soc.*, 1994, **116**, 7581.
17. J. Gloede, *Phosphorus, Sulphur & Silicon* , 1993, **82**, 211.
18. R. R. Holmes, T. K. Prakasha and R. O. Day, *Inorg. Chem.*, 1993, **32**, 4360.
19. T. K. Prakasha, R. O. Day and R. R. Holmes, *Inorg. Chem.*, 1994, **33**, 93.
20. S. Kojima, K. Kajiyama and K. Akiba, *Tet. Lett.*, 1994, **35**, (38), 7037.
21. T. Kawashima, H. Takami and R. Okazaki, *J. Am. Chem. Soc.*, 1994, **116**, 4509.
22. D. Houalla, Z. Bounja, M.-C. Monje and S. Skouta, *Phosphorus, Sulfur and Silicon*, 1993, **81**, 1.
23. M.-R. Mazieres, D. Houalla, M.-C. Monje and M. Sanchez, *Phosphorus, Sulfur & Silicon*, 1993, **83**, 157.
24. D. Houalla, L. Moureau, S. Skouta and R. Wolf, *Phosphorus, Sulfur & Silicon*, 1994, **88**, 83.
25. D. Houalla, L. Moureau, S. Skouta and R. Wolf, *Phosphorus, Sulfur & Silicon*, 1994, **90**, 131.
26. L. Liu, G. Li, Z. Zhang, R. Cao and S. Zhang, *Phosphorus, Sulfur & Silicon*, 1993, **84**, 1.
27. Y. Huang, A. E. Sopchik, A. M. Arif, and W. G. Bentrude, *Heteroatom Chemistry*, 1993, **4**, (2/3), 271.
28. F. U. Seifert and G.-V. Röschenthaler, *Phosphorus, Sulfur & Silicon*, 1994, **86**, 157.
29. I. Neda, T. Kaukorat and R. Schmutzler, *Phosphorus, Sulfur & Silicon*, 1993, **80**, 173.
30. M. A. Pudovik, S. A. Terent'eva, I. V. Litvinov, O. N. Kataeva, V. A. Naumov, and A. N. Pudovik, *Russian Chem. Bull.*, 1993, **42**, (10), 1706.

31. D. Katalenic, V. Skaric and B. Klaic, *Tet. Lett.*, 1994, **35**, (17) 2743.
32. C. D. Montgomery, *Phosphorus, Sulfur and Silicon*, 1993, **84**, 23.
33. D. V. Khasnis, J. M. Burton, J. D. McNeil, C. J. Santini, H. Zhang and M. Lattman, *Inorg. Chem.*, 1994, **33**, 2657.
34. R. Fernandez-Galan, B. R. Manzano, A. Otero, M. Lanfranchi and M. A. Pellinghelli, *Inorg. Chem.*, 1994, **33**, 2309.
35. C.Chuit, R. J. P. Corriu, P. Monforte, C.Reyé, J.-P. Declercq and A. Dubourg, *Angew. Chem,. Int. Ed. Engl.*, 1993, **32**, (10) 1430.

3
Tervalent Phosphorus Acid Derivatives

BY O. DAHL

1 Introduction

Two books have been published which contain much information on the use of tervalent phosphorus acid derivatives for the synthesis of oligonucleotide derivatives and analogues.[1, 2] A review on the synthesis of modified oligonucleotides by the phosphoramidite approach and their applications,[3] and another review on the synthesis of specific ribonucleotides and unrelated phosphorylated biomolecules by the phosphoramidite method [4] have appeared. A review on anellated heterophospholes gives an account on the preparation, characterization, and reactivity of such two-co-ordinate phosphorus compounds, e.g. (1).[5]

The previous section 3.3 on the use of tervalent phosphorus acid derivatives for synthesis of compounds of biological relevance has this year been divided into two parts, 3.3 use for nucleotide synthesis, and 3.4 use for sugar phosphate, phospholipid, or phosphopeptide synthesis.

2 Nucleophilic Reactions

2.1 Attack on Saturated Carbon. - Ethylene dicarboxylic diphosphonic acid (EDCP, 2) has been prepared in 70% yield from 2,3-dichlorosuccinic anhydride (3) and trimethyl phosphite, followed by hydrolysis of the Arbuzov product.[6] Tris(trimethylsilyl) phosphite, in contrast to trialkyl phosphites, attacks an oxirane carbon of epihalohydrins (4) to give the phosphonates (5).[7] Bis(trimethylsilyl) phosphonite (6) has previously been prepared *in situ* and used to obtain γ-ketophosphinic acids; similar reactions with simple alkyl halides to give alkylphosphinic and dialkylphosphinic acids acids in high yields have now been described.[8]

The new trinorborn-1-yl phosphite (7) has been prepared and shown to give quasi-phosphonium salts which are even more stable than those derived from trineopentyl phosphite;[9] thus the salt with methyl iodide had a half-life of more than 100 h in deuterochloroform at 150 ^{o}C, compared to less than 2 h at 33 ^{o}C for the neopentyl analogue. Although phenacyl chloride gave mainly the Arbuzov intermediate (8), with p-nitrophenacyl chloride a stable Perkow intermediate (9) could be isolated and its X-ray crystal structure studied. Trifluoroiodomethane with tris(dialkylamino)phosphines gave the normal trifluoromethylphosphonium salts (10) in the absence of a solvent, but iodophosphonium salts (11) in solvents that can deliver a proton to the intermediate trifluoromethyl anion.[10]

2.2 Attack on Unsaturated Carbon.- Tervalent phosphorus compounds are effective catalysts for the polymerization of acrylic acid derivatives, and the monomeric adducts which initiate the polymerization usually cannot be isolated. However, by addition of ethyl 2-cyanoacrylate to strongly nucleophilic tervalent phosphorus compounds it was possible to isolate monomeric adducts, e.g. (12).[11] Arbuzov-type reactions at sp^2 carbon atoms, like other nucleophilic substitution reactions, are possible via an addition-elimination reaction. This has been realized in the preparation of some phosphono-cephalosporin analogues (13),[12] and in the reactions of several alkyl diphenylphosphinites with 10-methylacridinium salts (14).[13] In the latter reactions the intermediate phosphonium salts (15) were quite stable, which allowed a kinetic study of the two steps of this Arbuzov-type reaction as a function of the alkyl group.

A full paper has appeared on the stereoselective phosphonylation of aldehydes with the 1,3,2-oxazaphospholans (16) derived from (1R,2S)-ephedrine.[14] From NOE and other n.m.r. data on one of the products the main isomer formed was assigned the structure (17). Similar highly stereoselective Abramov reactions have been studied using (18), and the X-ray structure of the main product determined in one case.[15] New chiral 1,3,2-dioxaphospholans (19) [16] and a cyclic phosphite derived from 1,1'-binaphth-2-ol [17] prepared by the same group were much less reactive towards aldehydes than (16) or (18).

Salicylaldehyde and triethyl phosphite upon heating gave, instead of the expected diphosphonate (20), the rearranged phosphate-phosphonate (21).[18] A new method to obtain 1-aminoalkylphosphinic acids (22) has been published;[19] it involves addition of bis(trimethylsilyl) phosphonite to *N*-trityl imines, followed by hydrolysis. A similar reaction of bis(trimethylsilyl) phosphonite with 1-pyrroline trimer gave 2-pyrrolidinylphosphinic acid in 90% yield.[20] Bis(trimethylsilyl) phosphonite with carbamoyl chlorides and triethylamine gave fair yields of the carbamoylphosphonites (23).[21]

2.3 Attack on Nitrogen, Chalcogen, or Halogen.- Some tertiary alkyl azides and tris(dimethylamino)phosphine gave thermally stable phosphazides (24), allowing the X-ray crystal structure of one to be determined.[22]

The reactions of bis(phosphoryl) disulphides with a variety of tervalent phosphorus acid derivatives or triphenylphosphine have been studied by means of low-temperature ^{31}P n.m.r.[23] Intermediates were detected which indicated initial attack on one sulphur atom followed by substitution reactions at one of the phosphorus atoms, Arbuzov dealkylations, or phosphorane formation. An example is shown for the reaction of tributyl phosphite with bis(diethoxyphosphoryl) disulphide (25). In order to obtain solely thiophosphate products it is necessary to use bis(thiophosphoryl) disulphides, and one such compound (26) has been proposed as a highly effective reagent for the oxidation of oligonucleoside phosphites to phosphorothioates.[24] A comparison of the efficiency of (26) with that of four other sulphurizing agents for use in

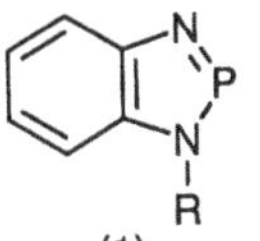

(1)

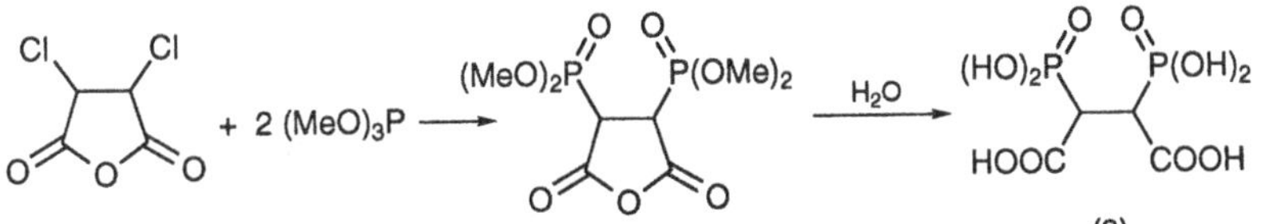

Cl, Cl (3) + 2 $(MeO)_3P$ ⟶ $(MeO)_2P(O)$, $P(O)(OMe)_2$ $\xrightarrow{H_2O}$ $(HO)_2P(O)$, $P(O)(OH)_2$, HOOC, COOH (2)

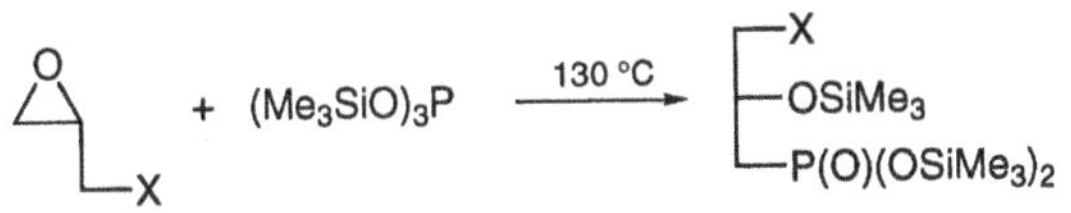

(4) X = Cl or Br + $(Me_3SiO)_3P$ $\xrightarrow{130\ °C}$ X, $OSiMe_3$, $P(O)(OSiMe_3)_2$ (5)

$(Me_3SiO)_2PH$ (6)

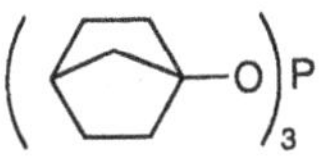

(7)

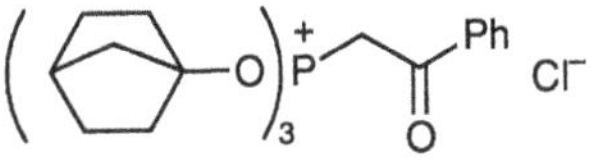

Cl^- (8)

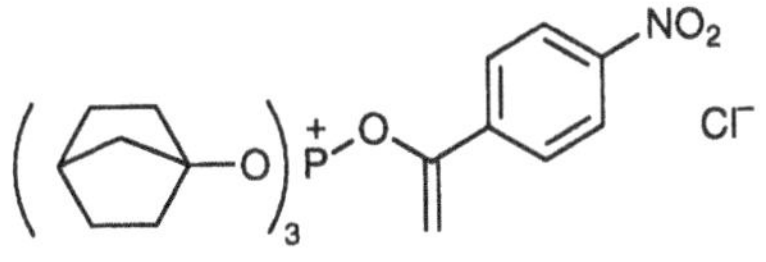

NO_2 Cl^- (9)

$(R_2N)_3\overset{+}{P}–CF_3$ I^- $\xleftarrow{neat}$ $(R_2N)_3P + CF_3I$ $\xrightarrow{CH_3CN}$ $(R_2N)_3\overset{+}{P}–I$ X^- + HCF_3

(10) R = Me or Et (11)

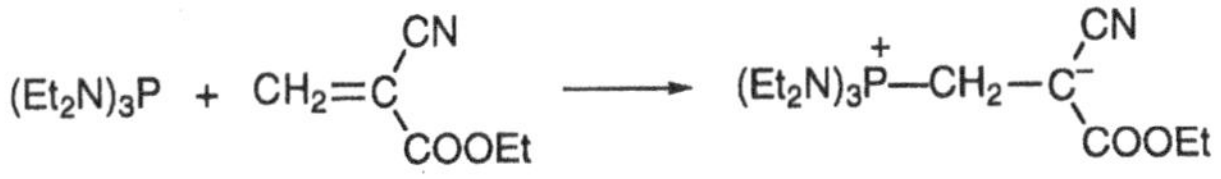

$(Et_2N)_3P$ + $CH_2{=}C(CN)(COOEt)$ ⟶ $(Et_2N)_3\overset{+}{P}–CH_2–\overset{-}{C}(CN)(COOEt)$

(12)

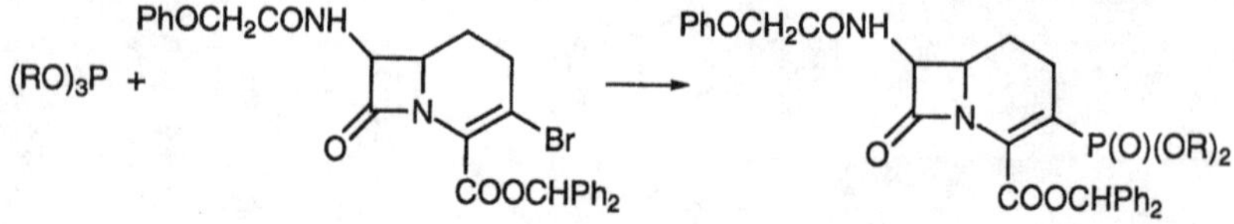

$(RO)_3P$ + PhOCH$_2$CONH … N, O, Br, COOCHPh$_2$ → PhOCH$_2$CONH … N, O, P(O)(OR)$_2$, COOCHPh$_2$

R = Me, Et, or Bu

(13)

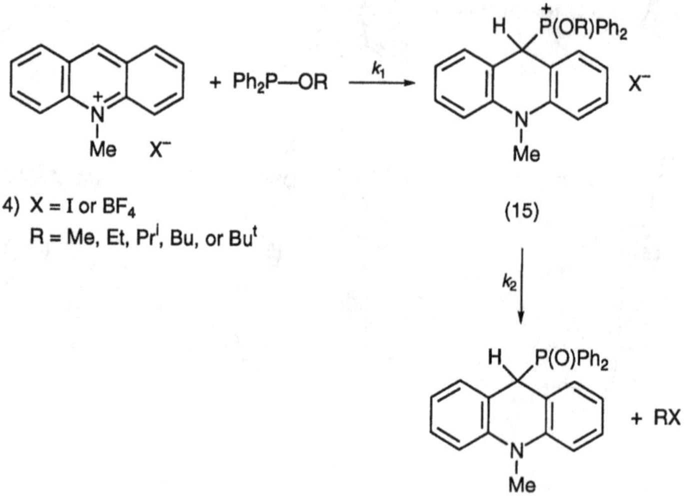

N$^+$, Me, X$^-$ + Ph$_2$P—OR $\xrightarrow{k_1}$ H, $\overset{+}{P}$(OR)Ph$_2$, N, Me, X$^-$

(14) X = I or BF$_4$
R = Me, Et, Pr^i, Bu, or Bu^t

(15)

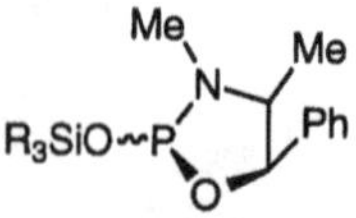

k_2

H, P(O)Ph$_2$, N, Me + RX

Me, N, Me, Ph, R$_3$SiO~P, O

(16)

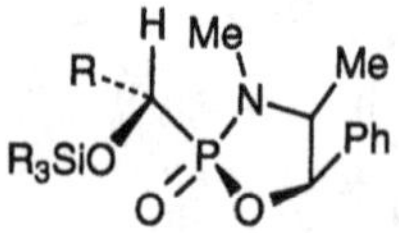

H, Me, R, N, Me, Ph, R$_3$SiO, O=P, O

(17)

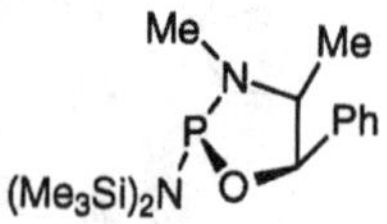

Me, N, Me, Ph, P, (Me$_3$Si)$_2$N, O

(18)

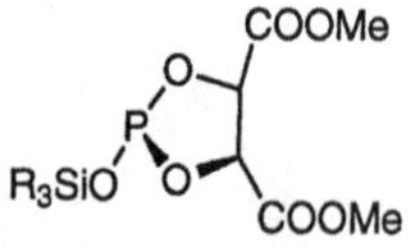

COOMe, O, P, R$_3$SiO, O, COOMe

(19)

2(EtO)$_3$P + CHO, OH → [CH[P(O)(OEt)$_2$]$_2$, OH (20)] → CH$_2$–P(O)(OEt)$_2$, O–P(O)(OEt)$_2$

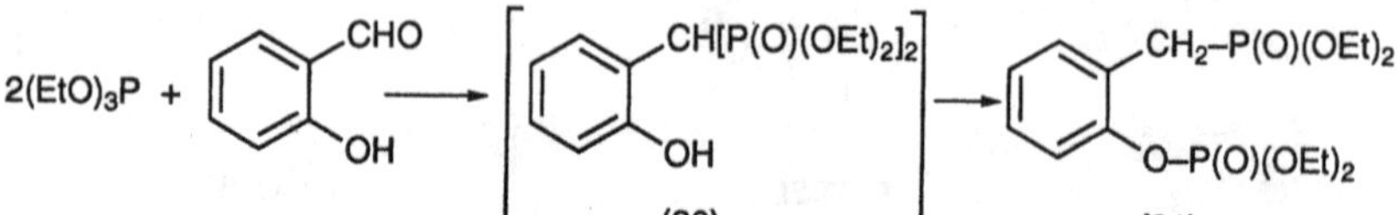

(20)

(21)

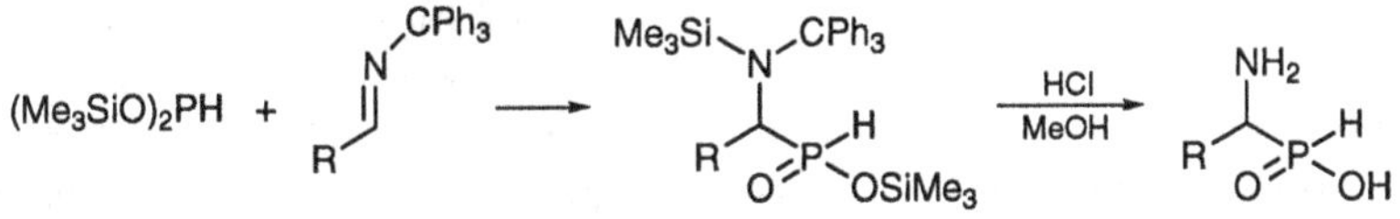

$(Me_3SiO)_2PH$ + $RCH{=}NCPh_3$ ⟶ $RCH(N(SiMe_3)CPh_3)P(O)(H)OSiMe_3$ —HCl, MeOH⟶ $RCH(NH_2)P(O)(H)OH$

(22)

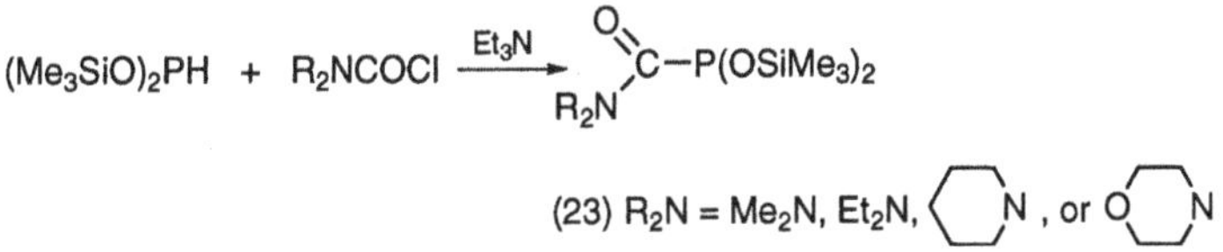

$(Me_3SiO)_2PH$ + R_2NCOCl —Et_3N⟶ $R_2NC(O){-}P(OSiMe_3)_2$

(23) R_2N = Me_2N, Et_2N, piperidino, or morpholino

$(Me_2N)_3P$ + RN_3 ⟶ $(Me_2N)_3P{=}N{-}N{=}N{-}R$

(24) R = Bu^t, 1-adamantyl, or Ph_3C

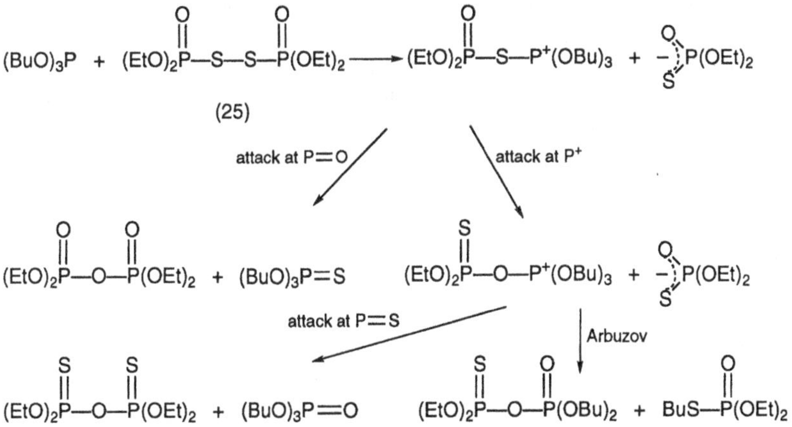

$(BuO)_3P$ + $(EtO)_2P(O){-}S{-}S{-}P(O)(OEt)_2$ ⟶ $(EtO)_2P(O){-}S{-}P^+(OBu)_3$ + $^-[O{\cdots}P(OEt)_2{\cdots}S]$

(25)

attack at P=O ⟶ $(EtO)_2P(O){-}O{-}P(O)(OEt)_2$ + $(BuO)_3P{=}S$

attack at P^+ ⟶ $(EtO)_2P(S){-}O{-}P^+(OBu)_3$ + $^-[O{\cdots}P(OEt)_2{\cdots}S]$

attack at P=S ⟶ $(EtO)_2P(S){-}O{-}P(S)(OEt)_2$ + $(BuO)_3P{=}O$

Arbuzov ⟶ $(EtO)_2P(S){-}O{-}P(O)(OBu)_2$ + $BuS{-}P(O)(OEt)_2$

$(Pr^iO)_2P(S){-}S{-}S{-}P(S)(OPr^i)_2$

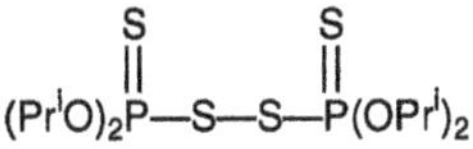

(26)

$RO{-}P(NEt_2)_2$ + CF_3SSCF_3 ⟶ $RSCF_3$ + $CF_3S{-}P(O)(NEt_2)_2$

(27) R = CH_2Ph, CH_2COOEt, or $CH(Me)COOEt$

oligonucleoside phosphorothioate synthesis has indeed shown that (26) is as effective as the best of the previous known compounds.[25] Alkyl trifluoromethyl sulphides are conveniently made by the treatment of phosphorodiamidites (27) with bis(trifluoromethyl) disulphides.[26] 1,3-Dithiol-2-thiones (28) when heated in neat triethyl phosphite normally give good yields of tetrathiafulvalenes (29), but with low concentrations of (28) the main product is the diethyl phosphonate (30).[27] A new selenium-transfer reagent (31) has been shown to oxidize triethyl phosphite to *O, O', O''*-triethyl selenophosphate in acetonitrile-triethylamine at rt in 5 min.[28]

Trialkyl phosphites have been converted to dialkyl chlorophosphates (32) in a mild and efficient way by tellurium tetrachloride.[29] Tris(dimethylamino)phosphine abstracts iodine from 1-iodoperfluoroalkanes to give metal-free perfluoroalkyl anions which have basic but poor nucleophilic properties.[30] The bicyclic phosphite (33) with chlorine gave a sole product (34) resulting from opening of the five-membered and not the seven-membered ring.[31]

3 Electrophilic Reactions

3.1 Preparation.- The short-lived and previously unknown chlorophosphine (35) has been generated from the slightly more stable fluoro analogue as shown and characterized by MS and gas phase i.r. spectroscopy.[32] Compounds with three acylamino groups bound to phosphorus, e.g. (36) and (37), have been prepared from phosphorus trichloride and a silylated amide.[33] Other aminophosphines with unusual *N*-substituents prepared this year are the tetramethylguanidinylphosphines (38) and (39);[34] although the guanidino group is usually quite basic the reactions of (38) with methyl iodide gave only the quaternary phosphonium salts. A convenient, high yield synthesis of the bis(dichlorophosphino)hydrazine (40) has been published.[35] Alkyldichlorophosphines with methylhydrazine gave the new cyclic hydrazinophosphines (41).[36]

A series of dialkylamino substituted triphosphines (42) has been prepared as shown, and the X-ray crystal structure of one determined;[37] upon treatment with methanol a silyl group on the central phosphorus atom could be replaced by hydrogen to give a rather unstable triphosphine (42, $R^1 = Pr^i$, $R^2 = H$). The diastereoselectivity of the formation of phosphazanes, e.g. (43) from a benzo-1,3,2-diazaphospholan (44) and a chlorophosphine, has been studied;[38] the main isomer is the one shown, as seen from an X-ray crystal structure determination. Some new 1,3,2,4-diazadiphosphetidines (45) with RNH substituents on phosphorus have been prepared as a mixture of cis and trans isomers;[39] the cis isomers are the thermodynamically most stable. A series of chloro(dimethylamino)phosphines (46) has been prepared from chlorobis(dimethylamino)phosphine and used to prepare 1,3,2-oxazaphospholans (47) derived from (-)-ephedrine.[40] This method gave much higher yields of (47), isolated as the borane adducts (72-82%), than previous one-step methods involving coupling of bis(dialkylamino)phosphines with ephedrine.

In a series of papers R. Schmutzler *et al.* have studied reactions of tervalent chlorophosphorus compounds with *N,N'*-dimethylthiourea in the presence of a base. With one equivalent of monochlorophosphines and dialkyl phosphorochloridites the products normally were *N*-phosphinoylated

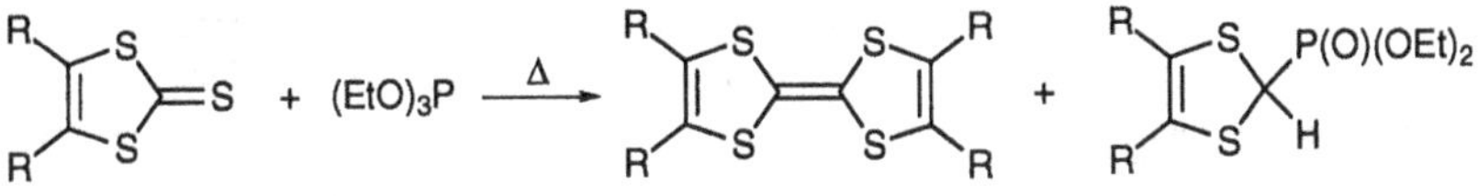

(28) + $(EtO)_3P$ —Δ→ (29) + (30)

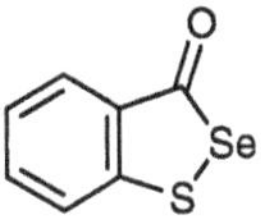

(31)

$$2\ (RO)_3P + TeCl_4 \xrightarrow{\text{r.t.}} 2\ (RO)_2P(O)Cl + 2\ RCl + Te$$

(32)

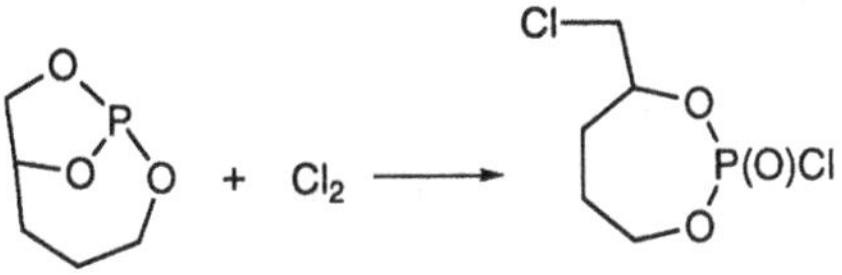

(33) + Cl_2 → (34)

$$H_3PF_2 \xrightarrow{AlCl_3} H_2PF \xrightarrow{BCl_3} H_2PCl$$

(35)

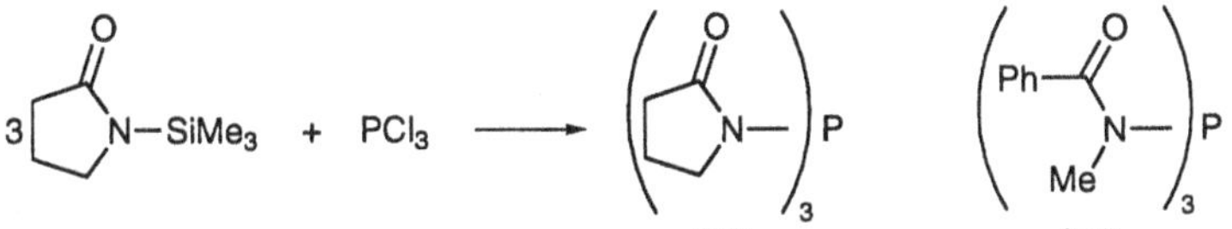

3 (N-SiMe$_3$ pyrrolidinone) + PCl_3 → (36)

(37)

$R-P[-N=C(NMe_2)_2]_2$

(38) R = Me, Bu^t, or Ph

$Z_2P-N=C(NMe_2)_2$

(39) Z = Pr^i, Ph, NEt_2, or Z_2 =

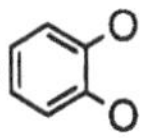

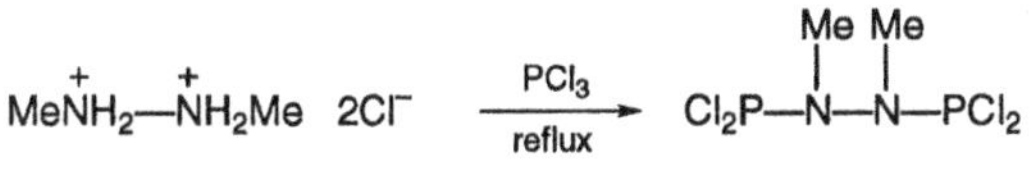

$$Me\overset{+}{N}H_2-\overset{+}{N}H_2Me\ \ 2Cl^- \xrightarrow[\text{reflux}]{PCl_3} Cl_2P-N(Me)-N(Me)-PCl_2$$

(40)

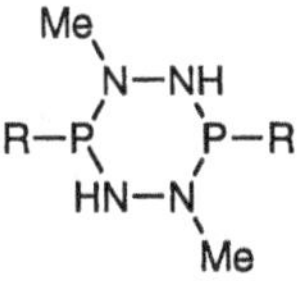

(41) R = Et, Bu^t, or Ph

compounds, e.g. (48), although chlorodi-t-butylphosphine gave exclusively the *S*-phosphinoylated (49).[41] With two equivalents *N,N'*-diphosphinoylated products (50) were obtained, but chlorodiisopropylphosphine initially gave the *N,S*-disubstituted compound (51).[42] Some unsymmetrical derivatives, e.g. (52), were also prepared;[43] the labile, cyclic compound (53) was formed from (52) upon treatment with phosphorus pentafluoride. Lithiated *N,N'*-dimethylthiourea and tervalent dichlorophosphorus compounds gave 1,3,2-diazaphosphetidin-4-thiones (54).[44] The 1,3,2-oxathiaphospholan (55) has been prepared from the mercapto-alcohol and phenyldichlorophosphine and the sulphide used for enantioselective synthesis of *P*-chiral phosphines.[45]

Phosphorus trichloride with 3-aminopropan-1-ol gave the 1,3,2-oxazaphosphorinanes (56) and (57), the latter being very insoluble and therefore not further phosphinoylated.[46] Similar reactions with 1,3-propanediamine or the *N*-methyl analogue gave (58),[47] and the chloro groups in (56) and (58) could be substituted in a stepwise manner with amino, alkoxy, or fluoro groups to give e.g. (59) and (60).[48, 49] A series of 1,3,2-oxazaphosphorinanes (61)-(63) has been prepared for conformational studies;[50] the majority of the compounds exist in a conformation with the phosphorus substituent in an axial position. Further studies have appeared on the preparation of differently *P*-substituted benzo-1,3,2-oxazaphosphorinan-6-ones (64),[51] and the benzo-1,3,2-diazaphosphorinan-4-ones (65),[52] (66),[53] and (67).[54]

The 2-hydroxybenzyl alcohol (68) with propyl phosphorodichloridite gave the first isolable tervalent benzo-1,3,2-dioxaphosphorinan (69).[55] 2-Hydroxybenzyl alcohols without the trifluoromethyl groups form phosphonates because of the easy formation of benzyl cations. Naphthalene-1,8-diamine with phosphorodiamidites gave the new 1,3,2-diazaphosphorinanes (70);[56] the X-ray crystal structure was determined for (70, R = Et). Several new 1,3,5,2-triazaphosphorinan-4,6-diones (71) have been prepared from 1,3,5-trimethylbiuret, and the X-ray crystal structure of the previously prepared (72) determined.[57, 58, 59] Tris(dimethylamino)phosphine with two equivalents of benzamidine gave the dihydro-1,3,5,2-triazaphosphorine (73) which did not form (74) upon further heating.[60] One trialkoxy- and several triaryloxy-λ^3-cyclotriphosphazanes (75) have been prepared from the known trichloro analogue, and the X-ray crystal structure of cis- and trans-(75, R = 4-BrC_6H_4) determined.[61]

3.2 Mechanistic Studies.- A study has been published on the alcoholysis of a phosphonamidite (76) containing a properly positioned amino group which results in intramolecular catalysis upon protonation.[62] The reaction of (76) with methanol was 150-300 times faster than analogous reactions of (77), and the catalytic rate constant k_c was independent of the pK_a of the acid catalyst used (amine hydrochlorides), indicating that (78) is the actual catalyst. A similar chiral phosphonite (79) (mixture of two diastereomers) reacted with optically active alcohols to give phosphonites with a maximum stereoselectivity of 3:1 when an achiral amine hydrochloride was used as the catalyst.

The rates of methanolysis of several benzimidazolides (80),[63] cyclic amidine derivatives (81)-(83),[64] 2-aminopyridine derivatives (84)-(86),[65] and the guanidine derivative (87) [66] have been

$$2\ (R^1{}_2N)_2PCl + R^2PLi_2 \longrightarrow (R^1{}_2N)_2P{-}P(R^2){-}P(NR^1{}_2)_2$$

R^1 = Et or Pr^i R^2 = Bu^t, Mes, or R_3Si (42)

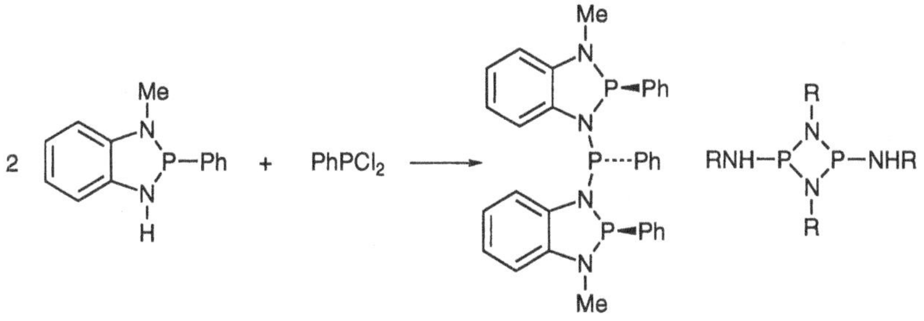

2 [Me, N, P–Ph, N, H] + $PhPCl_2$ ⟶ [Me, N, P◄Ph, N, P····Ph, N, P◄Ph, N, Me] RNH–P[N(R)]₂P–NHR

(44) (43) (45) R = Me , Et, or Pr^i

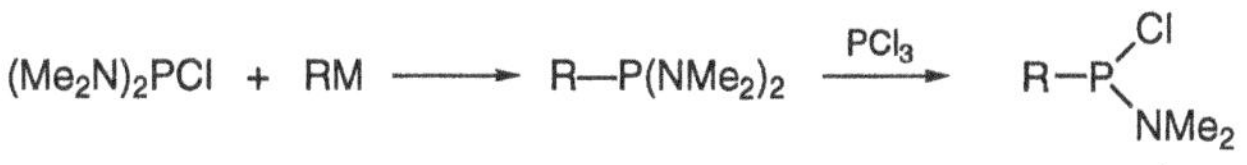

$$(Me_2N)_2PCl + RM \longrightarrow R{-}P(NMe_2)_2 \xrightarrow{PCl_3} R{-}P(Cl)NMe_2$$

(46)

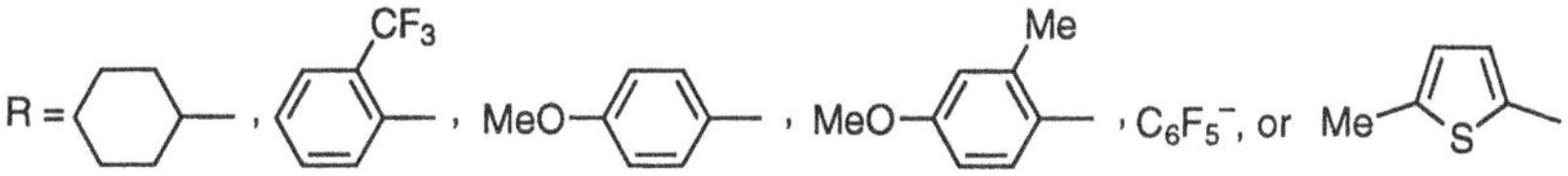

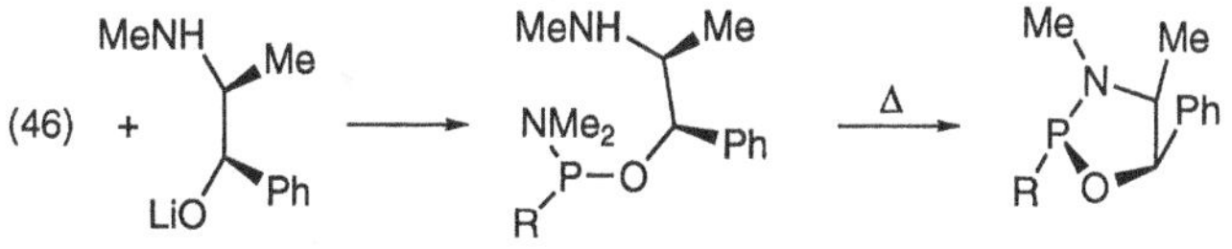

R = [cyclohexyl] , [CF_3-phenyl] , MeO–[phenyl] , MeO–[Me-phenyl] , $C_6F_5^-$, or Me–[thienyl, S]

(46) + MeNH, Me, LiO, Ph ⟶ MeNH, Me, NMe_2, P–O, R, Ph $\xrightarrow{\Delta}$ Me, N, Me, Ph, P, R, O

(47)

S, Me, N, N, Me, Z_2P, H S–PBu^t_2, Me, N, N, Me, H S, Me, N, N, Me, Z_2P, Z_2P S–PPr^i_2, Me, N, N, Me, Pr^i_2P

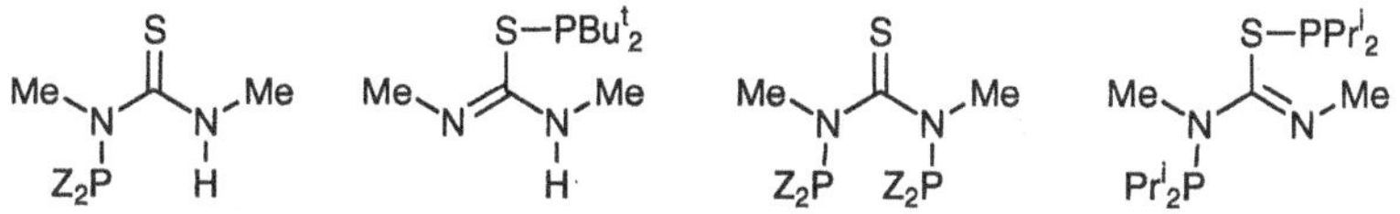

(48) Z = Me, Pr^i, Ph, or OMe (49) (50) Z = Pr^i, Ph, or OMe (51)

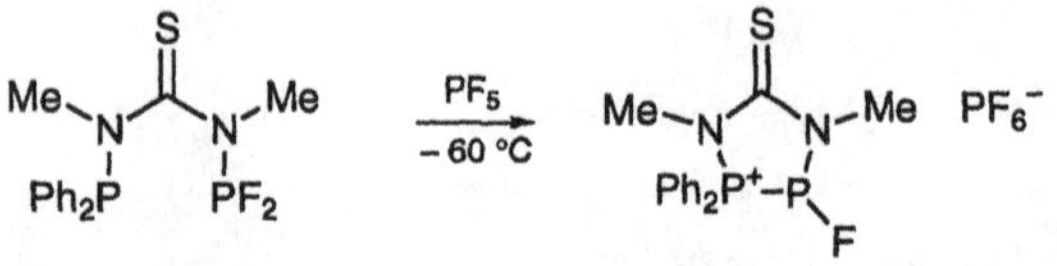

(52) $\xrightarrow[-60\ ^\circ C]{PF_5}$ (53) PF_6^-

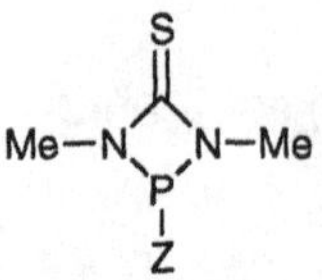

(54) Z = NMe_2, NEt_2, NCy_2, or CCl_3

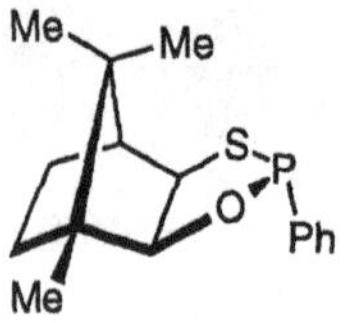

(55)

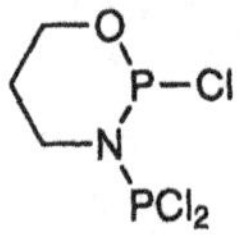

(56)

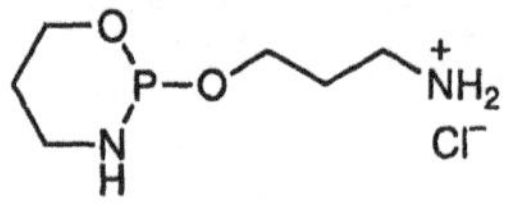

(57)

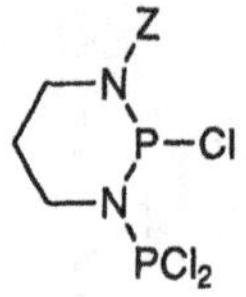

(58) Z = Me or PCl_2

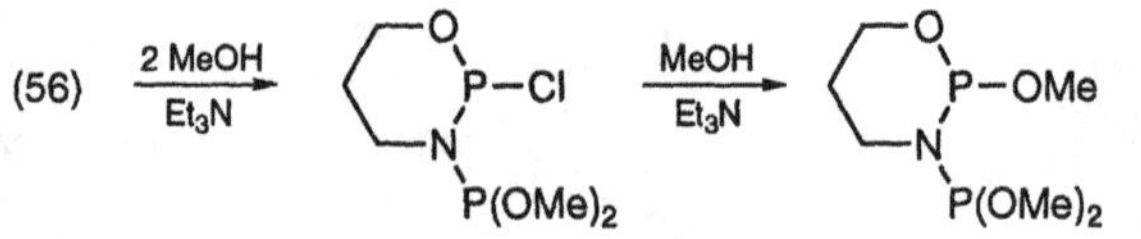

(56) $\xrightarrow[Et_3N]{2\ MeOH}$ (59) $\xrightarrow[Et_3N]{MeOH}$ (60)

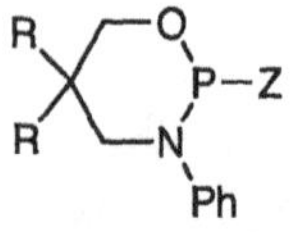

(61) R = H or Me
Z = Ph, NHMe, NMe_2, OMe, or $OCH(CF_3)_2$

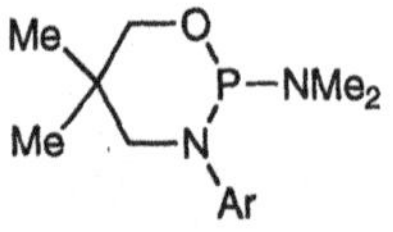

(62) Ar = 3-$CF_3C_6H_4$ or 4-$Me_2NC_6H_4$

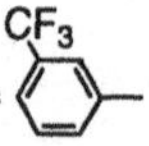

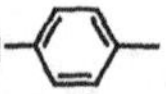

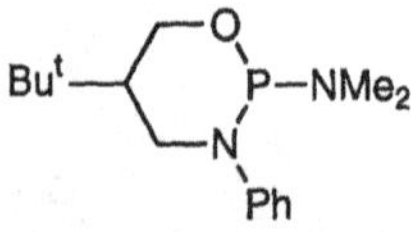

(63)

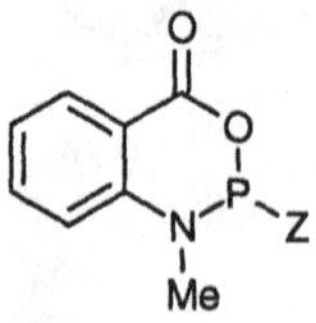

(64) Z = halogen NR_2, or CN

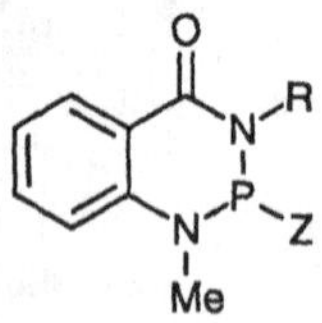

(65) R = Me, Z = Cl or $NMeCH_2CH_2NMe_2$

(66) R = CH_2CH_2Cl, Pr^i, or CH_2Ph
Z = Cl, NMe_2, $NHCH_2CH_2Cl$, or $N(CH_2CH_2Cl)_2$

(67) R = $(CH_2)_3NR^1{}_2$ or 2-picolyl
Z = Cl, NMe_2, or $N(CH_2CH_2Cl)_2$

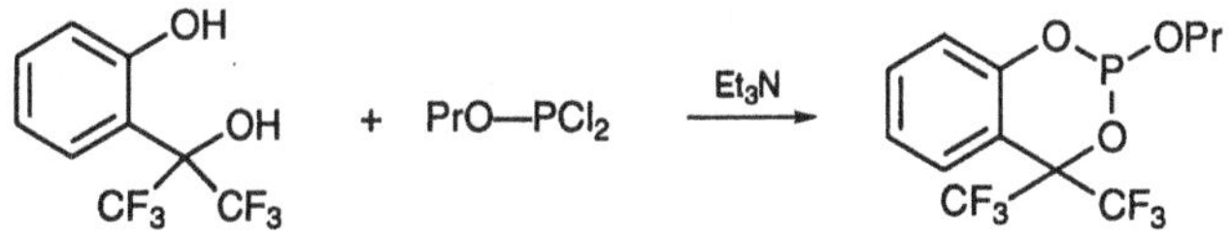

(68) + $PrO—PCl_2$ $\xrightarrow{Et_3N}$ (69)

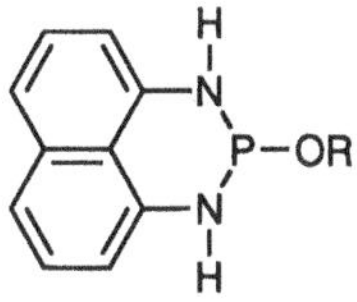

(70) R = Et or Ph

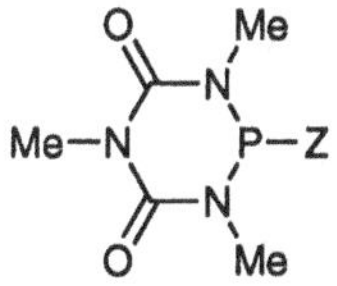

(71) Z = OMe, $OSiMe_3$, NCy_2, N-pyrrolidinyl, $NHCH_2CH_2Cl$, or $N(CH_2CH_2Cl)_2$

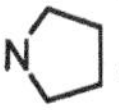

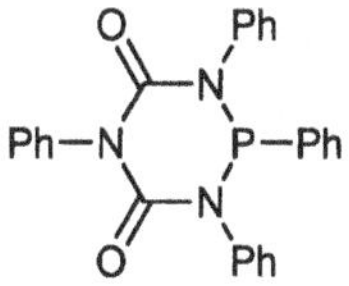

(72)

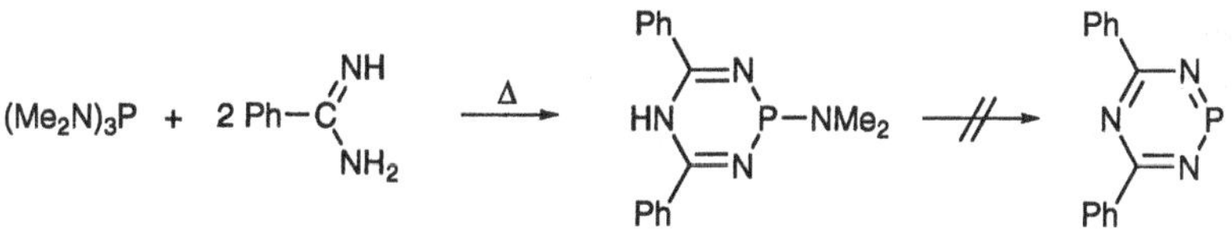

$(Me_2N)_3P$ + 2 $Ph—C(=NH)NH_2$ $\xrightarrow{\Delta}$ (73) $\not\rightarrow$ (74)

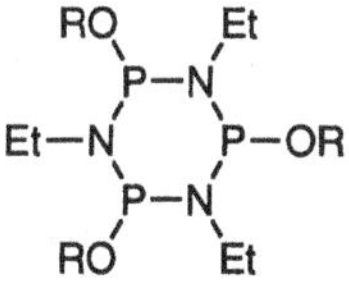

(75) R = CH_2CF_3, Ph, 4-Br-phenyl, 2,4,6-trimethylphenyl (mesityl), or 2,6-dimethylphenyl

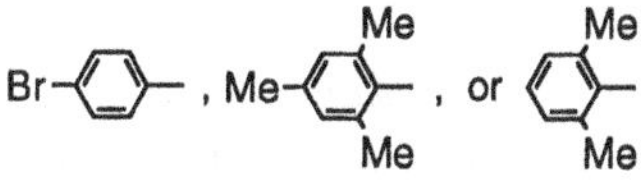

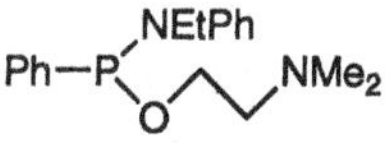

(76)

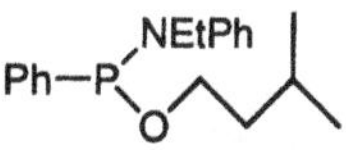

(77)

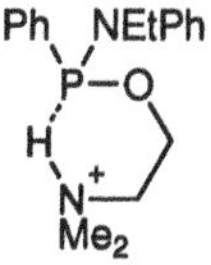

(78)

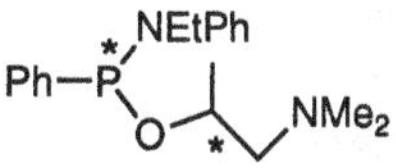

(79)

measured. The benzimidazolides (80), including the tris-benzimidazolide, can be isolated and stored as pure compounds in contrast to the corresponding imidazolides, but are still quite reactive. For the remaining compounds the reactivity order was (81) > (82) > (83), (84) > (85) > (86), with (87) being very unreactive. Apparently phosphoramidites containing a fixed double bond to nitrogen are less reactive than normal amidites.

3.3 Use for Nucleotide Synthesis.- Two books, edited by S. Agrawal, have been published containing protocols for the preparation of oligonucleotides and analogues,[1] and the preparation of oligonucleotide conjugates.[2] The first contains chapters on the preparation, using tervalent phosphorus acid derivatives, of unmodified oligodeoxyribo- and oligoribonucleotides, 2'-*O*-alkyloligoribonucleotides, α-oligodeoxyribonucleotides, and oligodeoxyribonucleoside methylphosphonates, phosphorothioates, phosphorodithioates, phosphotriesters, boranophosphates, and phosphorofluoridates. The second book has chapters on the preparation of oligonucleotides with reporter groups or enzymes covalently bound to the 5'-end, the 3'-end, or at internucleotidic positions, via amino or thiol linkers. Two reviews by Beaucage and Iyer cover exhaustively the synthesis of modified oligonucleotides [3] and the synthesis of oligoribonucleotides [4] by the phosphoramidite approach.

The phosphoramidite (88) with orthogonal diisopropylamino and 4-nitrophenoxy leaving groups has been prepared and used to obtain thymidine dimers or trimers containing phosphate, phosphorothioate, or phosphoroselenoate linkages.[67] More versatile is the phosphoramidite (89), which gave phosphorodithioates or phosphoroselenothioates (90) as shown.[68] The catalyst, DBU, used here for the displacement of 4-nitrophenoxy groups is certainly an improvement over the previously recommended sodium hydride.[67] A cyclic selenophosphoramidite (91) has been used as a precursor for monomers (92) which by analogy with the known sulphur analogues can lead to stereoselective syntheses of nucleoside phosphorothioates.[69] Although (92) gave a higher coupling yield (98%) than the sulphur analogue (90-95%), weaker bases than DBU were not effective catalysts in spite of the presence of the weak P-Se bond. A full paper has appeared on attempts to prepare diastereomerically pure nucleoside phosphonamidites as potential monomers for stereoselective synthesis of nucleoside methylphosphonates.[70] The best result was obtained for (93) which could be prepared with a diastereomeric excess of 81%; tetrazole-catalyzed coupling reactions, however, were not significantly stereoselective.

β-Silylethyl protecting groups for nucleoside phosphoramidites continue to be studied. One group examined the phosphorodiamidite (94), which was deemed too labile, whereas (95) was found to have favourable properties and gave good yields of nucleoside phosphoramidites (96).[71] However, another group had no problems using (94) to prepare (97).[72] After coupling and oxidation the β-silylethyl group was removed with conc. aqueous ammonia [71] or tetrabutylammonium fluoride in THF.[72] A phosphoramidite reagent (98) has been developed which after basic deblocking yields two oligonucleotides when built into one longer oligonucleotide during solid phase synthesis;[73] this could

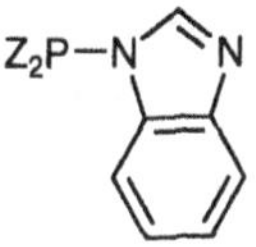

Z_2P—N (benzimidazolyl)

(80) Z = Pr^i, RO, or R_2N

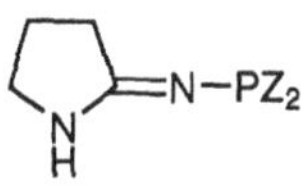

(81)

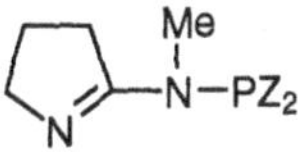

(82)

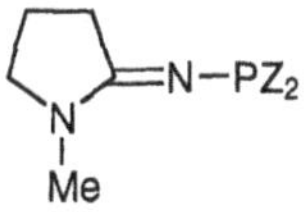

(83)

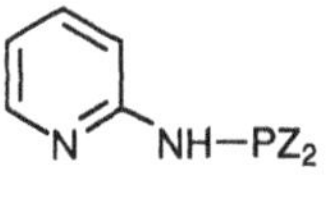

(84)

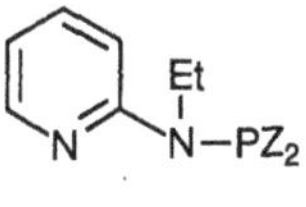

(85)

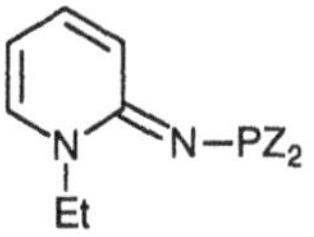

(86)

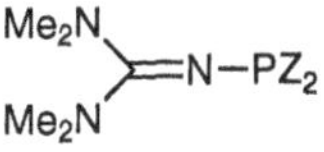

(87)

(81)-(87) PZ_2 = 5,5-dimethyl-1,3,2-dioxaphosphorinan-2-yl

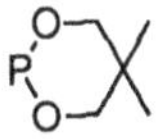

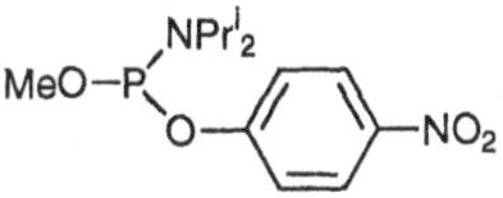

(88) $MeO-P(NPr^i_2)-O-C_6H_4-NO_2$

(89) $(O_2N-C_6H_4-O)_2P-NPr^i_2$

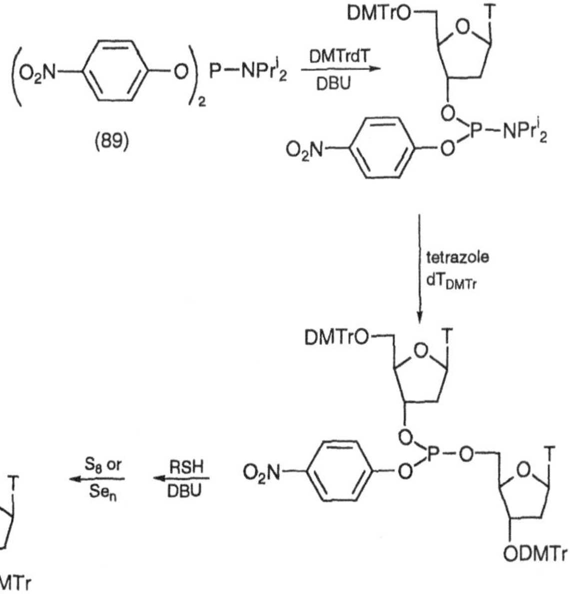

DMTrdT, DBU

tetrazole, dT_{DMTr}

RSH, DBU

S_8 or Se_n

(90) X = S or Se

be useful e.g. when large numbers of pairs of primers are wanted for PCR priming. The removal of phosphate groups at the 3'- and 5'-end is secured during deblocking by a neighbouring group attack of the 2-hydroxyalkyl group on the phosphate as shown in (99). A similar reagent (100) was also developed which leaves one of the two oligonucleotides with a 5'-phosphate group.[73]

Two types of phosphate-modified oligonucleotides have been prepared from new phosphoramidite precursors. The boron-rich phosphoramidite (101) coupled normally during solid phase synthesis to give compounds useful for boron neutron capture therapy of cancer.[74] The trifluoroethyl phosphoroamidochloridite (102) gave nucleoside phosphoramidites which lead to trifluoroethyl phosphotriester analogues that could be useful for ^{19}F n.m.r. imaging of the biodistribution of antisense probes.[75] A pentacarbonyl tungsten dinucleoside phosphite (103) has been prepared for potential heavy metal labeling of oligonucleotides;[76] however, an attempted preparation of a similarly labeled oligonucleotide failed.

Several new phosphoramidite reagents have been developed this year to enable the introduction of bio-active groups in oligonucleotides. These include the cholesterol derivative (104),[77] the biotin derivatives (105),[78] the linkage phosphoramidite (106),[79] and two thiol-linker phosphoramidites (107) [80] and (108) [81] with base-labile *S*-protecting groups. Imidazole oligonucleotide conjugates are of interest as artificial endonucleases, and two papers have appeared on the preparation of such compounds. One described some imidazole-alkyl phosphoramidites, e.g (109), and their coupling products with thymidine;[82] another some nucleoside phosphoramidites with imidazole attached to *N*-2 of 2-amino-A or G, e.g. (110), which when built into oligonucleotides resulted in increased binding to complementary DNA and RNA.[83]

A large number of papers continue to appear on the preparation of base or sugar modified oligonucleotides by more or less standard phosphoramidite methods. Among base modified monomers prepared this year are the phosphoramidites of 2-amino-2'-deoxyadenosine (111),[84] 1,*N*-6-etheno deoxy and ribo adenosine (112) and 3,*N*-4-etheno deoxy and ribo cytidine (113),[85] 2-thiothymidine and 2-thiodeoxyuridine (114), [86] nucleoside analogues with long alkyl chains instead of the base, e.g. (115),[87] and 3-nitropyrrole (116) [88, 89] and 4-, 5-, or 6-nitroindole analogues (117).[89] The dA analogue (111) gave increased binding to both DNA and RNA in several mixed-base sequences, the etheno analogues (112) and (113) gave fluorescent oligonucleotides, the 2-thio analogues (114) gave photosensitive probes, the long chain analogues (115) increased the binding to DNA, probably because of a better screening of the aliphatic chain from water in the duplex than in the single-stranded oligomer, and the nitropyrrole and nitroindole analogues (116) and (117) were universal base analogues, the best being the 5-nitroindole analogue.

Among sugar modified phosphoramidite derivatives prepared this year are 3',5'-ethano-2'-deoxyribonucleoside derivatives (118),[90, 91] the 4',6'-methano carbocyclic thymidine derivative (119),[92] a 2'-*O*-methyl-6,3'-ethanouridine derivative (120),[93] (R)- and (S)-3',4'-*seco*-thymidine derivatives, e.g. (121),[94] and arabino derivatives (122).[95] The first three are conformationally rigid nucleoside analogues, and were made in order to prepare oligonucleotide analogues which bind better

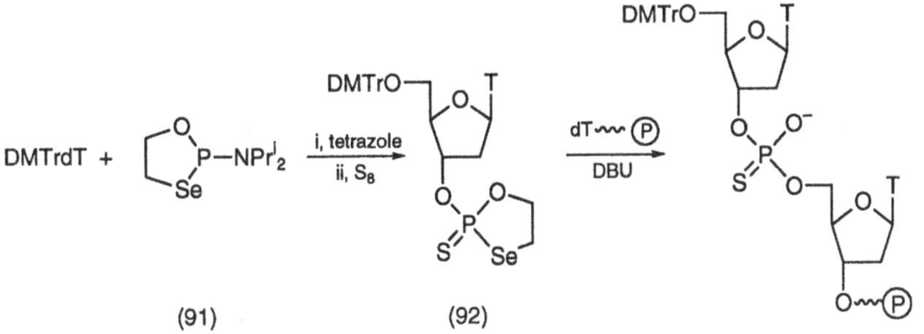

DMTrdT + (91) $\xrightarrow{\text{i, tetrazole; ii, } S_8}$ (92) $\xrightarrow[\text{DBU}]{\text{dT}\sim\text{(P)}}$

DMTrO T O P NPr$^i_2$ Se S O$^-$ O~(P)

(91) (92)

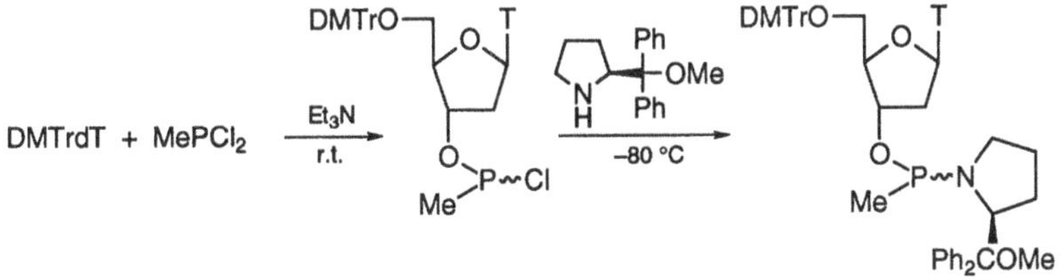

DMTrdT + $MePCl_2$ $\xrightarrow[\text{r.t.}]{Et_3N}$ DMTrO T O Me P~Cl $\xrightarrow{-80\ °C}$ Ph OMe Ph N H DMTrO T O Me P~N Ph_2COMe

(93) R_p: S_p = 9.7:1

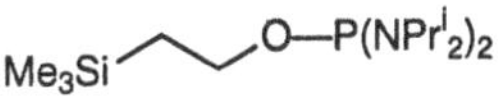

Me_3Si O—P(NPr$^i_2$)$_2$ Ph_2MeSi O—P(NPr$^i_2$)$_2$ DMTrO T O R_2MeSi O P—NPr$^i_2$

(94) 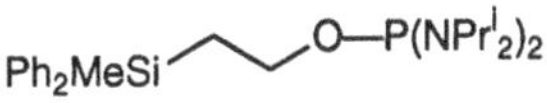 (95)

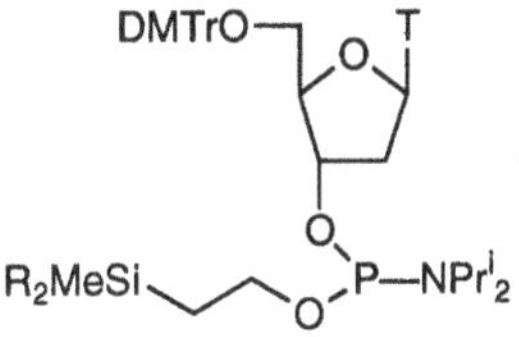

(96) R = Ph
(97) R = Me

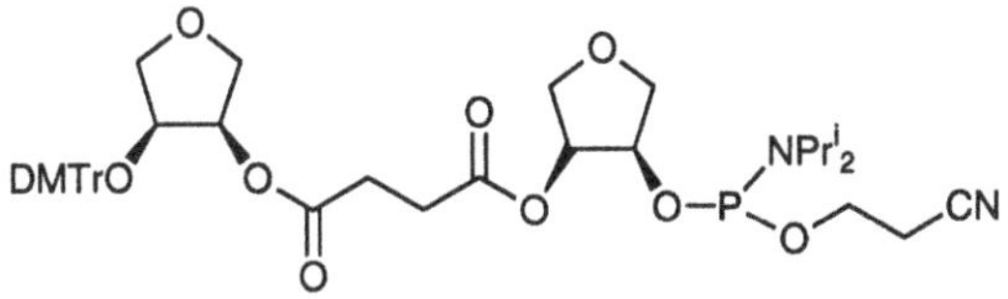

DMTrO O O O O O O—P NPr$^i_2$ O CN DNA—O—P O O$^-$ O OH O

(98) 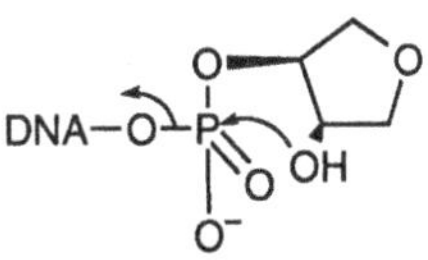 (99)

DMTrO O O O O O S O O O P NPr$^i_2$ O CN

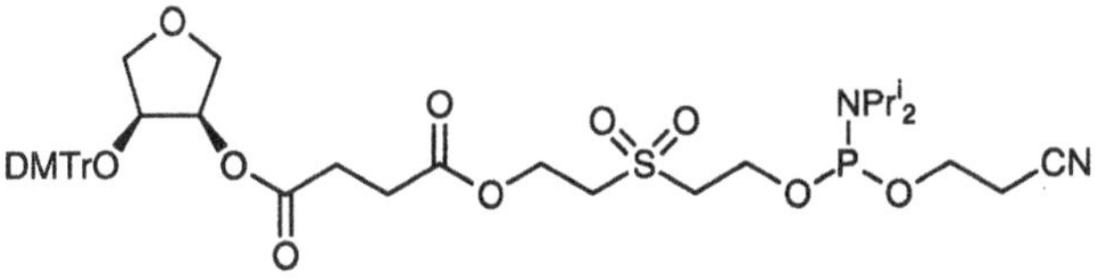

(100)

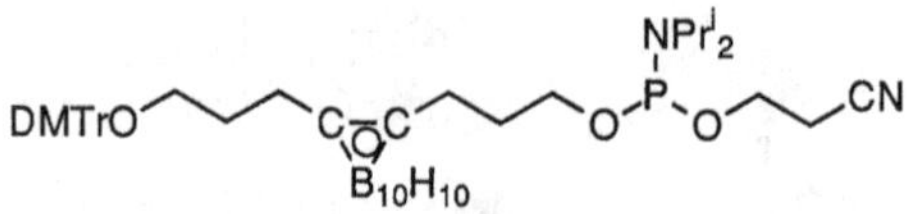

NPr^i_2
DMTrO C C O P O CN
$B_{10}H_{10}$

(101)

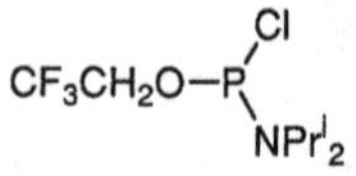

Cl
CF_3CH_2O—P
NPr^i_2

(102)

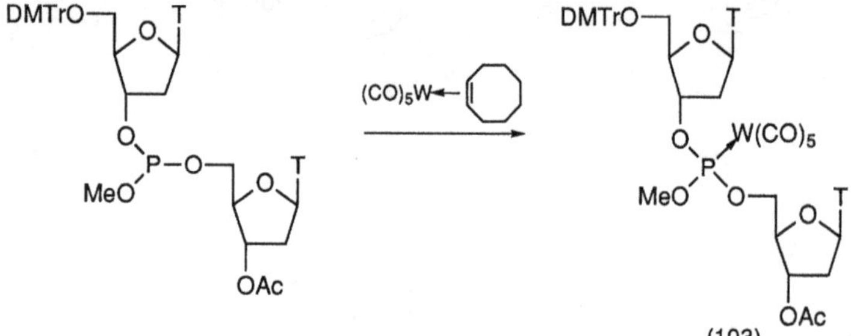

DMTrO O T
O P—O O T
MeO
OAc

$(CO)_5W$←

DMTrO O T
O $W(CO)_5$
P
MeO O O T
OAc

(103)

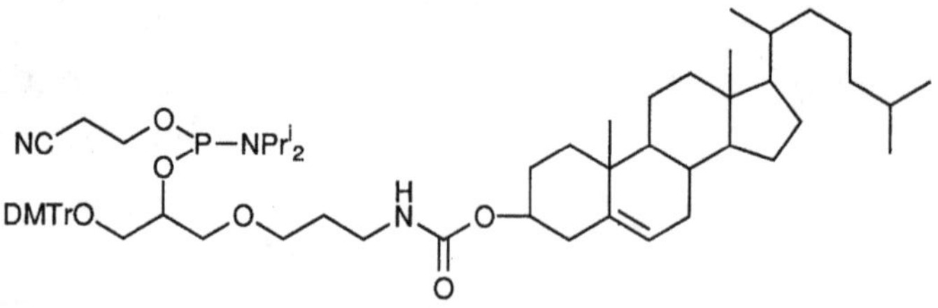

NC O P—NPr^i_2
O
DMTrO O H N O
O

(104)

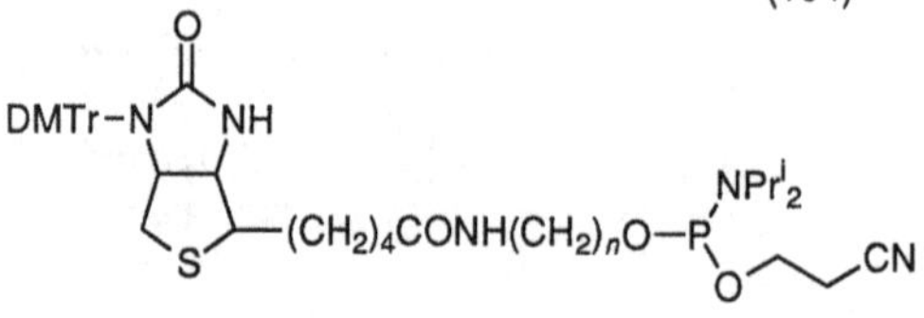

O
DMTr—N NH
S —$(CH_2)_4CONH(CH_2)_nO$—P NPr^i_2
O CN

(105) $n = 3$ or 5

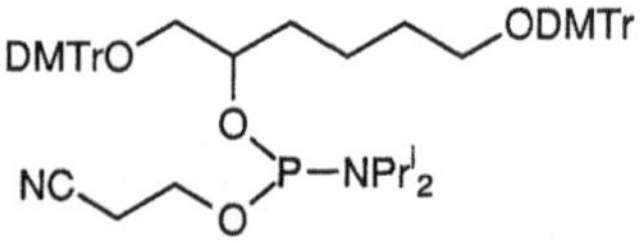

DMTrO ODMTr
O
NC O P—NPr^i_2

(106)

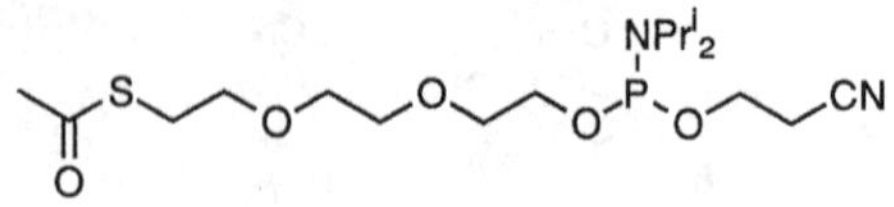

NPr^i_2
S O O O P O CN
O

(107)

NO_2 NO_2
S O
NC O P—NPr^i_2

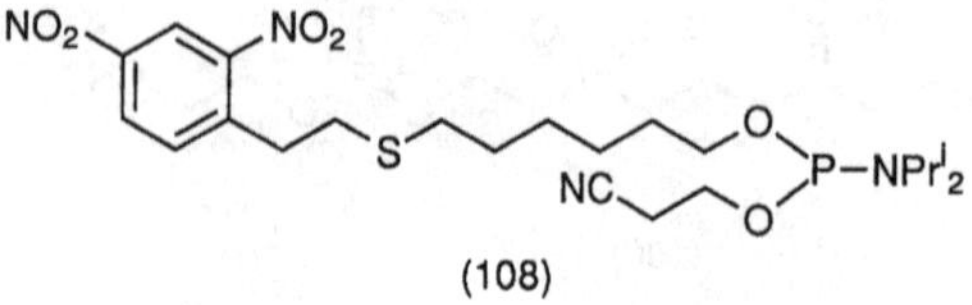

(108)

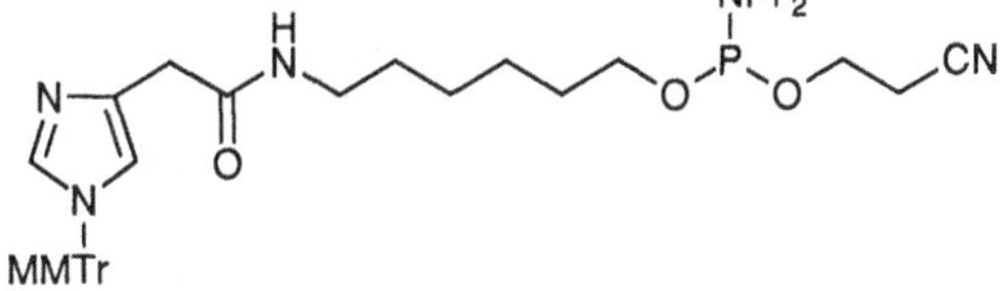

(109)

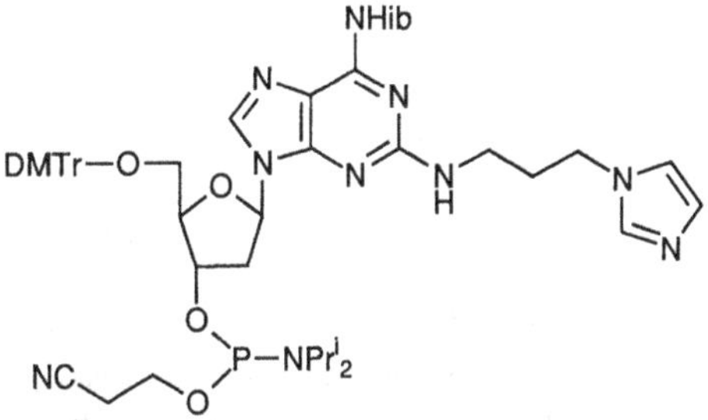

(110)

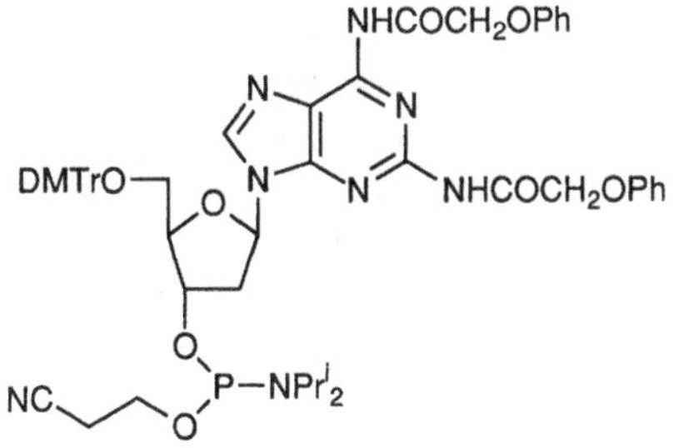

(111)

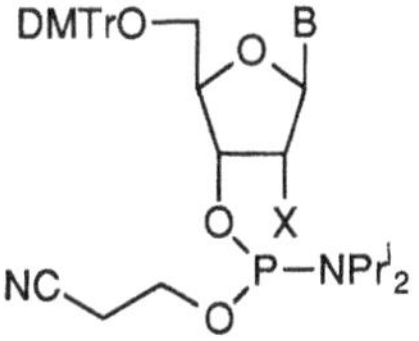

X = H or $OSiBu^tMe_2$

(112) B =

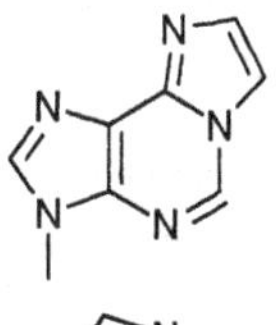

(113) B =

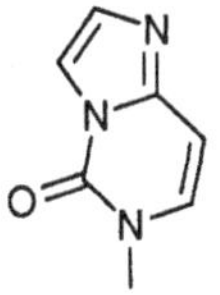

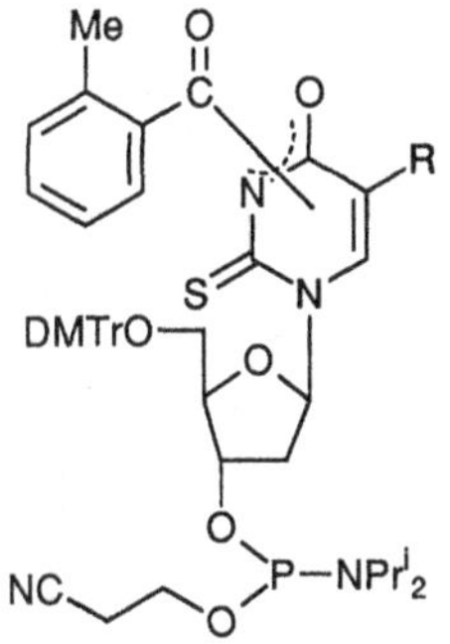

(114) R = H or Me

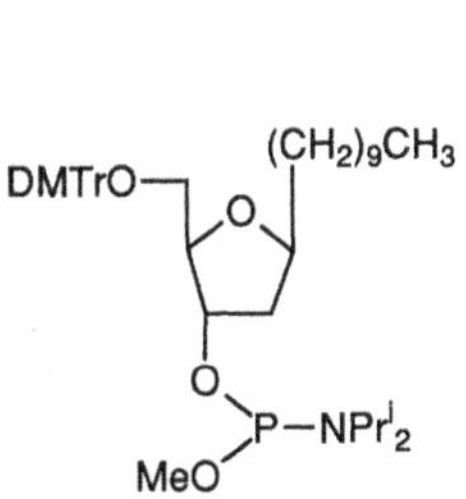

(115)

to complementary DNA or RNA for entropic reasons. This is the case for (118, B = A^{bz}) and (119), but not for (118, B = T) and (120). The flexible analogue (121) gave oligonucleotides with decreased binding in accordance with results for other *seco*-nucleosides. The arabino oligonucleotides prepared from (122) bound to DNA and RNA with an affinity comparable to unmodified oligodeoxyribonucleotides.

Two papers have been published on recovery of nucleoside phosphoramidites used in excess during solid support synthesis of oligonucleotides. This could be important in large-scale synthesis to reduce the costs. In one study the phosphoramidites were collected in bottles containing an excess of diisopropylamine which regenerated the phosphoramidite from the tetrazolide formed during coupling.[96] The collected products were more than 85% pure and could be reused after silica column chromatography. In another study the excess of phosphoramidites were hydrolysed, and the H-phosphonate group cleaved off by transesterification with methanol, catalysed by imidazole, to give the 3'-OH nucleosides in 71-100% yield.[97] A similar procedure was found to regenerate nucleosides from nucleoside methylphosphonamidites.

Nucleoside phosphoramidites or methylphosphonamidites with protecting groups on the bases that are removed faster or milder than the conventional benzoyl and isobutyryl groups have been developed, and several are commercially available. Papers in this area include one on the preparation and use of dialkylformamidine protected dA and dG phosphoramidites, which together with dC^{bz} yielded oligonucleotides that were completely deprotected by conc. aqueous ammonia in 1 h at 65 ^{o}C;[98] one paper on the use of acetyl protected dC phosphoramidites, allowing deprotection in 5 min at 65 ^{o}C with a mixture of methylamine and aqueous ammonia, without the modification of dC seen when dC^{bz} derivatives are treated with amines;[99] and three papers on the preparation of the base-sensitive methylphosphonates employing methylphosphonamidites with 2-acetoxymethyl-benzoyl,[100] formamidine on dA and dG plus isobutyryl on dC,[101] or 4-t-butylphenoxyacetyl [102] *N*-protecting groups.

3.4 Use for Sugar Phosphate, Phospholipid, or Phosphopeptide Synthesis.- A review has been published on the synthesis of phosphorylated biomolecules by the phosphoramidite approach.[4] It contains chapters on sterol-mononucleotide conjugates, mononucleotide glycoconjugates, sugar phosphates, phospholipids, phosphopeptides, and nucleopeptides.

A simple, one pot synthesis has been developed to prepare sugar phosphodiesters from the phosphoramidochloridites (123).[103] Hydrolysis of the intermediate phosphoramidite is apparently catalysed by the pyridine hydrochloride formed in the first step. Two sugar hydroxycholesterol phosphodiesters, needed for targeting of oxysterol drugs to defined organs, were prepared from the hydroxycholesterol phosphoramidite (124).[104] A glycolipid lipoteichoic acid was prepared from a lipid phosphoramidite (125).[105] Phospholipid analogues (126) containing phosphorothioate groups have been obtained from (127) via a phosphite coupling as shown.[106] Phospholipid analogues (128) containing phosphorodithioate groups have been prepared in high yields from 2-chloro-1,3,2-

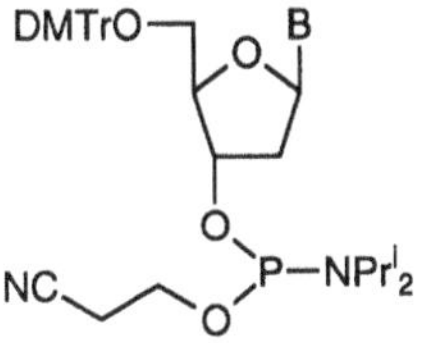

(116) B =

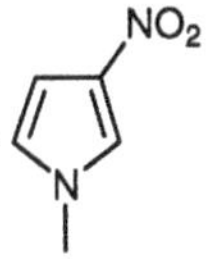

(117) B =

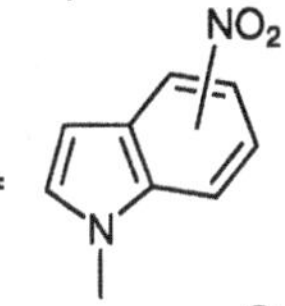

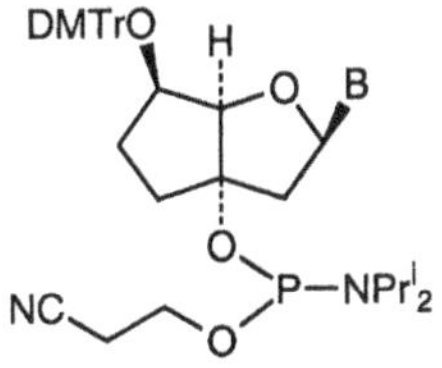

(118) B = A^{bz} or T

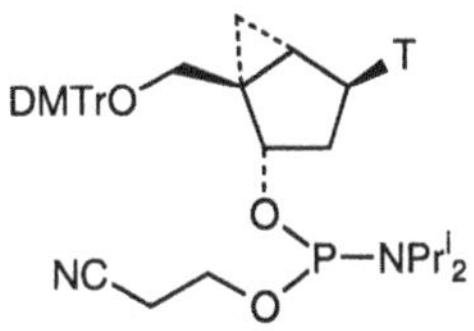

(119)

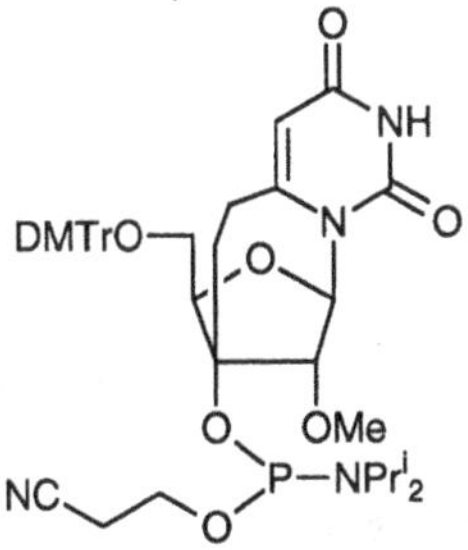

(120)

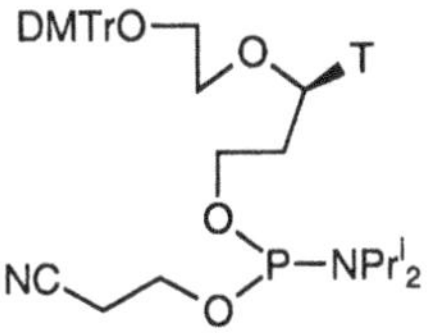

(121)

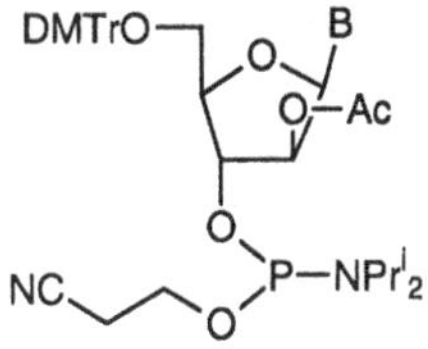

(122)

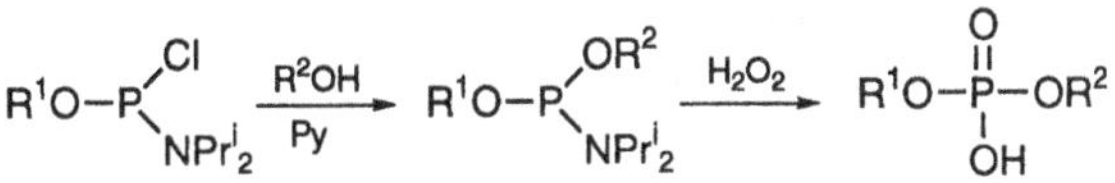

(123) R^1 = Me or octyl, R^2 = protected sugar alcohol

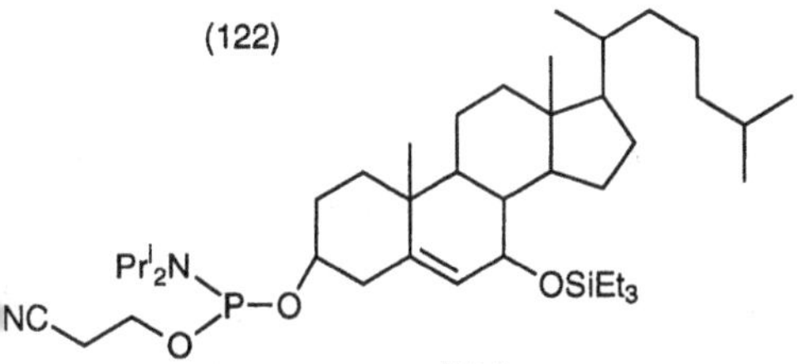

(124)

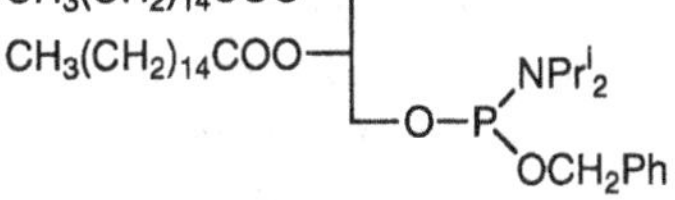

(125)

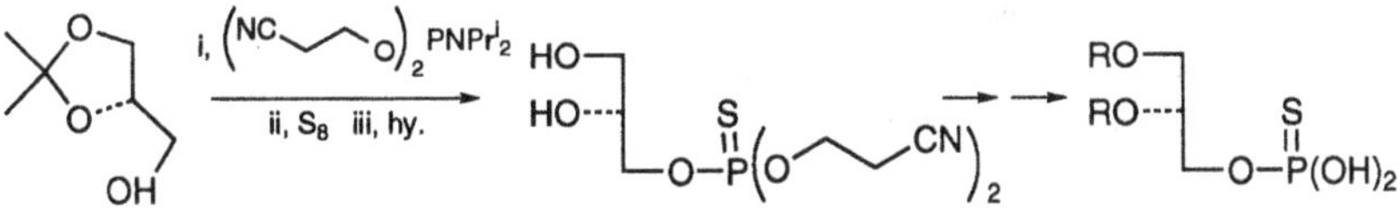

(127)

(126) R = H or oleoyl

dithiaphospholan (129).[107] A phosphatidylinositol analogue (130) likewise modified with a phosphorodithioate linkage was prepared from the thiophosphoramidite (131).[108]

Some cyclic phosphopeptides (132) were prepared in surprisingly high yields for the formation of a 12-membered ring (62-70%) using the phosphorodichloridite (133);[109] similar reactions with the phosphorodiamidite (134) gave lower yields. Synthesis of cryptands containing phosphodiester linkages between amino acids has been realised by the same approach using (133) as the reagent of choice.[110] A high yield synthesis of *O*-phosphohomoserine has been developed, where the phosphate group was introduced by phosphitylation with the phosphoramidite (135).[111] Protected serine has been phosphitylated with the phosphoramidite (136) which after oxidation and cleavage of the trichloroethyl group gave the benzyl serine phosphodiester (137);[112] the yields of phosphorylated peptides were higher using (137) than similar phosphoserine derivatives with none or two benzyl groups. Phosphitylation of resin-bound peptides with the phosphoramidite (138) followed by oxidation with t-butylhydroperoxide or dibenzoyl tetrasulphide gave phosphorylated or thiophosphorylated peptides.[113]

Three papers have appeared on the solid support synthesis of oligonucleoside-peptide hybrids, using different solid supports and protection group strategy.[114, 115, 116]

3.5 Miscellaneous.- A simple method to resolve racemic 1,1'-binaphthalene-2,2'-diol involves reaction with menthyl phosphorodichloridite (139) and separation of the diastereomeric phosphites by crystallisation from ether.[117] The new cyclic chlorodiaminophosphine (140), prepared *in situ*,[118] and two new cyclic triaminophosphines (141) [119] have been used to determine the enantiomeric purity of chiral alcohols, thiols, and amines by means of ^{31}P n.m.r. New optically pure phosphites used for Rh catalysed asymmetric hydroformylation reactions are (142),[120] two diastereomers of (143),[121] and (144);[122] the diphosphinite (145) was similarly employed.[123] Some new cyclic aminophosphines (146) were prepared for use as ligands in asymmetric Pd catalysed allylic substitution reactions.[124]

4 Reactions involving Two-co-ordinate Phosphorus

The first thioxophosphines (147) which are stable at room temperature have been prepared.[125] Their physical properties and an X-ray crystal structure determination point to a considerable contribution from the dipolar resonance formula shown. The amino-oxophosphine (148) has been generated from the reaction of a rhodium complex containing the diphosphoxane (149) with carbon monoxide.[126] The generation of an aryl phosphenite (150) from a 7-phospha-norbornene derivative (151) by thermolysis or photolysis has been investigated.[127] Thermolysis gave a condensate containing some (150) besides dimers and trimers; photolysis, however, gave no fragmentation products from (151) or from similar phospha-norbornenes unless alcohols were present. Photolysis of 7-phospha-norbornenes is therefore not a route to two-co-ordinate phosphorus compounds, as assumed earlier.

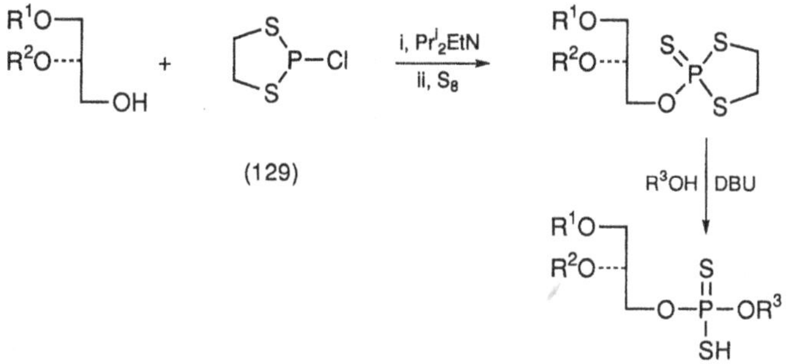

i, Pr^i_2EtN
ii, S_8

(129)

R^3OH | DBU

(128)

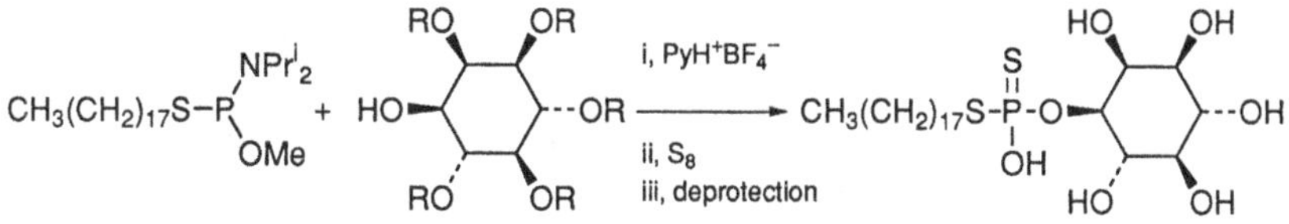

i, $PyH^+BF_4^-$
ii, S_8
iii, deprotection

(131) $R = CH_2Ph$ (130)

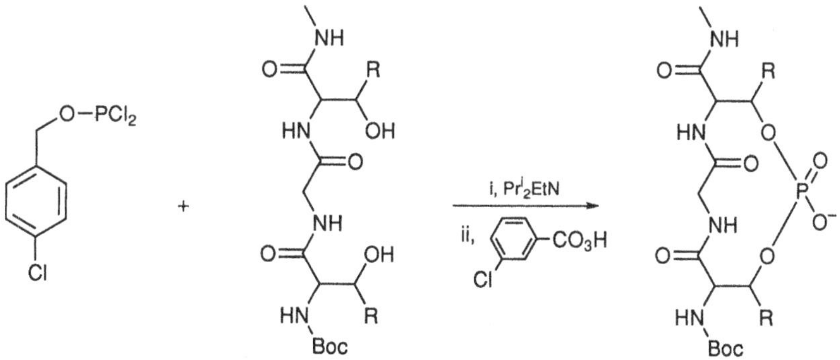

i, Pr^i_2EtN
ii, $ClC_6H_4CO_3H$

(133) (132) R = H or Me

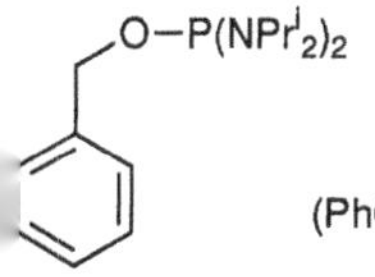

(134) (135) (136) 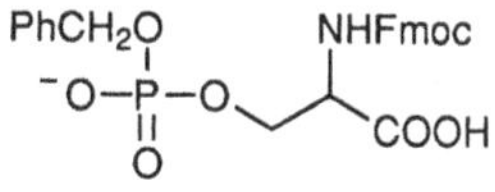 (137)

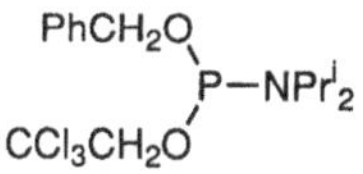

$(Bu^tO)_2P—NEt_2$

O—PCl_2

CF_3 CF_3 Me N P—Cl N Me

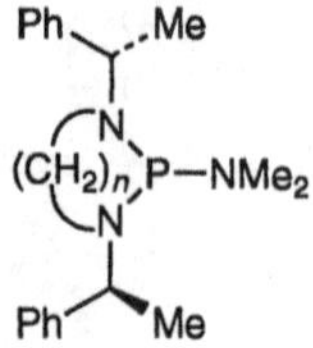

Ph Me N $(CH_2)_n$ P—NMe_2 N Ph Me

(138)

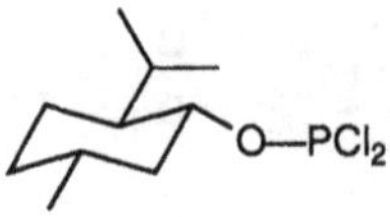

(139)

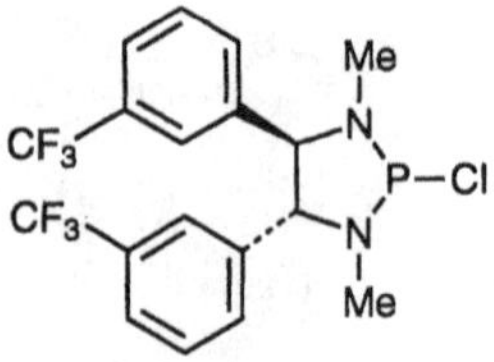

(140)

(141) n = 2 or 3

PAr_2 O P O O

Me Cl Me PPh_2 Me Cl Me O P O O

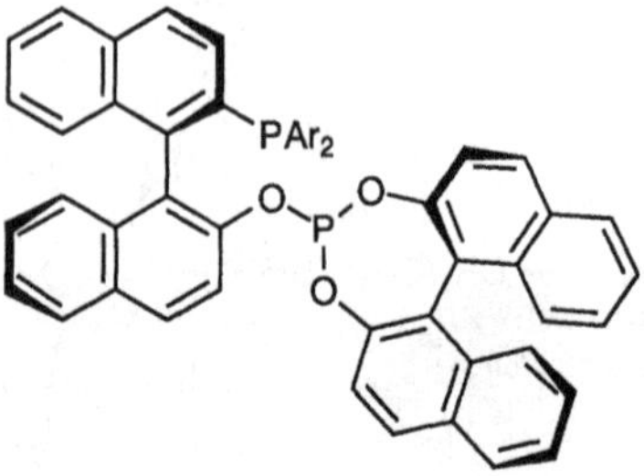

(142) Ar = Ph or 3, 5-dimethylphenyl

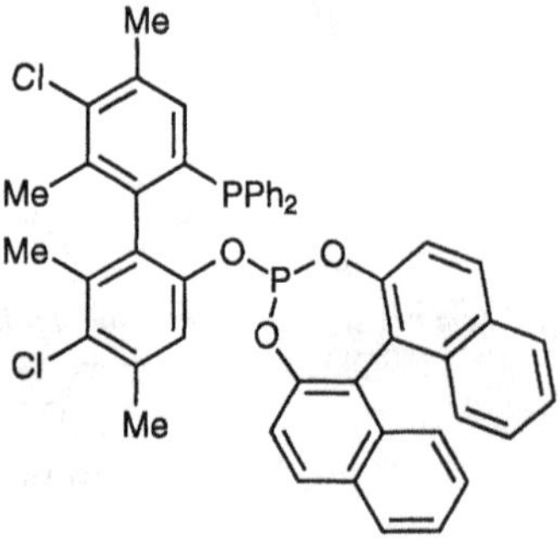

(143)

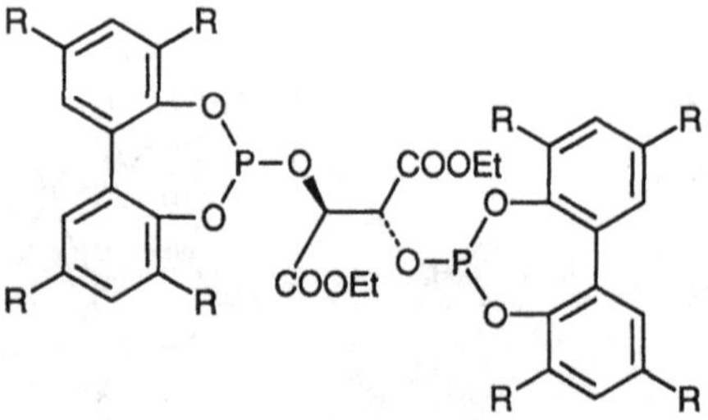

R R O P—O O R R COOEt COOEt O—P O O R R R R

O—PPh_2 O—PPh_2

(144) R = H or Bu^t

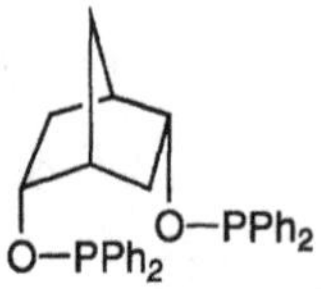

(145)

Me N—R P Ph

$$Ph_3P{=}C(R)PCl_2 \xrightarrow{Na_2S} Ph_3P{=}C(R){-}P{=}S \longleftrightarrow Ph_3\overset{+}{P}{-}C(R){=}P{-}S^-$$

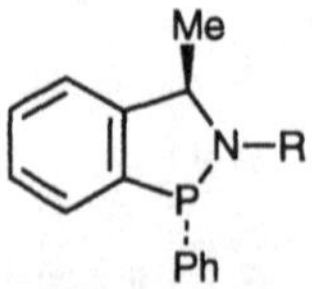

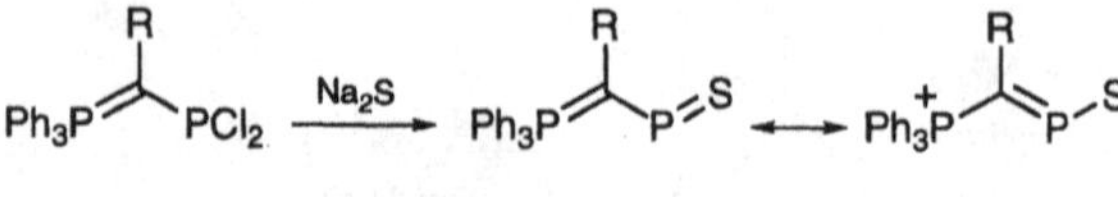

(146) R = CH_2CH_2–N(pyrrolidinyl) or $SiBu^tPh_2$

(147) R = Me or Et

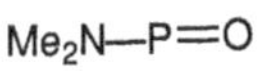

$Me_2N—P=O$

(148)

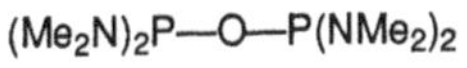

$(Me_2N)_2P—O—P(NMe_2)_2$

(149)

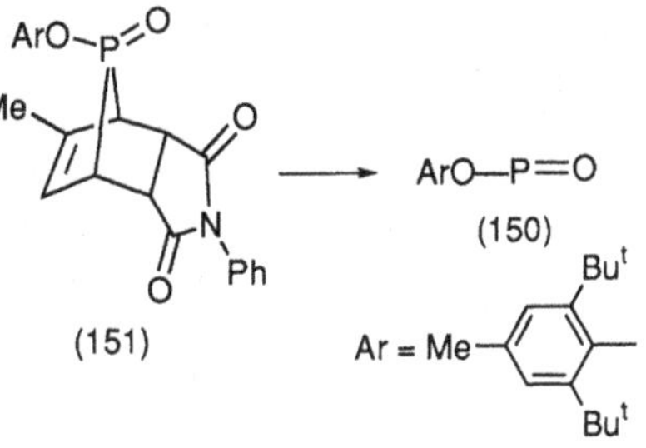

ArO—P=O, Me, O, N, Ph, O

(151)

→ ArO—P=O

(150)

Ar = Me—C_6H_2(Bu^t)(Bu^t)—

R^1—P=N—R^2

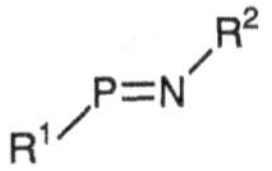

(152) R^1 = Bu^t, $EtMe_2C$, Et_2MeC, or Et_3C
R^2 = Bu^t or Et_3C

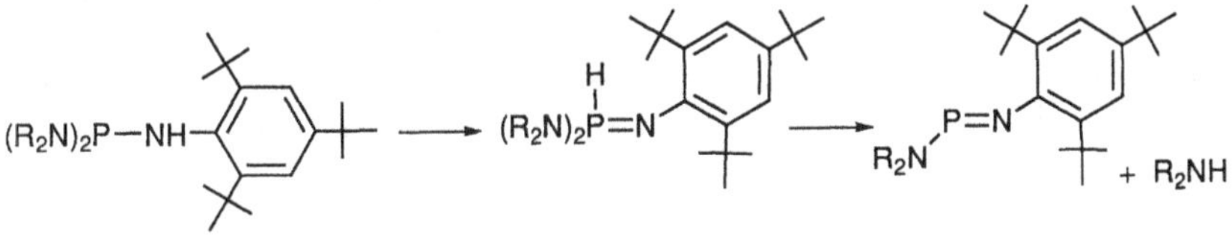

$(R_2N)_2P—NH$— → H, $(R_2N)_2P=N$— → R_2N—P=N— + R_2NH

(154) R = Me, Et, Pr^i, or R_2N = N

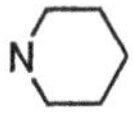

(153)

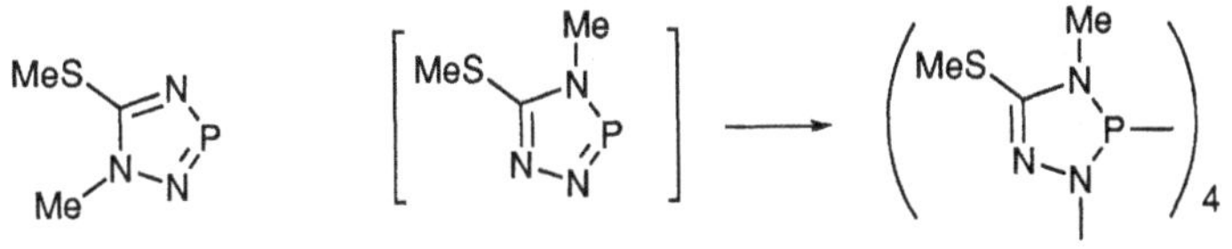

MeS, N, P, N, N, Me

(155)

[Me, MeS, N, P, N, N]

(156)

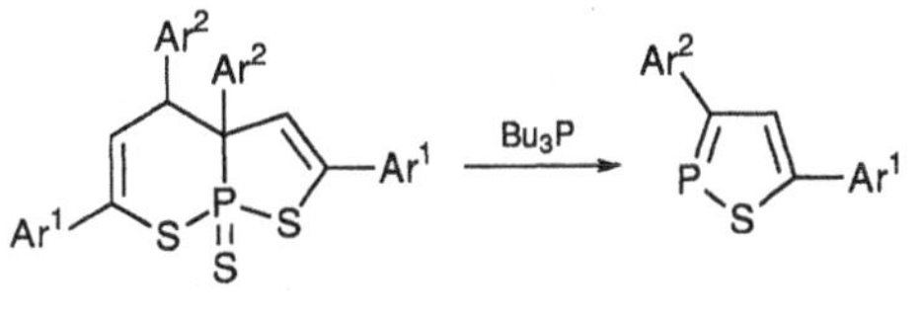

→ (Me, MeS, N, P—, N, N)$_4$

Ar^2, Ar^2, Ar^1, Ar^1, S, P, S, S

(158)

$\xrightarrow{Bu_3P}$ Ar^2, P, S, Ar^1

(157)

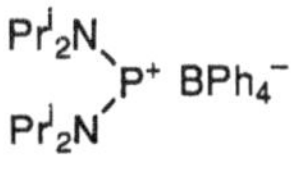

Pr^i_2N, P^+ BPh_4^-, Pr^i_2N

(159)

Me, N, +, Ph, P, −, N, BPh_3, Me

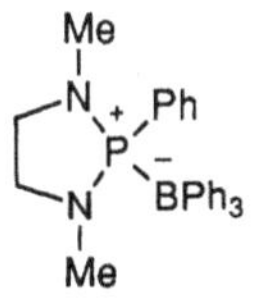

(160)

A series of alkyl substituted iminophosphines (152) have been prepared and their tendency towards (2+1) cyclodimerization studied.[128] Not surprisingly, the stability of the iminophosphines increased with increasing size of the substituents, with (152, R = R' = Et_3C) being unchanged after several days at 40 °C. The first examples of spontaneous formation of aminoiminophosphines (153) from triaminophosphines (154) have been published.[129] This is a remarkable reaction which could be general for aminophosphines containing an NH group, although the aminoiminophosphines would not be stable without the large NH substituent present here.

A review has appeared on the preparation and properties of anellated heterophospholes, all of which contain two-co-ordinated phosphorus atoms.[5] Some new 5-methylthio-1,2,4,3-triazaphospholes have been prepared;[130] the 1-methyl derivative (155) was stable whereas the 2- or 4-substituted derivatives, e.g. (156), gave cycloaddition products. Several 3,5-diaryl-1,2-thiaphospholes (157) were obtained in good yields by treatment of the cyclic phosphonotrithioates (158) with tributylphosphine.[131]

A full paper has been published on the anionic and steric factors responsible for the stabilization of phosphenium ions.[132] It was found that (159) was ionic while a corresponding cyclic compound with less steric hindrance had the covalent structure (160), and that the tetrachlorogallate of both structures was ionic. New phosphenium ions prepared this year are (161) and (162) made by the routes shown,[133] and (163).[134] The compound (164) can be regarded as a phosphenium ion stabilized by two N->P donor-acceptor bonds; its structure and some reactions have been studied.[135] The properties of the 2-chloro-1,3,2-diazaphospholene (165) indicate that it is in equilibrium with a phosphenium chloride in polar solvents;[136] this explains the unprecedented exchange of (165) with alkyl halides to give (166).[137]

The unsymmetrical diphosphene (167) and its reaction product (168) with sulphur have been described.[138] A corresponding 1,2,3-selenadiphosphirane (169) has been prepared from the very stable bis(2,4,6-tris(trifluoromethyl)phenyl)diphosphene and its X-ray crystal structure determined.[139] Addition of chlorine to a diphosphene has been realised for the first time in the reaction of (170) with phosphorus pentachloride to give (171);[140] cyclization of (171) to (172) followed by substitution of the chlorine atom gave interesting ring-opened compounds, e.g. (173).

5 Miscellaneous Reactions

Dialkyl trimethylsilyl phosphites (174) are conveniently prepared from sodium salts of dialkyl phosphites and trimethylsilyl chloride. It has now been found that (174) prepared in this way rearrange to silylphosphonates (175) upon heating.[141] Azidobis(diisopropylamino)phosphine (176) and an alkyne gave the 1,2,3,4-triazaphosphinine (177) at room temperature, and the 1,2-azaphosphete (178) upon reflux in toluene.[142] Photoinduced rearrangement of the allylic phosphite (179) to (180) has been shown to proceed much more efficiently when run under single electron transfer conditions than under triplet forming conditions.[143] Photoinduced rearrangements of benzylic phosphites continue to be used to prepare acyclic nucleoside phosphonate analogues like

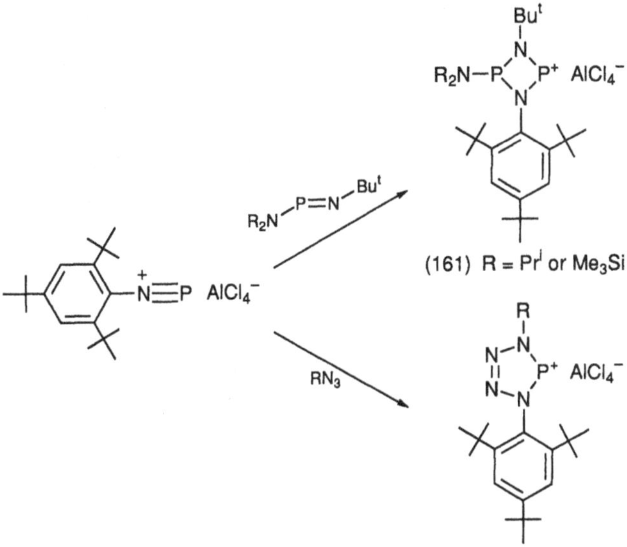

Bu^t
N
R_2N-P P^+ $AlCl_4^-$
N

$R_2N-P=N-Bu^t$

$\overset{+}{N}\equiv P$ $AlCl_4^-$

(161) $R = Pr^i$ or Me_3Si

RN_3

R
N–N
N P^+ $AlCl_4^-$
N–N

(162) $R = Bu^t$ or Et_3C

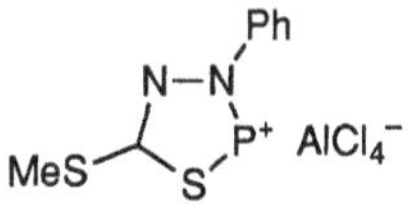

Ph
N–N
MeS P^+ $AlCl_4^-$
S

(163)

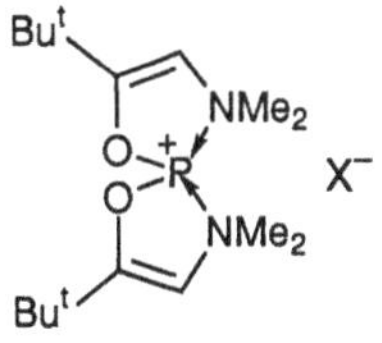

Bu^t
NMe_2
O $\overset{+}{P}$ X^-
O
NMe_2
Bu^t

(164) X = Cl or BPh_4

Bu, Me, N, P–Cl, N, Me, Bu ⇌ Bu, Me, N, P^+ Cl^-, N, Me, Bu $\xrightarrow[r.t.]{RX}$ Bu, Me, N, P–X + RCl, N, Me, Bu

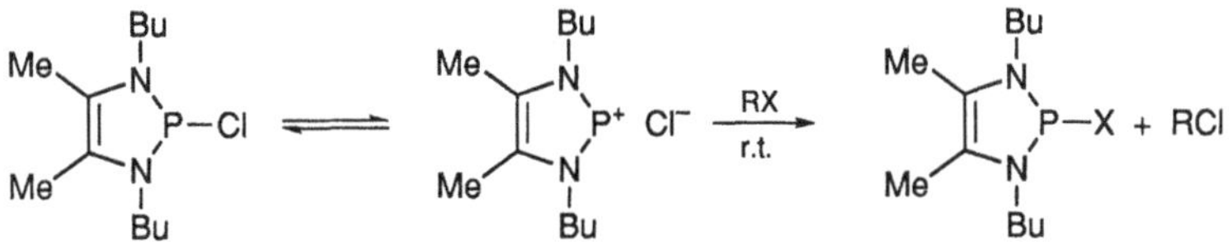

(165)

(166) X = Br or I
R = Me, Et, or Bu

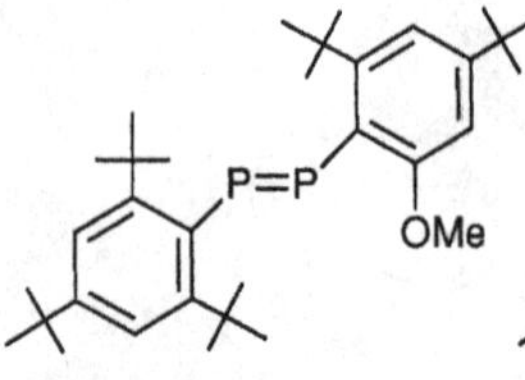

P=P OMe CF3 CF3 CF3 P–P S OMe P–P Se CF3 CF3 CF3

(167) 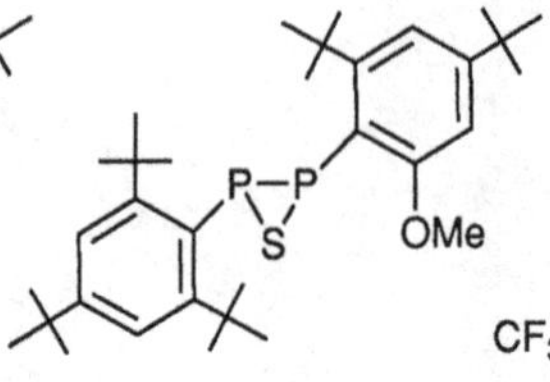 (168) 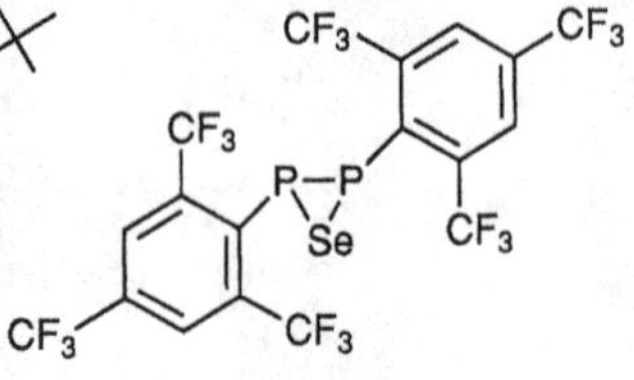 (169)

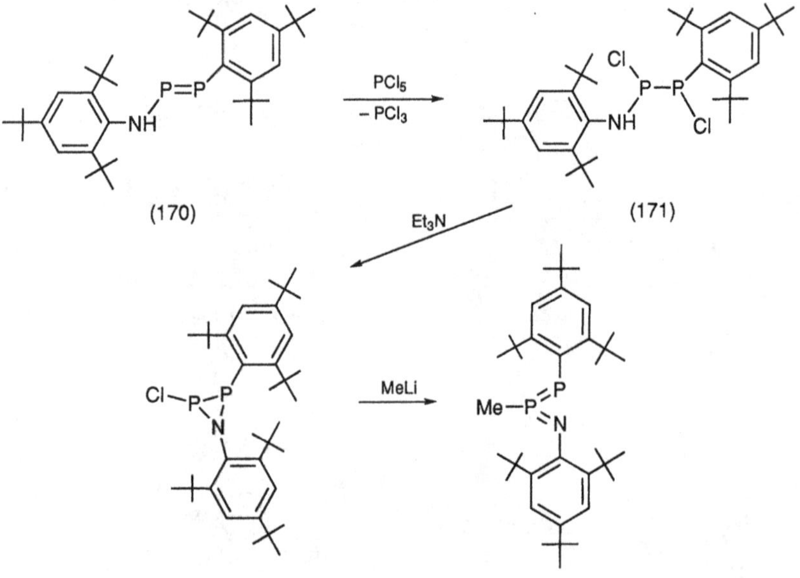

P=P NH PCl_5 $-PCl_3$ Cl P–P NH Cl

(170) (171)

Et_3N

Cl–P–P N MeLi Me–P=P =N

(172) (173)

$(RO)_2P$—$OSiMe_3$ 80 °C $(RO)_2P(=O)SiMe_3$

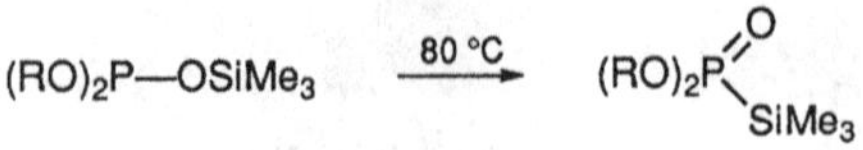

(174) (175)

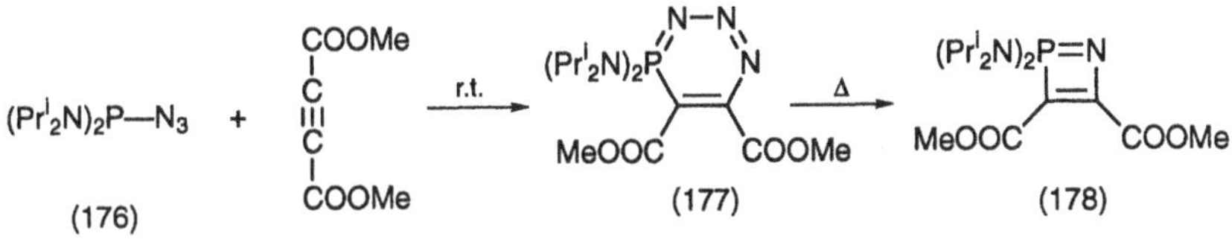

$(Pr^i_2N)_2P—N_3$ + COOMe–C≡C–COOMe →(r.t.) $(Pr^i_2N)_2P$ (N–N, N ring) MeOOC COOMe →(Δ) $(Pr^i_2N)_2P=N$ MeOOC COOMe

(176) (177) (178)

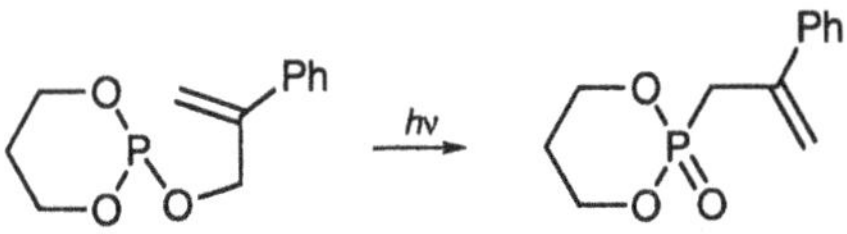

Ph →(hν) Ph B P(O)(OH)$_2$ Ph HO

(179) (180) 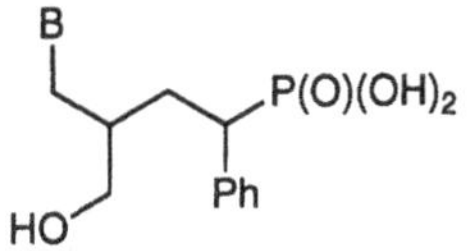 (181) B = A, C, or G

(181).[144] Glycosylation reactions with $(RO)_2P$-O- as the leaving group are effective with Lewis acid catalysis, and the mechanism and scope of triflate catalysed reactions has been studied,[145] as well as those catalysed by zinc chloride.[146]

References

1. S. Agrawal (ed.), *Protocols for Oligonucleotides and Analogs, Synthesis and Properties*, *Methods in Molecular Biology*, **20**, Humana Press Inc., Totowa, New Jersey, 1993.
2. S. Agrawal (ed.), *Protocols for Oligonucleotide Conjugates, Synthesis and Analytical Techniques, Methods in Molecular Biology*, **26**, Humana Press, Inc., Totowa, New Jersey, 1994.
3. S.L. Beaucage and R.P. Iyer, *Tetrahedron*, 1993, **49**, 6123-6194.
4. S.L. Beaucage and R.P. Iyer, *Tetrahedron*, 1993, **49**, 10441-10488.
5. R.K. Bansal, K. Karaghiosoff, and A. Schmidpeter, *Tetrahedron*, 1994, **50**, 7675-7745.
6. S. Andreae and U. Pieper, *J. Prakt. Chem.*, 1994, **336**, 75.
7. S.V. Serves, A.G. Teloniati, D.N. Sotiropoulos, and P.V. Ioannou, *Phosphorus, Sulfur, and Silicon*, 1994, **89**, 181.
8. E.A. Boyd, A.C. Regan, and K. James, *Tetrahedron Lett.*, 1994, **35**, 4223.
9. H.R. Hudson, R.W. Matthews, M. McPartlin, M.A. Pryce, and O.O. Shode, *J. Chem. Soc., Perkin Trans. 2*, 1993, 1433.
10. L. Riesel, K. Lauritsen, and H. Vogt, *Z. Anorg. Allgem. Chem.*, 1994, **620**, 1099.
11. Y.G. Gololobov, G.D. Kolomnikova, and T.O. Krylova, *Tetrahedron Lett.*, 1994, **35**, 1751.
12. R.J. Ternansky and A.J. Pike, *Bioorg. Med. Chem. Lett.*, 1993, **3**, 2237.
13. S. Yasui, K. Shioji, M. Yoshihara, T. Maeshima, and A. Ohno, *Bull. Chem. Soc. Japan*, 1993, **66**, 2077.
14. V. Sum and T.P. Kee, *J. Chem. Soc., Perkin Trans. 1*, 1993, 2701.
15. V. Sum, T.P. Kee, and M. Thornton-Pett, *J. Chem. Soc., Chem. Commun.*, 1994, 743.
16. I.F. Pickersgill, P.G. Devitt, and T.P. Kee, *Synth. Commun.*, 1993, **23**, 1643.
17. N. Greene and T.P. Kee, *Synth. Commun.*, 1993, **23**, 1651.
18. H. Gross, B. Costisella, S. Ozegowski, I. Keitel, and K. Forner, *Phosphorus, Sulfur, and Silicon*, 1993, **84**, 121.
19. X.-Y. Jiao, C. Verbrugge, M. Borloo, W. Bollaert, A. De Groot, R. Dommisse, and A. Haemers, *Synthesis*, 1994, 23.
20. X.-Y. Jiao, M. Borloo, C. Verbruggen, and A. Haemers, *Tetrahedron Lett.*, 1994, **35**, 1103.
21. A.A. Prishchenko, D.A. Pisarnitskii, M.V. Livantsov, and V.S. Petrosyan, *Russ. J. Gen. Chem.*, 1993, **63**, 167.
22. J.R. Goerlich, M. Farkens, A. Fischer, P.G. Jones, and R. Schmutzler, *Z. Anorg. Allgem. Chem.*, 1994, **620**, 707.
23. E. Krawczyk, A. Skowronska, and J. Michalski, *J. Chem. Soc., Perkin Trans. 1*, 1994, 89.
24. W.J. Stec, B. Uznanski, A. Wilk, B.L. Hirschbein, K.L. Fearon, and B.J. Bergot, *Tetrahedron Lett.*, 1993, **34**, 5317.
25. T.K. Wyrzykiewicz and V.T. Ravikumar, *Bioorg. Med. Chem. Lett.*, 1994, **4**, 1519.
26. A.A. Kolomeitsev, K.Y. Chabanenko, G.-V. Röschenthaler, and Y.L. Yagupolskii, *Synthesis*, 1994, 145.
27. R.P. Parg, J.D. Kilburn, and T.G. Ryan, *Synthesis*, 1994, 195.
28. J. Stawinski and M. Thelin, *J. Org. Chem.*, 1994, **59**, 130.
29. Y.J. Koo and D.Y. Oh, *Synth. Commun.*, 1993, **23**, 1771.
30. S. Thiebaut and C. Selve, *New J. Chem.*, 1993, **17**, 595 (Chem. Abstr., 1994, **121**, 8390q).
31. A.M. Koroteev, M.P. Koroteev, A.R. Bekker, M.Y. Antipin, B.K. Sadybakasov, Y.T. Struchkov, and E.E. Nifant'ev, *Russ. J. Gen. Chem.*, 1993, **63**, 49.
32. H. Beckers, *Z. Anorg. Allgem. Chem.*, 1993, **619**, 1880.
33. D.M. Malenko, L.I. Nesterova, S.N. Lukyanenko, L.V. Randina, and A.D. Sinitsa, *Russ. J. Gen. Chem.*, 1993, **63**, 1171.

34. J. Münchenberg, O. Böge, A.K. Fischer, P.G. Jones, and R. Schmutzler, *Phosphorus, Sulfur, and Silicon*, 1994, **86**, 103.
35. V.S. Reddy and K.V. Katti, *Inorg. Chem.*, 1994, **33**, 2695.
36. V.S. Reddy, K.V. Katti, and C.L. Barnes, *Chem. Ber.*, 1994, **127**, 979.
37. H.R.G. Bender, M. Nieger, and E. Niecke, *Z. Naturforsch.*, 1993, **48b**, 1742.
38. S.A. Katz, V.S. Allured, and A.D. Norman, *Inorg. Chem.*, 1994, **33**, 1762.
39. T.G. Hill, R.C. Haltiwanger, M.L. Thompson, S.A. Katz, and A.D. Norman, *Inorg. Chem.*, 1994, **33**, 1770.
40. S.K. Sheehan, M. Jiang, L. McKinstry, T. Livinghouse, and D. Garton, *Tetrahedron*, 1994, **50**, 6155.
41. M. Gruber and R. Schmutzler, *Phosphorus, Sulfur, and Silicon*, 1993, **80**, 181.
42. M. Gruber, P.G. Jones, and R. Schmutzler, *Phosphorus, Sulfur, and Silicon*, 1993, **80**, 195.
43. M. Gruber, R. Schmutzler, and D. Schomburg, *Phosphorus, Sulfur, and Silicon*, 1993, **80**, 205.
44. M. Gruber and R. Schmutzler, *Phosphorus, Sulfur, and Silicon*, 1993, **80**, 219.
45. E.J. Corey, Z. Chen, and G.J. Tanoury, *J. Am. Chem. Soc.*, 1993, **115**, 11000.
46. C. Mundt and L. Riesel, *Phosphorus, Sulfur, and Silicon*, 1994, **88**, 179.
47. C. Mundt and L. Riesel, *Phosphorus, Sulfur, and Silicon*, 1994, **88**, 75.
48. C. Mundt and L. Riesel, *Phosphorus, Sulfur, and Silicon*, 1994, **88**, 169.
49. C. Mundt and L. Riesel, *Phosphorus, Sulfur, and Silicon*, 1994, **89**, 133.
50. Y. Huang, A.M. Arif, and W.G. Bentrude, *J. Org. Chem.*, 1993, **58**, 6235.
51. A. Fischer, I. Neda, P.G. Jones, and R. Schmutzler, *Phosphorus, Sulfur, and Silicon*, 1993, **83**, 135.
52. I. Neda, T. Kaukorat, and R. Schmutzler, *Phosphorus, Sulfur, and Silicon*, 1993, **80**, 241.
53. I. Neda, T. Kaukorat, and R. Schmutzler, *Phosphorus, Sulfur, and Silicon*, 1993, **84**, 205.
54. R. Sonnenburg, I. Neda, A. Fischer, P.G. Jones, and R. Schmutzler, *Z. Naturforsch.*, 1994, **49b**, 788.
55. E.E. Nifant'ev, T.S. Kukhareva, V.I. Dyachenko, and A.F. Kolomiets, *Russ. J. Gen. Chem.*, 1993, **63**, 498.
56. E.E. Nifant'ev, A.I. Zavalishina, V.G. Baranov, E.I. Orzhekovskaya, L.K. Vasyanina, M.Y. Antipin, B.K. Sadybakasov, and Y.T. Struchkov, *Russ. J. Gen. Chem.*, 1993, **63**, 1220.
57. M. Farkens, T.G. Meyer, I. Neda, R. Sonnenburg, C. Müller, A.K. Ascher, P.G. Jones, and R. Schmutzler, *Z. Naturforsch.*, 1994, **49b**, 145.
58. I. Neda, M. Farkens, and R. Schmutzler, *Z. Naturforsch.*, 1994, **49b**, 165.
59. M. Farkens, I. Neda, and R. Schmutzler, *Z. Naturforsch.*, 1994, **49b**, 445.
60. M. Haddad, L. Lopez, J. Barrans, Y.K. Rodi, and E.M. Essassi, *Phosphorus, Sulfur, and Silicon*, 1993, **80**, 37.
61. R. Murugavel, S.S. Krishnamurthy, J. Chandrasekhar, and M. Nethaji, *Inorg. Chem.*, 1993, **32**, 5447.
62. E.E. Nifant'ev, M.K. Grachev, and L.K. Vasyanina, *Russ. J. Gen. Chem.*, 1993, **63**, 403.
63. V.Y. Iorish, M.K. Grachev, A.R. Bekker, and E.E. Nifant'ev, *Russ. J. Gen. Chem.*, 1993, **63**, 551.
64. E.E. Nifant'ev, M.K. Grachev, G.I. Kurochkina, and L.K. Vasyanina, *Russ. J. Gen. Chem.*, 1993, **63**, 660.
65. M.K. Grachev, G.I. Kurochkina, A.R. Bekker, L.K. Vasyanina, and E.E. Nifant'ev, *Russ. J. Gen. Chem.*, 1993, **63**, 240.
66. M.K. Grachev, G.I. Kurochkina, A.R. Bekker, L.K. Vasyanina, and E.E. Nifant'ev, *Russ. J. Gen. Chem.*, 1993, **63**, 661.
67. J. Helinski, W. Dabkowski, and J. Michalski, *Nucleosides and Nucleotides*, 1993, **12**, 597; *Phosphorus, Sulfur, and Silicon*, 1993, **76**, 395.
68. J. Helinski, W. Dabkowski, and J. Michalski, *Tetrahedron Lett.*, 1993, **34**, 6451.
69. K. Mitsiura and W.J. Stec, *Bioorg. Med. Chem. Lett.*, 1994, **4**, 1037.
70. P. Rosmanitz, S. Eisenhardt, J.W. Bats, and J.W. Engels, *Tetrahedron*, 1994, **50**, 5719.
71. V.T. Ravikumar, H. Sasmor, and D.L. Cole, *Bioorg. Med. Chem. Lett.*, 1993, **3**, 2637.
72. T. Wada and M. Sekine, *Tetrahedron Lett.*, 1994, **35**, 757.
73. P.M. Hardy, D. Holland, S. Scott, A.J. Garman, C.R. Newton, and M.J. McLean, *Nucleic Acids Res.*, 1994, **22**, 2998.

74. R.R. Kane, K. Drechsel, and M.F. Hawthorne, *J. Am. Chem. Soc.*, 1993, **115**, 8853.
75. W.D. Luo, E. Atrazheva, N. Fregeau, W.H. Gmeiner, and J.W. Lown, *Can. J. Chem.*, 1994, **72**, 1548.
76. J.M.D.R. Toma and D.E. Bergstrom, *J. Org. Chem.*, 1994, **59**, 2418.
77. H. Vu, P. Singh, L. Lewis, J.G. Zendegui, and K. Jayaraman, *Nucleosides Nucleotides*, 1993, **12**, 853.
78. P. Kumar, D. Bhatia, B.S. Garg, and K.C. Gupta, *Bioorg. Med. Chem. Lett.*, 1994, **4**, 1761.
79. S. Teigelkamp, S. Ebel, D.W. Will, T. Brown, and J.D. Beggs, *Nucleic Acids Res.*, 1993, **21**, 4651.
80. W.H.A. Kuijpers and A.A. van Boeckel, *Tetrahedron*, 1993, **49**, 10931.
81. B.G. de la Torre, A.M. Avino, M. Escarceller, M. Royo, F. Albericio, and R. Eritja, *Nucleosides Nucleotides*, 1993, **12**, 993.
82. N.N. Polushin, B. Chen, L.W. Anderson, and J.S. Cohen, *J. Org. Chem.*, 1993, **58**, 4606.
83. K.S. Ramasamy, M. Zounes, C. Gonzalez, S.M. Freier, E.A. Lesnik, L.L. Cummins, R.H. Griffey, B.P. Monia, and P.D. Cook, *Tetrahedron Lett.*, 1994, **35**, 215.
84. S. Gryaznov and R.G. Schultz, *Tetrahedron Lett.*, 1994, **35**, 2489.
85. S.C. Srivastava, S.K. Raza, and R. Misra, *Nucleic Acids Res.*, 1994, **22**, 1296.
86. R.G. Kuimelis and K.P. Nambiar, *Nucleic Acids Res.*, 1994, **22**, 1429.
87. P. Francois, P. Muzzin, M. Dechamps, and E. Sonveaux, *New J. Chem.*, 1994, **18**, 649.
88. R. Nichols, P.C. Andrews, P. Zhang, and D.E. Bergstrom, *Nature*, 1994, **369**, 492.
89. D. Loakes and D.M. Brown, *Nucleic Acids Res.*, 1994, **22**, 4039.
90. M. Tarköy and C. Leumann, *Angew. Chem., Int. Ed. Engl.*, 1993, **32**, 1432.
91. M. Tarköy, M. Bolli, and C. Leumann, *Helv. Chim. Acta*, 1994, **77**, 716.
92. K.-H. Altmann, R. Kesselring, E. Francotte, and G. Rihs, *Tetrahedron Lett.*, 1994, **35**, 2331.
93. M.-O. Bévierre, A. De Mesmaeker, R.M. Wolf, and S.M. Freier, *Bioorg. Med. Chem. Lett.*, 1994, **4**, 237.
94. P. Nielsen, F. Kirpekar, and J. Wengel, *Nucleic Acids Res.*, 1994, **22**, 703.
95. P.A. Giannaris and M.J. Damha, *Can. J. Chem.*, 1994, **72**, 909.
96. C.L. Scremin, L. Zhou, K. Srinivasachar, and S.L. Beaucage, *J. Org. Chem.*, 1994, **59**, 1963.
97. W.K.-D. Brill, *Tetrahedron Lett.*, 1994, **35**, 3041.
98. P. Theisen, C. McCollum, and A. Andrus, *Nucleosides Nucleotides*, 1993, **12**, 1033.
99. M.P. Reddy, N.B. Hanna, and F. Farooqui, *Tetrahedron Lett.*, 1994, **35**, 4311.
100. W.H.A. Kuijpers, E. Kuyl-Yeheskiely, J.H. van Boom, and C.A.A. van Boeckel, *Nucleic Acids Res.*, 1993, **21**, 3493.
101. E. Kuyl-Yeheskiely, N.J. Meeuwenoord, G.A. van der Marel, and J.H. van Boom, *Recl. Trav. Chim. Pays-Bas*, 1993, **113**, 40.
102. N.D. Sinha, D.P. Michaud, S.K. Roy, and R.A. Casale, *Nucleic Acid Res.*, 1994, **22**, 3119.
103. Y. Lu and F. Le Goffic, *Synth. Commun.*, 1993, **23**, 1943.
104. X. Pannecoucke, G. Schmitt, and B. Luu, *Tetrahedron*, 1994, **50**, 6569.
105. K. Fukase, T. Yoshimura, S. Kotani, and S. Kusumoto, *Bull. Chem. Soc. Japan*, 1994, **67**, 473.
106. N.V. Heeb and K.P. Nambiar, *Tetrahedron Lett.*, 1993, **34**, 6193.
107. S.F. Martin and A.S. Wagman, *J. Org. Chem.*, 1993, **58**, 5897.
108. M.A. Alisi, M. Brufani, L. Filocamo, G. Gostoli, L. Cellai, M.A. Iannelli, G. Melino, M.C. Cesta, and S. Lappa, *Bioorg. Med. Chem. Lett.*, 1993, **3**, 1931.
109. A.H. van Oijen, S. Behrens, D.F. Mierke, H. Kessler, J.H. van Boom, and R.M.J. Liskamp, *J. Org. Chem.*, 1993, **58**, 3722.
110. A.H. van Oijen, N.P.M. Huck, J.A.W. Kruijtzer, C. Erkelens, J.H. van Boom, and R.M.J. Liskamp, *J. Org. Chem.*, 1994, **59**, 2399.
111. F. Barelay, E. Chrystal, and D. Gani, *J. Chem. Soc., Chem. Commun.*, 1994, 815.
112. T. Wakamiya, K. Saruta, J. Yasuoka, and S. Kusumoto, *Chem. Lett.*, 1994, 1099.
113. W. Tegge, *Int. J. Peptide Protein Res.*, 1994, **43**, 448.
114. J.-C. Truffert, O. Lorthioir, U. Asseline, N.T. Thuong, and A. Brack, *Tetrahedron Lett.*, 1994, **35**, 2353.

115. B.G. de la Torre, A. Avino, G. Tarrason, J. Piulats, F. Albericio, and R. Eritja, *Tetrahedron Lett.*, 1994, **35**, 2733.
116. J. Robles, E. Pedroso, and A. Grandas, *Tetrahedron Lett.*, 1994, **35**, 4449.
117. J.-M. Brunel and G. Buono, *J. Org. Chem.*, 1993, **58**, 7313.
118. A. Alexakis, J.C. Frutos, S. Mutti, and P. Mangeney, *J. Org. Chem.*, 1994, **59**, 3326.
119. R. Hulst, N.K. de Vries, and B.L. Feringa, *Tetrahedron: Asymmetry*, 1994, **5**, 699.
120. N. Sakai, S. Mano, K. Nozaki, and H. Takaya, *J. Am. Chem. Soc.*, 1993, **115**, 7033.
121. T. Higashizima, N. Sakai, K. Nozaki, and H. Takaya, *Tetrahedron Lett.*, 1994, **35**, 2023.
122. G.J.H. Buisman, P.C.J. Kamer, and P.W.N.M. van Leeuwen, *Tetrahedron: Asymmetry*, 1993, **4**, 1625.
123. K. Yamamoto, S. Momose, M. Funahashi, S. Ebata, H. Ohmura, H. Komatsu, and M. Miyazawa, *Chem. Lett.*, 1994, 189.
124. G. Brenchley, E. Merifield, M. Wills, and M. Fedouloff, *Tetrahedron Lett.*, 1994, **35**, 2791.
125. G. Jochem, H. Nöth, and A. Schmidpeter, *Angew. Chem., Int. Ed. Engl.*, 1993, **32**, 1089.
126. K. Wang, T.J. Emge, and A.S. Goldman, *Organometallics*, 1994, **13**, 2135.
127. L.D. Quin, S. Jankowski, J. Rudzinski, A.G. Sommese, and X.-P. Wu, *J. Org. Chem.*, 1993, **58**, 6212.
128. D. Barion, C. Gärtner-Winkhaus, M. Link, M. Nieger, and E. Niecke, *Chem. Ber.*, 1993, **126**, 2187.
129. N. Burford, J.A.C. Clyburne, S. Mason, and J.F. Richardson, *Inorg. Chem.*, 1993, **32**, 4988.
130. Y.K. Rodi, L. Lopez, C. Malavaud, M.T. Boisdon, and J.P. Fayet, *Can. J. Chem.*, 1993, **71**, 1200.
131. S. Motoki, T. Sakai, I. Shinoda, T. Saito, and T. Uchida, *Chem. Lett.*, 1993, 1563.
132. N. Burford, P. Losier, C. McDonald, V. Kyrimis, P.K. Bakshi, and T.S. Cameron, *Inorg. Chem.*, 1994, **33**, 1434.
133. G. David, E. Niecke, M. Nieger, V. von der Gönna, and W.W. Schoeller, *Chem. Ber.*, 1993, **126**, 1513.
134. T.B. Huang, J.L. Zhang, C.G. Zhan, and Y.J. Zhang, *Sci. China, Ser. B*, 1993, **36**, 1993 (Chem. Abstr., 1994, **121**, 35689a).
135. S.E. Pipko, Y.V. Balitzky, A.D. Sinitsa, and Y.G. Gololobov, *Tetrahedron Lett.*, 1994, **35**, 165.
136. A.M. Kibardin, T.V. Gryaznova, T.A. Zyablikova, S.K. Latypov, R.Z. Musin, and A.N. Pudovik, *Russ. J. Gen. Chem.*, 1993, **63**, 22.
137. A.M. Kibardin, T.V. Gryaznova, R.Z. Musin, and A.N. Pudovik, *Russ. J. Gen. Chem.*, 1993, **63**, 27.
138. M. Yoshifuji, D.L. An, K. Toyota, and M. Yasunami, *Chem. Lett.*, 1993, 2069.
139. H. Voelker, U. Pieper, H.W. Roesky, and G.M. Sheldrick, *Z. Naturforsch.*, 1994, **49b**, 255.
140. E. Niecke, B. Kramer, M. Nieger, and H. Severin, *Tetrahedron Lett.*, 1993, **34**, 4627.
141. Z. Li, C. Zhu, and Y. Zhao, *Phosphorus, Sulfur, and Silicon*, 1994, **86**, 229.
142. J. Tejeda, R. Réau, F. Dahan, and G. Bertrand, *J. Am. Chem. Soc.*, 1993, **115**, 7880.
143. S. Ganapathy, K.P. Dockery, A.E. Sopchik, and W.G. Bentrude, *J. Am. Chem. Soc.*, 1993, **115**, 8863.
144. K.B. Mullah and W.G. Bentrude, *Nucleosides Nucleotides*, 1994, **13**, 127.
145. H. Kondo, S. Aoki, Y. Ichikawa, R.L. Halcomb, H. Ritzen, and C.-H. Wong, *J. Org. Chem.*, 1994, **59**, 864.
146. Y. Watanabe, C. Nakamoto, T. Yamamoto, and S. Ozaki, *Tetrahedron*, 1994, **50**, 6523.

4
Quinquevalent Phosphorus Acids

BY R. S. EDMUNDSON

The low activity in phosphoric acid chemistry, particularly in the specific area of synthesis, coupled with an increase in interest in phosphonic and phosphinic acid chemistry, follows the trend observed during recent years.

1. Phosphoric Acids and their Derivatives

1.1 Syntheses of Phosphoric Acids and their Derivatives.-Two recent reviews, whilst concentrating on nucleoside phosphates and phosphorothioates, contain much of general interest with regard to phosphoric acid chemistry.[1,2]

Interest has continued in the preparation of phosphate esters from fundamental materials. Studies have been carried out on the formation of simple trialkyl phosphates from a zinc phosphide in the presence (equation 1) or absence (equation 2) of oxygen;[3] through the oxidation of white phosphorus in the presence of alcohols and Cu(II) (equation 3);[4] by reactions between chlorine and phosphine (each diluted with an inert gas) in the presence of an alcohol (equation 4);[5] and by the interaction of phosphine with alcohols in the presence of Pt(IV) halides, a system (Scheme 1) in which trialkyl phosphites are formed, but which can be oxidized to trialkyl phosphates when the reaction mixture is chlorinated.[6] More data have been presented on the electrochemical conversion of white phosphorus, in the presence of tetraethylammonium iodide, into trialkyl or triaryl phosphate esters; in the case of the latter esters, the process is thought to occur *via* $(ArO)_5P$ species.[7] The subject of oxidative *O*, *N*, and *C*-phosphorylations of organic compounds by elemental phosphorus, or phosphides, in the presence of metal catalysts, has been reviewed.[8]

Simple triaryl phosphates have been obtained in yields higher than 90% by the interaction of $POCl_3$ and polymer-supported ArO^- in benzene at room temperature.[9] Other phosphoric triesters have been prepared by the use of $(R^1O)(Pr^i{}_2N)PCl$/tetrazole as phosphitylation reagents with the alcohols R^2OH and R^3OH, with subsequent oxidation;

the removal of one alkyl group (2-cyanoethyl or benzyl) led to dialkyl hydrogen phosphates.[10]

Titanium tetrachloride is an efficient reagent for the conversion of trialkyl phosphites and dialkyl hydrogen phosphonates into dialkyl phosphorochloridates.[11] Imidazolides and dialkyl or diaryl phosphoric acids react with acyl fluorides - benzoyl fluoride and oxalyl difluoride being the reagents of choice - to give quantitative yields of the phosphoryl fluorides. The procedure is adaptable to the preparation of fluorides of carbohydrate phosphates; in this field, the reaction between the carbohydrate and tris-1*H*-imidazolylphosphine oxide or sulphide with the replacement of one imidazole nucleus, is followed by a reaction with PhCOF to give the phosphoryl or thiophosphoryl difluoride.[12] 1,1,2,3,3,3-Hexafluoropropyl azide is a potent fluorinating agent for many organic phosphorus compounds; phosphorus(III) triesters and dialkyl hydrogen phosphonates each yield dialkyl phosphorofluoridates, phosphorodiamidites yield phosphorodiamidic fluorides, and other phosphorus(III) halides undergo valence expansion with, where possible, halogen exchange.[13]

Interest is growing in the chemistry of phosphorus-based derivatives of the calixarenes. The reaction between the calix[4]arene (1)(R = Bu^t) (because of folding in the molecule, an alternative representation is (2)) and diethyl phosphorochloridate under phase transfer conditions yields the tristriester (3)(this has a partial cone conformation) together with the bridged ester (4); the latter derivative also has a distorted cone conformation with the unusual orientation of one Bu^t group into the cone cavity.[14] A reaction between the same substrate and PCl_5 affords the hygroscopic salt (5), further characterized as hexachloroantimonate salts, and which, on ethanolysis, yields the bistriester (6), characterized by X-ray analysis.[15] Another paper, concerned primarily with compounds which have six-, five-, or three-coordinated phosphorus, also describes some with the core element in the four-coordinate state, including compound (7), obtained by the oxidation of the corresponding phosphite triester with *tert*-butyl hydroperoxide.[16] The tetramethylcalix[4]arenoctol (8), from its reactions with alkyl phosphorodichloridates, affords the products (9) as mixtures, each of six diastereoisomers, the structures of which (the phosphoryl groups point inwards or outwards with respect to the cavity) were ascertained from ^{1}H and ^{31}P n.m.r. spectroscopic data which reflect the molecular symmetry.[17]

Further developments in the synthesis of phosphate and thiophosphate esters from *myo*-inositol (10) have been widely described. By and large, these syntheses have followed the procedures and used reagents (particularly with regard to the nature of protection

$$Zn_3P_2 + 10\,ROH + 4\,O_2 + 4\,HCl \xrightarrow{CuCl_2} 2\,(RO)_3P(O) + 4\,RCl + 3\,ZnO + 7\,H_2O \quad (1)$$

$$Zn_3P_2 + 16\,CuCl_2 + 11\,ROH \longrightarrow 2\,(RO)_3P(O) + 5\,RCl + 16\,CuCl + 11\,HCl + 3\,ZnO \quad (2)$$

$$P_4 + 16\,ROH + 5\,O_2 + 4\,HCl \xrightarrow{CuCl_2} 4\,(RO)_3P(O) + 4\,RCl + 10\,H_2O \quad (3)$$

$$PH_3 + 4\,Cl_2 + 4\,ROH \longrightarrow (RO)_3P(O) + RCl + 7\,HCl \quad (4)$$

$$PH_3 + 3\,ROH + 3\,Na_2PtCl_6 \longrightarrow (RO)_3P + 3\,Na_2PtCl_4 + 6\,HCl$$

$$(RO)_3P + Cl_2 + ROH \longrightarrow (RO)_3P(O) + RCl + HCl$$

$$Na_2PtCl_4 + Cl_2 \longrightarrow Na_2PtCl_6$$

Scheme 1

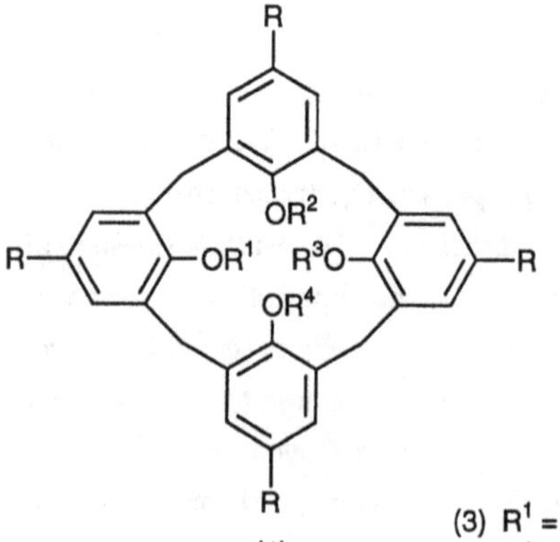

(1)

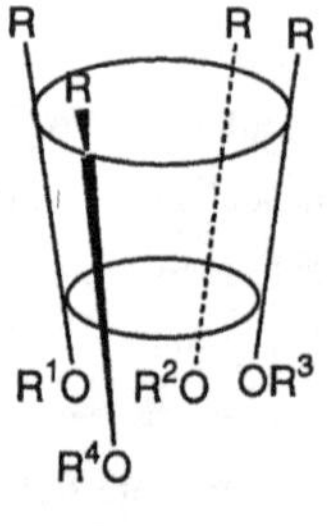

(2)

(3) $R^1 = R^2 = R^3 = (EtO)_2P(O)$, $R^4 = H$
(4) $R^1 = R^2 = {=}P(O)OEt$, $R^3 = (EtO)_2P(O)$, $R^4 = H$
(5) $R^1 = R^2 = R^3 = {\equiv}P^+Cl\ PCl_6^-$, $R^4 = PCl_4$
(6) $R^1 = R^2 = R^3 = {\equiv}P(O)$, $R^4 = (EtO)_2P(O)$
(7) $R^1 = R^2 = R^3 = {\equiv}P(O)$, $R^4 = H$

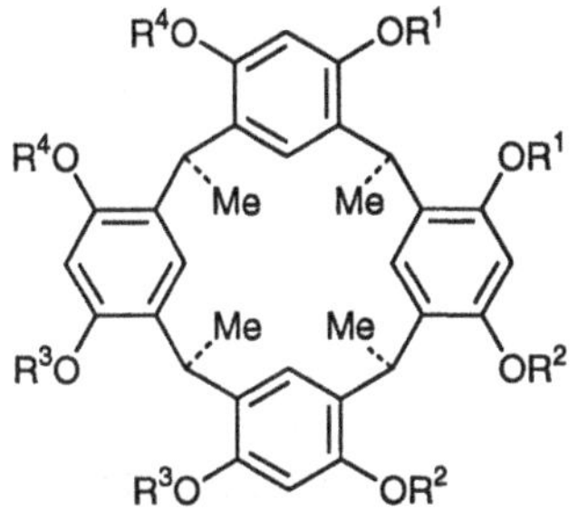

(8) $R^1 = R^2 = R^3 = R^4 = H$
(9) $R^1 = R^2 = R^3 = R^4 = EtO(O)P=$

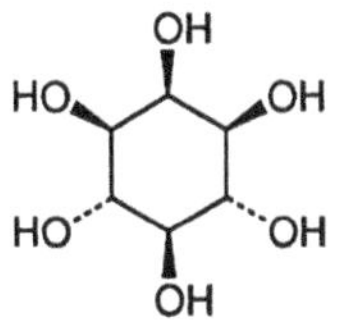

(10)

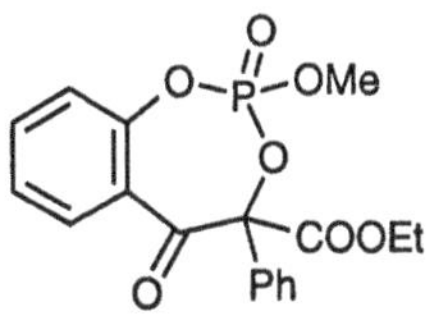

(11)

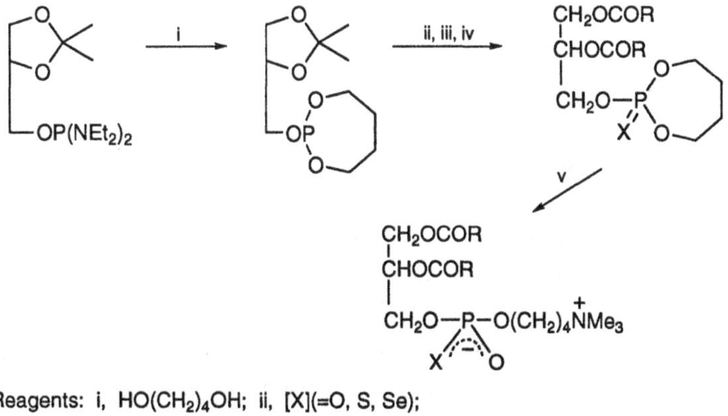

Reagents: i, $HO(CH_2)_4OH$; ii, [X](=O, S, Se); iii, H_3O^+; iv, RCOCl, base; v, Me_3N

Scheme 2

groups) summarized in recent reviews.[18] Recent papers have described preparations of *myo*-inositol 4- and 6- monophosphates[19]; derivatives of the 1,4,5-trisphosphate[20,21]; the 1,4,6-trisphosphate;[22] the 1,5,6-trisphosphate;[23] the 1,2,4,5-tetrakis-,[24] 2-substituted 1,3,4,5-tetrakis- and 1,3,4,5,6-pentakis- phosphates[25] as well as the 1,4,5-trisphosphate 3-phosphorothioate.[26] A synthesis of L-*chiro*-inositol 1,2,3,5-tetrakisphosphate from L-quebrachitol has also been described.[21]

Dialkyl 2-bromoethenyl phosphates are prepared by a reaction sequence which commences with the bromination of dialkyl ethenyl phosphates, and this is followed by dehydrobromination of the dibromides with $LiNPr^{i}_{2}$ in toluene (best) but similar results can be obtained with $MN(SiMe_3)_2$ (M = Li, Na, or K).[27] A successful synthesis of trisketol phosphates (with yields of 60-75%) involves the interaction of enol trimethylsilyl ethers with phosphoric acid in the presence of 4-(difluoroiodo)toluene; a further reaction with LiBr in acetone results in the formation of bisketol hydrogen phosphates in yields of from 80% to greater than 90%. [28] Other examples of dialkyl monoketol phosphates have been prepared, within certain structural restrictions, from diazomethyl ketones and dialkyl hydrogen phosphates.[29]

Phosphoric triesters have been prepared by the direct alkylation or acylation of phosphodiesters in the presence of Ag_2O in MeCN; a variety of alkyl, aralkyl, aroyl, and (diethoxyphosphinoyl) phosphates was so prepared in good to excellent yields, and the reaction was also applicable to the preparation of esters of diphenylphosphinic acid.[30] Examples of the Baeyer-Villiger oxidation of acylphosphonamidic esters to alkyl acyl phosphoramidates have been reported;[31] this work will be considered more fully in Section 2.2. The reaction between ethyl benzoylformate and 2-methoxy-1,3,2-benzophosphorin-4-one is said to give the product (11).[32]

Diethyl phosphorisothiocyanatidate is obtainable from $(EtO)_2P(O)Cl$ and Me_3SiNCS in acetonitrile in the presence of 4-methylmorpholine, a procedure also applicable to a phosphonic chloride.[33]

Phosphitylation methodology has been involved in the synthesis of phosphopeptides[34] and thiophosphorylated peptides.[35] The same principle was utilized in the preparation of phosphatidylcholine analogues (Scheme 2)[36] and phosphorodithioate esters (Scheme 3)[37] and other 2,3-dihydroxyalkyl phosphate esters.[38] Sulphur adds to the product from the interaction of 1,2-diacylglycerols and (12) in the presence of a tertiary amine HCl acceptor to give the trithioates (13); these undergo ring opening to give the dithioate esters (14) when acted upon with R^3OH in the presence of DBU. Ring opening also occurs, to give the choline derivatives (15), when the 1,3,2-dithiaphosph(V)olanes

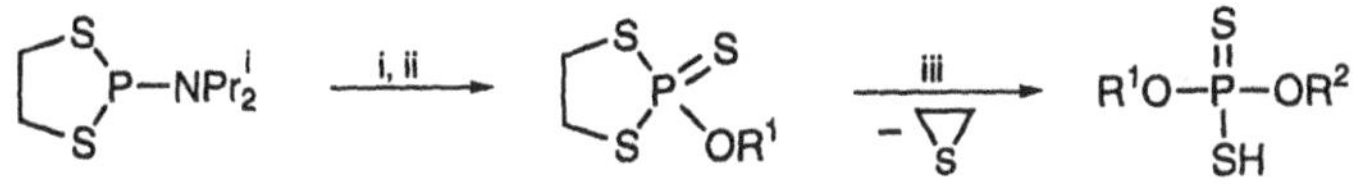

Reagents: i, R^1OH, tetrazole; ii, S ; iii, R^2OH

Scheme 3

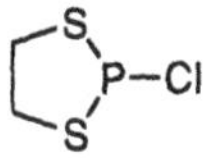

(12) 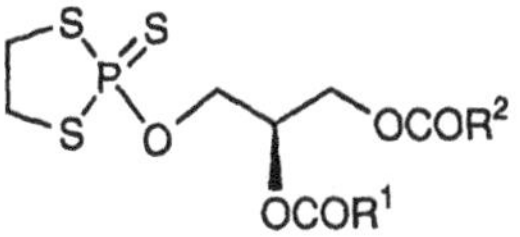 (13) 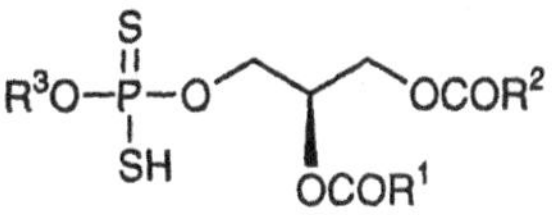 (14)

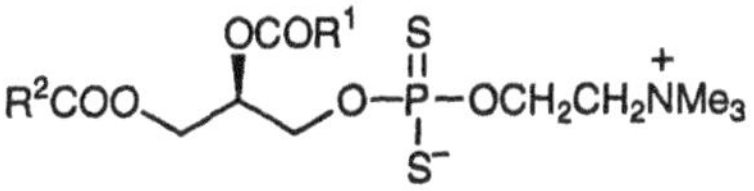

(15)

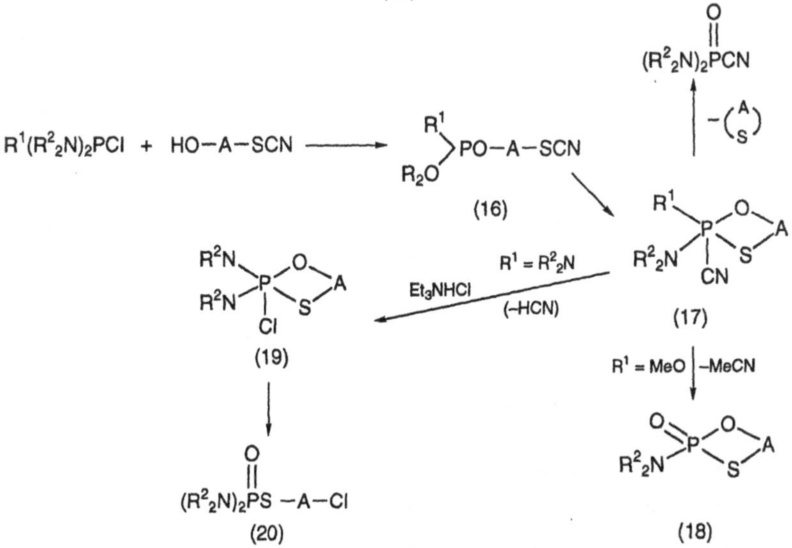

(16)

(17)

(19)

(18)

(20)

(13) react with choline *p*-toluenesulphonate.[37]

The phosphitylation, by the chlorides $R^1(R^2{}_2N)PCl$, (R^1 = OMe, R^2 = Et)[39] or (R^1 = $R^2{}_2N$, R^2 = Et),[39,40] of substrates which have the general structure HO-A-SCN, (where A is a two or three carbon chain, with or without Me substituents), is complex. With both reagents, the initial reaction affords the esters (16), but even at room temperature, the use of the first reagent leads to the liberation of MeCN, presumably from a pentacoordinate intermediate, and the formation of the cyclic products (18). With the second reagent, it would appear that the loss of CN also occurs by the elimination of the cyanidate from the pentacoordinate intermediate (17), but in a different manner; at the same time, further reaction also leads to the esters (20) through another intermediate (19).

The interaction of benzopentathiepin (21) with a trialkyl phosphite yields a thioether (22)(R^1 = R) when the reaction is carried out in boiling dichloromethane or, if the reaction is performed at -15°, the product is a thiol, (22)(R^1 = H); the pentathiepin is unreactive towards triphenyl phosphite.[41]

S,S-Dialkyl phosphorochloridodithioates (23) are formed in the reactions between *O,O,S*-trialkyl phosphorodithioates and $POCl_3$ at 60-100°.[42,43] An attempt to prepare 2-chloro-4,5-benzophosphorin-4-one 2-sulphide by the thiation of the corresponding phosphorus(III) chloride with $PSCl_3$ resulted in a complex mixture containing the desired compound, but also others in which one or more of the ring oxygen atoms was replaced by sulphur.[44] The treatment of *O,O*-dimethyl *N*-(1-phenylethyl) phosphoroselenoic amide, derived from the chiral amine, with MeI produced a mixture of diastereoisomeric and separable *O,Se*-dimethyl isomers.[45]

Reactions between tetraphosphorus decasulphide and silyl ethers (equation 5; Z = Si)[46] or their tin analogues (equation 5; Z = Sn) yield *O,O*-dialkyl phosphorodithioate *S*-silyl or *S*-stannyl esters;[47] that with diol silyl ethers leads to *O,O*-alkylene *S*-trimethylsilyl phosphorodithioic esters[46] and the use of the reagents $RSZMe_3$ affords phosphorotetrathioic esters;[47,48] the stannyl phosphorotetrathioic esters (24) disproportionate when distilled (equation 6).[47] Tetraphosphorus decasulphide and dialkyl disulphides together yield the pentathio esters (25)(equation 7), which were characterized spectroscopically and which, when distilled, yield the known trialkyl phosphorotetrathioates, $(RS)_3P(S)$.[49] The same products, (25) are also obtainable from dialkyl disulphides and the bisanhydrosulphides (26).[49] Reactions between P_4S_{10} and dithioacetals produce mixtures of esters of the types (27) and (28)(equation 8).[49]

The scope of the Todd-Atherton reaction has been extended to include reactions with mono- and with bis-(trimethylsilyl) hydrogen phosphonates.[50] The acylation of

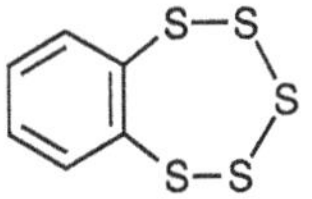

(21)

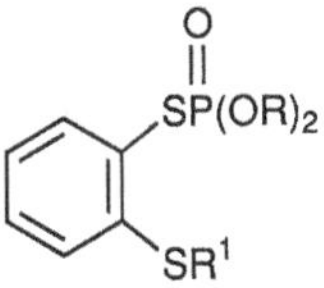

$SP(O)(OR)_2$, SR^1

(22)

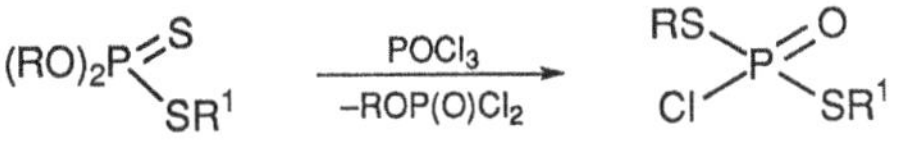

$$(RO)_2P(=S)SR^1 \xrightarrow[-ROP(O)Cl_2]{POCl_3} (RS)(Cl)P(=O)SR^1$$

(23)

$$P_4S_{10} + 8\ ROZMe_3 \longrightarrow 4\ (RO)_2\overset{\overset{S}{\|}}{P}SZMe_3 + 2\ (Me_3Z)_2S \quad (5)$$

$$(R^1S)_2\overset{\overset{S}{\|}}{P}SSnR^2{}_3 \longrightarrow R^1S\overset{\overset{S}{\|}}{P}(SSnR^2{}_3)_2 + (R^1S)_3P(S) \quad (6)$$

(24)

$$P_4S_{10} + 6\ RSSR \longrightarrow 4\ (RS)_2\overset{\overset{S}{\|}}{P}SSR + {}^1/_4\ S_8 \quad (7)$$

(25)

RS–P(=S)(S)(S)P(=S)–SR

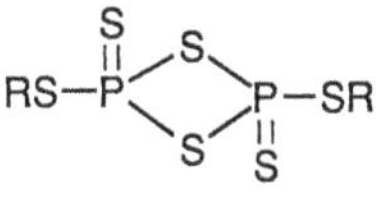

(26)

$$P_4S_{10} + 6\ (R^1S)_2CHR^2 \longrightarrow 2\ (R^1S)_2\overset{\overset{S}{\|}}{P}SCH(SR^1)R^2 + 2\ R^1S\overset{\overset{S}{\|}}{P}[SCH(SR^1)R^2]_2 \quad (8)$$

(27) (28)

diethyl phosphoramidate is achieved indirectly through an initial reaction with an orthoalkanoic ester, followed by acid hydrolysis of the product (29).[51] Diethyl phosphoramidate and oxalyl chloride together yield $(EtO)_2P(O)NCO$.[51] Macrocyclic phosphoramides (30) have been prepared conventionally and characterized by X-ray analysis (R = Ph or 1-adamantyl).[52]

The nitrogen-containing systems (31)(Z = N, CHCN, or CHCOOEt) have been prepared as indicated[53] and new cyclophosphoramide derivatives in the carbohydrate[54,55] and steroid[56] fields have been described. The novel compounds (32) (Z = O or S)[57] and (33)(R = MeO or Me_2N)[58] have been characterized. Further macrocyclic systems have been prepared from the bishydrazides (34)(R = PhO, or Ph) and the dialdehydes (35)(R = H or Me) and contain two, three, or four, of each of the phosphoryl and pyrrole moieties.[59] A phosphoramidate analogue of phosphoenolpyruvate, (36), reasonably stable only in fully ionized form (with a half-life of about four days at pH 9 and room temperature), has been prepared from D-serine in a two-stage sequence.[60]

N-(Dialkoxyphosphinoyl)-*O*-(4-nitrobenzenesulphonyl)hydroxylamines have been prepared as potential reagents for electrophilic amination reactions.[61]

1.2. Reactions of Phosphoric Acid Derivatives.- Polar substituent effects in 4-substituted-2,6,7-trioxa-1-phosphabicyclo[2.2.2]octanes, monohexyl arylphosphonates, and diethyl arylphosphonates, as indicated by, for example, pK_a values and ^{31}P chemical shifts, have been found to parallel those for carbon compounds.[62]

Two long and detailed papers consider the reactivity of acylic phosphate esters in terms of intramolecular catalysis. The dianion of the ester (37)(the trifluoromethyl groups are present to minimize any competitive dephosphorylation by an S_N1 process) hydrolyses rapidly in water at 50°, with a half-life of less than two minutes. In a detailed analysis of the kinetics results, the case was argued for hydrolysis through intramolecular nucleophilic catalysis by the phenol OH group over the whole of the pH range. In water at 50° or less, the sole product from the dianion (37) is the diester (38), and further reaction occurs only at a higher temperature or if the medium is strongly alkaline; under the latter circumstances ring opening gives rise to the two products indicated. No evidence for P-O bond cleavage was presented. Phosphoranes were considered to act as intermediates in the reactions depicted in Scheme 4.[63] The methylation of phenyl 1,2-isopropylidene- β-D-xylofuranose 3'-phosphate, to give the diastereoisomeric (39) (R = Me) increases the rate of intramolecular cyclization by a factor of $>10^5$. The release of

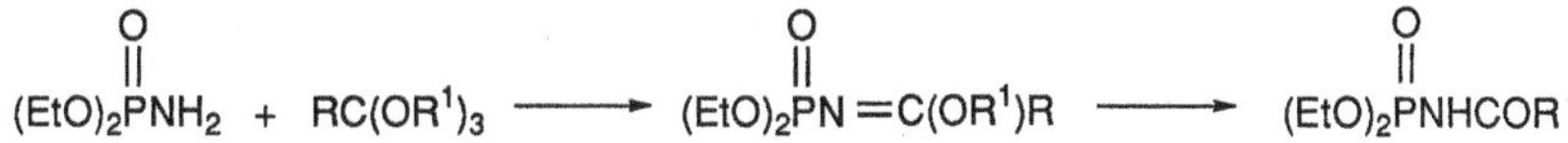

$(EtO)_2P(O)NH_2 + RC(OR^1)_3 \longrightarrow (EtO)_2P(O)N{=}C(OR^1)R \longrightarrow (EtO)_2P(O)NHCOR$

(29)

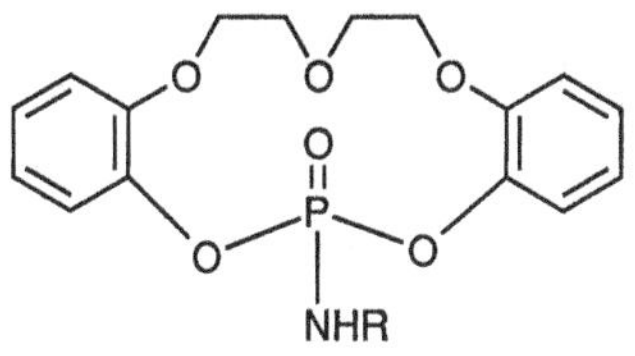

NHR

(30)

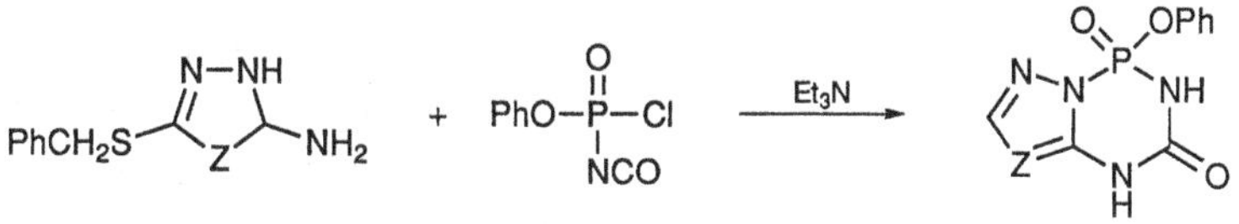

$PhCH_2S$ N–NH Z NH_2 + PhO–P(O)–Cl NCO $\xrightarrow{Et_3N}$ O OPh N N P NH Z N H O

(31)

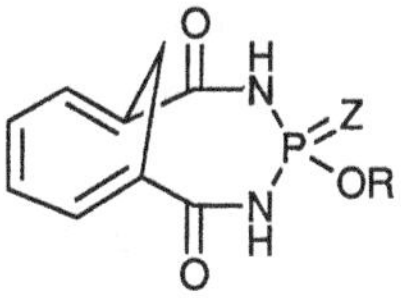

O H N P Z OR N H O

(32)

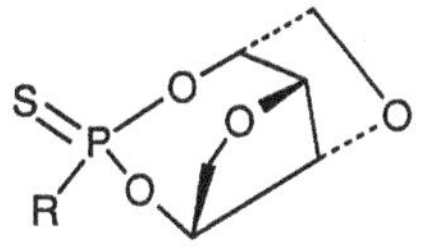

S P R O O O O

(33)

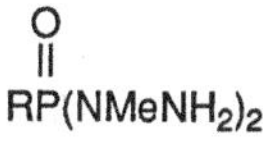

$RP(O)(NMeNH_2)_2$

(34)

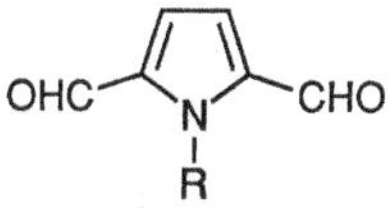

OHC N R CHO

(35)

$H_2C{=}C(COOH)NHPO_3H_2$

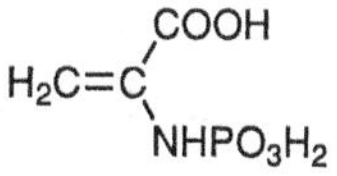

(36)

MeOH leads to the 1,3,2-dioxaphosphorinanes (40a) and (40b) with the conversion of (39a) into (40b) a process of minor importance.[64]

The base-catalysed methanolysis of 4-methyl-2,6,7-trioxa-1-phosphabicyclo[2.2.2]-octane 1-oxide and 1-sulphide (41)(X = O or S) (Scheme 5) initially yields *trans* forms of the 5-hydroxymethyl-2-methoxy-5-methyl-1,3,2-dioxaphosphorinanes (42)(X = O or S). The progress of the reactions was followed by ^{31}P n.m.r. spectroscopy. With the passage of time, the *trans*-1,3,2-dioxaphosphorinanes equilibrate with the corresponding *cis* forms (44) *via* the acyclic esters (43). During 2 h of contact, (41) affords only (42), but within 13 h, the starting material disappears and a second product is the acyclic phosphate (43); after 180 h, all three compounds are present. Under the same conditions, the methanolytic behaviour of all three products, (42)-(44), is consistent with the mechanism depicted in Scheme 5.[65]

The importance of 2-iodosobenzoic acid for the rapid cleavage of reactive phosphate esters is widely recognized; in the cleavage of diphenyl 4-nitrophenyl phosphate, for example, it has been suggested that the catalytic activity results through the formation of the compound (45), although no evidence has been forthcoming for the existence of such a species. However, the compound (45) has now been prepared from (46)(X = Cl) (Scheme 6) and shown to be degradable to diphenyl phosphate and other species through the action of a variety of nucleophiles, particularly water.[66]

Photo-labile triesters include diethyl pyren-1-ylmethyl phosphate (47), diethyl [(7-methoxy-2-oxobenzo[b]pyran-4-yl)methyl] phosphate (48), and diethyl 1-(2-nitrophenyl)ethyl phosphate.[30] Under conditions of photolysis in MeOH solution, the aforementioned esters fragment to diethyl phosphate and fractions which can be characterized as methyl ethers, for examples, 1-methoxymethylpyrene from (47).[67] Two other groups of phosphate esters whose photochemistry has been studied are the desyl phosphates (49) (prepared as indicated in Scheme 7) and the 4-methoxybenzoylmethyl phosphates (50); the reactivities of both groups are of potential importance from the viewpoint of protection, the particular P-O-C bonding being sensitive under photolysis conditions. When irradiated at 300 nm in a variety of solvents, fragmentation of the desyl esters yields the dialkyl hydrogen phosphate together with 2-phenylbenzo[b]furan. The irradiation of a methanolic solution of 4-methoxyphenacyl diethyl phosphate (50)(R = Et) releases diethyl hydrogen phosphate, 4-methoxyacetophenone and methyl 4-methoxyphenylacetate. Mechanisms for the formation of the noted dephosphorylated products were discussed.[68]

Isotope(^{17}O)-labelling has been used to demonstrate that, when a 1:1 mixture of the

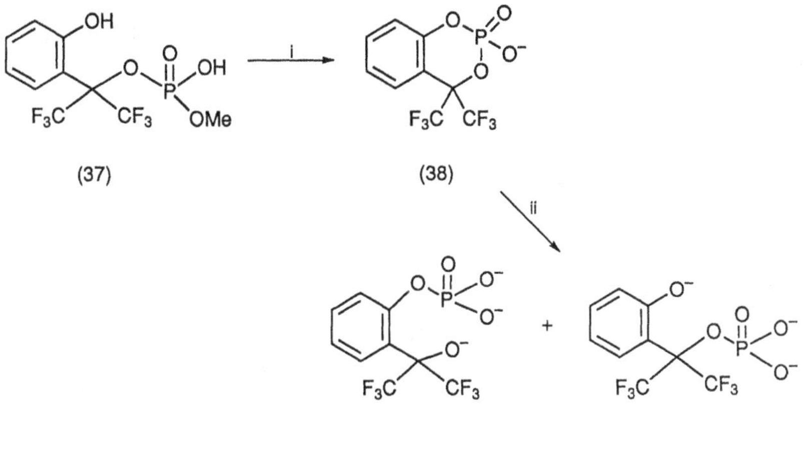

(37) (38)

Conditions: i, 50 °C, pH < 11; ii, 50 °C, pH > 11

Scheme 4

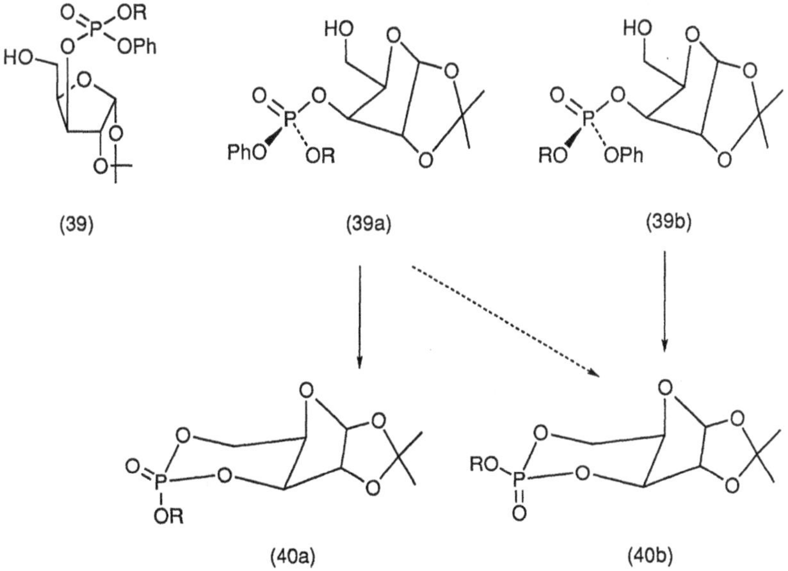

(39) (39a) (39b)

(40a) (40b)

Scheme 5

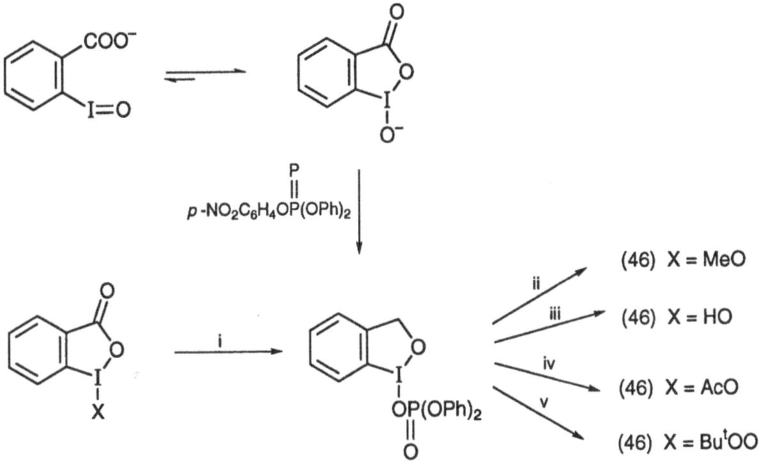

COO^-

I=O

O

O

I

O^-

P

p-$NO_2C_6H_4OP(OPh)_2$

O

O

I

X

O

I

$OP(OPh)_2$

O

ii

iii

iv

v

(46) X = MeO

(46) X = HO

(46) X = AcO

(46) X = Bu^tOO

(46)

(45)

Reagents: i, $(PhO)_2PO_2^-Ag^+$, DMSO ; ii, MeOH; iii, $H_2O(H^+)$; iv, HOAc; v, Bu^tOOH

Scheme 6

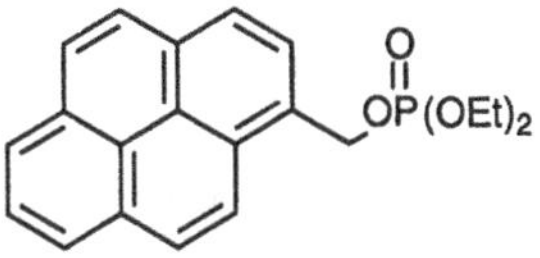

O

$OP(OEt)_2$

(47)

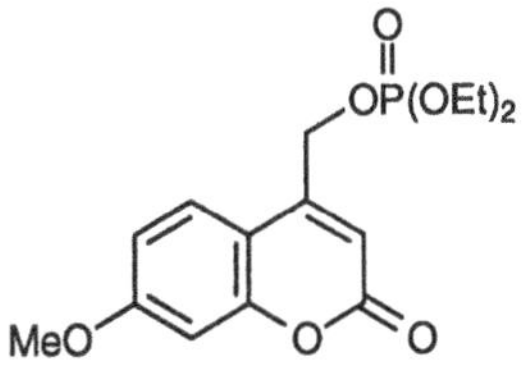

O

$OP(OEt)_2$

MeO

O

O

(48)

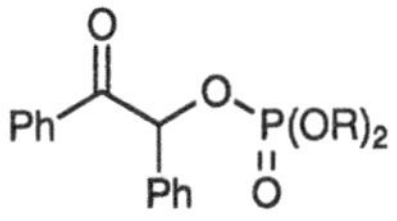

O

Ph

O

$P(OR)_2$

Ph

O

(49)

MeO

O

O

O

O–$P(OR)_2$

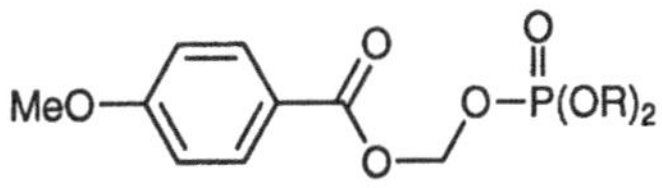

(50)

two esters, (51) (R^1 = Ph) and (52) (R^2 = CF_3CH_2) is heated for 10 h at 150°, the ensuing transalkylation reaction involves the attack by the phosphoryl group of one ester on to the carbon of the OCH_2Si group of the other; the product consists of a mixture of an equal amount of each of two transalkylated esters, (51)(R^1 = CF_3CH_2) and (52) (R^2 = Ph) together with an equilibrated amount of the two initial esters.[69]

The reactions between allylic magnesium chlorides and diphenyl alkyl phosphates $(PhO)_2P(O)OR$, where R = Me, CH_2Ph, or $CH_2CH{=}CHR'$ (R' = Pr or heptyl, for example) are remarkably regiospecific. The reaction is exemplified in equation 9, and the yields are reported to be within the range 60-99% with the γ:α ratios between 92:8 and 99:1.[70]

The products from a treatment of the enol phosphate esters (53)(R^1, R^2, and R^3 = H or Me), prepared *in situ*, with the nucleophiles (MeOH, PrOH, Bu^tOH, EtSH, MeNHOMe) in the presence of Et_3N and with cooling, are of type (55); these result from the acylation of the nucleophile by the mixed anhydrides (54), evidently formed as the result of Claisen rearrangements.[71]

Aspects of the chemistry of carbohydrate bicyclic (thio)phosphate chemistry have been revewed[72] and an example has been given of the cleavage of a P-O bond in a bicyclic thio(no)phosphate by LiBr, thus allowing, after further reaction, the re-formation of the bicyclic geometry but with an endocyclic P-S bond, effectively the product of a 'thione-thiol' isomerization.[73] Cleavage of the P-S bond takes place during the acid hydrolysis (with 0.1-7.0 M HCl in aqueous dioxan) of $(2,4,6\text{-}Br_3C_6H_2S)_3P(O)$ [74] and of $2,4,6\text{-}Br_3C_6H_2SPO_3H_2$.[75] In the enzymic hydrolysis of enantiomers of *O,S*-dimethyl *N*-acetylphosphoramidothioate (acephate) with a bacterial phosphotriesterase, the (S_P) enantiomer is degraded 100x faster than the (R_P) antipode.[76] New observations have been reported on the ring-opening of epoxides in the system (56) by phosphorodithioate anions; the compounds (57) and (58) are intermediates in the ultimate formation of (59), the transition probably occurring through a pentacoordinate intermediate (Scheme 8) (for an earlier report, see *Organophosphorus Chemistry*, 1993, **24**, 106, references 88, 89).[77]

(*S*)-*O*-Ethyl *O*-phenyl phosphorochloridothioate reacts with amines with inversion of configuration at phosphorus; the resultant phosphorothioic amides undergo acid hydrolysis (P-N bond cleavage) or alkaline hydrolysis (P-O bond cleavage), and both reactions yield products with inverted configuration.[78] *O*-Methyl *O*-phenyl hydrogen phosphorothioate has been resolved with brucine; through sequential chlorination (with PCl_5) and reaction with EtO^-, the (+)-acid affords (+)-*O*-ethyl *O*-methyl *O*-phenyl phosphorothioate of known (*S*) configuration; since both steps are known to occur with inversion, the configuration of

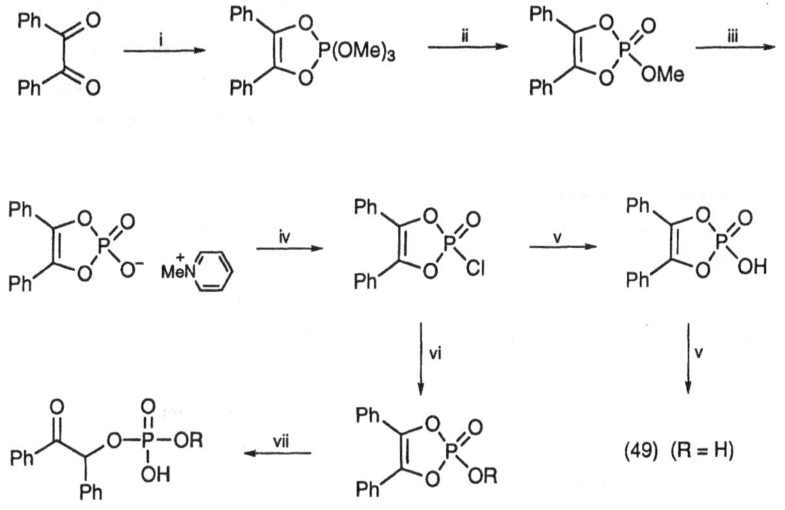

Reagents: i, $(MeO)_3P$, MeCN; ii, MeCOBr, MeCN;
iii, py, C_6H_6 ; iv, Cl_2CO, C_6H_6 ; v, H_2O (1 equiv.), THF ;
vi, ROH, Et_3N, THF ; vii, H_2O, THF.

Scheme 7

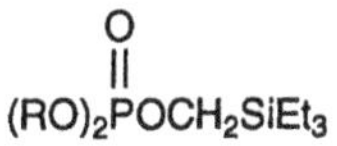

(49) (R = H)

$(RO)_2P(O)OCH_2SiEt_3$

(51)

$(R^1O)_2P(O)OCH_2SiMe_3$

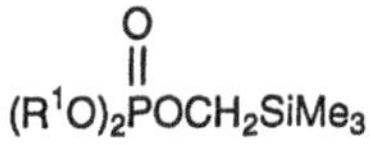

(52)

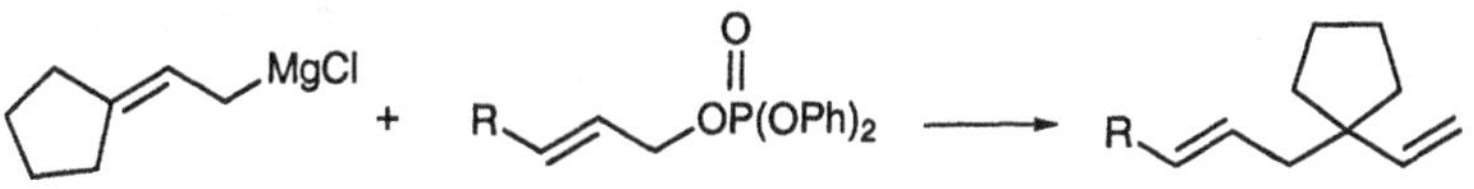

(9)

the acid must be (*R*).[79] The additions of dialkoxyphosphinoylsulphenyl chlorides to ethyl 2-cyanopropenoate have been reported to give the esters (60) and not (61).[80] When heated, the *syn* or *anti* forms of the product (62), from *O,O*-diethyl phosphorochloridothioate and *N*-acetyl-*O*-methyl hydroxylamine each afford a 1:1 mixture of the *syn* and *anti* forms of the thiolo isomer.[81]

Following a treatment with one equivalent of BuLi, the phosphorodiamidate (63) (R = PhO) undergoes dehydrochlorination to give the mono-aziridide (64); neither a further reaction to give the bis-1-aziridide, (65), nor cyclization to give the 1,3,2-diazaphospholidine (66), is observed, even in the presence of an excess of BuLi. Compound (64) is thus isolable in pure form. This behaviour is in contrast to that demonstrated by (63)(R = Ph) for which the conversion of (64) into (65) is faster than that of (63) into (64). (It might be noted that the P(O)-N bond is not immune to attack by BuLi; 1-diphenylphosphinoylaziridine reacts with BuLi in THF to give butyldiphenylphosphine oxide).[82]

Nitrosoalkanes condense with diethyl phosphoramidate in the presence of Bu^tOCl, *N*-bromosuccinimide, or more particularly, of dibromoisocyanurate, to give (diethoxyphosphinoyl)diazene oxides; it is thought that this reaction proceeds through the initial dihalogenation of the amide, an explanation also offered for the analogous behaviour of phosphonic amides (see reference 303).[83] Michael type additions of diethyl phosphoramidate to α,β-unsaturated esters have been carried out in boiling toluene with K_2CO_3 and a quaternary ammonium salt.[84] A novel cyclization occurs when alkyl acetonyl phosphoramidates are treated with $BF_3.Et_2O$; spectroscopic evidence has been advanced to support the proposed structure (67) for the product.[85] The synthesis and phosphorylating properties of several 2-chloro-2,3-dihydro-3-methanesulphonyl-1,3,2-benzoxazaphosph(V)oles (68) have been examined.[86]

Diphenyl pentafluorophenyl phosphate has been recommended as a peptide coupling reagent which produces higher yields than DCC and with a lower tendency to cause racemization.[87] Dialkyl 3-nitro-2-pyridinyl phosphates bring about lactam formation from β-amino carboxylic acids.[88] The menthyl ester (69)(R = (-)-menthyl) has been recommended for the resolution of the cyclic phosphoric acid derived from 1,1'-binaphthalene-2,2'-diol, to which the acid is readily cleaved by the action of $LiAlH_4$.[89]

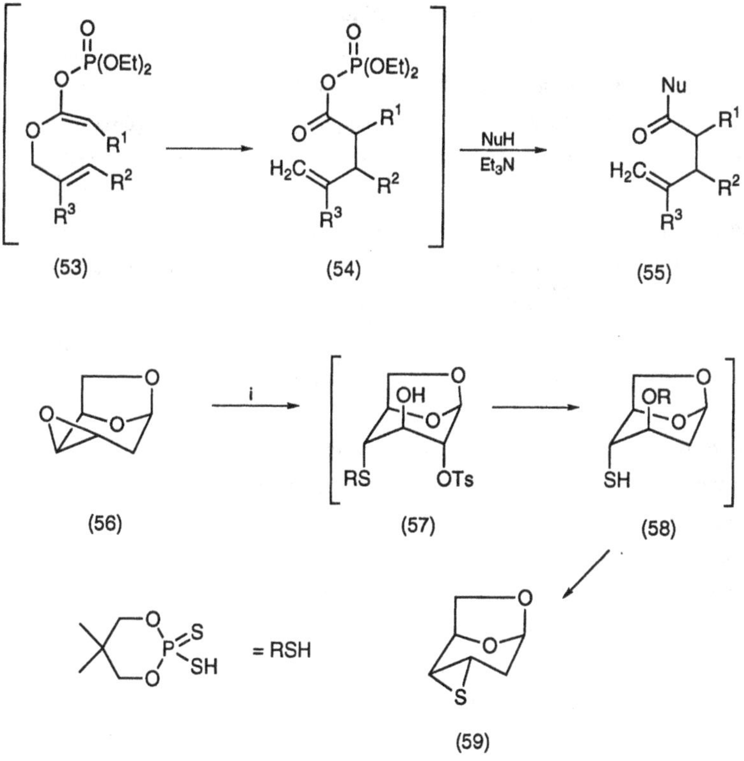

(53) (54) (55)

NuH, Et_3N

(56) (57) (58) (59)

= RSH

Reagents: i, RSH, Et_3N

Scheme 8

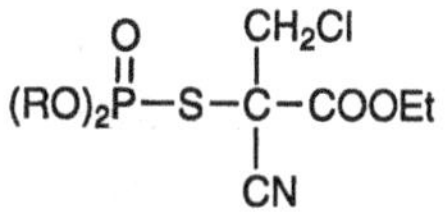

(60) 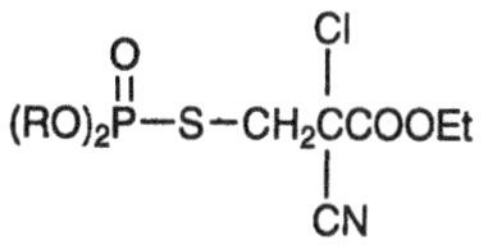 (61)

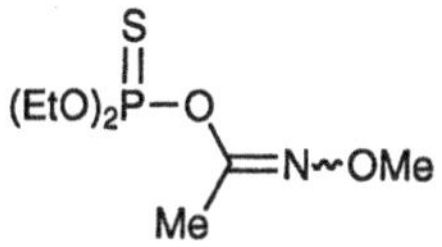

(62)

2. Phosphonic and Phosphinic Acids and their Derivatives

2.1. Syntheses of Phosphonic and Phosphinic Acids and their Derivatives.-

Aspects of the chemistry of hydrogen phosphonates have been discussed in two reviews.[90,91] A review on diphosphonic acids is concerned mainly with methylenebisphosphonic acid, hydroxyethylidenebisphosphonic acid, and some other *vic*-bisphosphonic acids.[92]. The scope and mechanistic aspects of the photochemical cleavage of the phosphorus-carbon bond in alkylphosphonic acids has been discussed (with 34 references),[93] and a review concerned with six- and seven-membered phosphorus-containing ring systems, in general, includes much information on phosphinic acids based on the phospholane and similar ring systems.[94]

2.1.1. *Phosphonic and Phosphinic Halides and related compounds.*- A mixture of 2.5 eq. PCl_5 and 1.3 eq. $POCl_3$ converts simple or functionalized phosphonic diethyl esters into the phosphonic dichlorides.[95]

Compounds (71)-(73) are the potential products from the oxidative chloro-phosphonation of the alkenes (70); in practice, compounds (70a) and (70b) with PCl_3/O_2 yield mainly (71a) and (71b), respectively. The main difference between these examples and (70c) is that in this case the oxidative chlorophosphonation process is accompanied by the breakdown of the hydrocarbon chain. Although the main reaction pathway leads to the formation of a phosphorus-carbon bond, the chemoselectivity of the process is not as high as for other examples in which, for example, R^1 and R^2 are H or Me. The dehydrochlorination of certain of the products was investigated. With one equivalent of Et_3N, (71a) loses one HCl, and the second is removable with $KOBu^t$ after esterification with $EtOH/Et_3N$; (73c) is fully dehydrochlorinated with two equivalents of Et_3N without prior esterification.[96] The phosphorylation of an α,β-unsaturated phosphonic dichloride (74) has been recorded; the nature of the product indicates a migration of the carbon-carbon double bond, and although this is not an entirely new phenomenon, this is a slightly unusual example of the reaction.[97] 1-Adamantylphenyl- and di-1-adamantyl-phosphinic acids have been prepared from 1-hydroxyadamantane and the appropriate phosphonous dichloride in sulphuric acid.[98]

The reaction between $Me_2Si(OMe)NCS$ and $PhPF_4$ yields $PhP(O)F_2$ in 90% yield, together with MeF.[99] The imidazolides of dimethyl- and diphenyl-phosphinic acids, and also the same derivatives of diphenylphosphinothioic and *tert*-butylphenyl-phosphinoselenoic acid, yield the respective fluorides when treated with benzoyl

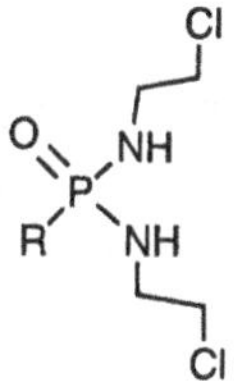

(63) 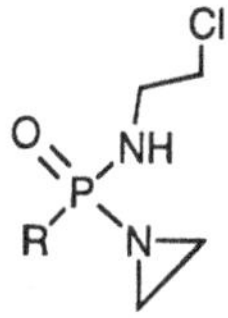 (64) 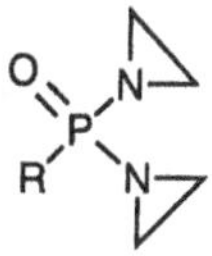 (65) 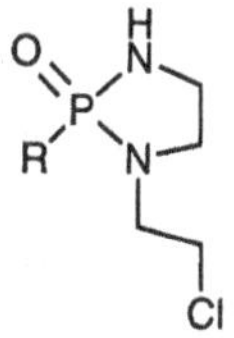 (66)

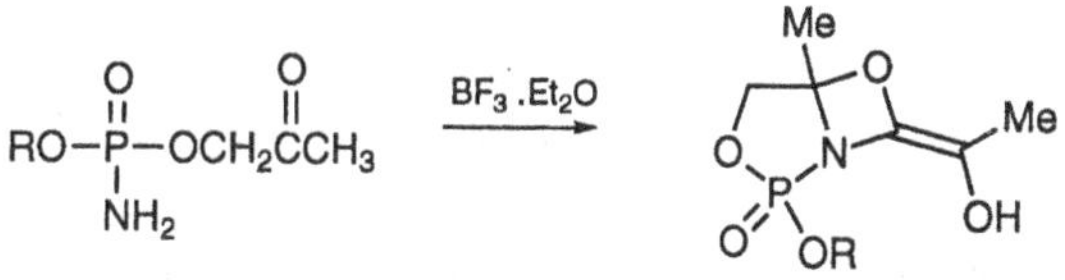

$RO-P(O)(NH_2)-OCH_2C(O)CH_3$ $\xrightarrow{BF_3.Et_2O}$ (67)

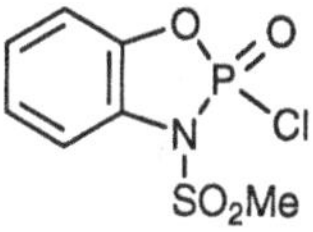

(68)

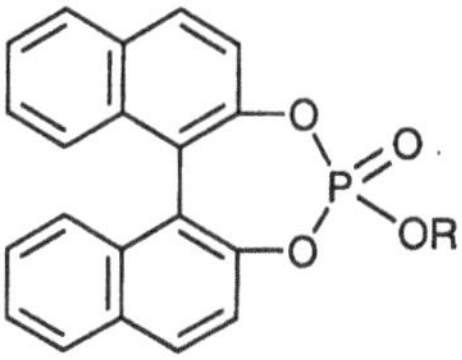

(69)

$H_2C{=}CHCH_2CClR^1R^2$

(70)

	R^1	R^2
(a)	H	Ph
(b)	Me	Ph
(c)	Me	Bu^t

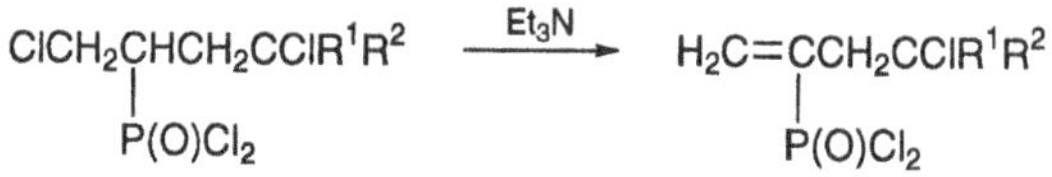

$ClCH_2CH(P(O)Cl_2)CH_2CClR^1R^2 \xrightarrow{Et_3N} H_2C{=}C(P(O)Cl_2)CH_2CClR^1R^2$

(71)

$Cl_2P(O)CH_2CHClCH_2CClR^1R^2$

(72)

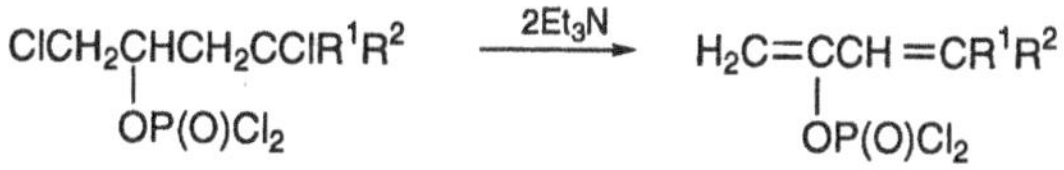

$ClCH_2CH(OP(O)Cl_2)CH_2CClR^1R^2 \xrightarrow{2Et_3N} H_2C{=}C(OP(O)Cl_2)CH{=}CR^1R^2$

(73)

$Pr^iOCMe{=}CHP(O)Cl_2 \xrightarrow{PCl_5} \xrightarrow{2PCl_5} Pr^iOC(O)CHClP(O)Cl_2 \;(=\bar{C}HPCl_3.\;\overset{+}{P}Cl_6) \longrightarrow Pr^iOC(=CHP(O)Cl_2)CHClP(O)Cl_2$

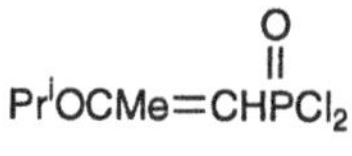

(74)

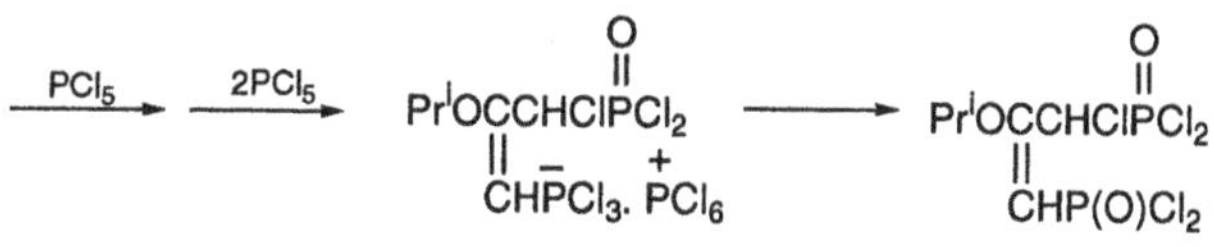

fluoride.[12] Alkyl and cycloalkyl phosphorodifluoridites, $ROPF_2$, react with 1,1,2,3,3,3-hexafluoropropyl azide to yield phosphonic difluorides, $R'P(O)F_2$, with, in certain cases, an isomerized alkyl group.[13]

2.1.2. *Alkyl, Aralkyl, and Cycloalkyl Acids.* In this area, pride of place should arguably go to the syntheses of the C_{60} and C_{70} fullerenylphosphonic acids (as esters) through the irradiation of the hydrocarbon with $Hg[P(O)(OR)_2]$ in toluene.[100,101] Bis(trimethylsilyl) phosphonite is the starting point for the preparation, through successive alkylation steps, of either symmetrical or unsymmetrical phosphinic acids (Scheme 9).[102] Other newly recorded examples of well-established synthesis procedures include: the addition of dialkyl hydrogen phosphonates to alkenes,[103-105] and the use of the Kinnear-Perren reaction to prepare higher esters of (1-methylpropyl)phosphonic acid.[106] More unusual has been the formation of cyclopropyl-1,1-bisphosphonic acids (as their esters) through the Al_2O_3/KF-catalysed additions of tetraethyl methylenebisphosphonate to electron-defficient alkenes.[107] The bisphosphonic esters (75), obtainable in two steps from methylenebisphosphonic ester, undergo an enyne cycloisomerization process in the presence of Pd(0) to give derivatives of cyclopentyl-1,1-bisphosphonic acid, (76) or the related (77).[108] The reagent (78) was employed in a synthesis of the methylphosphonic mono ester of *myo*-inositol (79).[109] Alkylation reactions between hydrogen phosphonates and diazoalkanes have provided di-*tert*-butyl esters of phosphonic acids[110] and a similar reaction between diethyl hydrogen phosphonate and tetraethyl diazomethylenebis- phosphonate in the presence of $Cu(acac)_2$ yielded hexaethyl methanetrisphosphonate, not obtainable by some other conventional procedures.[111] Details have been presented for the conversion of trialkyl phosphates into dialkyl alkylphosphonates by the action of two equivalents of alkyllithium reagents at -78°.[112]

Several products have been identified as deriving from the interaction of bis(trimethylsilyl) phosphonite and 2-(2-benzoyl-1-phenyl-2-propen-1-yl)cyclohexanone (80); they include at least two phosphorinane derivatives (81) and (82)(R = H or Me_3Si) and various acyclic phosphinic esters (83)-(85)(R = H or Me_3Si).[113] A series of cyclic phosphinic acids (86)(n = 1-3; R = H) have been prepared, *via* their ethyl esters; the latter were, in turn, prepared from di-Grignard reagents and $EtOP(O)Cl_2$.[114]

1,3-Dithiol-2-phosphonate esters (88) are already known, and have been prepared previously from (87) as the result of several steps, but a new method of preparation involves the direct interaction of (87)(X = O or S) with a trialkyl phosphite; the yields may be as low as 5% (R = Me or Et, X = O for R = Ph) or up to 75%.[115]

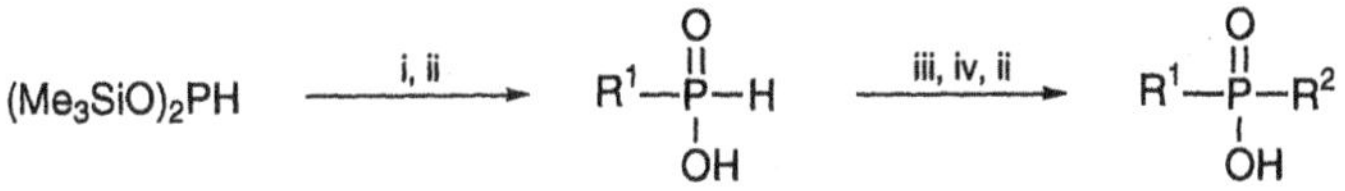

$$(Me_3SiO)_2PH \xrightarrow{\text{i, ii}} R^1-P(=O)(OH)-H \xrightarrow{\text{iii, iv, ii}} R^1-P(=O)(OH)-R^2$$

Reagents: i, R^1X, CH_2Cl_2; ii, MeOH; iii, $(Me_3Si)_2NH$; iv, R^2X

Scheme 9

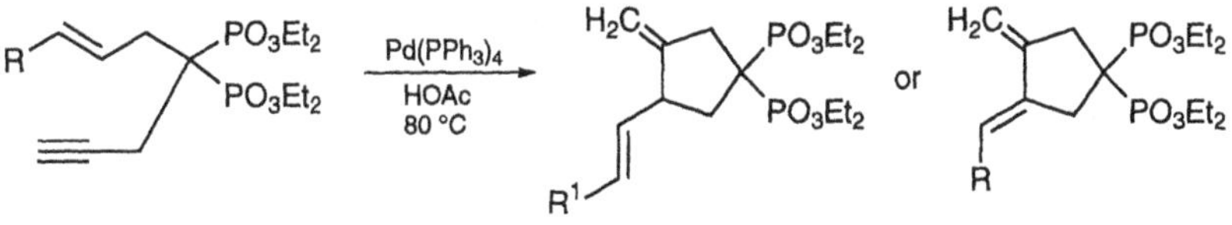

PO_3Et_2 PO_3Et_2 R — $Pd(PPh_3)_4$, HOAc, 80 °C → H_2C PO_3Et_2 PO_3Et_2 R^1 or H_2C PO_3Et_2 PO_3Et_2 R

(75) (76) (77)

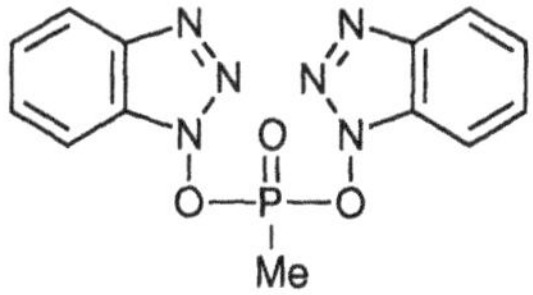

N N N N N N O O P O Me

OH HO Me O P OH O H_2O_3PO OH OPO_3H_2

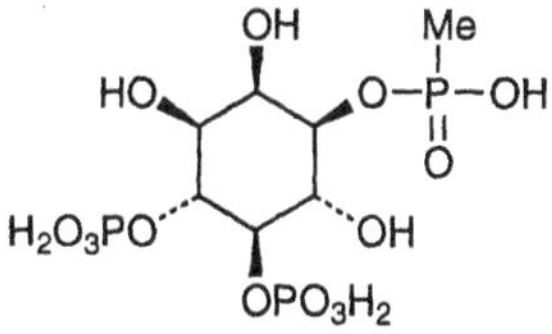

(78) (79)

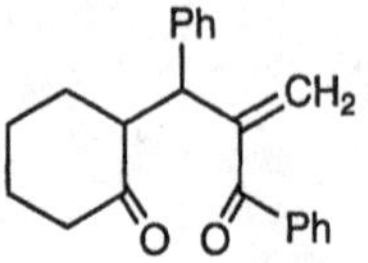

(80) 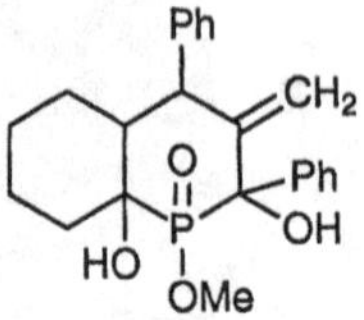 (81) 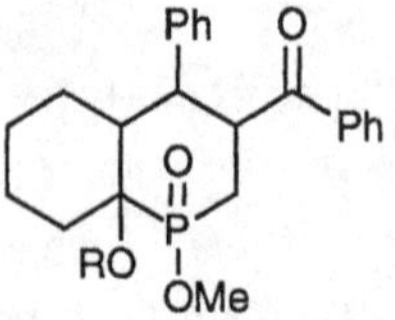 (82)

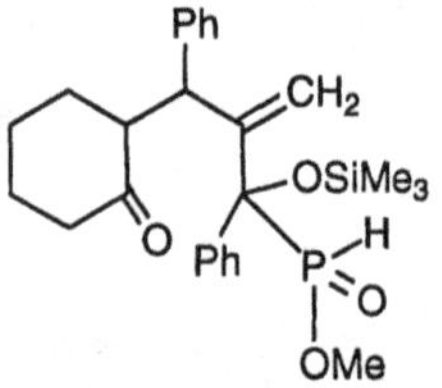

(83) 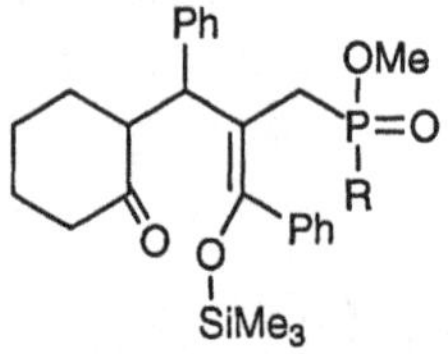 (84) 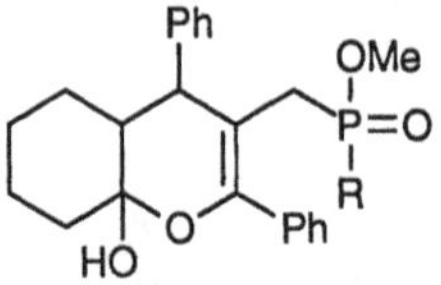 (85)

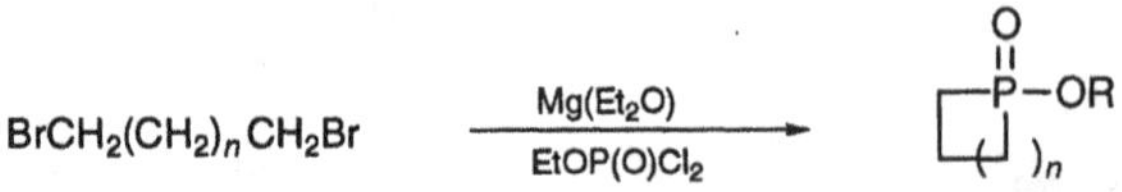

$BrCH_2(CH_2)_nCH_2Br$ $\xrightarrow[EtOP(O)Cl_2]{Mg(Et_2O)}$ (86)

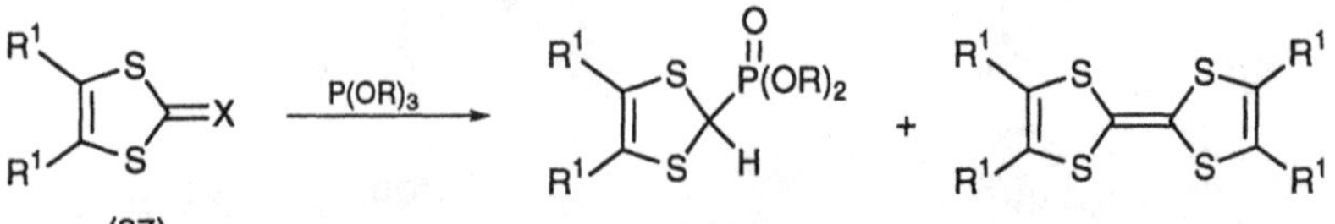

(87) $\xrightarrow{P(OR)_3}$ (88) +

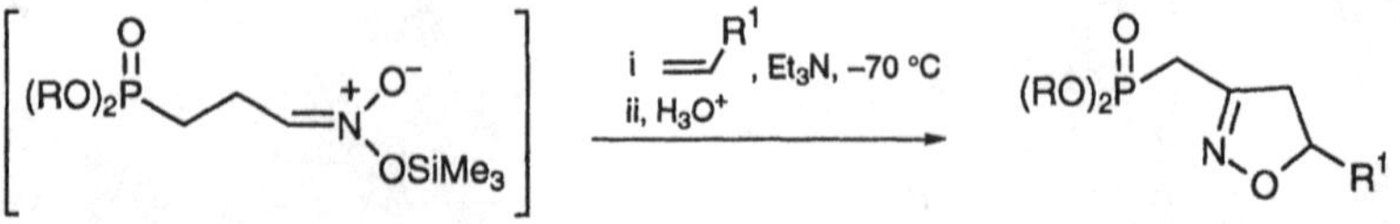

(89) $\xrightarrow[\text{ii, } H_3O^+]{\text{i } CH_2{=}CHR^1\text{, } Et_3N\text{, } -70\ ^\circ C}$ (90)

Dialkyl methylphosphonate carbanions have a multitude of uses; their Michael addition to nitroethene and silylation yield the intermediates (89) which themselves undergo additions to appropriate alkenes (R^1 = COMe or COOMe) to give the 4,5-dihydroisoxazoles (90).[116] The deuteriolysis of α-silylated α-phosphonylated carbanions with D_2O has furnished a range of α,α-dideuteriated phosphonic diesters.[117]

2.1.3. *Alkenyl, Alkynyl, Aromatic, Heterocyclic, and related acids.*-(1-Alkenyl)phosphonic diesters have been prepared through the DBU-catalysed isomerization of the (2-alkenyl)phosphonic isomers, and equally by the DBU-initiated dehydrobromination of (2-bromoalkyl)phosphonic diesters.[118] The Pd(0) catalysed coupling of dialkyl hydrogen phosphonates and 1-bromo-2-(trimethylsilyl)ethene (as a 1:9 (*Z*):(*E*) mixture) afforded high yields of pure (*E*) stereoisomers of dialkyl [(2-trimethylsilyl)ethenyl]phosphonates, the Diels-Alder reactions of which (with cyclopentadiene) were investigated.[119] The esters (91), in pure (*E*) form, are obtainable from aromatic aldehydes and the triester (92) in the presence of tributylarsine at 70°.[120] Condensations between tetraethyl methylenebisphosphonate (activated with NaH) and carbonyl compounds (isobutyraldehyde, heptanal, and phenylacetaldehyde gave 90%+ yields, and cyclohexanone from which the yield was about 50%) give only the diethyl (*E*)-(1-alkenyl)phosphonate; for the reaction with isobutyraldehyde, NaH proved to be superior (THF solvent) to NaOEt, Bu^tOK, or K_2CO_3. Isomerization of the products to the dialkyl (2-alkenyl)phosphonates was best achieved with $KOBu^t$ in DMSO.[121]

The Michaelis-Arbuzov-like coupling of aryl bromides and tris(trimethylsilyl) phosphite occurs quite quickly in the presence of $NiCl_2$ at 150°; for pentachloro- and pentafluoro-pyridines, the coupling takes place at the 4-position, but the presence of a CN group results in the loss of chlorine from position 6 in 2,4,5,6-tetrachloro-3-cyanopyridine.[122] Iodoaromatics couple with esters of (2-propenyl)phosphonic acid in the presence of $Pd(OAc)_2/Et_3N$ to give the dialkyl (3-aryl-2-propenyl)phosphonates.[123] In the calix[4]arene series, the tetrabromo derivative (93) reacts with triisopropyl phosphite to give the phosphonate (94)(and also with isopropyl diphenylphosphinite to give the phosphine oxide (95)).[124] Further examples of the rearrangement of aryl phosphate esters to (2-hydroxyphenyl)phosphonic esters have been noted[125] and pyridinyl and other heteroaryl phosphonic diesters have also been recorded as the products of similar rearrangement reactions.[126,127] Phospha-crown ethers based on phenylphosphonic acid have been prepared by conventional methods.[128] Phosphonoylated pyrazoles[129] and pyrroles[130,131] have been prepared through reactions of

C-phosphorylated carbonyl compounds. The reactions between benzopyrylium salts and phosphorus(III) triesters are of the Michaelis-Arbuzov type (overall) and yield the phosphonates (96).[132]

The thermal decomposition of a variety of dimethyl (1-diazo-2-oxo-3-alkenyl)phosphonates in refluxing benzene (during 30 h) or toluene (3 h) has provided a selection of phenolic phosphonic derivatives and cyclobutenones according to Scheme 10.[133]

Knoevenagel reactions between triethyl phosphonoacetate and salicylaldehyde or its derivatives (Scheme 11)(R^1 = H or Br, R^2 = H, OMe or OEt) in the presence of piperidine acetate/β-alanine in refluxing toluene resulted in condensation followed by cyclization to give 3-coumarinylphosphonic esters, reducible to the dihydro derivatives.[134]

2.1.4. *Halogenoalkyl and related acids.*-The additions of perfluoroalkyl iodides to diethyl 2-propenylphosphonate are free-radical in nature, and have been achieved in a water/CH_2Cl_2 system containing sodium dithionite; diethyl (3-R_f-2-iodopropyl)-phosphonates are obtained in excellent yields, and may be de-iodinated to the esters (97) with zinc.[135] The similar addition of some iodoperfluoroalkylsulphonyl fluorides to diethyl [2-propenyloxy)methyl]phosphonate has been noted; hydrolysis of the products affords the phosphonic-sulphonic diacids (98)(Z = OH) as hydrates.[136] Triflates offer distinct advantages in reactions with halogenated phosphoryl carbanions: thus, the compounds (99)(X = I or OSO_2Me) do not react with $(EtO)_2P(O)CF_2Li$, but with X = OSO_2CF_3, a reaction occurs readily (during 5-10 m in THF at -78°) to give (99)(X = $CF_2PO_3Et_2$). Additionally, no reaction occurs with the corresponding Na or K complexes, and even for the Li salt, the yields can depend on the manner and timing of the anion formation, as well as on the presence of other solvent in mixtures.[137]

A Michaelis-Arbuzov reaction between triethyl phosphite and tribromofluoromethane in boiling THF yields diethyl (dibromofluoromethyl)phosphonate; when this is acted on by BuLi followed by Me_3SiCl, there results the Li complex of diethyl [fluoro(trimethylsilyl)methyl]phosphonate; this, on ethanolysis followed by acidification, gives diethyl (fluoromethyl)phosphonate. Also, an initial reaction of the compound with RI followed by an identical workup, produces a diethyl (1-fluoroalkyl)phosphonate.[138,139] In the same way, it is known that triethyl phosphite and dibromodifluoromethane yield diethyl (bromodifluoromethyl)phosphonate and, whilst this may be converted into an organozinc reagent for further reaction, it is also reducible to diethyl (difluoromethyl)phosphonate by Bu_3SnCl.[140]

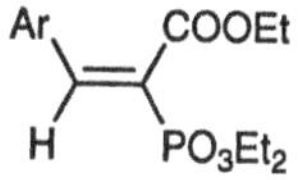

Ar, COOEt, H, PO_3Et_2

(91)

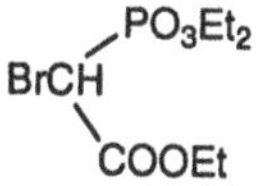

PO_3Et_2, BrCH, COOEt

(92)

(93) = (1), R = Br
(94) = (1), R = $P(O)(OPr^i)_2$
(95) = (1), R = $P(O)Ph_2$

$R^1 = R^2 = R^3 = R^4 = Me$

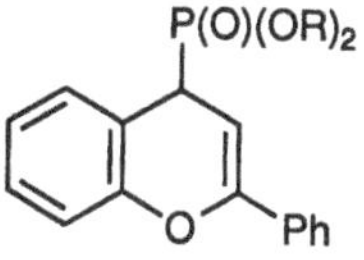

$P(O)(OR)_2$, O, Ph

(96)

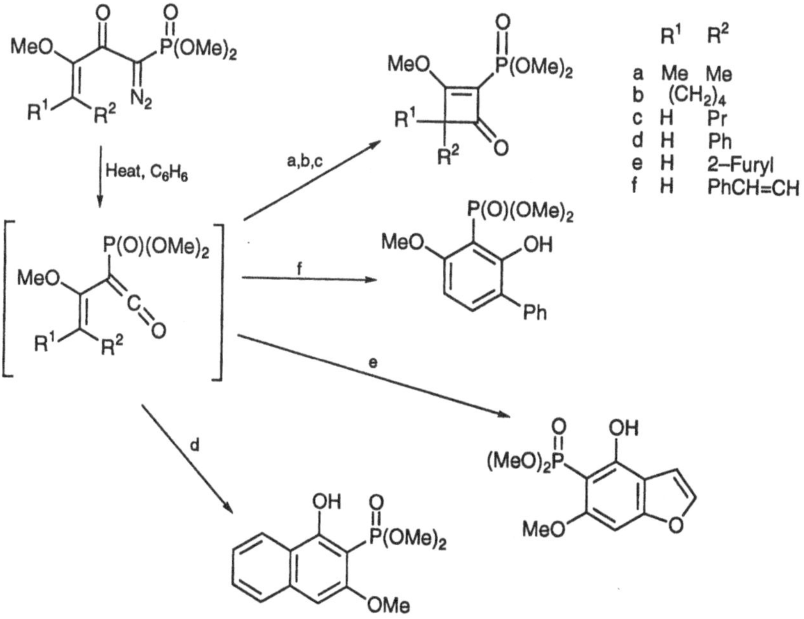

MeO, O, O, $P(OMe)_2$, N_2, R^1, R^2

Heat, C_6H_6

$P(O)(OMe)_2$, MeO, C, O, R^1, R^2

a,b,c

MeO, O, $P(OMe)_2$, R^1, R^2, O

f

$P(O)(OMe)_2$, MeO, OH, Ph

e

O, OH, $(MeO)_2P$, MeO, O

d

OH, O, $P(OMe)_2$, OMe

	R^1	R^2
a	Me	Me
b	$(CH_2)_4$	
c	H	Pr
d	H	Ph
e	H	2–Furyl
f	H	PhCH=CH

Scheme 10

The reactivity of phosphorus-containing nucleophiles towards polyfluoro organic compounds has been reviewed (with 120 references).[141]

Burton and his coworkers have presented an account of the syntheses of several fluorinated bisphosphonic acids which possess two, three, four, or six difluoromethylene groups. The conversion of diethyl or diisopropyl (bromodifluoromethyl)phosphonate into the corresponding (iododifluoromethyl)phosphonate by the initial generation of the zinc reagent and its subsequent iodination, was followed by a reaction with cadmium (but not Zn) in boiling CH_2Cl_2 to give good yields of the tetraalkyl (tetrafluoroethane-1,2-diyl)-bisphosphonate. In the same way diethyl (1,1,2,2-tetrafluoro-2-iodoethyl)phosphonate afforded tetraethyl (octafluorobutane-1,4-diyl)bisphosphonate, and the procedure was also adaptable to the preparation of the (dodecafluorohexane-1,6-diyl)bisphosphonic ester. A second approach to the synthesis of these compounds is illustrated by the initial reaction between 1,1,2,2,3,3-hexafluoro-1,3-diiodopropane and tetraethyl diphosphite; subsequent oxidation of the product with Bu^tOOH yielded the (perfluoropropane-1,3-diyl)-bisphosphonic acid ester. In all cases, reactions between the tetraalkyl esters and Me_3SiCl, followed by hydrolysis of the resulting silyl esters, yielded the free acids.[142]

The hydrogenolysis of diethyl (1-fluoro-2-propynyl)phosphonate (obtained by the fluorination of diethyl (1-hydroxy-2-propynyl)phosphonate with Et_2NSF_3 (DAST)) resulted only in the formation of diethyl (1-fluoro-2-propenyl)phosphonate and a loss of fluorine was not observed.[143] The coupling of 1-fluoro-1-iodoalkenes with dialkyl hydrogen phosphonates occurs slowly (20 h at 90°) in the presence of $PdCl_2(PPh_3)_2$ to give the products (100) with retention of geometry.[144] Finally, dialkyl (perfluoro-1-alkynyl)phosphonates (102) result from the (perfluoroacyl)ation of dialkyl methylphosphonate and subsequent dehydration through the enol form (101) of the (acylmethyl)phosphonate by the action of trifluoromethanesulphonic anhydride in the presence of $EtNPr^i_2$.[145]

2.1.5. *Hydroxyalkyl and Epoxyalkyl Acids*.-Some worthwhile developments have occurred in the synthesis of (1-hydroxyalkyl)phosphonic acid derivatives through the so-called Abramov reaction. The esters (103) have been prepared from dimethyl hydrogen phosphonate and R_fCHO [146] and the same esters have also been obtained from the hydrogen phosphonate and $F_3CCH(OH)(OEt)$ at room temperature in the presence of K_2CO_3 or a tertiary amine base; the reaction with dimethyl hydrogen phosphonate produced a yield much lower than for other dialkyl hydrogen phosphonates during the normal reaction times of 5-16 h.[147] Abramov reactions with 4-selenanone have been

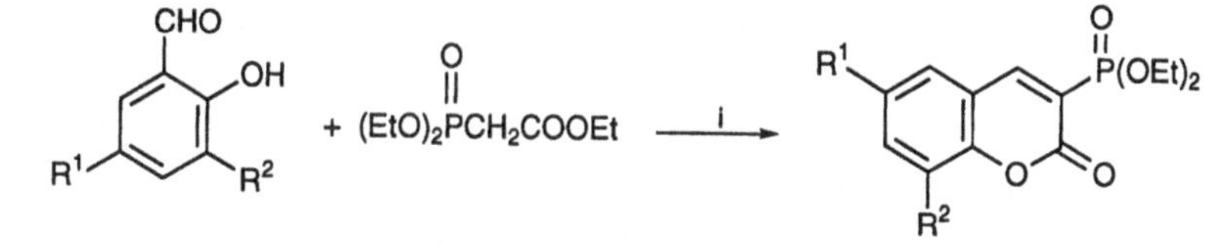

$$\text{(salicylaldehyde, } R^1, R^2\text{)} + (EtO)_2P(O)CH_2COOEt \xrightarrow{i} \text{3-}P(O)(OEt)_2\text{-coumarin } (R^1, R^2)$$

Reagents: i, piperidine acetate, β-alanine, PhMe, 110 °C

Scheme 11

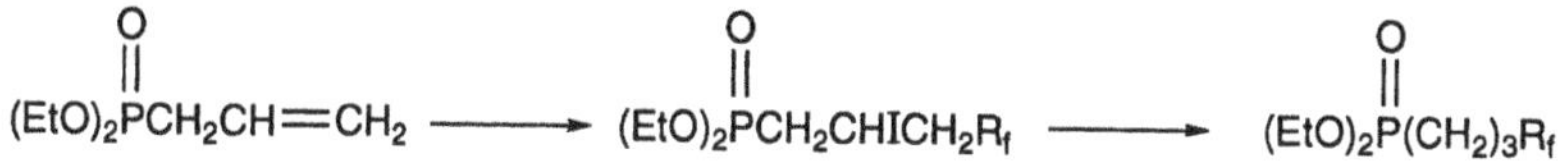

$$(EtO)_2P(O)CH_2CH{=}CH_2 \longrightarrow (EtO)_2P(O)CH_2CHICH_2R_f \longrightarrow (EtO)_2P(O)(CH_2)_3R_f$$

(97)

$$(EtO)_2P(O)CH_2OCH_2CH{=}CH_2 \xrightarrow[\text{ii, } H_2O]{\text{i, } R_fI\ (Z{=}F)} H_2O_3PCH_2O(CH_2)_3R_f$$

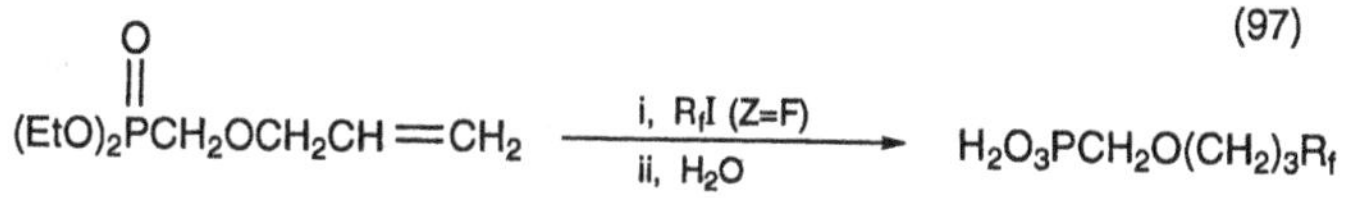

(98)

(a) $R_f = (CF_2)_4SO_2Z$
(b) $R_f = (CF_2)_2O(CF_2)_2SO_2Z$

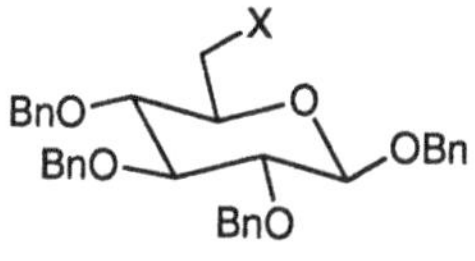

(99) [BnO, BnO, BnO, OBn, X]

$$R_2C{=}C(I)F \longrightarrow R_2C{=}C(F)PO_3Et_2$$

(100)

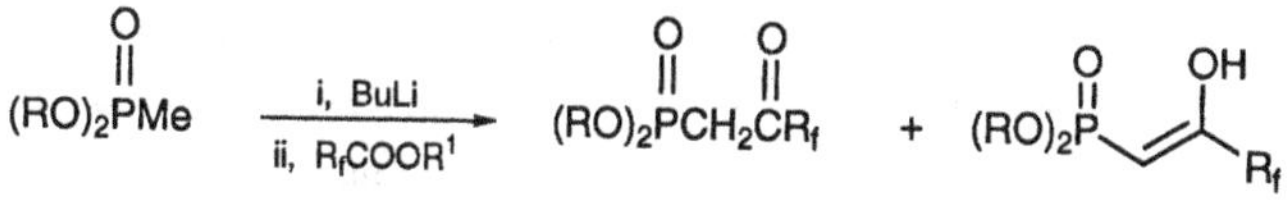

$$(RO)_2P(O)Me \xrightarrow[\text{ii, } R_fCOOR^1]{\text{i, BuLi}} (RO)_2P(O)CH_2C(O)R_f + (RO)_2P(O)CH{=}C(OH)R_f$$

(101)

$$(RO)_2P(O)C{\equiv}CR_f$$

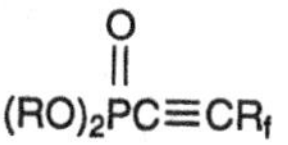

(102)

reported.[148] The reaction between salicylaldehyde and triethyl phosphite has been investigated in some detail, and a complex sequence of reactions has been proposed to account for the formation of the products (104) and (105) through the betaine (106).[149] Triethyl phosphite and 3,5-di-*tert*-butyl-4-hydroxybenzaldehyde in the molecular ratio 2:1 gave the ether (107) instead of the expected (108), but if the reactants are in the ratio 3:1, the product is (109), products consistent with those found in the study with salicylaldehyde.[150]

Most of the efforts within this area have been directed towards the achievement of enantioselectivety in the Abramov process. Thus, in the reactions between ArCHO and diethyl hydrogen phosphonate, the presence of 20% of a catalyst derived from $LaCl_3.7H_2O$ and dilithium (*R*)-binaphthoxide, in THF at temperatures lower than -20°, results in the formation of (*S*)-(-)-(α-hydroxybenzyl)phosphonic diesters (110) in >90% yields at -40° - the most effective temperature. However, the enantioselectivity does depend on the individual aromatic substituent, being 82% when Ar is 4-methoxyphenyl, but <50% for Ar = Ph or 4-chlorophenyl.[151] Similar results were later reported for reactions which involved cinnamaldehyde; the enantioselectivity, described as modest, was also dependent on temperature, timing, and ratio of reactants; the enantiomeric forms of the product were separated by HPLC of the (*R*)-mandelate esters, and the minor, but crystalline, diastereoisomer was shown by X-ray analysis to have the (*R,R*) configuration (111). In this case, also, it was demonstrated that the presence of $LaCl_3$ was necessary for enantiomeric preferment to occur.[152]

In reactions between the chiral phosphonic diamide (112), $LiNPr^i_2$, and the aldehydes RCHO, the ratios of diastereoisomeric α-hydroxy phosphonic diamides ranged from 29:1 (R = 1-naphthalenyl) to as low as 3.4:1 (with isovaleraldehyde), with the (*S*) forms being the major components; in all cases but the last, the products could be crystallized to yield a single diastereoisomer.[153]

Also of interest are the results obtained through the use of perhydro-1,3,2-oxazaphosphorines; the reactions between the 2-methoxy derivative (113) (Z = OMe) and RCHO in the presence of $BF_3.Et_2O$/LiI in THF at -78° gave mixtures of the (R_C) (114) and (S_C)(115) products; diastereoisomeric excesses were rather poor, being in the range 8-18%, increasing with a gradual increase in the size of R from Me to Bu^t. In the reactions between (113)(Z = OLi) and ArCHO with the aromatic substituents H, 4-Me, 4-Cl, and 4-NO_2, the diastereoisomeric excesses were then 32, 40, 22, and 4%; the configuration of (114)(R = Ph) was confirmed by X-ray analysis.[154]

The results obtained for the hydrophosphorylation of benzaldehyde with the

$(MeO)_2P(O)CH(OH)R_f$

(103)

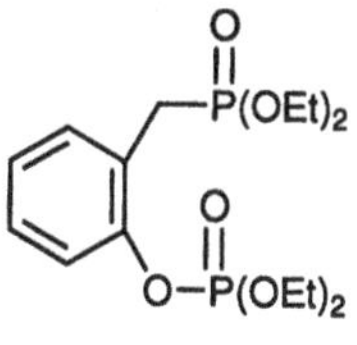

(104)

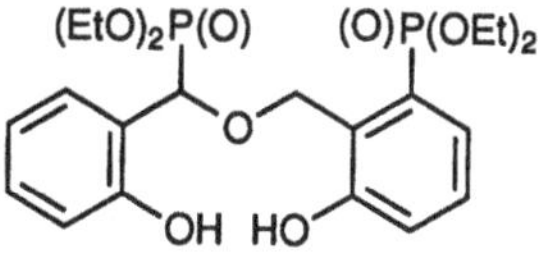

(105)

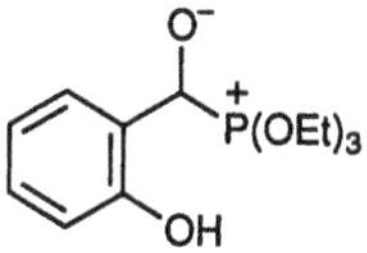

(106)

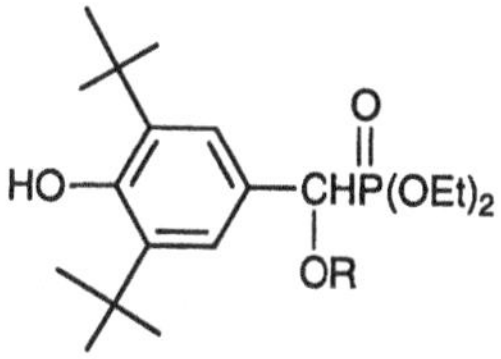

(107) R = Et
(108) R = H

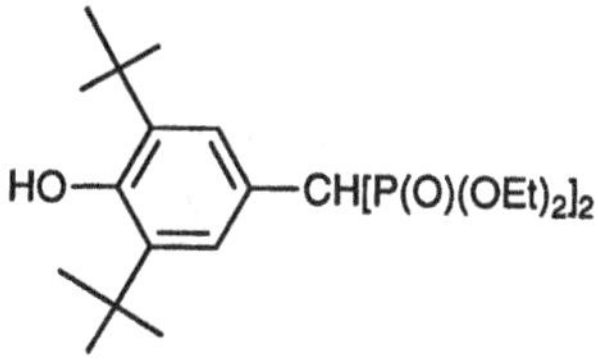

(109)

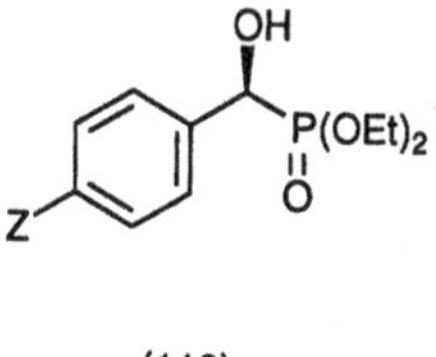

(110)

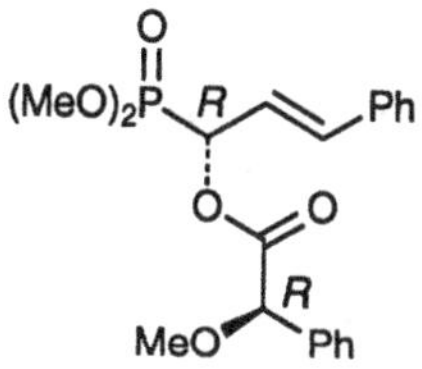

(111)

phosphite ester (116) in the presence of the chiral titanium-containing catalysts (117) (R = Me or Ph) (and from which the product was a mixture of (118) and (119)) were excellent from the point of view of yield, but very disappointing from that of enantioselectivity.[155] The formation of ethers of α-hydroxyalkylphosphonic diesters from aldehydes and phosphorus(III) triesters has already been reported on several occasions during past decades, but a particularly detailed study has been made of Abramov reactions between the chiral 2-silyloxy-1,3,2-oxazaphosph(III)olidines (120) and benzaldehyde or 2,2-dimethylpropanal; the products are the corresponding silyl ethers (121) (R' = Ph or Me_3C) obtained in high yields and with good stereoselectivities at room temperature. Evidence was presented to support the conclusion that the main products are formed from the major epimers of compounds (120) and R'CHO with retention of configuration at phosphorus, and the absolute configurations at phosphorus and the α-carbon in the major epimer of product (121c) is (S_P,S_C). The reactions of the series (120) with ketones occur much more slowly.[156] These results were presented in preliminary form in last year's report *Organophosphorus Chemistry*, 1994, **25**, 111, reference 173). Abramov reactions for the 1,3,2-diazaphosph(III)olidine (122) have also been described.[157]

1-(1-Naphthalenyl)ethylamine has been recommended as a chiral agent for the direct ^{31}P n.m.r. spectroscopic determination of enantiomer composition of (1-hydroxyalkyl)-phosphonic derivatives.[158]

Some unusual results have been observed in a study of the reactions between the 1,4-benzoquinonimines, (123) and (124), and dialkyl hydrogen phosphonates $(R^2O)_2P(O)H$, the outcome of which depends on the natures of both R^1 and R^2. Thus, in the reaction between (123) and diethyl hydrogen phosphonate, the two products are (125) and (126)(R^2 = Et); comparable reactions with dimethyl or diisopropyl hydrogen phosphonates yield compounds of type (126) only. On the other hand, the interaction of diethyl hydrogen phosphonate and (124) yields no phosphorus-containing product, and that from dimethyl hydrogen phosphonate is (127).[159]

[2,3]-Wittig rearrangements of the ethers (128) occur when they are treated with two equivalents of $LiNPr^i_2$. Yields of the isolated α-hydroxy compounds (129) (Scheme 12) are 60-72% when R^1 = H, R^2 = H, Me, or Ph, but only about 30% when R^1 = R^2 = Me. The reactions are slow at -70°, and the yields of products are enhanced when the product is rapidly quenched at -70° with the minimum of aqueous acid.[160]

In the additions of an aldehyde ArCHO to the carbanion from diethyl 2-propenylphosphonate - a fully reversible process - kinetic control leads, almost exclusively, to α-adducts, whereas the thermodynamic control affords the γ-adducts.[161]

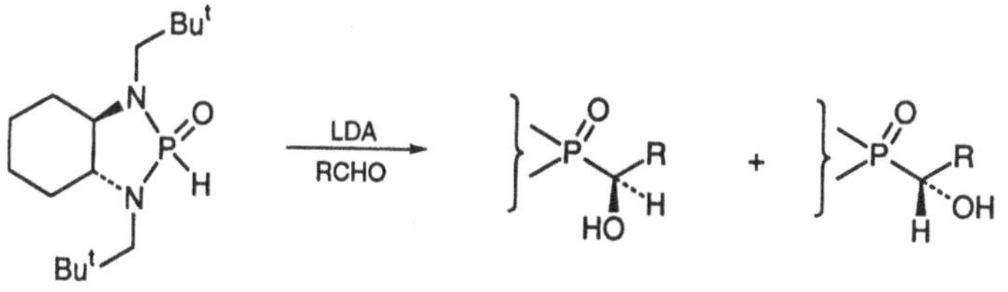

LDA, RCHO

(112)

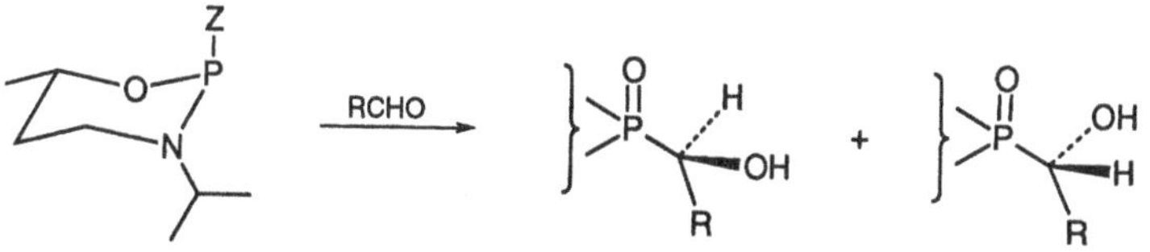

RCHO

(113) (114) (115)

$(EtO)_2POSiMe_2Bu^t$

(116)

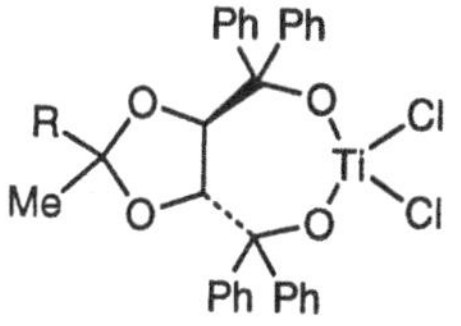

(117)

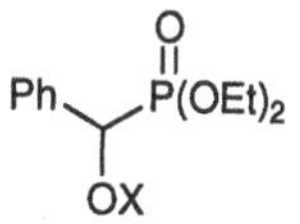

(118) X = H
(119) $X = SiMe_2Bu^t$

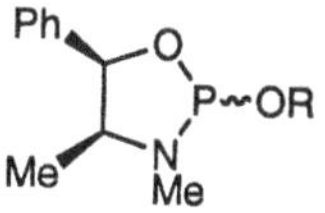

(120)

(a) $R = SiMe_3$
(b) $R = SiEt_3$
(c) $R = SiPh_3$
(d) $R = SiMe_2Bu^t$

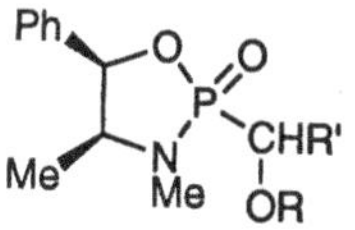

(121)

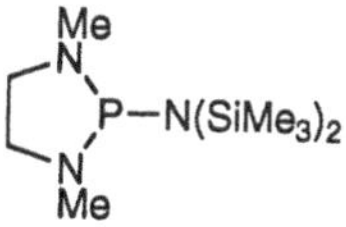

$P{-}N(SiMe_3)_2$

(122)

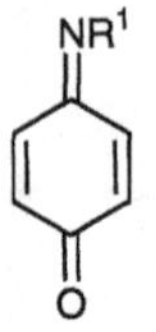

NR^1

(123) $R^1 = PhSO_2$
(124) $R^1 = MeSO_2$

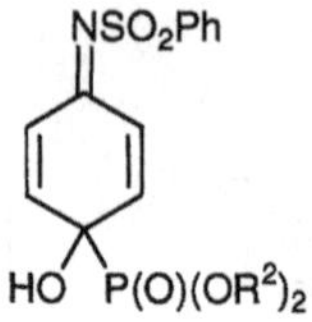

NSO_2Ph HO $P(O)(OR^2)_2$

(125)

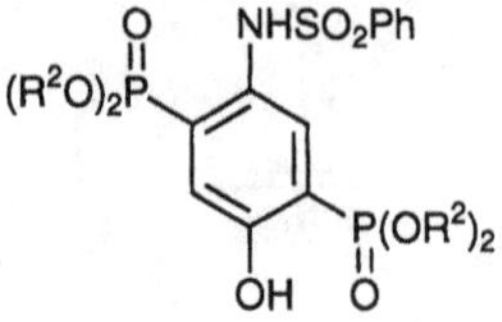

$(R^2O)_2P(O)$ $NHSO_2Ph$ $P(OR^2)_2$ OH

(126)

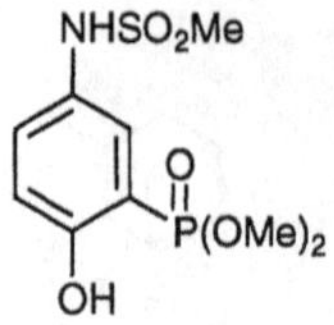

$NHSO_2Me$ $P(OMe)_2$ OH

(127)

$(EtO)_2P(O)$ O R^1 H R^2 → i → [$(EtO)_2P(O)$ O R^1 H R^2 → $(EtO)_2P(O)$ O^- R^1 R^2 CH_2 H] → ii → $(EtO)_2P(O)$ OH R^1 R^2 CH_2 H

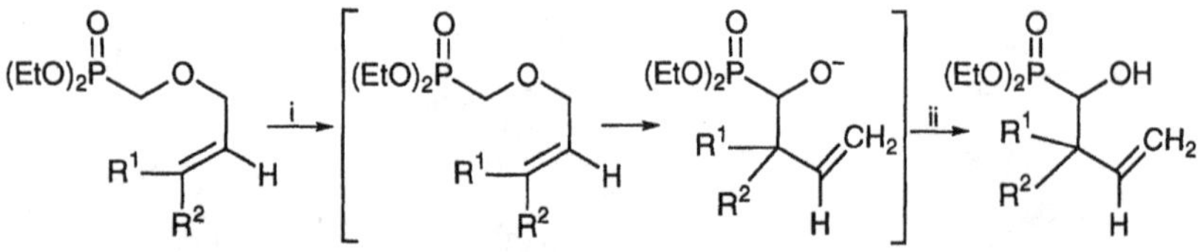

(128) (129)

Reagents: i, 2 equiv. $LiNPr^i_2$; ii, H_3O^+

Scheme 12

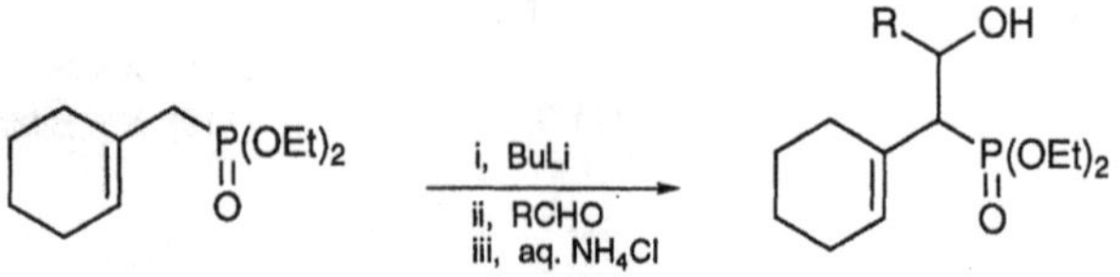

$P(OEt)_2$ → i, BuLi; ii, RCHO; iii, aq. NH_4Cl → R OH $P(OEt)_2$

(130)

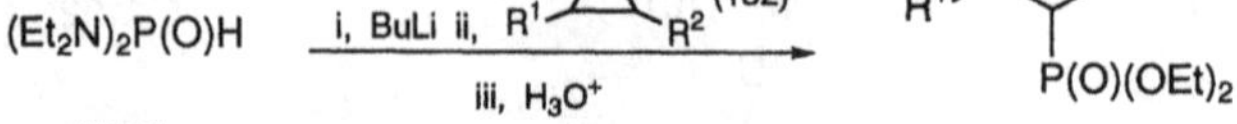

$(Et_2N)_2P(O)H$ (131) → i, BuLi ii, R^1–epoxide–R^2 (132); iii, H_3O^+ → HO R^1 R^2 $P(O)(OEt)_2$

The solution and solid-state structures of α-adduct single stereoisomers (130) from diethyl [(1-cyclohexenyl)methyl]phosphonate and three aldehydes (R = Pr^i, Ph, and CH_2Ph) have been determined by ^1H, ^{13}C, and ^{31}P n.m.r. spectroscopy and X-ray crystallography.[162] Conformational analyses have been carried out on other (2-hydroxypropyl)phosphonic dimethyl esters.[163]

In their reactions with the lithium salts of phosphonic diamides (131), the epoxides (132)(R^2 = H) undergo ring opening under mild conditions with full regiospecificity, the phosphorus attacking the less hindered carbon atom. Epoxides with $R^2 \neq$ H are less reactive, and some, e.g. *exo*-2,3-epoxynorbornane, may be completely unreactive; in general, however, for this group, the ring opening step is *trans* selective, and only one diastereoisomeric product is formed. In the reaction between (133) and 1,2-epoxybutane, ring opening gives the products (134) as a 2:1 mixture of diastereoisomers.[164]

The potential for the synthesis of hydroxyalkyl phosphonic diesters by means of epoxide ring cleavage has been well demonstrated in the reactions of (135)(Scheme 13), a preliminary notification of which was made last year (*Organophosphorus Chemistry*, 1994, **25**, 111, reference 174). Ring opening is achieved by a reaction with a phosphorus(III) triester, a dialkyl hydrogen phosphonate, or with a dialkyl methylphosphonate carbanion, to produce compounds with n = 1 or 2, and with R = H or OR', and it is thus possible to obtain phosphonate esters with two or three hydroxy functions in the carbon chain. Moreover, benzoate esters survive attack by nucleophilic phosphorus, but are cleaved by phosphonate carbanions, thus widening the scope still further in allowing the preparation of the compounds (136)(X = OBz or OH).[165] Regioselectivity in the ring opening of cyclohexene oxides by diisopropyl (lithiomethyl)phosphonate/BF_3 is quite modest, but the procedure was adapted to the preparation of an ester of a [(pentahydroxycyclohexyl)methyl]phosphonic acid.[166]

Polyhydroxy(cyclo)alkyl phosphonic esters in the *myo*-inositol series have been prepared through a simple application of the Michaelis-Arbuzov reaction[167], and the displacement of methanesulphonyl groups from (137) occurs in reactions with simple trialkyl phosphites to give, for example, (138)[168]. Interest continues in carbohydrate-like substances which possess a ring phosphorus atom, and are thus fundamentally derivatives of phosphole or phosphorin, but at the same time are also phosphinic acids. This work also incorporated examples of Michaelis-Arbuzov reactions at exocyclic sites, as in the conversion, (139) -> (140).[169]. By means of the reaction sequence abbreviated in Scheme 14, a series of dimethyl phosphonates (141)/(142)(R = a carbohydrate nucleus) was obtained; typically, (143) was then convertible into the L-galactopyranose phosphorinane

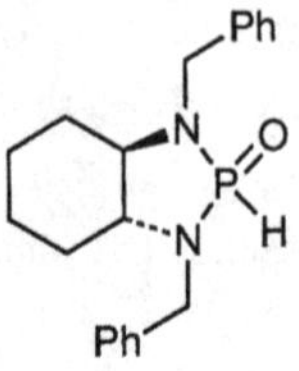

(133) 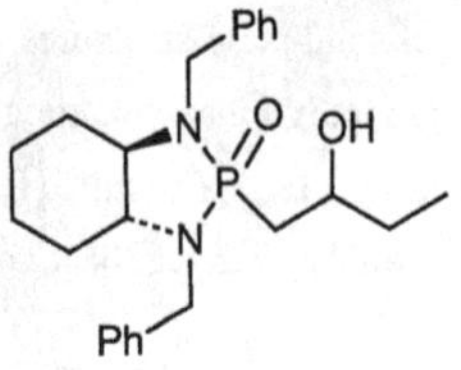 (134)

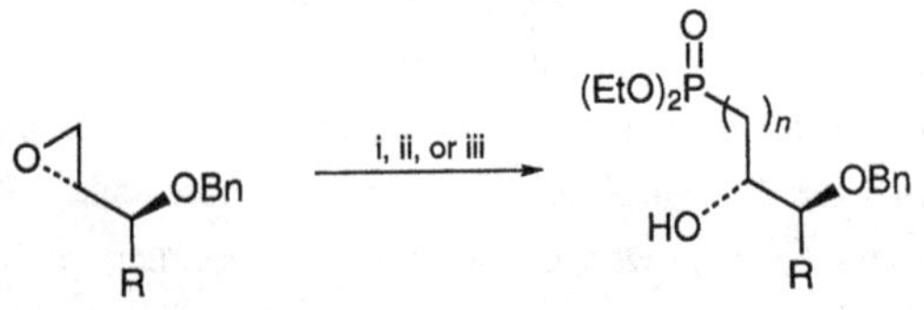

(135)

Reagents: i, $(EtO)_2POSiMe_3$, ZnI_2 or Et_2AlCl, 140 °C
ii, $(EtO)_2P(O)H$, NaH, BF_3. Et_2O, C_6H_6, 0 °C
iii, $(EtO)_2P(O)Me$, BuLi, BF_3. Et_2O, THF, –78 °C

Scheme 13

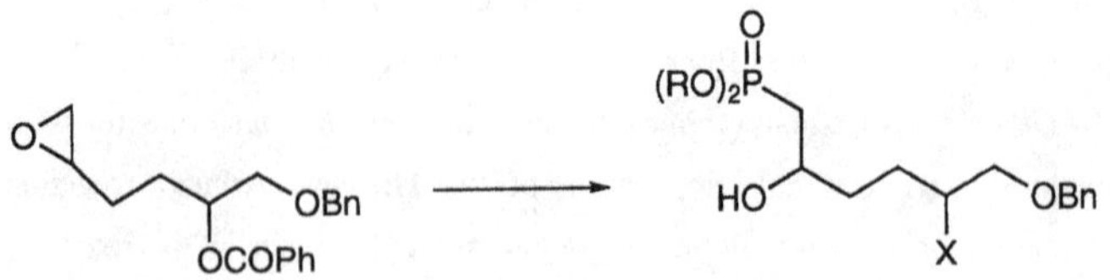

(136)

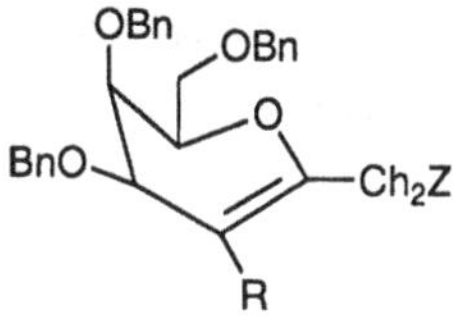

(137) R = H or SO_2Ph, Z = OSO_2Me

(138) R = H or SO_2Ph, Z = $(MeO)_2P(O)$

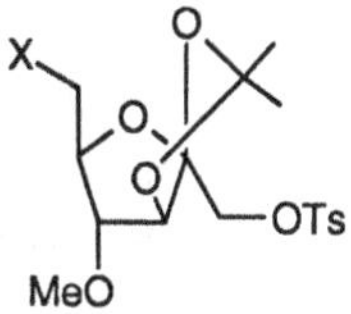

(139) X = I

(140) X = P(O)(OEt)Ph

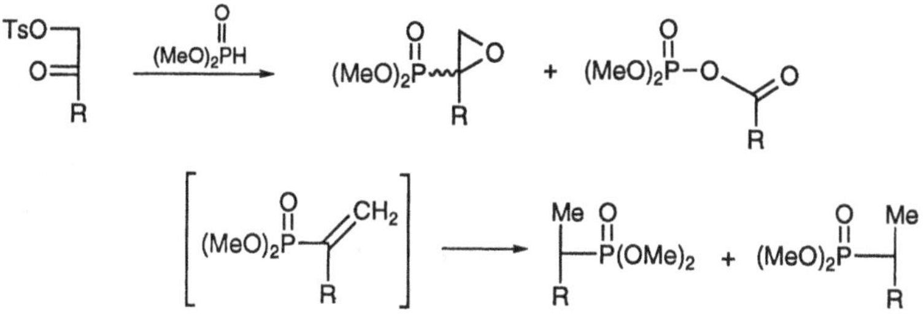

(141)

(142)

Scheme 14

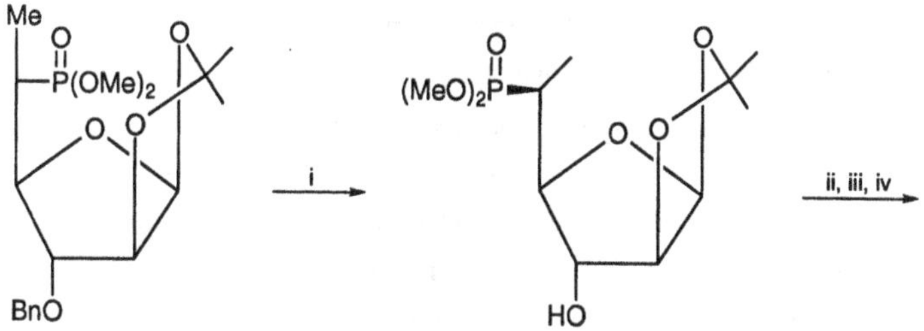

(143)

(144)

(145) through the steps indicated in Scheme 15. The tetra-*O*-acetyl methyl ester (146) was separable chromatographically into the (*S*)$_P$-α-L-galactopyranose (6% overall yield from (144), its β-anomer (5%), the (*R*)$_P$-α-isomer (3%), and its β-anomer (12%).[170] The conversion of (147)(R, R' = H or NO_2) into (148)(and thence to (149)) requires four steps of a similar nature.[171]

The structure (150) was confirmed through an X-ray analysis of the isopropylidene derivative (151)(R = H). The reaction between (151)(R = H) and 6-chloropurine under Mitsunobu conditions provided a mixture of (152) and (153). The action of potassium phthalimide upon (151)(R = methanesulphonyl) led to (154)(24%) and (155)(29%) together with smaller amounts of other products.[172]

The Abramov reaction has been applied in the synthesis of cyclic phosphonic analogues of hexopyranoses. In preliminary experiments, no stereoselectivity was observed in the addition of dimethyl or diethyl hydrogen phosphonate to di-*O*-isopropylidene-D-arabinose to give (156)(R^1, R^2 = H or OH), although fractional crystallization allowed the isolation of much purified diastereoisomers, and the structures of the products were ascertained by acid hydrolysis to the linear (pentahydroxyalkyl)phosphonate esters and an n.m.r. spectroscopic examination of the penta-*O*-acetyl derivatives. A successful approach to the preparation of the cyclic carbohydrate analogues is indicated in Scheme 16, in which the Abramov reaction between the aldehyde (157) and trimethyl phosphite furnished the diastereoisomeric adducts (158)(R^1, R^2 = H or OH) in ca. 1:1 ratio. Basic methanolysis of the mixture led to (159) as a mixture of four diastereoisomeric 1,2-oxaphosphorinanes; NaI-demethylation of this mixture provided only two products, namely the cyclic phosphonic acids as their sodium salts (160). A partial separation of the compounds (159) was achieved by reversed-phase HPLC. A combination of hydrogenolytic debenzylation and ester hydrolysis experiments coupled with a detailed ^1H n.m.r. spectroscopic examination of the various products allowed structural assignments to be made.[173] Abramov reactions have also been carried out on keto nucleosides.[174]

The synthesis of dialkyl (1,2-epoxyalkyl)phosphonates from dialkyl hydrogen phosphonates and α-chlorocarbonyl compounds is possible under phase-transfer conditions.[175] Phosphorylated oxiranes have also been obtained from trialkyl phosphites and silylated α,α-dichloroketones.[176] The epoxidation of tetraalkyl ethylidene-1,1-bisphosphonates to give tetraalkyl 2,2-oxiranylbisphosphates is readily accomplished by the action of alkaline 30% H_2O_2.[177]

2.1.6. *Oxoalkyl Acids*.-Surprisingly little activity in this area has been apparent during the

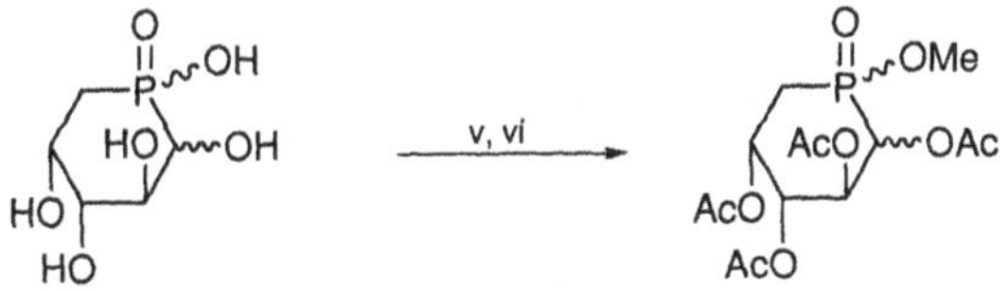

(145) $\xrightarrow{\text{v, vi}}$ (146)

Reagents: i, H_2, $Pd(OH)_2$; ii, $(MeOC_2H_4O)_2AlH_2Na$;
iii, H_3O^+ ; iv, H_2O_2 ; v, Ac_2O, py ; vi, CH_2N_2

Scheme 15

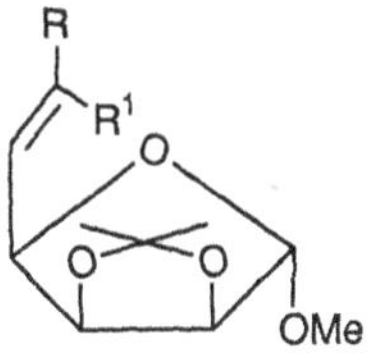

(147)

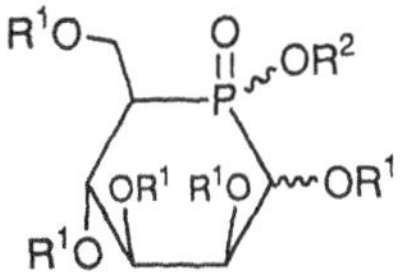

(148) $R^1 = R^2 = H$
(149) $R^1 = Ac$, $R^2 = Me$

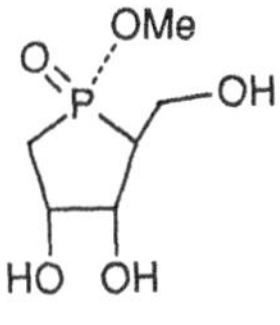

(150) 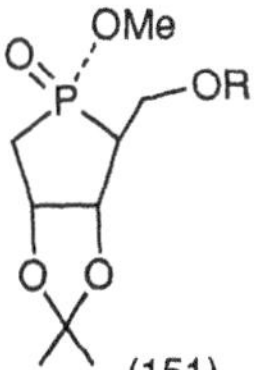 (151) 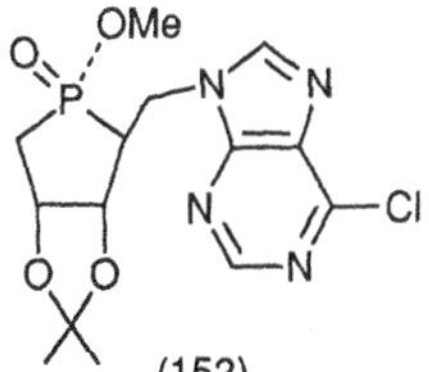 (152)

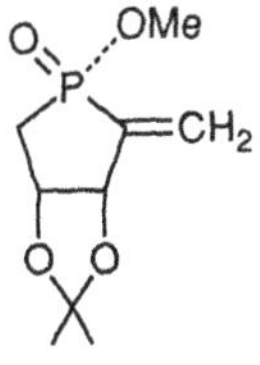

(153) 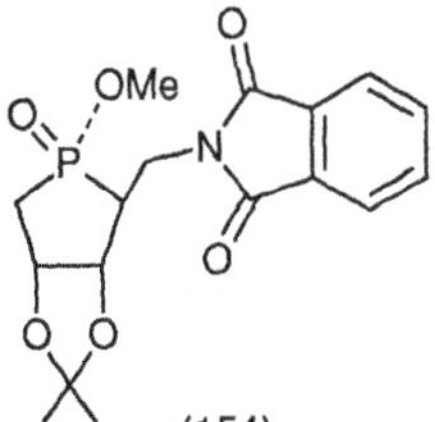 (154) 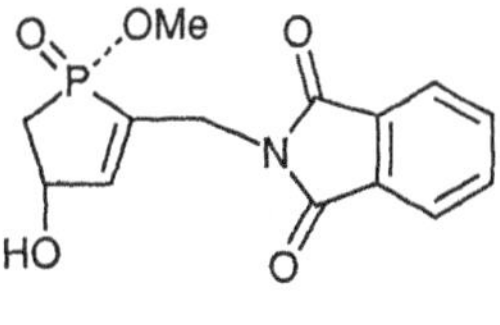 (155)

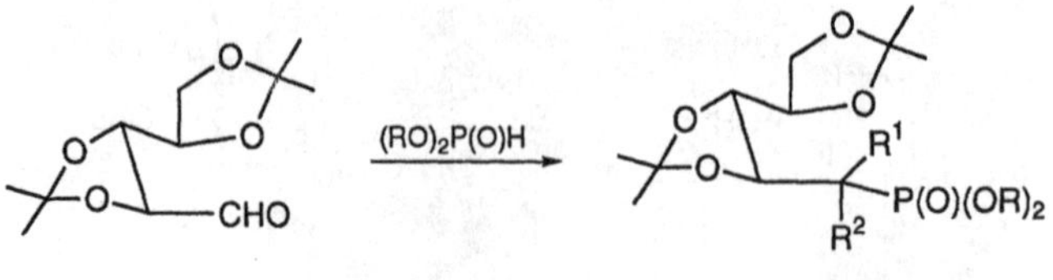

CHO

(RO)$_2$P(O)H

R^1 P(O)(OR)$_2$ R^2

(156)

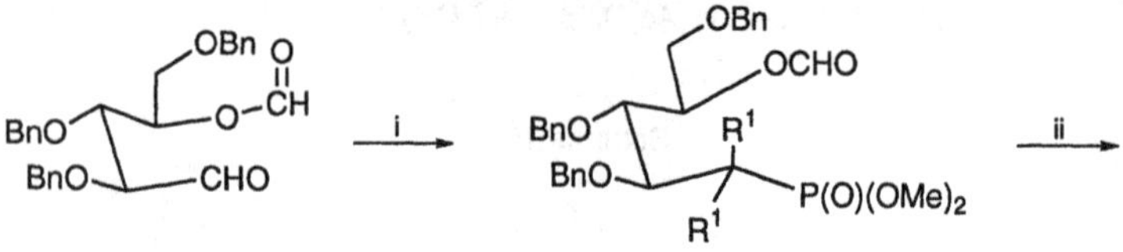

OBn O O–CH BnO BnO CHO

(157)

i

OBn OCHO BnO BnO R^1 P(O)(OMe)$_2$ R^1

(158)

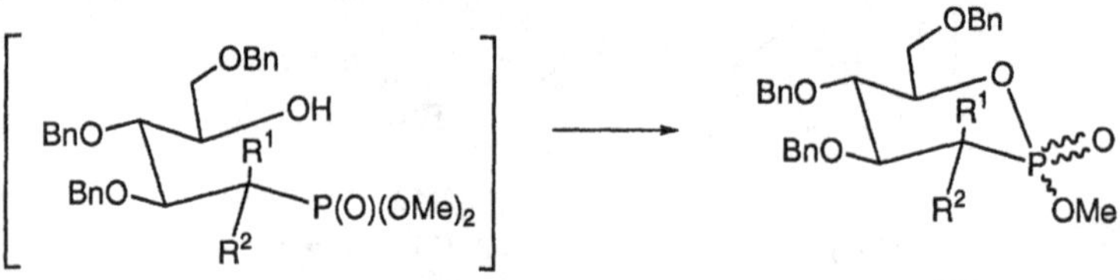

ii

OBn OH BnO BnO R^1 P(O)(OMe)$_2$ R^2

OBn BnO BnO O R^1 P O R^2 OMe

(159)

OBn BnO BnO O R^1 P–O R^2 O – M$^+$

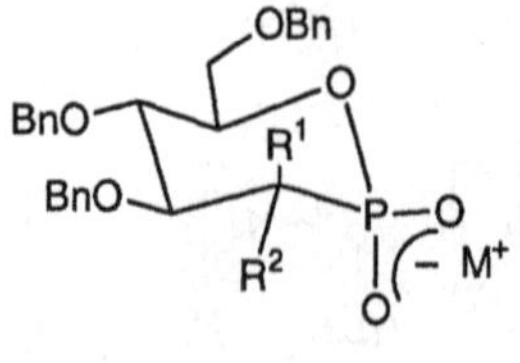

(160)

Reagents: i, P(OMe)$_3$, HOAc ; ii, MeO$^-$, MeOH

Scheme 16

year. The predictably straightforward reaction between trialkyl phosphites and 1-aryloxyalkanoic chlorides has been shown not to be so; rather than (161), the products have been shown to be the esters (162) and dialkyl (chloroacetyl)phosphonate.[178] The course of the reaction between 4,4-dichloro-4*H*-benzopyran-2-carbonyl chloride (163) and a trialkyl phosphite is complex, it evidently includes the initial formation of the expected phosphonic diesters (164); thereafter, the reaction is completed with the formation of the phosphate-phosphonate tetra-esters (165),[179] the structure of the tetramethyl ester being confirmed by X-ray analysis.[180] The suggested mechanism for the formation of the esters (165) parallels that proposed for the origin of similar compounds from other heterocyclic systems (see *Organophosphorus Chemistry*, 1994, **25**, 111) and involves a combination of Michaelis-Arbuzov and Perkow steps. A 92:8 *trans*:*cis* mixture of the phosphorus(III) ester amides (113)(Z = OMe) reacts with propanoyl chloride to give a mixture of the phosphonates (166) of similar composition and with retention of configuration at phosphorus.[181]

In the reaction between the (2-oxopropyl)phosphonic diesters and the aminals (167) in the presence of monochloroacetic or trifluoroacetic acid, the Wittig-Horner condensation is entirely suppressed and the (acetylstyryl)phosphonic diesters (168) can be obtained in good to excellent yields. The use of acetic acid as the catalyst led to poor results. Variations in the yields of products are to be found associated with the nature of group R^1; NO_2, CN and CF_3 groups favour the reaction, but reduced yields were found for 2-NO_2 and 2- and 4-MeO substituents. The products (168) consist of *(E)*:*(Z)* mixtures of isomers, this ratio being commonly 80-90:20-10, and only rarely is the (*Z*) isomer in excess.[182]

The *cis* and *trans* forms of the enols from the (diethoxyphosphinoyl)acetaldehydes $(EtO)_2P(O)CHRCHO$ (R = Me or Ph) have been isolated as pure lithium complexes.[183] Although the mono- and di-chloro derivatives of (diethoxyphosphinoyl)acetaldehyde have been obtained, in the past, by a direct chlorination step, an alternative route has now been proposed which involves, as the initial step, the chlorination of a dialkyl (2-ethoxyethenyl)phosphonate in an aqueous medium; this results in the formation of the aldehyde hydrate (169). Yet a further variation consists in the chlorination of a dialkyl (2,2-diethoxyethyl)phosphonate followed by hydrolysis of the resultant dichlorodiethylacetal.[184] (Diethoxyphosphinoyl)methanal was also prepared as its hydrate by the reaction between diethyl (diazomethyl)phosphonate and dimethyldioxirane; it acts as a formylating agent for secondary amines, and also reacts with a primary amine with the formation of a *C*-phosphorylated imine.[185] Diethyl (3-oxo-1-butenyl)phosphonate and

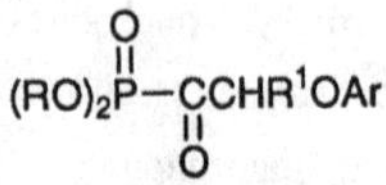

$(RO)_2P(O)-C(O)CHR^1OAr$

(161)

$ArOCH_2COOAr$

(162)

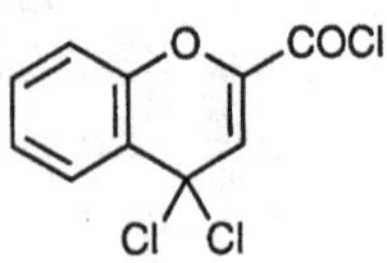

COCl

Cl Cl

(163)

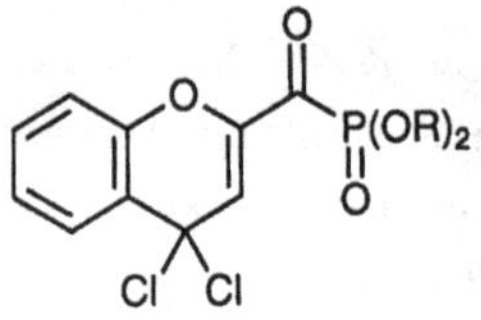

$P(OR)_2$

Cl Cl

(164)

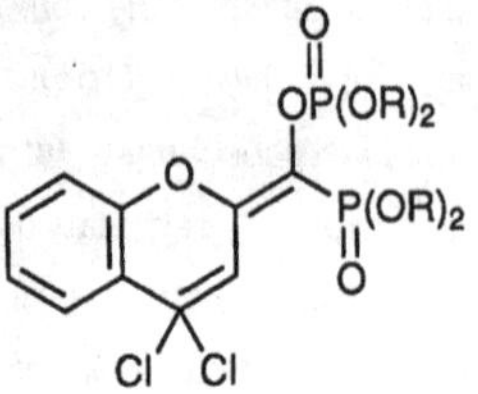

$OP(OR)_2$

$P(OR)_2$

Cl Cl

(165)

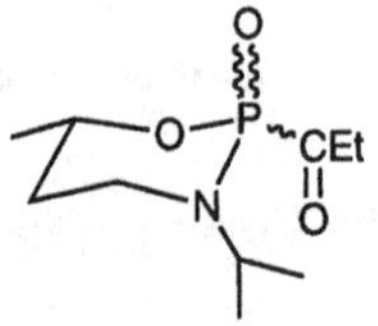

CEt

(166)

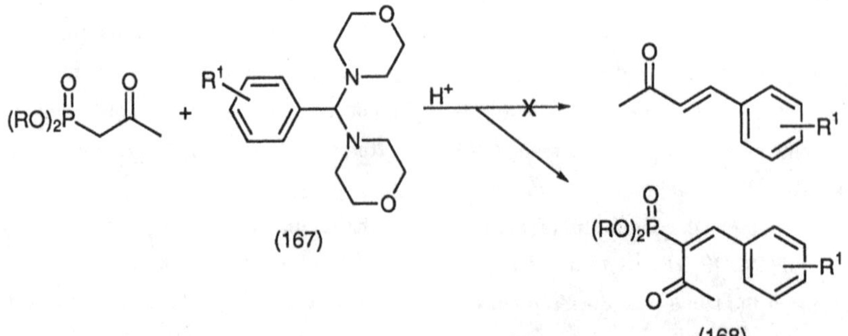

$(RO)_2P$

R^1

H^+

R^1

(167)

$(RO)_2P$

R^1

(168)

$(RO)_2P$

OEt

Cl_2/H_2O

$(RO)_2P$

OH

OH

Cl Cl

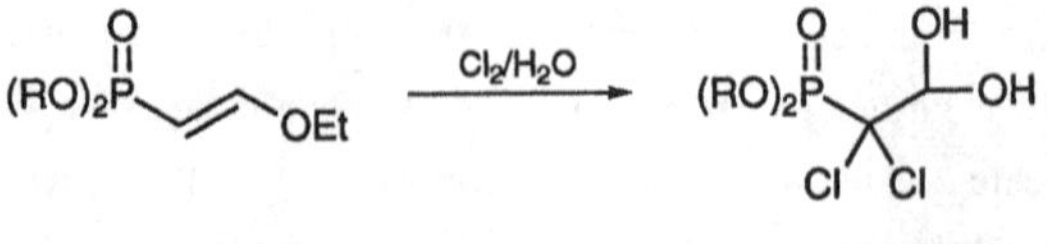

(169)

derived acetals have been synthesized through Horner-Wittig reactions.[186]

A highly enantioselective synthesis of (3-hydroxy-2-oxoalkyl)phosphonic diesters (173) has been designed, in which the important step consists in a reaction between the aldehyde (170) and a dialkyl hydrogen phosphonate, with the last (practical) step comprising ring opening with tetrabutylammonium fluoride.[187] A new route to β-oxoalkyl phosphonic esters (Scheme 17) involves the interaction of a trialkyl phosphite with a nitroalkene in the presence of $TiCl_4$ in CH_2Cl_2; the intermediate complex is then treated with mCPBA when yields of the desired compounds (174)(R^1 = aryl, R^2 = Me or Et; R^1 = H, R^2 = Ph) lie in the range 72-93%.[188]

2.1.7. *Nitroalkyl and Aminoalkyl Acids*.- The chemistry of nitro aliphatic organophosphorus compounds has been reviewed,[189] as has that of phosphorus-containing analogues of amino acids.[190]

The year's literature has provided more examples of the synthesis of derivatives of (1-aminoalkyl)phosphonic acids by the standard procedures, including the addition of hydrogen phosphonates to imines.[191-195] Amongst the results presented are the use of the imines (175)(R^2 = $CHPh_2$ or CPh_3) which allow hydrogenolytic or acidolytic removal of *N*-protection in the final products;[192,196,197] and the observation that the addition of diethyl hydrogen phosphonate to the diimine (176) yields a single diastereoisomeric product displaying only one signal in its ^{31}P n.m.r. spectrum, with implications for its manner of formation.[195]

Dialkyl [(diethylamino)alkyl] phosphonates have been prepared by the Kabachnik-Medved'-Fields reaction[198] and further discussion has centred on the nature of unexpected products observed as a result of that reaction. It has been known for many years that aromatic ketones such as benzophenone and fluorenone do not participate in this reaction, or at best, do so very reluctantly; in mixtures of these or similar ketones, primary amine, and dialkyl phosphonate, the formation of α-hydroxy phosphonate takes place 100x faster than fission of the latter in the reverse reaction; meanwhile, the α-hydroxy phosphonate rearranges into isomeric phosphate 6x faster than that reverse reaction. Thus, in such cases, the main reaction product is a phosphate ester.[199] The Kabachnik-Medved'-Fields procedure was employed in the first step of sequences which led, ultimately, to derivatives of 3-amino-1,2-oxaphospholane and 3-amino-1,2-oxaphosphorinane, albeit obtainable in only impure form (Scheme 18);[200] the ring opening of one of these yielded an ester of (1-amino-3-hydroxy-1-methylpropyl)-phosphonic acid. Further examples have also been presented of the Oleksyszyn

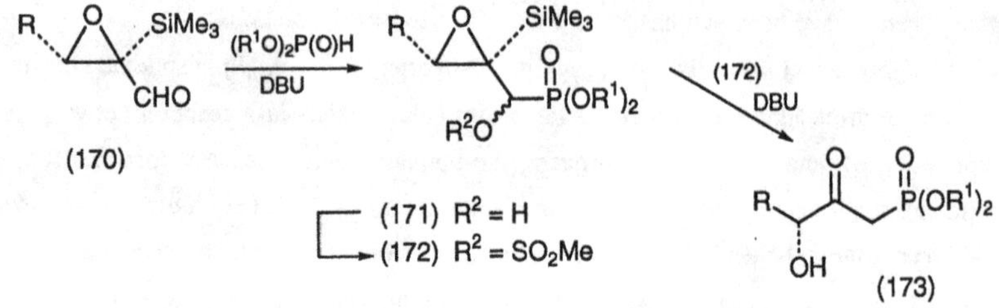

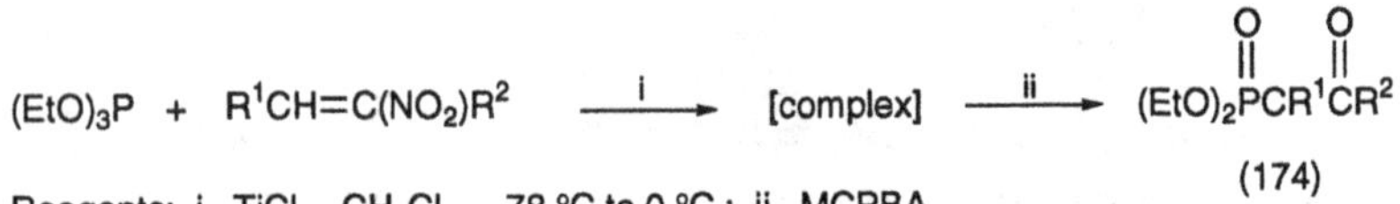

(170) $\xrightarrow[\text{DBU}]{(R^1O)_2P(O)H}$ (171) $R^2 = H$ → (172) $R^2 = SO_2Me$ $\xrightarrow[\text{DBU}]{(172)}$ (173)

$(EtO)_3P + R^1CH{=}C(NO_2)R^2 \xrightarrow{i} [complex] \xrightarrow{ii} (EtO)_2P(O)CR^1C(O)R^2$ (174)

Reagents: i, $TiCl_4$, CH_2Cl_2, –78 °C to 0 °C ; ii, MCPBA

Scheme 17

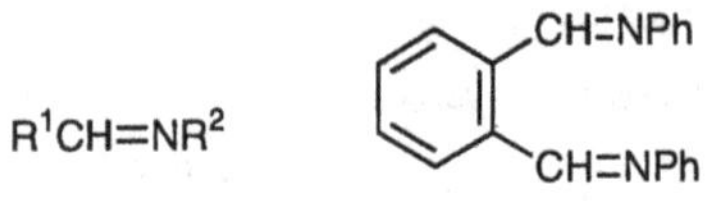

$R^1CH{=}NR^2$ (175)

(176)

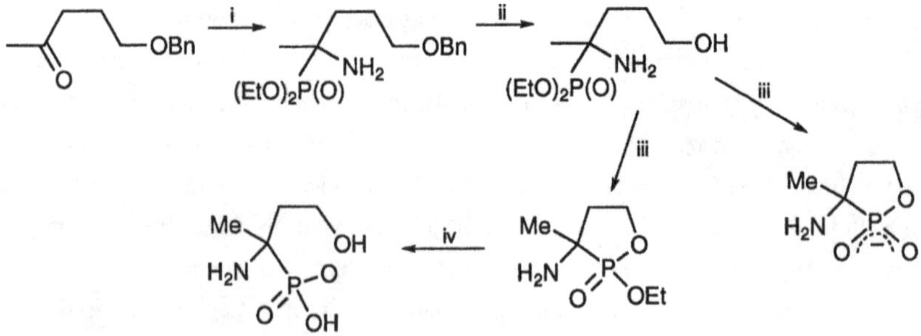

Reagents: i, $(EtO)_2P(O)H$, NH_3, 60 °C; ii, H_2, Pd/C, EtOH, HCl; iii, NaH, DMF; iv, EtOH

Scheme 18

reaction[201-203] and of the use of the Gabriel reaction in the preparation of (ω-aminoalkyl)phosphonic acids.[202]

Tetraethyl (aminomethylene)bisphosphonate has been prepared by the amination of methylenebisphosphonic ester carbanion with *O*-diphenylphosphinoyl hydroxylamine[204] and the *S*-oxides and dioxides of phosphonocysteine, its homologues, and *S*-substituted derivatives have been obtained by H_2O_2 oxidations of the corresponding sulphides in dimethylsulphoxide.[205] A series of acids (177)(X = O, S, or NH) has been obtained through standard procedures exemplified in Scheme 19.[206] Phosphinopeptides have been derived from (aminomethyl)methylphosphinic acid through the *N*-protected acid chloride.[207]

As in the case of the α-hydroxyalkyl phosphonic acids, much effort continues to be expended on the design of reactions with a potential for asymmetric synthesis. Mention should be made of the use of the imine (178); here, an increase in the size of the alkyl group R^2 leads to an increase in the (*R*)(at the centre C*) content of the product (179).[194] The two diastereoisomeric phosphonic amides (180a,b) are alkylated (BuLi, RX) to (181a,b) and the separated products made to react with potassium phthalimide to give (182), and these are then worked up in the usual way.[208] Chinese workers employed the diastereoisomeric 1,3,2-diazaphospholidines (183), obtained conventionally from (*S*)-2-anilinomethylpyrrolidine; here, however, the separated compounds were alkylated, and the products acted upon by NaN_3 in DMF to give the diastereoisomeric azides (184); the conversion of these into the amino compounds (186) was achieved by an initial reaction with Ph_3P, to give the respective (185), followed by acid hydrolysis. Finally, the diazaphospholidine chiral template was removed with aq. HCl.[209] A complete lack of diastereoselectivity was observed in additions to the salt (187) to give (188). However, a modest diastereoselectivity was reported for the addition of ethyl methylphosphinate to the quinonemethide (189); the two diastereoisomeric products (190) and (191)(Ar = 4-hydroxyphenyl) were obtained in ca. 1:2 ratio, the minor isomer being isolable in pure form.[210] Dialkyl (1-aminoalkyl)phosphonates are formed by the reduction of the oximes of (1-oxoalkyl)phosphonates with $LiBH_4/Me_3SiCl$ in THF[211] and also through the interaction of a dialkyl alkylphosphonate carbanion with BocN=NBoc to give an adduct which is hydrogenolytically cleaved at the N-N bond.[212]

A further application of the so-called Schoellkopf synthesis has been reported. This employs the lithiated bislactim ether (192) of known chirality (*R*); this was condensed (Scheme 20) with 4-benzyloxy-3-(dimethoxyphosphinoyl)benzaldehyde to form a second centre of chirality (*S*) in the intermediate (193); further work-up according to the steps in

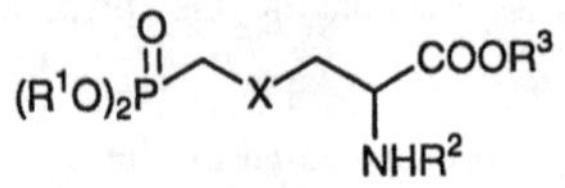

(177)

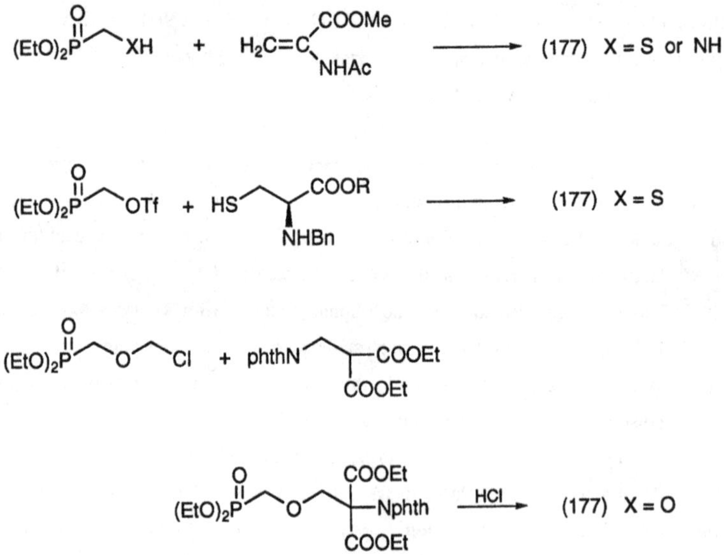

$(EtO)_2P(O)CH_2XH + H_2C{=}C(COOMe)NHAc \rightarrow$ (177) X = S or NH

$(EtO)_2P(O)CH_2OTf$ + HS–CH$_2$–CH(COOR)NHBn → (177) X = S

$(EtO)_2P(O)CH_2OCH_2Cl$ + phthN–CH$_2$–CH(COOEt)$_2$

$(EtO)_2P(O)CH_2OCH_2C(COOEt)_2$Nphth $\xrightarrow{HCl}$ (177) X = O

Scheme 19

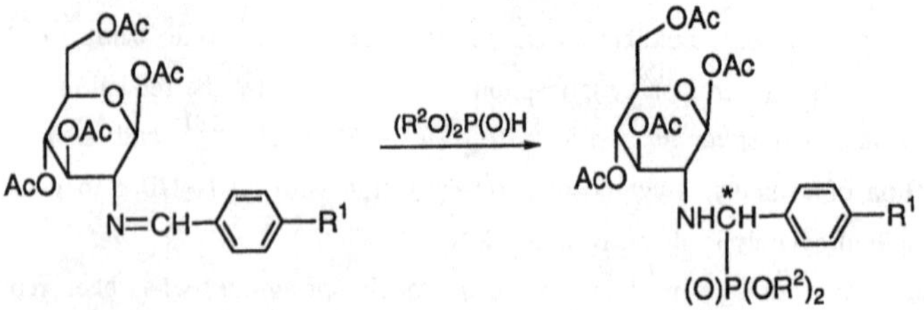

$(R^2O)_2P(O)H$

(178) (179)

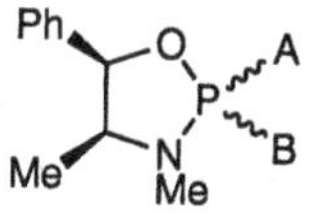

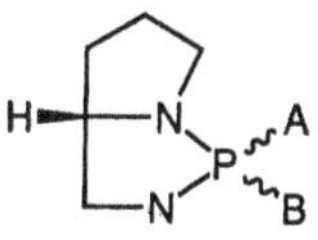

(180) (a) A = =O, B = CH_2Cl
(b) A = CH_2Cl, B = =O
(181) A, B = CHRCl, =O
(182) A, B = CHRphth, =O

(183) A, B = CH_2Cl, =O
(184) A, B = $CHRN_3$, =O
(185) A, B = $CHRN{=}PPh_3$, =O
(186) A, B = $CHRNH_2$, =O

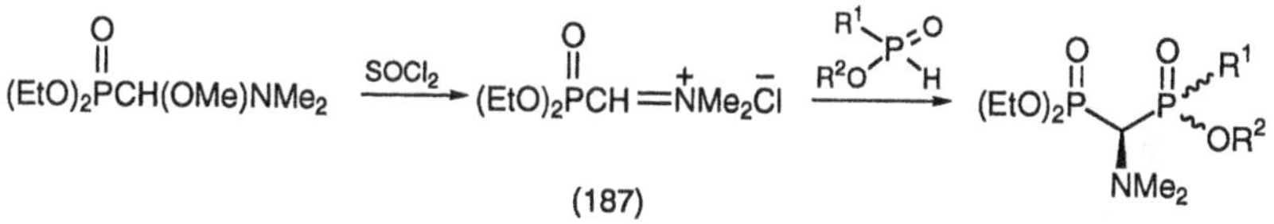

$(EtO)_2P(O)CH(OMe)NMe_2 \xrightarrow{SOCl_2} (EtO)_2P(O)CH{=}\overset{+}{N}Me_2\overset{-}{Cl}$ (187) → (188)

(187)

(188)

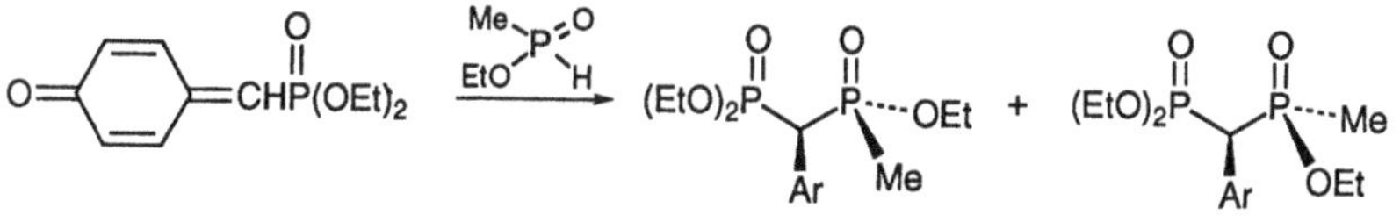

(189)

(190)

(191)

Scheme 20 yielded the target 3'-phosphono-L-tyrosine as its trimethyl ester benzyl ether (194). The stereoisomeric composition of the product was determined by the generation of the diastereoisomeric peptides (195), the h.p.l.c. of which indicated the (*S,S*)/(*R,S*) ratio as 87.4:12.6.[125]

In order to obtain examples of aminohydroxyalkyl phosphonic acids diastereoisomerically enriched at the hydroxy site, the aldehyde (196) (from L-phenylalanine) was subjected to reactions with phosphorus(III) nucleophiles (Scheme 21); with diethyl *tert*-butyldimethylsilylphosphite it gave the adducts (197) and (198) (X = Bn, Y = $SiMe_2Bu^t$), convertible conventionally into (197) and (198)(X = Bn, Y = H), and thence to (197) and (198)(X = Y = H). The diastereoisomeric composition of the mixture was determined through a conversion into the oxazolidin-2-ones (199). In this way it was found that the ratio of (197)/(198)(X = Y = H) was >98:<2, a degree of selectivity which was not approached when the nucleophile was either triethyl phosphite or diethyl hydrogen phosphonate.[213]

Oximes of dialkyl (1-oxoalkyl)phosphonates are reduced to dialkyl (1-hydroxyaminoalkyl)phosphonates with borane-pyridine,[214] and the oxime *O*-benzyl ethers are converted into dialkyl [(benzyloxyamino)alkyl]phosphonates through the use of Et_3SnH and trifluoroacetic acid.[215] [(Benzyloxyamino)methyl]phosphonic acid has been obtained by the interaction of PCl_3 and formaldoxime *O*-benzyl ether in acetic acid,[216] while reduction of a nitro group at a carbon atom β to the hydroxyamino function is brought about with Al/Hg or $SnCl_2$/HCl.[217] The alkylation of nitrones yields oximinium salts; these, for example, (200) from ArCHO, add dialkyl hydrogen phosphonates or alkyl phosphinates to give the alkoxyaminoalkyl compounds, (201) which are hydrogenolysed under acidic conditions to give the *N*-substituted (α-methylaminobenzyl)-phosphonic or -phosphinic derivative (202).[218,219]

The use of imines (175)(R^2 = CPh_3) in conjunction with bis(trimethylsilyl) phosphonite at room temperature, afforded adducts which, on methanolysis (to desilylate) in M HCl (to remove *N*-protection) yielded the phosphinic acids (203).[197] The pyrroline trimer (204) with the same phosphonite ester could be made to furnish 2-pyrrolidinephosphinic acid (205), a procedure which also provided the higher ring homologues - 2-piperidine- and 2-perhydroazepine-phosphinic acids.[220]

Novel results were observed in the attempted syntheses of some branched chain aminoalkyl phosphonic acids by the use of alcohol tosylates in the Michaelis-Becker reaction. In the reaction of (206)(R^1 = *p*-OTos; R^2 = Me) with sodium diethyl phosphite, the product was not the expected (207), but rather the phosphonic diester (208). Similar

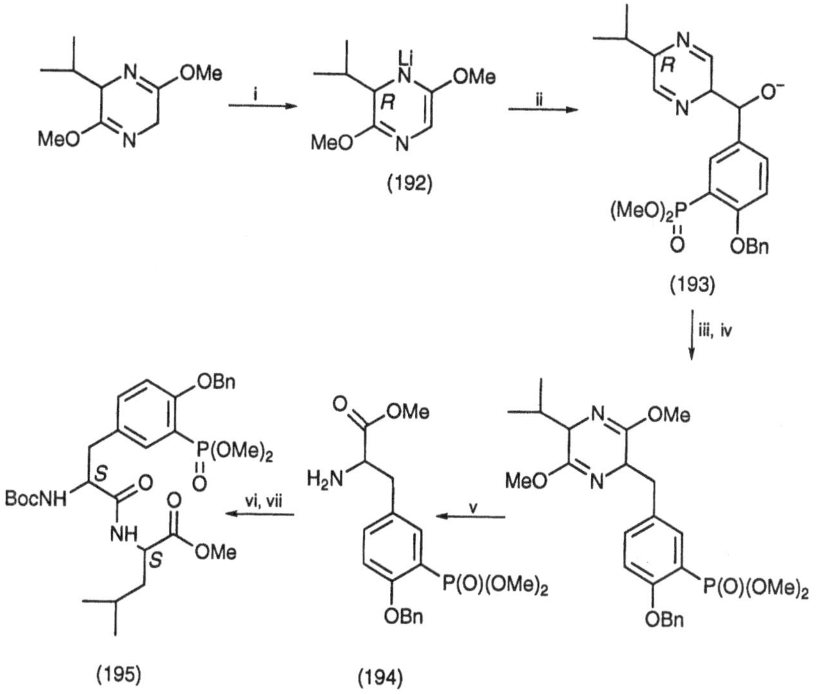

(192)

(193)

(195)

(194)

Reagents: i, BuLi ; ii, 4-benzyloxy-3-(dimethoxyphosphinoyl)benzaldehyde; iii, PhOCSCl; iv, Bu_3SnH, AIBN, toluene ; v, 0.25 M HCl ; vi, Boc_2O, Et_3N ; vii, Leu-OMe, DCC, BuOH, Et_3N

Scheme 20

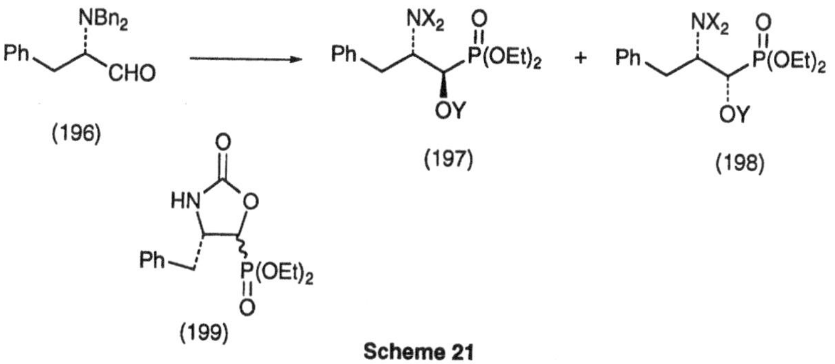

(196)

(197)

(198)

(199)

Scheme 21

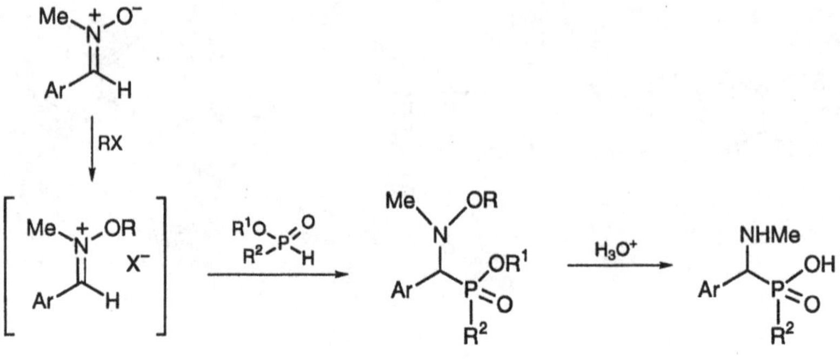

Me $\overset{+}{N}$ O⁻ Ar H

RX

Me $\overset{+}{N}$ OR X^- Ar H

R^1O P=O R^2 H

Me N OR Ar OR^1 P=O R^2

H_3O^+

NHMe Ar OH P=O R^2

(200) (201) (202)

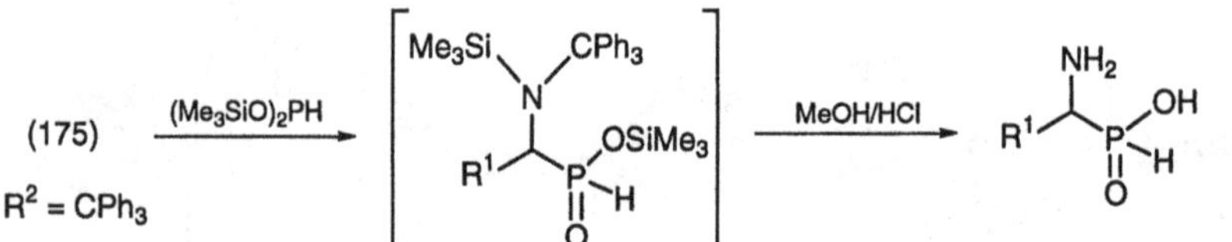

(175) $R^2 = CPh_3$

$(Me_3SiO)_2PH$

Me_3Si N CPh_3 R^1 P $OSiMe_3$ H O

MeOH/HCl

NH_2 R^1 P OH H O

(203)

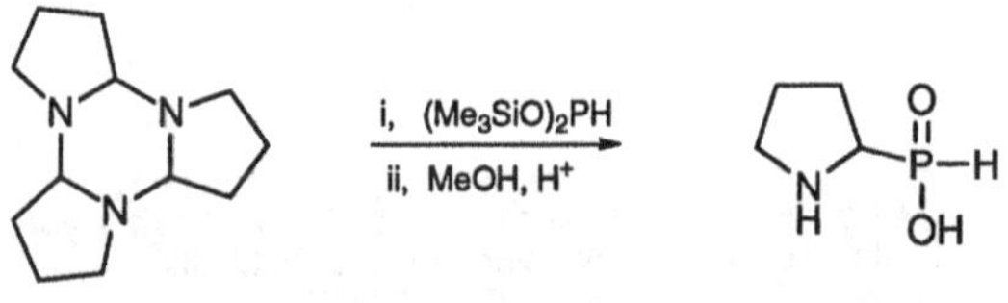

i, $(Me_3SiO)_2PH$
ii, MeOH, H^+

(204) (205)

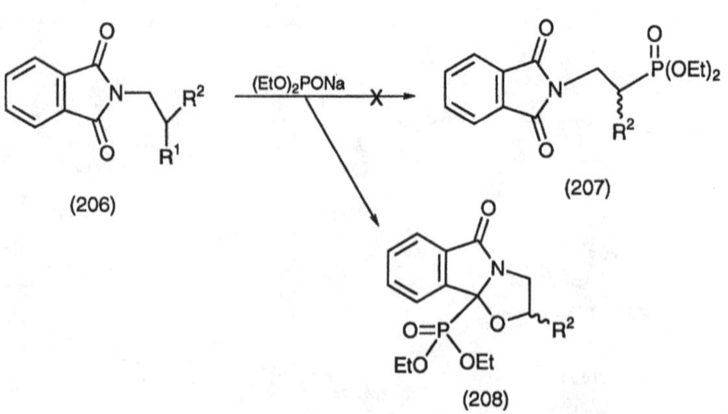

$(EtO)_2PONa$

O N R^2 O R^1

O $P(OEt)_2$ O N R^2 O

(206) (207)

O N O R^2 O=P EtO OEt

(208)

results were obtained with *N*-(2-bromoethyl)phthalimide and the tosylate in their reactions with butyl phenylsodiophosphinate.[221]

2.1.8. *Sulphur- and selenium -containing compounds*.-A selection of phosphonothioic and phosphinothioic esters have been prepared by the direct alkylation of $R^1R^2P(S)H$ (R^1 = R^2 = EtO, BuO; R^1 = Me, R^2 = alkoxy) under phase transfer conditions.[222] *O,O*-Dialkyl phenylphosphonothioates react with $POCl_3$ at 50-90° to give the phosphonothioic chlorides, Ph(RS)P(O)Cl, from which other esters and amides were obtained.[223,224] A convenient preparation of the chloride-esters PhP(S)(OAr)Cl consists in the reactions between $PhP(S)Cl_2$ and dicyclohexylammonium salts of phenols.[225]

A new and quite remarkable example of a dithioxophosphorane has come to light. By means of conventional steps, 2,4-di-*tert*-butyl-6-methoxyphenylphosphonous dichloride has been converted into the phosphorane (209), which is stable for several days in a nonpolar solvent but then dimerizes to a mixture of the *trans* (210) and *cis* forms of the dimer, 2,4-bis(2,4-di-*tert*-butyl-6-methoxyphenyl)-1,3,2,4-dithiadiphosphetane 2,4-disulphide; these revert to the monomer even in boiling benzene. When (209) is acted upon with benzophenone, thiobenzophenone (63%) is isolable; evidently a coproduct from this step then reacts with the dithioxophosphorane to give the 1,3,2,4-oxathiadiphosphetane (212) and so presumably has the mixed phosphorane structure (211). In a reaction carried out without the isolation of (209), the other isolable products are reported to be the *cis* and *trans* isomers of the system (213) (Ar = 2,4-di-*tert*-butyl-6-methoxyphenyl).[226] Other 2,4-diaryl-1,3,2,4-dithiadiphosphetane 2,4-disulphides (Aryl = 2,4,6-trisubstituted-phenyl) have been prepared and, along with certain monomeric dithioxophosphoranes, characterized by X-ray crystallography.[227] The equilibria between monomers and dimers was studied by ^{13}C and ^{31}P n.m.r. spectroscopy.[227] Cyclic esters of 4-methoxyphenylphosphonothioic acid have been prepared by reactions between dihydric phenols and Lawesson's reagent.[228] Other uses of Lawesson's reagent in synthesis consist in its reactions with dialkyl disulphides to give the polythiophosphonic esters (214)(Ar = 4-methoxyphenyl, R = Et or Bu^t)[229] and with organosilyl or -stannyl reagents to give (215) and (216) (Ar = 4-methoxyphenyl).[230]

It is already known that, in the 2,4-di-*tert*-butyl-6-dimethylaminophenyl series, the action of selenium or sulphur on the diphosphene (217) produces the monomeric diselenoxo- or dithioxo-phosphoranes (see *Organophosphorus Chemistry*, 1994, **25**, 111, reference 228). The diselenoxophosphorane (218) loses selenium when treated with tris(dimethylamino)phosphine to give the phosphinoselenoylidene (221). It has now been

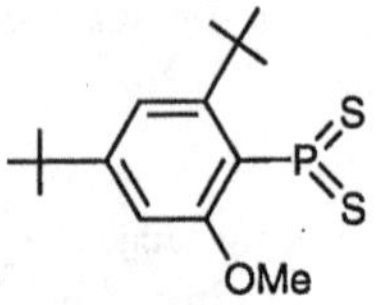

(209) 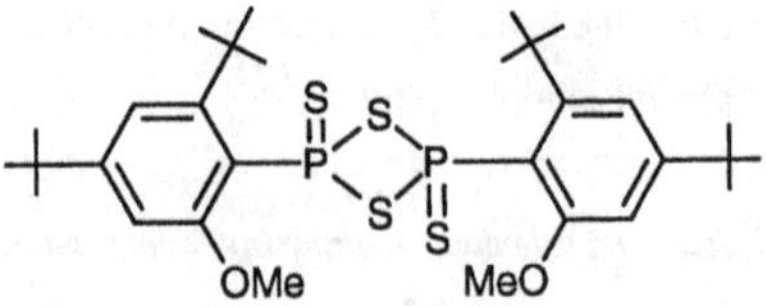 (210)

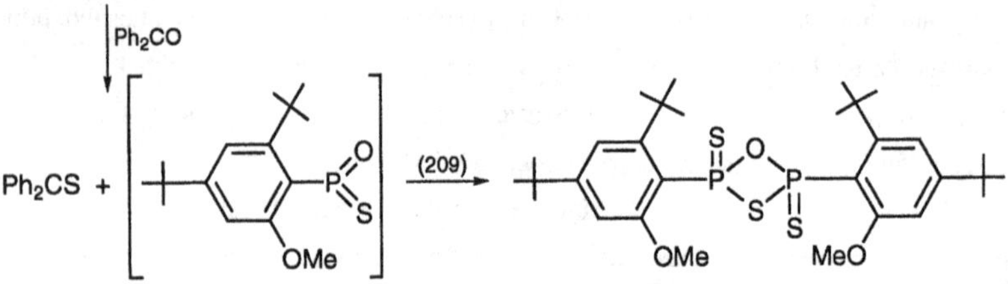

Ph_2CO

Ph_2CS + (211) —(209)→ (212)

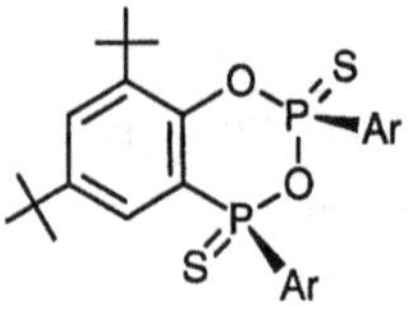

(213)

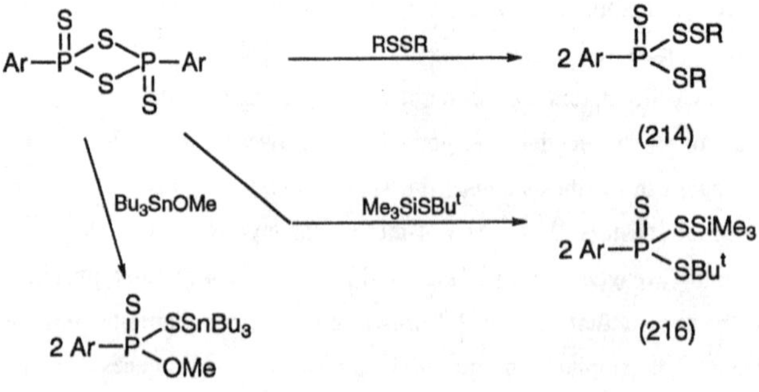

RSSR (214)

Bu_3SnOMe (215)

Me_3SiSBu^t (216)

shown that a similar desulphuration of the dithioxophosphorane (219) does not occur. Nevertheless, the action of sulphur on (218) generates the mixed phosphorane, (220) - not formed in a reaction between (219) and selenium. The selenoxothioxophosphorane with tris(dimethylamino)phosphine produces (222), thus far characterized spectroscopically.[231]

Further results have appeared in relation to the 1,2,4-thiadiphosphetane system (see *Organophosphorus Chemistry*, 1993, **24**, 106, reference 214). The addition of sulphur to 1,3-diaryl-1,3-diphosphaallene (aryl = 2,4,6-tri-*tert*-butylphenyl) in a water-toluene system containing DBU, not only yields the stereoisomers of 2,4-diaryl-1,2,4-thiadiphosphetane 2,4-disulphide, but also the same derivative of the analogous 1,2,4-oxadiphosphetane system, the *cis* geometry of which was confirmed by X-ray crystallography.[232]

2.1.9. *Phosphorus-nitrogen bonded compounds*.-Little activity has been noted in this area.

N-Diphenylphosphinothioylhydroxylamine has been prepared *via* its *O*-trimethylsilyl derivative; when treated with NaOMe/MeOH, its *O*-benzoyl derivativative rearranges to a product tentatively identified by spectroscopic means as (223), and which bears comparison with material obtained as indicated from (224).[233]

Diastereoisomeric derivatives of the dihydrobenzazaphosphole system have been recorded; the synthesis of a mixture of the (S_P,R_C)- (illustrated) and (R_P,R_C) forms of the phosphinic amide (225), starting with the silylamine (226), has been noted; the structure of the former isomer has been determined by X-ray crystallography.[234]

Examples of organic derivatives of the new and unusual inorganic ring system (227)(R = Bu^t or Ph) have been obtained in reactions between the phenylphosphonic diamides, $PhP(O)(NHR)_2$ and S_2Cl_2, and the products thoroughly characterized by methods which included X-ray analysis.[235]

Reactions between the bishydrazides, $RP(Z)(NMeNH_2)_2$ (Z = O or S; R = Ph or N_3) and various dialdehydes of the general type OHC-Y-CHO where Y = 1,3-phenylene or $N_3(X)P(OC_6H_4\text{-})_2$ (X = l.p. or S), yield products with 16-membered rings which, through further reactions with the diphosphines, $Ph_2P\text{-}A\text{-}PPh_2$, where A = $(CH_2)_6$ or $(CH_2)_2\text{-}B\text{-}(CH_2)_2\text{-}B\text{-}(CH_2)_2$ (B = O or S), yield bridged structures which possess 29 or 31-membered rings connected through moieties of type (228).[236]

The crystal structure and absolute configuration of one diastereoisomer of the methylphosphonic diamide (229) have been determined.[237]

2.1.10. *Compounds of potentional pharmacological interest*.-Work referred to in last year's report on the synthesis of the enantiomers of the ether linked antitumour agent

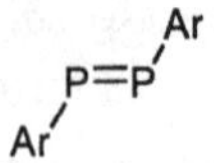

Ar–P=P–Ar

(217)

Ar–P(=X)(=Y)

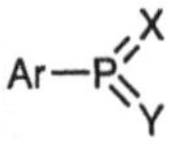

(218) X = Y = Se
(219) X = Y = S
(220) X = Se, Y = S

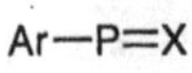

Ar–P=X

(221) X = Se
(222) X = S

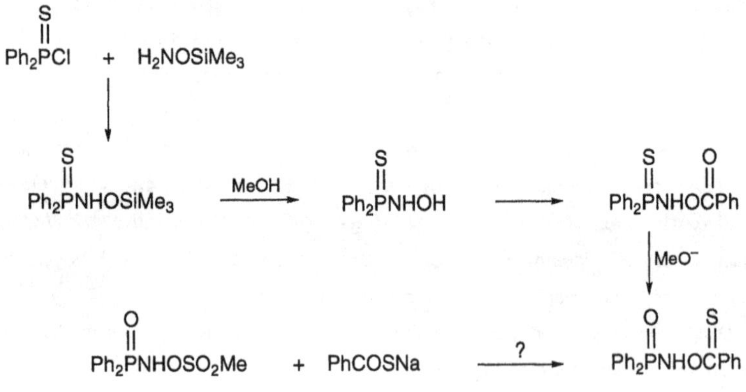

$Ph_2P(S)Cl$ + $H_2NOSiMe_3$ → $Ph_2P(S)NHOSiMe_3$ —MeOH→ $Ph_2P(S)NHOH$ → $Ph_2P(S)NHOC(O)Ph$ —MeO^-→ $Ph_2P(O)NHOC(S)Ph$

$Ph_2P(O)NHOSO_2Me$ + PhCOSNa —?→ $Ph_2P(O)NHOC(S)Ph$

(224) (223)

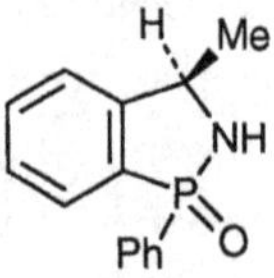

H, Me, NH, P, Ph, O

(225)

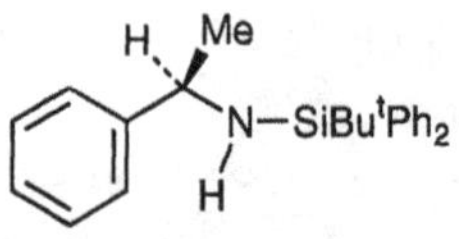

H, Me, N–SiBu^{t}Ph$_2$, H

(226)

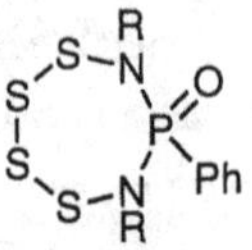

R, S, N, O, S, P, S, Ph, S, N, S, R

(227)

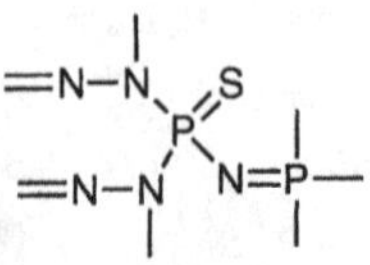

=N–N, S, P, =N–N, N=P–

(228)

$MeP(O)(NHPh)(NHCHMePh)$

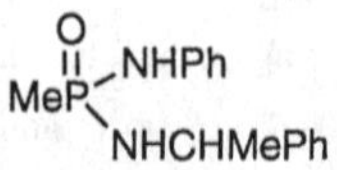

(229)

1-*O*-octadecyl-2-*O*-methyl-*sn*-glycero-3-phosphocholine (230)(X = O) has now been extended to include the 1-*S*-hexadecylthio analogue, the essential skeleton being constructed through the steps given in Scheme 22.[238] A series of *β*-substituted phosphonate diesters (231) and (232) was prepared as potential calcium antagonists through the standard routes indicated briefly in Scheme 23; amongst compounds which showed high activity were (231)(n = 2, R = Me, pentyl) but the aminophosphonic diesters (232) showed much reduced activity and neither series demonstrated the degree of activity shown by belfosdil (233).[239]

A series of novel renin inhibitors based on 2-{[(3-phenylpropyl)phosphoryl]oxy} alkanoic acid moieties (234)(R = Me, Et, Pr or Bu) and (235)(R' = Pr or Bu; R as in 234) were prepared as indicated in Scheme 24; diastereoisomers were separated by chromatography on silica gel. The syntheses of, and biological results for, a series of compounds of the general structure (236)(R and R' with the same significance as before) are also reported.[240] The synthesis of the farnesylamine derivative (237) has been described (Scheme 25) and its squalene synthetase inhibitory activity determined.[241] The 2-oxoalkyl phosphonates (238)(R = Bu^tMe_2Si) have been prepared by a conventional acylation route and their activity in Wittig-Horner reactions examined.[242]

The syntheses of the adenine (239) and guanine (240) derivatives, both with anti-herpes virus activity, have been carried out through Mitsunobu reactions between appropriately protected purines and diethyl (3-hydroxy-2-propynyl)phosphonate; it was noted that a reaction between the chloroalkyne (241) and triethyl phosphite gave the (*Z*)-(2-chloroalkene)phosphonate (242).[243] Compounds having the general structure (243)(B = purine or pyrimidine base) have been prepared by several groups. Such compounds include those with R = CH_2F,[244] CH_2OH,[245] and CH_2R' (R' = OH, N_3, F, Cl, CH_2OH, CH_2N_3, etc).[246] In these syntheses, the introduction of the $CH_2P(O)(OR)_2$ moiety occurred through Michaelis-Arbuzov reactions as, for example, in the conversion of (244) into (245) with $P(OPr^i)_3$, which led to the preparation of (243)(R = CH_2OH), or through the widespread use of esters of the type $RSO_2OCH_2P(O)(OPr^i)_2$ (R = Me or *p*-tolyl), from which the sulphonate grouping may be displaced to link up the phosphonic moiety with the purine or pyrimidine base. Many of the compounds so produced have been evaluated *in vitro* for HIV activity.

9-[(Phosphonoalkyl)benzyl]guanines (246)(typical substituents R being those indicated) were synthesized by a reaction between the substituted benzyl chloride (R' = Pr^i) and 2-amino-6-chloropurine in the presence of Cs_2CO_3 and examined for inhibitory properties against human erythrocyte purine nucleoside phosphorylase. The potency of the

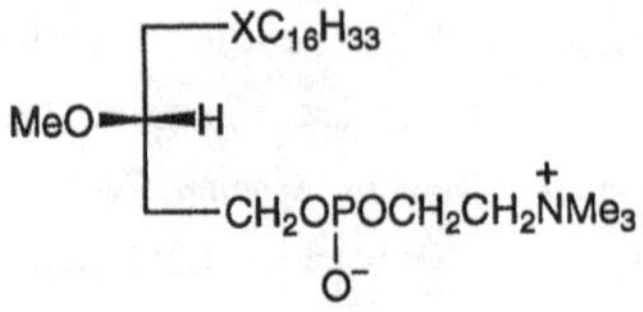

$XC_{16}H_{33}$
MeO–H
$CH_2OPOCH_2CH_2\overset{+}{N}Me_3$
O^-

(230)

$C_{16}H_{33}S$ (epoxide) —i, ii→ MeO–H ($SC_{16}H_{33}$), $P(O)(OMe)_2$ —iii, iv→ (230) (X = S)

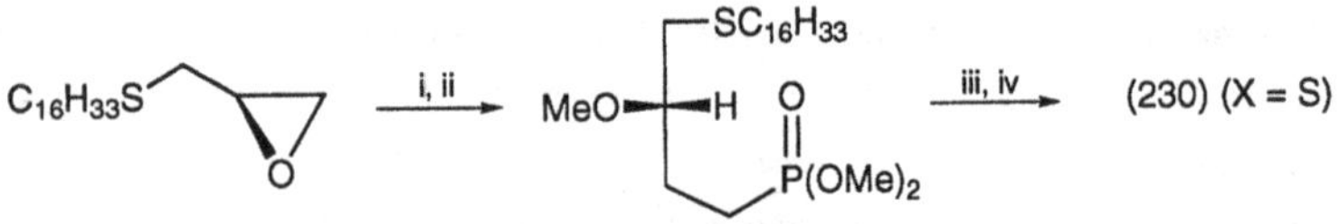

Reagents: i, $LiCH_2P(O)(OMe)_2$, $BF_3.Et_2O$, THF, –78 °C to –20 °C; ii, CH_2N_2, SiO_2, 0 °C; iii, (a) Me_3SiBr, CH_2Cl_2 (b) THF–H_2O; iv, choline tosylate, Cl_3CCN, py, 50 °C

Scheme 22

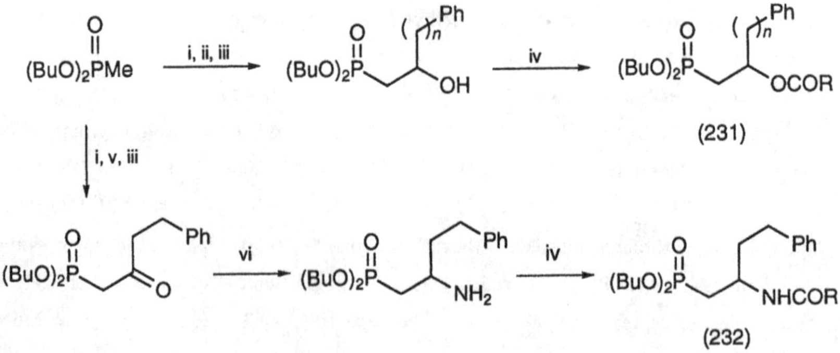

$(BuO)_2P(O)Me$ —i, ii, iii→ $(BuO)_2P(O)$, $(\ )_n$Ph, OH —iv→ $(BuO)_2P(O)$, $(\ )_n$Ph, OCOR

(231)

i, v, iii

$(BuO)_2P(O)$, Ph, O —vi→ $(BuO)_2P(O)$, Ph, NH_2 —iv→ $(BuO)_2P(O)$, Ph, NHCOR

(232)

Reagents: i, BuLi, THF, –78 °C; ii, $Ph(CH_2)_nCHO$; iii, aq. NH_4Cl ; iv, $(RCO)_2O$, R'_3N ; v, $PhCH_2CH_2COOEt$, THF, –78 °C ; vi, $NaBH_4$, NH_4OAc, MeOH

Scheme 23

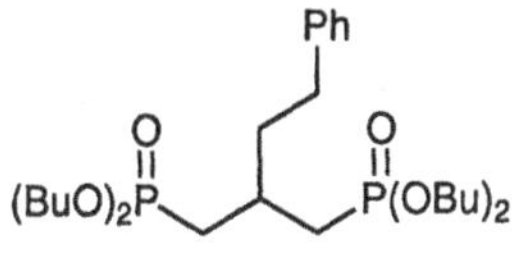

Ph
O O
(BuO)$_2$P P(OBu)$_2$

(233)

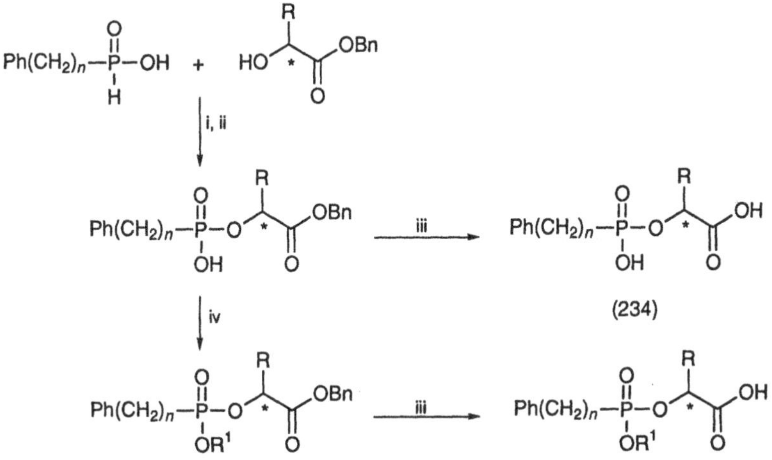

R
O
Ph(CH$_2$)$_n$—P—OH + HO * OBn
H O

i, ii

R
O
Ph(CH$_2$)$_n$—P—O * OBn
OH O

iii

R
O
Ph(CH$_2$)$_n$—P—O * OH
OH O

(234)

iv

R
O
Ph(CH$_2$)$_n$—P—O * OBn
OR[1] O

iii

R
O
Ph(CH$_2$)$_n$—P—O * OH
OR[1] O

(235)

Reagents: i, DCC, 4-dimethylaminopyridine, THF; ii, $NaIO_4$, dioxane;
iii, 1 M LiOH, dioxane; iv, EtI, or BnBr, K_2CO_3, DMF

Scheme 24

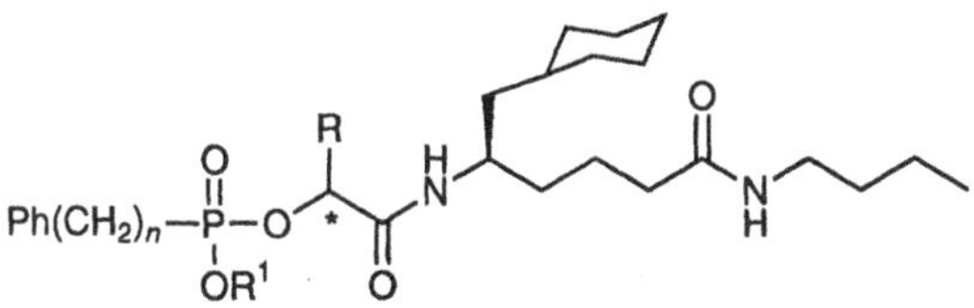

R O
O H
Ph(CH$_2$)$_n$—P—O * N N
OR[1] O H

(236)

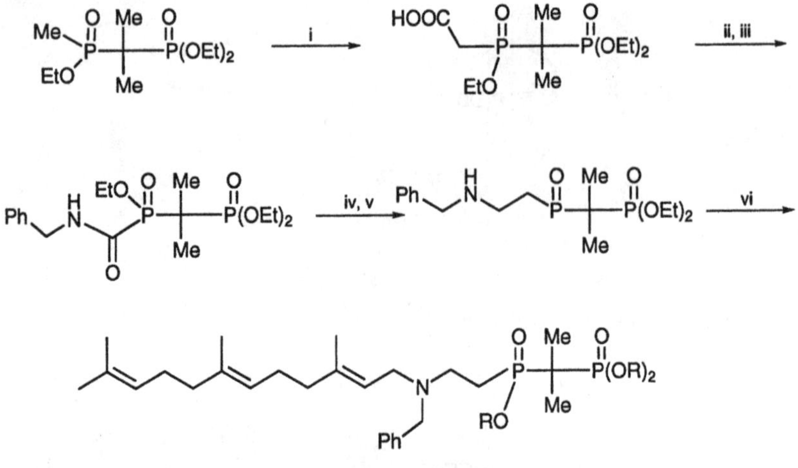

O Me O
Me P P(OEt)$_2$
EtO Me
i
HOOC O Me O
P P(OEt)$_2$
EtO Me
ii, iii

EtO O Me O
Ph H N P P(OEt)$_2$
O Me
iv, v
Ph H N O Me O
P P(OEt)$_2$
Me
vi

O Me O
N P P(OR)$_2$
Ph RO Me

(237)

vii R = Et
viii R = K

Reagents; i, BuLi, CO_2, –78 °C to 25 °C ; ii, $(COCl)_2$, DMF, C_6H_6, 0 °C ; iii, $PhCH_2NH_2$, Et_3N, CH_2Cl_2; iv, $Et_3O^+ BF_4^-$, CH_2Cl_2; v, $NaBH_4$, EtOH ; vi, farnesylamine, K_2CO_3, DMF; vii, Me_3SiBr, 2,4,6-collidine, CH_2Cl_2 ; viii, KOH, H_2O

Scheme 25

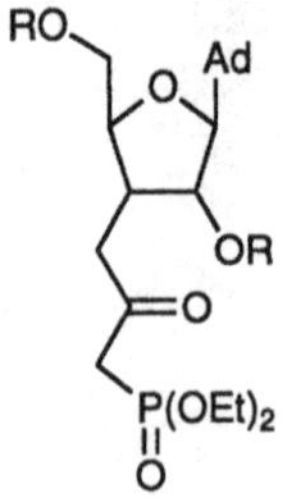

RO O Ad
OR
O
P(OEt)$_2$
O

(238)

R^1
N N
R^2 N N
O PO$_3$H$_2$

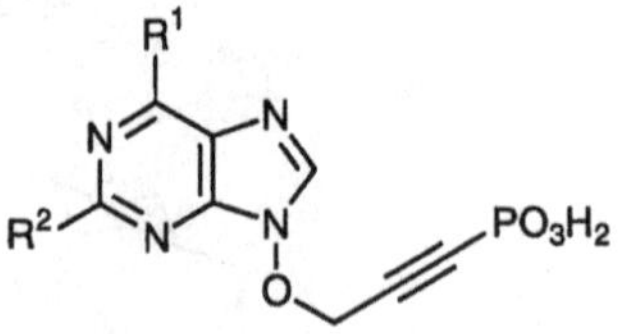

(239) $R^1 = NH_2$, $R^2 = H$
(240) $R^1 = HO$, $R^2 = NH_2$

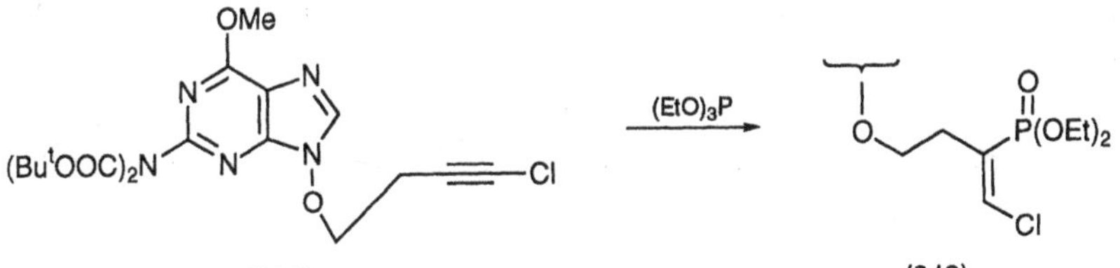

OMe N N N N O Cl $(Bu^tOOC)_2N$

(241)

$(EtO)_3P$

O O $P(OEt)_2$ Cl

(242)

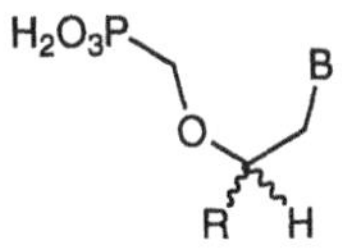

H_2O_3P O B R H

(243)

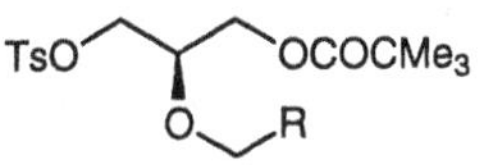

TsO $OCOCMe_3$ O R

(244) R = Br

(245) R = $P(O)(OPr^i)_2$

O HN N N N H_2N R

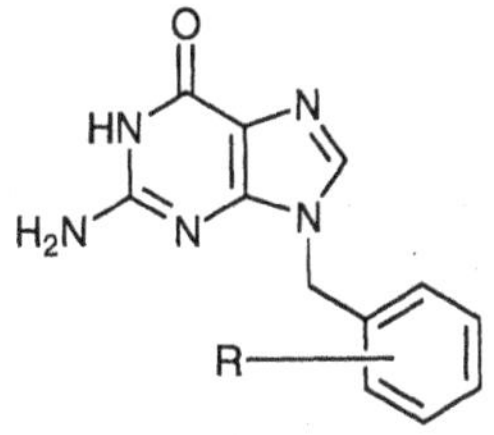

(246)

R = 2-,3-, or 4-$CH_2OCH_2P(O)(OPr^i)_2$

R = 3-$CH_2OCH_2CH_2P(O)(OPr^i)_2$

R = 3-$CH=CHCH_2P(O)(OR^1)_2$

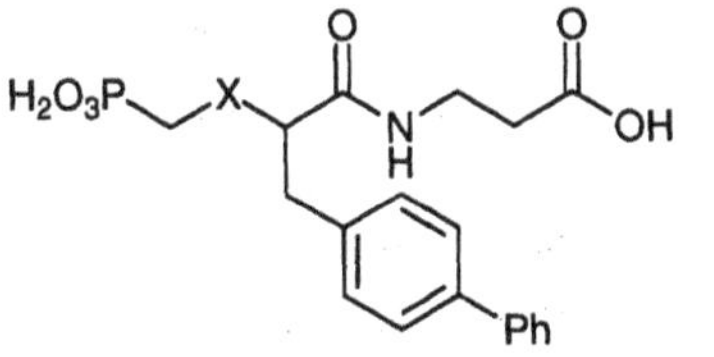

H_2O_3P X O N H O OH Ph

(247)

parent compound (246)(R = H) was increased more than 6000 fold with R = CH_2SCH_2.[123]

The phosphonic acids (247)(X = O, S, or SO_2), in racemic form, have been synthesized as inhibitors of neutral endopeptidase;[247] new phosphonopeptides[248,249] have been described.

2.2. The Reactions of Phosphonic and Phosphinic Acids. Exchange reactions between alkylphosphonic difluorides and HF in the presence of amines have been studied; the latter serve to generate F^-, and the ease of exchange appears to depend on the extent of electron withdrawal from the phosphorus atom, being easiest, in those cases examined, for $ClCH_2P(O)F_2$, with $Et_2NP(O)Cl_2$ at the other extreme.[250] The exchange of Cl for CN in phosphonic chlorides by HCN, and of CN by F, have also received attention; an exchange of radicals occurs when mixtures of phosphonic dihalides and silyl cyanides is warmed.[251]

The reactions between allylic alcohols and the species derived from diphenylphosphinic chloride and imidazole provide high yields of allylic esters of diphenylphosphinic acid; the reactions with primary alcohols are fast, and can take as little as 15 m, but those with secondary alcohols are much slower and can require up to 24 h. Alternatively, a procedure which involves diphenylphosphinic imidazolide and the alcohol in a solvent at ambient temperature provides lower yields but is suitable for hydrolytically sensitive esters.[252] Mixed diesters of [(1-amino-2-phenyl)ethyl]phosphonic acid, and derived peptides, are obtainable by the use of monoesters in combination with a primary or secondary alcohol in the presence of BOP reagent.[253] Aminoalkyl phosphonic diesters have also been prepared from the Et_3N salts of monoesters by means of a modified Mitsunobu reaction with tris(4-chlorophenyl)phosphine, rather than with Ph_3P, to reduce the reaction time.[254] DBU catalyses transesterification between a phosphonic bis(4-nitrophenyl) ester and a primary or secondary alcohol, other phenol, or an amine, leading to a mono(4-nitrophenyl) ester, and it is thus possible to replace the two aryl ester groups, the first by a primary alcohol (relatively fast) and the second by a secondary alcohol (the slower reaction).[255] The participation of a neighbouring group has been suggested to account for the ease of hydrolysis of bis(4-nitrophenyl) [(4-bromophenylsulphinyl)methyl]phosphonate.[256] Something of an analogy exists between *N,P*-diphenylphosphonamidic aryl esters and an aryl ester of [(methylsulphonyl)methyl]phenylphosphinic acid, with respect to *potential* mechanisms of hydrolysis. In actual fact, the former undergo hydrolysis by an E1cB mechanism, but the second group hydrolyse through an associative bimolecular process. It has now been shown that 2,4-dinitrophenyl esters of benzylphosphinic, benzylphosphonic, and benzylphosphonamidic acids do not undergo alkaline hydrolysis through the dissociative mechanism, but that the evidence suggests the associative process.[257]

C-Alkylated products are obtained when dicarbanions from nitroalkyl phosphonic diesters are treated with alkyl halides, but the regioselectivity of the process depends on the structure of the individual phosphonate. Surprisingly, the dicarbanion from an α-nitroalkyl phosphonate diester yields a β-alkylated product and, conversely, the alkylation of a β-nitroalkyl phosphonic diester occurs on the α-carbon centre.[258] In the alkylation of 2-(diethoxyphosphinoyl)cyclohexanone, when *C*- and *O*-alkylation are both possible under phase-transfer conditions, there is an increasing tendency for *O*-alkylation, with an increase in the steric bulk of the alkylating species.[259] Although the phosphonic diester (248) generates a carbanion, this cannot be alkylated by MeI or PrBr, and no deuterium is incorporated when the system is quenched by D_2O; on the other hand, the diester (249) generates a carbanion with BuLi (but not with NaH) which is readily alkylated at the α-carbon under very mild conditions, even by relatively bulky agents such as Pr^iBr and Me_3SiHal.[260] The reaction between the carbanion from diethyl [(methylthio)methyl]phosphonate and isocyanates or isothiocyanates affords the compounds (250)(X = O, R = Ph; X = S, R = alkyl, Ph or substituted phenyl); however, unusually in the reaction with 4-chlorophenyl isothiocyanate, the desulphurized compound (251) is formed in about 10% yield, but can be obtained in a higher yield from the isothiocyanate and diethyl methylphosphonate carbanion.[261]

The thiocyano group can be introduced into a phosphonic acid through the displacement of a tosyloxy group (generated from a hydroxyalkyl acid) with KCNS; alternatively, and for substrates of relatively high acidity , for example, triesters of phosphonoacetic acid, a reaction between ClSCN and a phosphonocarbanion is feasible, but the common use of $Al(OPr^i)_3$ as a catalyst sometimes suffers from the drawback of (carboxylic) ester exchange.[262]

An interesting alkylation study has centred around diastereoisomeric perhydro-1,3,2-oxazaphosphorines, in which structures were assigned by a comparison of ^1H n.m.r. spectroscopic data with those for (252)(R = Bu^t), a structure previously determined by X-ray analysis (S. E. Denmark *et al.* 1990). For the side-chain alkylation of the anions (generated with BuLi) from (252)(R = Me, Et, Pr^i, Bu^t or CEt_3), methylation (MeI) proceeded with (almost) uniformly high diastereoselectivity, with the composition of the products in the range 90/10 (R = Et_3C) to 97/3 (R = Et or Pr^i); benzylation (BnBr) was not so stereoselective, and diastereoselectivity ranged from 94/6 (R = Pr^i) to 80/20 (for R = CEt_3). By contrast, a change in the steric bulk of the group R in the series (253) has little influence on the diastereoselectivity (83/17 for R = Me or Bu^t and 85/15 for R = Et or Pr^i) although it does increase for R = CEt_3 (95/5).[263]

A reaction between diazomethane and carbohydrate-containing α-acyloximino phosphonates results in methylene group insertion between the phosphorus and (original) α-carbon atoms.[264]

The halogenation of (dialkoxyphosphinoyl)acetaldehydes involves a multistage sequence and there has been the recognition that enol forms are involved. Moreover, it has fairly recently been shown (Sokolov *et al.* 1990; Shagidullin *et al.* 1991) that the chlorination of (2-ethoxyethenyl)phosphonic acid derivatives leads to monochloro derivatives of the phosphinoylacetaldehydes in their enol forms. A detailed study of the halogenation of (2-ethoxyethenyl)phosphonic dimorpholide has now elucidated the overall sequence which leads to the dimorpholide of the halogenated (2-oxoethyl)phosphonic acid; these steps are indicated in Scheme 26 in which R = 4-morpholinyl.[265] The intermediates (254) are unstable, and if the halogenation (bromination) is carried out in a chloroform-water system, (254)(X = Br) is either hydrolysed rapidly to the monobromo derivative (255) which forms both (*E*) and (*Z*) oximes, and which is in equilibrium with its enol forms, or elimination of HBr occurs to give (256) which is isolable, and which can be brominated further and hydrolysed to give the dibromo(phosphinoyl)acetaldehyde (257); this last forms the expected hydrate and gives the expected oxime(s) but, under certain experimental circumstances, it is degraded to the dimorpholide of (dibromomethyl)phosphonic acid. In the case of chlorination, the intermediate (256) (X = Cl) is not isolable, and the monochloro derivative (255) actually exists in the (*Z*) enol form.[266] Dichloro(diethoxyphosphinoyl)acetaldehyde forms a hemithioacetal with Pr^iSH, the trimethylsilyl derivative of which, in boiling xylene, undergoes a transposition of the (original) carbonyl group to the α-position *via* a phosphinoylated oxirane.[267]

The nitrosation (propyl nitrite) of *N*-(diethoxyphosphinoyl)-*N'*-methylurea[268] and *N*-(diethoxyphosphinoyl)-*N''*-alkylbiurets[269] proceeds on the terminal nitrogen. The nitrosation (NOCl) of diethoxyphosphinoylacetyl chloride gives the oximes of the expected products following hydrolysis or methanolysis.[270]

The scope of the Schmidt reaction has been extended to include dialkyl acylphosphonates. Although acylphosphonic esters are known for their relative ease of cleavage at the P-C bond, it was found possible to carry out Schmidt reactions by using the diester in chloroform solution at 0° which is treated with an excess of HN_3 in the

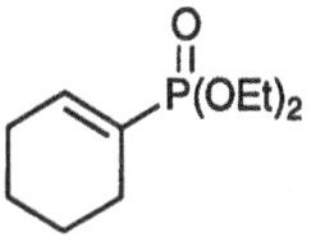

$P(O)(OEt)_2$

(248)

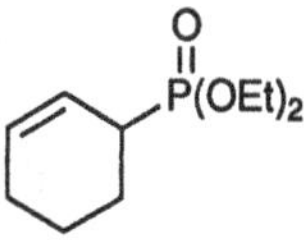

$P(O)(OEt)_2$

(249)

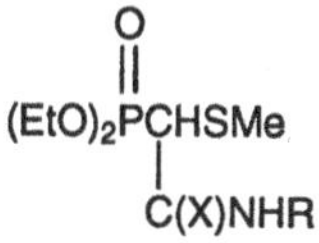

$(EtO)_2P(O)CH(SMe)C(X)NHR$

(250)

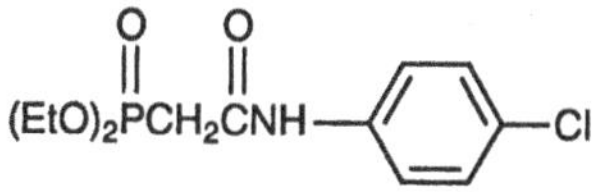

$(EtO)_2P(O)CH_2C(O)NH-C_6H_4-Cl$

(251)

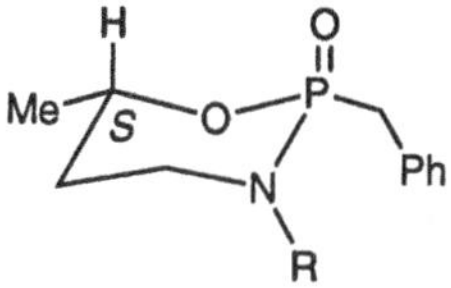

(252)

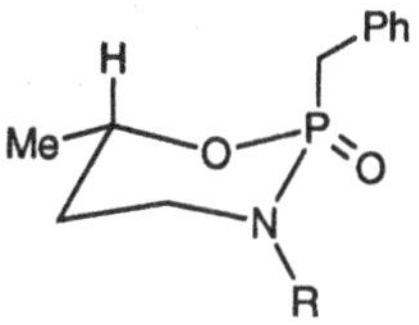

(253)

$R_2P(O)CH{=}CHOEt \xrightarrow{X_2} [R_2P(O)CHXCHX(OEt)]$ (254) $\xrightarrow{-HX} R_2P(O)CX{=}CHOEt$ (256)

(256) $\xrightarrow{H_2O} R_2P(O)CX_2CHX(OEt) \rightarrow R_2P(O)CX_2CHO$ (257)

(254) $\rightarrow R_2P(O)CHXCHO$ (255) $\rightleftharpoons R_2P(O)CX{=}CHOH \xrightarrow{X_2} R_2P(O)CX_2CHO$ (257)

(257) $\rightarrow R_2P(O)CX_2CH(OH)_2 \xrightarrow[X = Br]{H_2O} R_2P(O)CHBr_2$

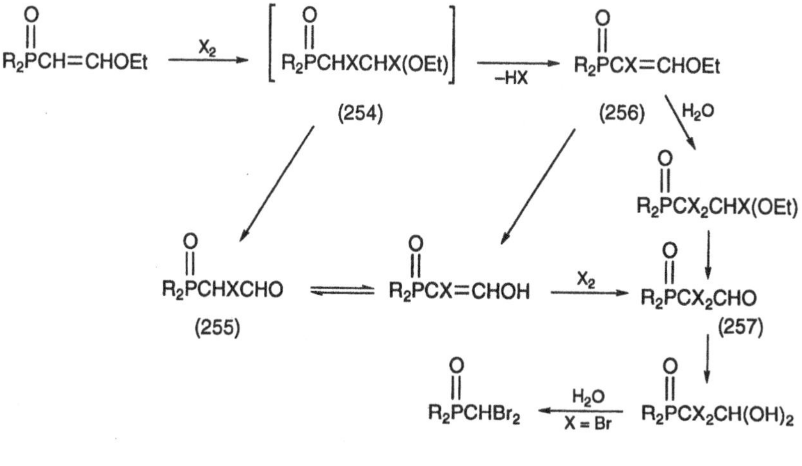

Scheme 26

presence of concentrated sulphuric acid. Several products from the rearrangement reactions have been recognized. When the acyl moiety is benzoyl, alone or carrying an electron-attracting, or only mild electron-releasing, substituent, the main product(s) result from the migration of the aryl group from *C* to *N;* the products are then the esters $ArNHCOP(O)(OR)_2$ together with *N*-arylformamides. If the substituent on the acyl group is strongly electron-releasing, migration of the phosphonate moiety from *C* to *N* takes place, the products then being the phosphoramidates, $ArCONHP(O)(OR)_2$ and the nitriles, ArCN. The authors considered mechanisms for such rearrangements and, it might be noted, felt that the results offered evidence for a possible alternative mechanism for the Schmidt reaction.[271]

Hammerschmidt has continued to report on wide-ranging aspects of the chemistry of hydroxyalkyl phosphonic acids. When (2-hydroxyethyl)phosphonic acid, labelled with ^{18}O in the alcohol-OH is fed to *Streptomyces fradiae*, the phosphomycin so formed (and isolated as (2-amino-1-hydroxy)phosphonic acid after ammonolysis) was found to be labelled to the extent of 50%, suggesting that the biological intermediate, phosphonoacetaldehyde, has a very short lifetime in the cell.[272] Hammerschmidt has also extended his earlier study (F. Hammerschmidt and H. Vollenkle, 1986) on the phosphonate-phosphate and phosphate-phosphonate rearrangements. The latest results indicate that diesters of (*R*)-(+)- and (*S*)-(-)-(1-hydroxy-1-phenylethyl)phosphonic acid, when in various solvents (which might contain up to 7% of water) containing $KOBu^t$, KOH, or DBU, rearrange to dialkyl (1-phenylethyl) phosphates with retention of configuration at phosphorus. The proposed mechanism for the rearrangement involves pentacoordinate phosphorus contained in an oxirane ring (Scheme 27). But, the story is rendered more complex by the finding that when the reactions are run in dry EtOH, racemic phosphate is formed together with acetophenone; thus, under these circumstances, complete fission of the substrate occurs before recombination to give the phosphate (Scheme 28).[273]

The enantioselective hydrolysis (de-*O*-acylation) of dialkyl [(1-acyloxy)alkyl]-phosphonates and dialkyl [(α-acyloxy)benzyl]phosphonates by esterolytic enzymes has been noted. The esters (259a-g) were examined as substrates for eight lipases as well as for pig liver esterase. The highest enantioselectivity was noted for a lipase towards the substrates (259a,b), when the (*S*) enantiomers were exclusively hydrolysed to give the optically pure alcohols (260a,b). Other alcohols, (260d,e), were obtained with enantiomeric excesses of 87% and 89% by the action of other enzyme systems.[274] (1-Hydroxyethylidene)bisphosphonic acid is thermally stable in quite acid solutions up to

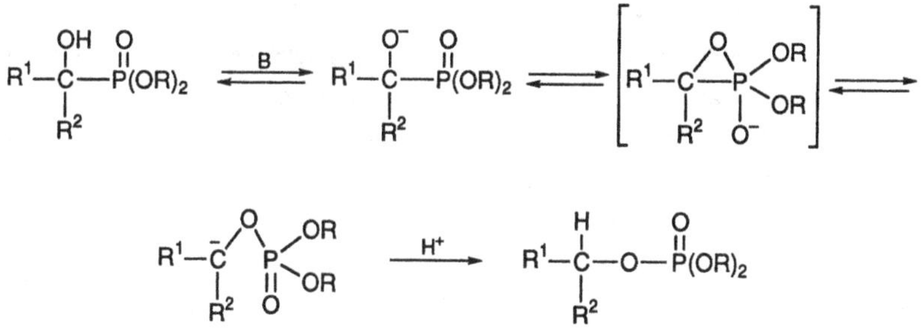

Scheme 27

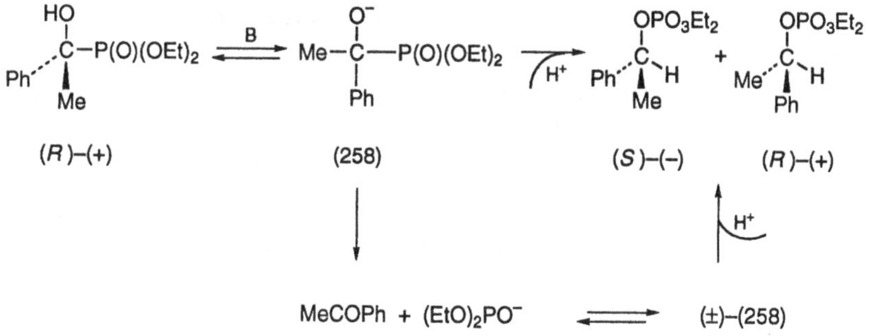

(R)-(+) (258) (S)-(−) (R)-(+)

$MeCOPh + (EtO)_2PO^-$ ⇌ (±)-(258)

Scheme 28

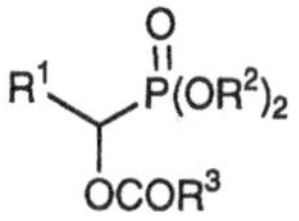

(259) 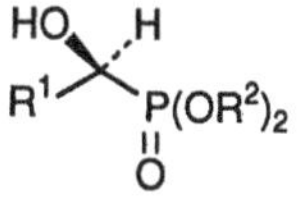 (260)

	R^1	R^2	R^3
(a)	Ph	Me	Me
(b)	Ph	Pr^i	Me
(c)	Me	Me	Me
(d)	Me	Pr^i	Me
(e)	(E)-MeCH=CH	Me	Me
(f)	Ph	Pr^i	CH_2Cl
(g)	Ph	Pr^i	Pr

125°, and in alkaline solutions up to 195°, when fission of the C-P bond occurs.[275]

A full paper on the pyrolysis of dialkyl [(2-acyloxy)alkyl]phosphonates has appeared (see *Organophosphorus Chemistry*, 1994, **25**, 111, reference 291).[276]

Wiemer and his coworkers have recorded further syntheses of α-phosphonolactones (261)(n = 1 or 2) and (262) and their reactions with aldehydes.[277]

The first example of a Baeyer-Villiger rearrangement involving a migrating phosphoryl moiety which, in turn, promotes the elimination of water from an intermediate α-hydroxy hydroperoxide, has been reported.[31] Experiments were conducted with the perhydro-1,3,2-oxazaphosphorine (263) and diethyl propanoylphosphonate (266). At -5° in a water/CH_2Cl_2 medium, a 92/8 mixture of (263) and its phosphorus epimer was treated with 30% H_2O_2; the insertion of oxygen, to give (265), proceeded in high yield and apparently, also, with retention of configuration at phosphorus, and similarly (266) gave (267) in 90% yield. In the light of data from experiments in which oxidations were carried out with ^{17}O-enriched hydrogen peroxide, a mechanism was proposed which is based on the movements depicted in structure (264).

Whilst the simple monoester monoanions (268) are stable for 30 h in refluxing MeCN, the corresponding dianions (269) display a tendency to fragment. The monoanions largely in the (*E*) form, and in which R is 2,2,2-trichloroethyl, 2,2,2-trifluoroethyl, or 2,2,2-trifluoro-1-(trifluoromethyl)ethyl, can act as phosphorylating agents through fragmentation (PhCN being eliminated) to a metaphosphate, e.g. (270)(X = F or Cl), which can be trapped in EtOH or Pr^iOH as the esters (271)(R = Et or Pr^i), but not by MeOH or water. The fragment (270)(X = F) was also trapped with styrene oxide, when a mixture of diastereoisomers of (272) was obtained.[278]

The kinetics of the additions of dialkyl hydrogen phosphites to enamines[279] and to conjugated imines[280] have been reported. Reactions have been carried out between ethene-1,1-diylbisphosphonic acid and nucleophiles - mostly amino acids, to give *N*-derivatives of (2-amino-1-ethylidene)bisphosphonic acid.[281] The addition reactions between activated methylenephosphonic esters and conjugated azoalkenes have already been noted earlier in this report;[130,131] the products, which consist of *N*-amino-pyrroles or dihydropyrroles, evidently result from an initial addition across the C=C bond followed, in certain cases, by loss of water (Scheme 29). The addition of hydrazones to enallene phosphonates leads to either indolines (by 1:2 addition) or dihydropyrazoles (by 1:4 addition) (Scheme 30).[282] (3-Alken-1-yl)phosphonic diesters add to α-nitroalkenes to give the phosphonic diesters (273).[283] In a reaction between a phosphonocarbanion from $(EtO)_2P(O)CH_2R$ (R = CN, COOMe, SO_2Me, or PO_3Et_2) and 2-nitro-1-phenylpropene,

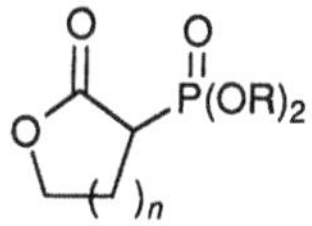

(261)

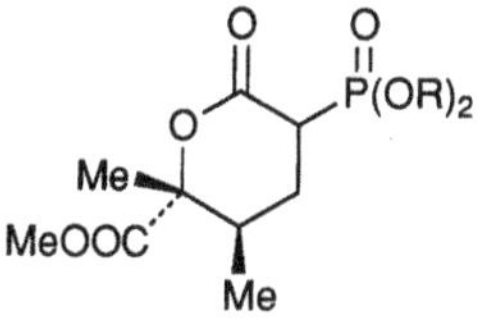

(262)

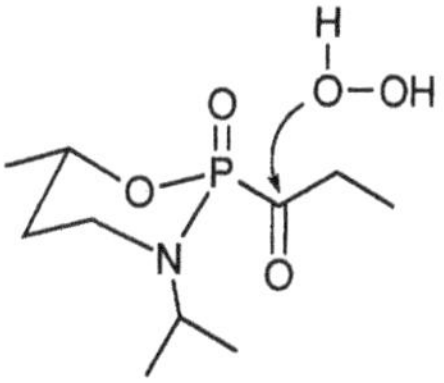

(263)

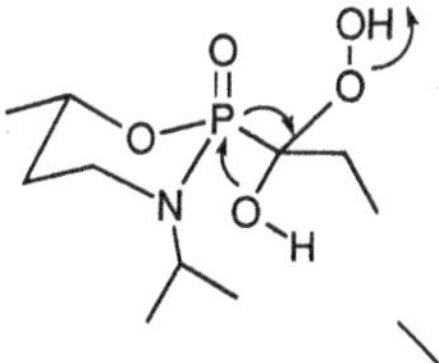

(264)

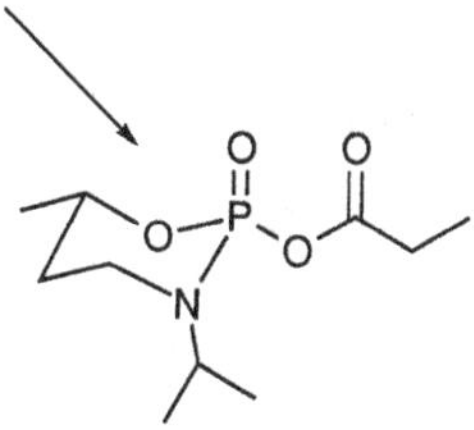

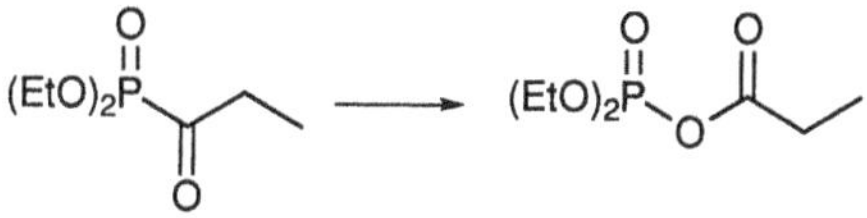

(266)

(267)

(265)

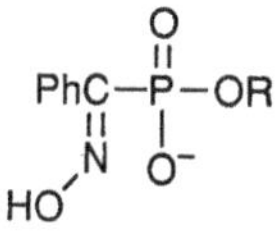

(268)

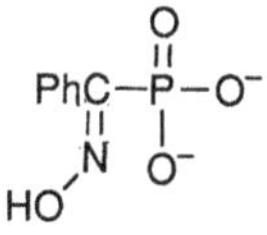

(269)

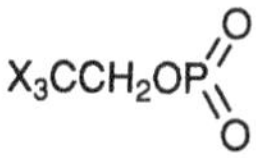

(270)

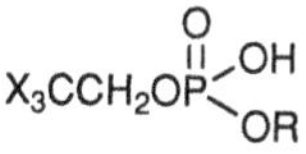

(271)

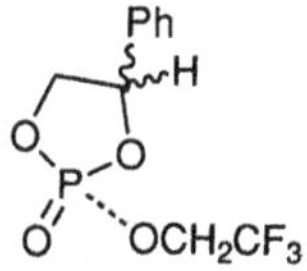

(272)

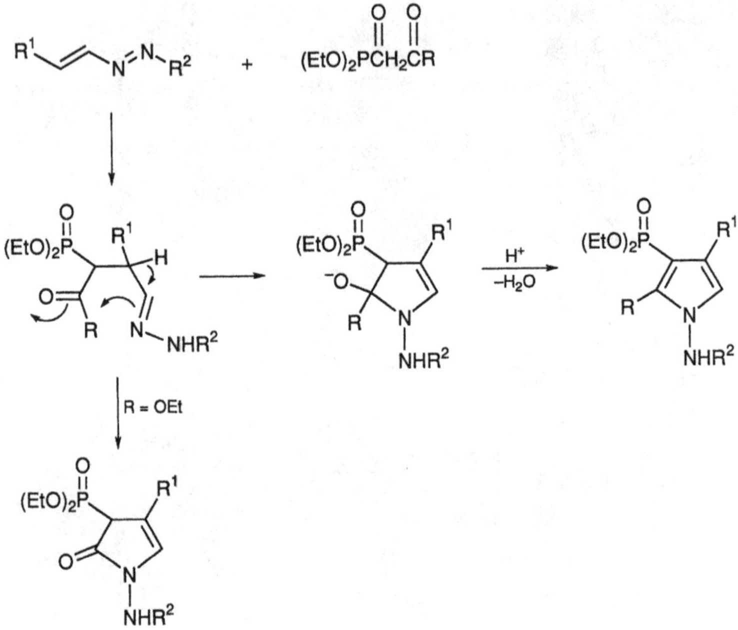

R^1–CH=CH–N=N–R^2 + $(EtO)_2P(O)CH_2C(O)R$

$(EtO)_2P(O)$ R^1 H O R N–NHR^2

$(EtO)_2P(O)$ R^1 ^-O R N NHR^2

H^+ / $-H_2O$

$(EtO)_2P(O)$ R^1 R N NHR^2

R = OEt

$(EtO)_2P(O)$ R^1 O N NHR^2

Scheme 29

$P(OR)_2$ (O) + $PhNHNH_2$ → (1:4) $(RO)_2P(O)$ N N Ph

1:2

N H $P(OR)_2$ O

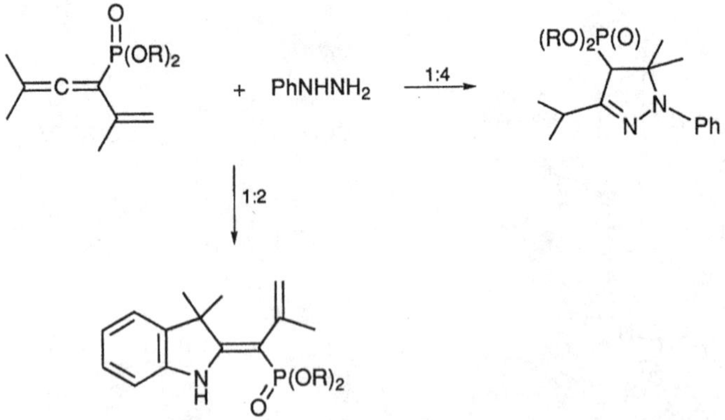

Scheme 30

the initial adduct evidently fragments (one fragment can be trapped by the addition of Me_3SiCl to the reaction mixture solution) before recombining as a dihydroisoxazole-phosphonic derivative (274).[284]

It is worthy of note that when diethyl vinylphosphonate is treated with $CHCl_3/HO^-$ in the presence of a phosphonium or ammonium salt or dibenzo-18-crown-6, the exothermic reaction provides about 50% of diethyl (3,3,3-trichloropropyl)-phosphonate[285,286] whereas an analogous reaction with $CHBr_3$ leads to diethyl (2,2-dibromocyclopropyl)phosphonate.[285] Diethyl (2-propenyl)phosphonate with CHX_3 (X = Br or Cl) gives the dihalogenocyclopropyl compound, identical to, for instance, the product from triethyl phosphite and 1-bromomethyl-2,2-dichlorocyclopropane.[285] Yet another difference to be observed is when dialkyl (1-methylethenyl)phosphonate is also treated with $CHCl_3/HO^-$ when the result is dichlorocyclopropyl phosphonate formation rather than chain extension.[286] Under phase transfer conditions, the esters of the cyclic phosphinic acids (275)(R = H) yield resolvable mixtures of stereoisomeric adducts (276) and (277).[287,288] Diethyl (trichloromethyl)phosphonate adds to 1-alkenes in the presence of Cu(I)-amine complexes to give derivatives of diethyl (1,1,3-trichloropropyl)-phosphonate.[289]

Dialkyl (1-diazoalkyl)phosphonates undergo a replacement of the diazo group by phenols, in a system containing $Rh_2(OAc)_4$, to give dialkyl (1-aryloxyalkyl)-phosphonates.[290] Cycloaddition reactions between dialkyl (1-azidoalkyl)phosphonates and alkynes to give dialkyl [(1,2,3-triazole)alkyl]phosphonates (278) as mixtures of isomers (depending on the relative positions of R^2 and R^3).[291,292] Further uses of (dialkoxyphosphinoyl)acetonitrile *N*-oxide (279) are noted in their reactions with enamines from ethyl acetoacetate, when the products are isoxazole derivatives (280).[293] Similarly formed are the phosphinoylformonitrile oxides (281)(R = 4-morpholinyl[294] or Pr^i [295]); the former compound is reasonably stable, not prone to dimerization, adds HCl to regenerate its precursor, and reacts with thiourea with subsequent elimination of urea to give the $R_2P(O)NCS$. The second formonitrile oxide undergoes [2 + 3] cycloaddition reactions with cyclopropenes to give the compounds (282).[295]

Further reactions of bis(diethoxyphosphinoyl)ketene dithioacetals with nucleophiles have been investigated - including their reactions with Grignard reagents, when the SR group is replaceable by Me or Ph.[296] An involved sequence is seen for the reactions between the readily available esters (283) and morpholine (MH; M = 4-morpholinyl); in general, when the compounds (283) have been derived from secondary amines, their further treatment with morpholine results in dehydrochlorination followed by cyclization

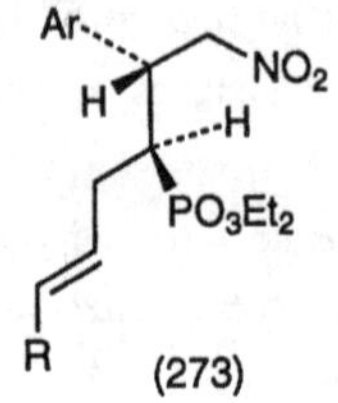

(273)

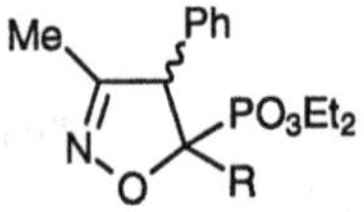

(274)

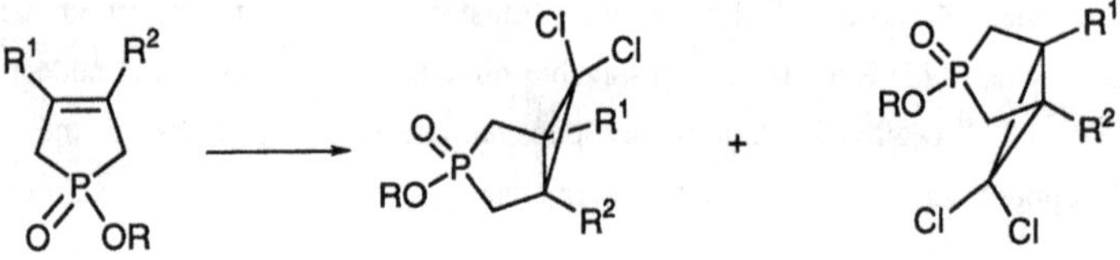

(275) (276) (277)

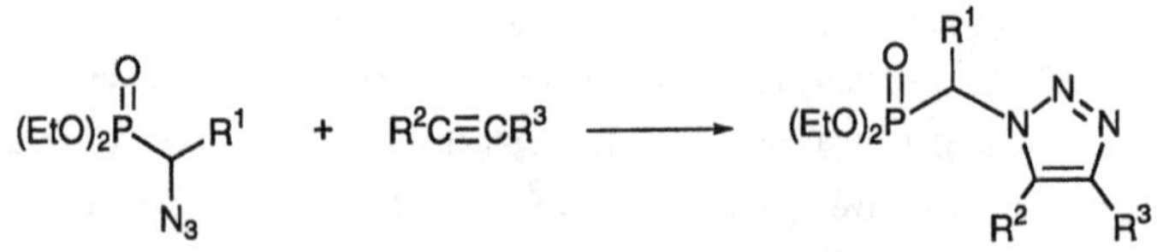

(278)

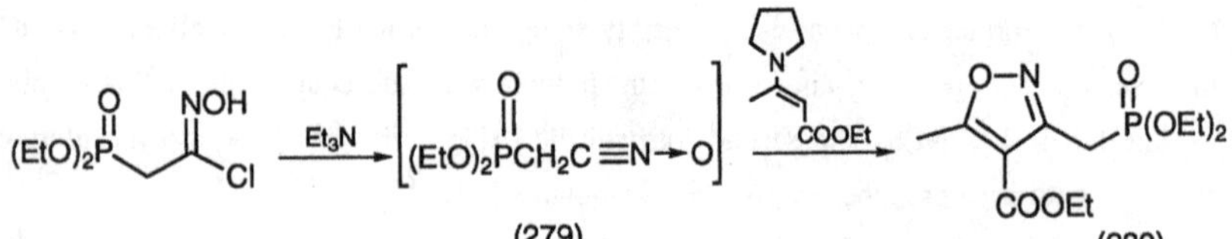

(279) (280)

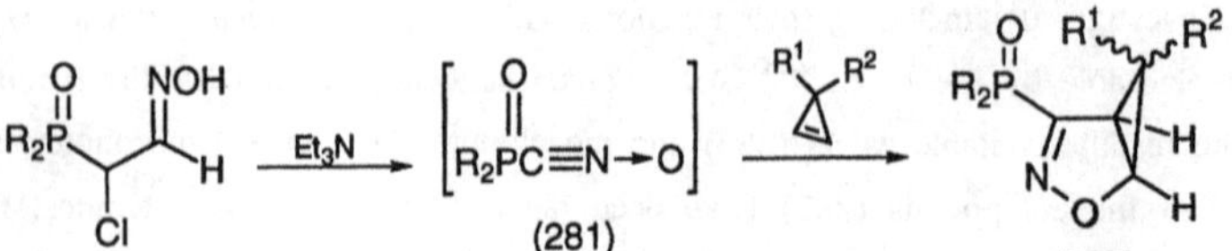

(281) (282)

to the 1,3-oxazoles (284), and further products (285) are the result of ring opening and reclosure when R^2 = H.[297]

In regard to the 2,3-oxaphosphabicyclo[2.2.2]octene system, the photolysis of (286)(R = neopentyl or 1-adamantyl) generates the metaphosphates $ROP(=O)_2$, (R = Me_3CCH_2 or 1-adamantyl) which could be trapped in water (as the phosphoric acid $ROP(O)(OH)_2$) or in EtOH, as the ethyl ester ROP(O)(OH)(OEt).[298] The UV irradiation of diaryl methylphosphonates leads to methylphosphonic acid and various biphenyl derivatives.[299]

The photolysis of *tert*-butylmesitylphosphinic azide generates the dihydrobenzazaphosphole (287), admixed with other rearrangement products if the azide is dissolved in MeOH. In an experiment designed to examine the possible role of a singlet nitrene in the formation of (287), the azide in MeCN was photolysed in the presence of MeOH or Me_2S. The normal rearrangement and insertion products were formed together with a further substance characterized as the sulphilimine (288), itself photolytically unstable. However, the increased formation of (288) had a much greater influence on that of (287)(with a marked reduction in this) than on that of the rearrangement products, (289) and (290)(Ar = 2,4,6-trimethylphenyl). This consideration, along with others on the relative proportions of insertion and rearrangement reactions, led to the conclusion that the Curtius-like photochemical rearrangement can proceed concertedly, without the intermediate formation of a nitrene.[300]

In a synthesis of 1-(diphenylphosphinoyl)aziridines, the intermediate formation of the *p*-toluenesulphonyl ester is now obviated by taking into account the marked fugacity of the diphenylphosphinoyl group itself (Scheme 31).[301]

An X-ray analysis of the diastereoisomeric (bromomethyl)phosphonamidic ester (291)(X = Br, R = (-)-menthyl), mp 155°, isolated from a 1:1 mixture, has the (S_P) configuration. This compound reacted with NaOMe to give (292), as the predominant of two diastereoisomers and shown to have (S_P) configuration, again by X-ray analysis; the minor product from the reaction, also predominantly as one diastereoisomer with (S_P) configuration was the phosphoramidate (293). Thus, the conversion of (291) into (292) occurs with inversion of configuration, but the last step, which involves phosphorus-carbon bond cleavage, proceeds with predominant retention (the stereochemistry of (293) was determined in a circuitous manner). The rational explanation for the changes requires that the formation of (293) must proceed through a permutational change (294) -> (295) in a true intermediate, and not within just a transition state.[302]

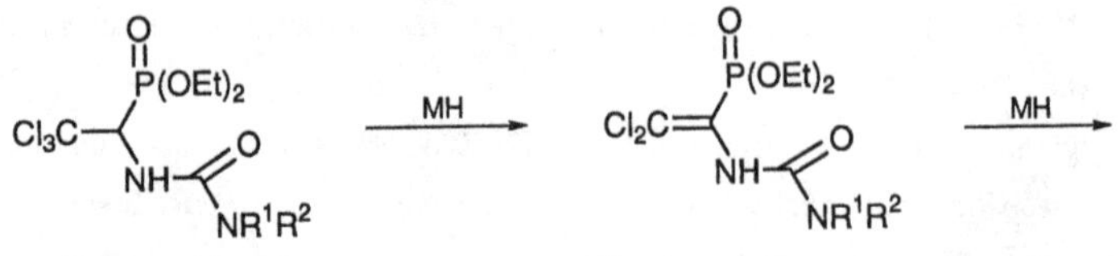

O
P(OEt)2
Cl_3C
NH
O
NR^1R^2

(283)

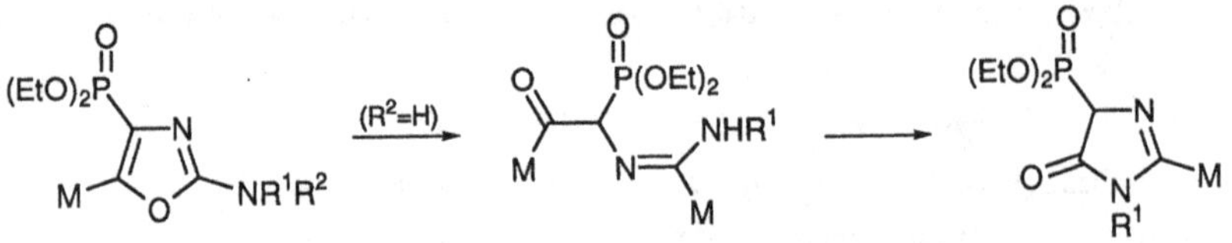

MH

O
P(OEt)2
Cl_2C
NH
O
NR^1R^2

MH

O
$(EtO)_2P$
N
M
O
NR^1R^2

(284)

(R^2=H)

O
O
P(OEt)2
NHR^1
M
N
M

O
$(EtO)_2P$
N
O
N
M
R^1

(285)

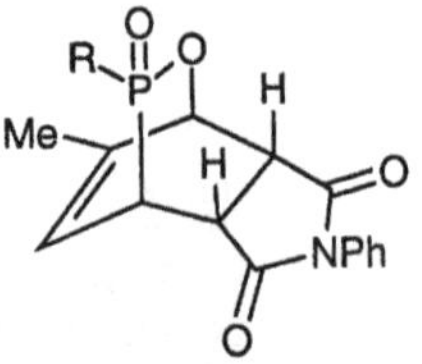

O
R
P
O
H
Me
H
O
NPh
O

(286)

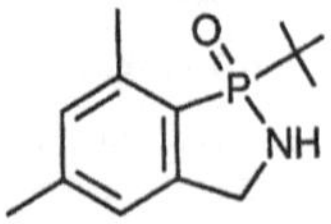

O
P
NH

(287)

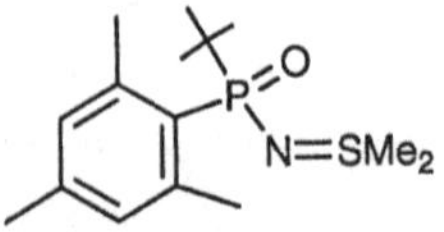

P
O
N=SMe_2

(288)

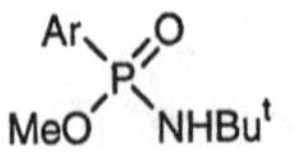

Ar
O
P
MeO
$NHBu^t$

(289)

MeO
O
P
Bu^t
NHAr

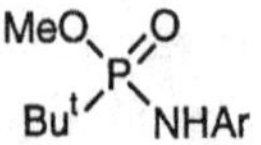

(290)

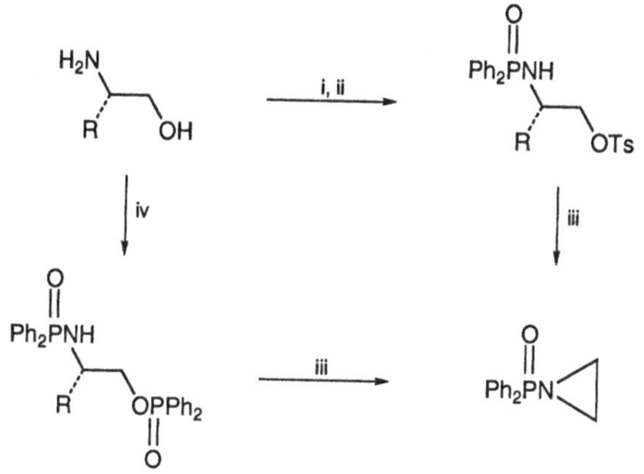

Reagents: i, 1 equiv. $Ph_2P(O)Cl, Et_3N$; ii, TosCl ; iii, NaH;
iv, 2 equiv. $Ph_2P(O)Cl$, Et_3N

Scheme 31

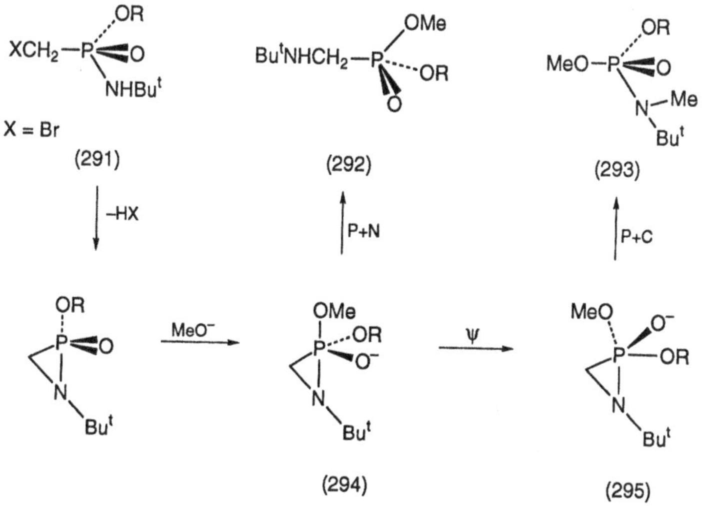

X = Br

(291)

(292)

(293)

(294)

(295)

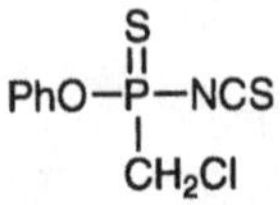

S
PhO—P—NCS
CH_2Cl

(296)

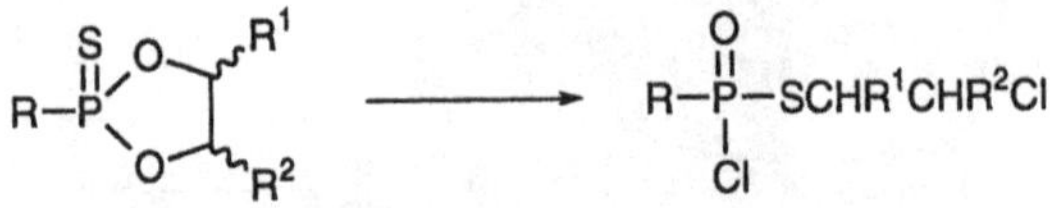

S O R[1]
R—P
O R[2]

(297)

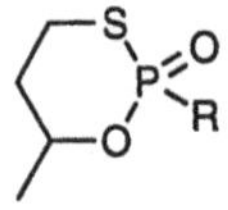

→ O
R—P—$SCHR^1CHR^2Cl$
Cl

S
P=O
O R

(298)

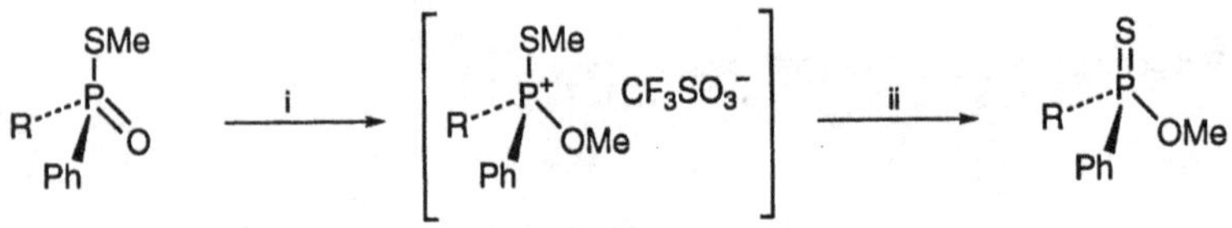

SMe
R····P=O
Ph

(299)

→ i

SMe
R····P+ $CF_3SO_3^-$
Ph OMe

→ ii

S
R····P
Ph OMe

Reagents: i, CF_3SO_3Me, CH_2Cl_2; ii, NaSH, Et_2O

Scheme 32

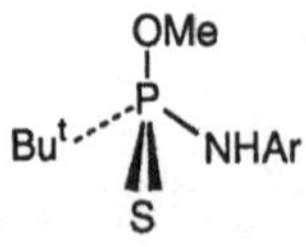

OMe
Bu[t]····P NHAr
S

(300)

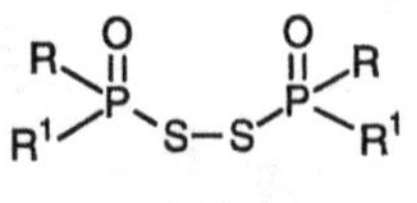

R O O R
P—S—S—P
R[1] R[1]

(301)

[>P(O)—S—$\overset{+}{P}$<] [$\bar{O}$– P(S)<] → [>P(S)—O—$\overset{+}{P}$<] [$\bar{O}$– P(S)<]

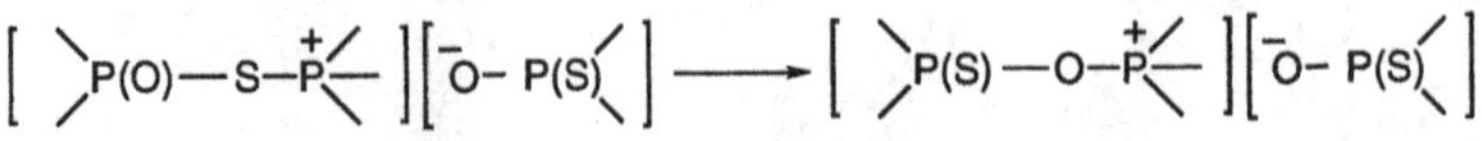

(302) (303)

Nitroso compounds condense with phosphonic amides in the presence of dibromisocyanurate to give phosphinoyldiazine *N*-oxides, probably *via* the *N*,*N*-dibromo amides (compare reference 83).[303]

Further reactions of the phosphonothioyl isocyanate (296), which lead to derivatives of the 1,3,4-thiazaphosph(V)ole system, have been described.[304] Phosphorus pentachloride reacts with the 1,3,2-dioxaphospholane 2-sulphides (297)(R = ArO or Me_2N) to give acylic phosphonic chloride esters, as it does also with the 1,3,2-dioxaphosphorinane (298)(R = Me_2N) but the latter reaction fails when R = ArO.[305]

The Pishchimuka reaction is a well established procedure to bring about the isomerization of *O*-alkyl esters of quinquevalent phosphorus acids into the *S*-alkyl esters. The reverse process, a 'retro-Pishchimuka reaction', has hitherto not been available, but has now been announced. The (*R*)-(+) esters (299)(R = Me or Bu^t) are converted into a quaternary salt on treatment with methyl triflate, and the products are then treated with NaSH (Scheme 32); no change occurs in the configurations at phosphorus.[306] At the same time, methyl triflate has been used in the normal Pishchimuka conversion of the phosphonothioic amide (300)(Ar = 2,4,6-tri-*tert*-butylphenyl) into its S-methyl isomer with retention of configuration.[307]

A detailed study of the desulphurization of the disulphides (301)(R,R^1 = dialkoxy, or alkyl(alkoxy)) by phosphorus(III) compounds, including $(Me_2N)_3P$, Ph_3P, or acyclic or cyclic (based on catechol) phosphotriesters or ester amides) considered the role of phosphonium and phosphorane intermediates. The isomerization of (302) into (303), and the decomposition (desulphurization, deoxygenation, or dealkylation) into end products, are influenced by electronic and steric factors.[308]

References

1. J. Baraniak and W. J. Stec, *Rev. Heteroat. Chem.*, 1993, **8**, 143.
2. W. J. Stec and A. Wilk, *Angew. Chem., Int. Ed. Engl.*, 1994, **33**, 709.
3. Ya. A. Dorfman, T. V. Petrova, R. R. Sagandykova, and D. M. Doroshkevich, *J. Gen. Chem. USSR (Engl. Transl.)*, 1992, **62**, 1667.
4. Ya. A. Dorfman and R. R. Abdreimova, *Russ. J. Gen. Chem. (Engl. Transl.)*, 1993, **63**, 206.
5. Ya. A. Dorfman, G. S. Polimbetova, E. Zh. Arbasov, A. K. Borangazieva, A. O. Kokpanbaeva, and F. Kh. Faizova, *J. Gen. Chem. USSR (Engl. Transl.)*, 1992, **62**, 1860.
6. Ya. A. Dorfman, L. V. Levina, and E. Zh. Arbasov, *J. Gen. Chem. USSR (Engl. Transl.)*, 1992, **62**, 1847.

7. Yu. G. Budnikova, Yu. M. Kargin, I. M. Zaripov, A. S. Romakhin, Yu. A. Ignat'ev, E. V. Nikitin, and A. P. Tomilov, *Bull. Acad. Sci. USSR., Div. Chem. Sci. (Engl. Transl.)*, 1992, 1585.
8. Ya. A. Dorfman, M. M. Aleshkova, G. S. Polimbetova, L. V. Levina, T. V. Petrova, R. R. Abdreimova, and D. M. Doroshkevich, *Russ. Chem. Rev. (Engl. Transl.)*, 1993, **62**, 877.
9. A. D. Sagar, M. T. Thorat, M. M. Salunkhe, R. S. Seiukar, and P. P. Wadgaonkar, *Synth. Commun.*, 1994, **24**, 2029.
10. Yang Lu and F. Le Goffic, *Synth. Commun.*, 1993, **23**, 1943.
11. Y. J. Koh and D. Y. Oh, *Synth. Commun.*, 1993, **23**, 1771.
12. W. Dabkowski, J. Michalski, J. Wasiak, and F. Cramer, *J. Chem. Soc. Perkin Trans. I*, 1994, 817.
13. S. A. Lermontov, I. I. Sukhojenko, A. V. Popov, A. N. Pushin, I. V. Martynov, and N. S. Zefirov, *Heteroat. Chem.*, 1993, **4**, 579.
14. L. T. Byrne, J. M. Harrowfield, D. C. R. Hockless, B. J. Peachey, B. W. Skelton, and A. H. White, *Aust. J. Chem.*, 1993, **46**, 1673.
15. J. Gloede, B. Costisella, M. Ramm, and R. Bienest, *Phosphorus, Sulfur, Silicon, Relat. Elem.*, 1993, **84**, 217.
16. D. V. Khasnis, J. M. Burton, J. D. McNeil, C. J. Santini, H. Zhang, and M. Lattman, *Inorg. Chem.*, 1994, **33**, 2657.
17. T. Lippmann, E. Dalcanale, and G. Mann, *Tetrahedron Lett.*, 1994, **35**, 1685.
18. R. S. Edmundson, in *Specialist Periodical Reports (Organophosphorus Chemistry)*, 1994, **25**, 111, and earlier volumes.
19. S. Iacobucci and M. d'Alarcao, *Synth. Commun.*, 1994, **24**, 809.
20. C. Liu, N. F. Thomas, and B. V. L. Potter, *J. Chem. Soc. Chem. Commun.*, 1993, 1687.
21. A. P. Kosikowski, V. I. Ognyanov, A. H. Fauq, R. A. Wilcox, and S. R. Nahorski, *J. Chem. Soc. Chem. Commun.*, 1994, 599.
22. S. J. Mills, J. Al-Hafidh, J. Westuick, and B. V. L. Potter, *Bioorg. Med. Chem. Lett.*, 1993, **3**, 2599.
23. G. M. Salamończyk and K. M. Pietrusiewicz, *Tetrahedron Lett.*, 1994, **35**, 4233.
24. S. J. Mills, S. T. Safrany, R. A. Wilcox, S. R. Nahorski, and B. V. L. Potter, *Bioorg. Med. Chem. Lett.*, 1993, **3**, 1505.
25. S. Ozaki, Y. Koga, L. Ling, Y. Watanabe, Y. Kimura, and M. Hirata, *Bull. Chem. Soc. Jpn.*, 1994, **67**, 1058.
26. C. Liu and B. V. L. Potter, *Tetrahedron Lett.*, 1994, **35**, 1605.
27. J.K. Stowell and T. S. Widlanski, J. Amer. Chem. Soc., 1994, **116**, 789.
28. G. F. Koser, K. Chen, Y. Huang, and C. A. Summers, *J. Chem. Soc. Perkin Trans. I*, 1994, 1375.
29. T. Gefflaut and J. Périe, *Synth. Commun.*, 1994, **24**, 29.
30. T. Furuta, H. Torigai, T. Osawa, M. Iwamura, *J. Chem. Soc. Perkin Trans. I*, 1993, 3139.
31. N. J. Gordon and S. A. Evans, *J. Org. Chem.*, 1993, **58**, 4516.
32. V. F. Mironov, L. A. Burnaeva, I. V. Konovalova, G. A. Khlopushina, and P. P. Chernov, *Zhur, Org. Khim.*, 1993, **29**, 639; *Chem. Abstracts*, 1994, **120**, 245331.
33. I. A. Rakhov, A. B. Uryupin, V. A. Kolesova, P. V. Petrovskii, and T. A. Mastryukova, *Russ. J. Gen. Chem.(Engl. Transl.)*, 1993, **63**, 998.
34. A. H. van Oijen, S. Behrens, D. F. Mierke, H. Kessler, J. H. van Boom, and R. M. J. Liskamp, *J. Org. Chem.*, 1993, **58**, 3722.
35. C. M. Thompson , A. I. Suarez, J. Lin, and J. A. Jackson, *Tetrahedron Lett.*, 1993, **34**, 6529.
36. É. E. Nifant'ev, D. A. Predvoditelev, and M. A. Malenkovskaya, *J. Gen. Chem. USSR (Engl. Transl.)*, 1992, **62**, 1636.

37. S. F. Martin and A. S. Wagman, *J. Org. Chem.*, 1993, **58**, 5897.
38. H. Saika, Th. Früh, G. Iwasaki, S. Koizumi, I. Mori, and K. Hayakawa, *Bioorg. Med. Chem. Lett.*, 1993, **3**, 2129.
39. O. N. Nuretdinova, V. G. Novikova, and L. B. Troitskaya, *Bull. Acad. Sci. USSR, Div. Chem. Sci. (Engl. Transl.)*, 1992, 2119.
40. O. N. Nuretdinova and V. G. Novikova, *Bull. Acad. Sci. USSR, Div. Chem. Sci. (Engl. Transl.)*, 1992, 2124.
41. R. Sato, T. Murata, S. Chida, and S. Ogawa, *Chem Lett.*, 1993, 1325.
42. C. Tang and G. Wu, *Chin. Chem. Lett.*, 1992, **3**, 783; *Chem. Abstracts*, 1993, **119**, 160398.
43. C. Tang and G. Wu, *Gaodeng Xuexiao Huaxue Xuebao*, 1993, **14**, 642; *Chem. Abstracts*, 1993, **119**, 250027.
44. L. V. Chvertkina, S. B. Smirnova, B. Ya. Chvertkin, and P. S. Khokhlov, *J. Gen. Chem. USSR (Engl. Transl.)*, 1992, **62**, 2078.
45. M. W. Wieczorek, J. Błaszczyk, and B. Krzyzanowska, *Heteroat. Chem.*, 1993, **4**, 79.
46. I. S. Nizamov, V. A. Kuznetsov, É. S. Batyeva, V. A. Al'fonsov, and A. N. Pudovik, *Phosphorus, Sulfur, Silicon, Relat. Elem.*, 1993, **79**, 179.
47. I. S. Nizamov, V. A. Kuznetsov, É. S. Batyeva, V. A. Al'fonsov, and A. N. Pudovik, *Heteroat. Chem.*, 1993, **4**, 379.
48. I. S. Nizamov, V. A. Kuznetsov, É. S. Batyeva, V. A. Al'fonsov, and A. N. Pudovik, *Bull. Acad. Sci. USSR, Div. Chem. Sci. (Engl. Transl.)*, 1992, 1517.
49. I. S. Nizaomov, L. A. Al'metkina, G. G. Garifzyanova, É. S. Batyeva, V. A. Al'fonsov, and A. N. Pudovik, *Phosphorus, Sulfur, Silicon, Relat. Elem.*, 1993, **83**, 191.
50. É. E. Nifant'ev, M. A. Kharstan, and S. A. Lysenko, *Russ. J. Gen. Chem. (Engl. Transl.)*, 1993, **63**, 547.
51. S. Bauermeister, T. A. Modro, and A. Zwierzak, *Heteroat. Chem.*, 1993, **4**, 11.
52. V. V. Tkachev, O. A. Raevskii, N. V. Luk'yanov, G. I. Vau'kin, R. I.. Yurchenko, V. G. Yurchenko, A. F. Solotnov, A. M. Pinchuk, T. G. Galenko, T. A. Ivanova, and L. O. Atmyan, *Bull. Acad. Sci. USSR, Div. Chem. Sci. (Engl. Transl.)*, 1992, 2211.
53. H. Yang and R. Lu, *Synth. Commun.*, 1994, **24**, 59.
54. C. Menneret, R. Cagnet, and J. C. Florent, *Carbohydrate Res.*, 1993, **240**, 313.
55. M. Yamashita, C. Takahashi, and K. Seo, *Heterocycles*, 1993, **36**, 651.
56. J. C. Kim, H. D. Pack, S. H. Moon, and S. H. Kim, *Bull. Korean Chem. Soc.*, 1993, **14**, 318; *Chem. Abstracts*, 1994, **120**, 77515.
57. C. S. Sarmah and J. C. S. Kataky, *Indian J. Heterocycl. Chem.*, 1993, 2, 165; *Chem. Abstracts* , 1993, **119**, 95650.
58. O. M. Plotnikova, N. S. Magomedova, N. G. Ruchkina, A. N. Sobolev, V. K. Bel'skii, and É. E. Nifant'ev, *Metalloorg. Khim.*, 1992, **5**, 1288; *Chem. Abstracts*, 1993, **119**, 250033.
59. B. Oussaid, B. Garrigues, J. Jaud, A. M. Caminade, and J. P. Majoral, *J. Org. Chem.*, 1993, **58**, 4500.
60. S. Sheffer-Dec-Noor and T. Baasov, *Bioorg. Med. Chem. Lett.*, 1993, **3**, 1615.
61. A. Koziara, M. Nowalinska, and A. Zwierzak, *Synth. Commun.*, 1993, **23**, 2127.
62. C. Yuan, S. Li, W. Hu, and H. Fen, *Heteroat. Chem.*, 1993, **4**, 23.
63. K. N. Dalby, A. J. Kirby, and F. Hollfelder, *J. Chem. Soc. Perkin Trans.* 2, 1993, 1269.
64. A. J. Chandler, F. Hollfelder, A. J. Kirby, F. O'Carroll, and R. Stromberg, *J. Chem. Soc. Perkin Trans.* 2, 1994, 327.
65. J. H. Kim, M. J. Gallagher, and R. F. Toia, *Aust. J. Chem.*, 1994, **47**, 715.
66. R. A. Moss and H. Zhang, *J. Amer. Chem. Soc.*, 1994, **116**, 4471.
67. T. Furuta, H. Torigai, T. Osawa, and M. Iwamura, *Chem. Lett.*, 1993, 1179.

68. R. S. Givens, P. S. Athey, B. Matuszowski, L. W. Kueper, J. Zue, and T. Fister, *J. Amer. Chem. Soc.*, 1993, **115**, 6001.
69. M.I. Kabachnik, L. S. Zakharov, G. N. Molchanova, E. I. Goryunov, P. V. Petrovsky, T. M. Shcherbina, and A. P. Larelina, *Dokl. Akad. Nauk, SSSR*, 1993, **330**, 582; *Chem. Abstracts*, 1994, **120**, 134698.
70. A. Yanagisawa, H. Hibino, N. Nomura, and H. Yamamoto, *J. Amer. Chem. Soc.*, 1993, **115**, 5879.
71. R. L. Funk, J. B. Stallman, and J. A. Wos, *J. Amer. Chem. Soc.*, 1993, **115**, 8847.
72. M. P. Koroteev and É. E. Nifant'ev, *Russ. J. Gen. Chem. (Engl. Transl.)*, 1993, **63**, 339.
73. M. P. Koroteev, N. M. Pugashova, É. E. Nifant'ev, and A. R. Bekker, *Russ. J. Gen. Chem. (Engl. Transl.)*, 1993, **63**, 337.
74. R. Patil, C. P. Shinde, and M. V. Sharma, *Asian J. Chem.*, 1993, **5**, 953; *Chem. Abstracts*, 1994, **120**, 217892.
75. R. Patil, C. P. Shinde and M. V. Sharma, *Asian J. Chem.*, 1993, **5**, 974; *Chem. Abstracts*, 1994, **120**, 217893.
76. M. Y. Chae, J. F. Postula, and F. M. Raushel, *Bioorg. Med. Chem. Lett.*, 1994, **4**, 1473.
77. M. Michalska, W. Kudelska, J. Pluskowski, A. E. Kozioł, and T. Lis, *J. Chem. Soc. Perkin Trans. I*, 1994, 979.
78. C. Tang, M. Zhang, and D. Zhang, *Gaodeng Xuexiao Huaxue Xuebao*, 1992, **13**, 1406; *Chem. Abstracts*, 1993, **119**, 49453.
79. C. C. Tang, M. J. Zhang, F. P. Ma, and Z. Li, *Chin. Chem. Lett.*, 1993, **4**, 949; *Chem. Abstracts*, 1994, **120**, 298749.
80. G. L. Kolomnikova, T. O. Krylova, and Yu. G. Gololobov, *Russ. J. Gen. Chem.,(Engl. Transl.)*, 1993, **63**, 506.
81. M. Sasaki, Y. Itoh, N. Yamamoto, and Y. Yamada, *Chem Lett.*, 1993, 1835.
82. C. le Roux, S. Bauermeister and T. A. Modro, *J. Chem. Res. (S)*, 1993, 372.
83. S. G. Zlotin, M. V. Sharashkina, Yu. A. Strelenko, O. A. Luk'yanov, *Bull. Acad. Sci. USSR, Div. Chem. Sci.*, 1992, 902.
84. K. Osowska-Pacewicka, S. Zowaszki, and A. Zwierzak, *Phosphorus, Sulfur, Silicon, Relat. Elem.*, 1993, **82**, 49.
85. Z. Liu and W. Chen., *Chin. Chem. Lett.*, 1992, **3**, 685; *Chem. Abstracts*, 1993, **119**, 95651.
86. S. Hünsch, W. Richter, I. Ugi, and J. Chattopadhyaya, *Justus Liebigs Ann. Chem.*, 1994, 269.
87. S. Q. Chen and J. C. Xu, *Chin. Chem. Lett.*, 1993, **4**, 303.
88. Y. H. Lee, C-H. Lee, and J. H. Lee, *Bull. Korean Chem. Soc.*, 1993, **14**, 415.
89. J-M. Brunel and G. Buono, *J. Org. Chem.*, 1993, **58**, 7313.
90. K. Troev, *Rev. Heteroat. Chem.*, 1993, **8**, 165.
91. P. N. Nagar, *Phosphorus, Sulfur, Silicon, Relat. Elem.*, 1993, **79**, 207.
92. M. M. Zolutukhin, V. I. Krutikov, and A. N. Lavrent'ev, *Russ. Chem. Rev. (Engl. Transl.)*, 1993, **62**, 647.
93. Y. Okamoto and S. Takamuku, *Rev. Heteroat. Chem.*, 1993, **8**, 183.
94. G. Keglevich, *Synthesis*, 1993, 931.
95. C. Patois, S. Berte-Verrando, and P. Savignac, *Bull. Soc. Chim. France*, 1993, **130**, 485.
96. I. E. Gurevich, A. V. Dogadina, B. I. Ionin, and A. A. Petrov, *Russ. J. Gen. Chem. (Engl. Transl.)*, 1993, **63**, 85.
97. V. E. Kolbina, G. V. Dolgushin, and V. G. Rozinov, *J. Gen. Chem. USSR (Engl. Transl.)*, 1992, **62**, 1179.
98. R. I. Yurchenko and L. P. Peresypkina, *J. Gen. Chem. USSR (Engl. Transl.)*, 1992, **62**, 1971.

99. A. A. Krolovets, G. A. Kirillov, and B. I. Martynov, *J. Gen. Chem. USSR (Engl. Transl.)*, 1992, **62**, 2296.
100. B. L. Tumanski, V. V. Bashilov, N. N. Bubnov, and S. P. Solodovnikov, *Bull. Acad. Sci. USSR. Div. Chem. Sci. (Engl. Transl.)*, 1992, 1519.
101. B. L. Tumanski, V. V. Bashilov, N. N. Bubnov, S. P. Solodovnikov, and V. I. Sokolov, *Bull. Acad. Sci. USSR., Div. Chem. Sci.*, 1992, 1521.
102. E. A. Boyd and A. C. Regan, *Tetrahedron Lett.*, 1994, **35**, 4223.
103. A. K. Brel', A. A. Gunger, and A. A. Ozerov, *Russ. J. Gen. Chem. (Engl. Transl.)*, 1993, **63**, 656.
104. G. Haegele, R. Peters, K. Kreidler, R. Boese, G. Grossmann, F. Steglch, *Phosphorus, Sulfur, Silicon, Relat. Elem.*, 1993, **83**, 77.
105. D. Villemin and A. Ben Alloum, *Phosphorus, Sulfur, Silicon, Relat. Elem.*, 1993, **79**, 33.
106. S. Jin and C. Yuan, *Huaxue Shiji*, 1993, **15**, 8; *Chem. Abstracts* , 1993, **119**, 49481.
107. D. Villemin, F. Thibault-Starzyk, and M. Hachemi, *Synth. Commun.*, 1994, **24**, 1425.
108. C. Li and C. Yuan, *Heteroat. Chem.*, 1993, **4**, 517.
109. L. Schmitt, B. Spiess, and G. Schlewer, *Tetrahedron Lett.*, 1993, **34**, 7059.
110. A. M. Polozov, and A. Kh. Mustafin, *Russ. J. Gen. Chem.,(Engl. Transl.)*, 1993, **63**, 169.
111. H. Gross, B. Costisella, I. Keitel, and S. Ozegowski, *Phosphorus. Sulfur, Silicon, Relat. Elem.*, 1993, **83**, 203.
112. C. Patois and P. Savignac, *Bull. Soc. Chim. France*, 1993, **130**, 630.
113. V. I. Vysotskii and A. G. Vilitkevitch, *Russ. J. Gen. Chem. (Engl.Transl.)*, 1993, **63**, 253.
114. S. C. Wong, N. I. Carruthers and T. M. Chan, *J. Chem. Res. (S)*, 1993, 268.
115. R. P. Parg, J. D. Kilburn, and T. G. Ryan, *Synthesis*, 1994, 195.
116. C. Li, D. B. Zhou, and C. Yuan, *Chin. Chem. Lett.*, 1993, **4**, 391; *Chem. Abstracts*, 1993, **119**, 271259.
117. S. Berte-Verrando, F. Nief, C. Patois, and P. Savignac, *J. Chem. Soc. Perkin Trans. I*, 1994, 821.
118. M. Ruiz, V. Ojea, G. Shapiro, H-P. Weber, and E. Pombo-Villar, *Tetrahedron Lett.*, 1994, **35**, 4551.
119. M. Maffei and G. Buono, *Phosphorus, Sulfur, Silicon, Relat. Elem.*, 1993, **79**, 297.
120. Y. Shen and B. Yang, *Synth. Commun.*, 1993, **23**, 3081.
121. J. J. Kiddle and J. H. Babler, *J. Org. Chem.*, 1993, **58**, 3572.
122. N. N. Demik, M. M. Kabachnik, Z. S. Novikova, and I. P. Beletskaya, *Bull. Acad. Sci. USSR, Div. Chem. Sci. (Engl. Transl.)*, 1992, 1913.
123. J.L. Kelley, J. A. Linn, E. W. McLean and J. V. Tuttle, *J. Med. Chem.*, 1993, **36**, 2355.
124. V. I. Kal'chenko, L. I. Atamas', V. V. Pirozhenko, and L. N. Markovskii, *J. Gen. Chem. USSR (Engl. Transl.)*, 1992, **62**, 2161.
125. J. Paladino, C. Guyard, C. Thurieau, and J. L. Fauchère, *Helv. Chim. Acta*, 1993, **76**, 2465.
126. S. Masson, J.-F. Saint-Clair, and M. Saquet, *Tetrahedron Lett.*, 1994, **35**, 3083.
127. P. P. Onys'ko, E. A. Suvalova, T. I. Chudakova, and A. D. Sinitsa, *Heteroat. Chem.*, 1993, **4**, 361.
128. M. El Malouli Bibout, A. Hannioui, G. Peiffer, A. Samat and K. H. Pannell, *Synth. Commun.*, 1993, **23**, 2273.
129. A. V. Khotinen and A. M. Polozov, *Phosphorus, Sulfur, Silicon, Relat. Elem.*, 1993, **83**, 53.
130. O. A. Attanasi, P. Filippone, D. Giovagnoli, and A. Mei, *Synthesis*, 1994, 181.
131. O. A. Attanasi, P. Filippone, D. Giovagnoli, and A. Mei, *Synth. Commun.*, 1994, **24**, 453.

132. H. Gross, B. Costisella, I. Keitel, H. Sonnenschein, and A. Kunath, *Phosphorus, Sulfur, Silicon, Relat. Elem.*, 1993, **84**, 129.
133. R. Andriamiadanarivo, B. Pujol, B. Chantegrel, C. Deshayes, and A. Doutheau, *Tetrahedron Lett.*, 1993, **34**, 7923.
134. G. Falsone, F. Cateni, M. M. De Nardo, and M. M. Darai, *Z. Naturforsch.* 1993, **B48**, 1391.
135. B. Huang, K. Wang, W. Huaang, and D. Prescher, *Chin. J. Chem.*, 1993, **11**, 169; *Chem. Abstracts*, 1994, **120**, 134701.
136. W. Qiu and D. J. Burton, *J. Fluorine Chem.*, 1993, **62**, 273.
137. D. B. Berkowitz, M. Eggen, Q. Shen, and D. G. Sloss, *J. Org. Chem.*, 1993, **58**, 6174.
138. C. Patois and P. Savignac, *J. Chem. Soc., Chem. Commun.*, 1993, 1711.
139. C. Patois and P. Savignac, *Synth. Commun.*, 1994, **24**, 1317.
140. S. Chen and C. Yuan, *Phosphorus, Sulfur, Silicon, Relat. Elem.*, 1993, **82**, 73.
141. G. G. Furin, *Russ. Chem. Rev. (Engl. Transl.)*, 1993, **62**, 243.
142. H. K. Nair, R. D. Guneratne, A. S. Modak, and D. J. Burton, *J. Org. Chem.*, 1994, **59**, 2393.
143. T. C. Sanders and G. B. Hammond, *J. Org. Chem.*, 1993, **58**, 5598.
144. R. S. Gross, S. Mehdi and J. R. McCarthy, *Tetrahedron Lett.*, 1993, **34**, 7197.
145. Y. Shen and M. Qi, *J. Chem. Soc. Perkin Trans. I*, 1993, 2153.
146. W. Cen, X. Dai, and Y. Shen, *J. Fluorine Chem.*, 1993, **65**, 49.
147. Y. Shen and M. Qi, *J. Chem. Soc. Perkin Trans. I*, 1994, 1179.
148. S. K. Tukanova, B. Zh. Dzhiembaev, S. F. Khalikova, and B. M. Butin, *J. Gen. Chem. USSR (Engl. Transl.)*, 1992, **62**, 2302.
149. H. Gross, B. Costisella, S. Ozegowski, and I. Keitel, *Phosphorus, Sulfur, Silicon, Relat. Elem.*, 1993, **84**, 121.
150. R. K. Ismagilov, V. V. Moskva, D. B. Bagautdinova, V.P. Arkipov, and L. Yu. Kopylova, *J. Gen. Chem. USSR (Engl. Transl.)*, 1992, **62**, 2054.
151. T. Yokomatsu, T. Yamagishi, and S. Shibaya, *Tetrahedron:Asymmetry*, 1993, **4**, 1783.
152. N. P. Rath and C. D. Spilling, *Tetrahedron Lett.*, 1994, **35**, 227.
153. V. J. Blazis, K. J. Koeller, and C. D. Spilling, *Tetrahedron: Asymmetry*, 1994, **5**, 499.
154. N. J. Gordon and S. E. Evans, *J. Org. Chem.*, 1993, **58**, 5293.
155. T. Yokomatsu, T. Yamagishi, and S. Shibuya, *Tetrahedron: Asymmetry*, 1993, **4**, 1779.
156. V. Sum and T. P. Kee, *J. Chem. Soc., Perkin Trans. I*, 1993, 2701.
157. V. Sum and T. P. Kee, *J. Chem. Soc., Perkin Trans., I*, 1993, 1369.
158. Z. Glowacki, M. Hoffmann, and J. Rachon, *Phosphorus, Sulfur, Silicon, Relat. Elem.*, 1993, **82**, 39.
159. L. S. Boulos and M. H. N. Arsanious, *Tetrahedron* , 1993, **49**, 4711.
160. M. Gulea-Purcarescu, E. About-Jaudet, and N. Collignon, *J. Organomet. Chem.*, 1994, **464**, C14.
161. E. L. Muller and T. A. Modro, *Bull. Soc. Chim. France*, 1993, **130**, 668.
162. K. P. Gerber, H. M. Roos, and T. A. Modro, *J. Mol. Struct.*, 1993, **296**, 85.
163. M-P. Belaug, T. A. Modro, and P. L. Wessels, *J. Phys. Org. Chem.*, 1993, **6**, 523.
164. C. P. de Jongh and T. A. Modro, *Phosphorus, Sulfur, Silicon, Relat. Elem.*, 1993, **79**, 161.
165. Z. Li, S. Racha, Li Dan, H. El-Subbagh, and E. Abushanab, *J. Org. Chem.*, 1993, **58**, 5779.
166. J-L. Montchamp, M. E. Migaud, and J. W. Frost, *J. Org. Chem.*, 1993, **58**, 7679.
167. A. P. Kozikowski, G. Powis, A. Gallegos, and W. Tüchmantel, *Bioorg. Med. Chem. Lett.*, 1993, **3**, 1323.
168. K. Frische and R. R. Schmidt, *Justus Liebigs Ann. Chem.*, 1994, 297.

169. T. Hanaya, R. Okamoto, Y. V. Prikhod'ko, M. Armour, A. M. Hogg, and H. Yamamoto, *J. Chem. Soc., Perkin Trans I*, 1993, 1663.
170. T. Hanaya, K. Yasuda, H. Yamamoto, and H. Yamamoto, *Bull. Chem. Soc. Jpn.*, 1993, **66**, 2315.
171. T. Hanaya, K. Hirose, and H. Yamamoto, *Heterocycles*, 1993, **36**, 2557.
172. M. Yamashita, A. Yabui, T. Oshikawa, and A. Kakehi, *Chem Lett.*, 1994, 23.
173. J. W. Darrow and D. G. Drueckhammer, *J. Org. Chem.*, 1994, **59**, 2976.
174. W. L. McEldoon, K. Lee, and D. F. Wiemer, *Tetrahedron Lett.*, 1993, **34**, 5843.
175. K. Kossev, K. Troev, and D. M. Roundhill, *Phosphorus, Sulfur, Silicon, Relat. Elem.*, 1993, **83**, 1.
176. D. M. Malenko, V. V. Simurova, and A. D. Sinitsa, *Russ. J. Gen. Chem. (Engl. Transl.)*, 1993, **63**, 657.
177. C. E. Burgos-Lepley, S. A. Mizsak, R. A. Nugent, and R. A. Johnson, *J. Org. Chem.*, 1993, **58**, 4159.
178. R. Y. Chen, H. Y. Li, and B. Z. Cai, *Chin. Chem. Lett.*, 1993, **4**, 403.
179. K. Kostka, R. Modranka, M. J. Grabowski, and A. Stepień, *Phosphorus, Sulfur, Silicon, Relat. Elem.*, 1993, **83**, 209.
180. T. A. Olszak, E. Stepień, M. Renz, M. J. Grabowski, K. Kostka, and R. Modranka, *Acta Crystallogr. Sec. C;Cryst. Struct. Commun.*, 1994, **50**, 284.
181. N. J. Gordon and S. A. Evens, *J. Org. Chem.*, 1993, **58**, 5295.
182. R. Sakoda, H. Matsumoto, and K. Seto, *Synthesis*, 1993, 705.
183. J. Petrova, Z. Zolravkova, J.C. Tebby, and E. T. K. Haupt, *Phosphorus, Sulfur, Silicon, Relat. Elem.*, 1993, **81**, 89.
184. F. I. Guseinov, V. V. Moskva, and V. M. Ismailov, *Russ. J. Gen. Chem. (Engl. Transl.)*, 1993, **63**, 66.
185. R. Hamilton, M. A. McKervey, M. D. Rafferty, and B. J. Walker, *J. Chem.Soc., Chem. Commun.*, 1994, 37.
186. R. Neidlein and B. Matuschek, *Monatsh. Chem.*, 1993, **124**, 789.
187. M. M. Kabat, *Tetrahedron Lett.*, 1993, **34**, 8543.
188. D. Y. Kim, J. Y. Mang, J. W. Lee, and D. Y. Oh, *Bull. Korean Chem. Soc.*, 1993, **14**, 526; *Chem. Abstracts*, 1994, **120**, 134638.
189. G. M. Baranov and V. V. Perekalin, *Russ. Chem. Rev. (Engl. Transl.)*, 1992, **61**, 1220.
190. V. P. Kuk'har, N. Yu. Zvistunova, V. A. Solodenko, and V. A. Soloshonok, *Russ. Chem. Rev. (Engl. Transl.)*, 1993, **62**, 261.
191. C. Hubert, B. Oussaid, G. Etemad-Moghadam, M. Koenig, and B. Garrigues, *Synthesis*, 1994, 51.
192. D. Green, G. Patel, S. Elgendy, J. A. Baban, G. Claeson, V. V. Kakkar, and J. Deadman, *Tetrahedron*, 1994, **50**, 5099.
193. C. Yuan, S. Li, G. Wang, and Y. Ma, *Phosphorus, Sulfur, Silicon, Relat. Elem.*, 1993, **81**, 27.
194. R-Y. Chen and X-R. Chen, *Phosphorus, Sulfur, Silicon, Relat. Elem.*, 1993, **83**, 99.
195. S. Failla, P. Finocchiaro, G. Haegele, and R. Rapisardi, *Phosphorus, Sulfur, Silicon, Relat. Elem.*, 1993, **82**, 79.
196. C. Y. Yuan, W. S. Huang, and Z. B. Yan, *Chin. Chem. Lett.*, 1993, **4**, 1047; *Chem. Abstracts*, 1994, **120**, 323709.
197. X. Jiao, C. Verbruggen, M. Barloo, W. Bollaert, A. De Groot, R. Dommisse, and A. Haemers, *Synthesis*, 1994, 23.
198. V. I. Krutikov and A. N. Lavrent'ev, *Russ. J. Gen. Chem. (Engl. Transl.)*, 1993, **63**, 73.
199. R. Gancarz, *Phosphorus, Sulfur, Silicon, Relat. Elem.*, 1993, **83**, 59.
200. J. P. Finet, C. Frejaville, R. Lauricella, F. Le Moigne, P. Stipa, and P. Tordo, *Phosphorus, Sulfur, Silicon, Relat. Elem.*, 1993, **81**, 17.

201. Y. H. Zhang, W. Q. Huang, A. J. Men, and B. L. He, *Chin. Chem Lett.*, 1993, **4**, 203.
202. D. G. Cameron, H. R. Hudson, and M. Pianka, *Phosphorus, Sulfur, Silicon, Relat. Elem.*, 1993, **83**, 21.
203. J. Oleksyszyn, B. Boduszek, C-M. Kam, and J. C. Powers, *J. Med. Chem.*, 1994, **37**, 226.
204. G. Sturtz and H. Couthon, *Comptes. Rendus Acad. Sci., Ser II*, 1993, **316**, 181.
205. Z. H. Kudzin, G. Andrjewski, and J. Drabowitz, *Heteroat. Chem.* 1994, **5**, 1.
206. C. Harde, K.-H. Neff, E. Nordhoff, K.-P. Gerbling, B. Laber, and H.-D. Pohlenz, *Bioorg. Med. Chem. Lett.*, 1994, **4**, 273.
207. W. J. Moree, G. A. van der Marel, J. H. van Boom, and R. M. J. Liskamp, *Tetrahedron*, 1993, **49**, 11055.
208. R. Jacquier, M. Lhassani, C. Petrus, and F. Petrus, *Phosphorus, Sulfur, Silicon, Relat. Elem.*, 1993, **81**, 83.
209. C. Yuan, S. Li, G. Wang, and H. Wu, *Chin. Chem. Lett.*, 1993, **4**, 753; *Chem. Abstracts*, 1994, **120**, 270544.
210. B. Costisella, I. Keitel, and S. Ozegowski, *Phosphorus, Sulfur, Silicon, Relat. Elem.*, 1993, **84**, 115.
211. D. Green, G. Patel, S. Elgendy, J. A. Baban, G. Claeson, V. V. Kakkar, and J. Deadman, *Tetrahedron Lett.*, 1993, **34**, 6917.
212. D. Maffre, P. Dumy, J-P. Vidal, R. Escale, and J-P.Girard, *J. Chem. Res. (S).*, 1994, 30.
213. T. Yokomatsu, T. Yamagishi, and S. Shibuya, *Tetrahedron: Asymmetry*, 1993, **4**, 1401.
214. C. Yuan, S. Chen, H. Zhou, and L. Maier, *Synthesis*, 1993, 955.
215. R. Neidlein and H. Keller, *Heterocycles*, 1993, **36**, 1925.
216. R. Jacquier and C. Petrus, *Phosphorus, Sulfur, Silicon, Relat. Elem.*, 1993, **78**, 71.
217. C. Yuan, G. Wang, H. Feng, J. Chen. and L. Maier, *Phosphorus, Sulfur, Silicon, Relat. Elem.*, 1993, **81**, 149.
218. S. Shatzmiller, R. Neidlein, and C. Weik, *Synth. Commun.*, 1993, **23**, 3009.
219. S. Shatzmiller, R. Neidlein, and C. Weik, *Justus Liebigs Ann. Chem.*, 1993, 955.
220. X-Y. Jiao, M. Barloo, C. Verbruggen, and A. Haemers, *Tetrahedron Lett.*, 1994, **35**, 1103.
221. R. Neidlein, P. Grenlich, and W. Kramer, *Helv. Chim. Acta*, 1993, **76**, 2407.
222. I. M. Aladzheva, I. L. Odinets, P. V. Petrovskii, T. A. Mastryukova, and M. I. Kabachnik, *Russ. J. Gen. Chem. (Engl. Transl.)*, 1993, **63**, 431.
223. C. C. Tang and G. P. Wu, *Chin. Chem. Lett.*, 1992, **3**, 967.
224. C. C. Tang, G. P. Wu, G. Y. Huang, Z. Li, and G. Y. Lin, *Phosphorus, Sulfur, Silicon, Relat. Elem.*, 1993, **84**, 159.
225. Purnanend and B. S. Batra, *Ind. J. Chem. Sec.B*, 1992, **31**, 778.
226. M. Yoshifuji, De-Lie An, K. Toyota, and M. Yasunami, *Tetrahedron Lett.*, 1994, **35**, 4379.
227. H. Beckmann, G. Grosmann, G. Ohms, and J. Sieler, *Heteroat. Chem.*, 1994, **5**, 73.
228. R. Shabana, F. H. Osman, and S. S. Atrees, *Tetrahedron*, 1994, **50**, 6975.
229. I. S. Nizamov, L. A. Al'metkina, É. S. Batyeva, V. A. Al'fonsov, and A. N. Pudovik, *Bull. Acad. Sci. USSR, Div. Chem. Sci.,(Engl. Transl.)* 1992, 1516.
230. I. S. Nizamov, V. A. Kuznetsov, É. S. Batyeva, and V. A. Al'fonsov, *Bull. Acad. Sci. USSR, Div. Chem. Sci.,(Engl. Transl.)*, 1992, 1936.
231. M. Yoshifuji, S. Sangu, M. Hirano, and K. Toyota, *Chem. Lett.*, 1993, 1715.
232. K. Toyota, Y. Ishikawa, K. Shirabe, M. Yoshifuji, K. Okada, and K. Hirotsu, *Heteroat. Chem.*, 1993, **4**, 279.
233. M. J. P. Harger, *Tetrahedron Lett.*, 1993, **34**, 7947.
234. B. Burns, E. Merifield, M. F. Mahon, K. C. Molloy, and M. Wills, *J. Chem. Soc., Perkin Trans I*, 1993, 2243.

235. S. E. Bottle, R. C. Bott, I. D. Jenkins, C. H. L. Kennard, G. Smith, and A. P. Wells, *J. Chem. Soc., Chem. Commun.*, 1993, 1684.
236. J. Mitjaville, A-M. Caminade, R. Mathieu, and J-P. Majoral, *J. Amer. Chem. Soc.*, 1994, **116**, 5007.
237. M. W. Wieczorek, J. Blaszczyk, and B. Krzyzanowska, *Heteroat. Chem.*, 1993, **4**, 399.
238. R. Bittman, H-S. Byun, B. Mercier, and H. Salari, *J. Med. Chem.* 1994, **37**, 425.
239. G. S. Poindexter, M. A. Foley, J. E. Macdonald, J. G. Sarmento, C. Bryson, G. D. Goggins, R. L. Cananagh, and J. P. Buyniski, *Bioorg. Med. Chem. Lett.*, 1993, **3**, 2817.
240. P. Raddatz, K-O. Minck, F. Rippmann, and C-J. Schmitges, *J. Med. Chem.*, 1994, **37**, 486.
241. M. Prashed, *Bioorg. Med. Chem. Lett.*, 1993, **3**, 2051.
242. K. Lee and D. F. Wiemer, *J. Org. Chem.*, 1993, **58**, 7808.
243. L. J. Jennings and M. J. Parratt, *Bioorg. Med. Chem. Lett.*, 1993, **3**, 2611.
244. J. Jindrich, A. Holý, and H. Dvořáková, *Coll. Czech. Chem. Comm.*, 1993, **58**, 1645.
245. P. Alexander and A. Holý, *Coll. Czech. Chem. Comm.*, 1993, **58**, 1157.
246. K-L. Yu, J. J. Bronson, H. Yang, A. Patick, M. Alam, V. Brankovan, R. Datema, M. J. M. Hitchcock, and J. C. Martin, *J. Med. Chem.*, 1993, **36**, 2726.
247. J. L. Stanton, G. M. Ksander, R. de Jesus, and D. M. Sperbeck, *Bioorg. Med. Chem. Lett.*, 1994, **4**, 539.
248. R-Y. Chen, Y-H. Zhang, and M-R. Chen., *Heteroat. Chem.*, 1993, **4**, 1
249. C. Yang, R. Qamar, S. J. Norton, and P. F. Cook, *Tetrahedron*, 1994, **50**, 1919.
250. Yu. I. Morozik and Yu. K. Knobel', *Russ. J. Gen. Chem.,(Engl. Transl.)*, 1993, **63**, 235.
251. A. A. Krolevets, V. V. Antipova, P. V. Petrovskii, and I. V. Martynov, *Bull. Acad. Sci. USSR., Div. Chem. Sci.,(Engl. Transl.)*, 1992, 153.
252. J. S. McCallum and L. S. Liebeskind, *Synthesis*, 1993, 819.
253. J-M. Campagne, J. Coste, and P. Jonin, *Tetrahedron Lett.* 1993, **34**, 6743.
254. D. A. Campbell and J. C. Bermak, *J. Org. Chem.*, 1994, **59**, 658.
255. D. S. Tawfik, Z. Eshhar, A. Bentolila, and B. S. Green, *Synthesis*, 1993, 968.
256. G. Cevasco, S. Penco, and S. Thea, *Phosphorus, Sulfur, Silicon, Relat. Elem.*, 1993, **84**, 257.
257. G. Cevasco and S. Thea, *J. Chem. Soc. Perkin Trans. 2*, 1994, 1103.
258. R. Zhang, J. Chen, and J. Qu, *Huaxue Xuebao*, 1993, **51**, 1016; *Chem. Abstracts*, 1994, **120**, 217894.
259. S. M. Ruder and V. R. Kulkarni, *Synthesis*, 1993, 945.
260. J. P. Gerber and T. A. Modro, *Phosphorus, Sulfur, Silicon, Relat. Chem.*, 1993, **84**, 107.
261. P. Bałczewski, P. P. Graczyk, W. Perlikowska, and M. Mikołajczyk, *Tetrahedron*, 1993, **49**, 10111.
262. B. A. Kashemirov, V. N. Osipov, N. F. Savenkov, B. Ya. Chvertkin, and P. S. Khokhlov, *J. Gen. Chem. USSR. (Engl. Transl.)*, 1992, **62**, 1042.
263. S. E. Denmark and C-T. Chen. *J. Org. Chem.*, 1994, **59**, 2922.
264. Z. I. Glebova and Yu. A. Zhdanov, *J. Gen. Chem. USSR (Engl. Transl.)*, 1992, **62**, 1972.
265. B. I. Buzykin, M. P. Sokolov, T. A. Zyablikova, and L. F. Chertanova, *J. Gen. Chem. USSR (Engl. Transl.)*, 1992, **62**, 1222.
266. L. F. Chertanova, B. I. Buzykin, A. A. Gazikasheva, and M. P. Sokolov, *Bull. Acad. Sci. USSR, Div. Chem. Sci., (Engl. Transl.)*, 1992, 2162.
267. F. I. Guseinov, G. Yu. Klimentova and V. V. Moskva, *Russ. J. Gen. Chem. (Engl.Transl.)*, 1993, **63**, 501.

268. A. D. Mikityuk, N. V. Kalinina, and P. S. Khokhlov, *Russ. J. Gen. Chem. (Engl. Transl.)*, 1993, **63**, 654.
269. P. S. Khokhlov, A. D. Mikityuk, and N. V. Kalinina, *Russ. J. Gen. Chem. (Engl. Transl.)* 1993, **63**, 655.
270. B. A. Kashemirov and P. S. Khokhlov, *Russ. J. Gen. Chem. (Engl. Transl.)*, 1993, **63**, 999.
271. M. Sprecher and D. Kost, *J. Amer. Chem. Soc.*, 1994, **116**, 1016.
272. F. Hammerschmidt, *Angew. Chem., Int. Ed. Engl.*, 1994, **33**, 341.
273. F. Hammerschmidt, *Monatsh. Chem.*, 1993, **124**, 1063.
274. Y-F. Li and F. Hammerschmidt, *Tetrahedron:Asymmetry*, 1993, **4**, 109.
275. N. A. Kaslina, I. A. Polyakova, A. V.Kessenikh, B. V. Zhadanov, L. V. Krinitskaya, and T. M. Balasheva, *J. Gen. Chem. USSR (Engl. Transl.)*, 1992, **62**, 1284.
276. C. P. de Jongh, T. A. Modro, and A. M. Modro, *Heteroat. Chem.*, 1993, **4**, 503.
277. K. Lee, J. A. Jackson, and D. F. Wiemer, *J. Org. Chem.*, 1993, **58**, 5967.
278. M. Mahajna and E. Breuer, *J. Org. Chem.*, 1993, **58**, 7822.
279. Yu. G. Safina, G. Sh. Malkova, and R. A. Cherkasov, *J. Gen. Chem. USSR, (Engl.Transl.)*, 1992, **62**, 1275.
280. V. V. Ovchinnikov, A. A. Sobanov, and A. N. Pudovik, *Doklady. Akad. Nauk. USSR*, 1993, **333**, 48; *Chem. Abstracts*, 1994, **120**, 217891.
281. I. S. Alfer'ev and N. V. Mikhalin, *Bull. Acad. Sci. USSR, Div. Chem. Sci. (Engl. Transl.)*, 1992, 1709.
282. M. R. Tirakyan, G. A. Panosyan, Yu. M. Dangyan, M. S. Aleksanyan, Yu. T. Struchkov, T. S. Kurtikyan, and Sh. O. Badanyan, *J. Gen. Chem. USSR, (Engl. Transl.)*, 1992, **62**, 2215.
283. C. Yuan and C. Li, *Tetrahedron Lett.*, 1993, **34**, 5959.
284. C. Yuan and C. Li, *Phosphorus, Sulfur, Silicon, Relat. Elem.*, 1993, **78**, 47.
285, S. Bieler and K. Kellner, *J. Organomet. Chem.*, 1993, **447**, 15.
286. O. A. Kolyamshin, V. V. Kormachev, and Yu. N. Mitrasov, *Russ. J. Gen. Chem. (Engl. Transl.)*, 1993, **63**, 750.
287. G. Keglevich, L. Tőke, A. Kovács, G. Tóth, and K. Újszászy, *Heteroat. Chem.*, 1993, **4**, 61.
288. G. Keglevich, A. Kovács, L. Tőke, and K. Újszászy , *Heteroat. Chem.*, 1993, **4**, 329.
289. D. Villemin, F. Sauvaget, and M. Hajèk, *Tetrahedron Lett.*, 1994, **35**, 3537.
290. D. Hough, *Tetrahedron*, 1994, **50**, 3177.
291. F. Palacios and A. M. Ochoa de Retana, *Heterocycles*, 1994, **38**, 95.
292. A. Elachqar, A. El Hallaoui, M. L. Roumestant, and P. Viallefont *Synth. Commun.*, 1994, **24**, 1279.
293. R. C. F. Jones, G. Bhalay, and P. A. Carter, *J. Chem. Soc. Perkin Trans. I*, 1993, 1715.
294. B. I. Buzykin and M. P. Sokolov, *J. Gen. Chem. USSR, (Engl. Transl.)*, 1992, **62**, 1868.
295. V. A. Pavlov, A. I. Kurdyukov, V. V. Plemenkov, R. R. Khaliullin, and V. V. Moskva, *Russ. J. Gen. Chem., (Engl. Transl.)*, 1993, **63**, 449.
296. D. Villemin and F. Thibault-Starzyk, *Phosphorus, Sulfur, Slicon, Relat. Elem.*, 1993, **80**, 251.
297. A. Loeckritz and M. Schnell, *Phosphorus, Sulfur, Silicon, Relat. Elem.*, 1993, **83**, 125.
298. L. D. Quin, X-P. Wu, N. D. Sadanani, I. Lukes, A.S. Ionkin, and R. O. Day, *J. Org. Chem.*, 1994, **59**, 120.
299. M. Nakamura, K. Sawasaki, Y. Okamoto, and S. Takamuku, *J. Chem. Soc. Perkin Trans. I*, 1994, 141.
300. M. J. P. Harger, *J. Chem. Res. (S)*. 1993, 334.

301. H. M. I. Osborn, A. A. Cantrill, and J. B. Sweeney, *Tetrahedron Lett.*, 1994, **35**, 3159.
302. J. Fawcett, M. J. P. Harger, D. R. Russell, and R. Sreedharan-Menou, *J. Chem. Soc. Chem. Comm.*, 1993, 1826.
303. S. G. Zlotin, M. V. Sharashkina, Yu. A. Strelenko, and O. A. Luk'yanov, *Bull. Acad. Sci. USSR, Div. Chem. Sci., (Engl. Transl.)*, 1992, 2096.
304. R. M. Kamalov, A. Schmidpeter, and K. Polborn, *Phosphorus, Sulfur, Silicon, Relat. Elem.*, 1993, **83**, 111.
305. O. N. Nuretdinova and V. G. Novikova, *Bull. Acad. Sci. USSR, Div. Chem. Sci. (Engl. Transl.)*, 1992, 2121.
306. J. Omelanczuk, *Tetrahedron*, 1993, **49**, 8887.
307. M. Mikołajczyk, J. Omelanczuk, W. Perlikowska, L. N. Markovski, V. D.Romanenko, A. V. Ruban, and A. B. Drapailo, *Phosphorus Sulfur*, 1988, **36**, 267.
308. E. Krawczyk, A. Skowronska, and J. Michalski, *J. Chem. Soc., Perkin Trans. I*, 1994, 89.

5
Nucleotides and Nucleic Acids

BY J. A. GRASBY AND D. M. WILLIAMS

Introduction

The main driving force behind the chemistry of nucleotides and nucleic acids continues to be the demand for novel chemotherapeutics. Over the past few years this motivation has extended from the traditional small nucleotide drugs to the bio-polymers DNA and RNA and more importantly their modified analogues. In addition, the understanding of recognition processes involving nucleic acids, with each other, proteins and small molecules is an area of considerable interest which spans from structural studies, such as NMR and crystallography, to the synthesis of modified nucleotides and nucleic acids. Since this review did not appear last year we have endeavoured to include some material form early 1993 and therefore cover publications from January 1993 to June 1994.

1. Mononucleotides

1.1 Nucleoside Acyclic Phosphates

There has been continuing interest in the development of nucleotide prodrugs which allow intracellular delivery of the nucleotides and their analogues to the target whereupon they are converted to their bioactive forms. An excellent review in this area has appeared.[1]

Several 2', 3'-dideoxynucleosides (ddN), in addition to others such as 3'-azido-3'-deoxythymidine (AZT), are potent inhibitors of HIV in cell culture due to the inhibition of HIV-reverse transcriptase (HIV-RT) by their 5'-triphosphate derivatives. The activity of the nucleosides depends upon the presence of cellular kinases which are often in low abundance or absent. Since the polar nucleoside phosphates are unable to penetrate the cell membrane much effort has been invested in recent years to 'deliver' them as lipophilic prodrugs which also alleviates the problem of kinase deficient viral strains. McGuigan and co-workers have shown that *bis*(2, 2, 2-trichloroethyl) phosphotriester analogues of several 3'-*O*-alkyl, sulphonyl and acyl modified thymidine analogues (**1**), which were previously devoid of anti-HIV activity, are consequently activated against the virus.[2, 3] The analogous phosphotriester of 2', 3'-dideoxyuridine (ddU) is also found to be active against HIV[4], whereas the parent nucleoside is essentially inactive. Moreover, this effect is retained in thymidine kinase deficient (TK^-) cells. The term 'kinase bypass' is used by these authors to describe the mode of action of these prodrugs and is an extension of earlier work by this group. The analogues are prepared by treating the respective nucleoside with *bis*(2, 2, 2-trichloroethyl) phosphorochloridate (**2**) in THF at room temperature in the presence of N-methyl imidazole.[4]

Specific carboxyesterase activation of 4-acyloxybenzyl phosphotri- and di-esters of AZT by cleavage of the acyloxy groups (Scheme 1) has been shown [5] but the compounds have comparable efficacies to AZT against HIV, but are more cytotoxic. Like AZT the compounds were inactive against thymidine kinase deficient cells despite evidence that some decomposition to the 5'-monophosphate of AZT occurs intracellularly.

The aryloxyphosphoramidates of AZT[6] and ddU[4], which contain amide-bound amino acid derivatives and ester-bound aryl alcohols which are cleaved by cellular non-specific esterases, are highly active against HIV in both TK^+ and TK^- strains. The derivative (**3**) displays a 50-fold lower ED_{50} than the parent nucleoside whereas the corresponding *bis*(2, 2, 2-trichloroethyl) phosphotriester is only 5-fold more active. Interestingly the corresponding ethyl triester of (**3**) is inactive.[4]

The *bis* (*o*, *m* and *p*-nitrophenyl)phosphate triesters of AZT[7, 8] have been found to be potent and non-toxic inhibitors of HIV in TK^+ strains, but display comparably poor activity in TK^- strains to that of AZT. Thus their mode of action differs from that of the phosphoramidates mentioned earlier in that they act as prodrugs of the free nucleoside rather than as nucleotide delivery forms. A study of several aryl phosphotriesters of AZT[8] has revealed that the anti-HIV activity is inversely related to the pKa of the respective phenol and probably therefore related to its leaving group ability. It has been shown that the triester is hydrolysed in human plasma to the corresponding diester[7] which is probably dephosphorylated extracellularly and thus enters the cell as the free nucleoside by diffusion.

A novel synthesis of the H-phosphonate derivative of AZT (**4**) has been described [9]which involves decarboxylation of a phosphonoformate intermediate (**5**). Preliminary results suggest that this is converted intracellularly to the monophosphate.

Several aza and deaza analogues of the anti-HIV agent 2', 3'-dideoxy-3'-oxoadenosine (isoddA) (**6**) have proved inactive against the virus.[10] The corresponding *bis*(2, 2, 2-trichloroethyl) phosphotriesters were equally inactive, but the phenylphosphoramidates of both 8-aza-isoddA (**7**) and isoddA were similar and more potent respectively than isoddA itself indicating the poor affinities of the nucleosides for cellular kinases. The 5'-triphosphates of 8-aza-isoddA and isoddA surprisingly showed the reverse trend when tested as inhibitors of HIV-RT.

In the absence of any phosphotriesterase activity inside cells, the use of phosphotriester prodrugs of nucleotides which rely on the spontaneous intracellular decomposition to the diester has been suggested. Ideally this process should preferentially occur subsequently rather than prior to cellular uptake. Imbach and co-workers [11] have described SATE (*S*-acetylthioethanol) and DTE (dithiodiethanol) phosphotriesters of ddU which it was envisaged would allow the formation of an unstable thioethanol phosphotriester by the action of cytosolic carboxyesterase or reductase activation respectively, which would then spontaneously decompose to the diester by elimination of episulphide (Scheme 2). Both triesters were active against two HIV-infected cell-lines, one being deficient in thymidine kinase. The free nucleoside (ddU) and its 5'-monophosphate were inactive. Decomposition studies of the two triesters in both cell extract and culture medium revealed approximately 100-fold reduced half-lives in the former. Extension of this work to AZT analogues has also been made.[12]

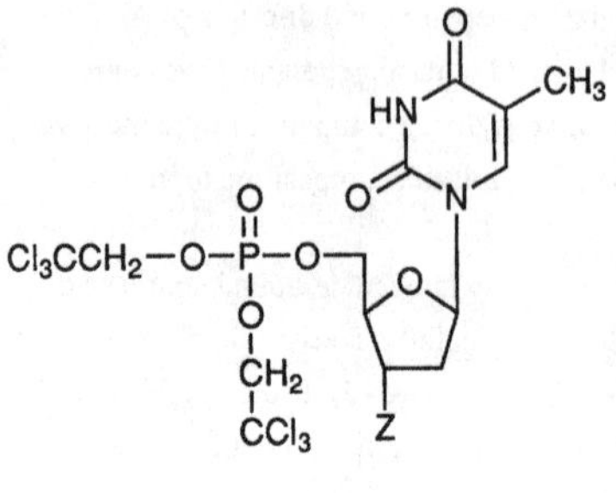

Z = OR, SO_2Me, OAc

(1)

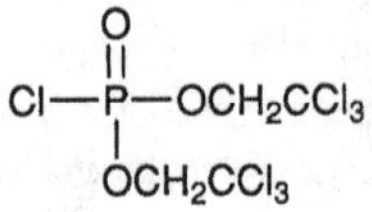

(2)

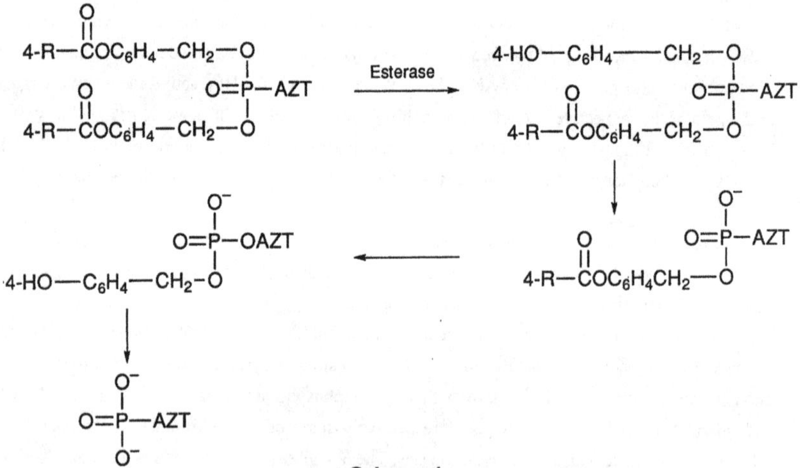

Esterase

Scheme 1

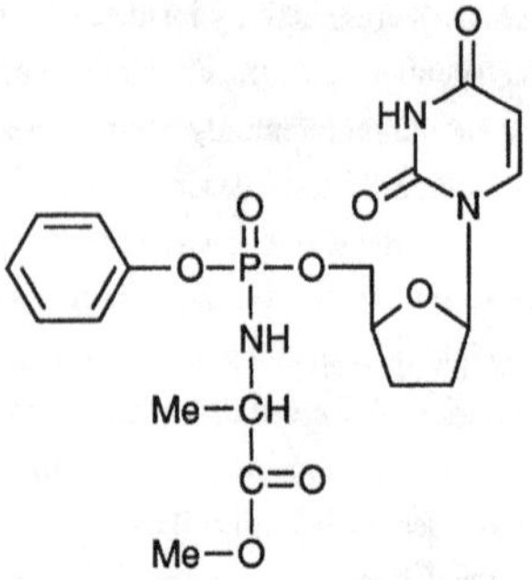

(3)

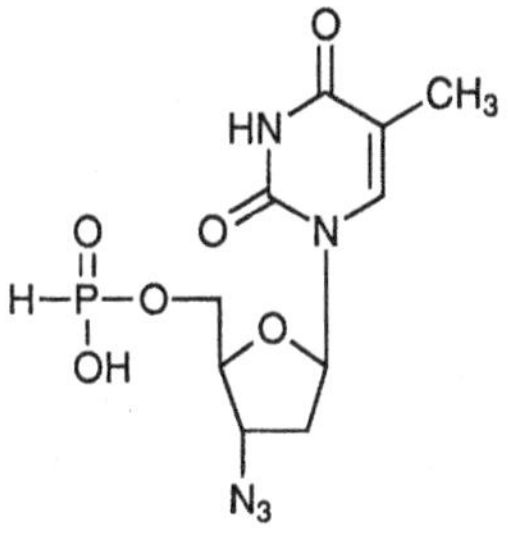

(4)

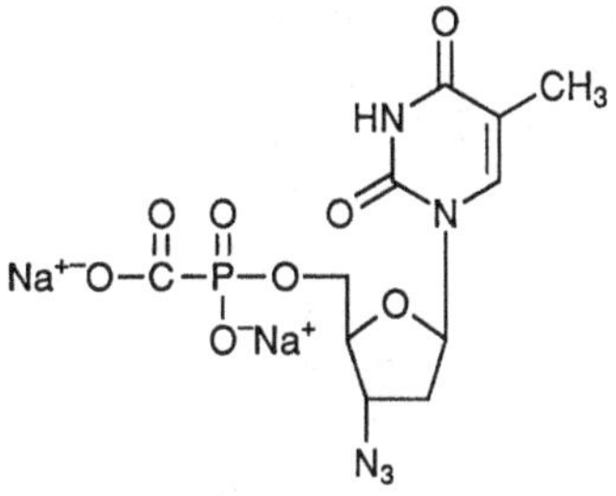

(5)

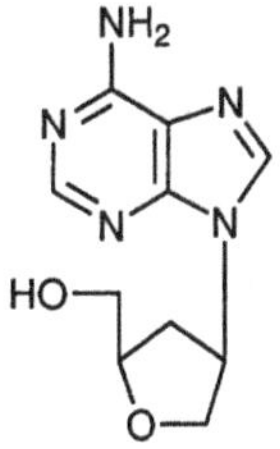

(6)

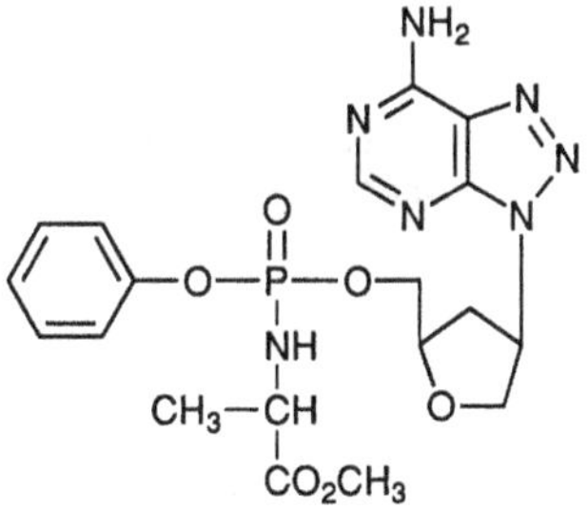

(7)

Alkyl steroidal phosphodiester[13] and triester[14] prodrugs of AZT have been prepared in an attempt to increase the metabolic stability and bioavailability of AZT whilst reducing its dose-related toxicity. The phosphotriesters (**8**) were prepared either by reaction of AZT with chloro (β-cyanoethyl) *N,N*-diisopropylphosphoramidite followed by treatment of the derived 5'-phosphoramidite with the respective steroid in the presence of tetrazole followed by iodine oxidation to the phosphate triester, or conversely by initial formation of the phosphoramidite derivatives of the respective steroids followed by reaction with AZT then oxidation. The latter procedure was found to be considerably more efficient. The individual diastereomers of the respective triesters were purified by chromatography and in the latter case one diastereomer was ten times more active than the other against HIV-infected cells. Both derivatives were however slightly less effective than AZT itself. The H-phosphonate (**9**), prepared from 7β-triethylsiloxy-cholesterol using salicyl chlorophosphite, reacted with AZT in the presence of pivaloyl chloride to give a good yield of the same phosphotriester after oxidation with iodine. In contrast, the reaction of AZT with cyanoethylphosphoramidite corresponding to (**9**) in the presence of tetrazole, followed by oxidation with mCPBA and desilylation gave the AZT phosphotriester in only 12% yield. Decyanoethylation with aqueous ammonia/methanol gave the AZT phosphodiester.

In order to combat progressive HIV infection in the CNS, often termed as "AIDS Dementia Complex" the design of anti-HIV agents able to penetrate the blood brain barrier (BBB) and suppress HIV replication is an attractive goal. The ability of the lipophilic glycosyl phosphodiester and triester prodrugs of AZT (**10**) to cross the blood brain barrier has been demonstrated and subsequent hydrolysis has resulted in significant levels of AZT monophosphate being found in the brain.[15]

The ampiphilic dinucleotide phosphates (**11**) and (**12**) have been prepared[16] as prodrugs bearing masked phosphate functionality. They display good water solubility and can be incorporated into liposomes.

Much work has concentrated on nucleoside phosphonate analogues as therapeutic agents since the phosphorus-carbon bond is resistant to enzymatic hydrolysis. The carbocyclic 5-bromovinyldeoxyuridine monophosphonate (**13**) has been prepared in 60% yield[17] by Mitsunobu condensation of 3-benzoyl-5-(2-bromovinyl)uracil and the alcohol (**14**) using triphenylphosphine and dimethylazodicarboxylate followed by removal of the protecting groups. The alcohol, which was obtained as a racemate in four steps from 6-oxabicyclo[3.1.0]hexan-3-*endo*-ol, was resolved after selective conversion to the 3-acetoxy derivatives with Amano lipase PS in vinyl acetate. The analogue (**13**) was unfortunately inactive against HSV *in vitro* presumably either due to lack of cell penetration or the absence of phosphorylation to the corresponding diphosphorylphosphonate.

Imbach and co-workers have synthesised the α (**15**) and β (**16**) 3'-C-phosphonate derivatives of both ribo and 2'-deoxyriboadenosine and thymidine[18] using 2-deoxy-*D*-ribose as starting material; both α and β nucleoside phosphonates are obtained, whilst 1, 2-isopropylidene-α-*D*-xylofuranose is converted exclusively to the β-anomers. The intermediates (**17**) and (**18**) in both routes were converted to the corresponding derivatives (**19**) and (**20**) *via* reaction with dimethylphosphite and triethylamine. The indicated stereochemistry of derivative (**20**), arising due to the presence of the isopropylidene group, allowed its use for the specific subsequent conversion to the β-anomer (**16**)

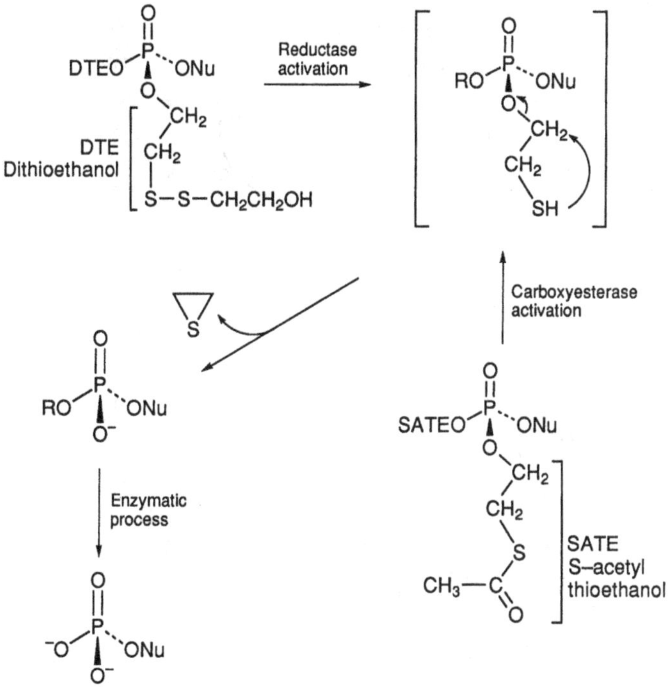

DTEO–P(=O)(ONu)–O–CH2–CH2–S–S–CH2CH2OH (DTE Dithioethanol) —Reductase activation→ [RO–P(=O)(ONu)–O–CH2–CH2–SH] ←Carboxyesterase activation— SATEO–P(=O)(ONu)–O–CH2–CH2–S–C(=O)–CH3 (SATE S–acetyl thioethanol)

[RO–P(=O)(ONu)–O–CH2–CH2–SH] → (episulfide) + RO–P(=O)(O⁻)(ONu) —Enzymatic process→ ⁻O–P(=O)(O⁻)(ONu)

Scheme 2

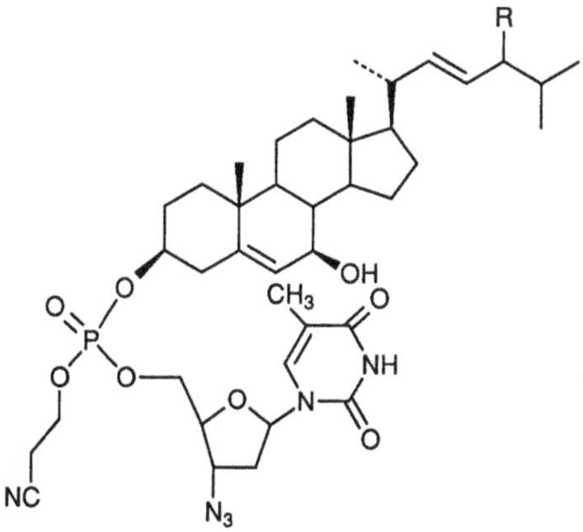

(8) R = H, C_2H_5

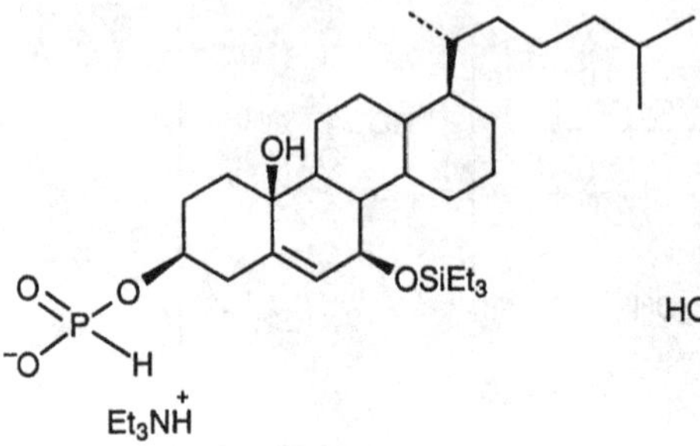

OH
O
P
O
$^{-}$O
H
OSiEt$_3$
Et$_3$NH$^+$

(9)

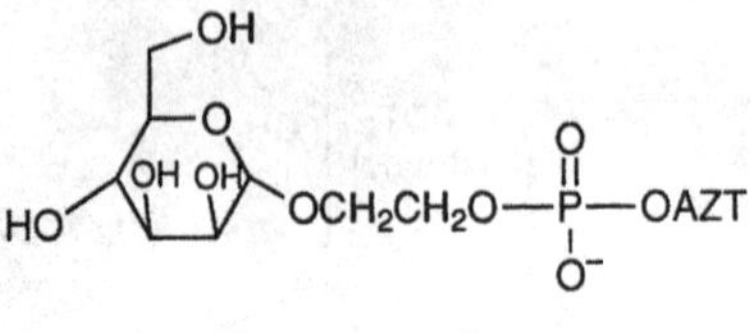

OH
O
OH OH
HO
OCH$_2$CH$_2$O—P—OAZT
O
O$^-$

(10)

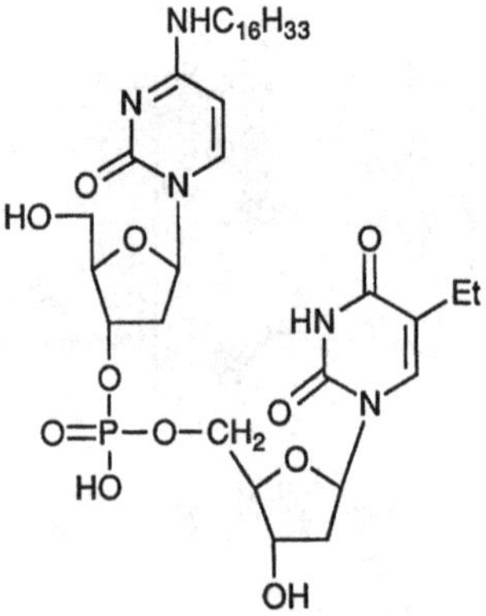

NHC$_{16}$H$_{33}$
N
O
N
HO
O
O
O=P—O—CH$_2$
HO
O
HN
Et
O
N
O
OH

(11)

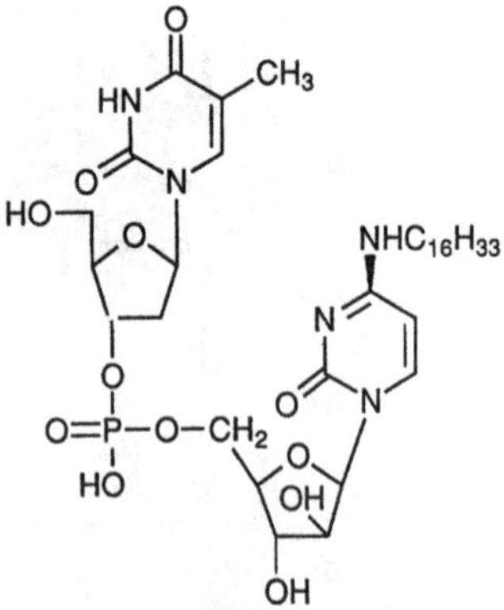

O
HN
CH$_3$
O
N
HO
O
O
O=P—O—CH$_2$
HO
NHC$_{16}$H$_{33}$
N
O
N
O
OH
OH

(12)

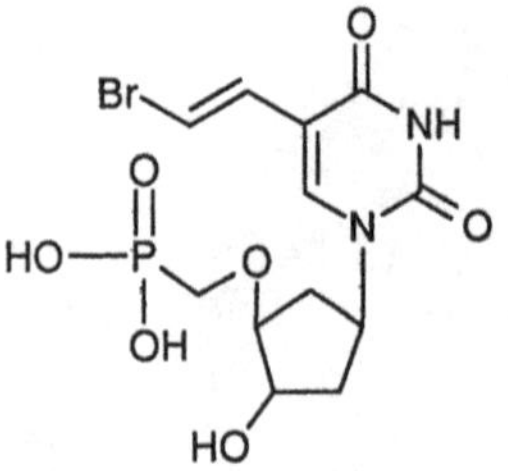

Br
O
NH
O
N
HO—P
O
OH
O
HO

(13)

(EtO)$_2$—P
O
O
OH
OH

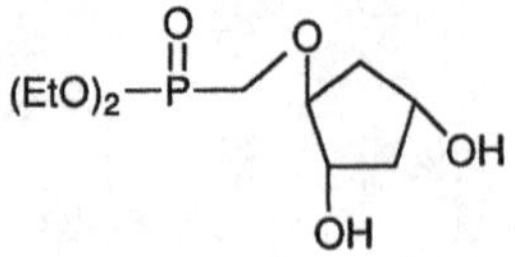

(14)

achieved by reduction and acetylation followed by glycosylation with the appropriate silylated nucleobase. None of the compounds displayed any significant anti-HIV activity.

The Wadsworth-Emmons condensation between the sodium enolate of the phosphono ester (**21**) and the 3'-ketone of 2', 5'-di-*O*-*tert*butyldimethylsilyladenosine gives (**22**) and (**23**).[19] Reduction of the alkene (**22**) followed by reaction with the lithio anion of diethylmethylphosphonate gave (**24**). A subsequent Wadsworth Emmons reaction of (**24**) with a nucleoside 5'-aldehyde offers a route to novel carbon bridged dinucleotide analogues.

The failure of carbocyclic-2'-deoxyguanosine (CDG) (**25**) to be phosphorylated in human cancer cells is probably related to its poor cytotoxicty. It is however a potent antiviral agent especially active against Herpes Simplex 1 (HSV-1) where it is known to be phosphorylated. Interestingly it has been found that both the D and L forms of the analogue are equally good kinase substrates, although only the former is subsequently converted to the triphosphate.[20] Recently the 5'-phosphonomethyl derivatives of CDG (**26**) and carbocyclic 6-(methylthio)purine 2'-deoxyriboside (**27**) have been prepared[21] *via* the phosphorylation of their respective 3'-*O*-benzyl-protected 5'-alkoxides (obtained in several steps) with diethyl *p*-(tolylsulfonyl)oxy)-methanephosphonate. Compound (**26**) was 40 times more cytotoxic to a human cancer cell line than the free nucleoside CDG, whereas (**27**) showed little activity. Efficient phosphorylation of (**26**) by guanylate kinase was observed, whilst (**27**) was assumed to be inactive due to its poor conversion to its triphosphate.

Despite its high antiviral activity against several retroviruses, the acyclic analogue 9-(2-phosphonomethoxyethyl)adenine (PMEA) (**28**) is limited in its therapeutic use due to its poor oral absorption. The *bis*-(pivaloyloxymethyl) ester (*bis*(pom) ester) of PMEA (**29**) was found to have increased HSV activity *in vitro* compared to PMEA.[22] This has been attributed to a combination of increased solubility and neutral charge, the latter facilitating cell permeation after which the acyloxy groups are rapidly cleaved by carboxyesterase resulting in the monoester and free phosphonate. A study of anti-HIV activity of several *bis*-(pivaloyloxymethyl) esters of acyclic nucleoside analogues[23] including 9-(2-phosphonylmethoxypropyl)adenine (PMPA) (**30**), 9-(2-phosphonylmethoxypropyl)2, 6-diaminopurine (PMPDAP) (**31**) and PMEA itself, has revealed 4-fold lower therapeutic indices (cytotoxicity / antiviral efficacy) for the latter two derivatives relative to the unmodified analogues, whilst that of PMPA is similar to its triester. Rapid cellular uptake was observed for the tritiated *bis*-(pivaloyloxymethyl) ester of PMEA with no detectable intracellular levels of the prodrug but significant amounts of PMEA phosphate and pyrophosphate. It has also been noted that the good acid stability of these *bis* (pom) esters may facilitate oral administration.

Extension of this work has generated an extensive range of prodrugs of PMEA in an attempt to obtain compounds with improved anti-HIV activities and increased bioavailabilities.[22] The dichlorophosphonate (**32**) obtained by treatment of PMEA with thionyl chloride and DMF was converted to several potential prodrugs as shown in Scheme 3. Several *bis*(acyloxy)alkyl phosphonate triesters, including the pivaloyl derivative were prepared from PMEA by reaction with the corresponding chloromethyl ester in the presence of the hindered base *N, N'*-dicyclohexylmorpholinecarboxamidine (**33**). The monopivaloyl ester was prepared by acylation of a benzyl protected PMEA diester. The glyoxamides (**34**) and (**35**) and difluoropropyl esters (**36**)

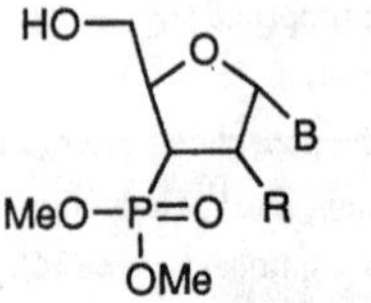

(15) B = Thymine
R = H, OH

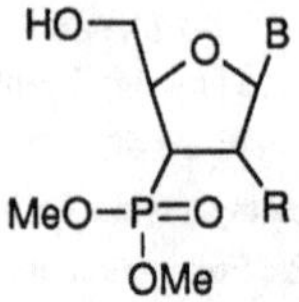

(16) B = Thymine or Adenine
R = H, OH

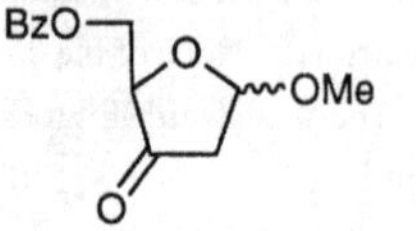

(17)

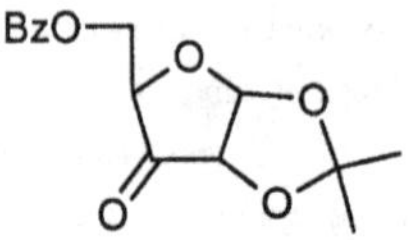

(18)

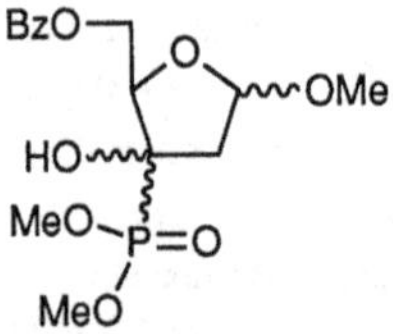

(19)

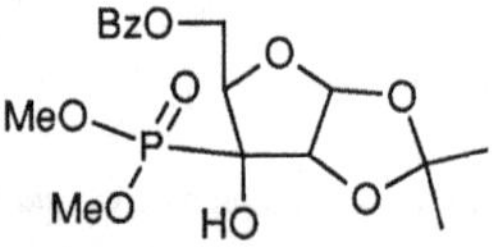

(20)

$(CH_3O)_2P(O)CH_2C(O)OCH_3$

(21)

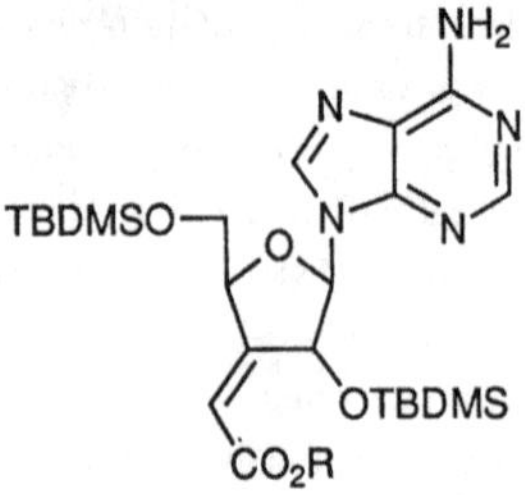

(22)

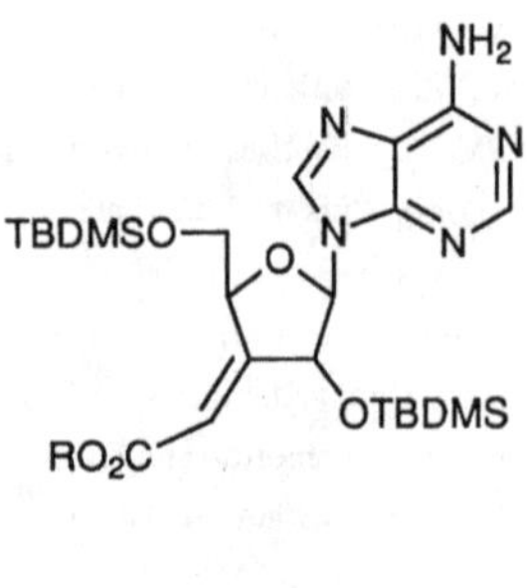

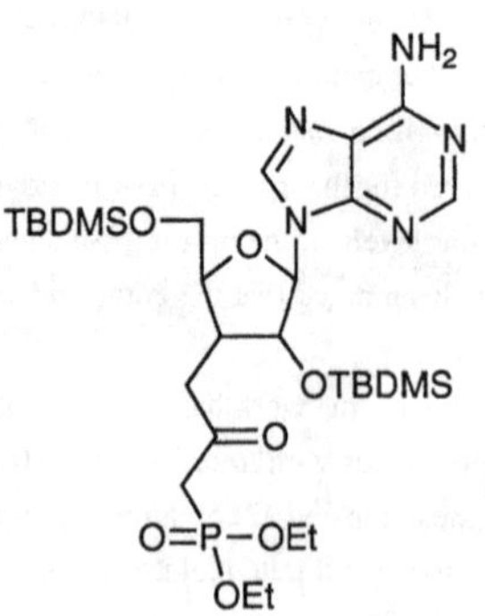

(23)

(24)

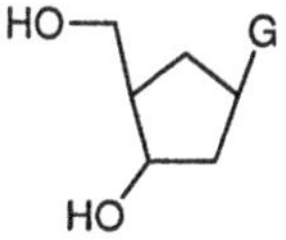

HO G
HO
G = guanine

(25)

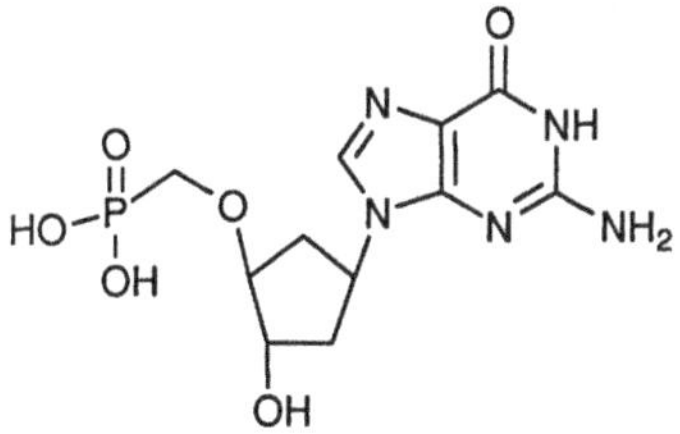

O
N NH
O
HO-P O N N NH$_2$
OH
OH

(26)

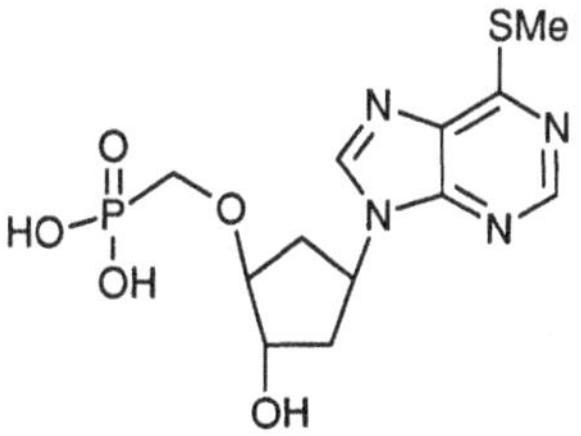

SMe
N N
O
HO-P O N N
OH
OH

(27)

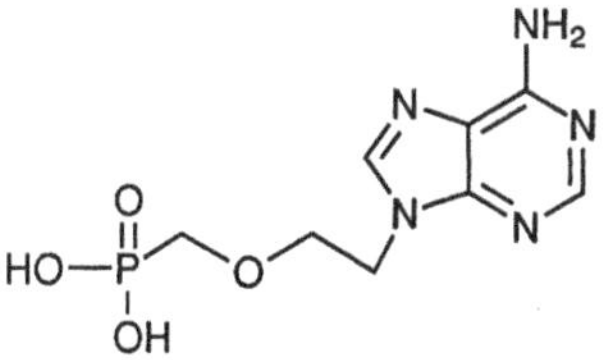

NH$_2$
N N
O N N
HO-P O
OH

(28)

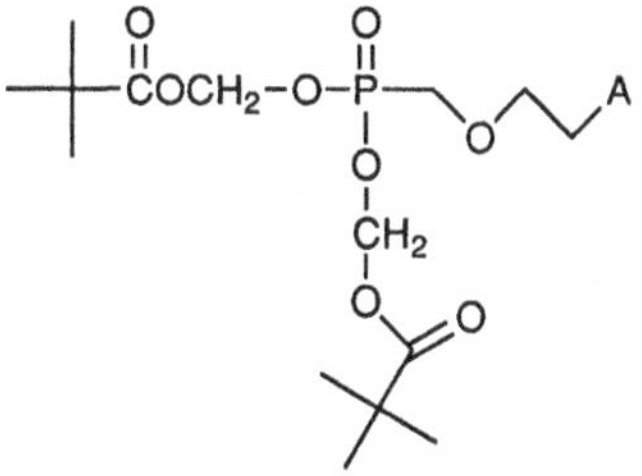

O O
COCH$_2$-O-P O A
O
CH$_2$
O O

A = adenine
(29)

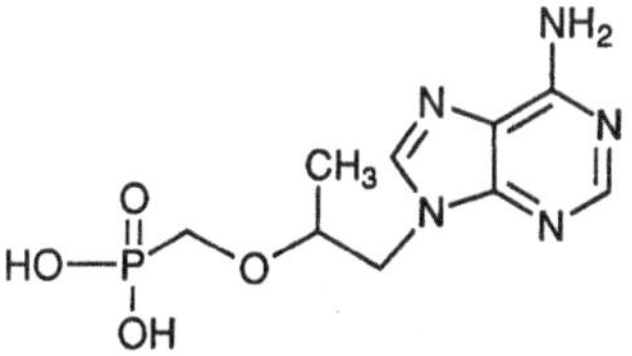

NH$_2$
N N
CH$_3$
O N N
HO-P O
OH

(30)

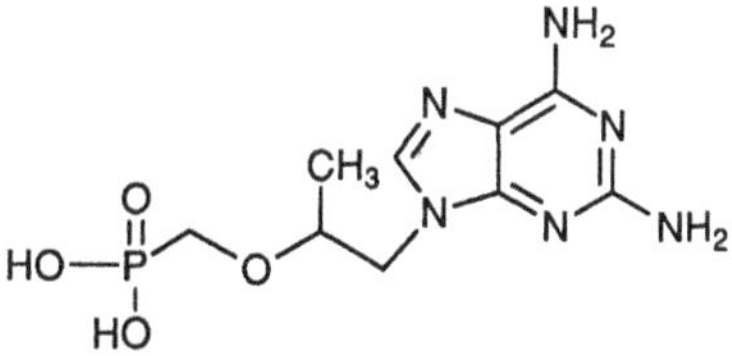

NH$_2$
N N
CH$_3$
O N N NH$_2$
HO-P O
HO

(31)

NH$_2$
N N
O N N
Cl-P O
Cl

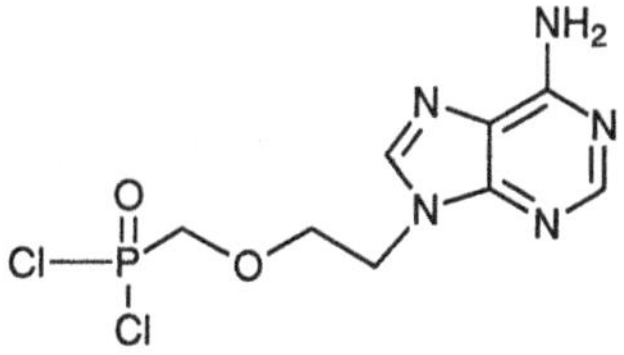

(32)

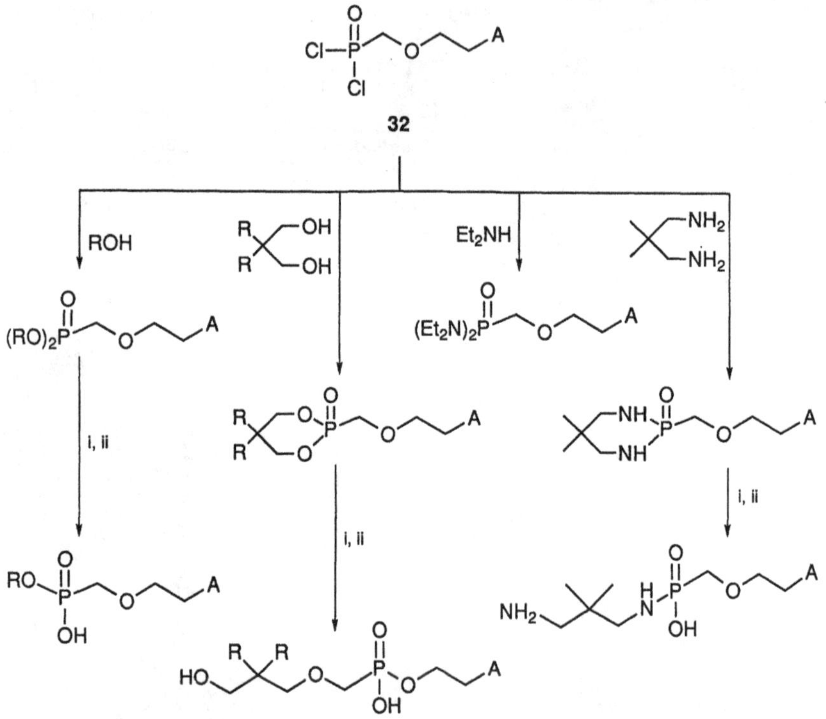

32

ROH

Et_2NH

i, ii

i, ii

i, ii

Reagents: i, NaOH; ii, HCl

Scheme 3

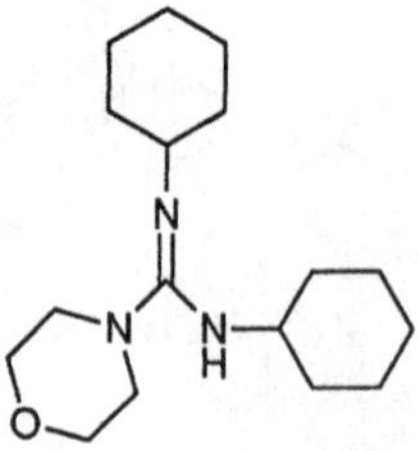

(33)

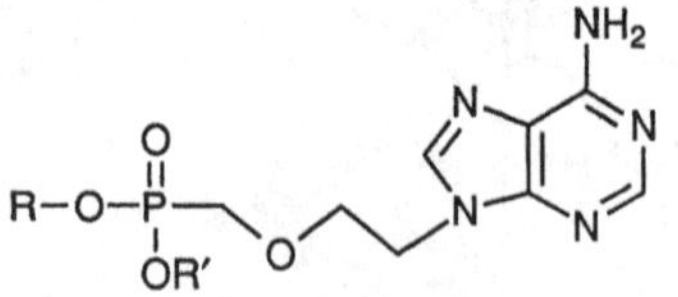

(34) R = Et_2N, R′ = Et_2N
(35) R = Et_2N, R′ = H

and (**37**) were prepared from (**32**) and the alcohols (**38**) and (**39**) respectively. The oral bioavailabilities of the analogues in rats were determined by measuring the concentration of PMEA in urine for 48 hours after administration of the prodrugs. The bioavailability of PMEA was found to be 7.8%. Absorption of the *bis* alkyl phosphonate triesters was observed but these were not effectively cleaved to liberate PMEA. The monoalkyl diesters and phosphonamides displayed similar bioavailabilities to PMEA and the latter compounds possessed poor acid stabilities. In contrast the *bis*[(acyloxy) alkyl] phosphonates demonstrated significantly improved bioavailabilities of between 14 .6 and 17.6%. In addition anti-HSV-2 activity of these analogues *in vitro* was more than 200-fold higher than that of PMEA. The analogous synthesis of (**32**), as an intermediate in the preparation of PMEA esters, has also been reported elsewhere.[24] In addition, the same article describes an alternative mild synthesis of PMEA ester analogues by reaction of PMEA with dimethylformamide dineopentyl acetal (**40**) in DMF in the presence of an appropriate alcohol. Transacetalisation affords the dimethylformamide dialkyl acetal of the more reactive alcohol *in situ* and makes this approach to obtaining derivatives of (**32**) highly versatile.

The novel acyclic pyrimidine analogues (**41**) and (**42**) of 9-[2-(phosphonomethoxy)ethyl]guanine (PMEG) have been obtained[25] from 2-amino-6-chloro-5-nitro-4-pyrimidone and the aminoalkyloxyphosphonate (**43**). Only the amino analogue was active against HSV-1 and displayed a comparable efficacy to that of PMEG.

The individual (*R*) and (*S*) enantiomers of 2-phosphonomethoxypropyl (PMP), 3-hydroxy-2-phosphonomethoxypropyl (HPMP) and 3-fluoro-2-phosphonomethoxypropyl (FPMP) derivatives (**44**) and (**45**) of adenine, 2-aminopurine, 2, 6-diaminopurine, and guanine have been prepared and tested against HIV and Moloney murine sarcoma virus (MSV).[26] Of these (*R*)-PMPDAP emerged as the most potent and selective *in vitro* and *in vivo* antiretroviral agent reported so far. Interestingly, (*S*) -FPMPA was far superior to (*R*) - FPMPA and (*R*) - PMPA far superior to (*S*) -PMPA in inhibiting MSV-induced cell-transformation and HIV replication, whereas for FPMPDAP, FPMPG and PMPG differences in activities of the two enantiomers were less apparent. This discrepancy in the adenine series of derivatives, notably of FPMPA, was attributed to the differential rates of phosphorylation of the two enantiomers by AMP kinase.

An enantiomeric synthesis of the human cytomegalovirus (HCMV) antiviral agent [(*R*) - 3-((2-amino-1, 6-dihydro-6-oxo-9*H*-purin-9-yl)methoxy)-4-hydroxybutyl]phosphonic acid (ganciclovir phosphonate) (**46**) has been described.[27] Comparison of the structures of the (*R*) (**46**) and(*S*) (**47**)-enantiomers shows that the former compound most closely resembles that of GMP and indeed it is this one which is the preferred substrate for GMP kinase. The acyclic precursor to (**46**) was coupled to persilylated N^2, N^7 - diacetylguanine to give (**48**) in 84% yield, but only 5% was the desired N-9 regioisomer. The mixture was converted from the chloro to the iodo derivative and after silylation with *N, O*-bis(trimethylsilyl)acetamide, reaction with diethylphosphite and potassium *tert*-butoxide in toluene gave exclusively (**49**) as the N-9 isomer. Presumably HI-catalysed isomerisation of the N-7 to N-9 isomer occurs during the reaction. The (*R*)- enantiomer was twice as active as the racemate against HCMV whilst both racemic and chiral compounds were less toxic than ganciclovir to bone marrow progenitor cells *in vitro*. Unfortunately, the main toxicity is also associated with the (*R*)- enantiomer.

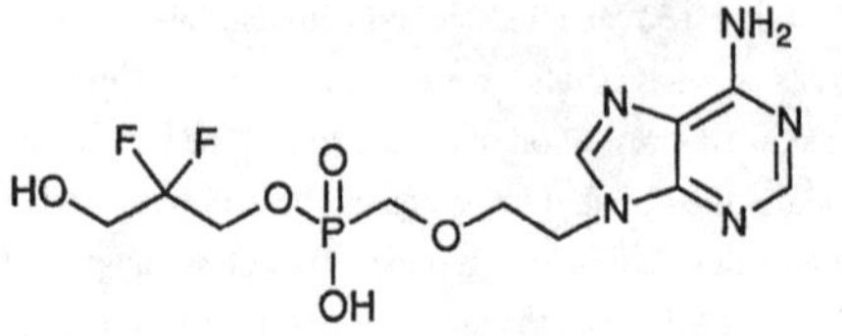

(36)

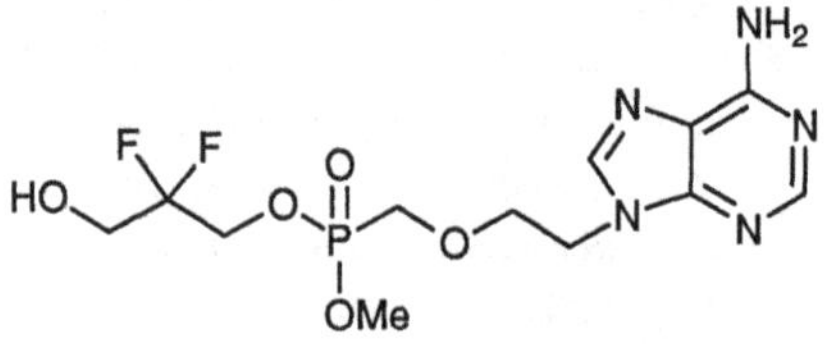

(37)

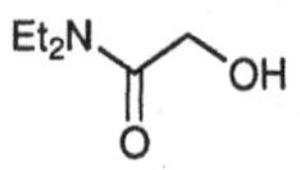

(38)

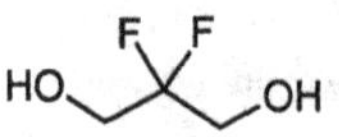

(39)

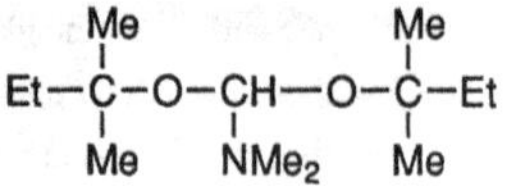

(40)

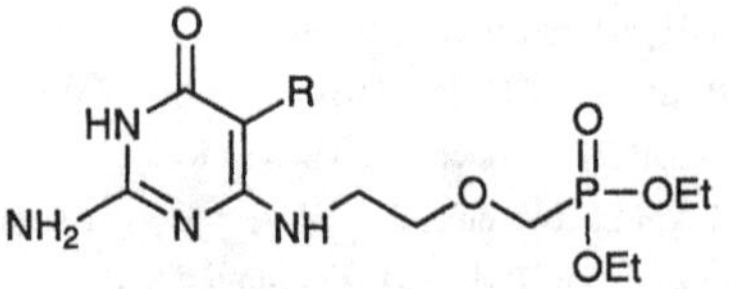

(41) R = NO_2
(42) R = NH_2

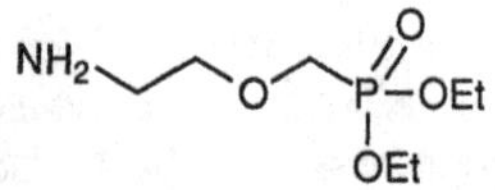

(43)

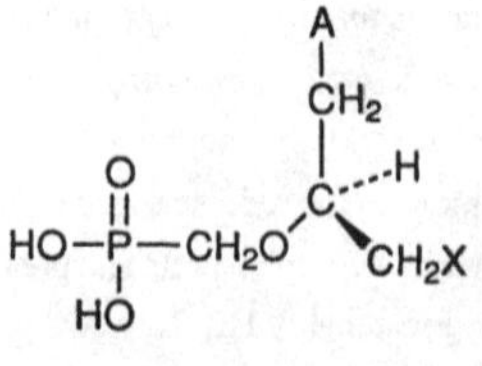

(*R*) X = H; (*S*) X = OH, X = CH_2F
(44)

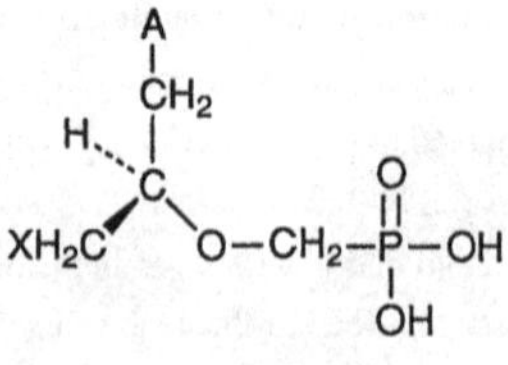

(*R*) X = OH, X = CH_2F; (*S*) X = H
(45)

The *S*-phosphates of several antiviral acyclonucleosides (**50**) have been prepared[28] in yields between 60 and 65% by reaction of the corresponding iodo derivatives with trisodium thiophosphate in water. The iodo derivatives were synthesised from the parent acyclonucleosides by initial formation of the *N, O*-bis acetyl derivatives (**51**) by reaction with triethylorthoformate followed by acid hydrolysis of the cyclic orthoester intermediate and then iodination with iodine/ trphenylphosphine/ imidazole. None of the derivatives were active against HSV-1, HSV-2 or CMV.

The 8-aza-analogues of of PMEA (**52**) (X=N) and its guanine counterpart PMEG (**53**) (X=N) have been synthesised[10] by alkylation of the sodium salt of azaadenine with the tosylate (**54**) in DMF or from azaguanine in DMSO in the presence of caesium carbonate. The use of the former procedure in the case of PMEA resulted in better yields and a more favourable regiocontrol of alkylation. The reaction of guanine with (**54**) using caesium carbonate in DMSO provided a novel synthesis of PMEG in 41% yield, the desired N-9 isomer predominating. Both 8-aza-PMEA and 8-aza-PMEG were less toxic for MT-4 cells but also less potent (notably for the PMEA derivative) than the parent compounds against HIV-1 and HIV-2. The inactivity of the regioisomers of (**52**) and (**53**) confirmed that only PME derivatives in which the phosphonate side chain is bonded at N-9 have affinity for both cellular kinases and HIV-RT.

A number of modifications to the 2-position of the acyclic portion of PMEA have been made by Holy and co-workers.[29] 2-Alkyl derivatives were made by reacting the tosylates of general structure (**55**) or oxiranes (**56**) with either the sodium salt of adenine in DMF or with adenine in DMF in the presence of caesium carbonate to give the corresponding alcohol precursors of PMEA. Analogously, compound (**57**) afforded the corresponding trifluoromethyl analogue. The corresponding alkoxides of these derivatives were subsequently reacted with the phosphonate synthon (**58**) followed by treatment with bromotrimethylsilane to give the desired analogues. The synthesis of the 2-trimethylsilylmethylene derivative employed (**59**) as alkylating agent. In addition to the adenine analogues, several other modified purine analogues of those described above were also synthesised. Preliminary antiviral assays provided negative results, suggesting that preservation of activity is not possible using substituents larger than a methyl or trifluoromethyl group.

The oxo-isostere of PMEG (**60**) has recently been synthesised[30] from the potassium salt of (**61**) and diethylchloromethoxymethanephosphonate in DMF. The alkylating agent diethyl chloromethoxymethanephosphonate was prepared *in situ* from diethyl hydroxymethane phosphonate by chloromethylation with paraformaldehyde and hydrogen chloride in dichloromethane. Testing of compound (**60**) for anti-HCMV activity in two different cell-lines revealed therapeutic indices equivalent or superior to ganciclovir.

Novel phosphonates of acyclovir (**62**) and ganciclovir (**63**), their ethyl diesters and the cyclic phosphonate ester (**64**) have been synthesised.[31] In each case 2-amino-6-chloropurine was used as the aglycon precursor, which was alkylated with either (**65**) or (**66**). Compound (**63**) was cyclised to give (**64**) using DCC. All analogues showed significant activity against several CMVs, the acyclovir analogues being more effective than acyclovir itself, whereas the activities of the ganciclovir analogues were comparable to the unmodified compound.

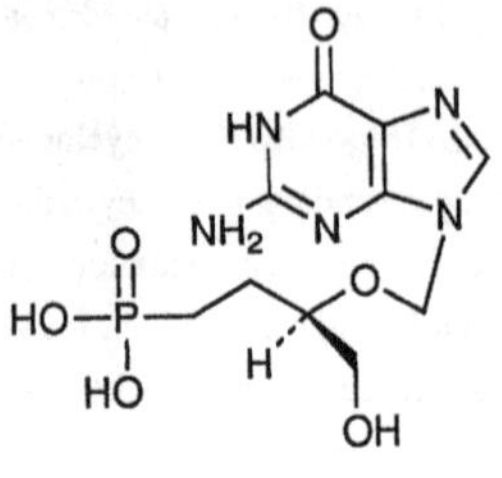

(46)

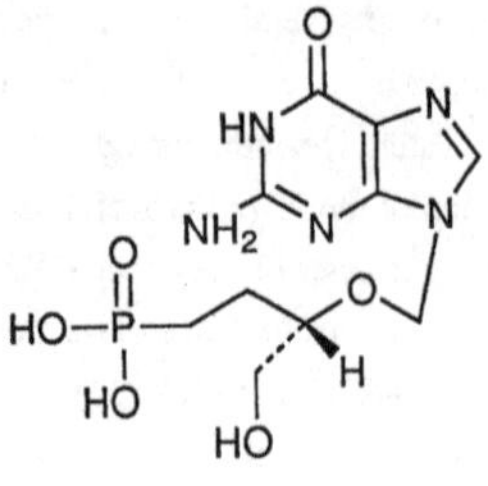

(47)

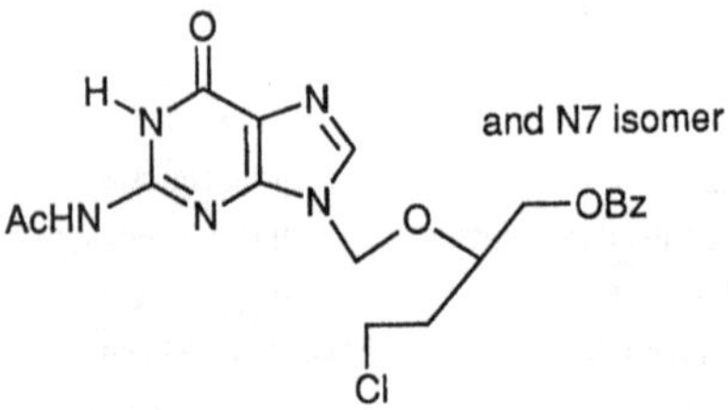

and N7 isomer

(48)

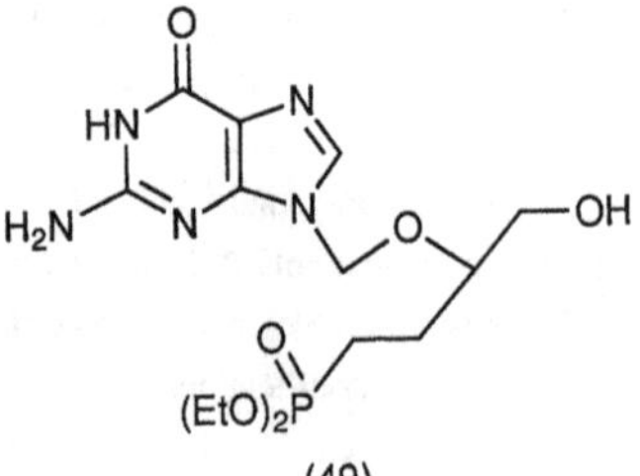

(49)

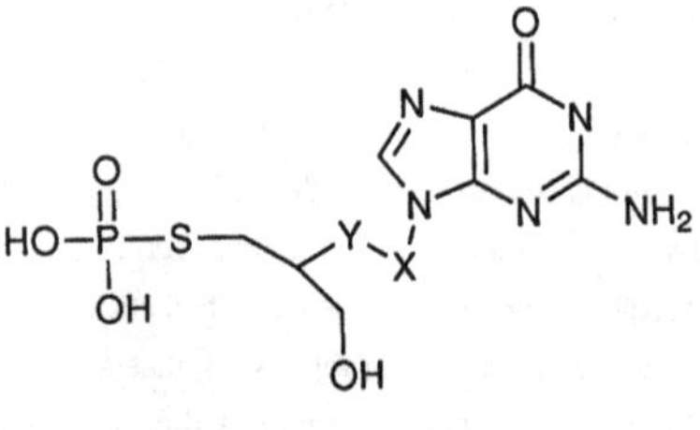

(50) Y, X = CH_2/O

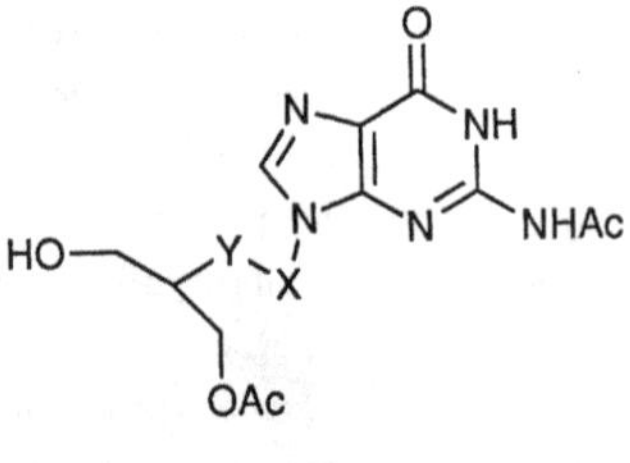

(51)

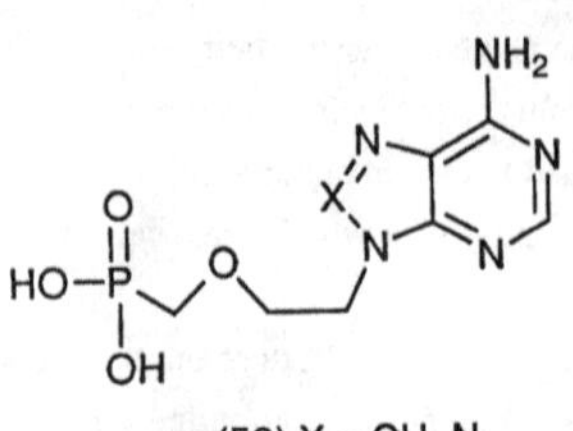

(52) X = CH, N

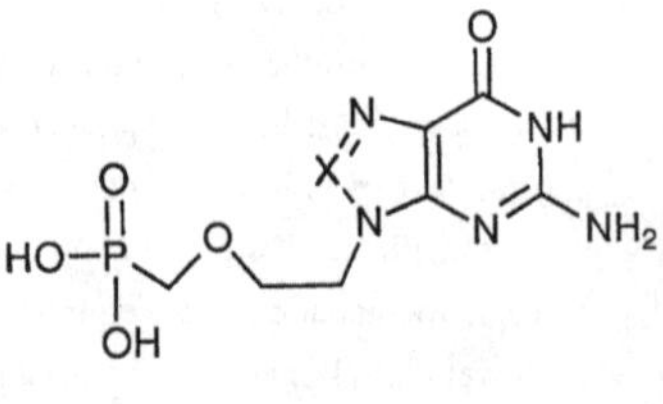

(53) X = CH, N

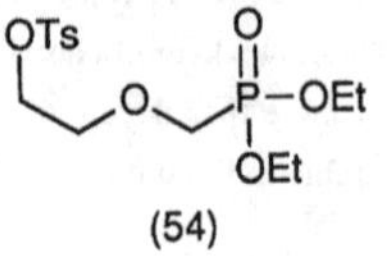

(54)

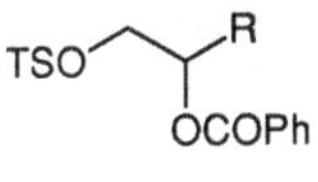

TSO R OCOPh

(55)

O R

(56)

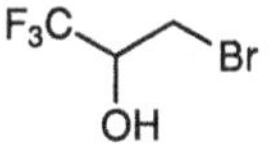

F_3C Br OH

(57)

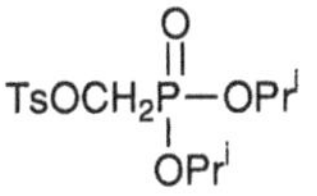

O TsOCH$_2$P—OPr^i OPr^i

(58)

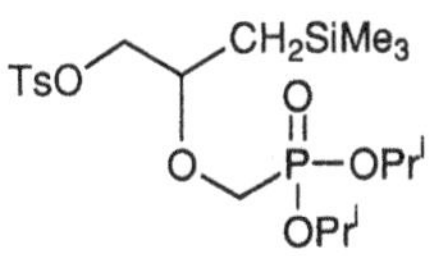

TsO CH_2SiMe_3 O P—OPr^i OPr^i

(59)

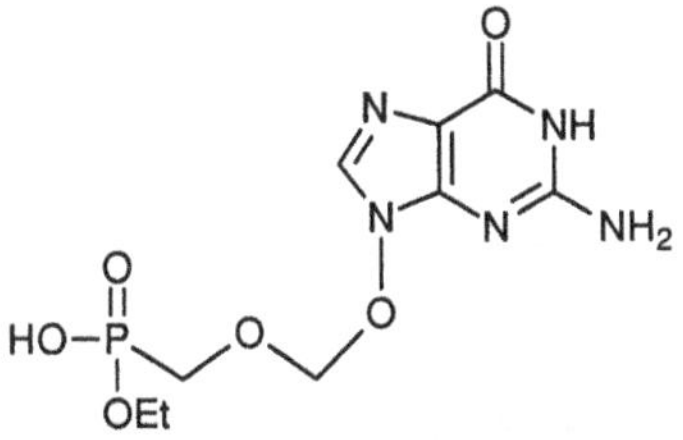

O N NH N N NH$_2$ HO—P O O OEt

(60)

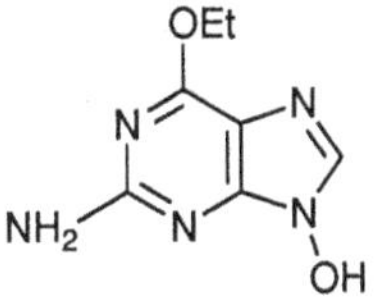

OEt N N NH$_2$ N N OH

(61)

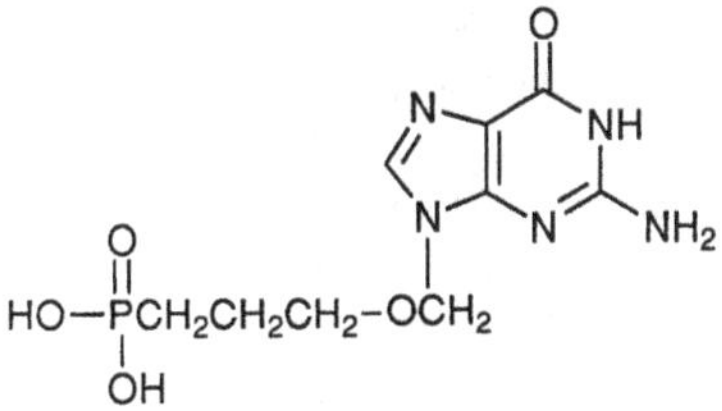

O N NH N N NH$_2$ HO—PCH$_2$CH$_2$CH$_2$-OCH$_2$ OH

(62)

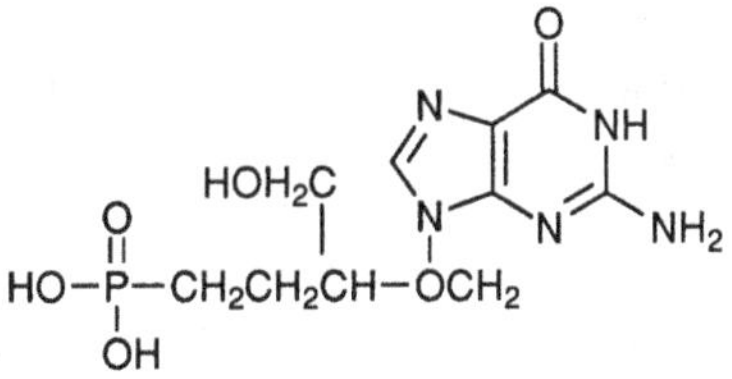

O N NH HOH$_2$C N N NH$_2$ HO—P—CH$_2$CH$_2$CH—OCH$_2$ OH

(63)

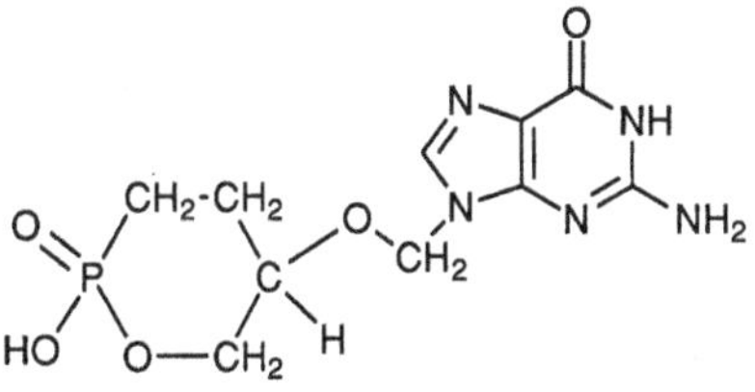

O N NH CH$_2$-CH$_2$ O N N NH$_2$ P C CH$_2$ HO O—CH$_2$ H

(64)

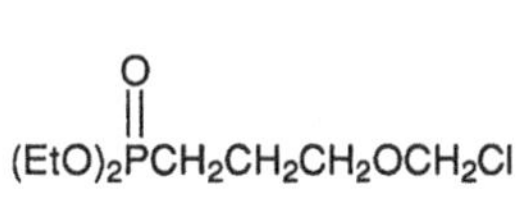

O (EtO)$_2$PCH$_2$CH$_2$CH$_2$OCH$_2$Cl

(65)

O (EtO)$_2$PCH$_2$CH$_2$ HCOCH$_2$Cl CH$_2$OCH$_2$Bz

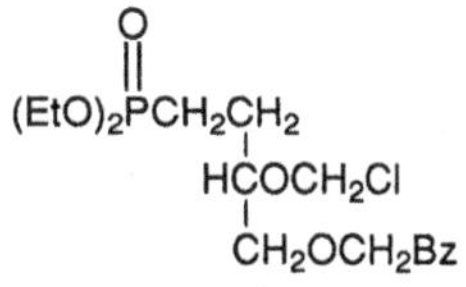

(66)

The enzyme purine nucleoside phosphorylase (PNP) catalyses the reversible conversion of both ribo- and 2'-deoxyribo-puranosides to the corresponding free base and ribose or deoxyribose-1-α-phosphate. Its absence is associated with T-cell immunodeficiency and thus inhibitors may be valuable in treating T-cell leukaemia. Of several recently prepared 9-[(phosphonoalkyl)benzyl]guanines of general structure (**67**)[32] those possessing X = CH_2OCH_2 and CH_2SCH_2 were excellent PNP inhibitors and displayed K_i values of 5.8 and 1.1 nM respectively. These drugs were approximately 6000-fold more effective than 9-benzylguanine stressing the importance of the alkylphosphonate side chain in binding.

A 5'-*P*-borane-substituted thymidine monophosphate (**68**) and triphosphate (see Section 2) have been prepared[33] from the boranophosphoramidate (**69**) which was obtained by reacting the thymidine 5'-β-cyanoethylphosphoramidite with borane-diisopropylethylamine complex followed by treatment with ammonia. Acid hydrolysis of (**69**) gave the thymidine monophosphate derivative (**68**).The compound shows similar stability to normal nucleoside monophosphates, hydrolysis proceeding by P-O rather than P-B cleavage. Acid phosphatase hydrolysed (**68**) to thymidine whilst the analogue was not a substrate for alkaline phosphatase.

Other novel analogues include the 5'-deoxy-5'-difluoromethylene nucleoside phosphonates (**70**) which have been obtained in good yield by the $SnCl_4$-catalysed glycosylation of the respective silylated nucleobases with glycosylating agent (**71**)[34] followed by cleavage of the ethyl triester groups with bromotrimethylsilane. The guanine derivative was not reported and N^6-benzoyladenine afforded an approximate 2:1 mixture of the N9 and N7 isomers respectively.

The analogues adenosine 5'-O-trifluoromethylphosphonate (**72**) and 2'-O-deoxythymidine 3'-O-trifluoromethylphosphonate have been prepared[35] from the bis-triazolide of trifluoromethylphosphinic acid which was obtained from trifluoromethylphosphorus dibromide.

The use of the Michaelis-Arbuzov reaction has been made in the synthesis of dinucleoside-3'-*S*-phosphorothiolates using disulphides (**73**) and nucleoside 5'-dialkylphosphite derivatives.[36] Particularly efficient reactions were achieved using a 5'-*bis*(trimethylsilyl)phosphite (**74**) which afforded the phosphorothiolate directly after aqueous work up. Derivative (**74**) was obtained by treating the corresponding nucleoside H-phosphonate with chlorotrimethylsilane in pyridine in the presence of triethylamine.

1.2 Nucleoside Cyclic Phosphates

Novel cyclic di- and triphosphate derivatives of 3'-amino-3'-deoxyadenosine-5'-diphosphate (**75**) and triphosphate (**76**) have been prepared[37] by cyclisation of the respective 5'-di- and triphosphate derivatives of 3'-*N*-Boc-3'-amino-3'-deoxyadenosine using the water-soluble carbodiimide 1-ethyl-3-(3-dimethylaminopropyl)carbodiimide hydrochloride (EDC), followed by removal of the Boc protecting group with TFA. The acyclic polyphosphate precursors were prepared according to Yoshikawa[38] and Ludwig[39]after first dissolving the nucleoside in hot triethylphosphate. The reaction proceeds via the intermediacy of the 5'-phosphorodichloridate derivative, partial hydrolysis of which gives the monophosphate, whilst traces of pyrophosphoryl chloride result in

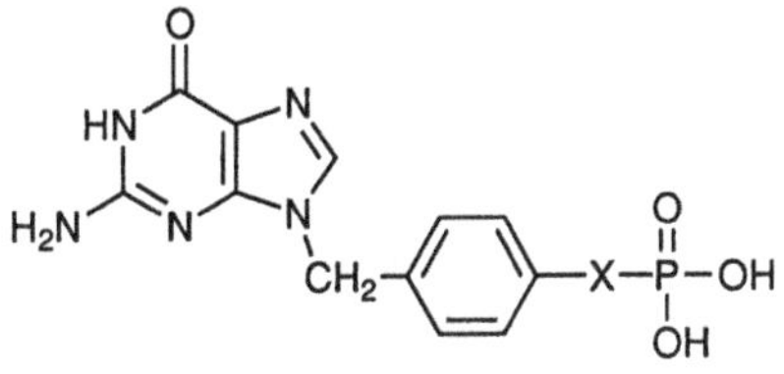

(67)

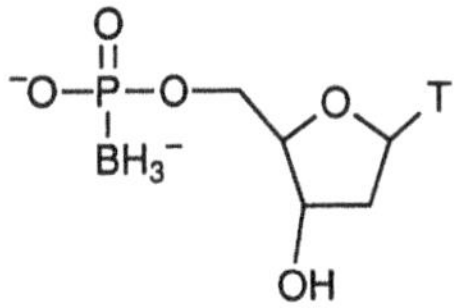

(68)

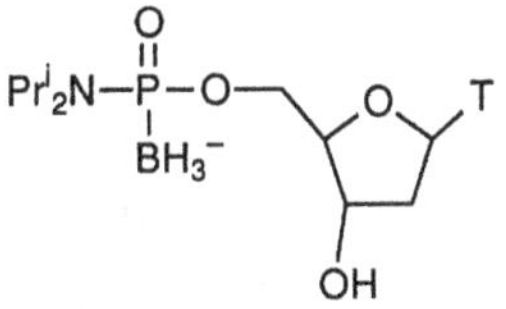

(69)

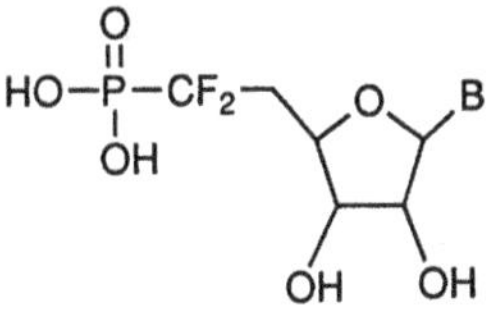

(70) B = U, C, A

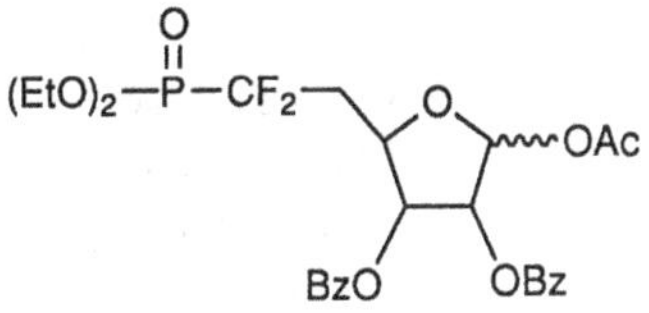

(71)

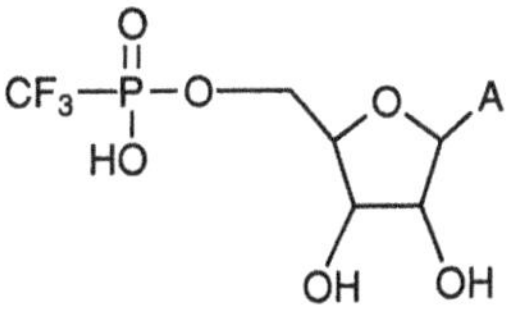

(72)

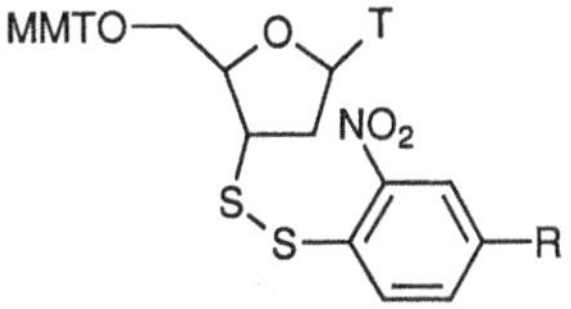

(73) R = H, NO_2

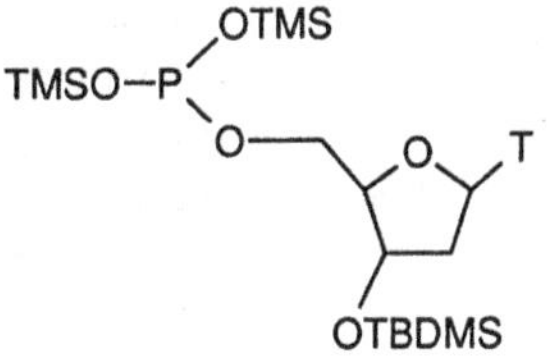

(74)

diphosphate formation.[38] Interestingly, phosphorylation of the nucleoside in a suspension of triethylphosphate produced almost exclusively the 5'-chloro-5'-deoxynucleoside, whereas phosphorylation in the absence of amino protection formed exclusively the cyclic phosphoramidate (**77**) upon alkaline hydrolysis. The compounds AMP, ADP, ATP and ATeP (adenosine tetraphosphate) do not undergo carbodiimide-mediated cyclisation[40] and (**76**) is the first example of a cATP derivative. Although the cyclic monophosphoramidate (**77**) is readily cleaved at pH 5, the di- and triphosphate derivatives display half lives of 18 and 30 hours respectively at pH 2 probably due to decreased ring strain. The cyclic methylene pyrophosphoramidate derivative 3'-amido-3'-deoxy-2'-*O*-tosyladenosine-3', 5'-cyclomethylene bis(phosphonate) (**78**) was also prepared by the same authors by displacement of a 5'-*O*-tosyl group with tris(tetra-*n*-butyl ammonium) methylene bis(diphosphonate) in acetonitrile followed by removal of the Boc group and subsequent cyclisation as described above. This compound also displays good stability in acid solution.

The cyclic dinucleotide phosphorothioate (**79**) has been prepared[41] stereoselectively and in good yield from the 3'-hydrogen phosphonate of the dinucleoside phosphorothioate (**80**) by standard activation with pivaloyl chloride, followed by sulphurisation with elemental sulphur. Interestingly the second stereocentre is formed with complete *Rp* stereoselectivity. This is attributed to the reduced conformational freedom associated with the formation of the cycle, since thiooxidation of H-phosphonate intermediates is known to be stereoselective.[42]

The disulphide (**81**) upon treatment with *N,N*-diisopropylaminodimethoxyphosphine forms the phosphite triester intermediate (**82**) which participates directly in a Michaelis-Arbuzov reaction to give the 3'-*S*, 5'-*O*-cyclic phosphorothiolate triester[43] (**83**) in 58% yield as a 20:1 mixture of two diastereomers, in which the OMe group predominantly has the axial configuration. In contrast, a previous synthesis of the analogous cyclic AMP derivative from 3'-thioadenosine and phosphorus oxychloride gave yields of only 2-4%.[44] The related analogue (**84**) was obtained in an analogous manner starting from 1-(2-deoxy-5-*O*-monomethoxytrityl-3-thio-β-*D*-xylofuranosyl)thymine.

2. Nucleoside Polyphosphates

Several novel nucleoside 5'-triphosphate derivatives have been described. Although these primarily possess either modified sugar or base residues, the novel 5'-*P*-borane-substituted thymidine triphosphate (**85**) has been prepared[33] as a 3:2 mixture of two diastereomers in a yield of 30% from the boranophosphoramidate (**69**) upon heating with *bis*(tributylammonium) pyrophosphate in DMF. The replacement of a non-bridging oxygen atom with the isoelectronic borane group and the resulting chirality introduced at the α-phosphate suggest that such derivatives will have widespread utility in biochemical and molecular biological investigations analogous to the nucleoside α-thiotriphosphates.[45] Indeed the analogue is a good substrate for Sequenase (modified T7 DNA polymerase).

Previous work has revealed the thymidine triphosphate analogue to be a substrate for human placenta DNA polymerase α, and AMV and HIV reverse transcriptases.[46] A one-pot procedure has been developed for the preparation of α-methylphosphonyl-β, γ-diphosphates of thymidine (**86**) and

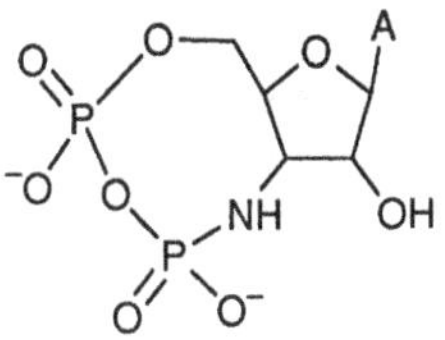

(75)

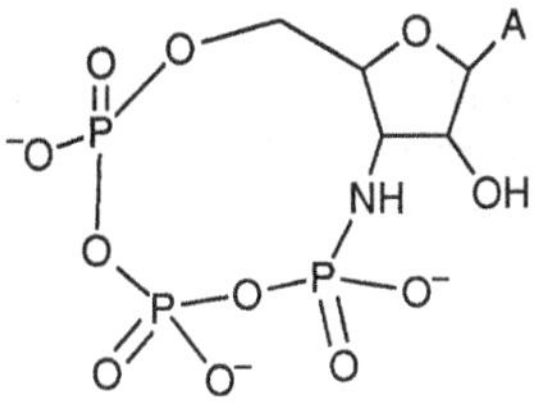

(76)

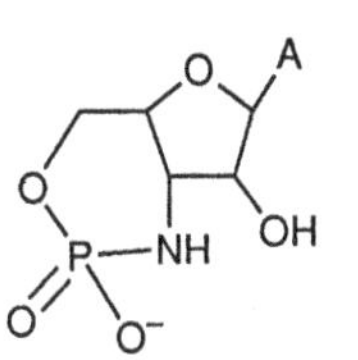

(77)

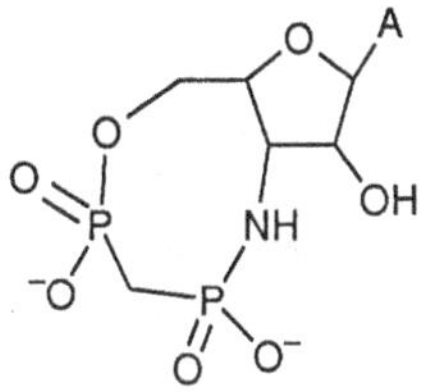

(78)

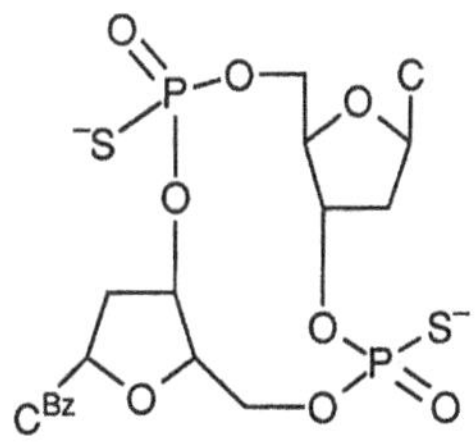

(79) Bz = Benzoyl

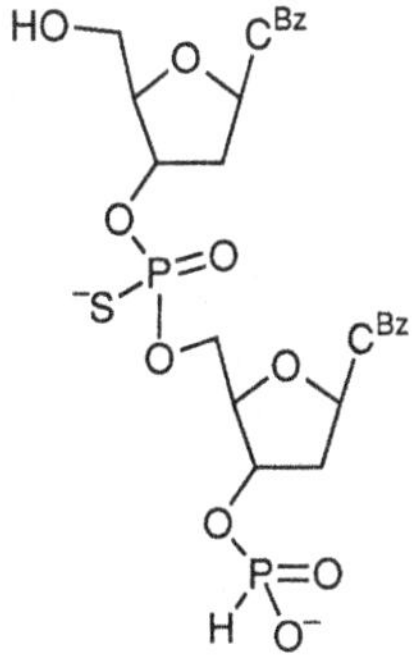

(80)

3'-azido-3'-deoxythymidine[47] whereby the nucleosides are treated with methylphosphonic *bis*(1, 2, 4-triazolide) followed by *bis*(tributylammonium) pyrophosphate. Yields of 26% are obtained, but in the case of thymidine, the 3'-hydroxyl function required protection with a *tert*butyldimethylsilyl group which was subsequently removed with *p*-toluenesulphonic acid in aqueous acetonitrile; the unusual base-lability of the α-methylphosphonyl-β, γ-diphosphates precluded the use of tetrabutylammonium fluoride for the deprotection. Although the use of methylphosphonic dichloride avoids this protection, lower yields are achieved. The procedure is considerably simpler than the previous method in which thymidine 5'-methylphosphonate was condensed with pyrophosphate in the presence of *N, N'*-carbonyl diimidazole.[46]

Several other modified nucleoside 5'-triphosphates have been prepared and evaluated as substrates for DNA polymerases, and in particular, novel, selective chain termination inhibitors of HIV-RT have been sought. Krayevsky *et al*. have shown that the O4'-nor-2', 3'-dideoxy-2', 3'-didehydronucleoside 5'-triphosphates (acyclo-d_4NTPs) (**87**) in common with their cyclic analogues are good substrates for HIV-RT, resulting in 50% inhibition of DNA synthesis at equimolar concentrations to the natural dNTPs[48]. Unfortunately they are also good substrates for the human DNA polymersases α and γ. In an attempt to increase their selectivity, the α-methylenephosphono isosteres (**88**) have been prepared[49] by condensation of the N-((Z)-4-hydroxybut-2-en-1-yl)purines and pyrimidines with ethyl [[(4-tolylsulphonyl)oxy]methyl]phosphonate. Subsequent treatment with *N, N'*-carbonyldiimidazole, followed by bis(tributylammonium) pyrophosphate produced the desired pyrophosphoryl (Z)-(phosphonomethoxy)but-2-enyl analogues. These analogues were equally effective chain-terminating substrates with HIV-RT and also of AMV-RT, but importantly, were not recognised by several other polymerases including human DNA polymerases α and ε, rat DNA polymerase β and *E.coli* DNA polymerase I. Of the 5'-phosphonate derivatives, only the adenine analogue was moderately effective against HIV infected MT-4 cells, displaying an ED_{50} of 9 μM. In contrast the nucleosides of the acyclo d_4NTPs are inactive against HIV in cell culture in concentrations up to 100 μM probably due to their inability to be phosphorylated in the cell.

Restricted glycon conformational flexibility and high coplanarity of C1', C2', C3' and C4' atoms of 2', 3'-dideoxy-2', 3'-didehydrothymidine 5'-triphosphate (d_4TTP) (**89**) have been suggested to account for the effectiveness of the analogue as a termination substrate for several DNA polymerases.[50] Extension of this work has shown that the 5'-triphosphates of the 2', 3'-epithio and anhydro derivatives of lyxothymidine (**90**) and ribothymidine (**91**) are excellent termination substrates for both AMV-RT and HIV-RT[51]; in the latter instance, K_m and V_{max} values differ only 2- to 3-fold compared to those for TTP. This provides further support for the idea that these modified substrates with a flattened sugar ring imitate the transition state conformation of the substrates in the polymerase-template-primer complex, the glycone performing a role for the correct positioning of the base and triphosphate residue. In addition, four diastereoisomers of the carbocyclic adenosine 5'-triphosphate analogues (**92**) - (**94**) have been prepared[52] and evaluated as substrates for AMV-RT and HIV-RT. Only the D-β and L-β ('cis-like') structures were recognised, the D-α (**93**) and L-α (**94**) derivatives being poor substrates.

The 5'-α-thiotriphosphate derivatives of 8-bromo-(**95**) and 8-azido-2'-deoxyadenosine (**96**) have been prepared[53] from the corresponding nucleosides using salicyl chlorophosphite followed by

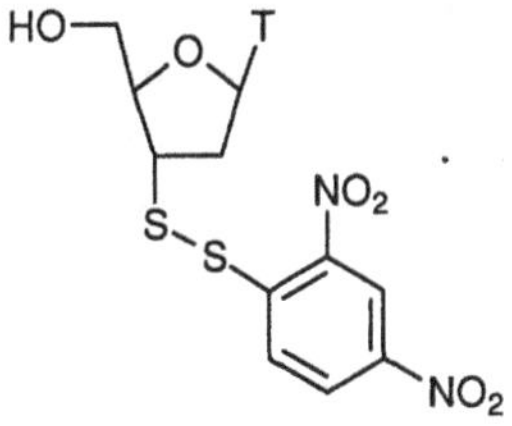

(81)

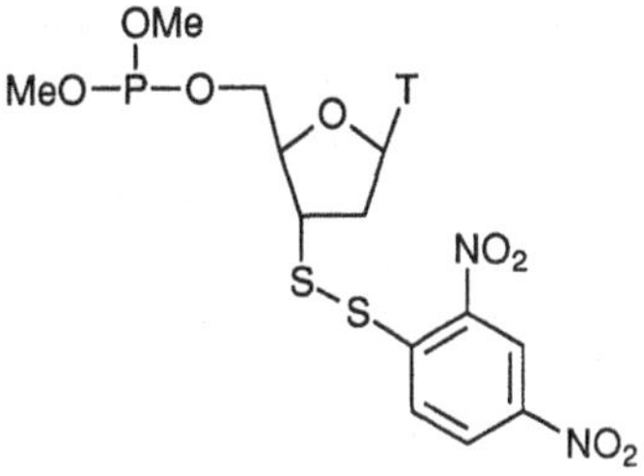

(82)

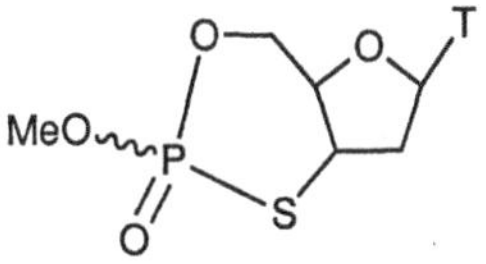

(83)

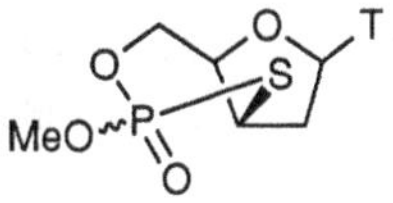

(84)

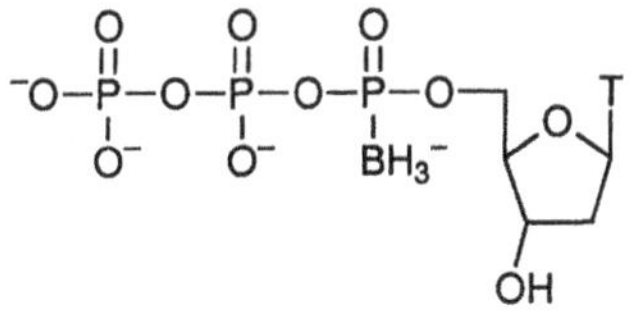

(85)

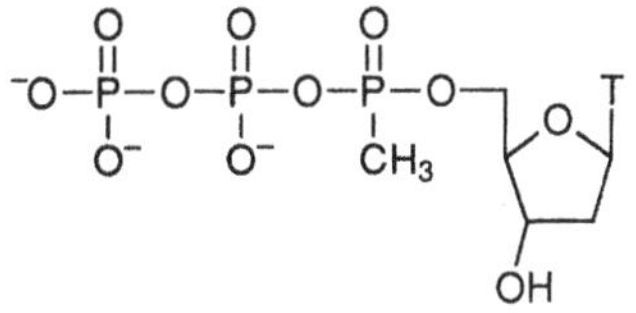

(86)

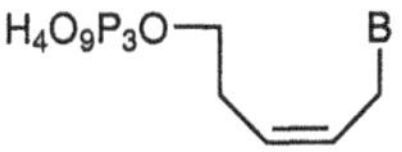

(87)

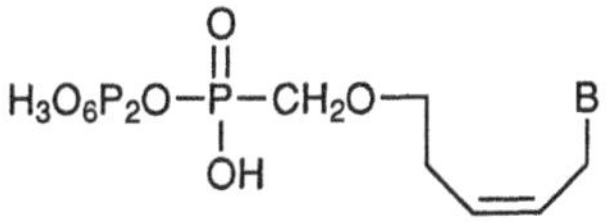

(88)

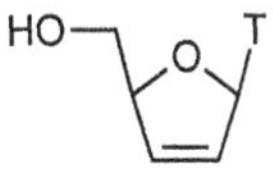

(89)

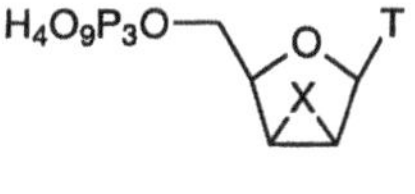

(90) X = O, S

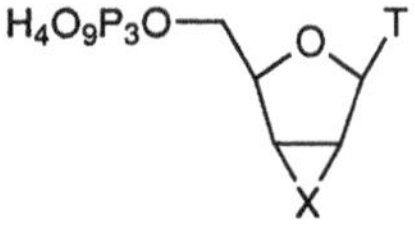

(91)

oxidation with sulphur[54] and by phosphorylation with thiophosphoryl chloride followed by reaction with *bis*(tributylammonium) pyrophosphate, respectively. Both derivatives were substrates for T7 DNA polymerase thereby allowing their incorporation into DNA. It was suggested that the resulting DNA by virtue of it containing 8-bromo- and 8-azido-adenine residues would offer the opportunity for combining interference and photo-crosslinking studies of sequence specific protein binding with the sequencing techniques based on phosphorothioate-containing DNA. Photochemical cross-linking studies of ribosomal RNA and a tRNA derivative have been achieved using the 5'-[γ-^{32}P]-labelled 3', 5'-bisphosphate of 2, 6-diazido-9-(β-D-ribofuranosyl)purine.[55] The compound was prepared by reaction of the corresponding nucleoside with pyrophosphoryl chloride followed by exchange of the 5'-phosphate with the γ-phosphate of [γ-^{32}P]ATP catalysed by T4 polynucleotide kinase. It was then enzymatically ligated to the 3'-terminus of a yeast tRNA^{Phe} molecule.

The recent trends in the use of fluorescent labelling techniques in DNA sequencing has employed either 5'-end labelled primers or base-labelled deoxynucleoside 5'-triphosphates. Since the latter are often poor substrates for DNA polymersases commonly used in sequencing, Engels and co-workers have prepared the 3'-amino-2', 3'-dideoxynucleoside 5'-triphosphates (**97**) to allow labelling via their modified sugars.[56] The triphosphates were prepared in yields of between 45 and 60% from their respective base-protected 3'-azido nucleosides by the procedure of Ludwig and Eckstein[54] employing salicyl chlorophosphite or via phosphorus oxychloride in triethylphosphate in the presence of a proton sponge,[57] followed by reaction with *bis*(tributylammonium) pyrophosphate. Reduction of the 3'-azido to the corresponding amino function was achieved using the Staudinger reaction using triphenylphosphine in pyridine and subsequent hydrolysis with aqueous ammonia. The 3'-amino-2', 3'-dideoxynucleoside 5'-triphosphates are themselves excellent chain terminators for DNA synthesis. Reaction with fluorescein 5-isothiocyanate (FITC) in pH 9.6 buffered aqueous DMF solution afforded the labelled 5'-triphosphate derivatives in yields of about 70%. Although the labelled triphosphates were good substrates in DNA sequencing reactions, low signal due to fluorescence quenching of the label suggested that an alkyl spacer between the dye molecule and the sugar may be required. A synthetic strategy for the preparation of 3'-O-(ω-aminoalkoxymethyl)thymidine 5'-triphosphates (**98**) has been described[58] and the potential of such analogues for use in DNA sequencing using fluorescent detection has been demonstrated. Phosphoryltris(triazole) in acetonitrile followed by treatment with *bis*(tributylammonium) pyrophosphate was used in place of phosphorus oxychloride to phosphorylate the 3'-trifluoroacetamido protected nucleosides due to the acid lability of the acetal linkage within the aminoalkyl chain.

Several 2'-fluoro- and 2'-amino-2'-deoxynucleosides (**99**) have been characterised as good T7 RNA polymerase substrates and thus permit the enzymatic synthesis of modified RNA polymers by transcription.[59]

An interesting enzymatic route for the synthesis of adenosine(5')tetraphospho(5')adenosine (Ap_4A) derivatives has recently been described.[60] The procedure involves the use of ATP[γS] (**100**) as adenylyl donor, an appropriate NTP and luciferin in a luciferase-catalysed reaction. Derivatives of Ap_4N are obtained in yields of about 50% and the procedure is appropriate for

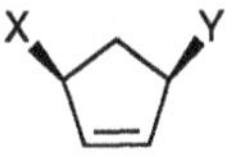

X Y

(92) X, Y = A, $H_3O_6P_2O-\overset{O}{\overset{\|}{P}}(OH)-CH_2^-$

$H_3O_6P_2O-P(=O)(OH)-CH_2-O$ A

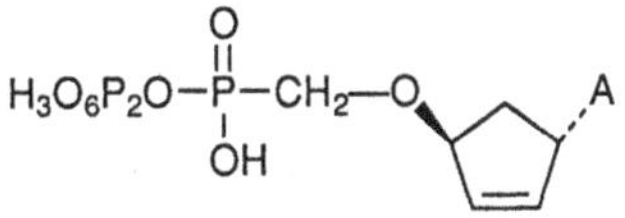

(93)

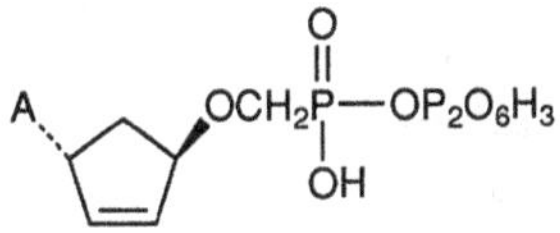

A $OCH_2P(=O)(OH)-OP_2O_6H_3$

(94)

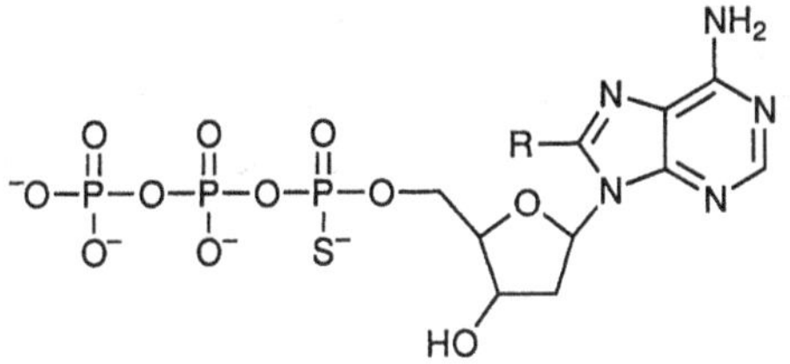

NH_2 N N R N N

$^-O-P(=O)(O^-)-O-P(=O)(O^-)-O-P(=O)(S^-)-O$ O HO

(95) R = Br
(96) R = N_3

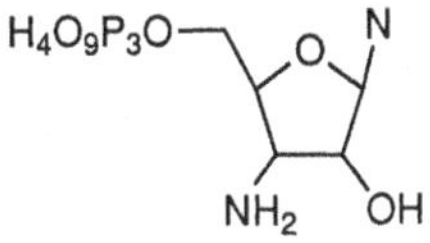

$H_4O_9P_3O$ O N NH_2 OH

N = A, C, G, T

(97)

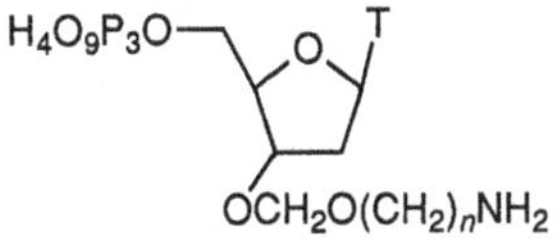

$H_4O_9P_3O$ O T $OCH_2O(CH_2)_nNH_2$

(98)

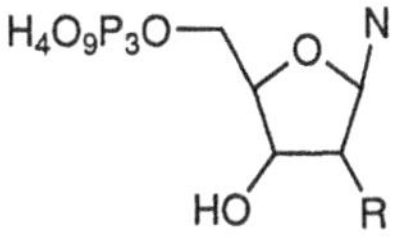

$H_4O_9P_3O$ O N HO R

(99) R = F, NH_2

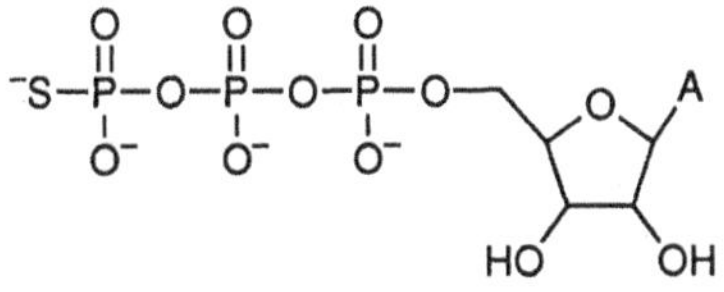

$^-S-P(=O)(O^-)-O-P(=O)(O^-)-O-P(=O)(O^-)-O$ O A HO OH

(100)

Cl N $O-P(=O)(O^-)-S$

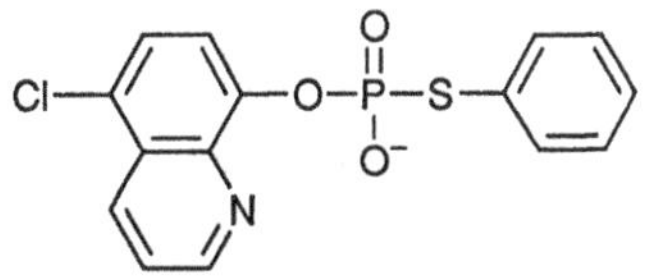

(101)

several nucleosides (N) including the fluorescent analogue N^6-ethenoadenosine. Between 1 and 5 mmols of the tetraphosphate derivative may be synthesised by this procedure.

A novel bifunctional phosphorylating agent has been reported which allows the preparation of the messenger RNA cap structure m^7G5'-pppG on a large scale and in good yield (55%) from the unprotected 5'-mononucleotides.[61, 62] The phosphorylating agent (**101**) is prepared by treating 5-chloro-8-quinolyl phosphate with diphenyl disulphide and tributylphosphine in pyridine. The synthesis involved the initial treatment of guanosine-5'-monophosphate in 1-methylpyrrolidone(MPD)-HMPA with the phosphorylating agent, the phenylthio group of which is activated by addition of silver nitrate. Subsequent addition of copper(II) chloride in MPD activated the 5-chloroquinolyl group which was displaced by the addition of *N7*-methylguanosine-5'-monophosphate to give the desired dinucleotide triphosphate derivative. In the same manner m^7G5'-pppA was obtained in 53% yield by using adenosine-5'-monophosphate in place of guanosine-5'-monophosphate.

Acyclovir diphosphate dimyristoylglycerol (**102**), prepared by coupling *sn*1, 2-dimyristoylglycerol-3-phosphomorpholidate with acyclovir monophosphate, has displayed good activity against a range of AZT resistant HSV strains.[63] Its activity appears to be due to the cellular metabolism of the prodrug, which gives rise to acyclovir monophosphate inside the target cell.

3. Oligo- and Poly-Nucleotides

3.1 DNA Synthesis

Methods for the synthesis of DNA are now routine and consequently very little literature appears on the synthesis of unmodified DNA. Advances in the synthesis of oligonucleotides by the phosphoramidite approach have been reviewed during 1992[64] and the following year a companion review dealt with the synthesis of modified oligonucleotides by the phosphoramidite approach.[65] In 1993 the Nobel Prize was awarded for two techniques which rely heavily on the ability to synthesise oligodeoxynucleotides, namely site directed mutagenesis and the polymerase chain reaction.[66]

An alternative very hydrophobic protecting group 4-(17-tetrabenzo[a,c,g,i]fluorenylmethyl)-4,'4"-dimethoxytrityl- (**103**) has been proposed for the 5'-hydroxyl group during oligodeoxynucleotide synthesis and purification.[67] The protecting group can be introduced to the nucleoside as the chloride but the coupling yields of the resultant phosphoramidites are lower than usual. Alternatively post oligonucleotide synthesis introduction is possible by treatment of the deprotected oligonucleotide with a 0.4M solution of the chloride in pyridine. The modified trityl is removed at twice the rate of 4,4'-dimethoxytrityl- with acid but is very much more hydrophobic. Oligonucleotides of 100 nucleotides can be purified by reverse phase hplc when protected with this group.

A novel support linkage for the synthesis of oligodeoxynucleotides, which is labile to photolysis, has been described.[68] 5'-*O*-dimethoxytrityl-3'-*O*-succinatothymidine was treated with the trichlorophenylester of 4-hydroxymethyl-3-nitrobenzoic acid in the presence of DCC. The activated ester is then linked to long chain alkyl controlled pore glass (LCAA-CPG) in the presence of 1-hydroxybenzotriazole. Photolysis in acetonitrile water using the $CuSO_4$ filtered output from a Hg/Xe lamp liberates oligonucleotides from the support (**104**).

The large scale synthesis of oligodeoxynucleotides has received some attention as the demand for oligonucleotides for antisense applications has increased. The recycling of nucleoside monomer reagents during oligonucleotide synthesis has been investigated.[69] An aminopolyethyleneglycol derivatised polystyrene support has been described.[70] The polyethylene linker has good solvation properties in acetonitrile and permits synthesis on a 1mmol scale.

Kumar *et al.* have investigated the effect of various parameters such as concentration of succinic anhydride, 4-dimethylamino pyridine(DMAP), temperature and solvent in the formation of nucleoside 4,4'-dimethoxytrityl-3'-succinates and their immobilisation onto solid supports.[71] 3'-*O*-Succinylation of protected 2'-deoxynucleosides is complete within 10 minutes using 1.5 equivalents of succinic anhydride, 0.5 equivalents of DMAP in 1,2-dichloroethane at 50°C. An alternative strategy for activation of the succinate and linkage to solid supports was devised. Appropriately protected 2'-deoxyribo- or 2'-*O*-*tert*butyldimethylsilyl-ribonucleosides-3'-*O*-succinates were reacted with one equivalent of toluene-2,4-diisocyannate or hexamethylene-1,6-diisocyanate in the presence of 1 equivalent of DMAP in dry dichloromethane to generate a monoisocyanate. This was then allowed to react without isolation with the amino groups of polymer supports in the presence of ethyldiisopropylamine. The resultant supports (**105a** and **105b**) were demonstrated to have comparable performance to those conventionally used in DNA synthesis. The reaction of 4,4'-dimethoxytrityl-2'-deoxynucleoside-3'-*O*-succinates with aminoethyl-polystyrene has also been studied by Montserrat *et al*.[72] The most reproducible reactions involved DCC and 1-hydroxybenzotriazole mediated coupling

Solution phase methods of synthesis are also of interest with a view to the production of large quantities of oligonucleotides. Bonora and co-workers have pioneered the use of polyethylene glycol as soluble supporting polymer for DNA synthesis. Cyclic oligonucleotides and oligonucleotides containing phosphorothioate linkages have also been synthesised with soluble polyethylene glycol supporting polymers.[73, 74] Nucleoside 3'-*N,N*-diisopropylphosphonamidates (**106**) are potentially useful in this context.[75] The phosphonamidates are stable to mild acid and mild oxidants and thus their presence in molecules undergoing coupling reaction at other sites is not prohibited so chain elongation in both a 3'-and 5'-direction is possible.

Trityl monitoring by conductivity offers the advantages of real time synthesis evaluation [76]and is now a feature of some DNA synthesisers. Alternative protection for guanine and thymine bases during oligodeoxyribonucleotide synthesis has been reported on. *O*-Allyl protecting groups are removed by Pd(0) but are stable to acid and fluoride ion and are easily introduced to the nucleosides by activation of the O^4- of protected thymidine or the O^6-of protected guanosine with 2-mesitylenesulphonylchloride.

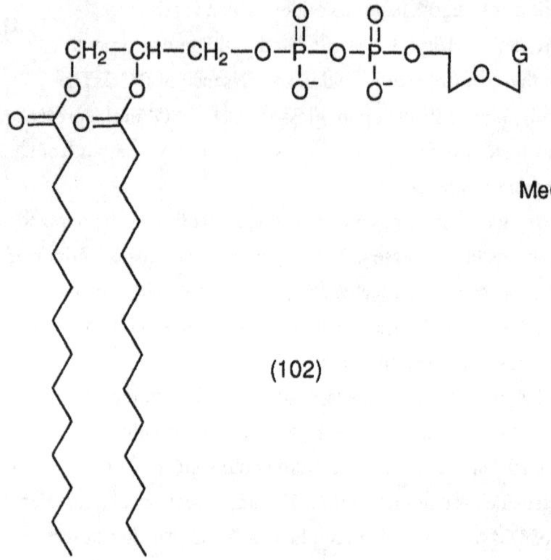

(102)

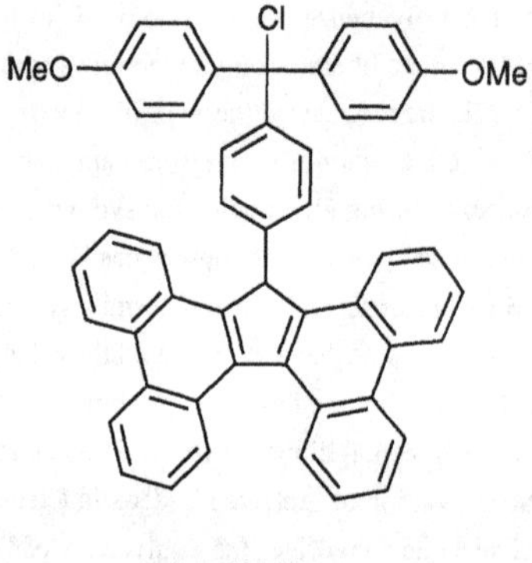

(103)

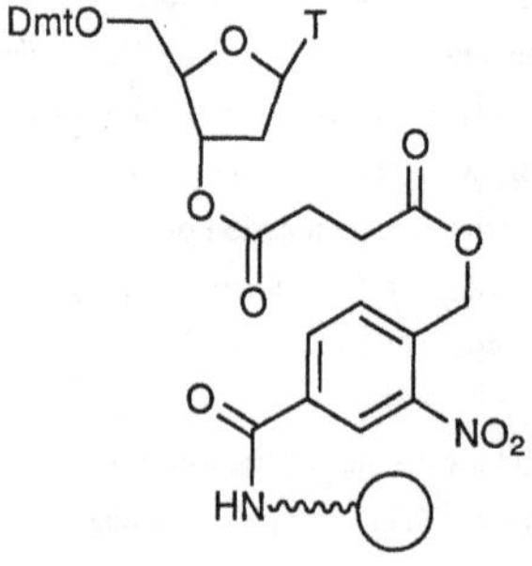

(104)

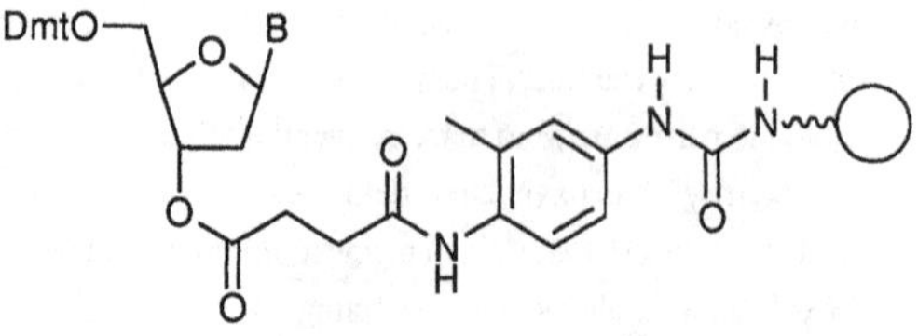

(105a)

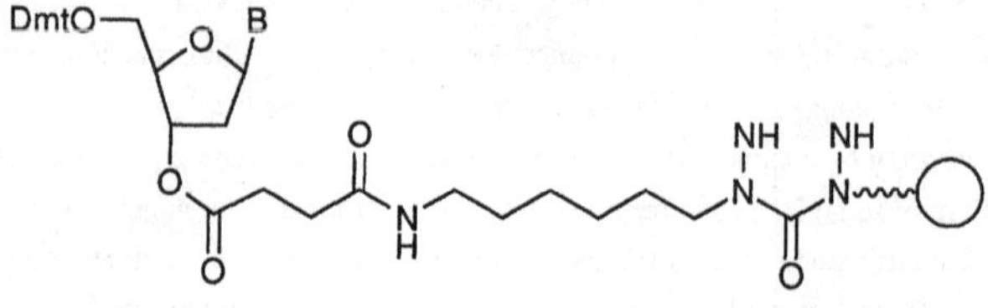

(105b)

Two linkers have been described which allow the preparation of two oligonucleotides per synthesis thus maximising the efficiency of use of a DNA synthesiser without intervention.[77] The linkers were prepared from mono-benzoyl-1,4-anhydroerythritol which was protected by dimethoxytritylation before removal of the benzoyl group with methylamine. The hemisuccinate ether was synthesised by treatment with succinic acid anhydride and DMAP. Carbodiimide mediated coupling of either 1,4-anhydroerythritol or 2,2-sulphonyldiethanol followed by phosphitilation yielded reagents (**107**) and (**108**). Upon ammonia treatment the succinate diester is hydrolysed and oligonucleotides released by attack of the vicinal alcohol on the terminal phosphate or β-elimination adjacent to the sulphone. Compound (**107**) yields two oligonucleotides whereas (**108**) yields one oligonucleotide bearing a 5'-phosphate function.

Several imidazole functionalised dimethoxytrityl protecting groups have been investigated for their ability to catalyse internucleoside bond formation using phosphate triester chemistry.[78] Diethyl oxamalonate has been suggested as an alternative detritylating agent for use in oligonucleotide synthesis.[79] An alternative phosphate protecting group for use in phosphoramidite chemistry, 2-(trimethylsilyl)ethyl-, has been reported.[80] The new protecting group is removed by treatment with 1.0 M tetrabutylammonium fluoride. H-phosphonate synthesis of oligodeoxynucleotides with dichloromethane rather than acetonitrile as solvent has been described.[81] Dipentafluorophenyl carbonate has been introduced as a condensing agent in the synthesis of oligonucleotides by the H-phosphonate approach.[82] An alternative protecting group, (2-pyridyl)methyl-, for the O6-function of 2'-deoxyguanosine has been investigated.[83]

3.2 RNA Synthesis

Unlike DNA synthesis the synthesis of RNA is much more challenging due to the presence of the 2'-hydroxyl group. Nevertheless reagents for the synthesis of RNA using phosphoramidite chemistry have become available over the last few years. The volume of literature relating to RNA synthesis is therefore greater than that for DNA as methods for synthesis are optimised and alternatives are suggested. A complete account of the synthesis of oligoribonucleotides using 2'-*O*-1-(2-fluorophenyl)-4-methoxypiperidin-4-yl (Fpmp) protection (**109**) has been reported[84] and oligoribonucleotide synthesis has been reviewed.[85] Commercially available reagents currently use either the *tert*-butyldimethylsilyl (**110**) or the Fpmp group for 2'-hydroxyl protection. Several variants on the former approach are available differing in the type of protection used for the exocyclic amino groups of the nucleobases.

Tert-butylphenoxyacetyl has been proposed as a protecting group for use in both DNA and RNA synthesis.[86] The group is more labile in ammonia solution than current alternatives and therefore minimises premature desilylation during ammonia deprotection in RNA synthesis and reduces depurination during DNA synthesis.

An alternative base labile 5'-protecting group, 2-dansylethoxycarbonyl (**111**) has been proposed for RNA synthesis utilising tetrahydro-4-methoxy-2H-pyran-4-yl for 2'-protection, 2-(4-nitrophenyl)ethyl and 2-(4-nitrophenyl)ethoxycarbonyl for phosphate and amino group masking, respectively.[87] 9-Fluorenylmethoxycarbonyl, conventionally associated with peptide synthesis, has previously been investigated as a potential protecting group for RNA synthesis. The utilisation of DBU to remove the F-moc function has led to problems with loss of functionalisation during chain assembly, in particular cleavage of the solid support linkage. The use of an acid labile alkoxybenzylidene linker (**112**) circumvents these problems and allows the use of piperidine to remove F-moc and methyl phosphoramidite protection which produces side reactions during DBU protection.[88] 5'-*O*-Fluorenylmethoxycarbonyluridine is treated with 4-formylacetic acid in the presence of an acid catalyst to introduce the linker and the resultant uridine derivative attached to LCAA-CPG using a DCC and 1-hydroxybenzotriazole mediated coupling.

The usual choice of reagent to remove the *t*-butyldimethylsilyl (*t*BDMS) protecting group in RNA synthesis is 1M tetrabutylammonium fluoride in tetrahydrofuran. Hogrefe *et al.* have studied the efficiency of desilylation of a model dimer in the presence of water.[89] Removal of the *t*-BDMS group from uridine and cytidine is greatly dependent on the water content of the TBAF reagent whereas the presence of water has no effect on the desilylation of purines. The authors recommend that TBAF is stored over molecular sieves. An alternative reagent to TBAF, triethylamine trihydrofluoride, suffers less from the problems of water presence and gives better results for the deprotection of long oligoribonucleotides.[90]

Stromberg and co-workers have studied a number of protecting groups with respect to their usefulness as a 2'-protecting group in RNA synthesis using the H-phosphonate approach.[91] The most successful syntheses were those where the *t*-butyldimethylsilyl group was used. Interestingly the Fpmp group was shown to be inadequate when this method of phosphodiester bond assembly was employed. The acid labile *tert*-butyloxycarbonyl group has also been investigated as a potential protecting group for the 2'-hydroxyl function during RNA synthesis using phosphotriester chemistry.[92]

The stability of inter-ribonucleoside phosphodiester linkages to acid has been investigated with a view to the use of acid labile 2'-hydroxyl protecting groups in RNA synthesis. Oligouridylic acids are susceptible to hydrolytic cleavage and migration at pH2, conditions that have previously been recommended for the deprotection of oligoribonucleotides bearing 2'-Fpmp and related acid labile protecting groups.[93] The authors therefore recommend deprotection of 2'-Fpmp bearing oligomers at pH 3.

The stability of the trisubstituted internucleotide bond in the presence of a vicinal 2'-hydroxyl group has been studied by Kierzek.[94] The acid treatment of model triesters with different protecting groups present on the internucleoside bond resulted in several different types of rearrangement reaction at the phosphoryl centre. These were removal of the phosphoryl protecting group with cleavage of the internucleoside bond to form the cyclic phosphate and the 3'-nucleoside, removal of the protecting group without cleavage but isomerisation of the internucleoside bond to produce a 2',5'-linked dimer. In some cases no further reaction was observed. For many of the protecting groups all these reactions occurred in parallel. As a result of these studies the synthesis of uridyl(2'-phosphate)-(3',5')uridine was undertaken without protection of the internucleoside phosphodiester.

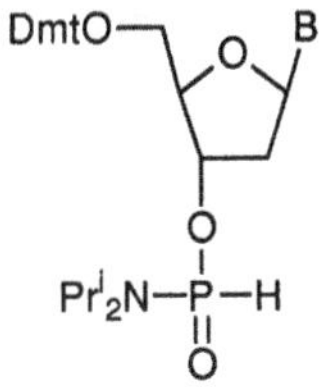

(106)

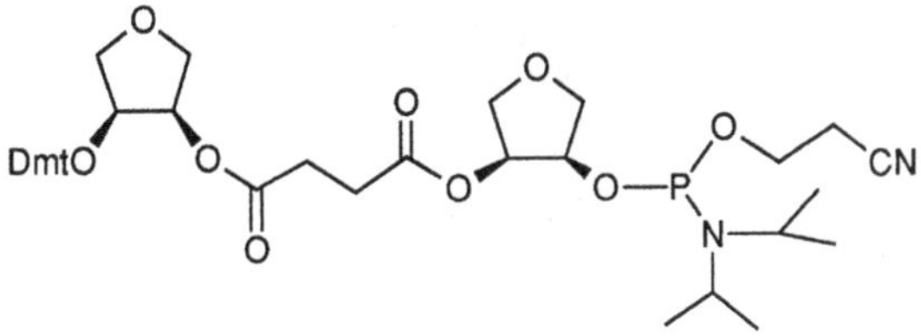

(107)

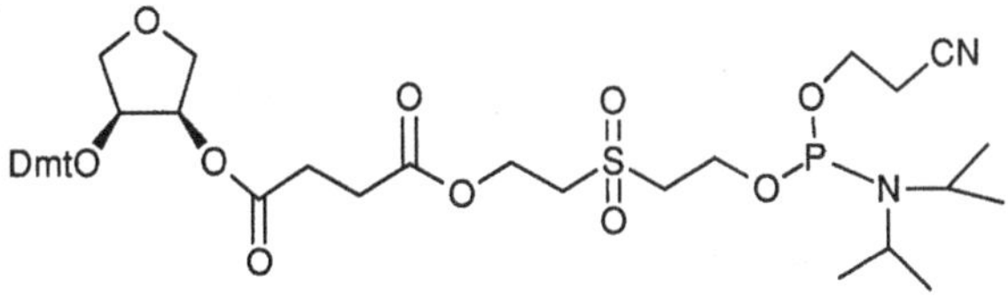

(108)

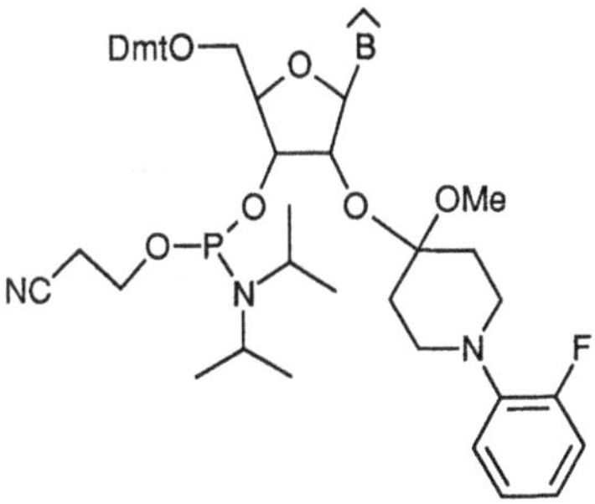

(109)

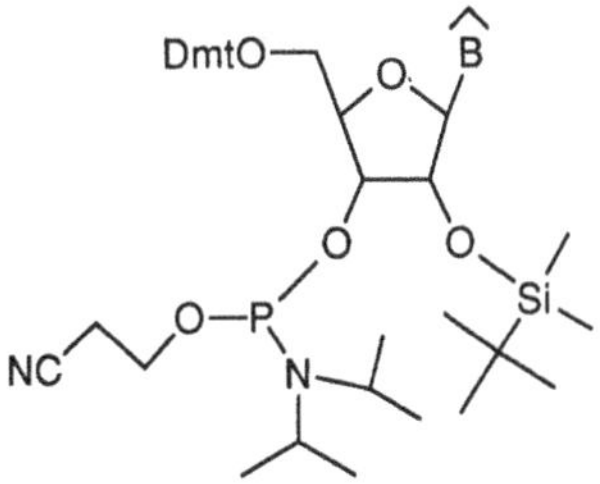

(110)

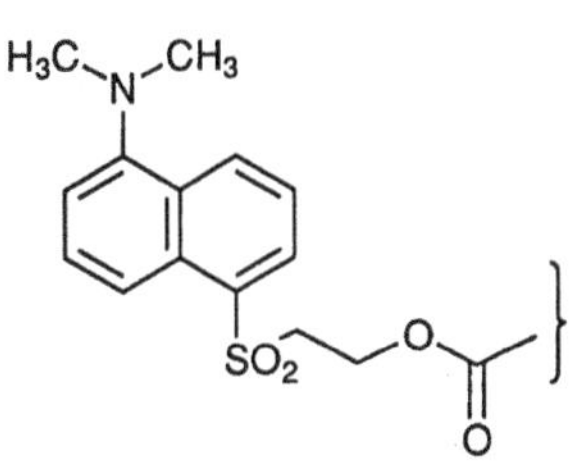

(111)

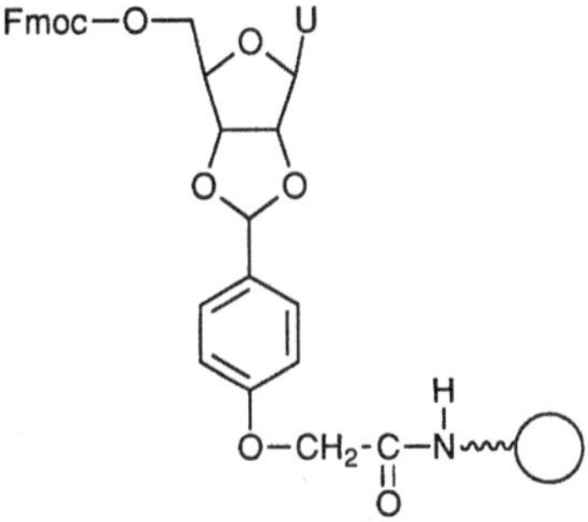

(112)

The enzymatic synthesis of milligram quantities of ribozymes using T7 RNA polymerase has been reported as a cheaper alternative to chemical synthesis.[95]

3.3. The Synthesis of Modified Oligodeoxynucleotides and Modified Oligoribonucleotides

3.3.1 Oligonucleotides Containing Modified Phosphodiester Linkages

The ability to synthesise oligonucleotides containing modified phosphodiester linkages continues to be of great interest in the therapeutic area. In particular it is desirable to confer through these modifications enhanced cell permeability properties and nuclease resistance. Several previous candidates that confer these properties have modified the phosphodiester linkage in such a way that creates a chiral centre at phosphorus. The resultant complex mixture of diastereoisomers all meet the above requirements to different extents. Over the past year efforts have therefore concentrated on the synthesis of achiral alternative linkages or stereospecific syntheses of chiral phosphate modifications. A number of literature summaries in this area have appeared recently.[96-98]

An alternative reagent for the formation of oligonucleotides containing phosphorothioate internucleoside linkages has been described by Stec and co-workers.[99] The oxidising solution used in phosphoramidite chemistry is replaced by a solution of *bis*(*O,O*-diisopropoxyphosphinothionyl) disulphide (**113**) in acetonitrile. An alternative protecting group strategy has been reported for use in the synthesis of oligonucleotides containing phosphorothioate linkages . Allyl and allyloxycarbonyl groups are removed from the phosphorothioate internucleoside linkages and the exocyclic amino groups by treatment with Pd^{0}.[100] A method has been described for the sequencing of oligonucleotide phosphorothioates. 2-Cyanoethyl groups can be removed selectively from unprotected solid-supported oligonucleotides by a solution of *tert*-butylamine in pyridine. Oxidation of the internucleoside phosphorothioate linkages with iodine, in the presence of water and *N*-methylimidazole, allows chemical sequencing of the oligonucleotide by Maxam and Gilbert methods.[101] The synthesis of phosphorothioate-containing oligonucleosides having abasic sites, utilising 5-*O*-(4,4'-dimethoxytrityl)-1,4-anhydro-2-deoxy-D-ribitol-3'-*O*-[2-diphenylmethylsilylethoxy)-*N,N*-diisopropylamino)phosphoramidite (**114**), has been reported by Ravikumar *et al.*[102]

Self-stabilised oligonucleotides containing phosphorothioate linkages have been produced.[103] These oligonucleotides have hairpin loop structures which leads to resistance to nucleases. The structure of the oligonucleotides does not, however, affect their ability to hybridise to complementary targets but the most nuclease-stable oligonucleotides are those with the maximum number of base pairs in the complementary stem region.

The synthesis of oligonucleotides containing internucleoside 3'-*S*-phosphorothiolate linkages using the Michaelis-Arbusov reaction, has been reported.[36] 5'-*O*-monomethoxytrityl-3'-*S*-(2,4-dinitrophenyldisulphanyl)-3'-deoxythymidine (**115**) was prepared by reaction of 5'-

monomethoxytrityl-3'-thiothymidine with the arene sulphenyl chloride. The *bis*-(*O*-trimethylsilyl) phosphite derivative of 3'-*O*-(*tert*-butyldimethylsilyl)thymidin-5'-yl triethylammoniumphosphonate reacts with the disulphide to produce the dithymidine phosphorothiolate diester (**116**). The ribonucleoside dimer analogue has also been reported.[104] The phosphonate may also be methylated prior to reaction with the disulphide to yield the corresponding methyl protected phosphorothiolate. An oligonucleotide with a phosphorothiolate linkage at the cleavage site of the *tetrahymena* ribozyme has been used to study the mechanism of this enzyme catalysed reaction.[105]

The synthesis of dimers containing phosphoroselonate internucleoside linkages has been described by Stawinski and Thelin.[106] Selenisation of H-phosphonate diesters or H-phosphonothioate diesters is accomplished using a selenium transfer reagent 3H-1,2-benzothiaselenol-3-one (**117**).

The synthesis of specific diastereoisomers of a DNA methylphosphonate heptamer has been reported. The synthesis involved the coupling of pure *Rp* or *Sp* dimers to assemble trimers and tetramers which were purified by hplc and then block coupled together to give diastereomerically pure samples. The stability of the isomers with a complementary unmodified oligonucleotide demonstrated that methylphosphonate oligonucleotides containing *Rp* linkages formed the most stable duplexes.[107-109]

The synthesis of oligodeoxynucleotides containing formacetal (**118**) and 3'-thioformacetal (**119**) linkages, has been described.[110] Formation of the formacetal linkage is achieved by reaction of 5'-(4,4'-dimethoxytrityl)-3'-[(methylthio)methyl]thymidine with 3'-*tert*-butyldimethylsilyl protected 2'-deoxynucleoside in the presence of bromine. After removal of the silyl protection, the H-phosphonate derivative of the dimer can be produced using conventional procedures. The thioformacetal linkage is assembled by chlormethylation of a 3'-phenoxyacetyl- protected nucleoside with paraformaldehyde and dry HCl gas. The resultant 5'-chloromethyl derivative is coupled with 5'-*O*-(4,4'-dimethoxytrityl)-3'-mercapto-3'-deoxythymidine in the presence of base. After removal of the 3'-protection the H-phosphonate derivatives of the dimers can be produced using conventional procedures. The resultant dimer blocks can be employed in DNA synthesis using H-phosphonate chemistry yielding oligonucleotides containing alternating phosphodiester and acetal linkages. Upon hybridisation with their complementary RNAs, the thioformacetal-containing oligomers proved to have higher melting temperatures than their unmodified counterparts, whereas those containing formacetal linkages have been demonstrated to have lowered stabilities. The reduced stability of DNA duplexes containing formacetal linkages has been investigated by computational techniques.[111]

Modified oligonucleotides containing sulphide linkages have been prepared by a dimer coupling procedure. In one approach the modified dimer is assembled from 3'-deoxy-3'-(2-hydroxyethyl)ribofuranosylthymine which is protected at the 5'- and 2'- positions, mesylated and then coupled to 2'-protected 5'-thio-2'-deoxythymidine in the presence of caesium carbonate in DMF.[112] After exchange of protecting groups to produce an appropriate derivatisation for oligonucleotide assembly, the 3'-*O*-phosphoramidite (**120**) of the dimer is formed. Interestingly 12mer oligonucleotides containing three sulphide linkages selectively bind to complementary RNA but not DNA. Oligodeoxynucleotides containing 3'-allylether, 3'-allylsulphide (**121**) and their

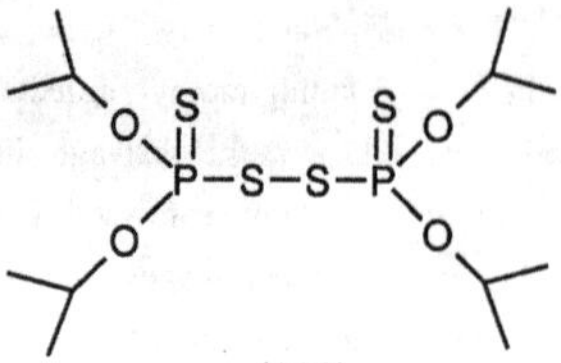

(113)

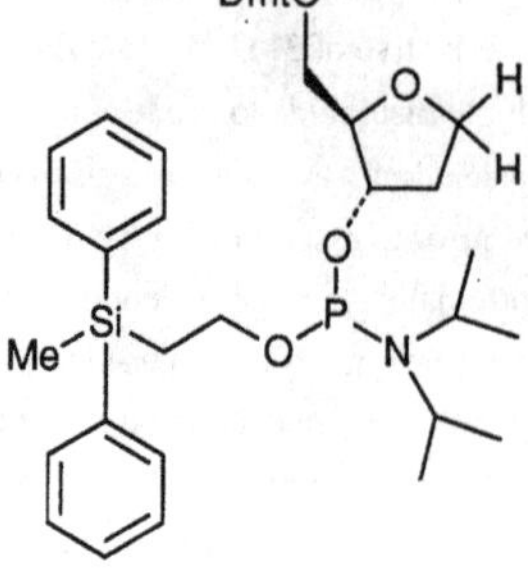

(114)

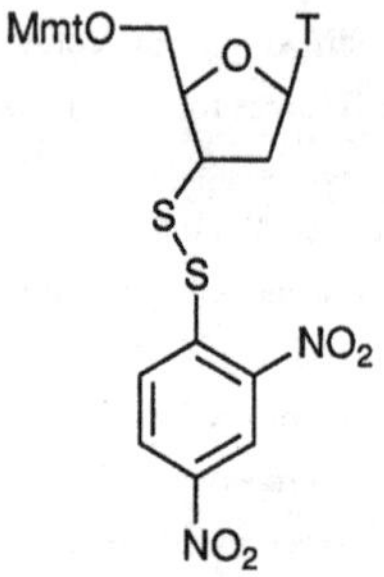

(115)

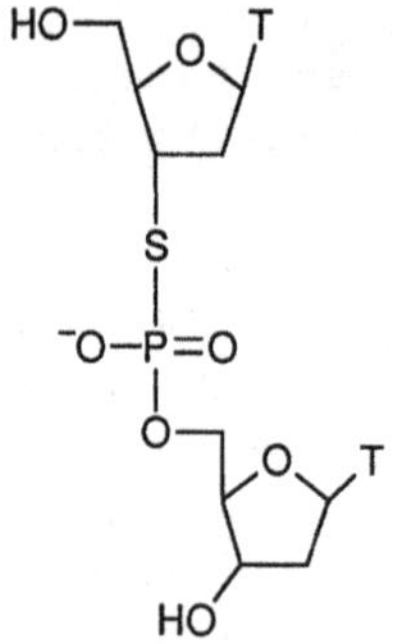

(116)

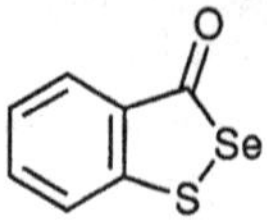

(117)

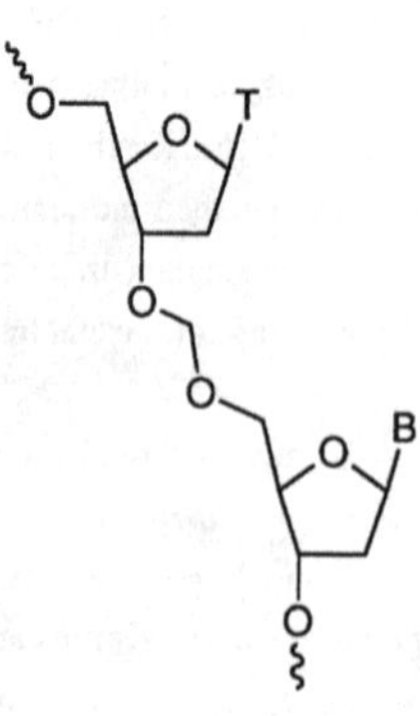

(118)

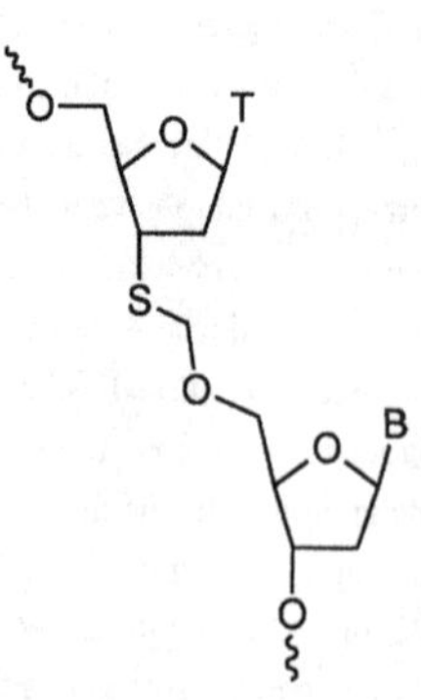

(119)

corresponding saturated derivatives (**122**) as phosphate mimics have been described. Dimers containing these internucleoside linkages were synthesised and then derivatised for H-phosphonate DNA synthesis.[113]

The synthesis and incorporation into oligonucleotides of thymidine dimers in which the phosphodiester internucleoside bond has been replaced with a carboxamide linkage has been reported by Chur *et al.*[114] The dimer (**123**) is assembled by the condensation of 5'-*O*-(4,4'-dimethoxytrityl)-3'-*O*-(2-aminoethyl)thymine with 1,2-dideoxy-1-thyminyl-β-D-erythro-petofuranuronic acid and then phosphitilated for use in oligonucleotide synthesis. Oligonucleotides containing the dimer exhibit modestly lowered melting temperatures on hybridisation with their complementary sequences.

A dimer block procedure has also been used for the synthesis of olithymidylates containing alternating (amino)ethyl phosphonate and phosphodiester linkages.[115] The compounds are synthesised from dimethyl(phthalimidomethyl)phosphonate which is deprotected using trimethylsilyl bromide to give phthalimidomethylphosphonate. This is treated with 5'-*O*-(4,4'-dimethoxytrityl)thymidine using triisopropylsulphonyl chloride as condensing agent. A second coupling with unprotected thymidine yields the dinucleotide phthalimidomethylphosphonate. Following separation of the diastereoisomers, phosphitilation under standard conditions yields the reagent (**124**), suitable for oligonucleotide synthesis. One isomer yields a duplex upon hybridisation with a complementary oligonucleotide which is more stable than the unmodified counterpart whereas the other isomer has a considerable destabilising effect. The linkage is stable to enzymatic degradation but not to hydrolysis at elevated temperatures.

The synthesis of oligonucleotides containing sulphonate linkages has been described. Thymidine is converted to the 5'-deoxy-5'-iodo- derivative using standard conditions and then protected at the 3'-position with *tert*-butyldimethylsilyl. The anion of isopropylmesylate is then utilised to displace the iodide. Deprotection of the sulphonate followed by formation of the acid chloride yields a thymidine derivative which can be coupled to 5'-monomethoxytrityl thymidine to yield the 3'-*O*-sulphonate ester linked dimer (**125**).[116] A series of *O*-(2,2,2-trifluoroethyl)-3',5'-dinucleoside phosphates (**126**) has been synthesised via preparation of a novel phosphitilating agent $CF_3CH_2OP(N^iPr_2)Cl$.[117] Thymidine(3',5')thymidine hydroxymethylphosphonate (**127**) has been synthesised by development of a novel phosphonylating agent methylacetoxymethylphosphonate.[118]

The synthesis of oligonucleotides which are derivatised with carboranes are of potential use in boron neutron-capture therapy. Thymidine(3',5')thymidine(*O*-carboran-1-ylmethyl)phosphonate (**128**) has been prepared via synthesis of the borophosphonylating agent 5'-*O*-(monomethoxytrityl)thymidine 3'-*O*-[methyl (*O*-carboran-1-ylmethyl)phosphonate].[119]

A comparative evaluation of a number of potential antisense oligonucleotides, summarised in figures (**129a** and **129b**) has been undertaken by Morvan *et al.*[120] Oligonucleotides containing 2'-*O*-methyl and 2'-*O*-allyl modification were shown to form the most stable hybrids with complementary RNA whereas alpha oligonucleotides and those containing phosphorothioate substitutions were the most stable in culture medium. However all the modified oligonucleotides studied exhibited sequence-non-specific inhibitory effects on HIV-replication.

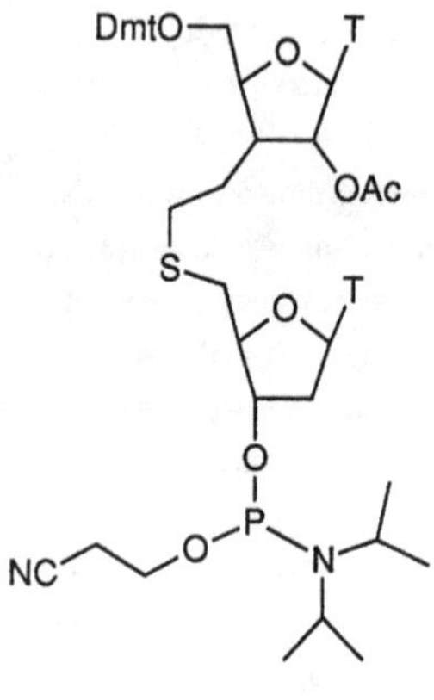

(120)

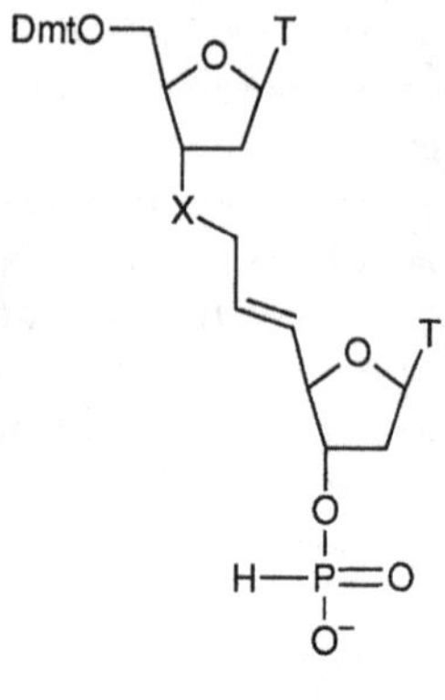

X = S or O
(121)

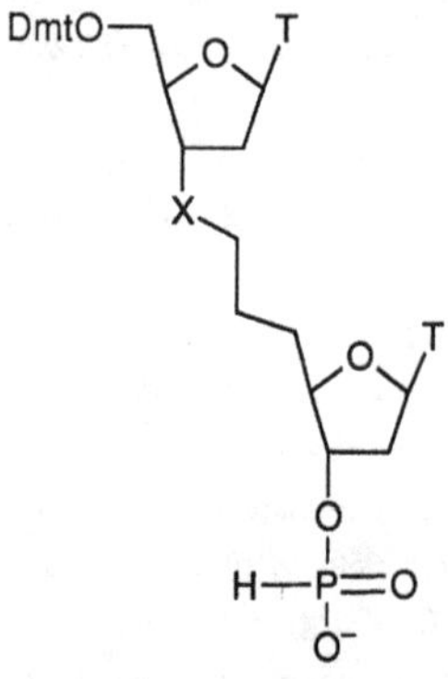

X = S or O
(122)

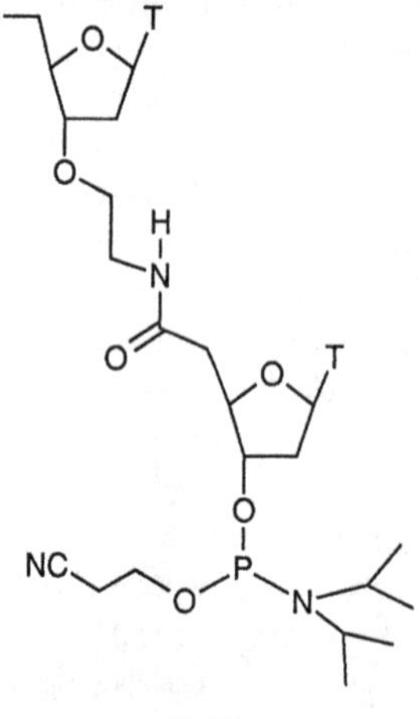

(123)

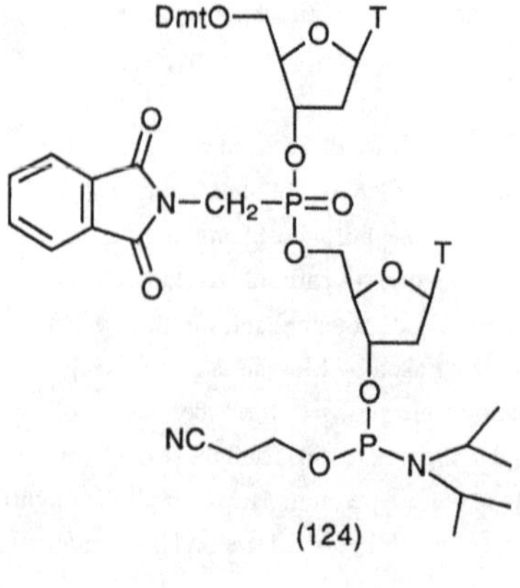

(124)

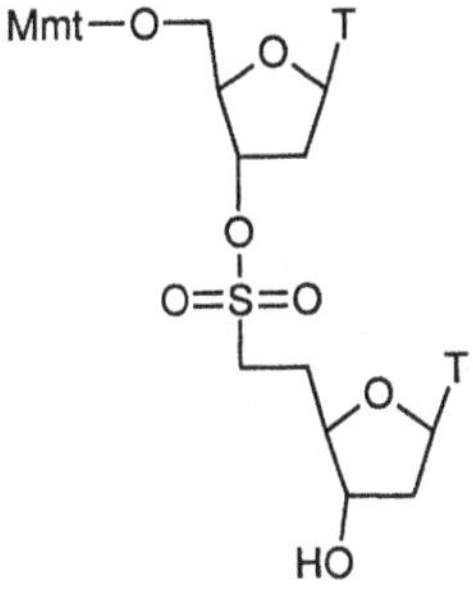

Mmt—O, T, O, O=S=O, T, HO

(125)

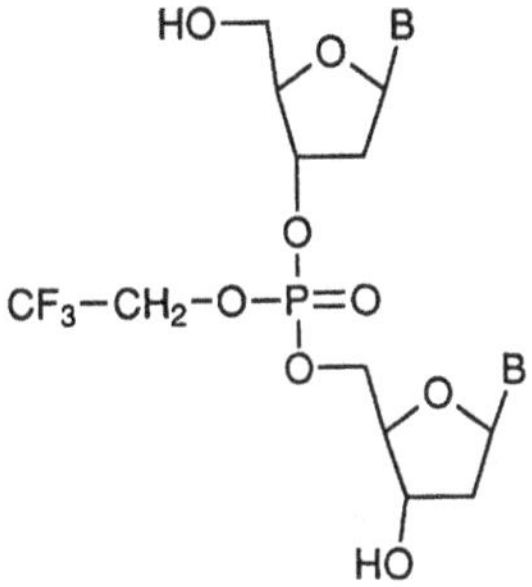

HO, B, O, $CF_3-CH_2-O-P=O$, O, B, HO

(126)

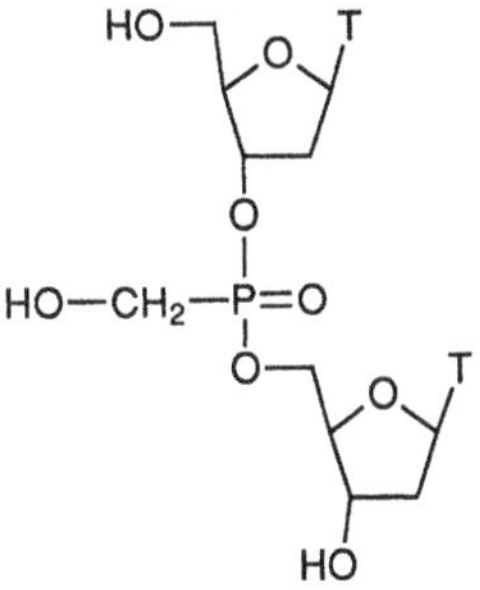

HO, T, O, $HO-CH_2-P=O$, O, T, HO

(127)

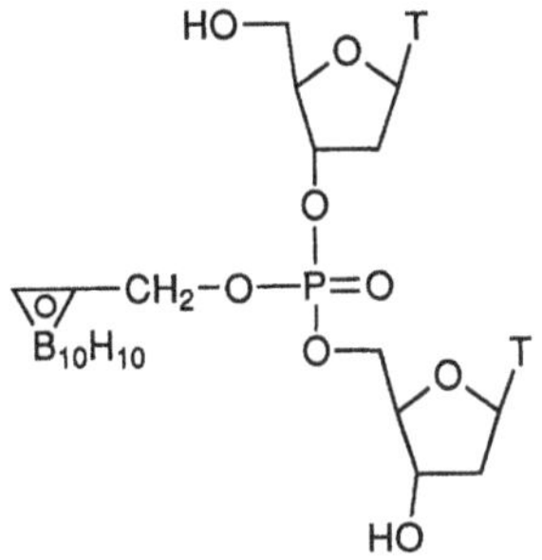

HO, T, O, $CH_2-O-P=O$, $B_{10}H_{10}$, O, T, HO

(128)

O, B, R, X—P=O, O

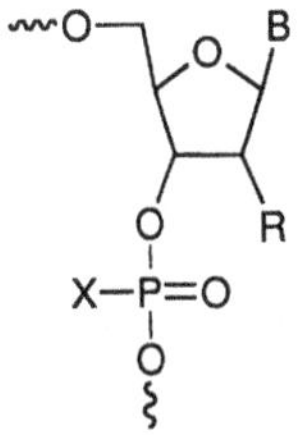

(129a)

β-DNA	X = O^-	R = H
β-2′-OMe-RNA	X = O^-	R = OCH_3
β-Me-P-DNA	X = CH_3	R = H
β-S-DNA	X = S^-	R = H

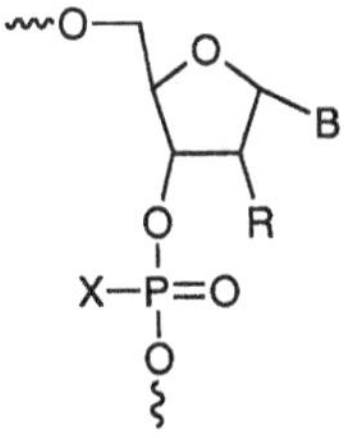

O, B, R, X—P=O, O

(129b)

α-DNA	X = O^-	R = H
α-RNA	X = O^-	R = OH
α-S-DNA	X = S^-	R = H

3.3.2 Oligonucleotides Containing Modified Sugars

The synthesis of oligodeoxyribonucleotides and oligoribonucleotides with modified sugars has been dominated by the alteration of ribose and deoxyribose moieties to produce oligonucleotides that have desirable properties with a view to chemotherapeutic applications. The interest in this area has thus been directed towards the production of oligomers with enhanced nuclease resistance and oligomers which form more stable hybrids with either DNA or RNA. These properties are also of interest when oligonucleotides are used in the detection of specific RNAs *in vivo* or in cell free replication or RNA processing systems.

Attempts to stabilise nucleic acid-RNA duplexes both in binding affinity and nuclease resistance continue to be been dominated by modifications to the 2'-hydroxyl substituent. In this regard the synthesis of the naturally occurring 2'-*O*-methyl ribonucleosides has historically received much attention. An efficient synthesis of 2'-*O*-alkyl ribonucleosides has been reported by Chanteloup *et al.*[121] Suitably protected 2-*O*-alkyl-β-ribosyl trichloroacetamide (**130**) can be synthesised from D-ribose in four steps and then utilised under Vorbrüggen conditions with appropriately protected nucleobases to synthesise the desired nucleoside. Introduction of a 5'-*O*-dimethoxytrityl and a 3'-phosphoramidite derivative thus yields monomers of 2'-*O*-methyl and 2'-*O*-propyl ribonucleosides suitable for solid phase synthesis strategies.

A phosphoramidite derivative of *N*4-isobutryl-2'-*O*-methylcytidine has been synthesised in order to decrease the risk of transamination reactions which occur at the C4-position of cytidine when alternatives to ammonia are used for oligonucleotide deprotection.[122] The monomer was prepared from cytidine protected with 1,1,3,3-tetraisopropyldisiloxane. Transient protection of the 2'-hydroxyl as the trimethylsilyl derivative followed by reaction with isobutyryl chloride gave the protected nucleoside which could be reacted with methyl iodide and silver oxide after removal of the transient protection. Suitable protection and derivatisation for synthesis yield the required synthon (**131**).

Oligodeoxynucleotides containing 3'-deoxy-3'-*C*-threo-hydroxymethylthymidine have been reported.[123] The sugar modified phosphoramidite building block (**132**)was obtained by phosphitilation of 1-(2,3-dideoxy-5-*O*-(4,4'-dimethoxytrityl)-3-*C*-hydroxymethyl-β-D-threo-pentofuranosyl)thymine which was synthesised in three steps from 5'-O-(4,4'-dimethoxytrityl)thymidine via hydroboration of the protected 3'-*C*-methylene nucleoside. 17mer Oligonucleotides containing one or three of the sugar modified thymidines hybridise to complementary DNA but with moderately lowered melting temperatures.

The synthesis of oligonucleotides containing 2'-deoxy-4'-thiothymidine have been reported by Walker and co-workers.[124] When this modified nucleoside is substiuted at the the cleavage site of the *Eco*RV resitriction endonuclease the resultant oligonucleotide is no longer a substrate for this enzyme. Replacement of other thymidine residues contained within the *Eco*RV recognition sequence with 2'-deoxy-4'-thiothymidine is severely inhibitory to the enzyme. These studies may shed some light on the toxic effects of the modified nucleoside in cell culture.

Two modifications which constrain the carbohydrate into a particular conformation have been reported in which the altered sugars mimic the conformation of deoxyribose in either DNA-DNA

[125]or DNA-RNA [126] hybrids. Potentially, such analogues of naturally occurring oligodeoxynucleotides could increase the affinity of a duplex by introducing favourable entropic terms in the energetics of hybrid formation. Tarkoy *et al.* have described the synthesis of 2'-deoxy-3',5'-ethano-D-ribonucleosides of adenine, guanine, cytosine and thymine (**133a** and **133b**). They differ from the natural 2-deoxyribose by an additional methylene bridge between 3'-C and 5'-C. The effect of this bridge is to confer on the sugar a preorganisation similar to that commonly found in DNA. In addition the torsion angles γ and δ are fixed in similar positions to that naturally found in a DNA-DNA duplex. Crystal structure studies confirm the assumptions made but the synthesis of oligonucleotides containing these modified sugars has not yet been reported. Altmann *et al.* have adopted a similar rationale in the synthesis of sugar modified deoxynucleosides which will stabilise an RNA-DNA duplex. Such hybrids exist predominately in the A-type conformation and so the stabilisation of a 3'-endo conformation is required. The bicyclo[3.1.0]hexane thymine derivative (**134**) was synthesised in 20 steps and then appropriately protected for DNA synthesis. Oligodeoxynucleotides of 14 and 15 residues containing a single modified thymidine increased the melting temperatures relative to the unmodified oligomer with a complementary RNA by 0.8°C and 2.1°C respectively.

The properties of oligonucleotides containing acyclic nucleosides where 3(*S*),5-dihydroxypentyl (**135**) or 4(*R*)-methoxy-3(*S*),5-dihydroxypentyl (**136**) replaces the sugar moiety have been investigated by Vandendrieessche *et al.*[127] Attachment of an acyclic nucleoside at the 3'-terminus of an oligonucleotide considerably enhances its resistance to enzymatic degradation. The hybridising properties of oligonucleotides containing multiple acyclic nucleosides are very much inferior to their unmodified counterparts and the modified nucleosides discriminate less than their 2'-deoxyribose counterparts in base-pairing. The latter property led the authors to suggest that these analogues may have application as universal nucleosides for use in probes and primers required to have ambiguous base pairing properties.

The synthesis of dimers (**137**) containing both sugar and linkage modifications has been described by Jenny *et al.*[128] Carbocyclic analogues of nucleosides which consist of cyclopentane bearing a 3-(hydroxymethyl) group, a 4-(2-hydroxyethyl) functionality and a nucleic acid base have been connected together via dimethylene sulphone linking groups.

Eschenmoser and co-workers have studied the relationship between the carbohydrate moiety of oligonucleotides and their resultant structures and properties. The properties of oligo(2'3'-dideoxy-β-D-glucopyranosyl)nucleotides have been summarised.[129] Model building studies suggested that oligonucleotides containing β-D-ribopyranosyl 4',2'-phosphodiester linkages would be capable of forming duplex structures analogous to those of natural RNA. When oligonucleotides of this sort derived from uracil and adenine were synthesised, they were found to form complementary base pairs which are stronger than those formed in corresponding natural RNA duplexes.[130]

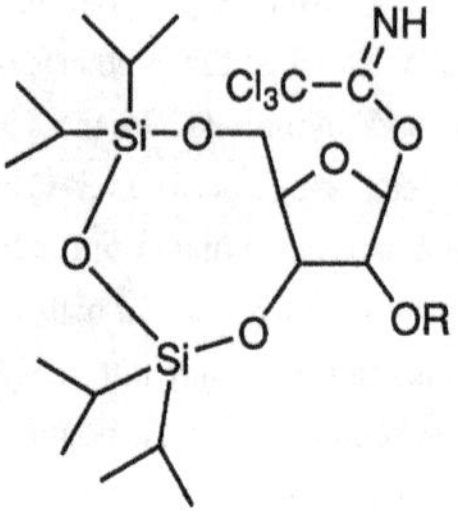

$R = Me$ or Pr^n

(130)

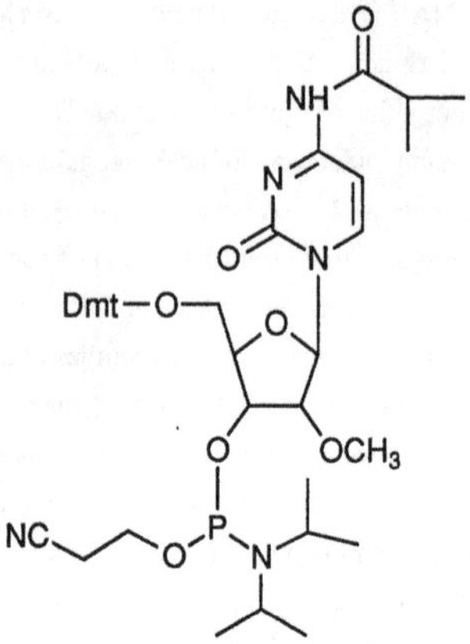

(131)

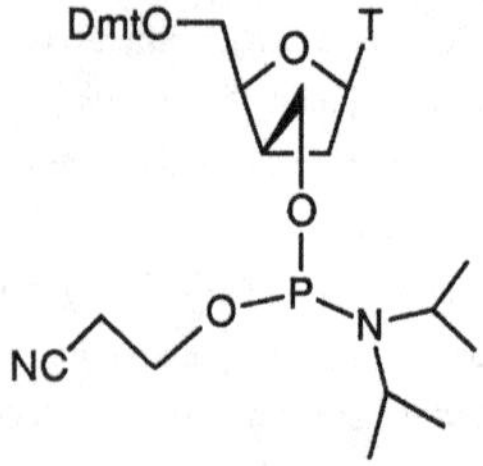

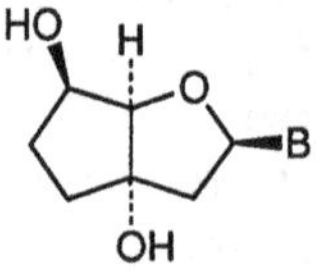

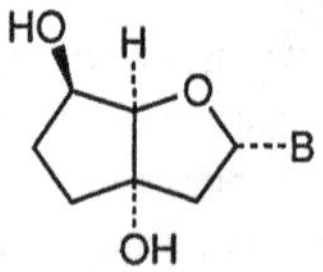

(132)

B = A, G, C, T

(133a)

(133b)

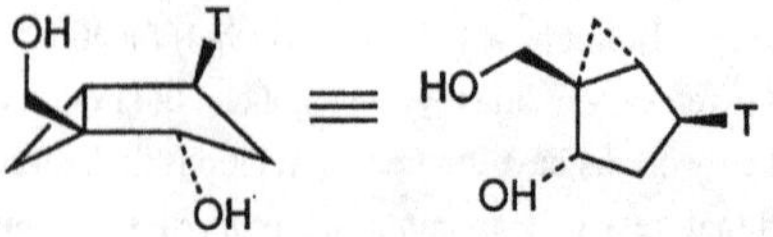

(134)

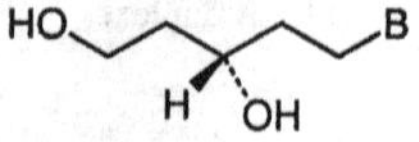

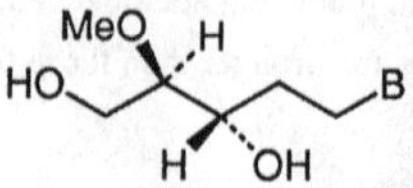

(135)

B = A, T, C, G, I

(136)

3.3.3 Oligonucleotides Containing Modified Bases

Several syntheses of deaza- or aza- nucleoside analogues and their subsequent incorporation into DNA have been described. Seela and Lampe have reported the incorporation of 8-aza-2'-deoxyguanosine (**138**) and its N-8 regioisomer (**139**) into DNA.[131] The (dimethylamino)methylidene group was used for exocyclic amino group protection and the phosphonate and phosphoramidites were prepared. Oligomers containing 8-aza-2'-deoxyguanosine had higher melting temperatures when hybridised to their complementary DNA strands than the corresponding unmodified duplexes whereas those containing the N-8 regioisomer were less stable. When 8-aza-2'-deoxyguanosine is incorporated into oligomers capable of forming guanosine tetrads disaggregation of the structure occurs. This behaviour was attributed to the change in pKa of the functions of the heterocyle interacting in the two structures. The pKa for protonation of N^7 (relevant to triple helix formation) in 8-aza-2'-deoxyguanosine is <1 whereas that of dG is 2.5. Deprotonation of N^1 in dG (relevant to base pairing in duplex strutures) has a pKa of 9.5 in contrast to the modified nucleoside which is 8.6.

A total synthesis of 2'-deoxy-9-deazaguanosine (**142**) and a protection strategy for DNA synthesis has been reported.[132] The ribonucleoside analogue is formed by standard C-glycosylation reactions, as following protection of the 2' and 3'-hydroxyl with Markieincz's reagent, a deoxygenation procedure converts the ribonucleoside to the corresponding 2'-deoxy equivalent. The exocyclic amino group of the guanosine analogue was protected with isobutyryl and then subsequently derivatised for DNA synthesis. When 2'-deoxy-9-deazaguanosine replaces 2'-deoxyguanosine in oligonucleotides self association and the ability to form triple helices is inhibited.

Removal of the N-1 nitrogen of adenosine prevents Watson -Crick hydrogen bonding in duplex structures. Alternative Hoogsteen or Reverse Hoogsteen pairing arrangements are however possible when polymers of 1-deaza-2'-deoxyadenosine (**143**) form duplexes with poly dT. The synthesis of oligonucleotides containing 1-deaza-2'-deoxyadenosine has been described by Seela and Wenzel.[133] The exocyclic amino group was protected with benzoyl although this proved difficult to remove and the authors suggest methoxyacetyl for further applications. H-Phosphonate chemistry was employed to synthesise oligonucleotides containing the modified base. Hybridisation to complementary oligonucleotides resulted in duplexes containing parallel DNA strands.

7-(2'-deoxy-β-D-erythro-pentofuranosyl)adenine (**144**), an analogue of 2'-deoxyadenosine in which the glycosidic bond is located between N^7- and the sugar, has been synthesised and incorporated into oligodeoxynucleotides.[134] The use of (dimethylamino)methylidene protection circumvents problems associated with removal or more traditional amino protection and protects the analogue from depurination during oligonucleotide synthesis. A 12mer containing 11 modified deoxyadenosines and one unmodified residue forms a duplex with dT_{12} with an antiparallel orientation and Watson-Crick type base pairing.

N^2-Isobutryl-7-(2'-deoxy-β-D-erthyro-pentofuranosyl)guanine (**140**) has been synthesised from the corresponding ribonucleoside derivative in a 4-step deoxygenation procedure.[135] Following suitable derivatisation for synthesis this was incorporated into oligonucleotides. A suitable protecting group for the amino functions of 8-amino-2'-deoxyguanosine (**141**) involves formation of the *bis-N*-(dimethylamino)methylene derivative.[135] Subsequent derivatisation for phosphoramidite DNA assembly techniques allows the synthesis of oligomers containing this modified base.

Benner and co-workers are interested in the base-pairing properties of unnatural 2'-deoxynucleosides with a view to extension of the genetic code. 2'-Deoxy-5-aza-7-deaza-isoguanine (**145**) is a nucleoside analogue which presents a hydrogen-bond donor-acceptor-acceptor pattern to a complementary pyrimidine analogue. A novel enzymatic synthesis of the analogue has been described in which 5-aza-7-deaza-isoguanine is enzymatically converted to the deoxynucleoside.[136] The synthesis of deoxynucleoside analogues via glycosylation procedures involving 2'-D-deoxyribose are less trivial than those involving ribosylation as they are not advantaged by the anomeric effect. In this procedure the selective formation of the β–D-anomer is afforded in a sugar exchange reaction catalysed by purine nucleoside phosphorylase. The deoxyribose sugar donor in this reaction is the zwitterionic 2'-deoxy-7-methylguanosine which is formed by neutralisation of 2'-deoxy-7-methylguanosine (**146**) with an aqueous solution of methylamine.

4-Pyrimidone nucleosides (**147**) are traditionally synthesised by desulphurisation of 2-thiopyrimidine (**148**) nucleosides using Raney Nickel based methods. The yield from these reactions, is however, poor. An improved synthesis of these compounds utilises *meta*-chloroperbenzoic acid to desulphurise the starting material.[137] The nucleosides were protected for DNA assembly and oligodeoxynucleotides were synthesised.

Synthesis of oligodeoxynucleotides containing nucleobases which represent the products of the action of mutagens on DNA is a subject of interest in order to facilitate the study of the physiochemical properties of the resultant oligomers and further the understanding of these potentially lethal events. The site specific incorporation of the alkaline labile oxidative stress product (5*R*)-5,6-dihydro-5-hydroxythymidine (**149**) into DNA has recently been reported.[138] This nucleoside is extremely sensitive to base and therefore usual strategies for DNA synthesis which include prolonged ammonia deprotection are unsuitable for the synthesis of oligonucleotides containing this modification. To circumvent these problems, bis-allyloxy protection was used to mask the reactive functionalities during synthesis. These protecting groups can be removed by treatment with Pd^0 at pH5.5. In addition, a novel linkage to the solid support which is cleaved by UV irradiation, was employed.

O^2-Methyl-2'-deoxythymidine (**150**) is one of the products of the action of methylating agents on DNA. A synthesis of this nucleoside analogue has been described in which the methyl group is introduced by reaction of methoxide ion with 2',5'-anhydrothymidine.[139] Standard protection for synthesis using phosphoramidite chemistry allows oligomer assembly. In order to avoid displacement of the O^2-methyl function by ammonia,.the oligomers were deprotected with DBU in methanol. When O^2-Methyl-2'-deoxythymidine is included in DNA duplexes opposite dA, the stability of the duplex is reduced whereas when the analogue appears opposite dG much more modest effects are observed.

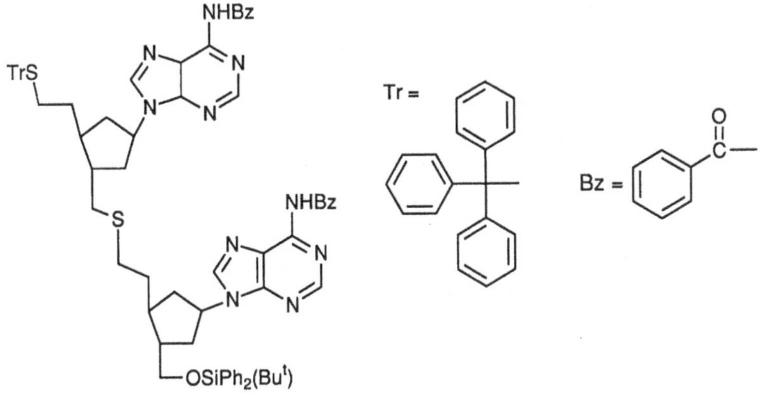

Tr =

Bz =

(137)

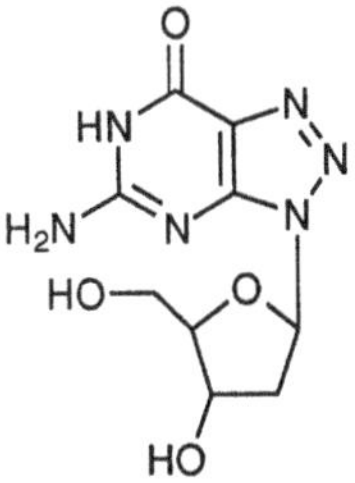

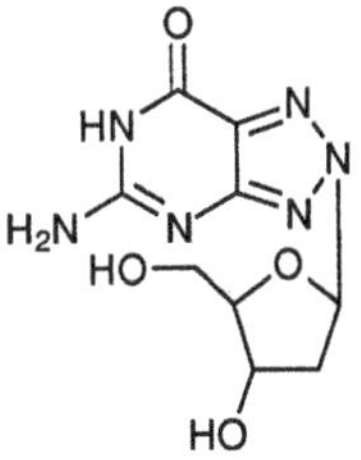

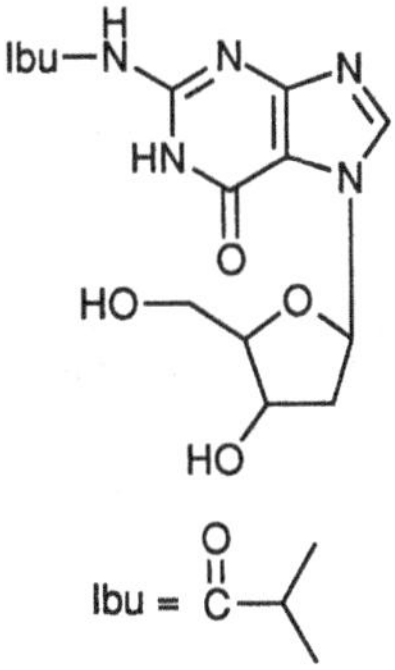

Ibu =

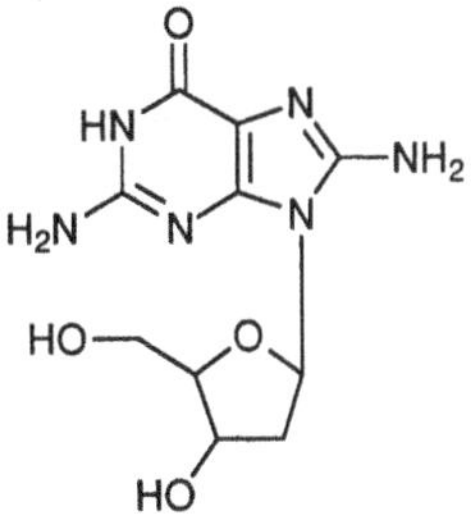

(138) (139) (140) (141)

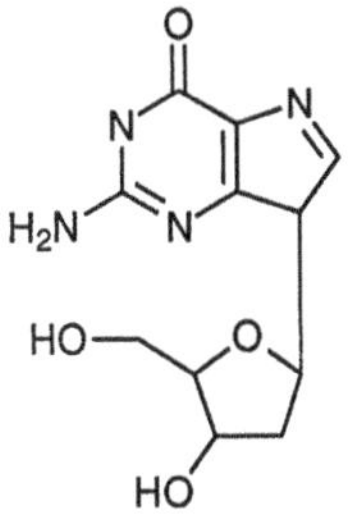

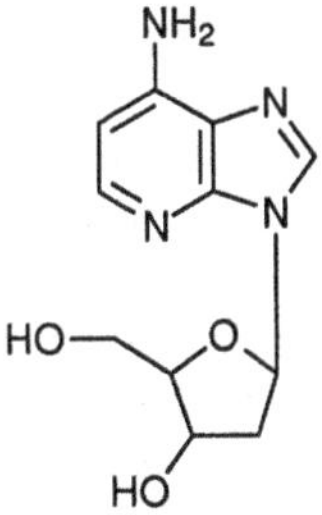

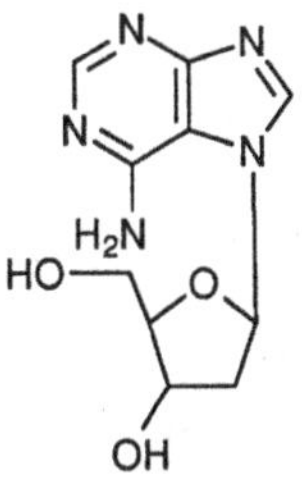

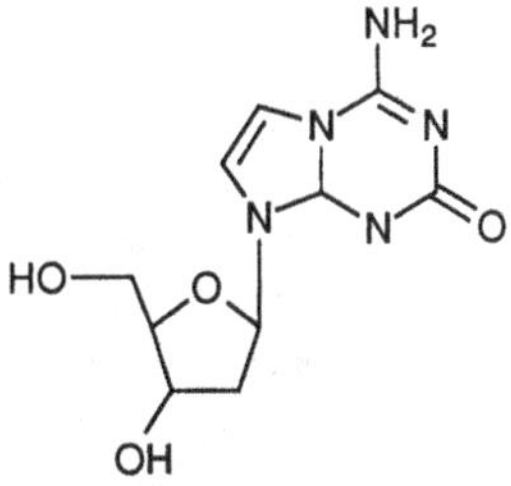

(142) (143) (144) (145)

5-Formyl-2'-deoxyuridine (**151**) is a product of DNA irradiation which has been reported as capable of cross linking to proteins via the formation Schiff base adducts. The synthesis of oligodeoxynucleotides containing 5-formyl-2'-deoxyuridine is complicated by the instability of this modified nucleoside, particularly to basic conditions. Ono *et al*. have developed a new protecting group, *N,N*-(3,5-dichlorophenyl)ethylenediamine, as in (**152**), which is suitable for protection of the formyl function during DNA synthesis.[140] Oligonucleotides containing the modified base were purified by reverse phase HPLC with a 5'-dimethoxytrityl and the formyl protecting group intact. Treatment with 80% acetic acid removes both these masking agents post purification. 5-Formyl-2'-deoxyuridine is degraded under the usual conditions used for enzymatic deoxynucleoside composition analysis. Oligonucleotides containing this modification were therefore treated with cyanomethylenetriphenylphosphorane to give a product containing (E) and (Z)-5- (2-cyanovinyl)-2'-deoxyuridine (**153a** and **153b**)which was then digested with enzymes allowing detection of the presence of 5-formyl-2'-deoxyuridine in the original oligomer. The melting temperature of a self-complementary oligonucleotide containing the analogue is higher than the corresponding unmodified compound. Klenow DNA Polymerase incorporates 2'-deoxyadenosine opposite the modified deoxyuridine but with decreased efficiency.

Another product formed from radical damage to thymidine residues in DNA is 2-deoxy-β-D-ribofuranosyl-*N*-formamide (**154**). The target modification was protected with a 5'-dimethoxytrityl residue and then included in oligonucleotides synthesised using phosphate triester chemistry.[141] The melting temperature of a duplex containing 2-deoxy-β-D-ribofuranosyl-*N*-formamide opposite dA was dramatically lowered by 30°C compared to that of the parent compound whilst the melting temperature decreased by 16°C compared to a duplex containing dC-dA mismatch.

The synthesis of 5-hydroxy-5-methylhydantoin nucleoside (**155**), which is a lesion induced by the action of ionising radiation of thymine residues, has been described.[142] The modified nucleoside is prepared by the action of permanganate in the presence of lead tetraacetate on 5'-monomethoxytritylthymidine. Careful phosphitilation under controlled conditions leads to selective reaction at the 3'-hydroxyl. The modified monomer was used in DNA synthesis with very base-labile protection (phenoxyacetyl for dA and dG, isobutyryl for dC) and a phenoxyacetic anhydride *N*-methylimidazole capping step.

8-Hydroxyguanosine (**156**) is generated by a variety of reagents that produce hydroxyl radicals. A new synthon for the incorporation of this mutagenic nucleoside into oligodeoxynucleosides has been reported by Koizume *et al.* [143] The protected nucleoside is synthesised from 8-bromodeoxyguanosine by treatment with methoxide ion to generate the 8-methoxy- compound. *N,N*-dimethylformamide dimethylacetal and monomethoxytrityl chloride were used to protect the amino and 5'-hydroxyl functions, respectively. Following removal of the methyl group with thiophenol in the presence of triethylamine, the 3'-*O*-phosphoramidite compound was synthesised. 8-Hydroxyguanosine was found to form stable base pairs with both G and T.

Nucleoside analogues with pendant imidazole heterocyles are of interest in the development of oligonucleotides with the ability to cleave RNA. This interest is inspired by the general base activity of the imidazole-containing amino acid histidine in ribonucleases. The synthesis of N^2-[3-(1H-imidazol-1-yl)propyl]-2'-deoxyguanosine (**157**) and N^2-[3-(1H-imidazol-1-yl)propyl]-2-

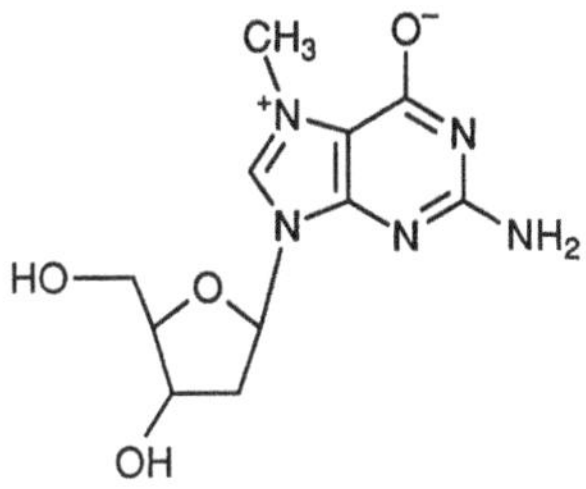

CH_3 O^- ^+N N N N NH_2 HO O OH

(146)

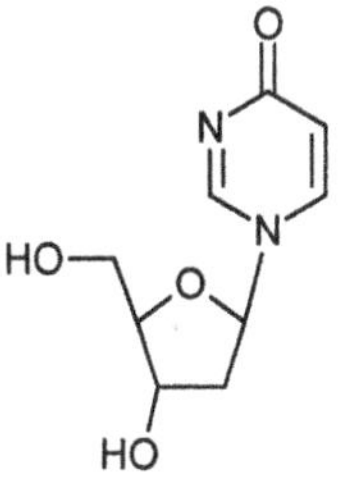

O N N HO O HO

(147)

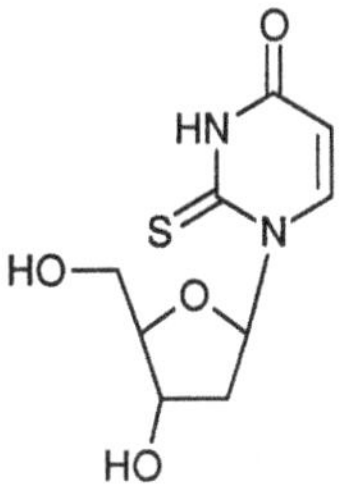

O HN S N HO O HO

(148)

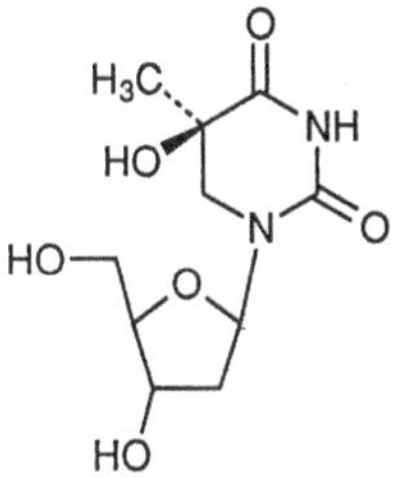

O H_3C NH HO N O HO O HO

(149)

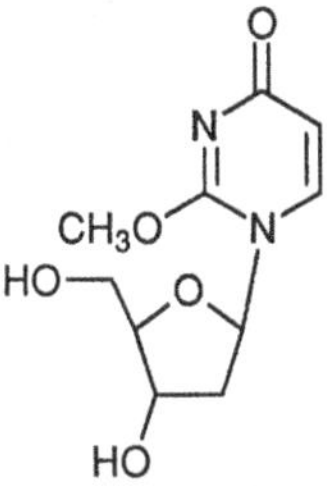

O N CH_3O N HO O HO

(150)

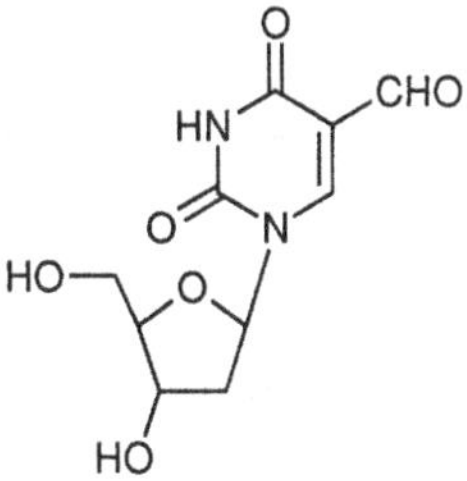

O CHO HN O N HO O HO

(151)

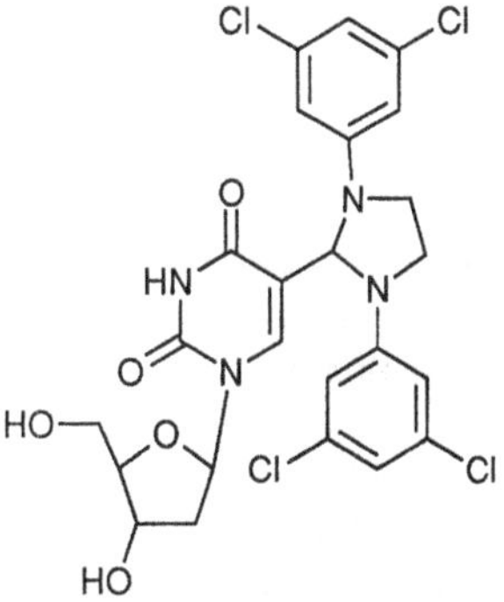

Cl Cl O N HN N O N HO O Cl Cl HO

(152)

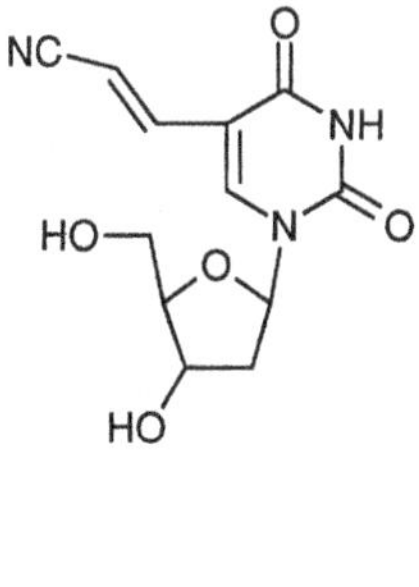

NC O NH HO N O O HO

E
(153a)

CN O NH HO N O O HO

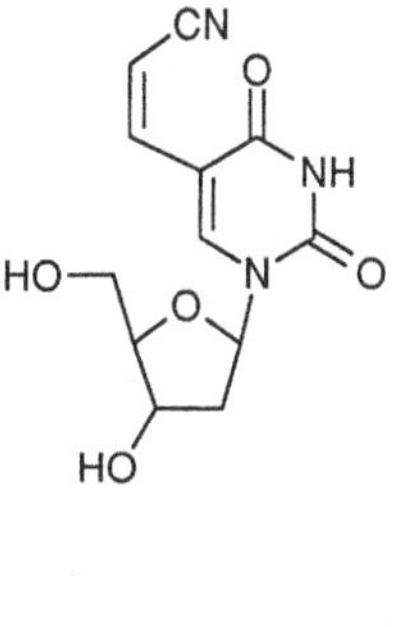

Z
(153b)

amino-2'-deoxyadenosine (**158**) and a suitable protection strategy for DNA synthesis has been described.[144] The modified nucleosides are synthesised from a protected deoxynucleoside 2,6-dichloropurine derivative. When incorporated into oligodeoxynucleotides, these modified bases did not effect cleavage upon hybridisation of the oligomers with a complementary RNA but the resultant duplexes had enhanced stability. Similar increases in melting temperature were also observed upon duplex formation with DNA. Model building studies suggest that the tethered imidazole is capable of binding in the minor groove close to the phosphodiester of B-form DNA duplexes but that in A form RNA-DNA duplexes this is not possible due to changes in the groove width. It is assumed that in the case of the heteroduplexes, the imidazole binds alternative hydrogen bond donor or acceptor groups in the minor groove.

An alternative approach to the introduction of pendant imidazole residues has been described by Wang and Bergstrom.[145] These authors synthesised and incorporated into oligodeoxynucleotides the modified deoxyguanosine nucleoside N^2-[2-(imidazol-4-ylacetamido)ethyl]-2'-deoxyguanosine (**159**). In this case, the N2-spacer arm contains an amide linker. The nucleoside was synthesised by displacement of a fluoro group at the 2-position of a protected purine derivative with 1,2-diaminoethane. An N-hydroxysuccinamide ester derivative of a substituted imidazole is coupled to the amine. The imidazole moiety is protected with fluorodinitrobenzene. Better yields in subsequent phosphitilation reactions were achieved when the O^6-position was protected as its *p*-nitrophenylethyl derivative. Post oligonucleotide synthesis, the *p*-nitrophenylethyl protecting group was removed by treatment with DBU in acetonitrile whilst the dinitrophenyl protection is removed during the standard ammonia deblock.

A related 2'-deoxyguanosine derivative (**160**) has also been described by Heeb and Benner.[146] An amide linkage likewise attaches the pendant imidazole to the N^2-position of guanosine but in this compound the amide linkage involves the exocyclic amino group of the nucleobase. This considerably simplifies the synthesis of the target modified nucleoside which is achieved by reacting 2',5'-protected deoxyguanosine with 3-imidazol-4-yl propionic acid in the presence of 1,1-carbonyldiimidazole. Again the imidazole residue and the O^6-function are protected with 2,4-dinitrophenyl and *p*-nitrophenylethyl and then suitably derivatised for DNA synthesis in the usual manner. The pKa of the imidazole moiety in the nucleoside was measured to be 7.1 and nmr studies of the association of the nucleoside with cytidine in chloroform suggested the presence of a hydrogen bond between the 2'-OH of cytidine and the imidazole residue. Oligodeoxynucleotides containing the modified nucleoside were not, however, capable of RNA cleavage upon hybridisation to an oligonucleotide in which the modification mispairs with adenine. Interestingly, the authors also observed an increase in melting temperature of the resultant hybrids.

Syntheses of oligonucleotides containing N^2-(5-carboxypentyl)-2'-deoxyguanosine (**161**) and 5-[2-(4'-methyl-2,2'-dipyrid-4-yl-carboxyamido)ethylthio]-2'-deoxyuridine (**162**) have also been described. The guanosine derivative offers an acidic function which may also be useful in catalysis of RNA cleavage whilst the latter modified nucleoside has a bipyridine residue capable of binding metal ions known to catalyse RNA hydrolysis such as zinc and copper. The guanosine derivative was synthesised from 4-desmethylwyosine which was alkylated with ethyl-6-bromohexanoate which following ring opening with N-bromosuccinimide and dilute ammonia, produces the desired

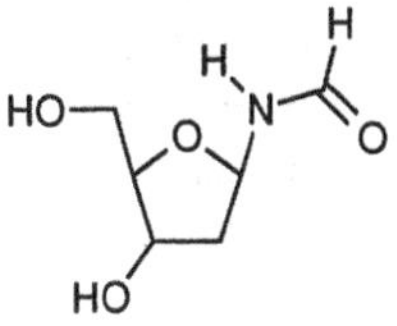

(154)

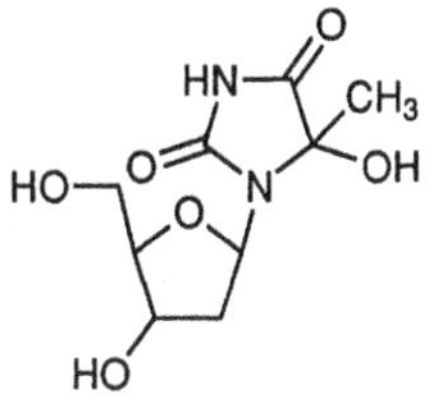

(155)

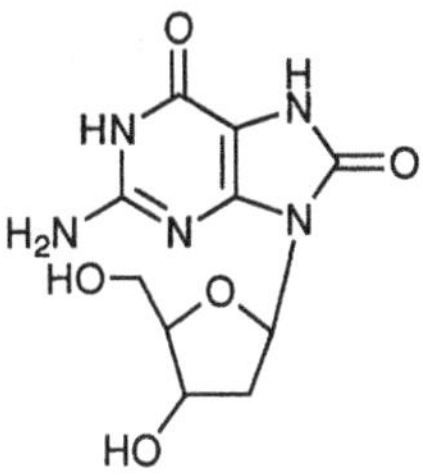

(156)

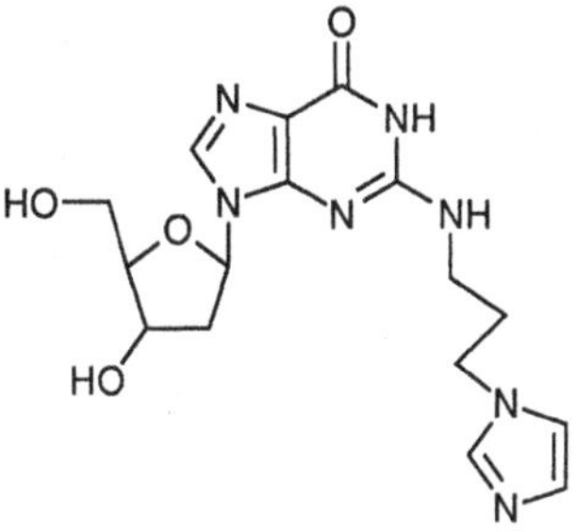

(157)

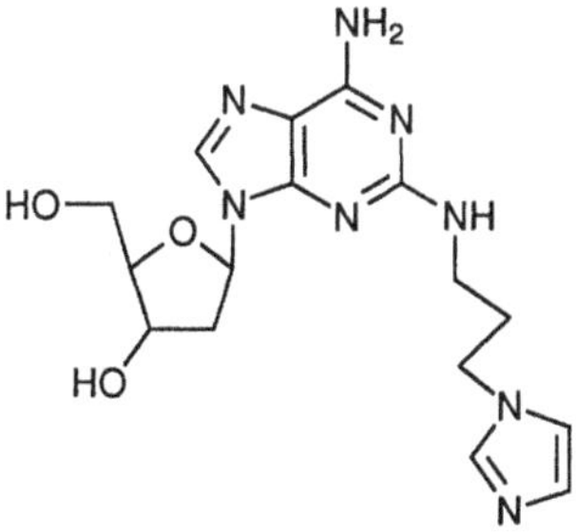

(158)

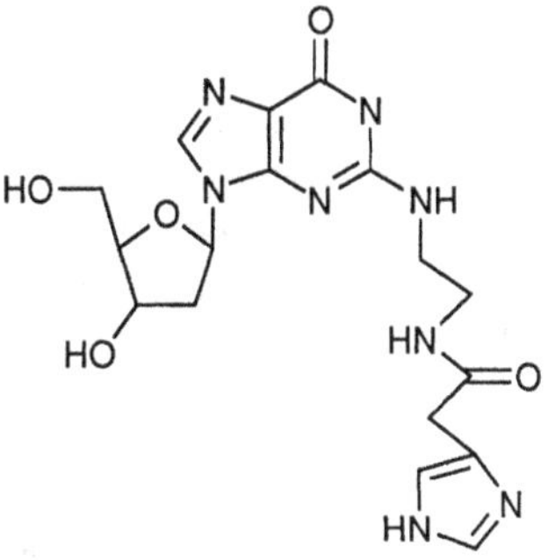

(159)

nucleoside. The incorporation into oligodeoxynucleotides was achieved after standard protection and functionalisation. Prior to ammonia treatment during deprotection the carboxylate was liberated by triethylammine in a mixture of ethanol and water in order to avoid amide formation. The bipyridine nucleoside derivative was synthesised from the known 5-thioethylamino derivative of uridine by reaction with a N-hydroxysuccinimide ester of a bipyridine derivative. No RNA cleavage data are reported in this manuscript.

In the past few years, considerable attention has been devoted to the synthesis of oligodeoxynucleotides and oligoribonucleotides containing 2-amino-2'-deoxyadenosine (**163**), and 2-aminoadenosine, respectively. This interest has focused on the ability to enhance the stability of DNA homo- and heteroduplexes with a view to the use of these modified nucleic acids as hybridisation probes and in antisense technologies. A straightforward synthesis of 2-amino-2'-deoxyadenosine has recently been described which involves treatment of 5'-O-(4,4'-dimethoxytrityl) guanosine with trifluoroacetic anhydride in pyridine followed by ammonia.[147] Subsequent protection of the exocyclic amino functions with phenoxyacetyl groups and phosphitilation produces the required monomer. Interestingly the authors report that oligodeoxynucleotides containing 2-amino-2'-deoxyadenosine in place of 2'-deoxyadenosine have enhanced melting temperatures upon duplex formation with RNA but not DNA.

Developments in reliable methods for the synthesis of RNA over the past two years have enabled investigations of the synthesis of oligoribonucleotides containing modified bases. These studies have been largely prompted by interest in RNA-protein interactions and catalytic RNAs. Consequently, many of the modifications introduced are analogues of the common naturally occurring ribonucleobases with a view to identification of important functional groups involved in RNA-RNA interactions or RNA-protein interactions.

Most attention has been devoted to the synthesis of oligoribonucleotides containing modified purines. The incorporation of N^7-deazaadenosine (**164**) (tubercidin) and N^7-deazaguanosine (**165**), into oligoribonucleotides has been described by several authors.[148-151] Very base labile exocyclic amino group protection of the formamidine type is preferable for these syntheses as difficulties are encountered with the more traditional benzoyl protection.[149, 150] Both phosphoramidite and H-phosphonate chemistries have been reported for chain assembly whilst silyl protection is employed in all cases These nucleoside analogues represent the replacement of the N7-nitrogen atom of adenosine and guanosine, respectively, with a C-H moiety and oligoribonucleotides containing these modifications have been used to study the importance of these functional groups to the action of the hammerhead ribozyme[148-150] and, in the case of N^7-deazaguanosine, the disaggregation of guanosine tetrads.[151]

Tuschl *et al.* have described the introduction of inosine (**166**), 2-amino purine riboside (**167**), xanthosine (**168**) and isoguanosine (**169**) into oligoribonucleotides with a view to the determination of important functional groups in the hammerhead cleavage reaction.[152] Inosine was protected with a *p*-nitrophenylethyl at the O^6-position in order to aid solubility but conventional isobutyryl /silyl protection strategies and phosphoramidite chemistry were employed for the other analogues. The exocyclic amino group of isoguanine was masked using N^6-(dimethylamino)methylene protection and further chain elongation after addition of modified nucleoside proved impossible.

Oligoribonucleotides were therefore assembled using an enzymatic procedure involving T4-DNA ligase and an oligodeoxynucleotide splint.

Grasby *et al.* have described the synthesis of oligoribonucleotides containing O^6-methyl guanosine (**170**).[153, 154] The methylated compound was generated using Mitsunobu chemistry using methanol and a protected guanosine derivative. Oligoribonucleotides were synthesised using acetyl protection for the N^2-amino group, silyl protection for the 2'-hydroxyl and phosphoramidite chemistry. Deprotection with DBU in methanol rather than the conventional ammonia leaves the O^6-methyl group intact during this procedure.

Four pyrimidine analogues have also been incorporated into RNA. Methods for the incorporation of 4-thiouridine (**171**) and 2-pyrimidone-1-β -D-riboside (**172**) using silyl protection and phosphoramidite chemistry have been reported by Adams *et al*. The thio base was protected with an S-cyanoethyl group which requires a modification to the deprotection protocol to include a treatment with DBU in acetonitrile to remove this group prior to ammonia treatment.[155] 2-Pyrimidone-1-β -D-riboside was obtained by coupling the nucleobase 2-hydroxypyrimidine with ribofuranose 1-acetate 2,3,5-tribenzoate in a Vorbrüggen coupling procedure.[156] During ammonia treatment of the resultant protected oligoribonucleotide, side reactions occur resulting in base modification and the authors therefore recommend the use of very base labile exocyclic amino group protection for the rest of the monomers used in assembly.

A strong interest in the naturally occurring modified nucleosides of RNA which result from post-translational processing remains. The literature in this area is vast and beyond the scope of this review. We confine our comments here to modified nucleosides that have been chemically synthesised and/or incorporated into synthetic oligoribonucleotides. A comprehensive listing of modified RNA nucleosides reported until mid 1994 appeared in Nucleic Acids Research.[157] Pseudouridine (**173**) is a naturally occurring modified base which is found in tRNA, rRNA, snRNA and other small RNAs. The synthesis of oligoribonucleotides containing pseudouridine has recently been described by Pieles *et al*.[158] Unusually, the authors chose to protect the 2'-hydroxy function with a 2'-*O*-[1-(2-fluorophenyl)-4-methoxypiperidin-4-yl] (Fpmp) group. The pivaloyloxymethyl (Pom) group was used to simultaneously protect both nitrogens of the heterocycle. A 2'-*O*-methyl nucleoside was also synthesised and incorporated into RNA using phosphoramidite chemistry. Interestingly the oligoribonucleotides containing the 2'-*O*-methyl pseudouridine derivative had higher melting temperatures with complementary oligonucleotides than those containing 2'-*O*-methyluridine in the same position.

A synthesis of 5'-*O*-(4,4'-dimethoxytrityl)-2'-*O*-(tert-butyldimethylsilyl)-5-fluorouridine 3'-(cyanoethyl N,N-diisopropylphosphoramidite) and its subsequent incorporation into RNA has been reported.[159] The resultant oligonucleotides containing 5-fluorouridine have similar properties to unmodified RNA when hybridised to a complementary RNA molecule except that the imino hydrogen NMR resonance is moved down field in an adenosine-5-fluorouridine base pair when compared to the unmodified parent compound.

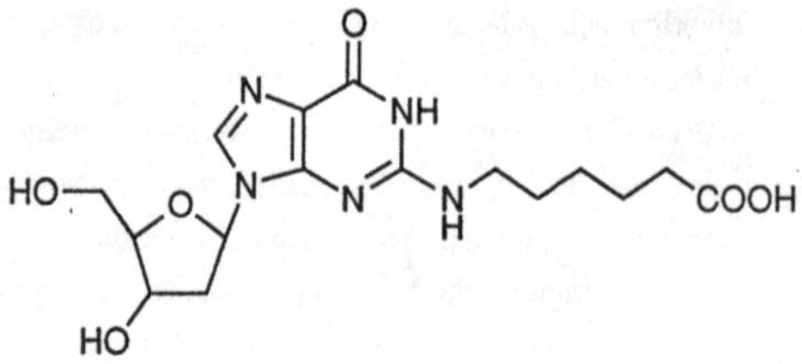

(160)

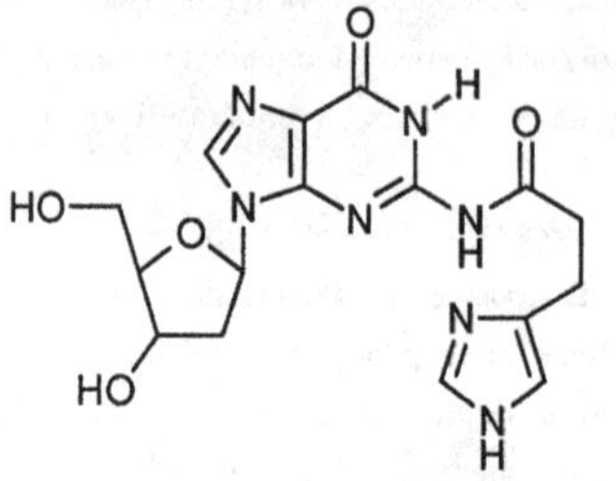

(161)

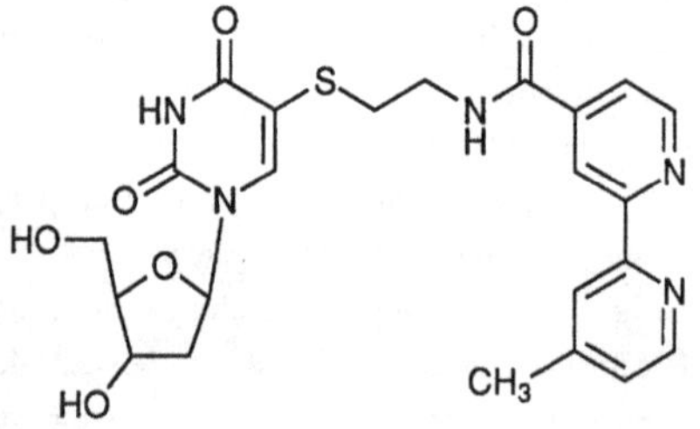

(162)

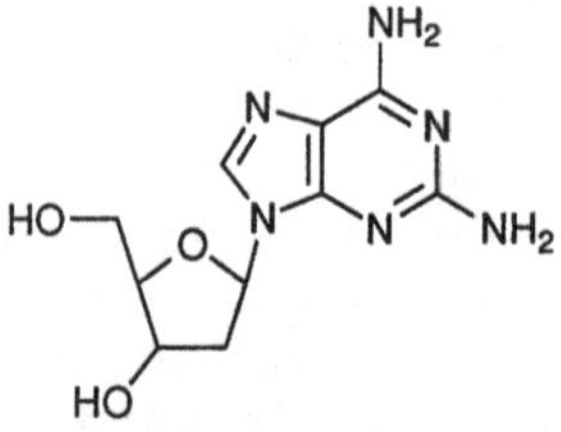

(163)

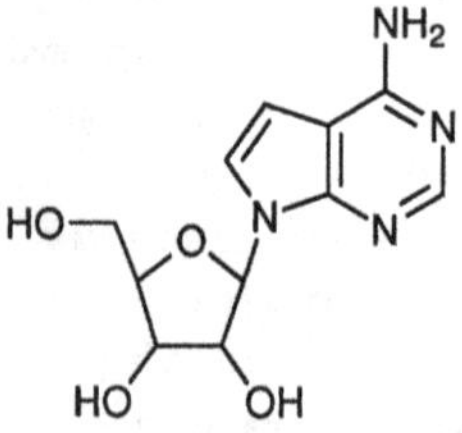

(164)

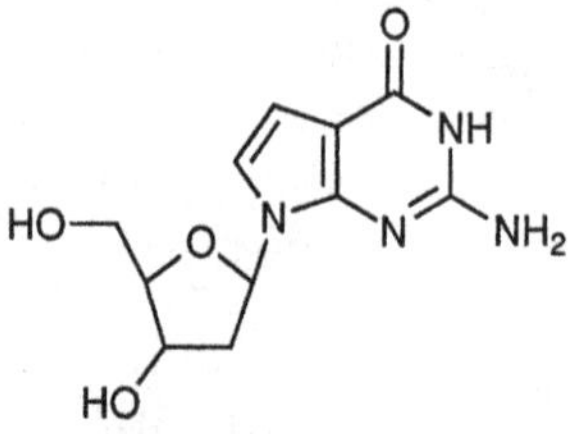

(165)

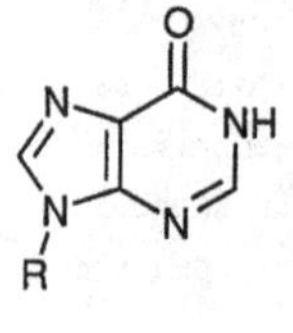

(166)

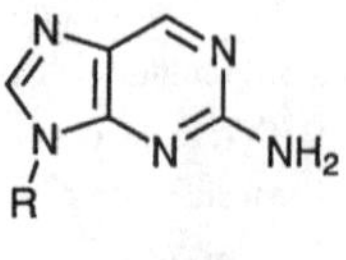

(167)

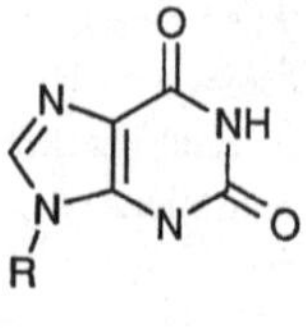

(168)

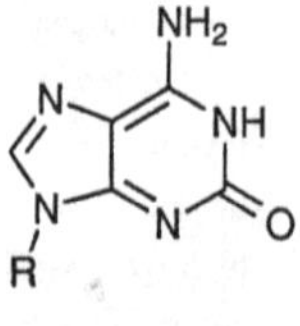

(169)

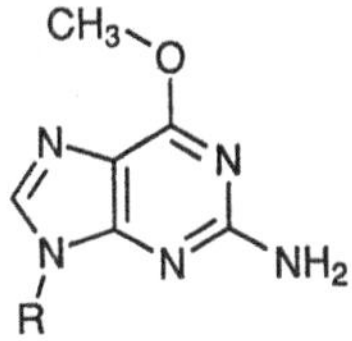

(170) 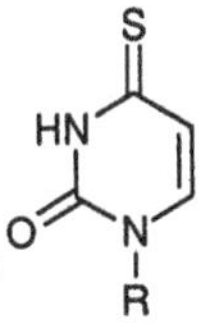 (171) 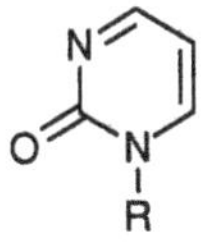 (172) 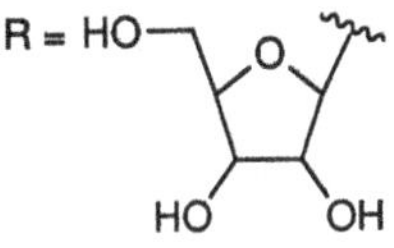 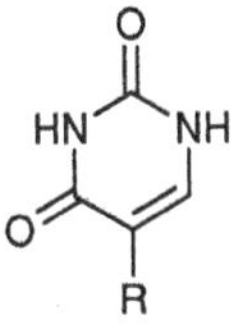 (173)

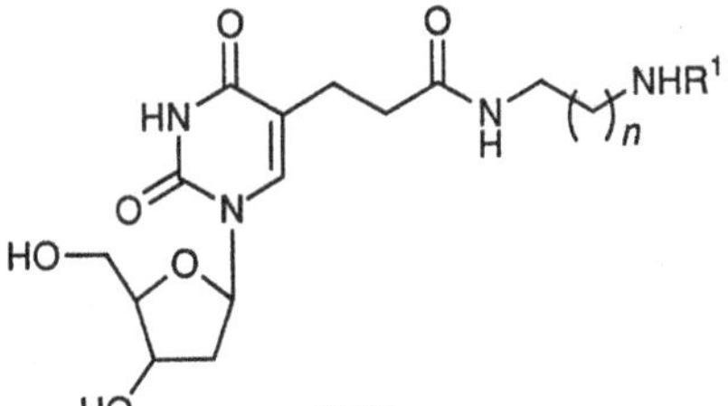

(174) 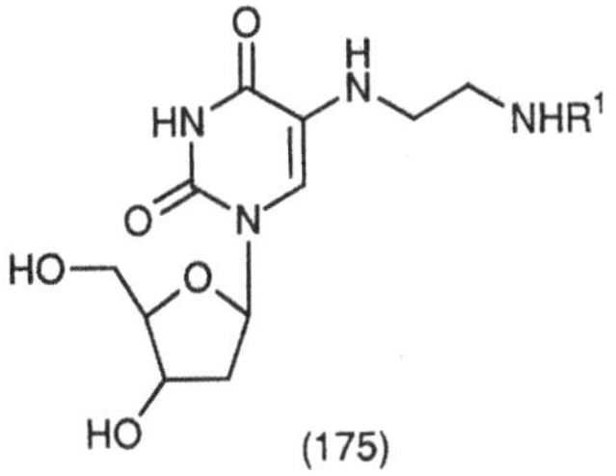 (175)

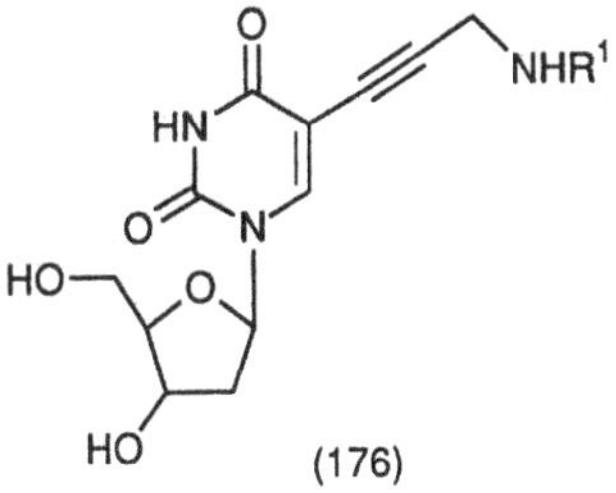

(176)

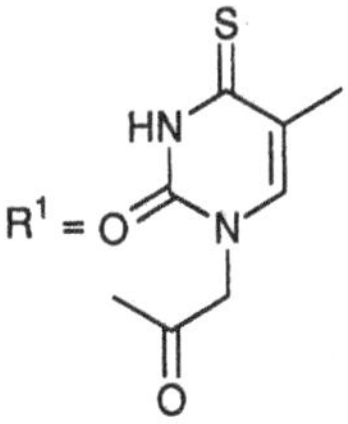

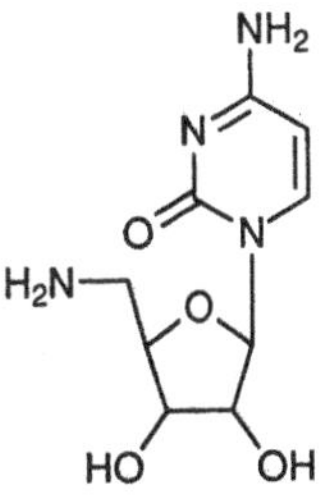

(177) 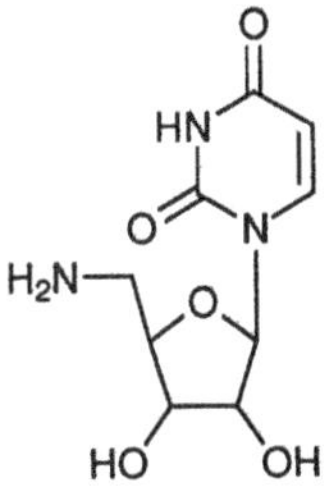 (178)

4. Linkers

Oligonucleotides containing 4-thiothymidine and 4-thiouridine have been previously used for UV cross-linking studies with a view to the study of nucleic acid structures and DNA-protein interactions. A recent publication describes the synthesis of a set of deoxyuridine derivatives (**174, 175, 176**) containing C5-pendant thiothymine residues suitable for incorporation into oligonucleotides using chemical methods.[160]

5'-Amino-cytidine (**177**) and 5'-amino-uridine(**178**) can be incorporated at the 5'-termini of oligoribonucleotides using F-moc protection for the amino group. The resulting oligomers can be labelled with the fluorescent compound pyrene.[161] Linking pyrene in this manner prevents intercalation of the dye and has allowed Turner and co-workers to study the interaction of the Sca-I ribozyme with an oligoribonucleotide, which binds to the 5'-exon site, by fluorescence. An alternative method for the introduction of a pyrene group to the termini of oligonucleotides has been described by Yamana *et al.* 5'-(1-pyrenylmethyl)thymidine has been converted to the phosphorobisdiethylamidite (**179**) and then coupled manually to a fully protected controlled pore glass bound oligonucleotide.[162] Alternatively a 3'-pyrenyl oligonucleotide is afforded by use of a solid support derivatised with an appropriately protected 3'-pyrenyl uridine (**180**).

One of the products of cleavage of DNA mediated by bleomycin and other related antibiotics is a series of oligonucleotides with a terminal 3'-phosphoglycolate group. Urat *et al*. have described the synthesis of oligonucleotides bearing this 3'-modification which can be used as an hplc standard in identifying products from these and related transition metal ion proton extraction reactions at the deoxyribose moiety of DNA.[163] The synthesis requires the production of a modified solid support (**181**) for DNA assembly. Mono-(4,4'-dimethoxytrityl) glycerol was immobilised on LCAA-CPG using conventional procedures. After DNA synthesis and deprotection, the resultant diol can be oxidised with $NaIO_4$ to produce the 3'-phosphoglycaldehyde. Treatment of the aldehyde with $NaClO_2$ affords the 3'-phosphoglycolate whereas reduction with $NaBH_4$ generates the corresponding primary alcohol.

Flavin derivatives can cleave DNA when subjected to photo irradiation. The synthesis of oligonucleotides bearing a 3'-deazaflavin derivative has been described utilising a 5-deazaflavin modified controlled pore glass support (**182**).[164] Normal flavin derivatives are unstable under basic conditions necessary for deprotection in DNA synthesis and consequently the 5-deaza compound bearing an electron withdrawing diethanolamine at the C-8 position was employed in the support derivatisation. Oligonucleotides were assembled using H-phosphonate chemistry.

General methods for the production of oligonucleotides bearing 3'-linkers have been investigated by Hovinen *et al*.[165, 166] A thymine derivatised controlled pore glass (**183**) bears a linker which may be cleaved using various reagents to generate oligonucleotides terminating with a variety of functional groups suitable for further derivatisation. 3'-*O*-methylthiomethyl-5'-5'-*O*-pivaloylthymidine was treated with ethyl-4-hydroxybutyrate in the presence of N-bromosuccinimide to produce the ethyl ester. After removal of the pivaloyl group and hydrolysis of the ester with sodium hydroxide, the 5'-hydroxyl of the nucleoside derivative was protected with 4,4'-dimethoxytrityl chloride. Esterification with 1,3-propanediol and then reaction with succinic

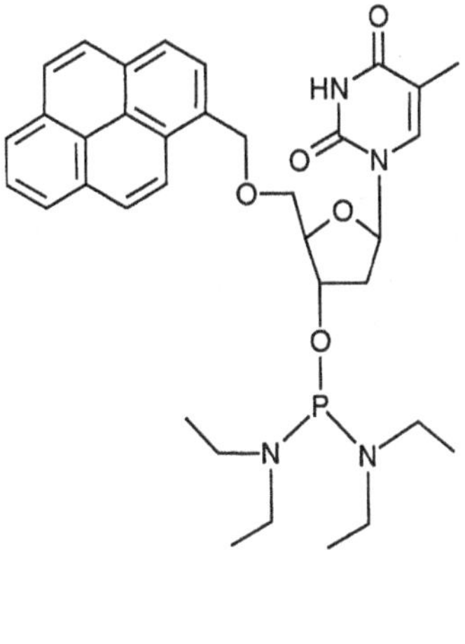

(179)

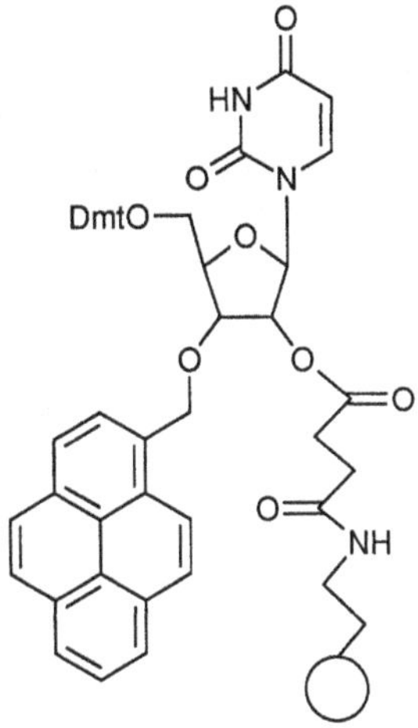

(180)

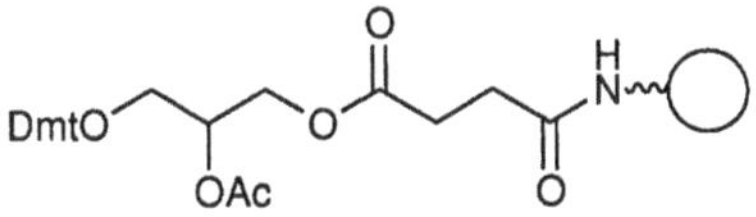

(181)

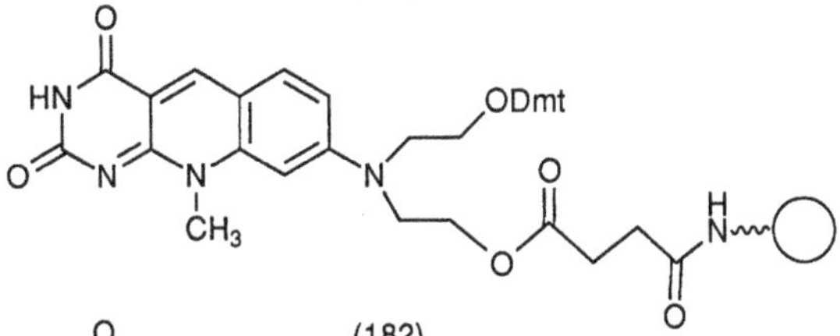

(182)

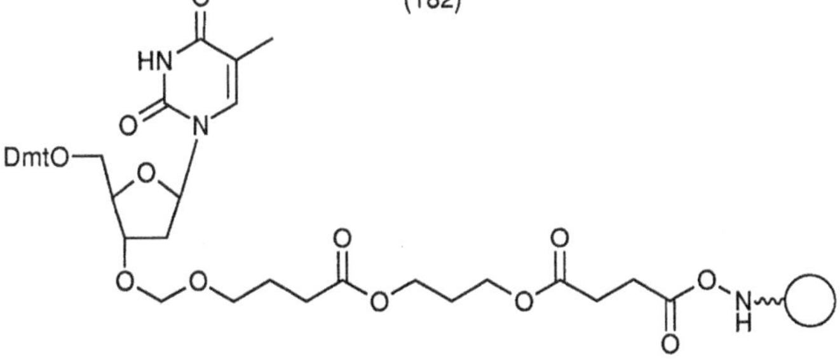

(183)

anhydride yielded a thymine derivative which could be immobilised on LCAA-CPG using conventional procedures. Following oligonucleotide synthesis, the resultant oligonucleotide could be deprotected with sodium hydroxide and then ammonia to yield the 3'-carboxylic acid derivative. An alternative method of deprotection involves the use of a mixture of hydrazine, acetic acid and pyridine to remove protecting groups from adenosine and cytidine residues, and then 1,3-diaminopropane to yield an oligonucleotide bearing a 3'-terminal amino group. The hydrazine treatment is required to avoid the risk of transamination reactions on exposure to the diamine. An extension of this work involves a universal support containing a 5-dimethoxytritylbutyric acid ester synthesised by coupling 4-(4,4'-dimethoxytrityl)butyric acid to LCAA-CPG. This support (**184**) may be utilised to synthesise oligonucleotides bearing 3'-terminal carboxy, amino, amidocarbonyl or mercaptoalkyl functions by treatment with the appropriate reagents.

The synthesis and properties of oligonucleotides bearing different functional groups at the 2'-position have been investigated by Keller and Haner.[167] Modified adenosine ribothymidine and cytidine derivatives (**185**) were prepared bearing n-octylcarbonyl, 2-aminoanthraquinoyl and alkyl-*N,N*-dimethylamino functions. Modification with n-octyl groups produced little effect on the stability of duplexes whereas the intercalator and N,N-dimethylamino residues stabilised duplex structures.

A series of non-nucleoside-based 2,4-dinitrophenyl phosphoramidites has been prepared and used in the multiple labelling of oligonucleotides during solid-phase synthesis.[168] The oligonucleotides bearing three or more dinitrophenyl labels can be detected in antibody binding assays at a comparable sensitivity to biotin. An example of a dinitrophenyl labelling reagent (**186**) was synthesised from 3-amino-1,2-propanediol. Reaction with 2,4-dinitrofluorobenzene followed by dimethoxytritylation and phosphitilation is used to synthesise the reagent. No significant change in antibody response was observed upon variation of the linker arm.

There is considerable interest in the synthesis of peptide-oligonucleotide conjugates.[169-171]

5. Nucleic Acid Triple Helices and Other unusual DNA and RNA Structures

Methods of stabilising unusual DNA and RNA structures by the introduction of cross-links between strands have been reported by several authors this year. Goodwin and Glick have described the synthesis of a uridine derivative to which alkyl thiol chains are attached at C5.[172] For 5-butylthiol or 5-propylthiol derivatives the synthesis begins with 5-iodouridine to which is coupled the corresponding acetylinic alcohol in a palladium catalysed coupling reaction. Following hydrogenation, the alcohol is converted to a thiol by activation as the mesylate and displacement with thiobenzoic acid. Finally an exchange reaction gives the t-butyldisulphide which serves as a protecting group during synthesis (**187**). The compounds with a three atom spacer arm are synthesised from 5-chloromercuri-2'-deoxyuridine by reactions with silyl chloride. Oxidation of the olefin to an aldehyde followed by reduction gives the C5-propanol derivative. Similar thiolation and protection of the other analogues yields the target compounds which are then derivatised for DNA synthesis using conventional procedures.

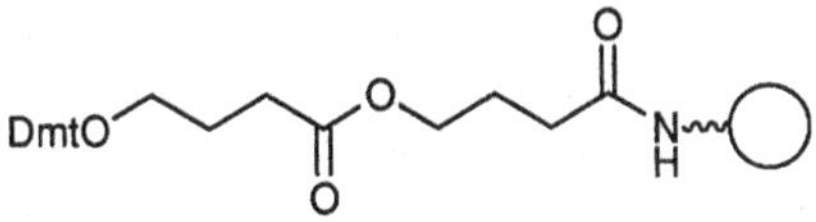

DmtO O O O N H

(184)

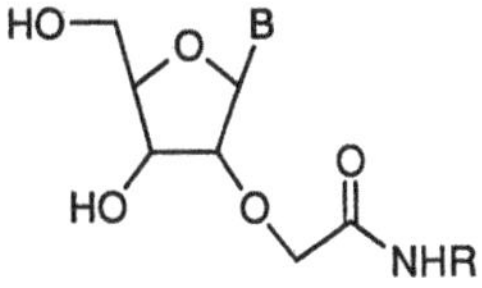

HO O B HO O O NHR

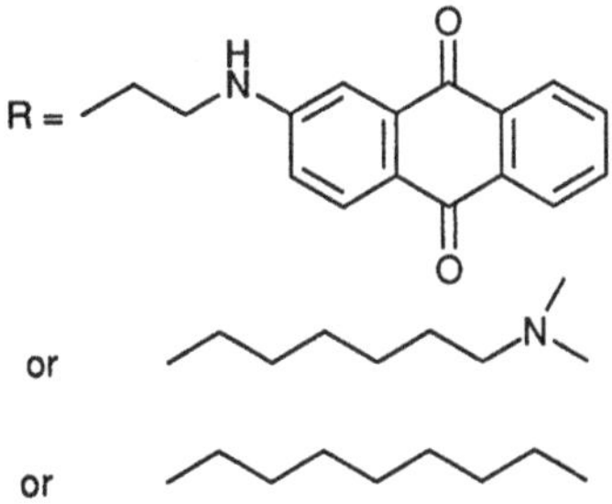

R = H N O O

or N

or

(185)

B = A, T, C

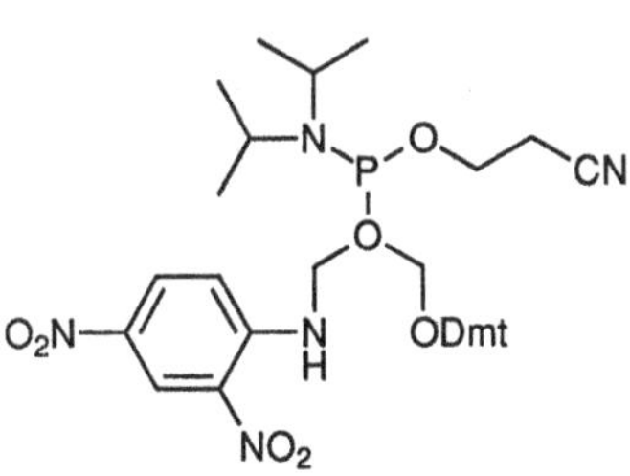

N P O CN O O_2N N H ODmt NO_2

(186)

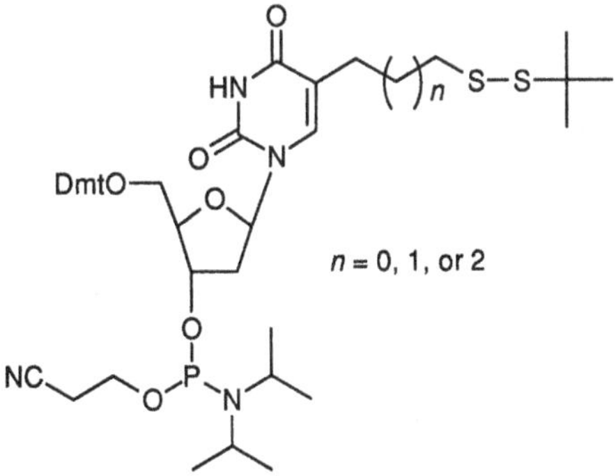

O HN O N S S n DmtO O O NC O P N

n = 0, 1, or 2

(187)

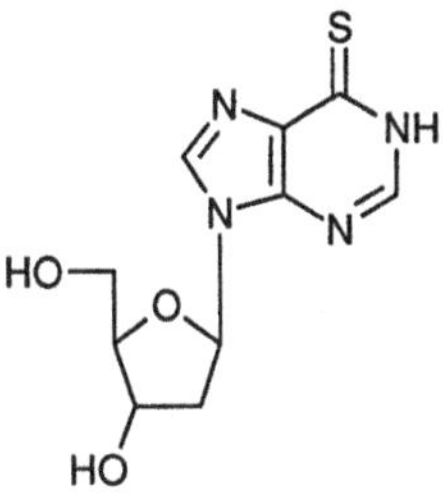

S N NH N N HO O HO

(188)

S HN O N HO O HO

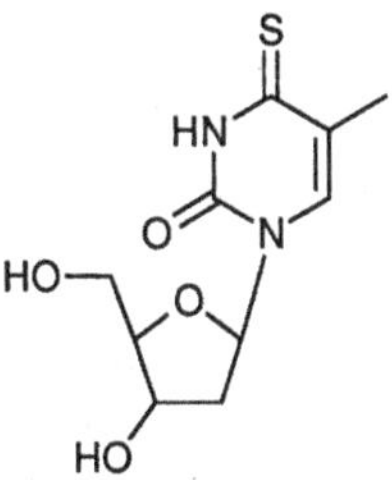

(189)

The synthesis of oligonucleotide duplexes containing site specific disulphide linkages has been reported.[173] Self complementary oligonucleotides were synthesised to present 6-thioinosine (**188**) and 4-thiothymidine (**189**) opposite one another upon duplex formation. The synthesis of the latter compounds has been described previously. 6-Thioinosine was protected for DNA synthesis with a 2-cyanoethyl group. Aerial oxidation afforded the disulphide.

Circular oligonucleotides have been demonstrated to be much more resistant to nucleases and there is consequently interest in the potential of these molecules as antisense agents. Gryaznov *et al.* have described the synthesis of circular oligonucleotides by the oligonucleotide template directed oxidation of oligonucleotides bearing 3' and 5'- terminal phosphorothioate groups to form diphosphorothioates using CHI_3 or $K_3Fe(CN)_6$.[174] An alternative strategy for circularisation involves templated formation of phosphodiester or pyrophosphate closed oligonucleotides by the action of cyanogen bromide or EDC with imidazole or N-hydroxybenzotriazole, respectively, on oligonucleotides bearing terminal 3' and 5' phosphate groups.[175]

Synthesis of an N^3-ethylthioluridine derivative (**190**) suitable for use in RNA synthesis has been described.[176] The nucleoside is synthesised from deoxyuridine by reaction with *S*-benzoyl-*O*-tosyl-3-mercaptopropanol in the presence of sodium hydride under transient protection conditions. Following exchange of the sulphur protecting group to give *t*-butyldisulphide and introduction of a 5'-dimethoxytrityl group, the nucleoside was treated with *t*-butyldimethylsilyl chloride. The resulting 3'-isomer was phosphitilated to yield a monomer suitable for use in RNA synthesis whereas the succinate ester of the 3'-compound was prepared and this isomer was attached to control pore glass. A 12mer with the tetraloop sequence UUCG containing 5'- and 3'-N^3- residues was synthesised from (**191**). The tert-butylthiol protection was removed with dithiothreitol and the liberated thiols allowed to oxidise. The resultant hairpin had a melting temperature which was approximately 20°C higher than the unmodified compound. The stabilisation of DNA triple helices has also been achieved by formation of a disulphide cross link, in this case between one strand of the duplex and DNA and the third strand of the triplex using C5- and N3- alkyl thiol derivatised deoxyuridine and thymidine compounds.[177]

There is considerable interest in the synthesis of modified nucleosides which are capable of stabilising triple helices or participating in such structures when the polypurine or polypyrimidine strand has been interrupted by the occurrence of a nucleobase from the alternative family. N^4-(6-aminopyridinyl)-2'-deoxycytidine (**192**) is a modified nucleoside capable of stabilising triple helices when it appears in the third stand interacting with A-T or C-G pairs. In contrast, the melting temperatures of the third strand, including the modified nucleoside with duplexes containing the reverse base pairs, are much lower. Based on this observation, it is proposed that the imino form of the nucleoside allows further interaction with the duplex in the major groove. N^4-(6-aminopyridinyl)-2'-deoxycytidine is prepared from 4-(1,2,4-triazol-1-yl)-(β-D-3,5-di-O-acetylribofuranosyl)pyrimidin-2-(1H)-one by reaction with 2,6-diaminopyridine.[178] The compound is protected with a benzoyl group during oligomer assembly.

Triplex formation has been observed between an oligonucleotide duplex and an oligopyrimindine strand containing N^4-(3-acetamidopropyl)-2'-deoxycytidine (**193**). This analogue selectively interacts with a G-C base pair in the target duplex. Oligomers containing N^4-(3-

acetamidopropyl)-2'-deoxycytidine were synthesised by treatment of an oligomer containing deoxycytidine residues with 1,3-diaminopropane in the presence of bisulphite.[179] Following transamination the N^4-(3-aminopropyl) functions were treated with *p*-nitrophenylacetate.

Triple helices formed with the third strand containing 7,8-dihydro-8-oxoadenine (**194**) as a replacement for cytosine are stable over a greater range of pH values as the requirement for protonation upon triple helix formation is eliminated.[180]

The non-natural pyrid[2,3-d] pyrimidine nucleoside (**195**) pairs preferentially with guanine and adenine in DNA but specifically recognises AT base pairs within the triple helix structure. It is thought that this sequence specificity may be caused by the ability of the nucleoside to adopt different tautomeric forms.[181]

An alternative approach to the formation of triple helices with sequences consisting of both purines and pyrimidines involves allowing the third strand to pair with purines on alternate strands of the duplex by cross-overs in the major groove. Each strand switch demands a reverse in the polarity of the third nucleic acid. Asseline and Thuong have described a linker in which two oligomers involved in a cross-over triplex are connected through their terminal nucleic acid bases via a short alkyl spacer.[182] The linked dimer (**196**) was synthesised from protected thymine which was activated at the 4-position by triazolide formation which upon treatment with 1,4-diaminobutane gave the target compound. After deprotection and formation of the succinate ester, the 3'-hydroxyl function was immobilised on LCAA-CPG using conventional procedures and any unreacted activated ester was capped with propylamine. The resultant solid-supported linked dimer was used in conventional DNA synthesis to yield oligonucleotides of opposite polarity tethered via nucleic acid bases. The modified support is currently capable of preparing only symmetrical oligonucleotides of this form but replacement of one dimethoxytrityl group with an alternative protecting group should lead to the synthesis of asymmetrical sequences.

Triple helices can be stabilised by the addition of spermine but not spermidine. Tung *et al.* have synthesised oligonucleotide-spermine conjugates which can significantly stabilise the binding of the third strand to the duplex and the resultant duplex after dissociation of this strand.[183] Synthesis of the activated spermine for attachment to the oligonucleotide is complicated by the necessity of maintaining all the nitrogens of spermine to preserve stabilising activity.

6. Interactions and Reactions of Nucleic Acids With Small Molecules

Potassium permanganate is classically used in experiments which relate the structure of DNA to reactivity of thymine residues. Hansler and Rokita have studied the oxidation rate of thymine as the free nucleoside and nucleotide, in single and double stranded DNAs and when thymidine residues are mispaired or unpaired.[184] The primary targets of modification were always unpaired and solvent accessible thymine residues. A drastic decrease in rate is observed upon conversion of thymidine to thymidine monophosphate. The rate of oxidation of thymine residues in single stranded oligonucleotides only approached that of thymidine residues when very high salt

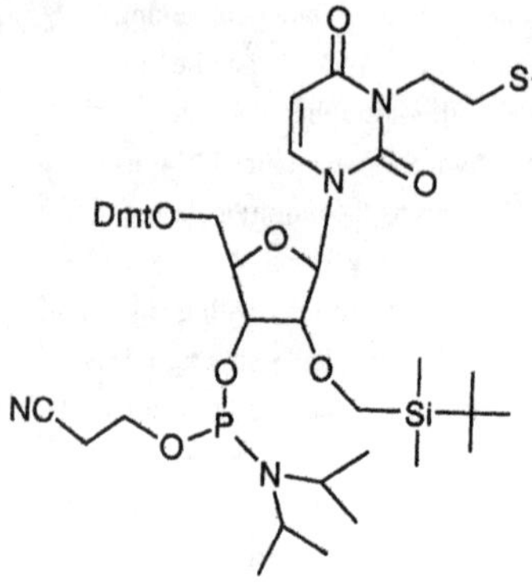

(190)

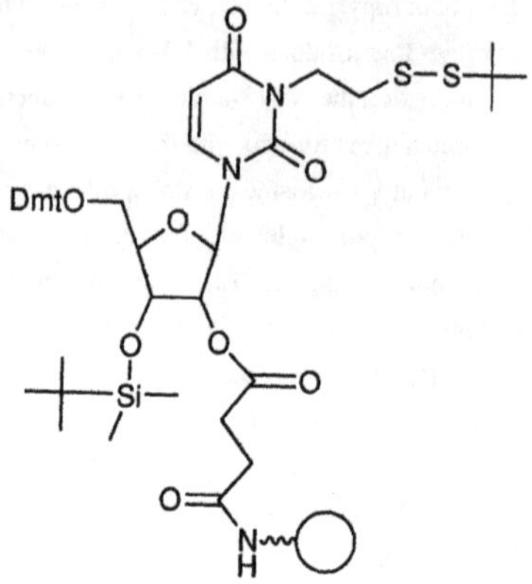

(191)

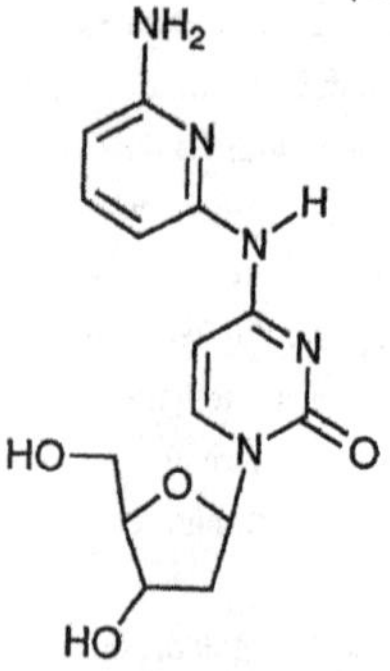

(192)

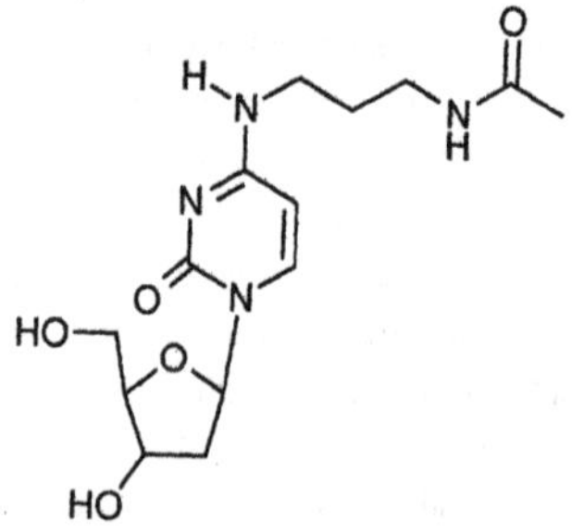

(193)

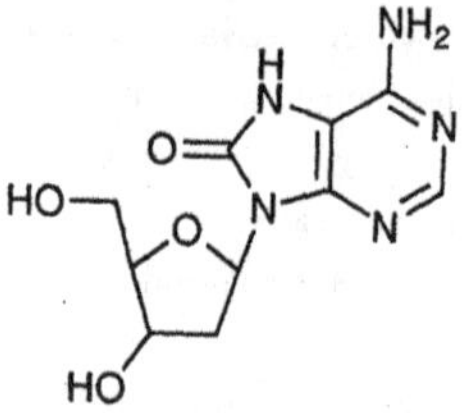

(194)

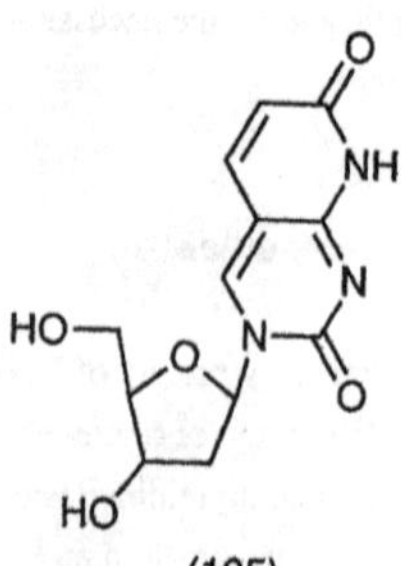

(195)

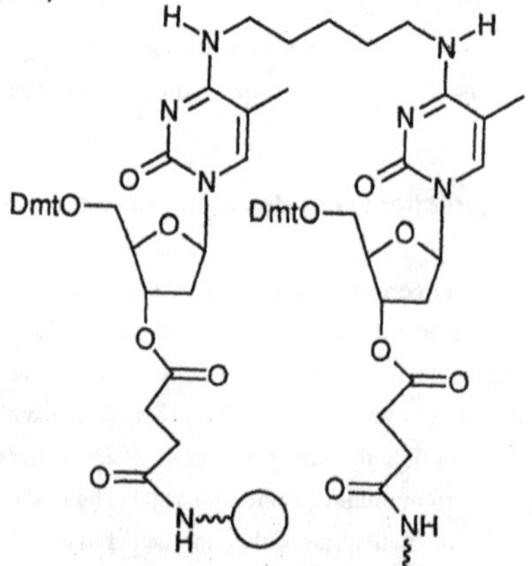

(196)

concentrations were employed. The effects of increasing salt concentration on the rate of reaction were much more dramatic than those associated with conformation.

Vinyl chloride is a potent mutagen which is metabolically converted to chloroethylene oxide. Rearrangement to chloracetaldeyde results in a bifunctional electrophile which forms an hydroxyethane bridge between the exocyclic amino groups and ring nitrogens of adenine and cytosine bases in DNA. The hydroxy ethane derivatives (**197**, **198**, **199a** and **199b**)are thought to dehydrate eventually to the etheno nucleoside derivatives. N^6-etheno-2'-deoxyadenosine converts to a bi-imidazole product ,4-amino-5-(imidazol-2-yl)imidazole (**200**), in alkaline solution. The mutagenic effects of these base modifications have been studied via synthesis of the etheno and bi-imidazole nucleosides and their incorporation into DNA.[185] The most mutagenic of these lesions was found to be the bi-imidazole derivative but the etheno adenine and cytosine derivatives were also demonstrated to be mutagenic.

The diol epoxide metabolites of polycyclic aromatic hydrocarbons react with the nucleobases of DNA and this selectivity is associated with the mutagenicity of these compounds. The effects of chirality on the structure of the resultant DNA lesions have been studied by Mao *et al.*[186] The (+) isomer of 7b,8a-dihydroxy-9a-10a-epoxy7,8,9,10-tetrahydrobenzo[a]pyrene ((+)-anti-BPDE) (**201**) is very much more mutagenic in mammalian cells. A 16mer containing one guanosine residue was modified at this position with each isomer of anti-BPDE. The cleavage rate of the resultant oligonucleotides with snake venom phosphodiesterase (a 3'-exonuclease) favours the (+) isomer adduct whilst the rate of cleavage with spleen phosphodiesterase (a 5'- exonuclease) is very much greater with the (-) adduct. It was concluded that in single stranded oligonucleotides containing the (-) adduct the pyrenyl must point towards 3'-end of the modified strand whilst in the (+) case the pyrenyl residue points to the opposite termini.

The mechanism of methoxyamine induced mutagenesis has been studied in synthetic oligonucleotides containing N^4-methoxycytosine (**202**) [187, 188]and its bicyclic analogue dihydropyrimido-4,5-c[1,2]oxazin-7-one (**203**)[187] The mutagenic effect of this base modification has been ascribed to the ability of the nucleoside to exist in two tautomeric forms, thus allowing base pair formation with both adenine and guanine. Crystallographic studies have, however, failed to confirm this explanation. The authors report observation of both tautomeric forms of these modified nucleosides in oligonucleotides by ^{1}H NMR.

The structure of the adducts of platinum derivatives with DNA continues to be of interest because of their use and potential as anticancer drugs. The specificity of DNA interstrand cross-links formed between the ineffective *trans*-diamminechloroplatinum(II) has been determined as being preferentially between complementary guanine and cytosine residues.[189]In addition, the rate of cross-linking with the *trans* isomer was found to be lower than that for the clinically active *cis* compound. The NMR solution structure of a DNA duplex containing the cis-platinumdiammine N7(G), N7(G) adduct has been determined.[190] A second NMR structure examines adduct formation in an intrastrand cross-link created in a self-complementary duplex. A pair of enantiomeric Pt(II) complexes has been synthesised and their cytotoxicity has been evaluated. [Pt(RR-eaP)Cl_2] (eap= N,N-diethyl-2,4-pentadiamine) was found to be more active in leukaemia cells but the difference was not as great as expected.[191]

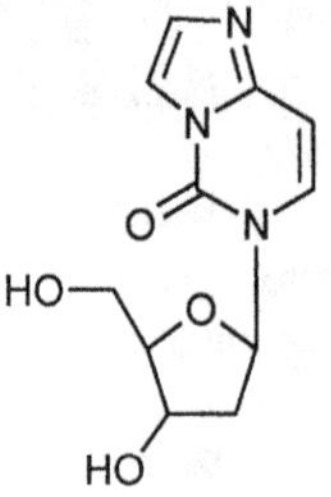

(197) 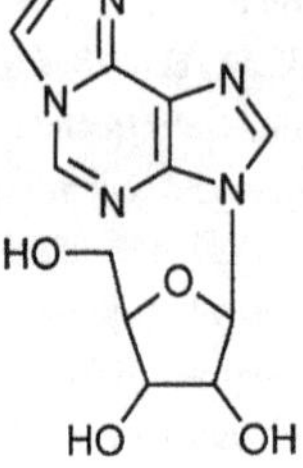 (198) 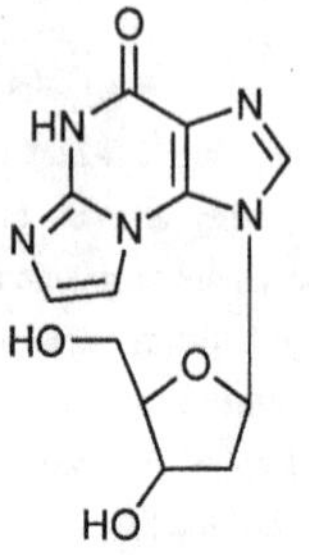 (199a) 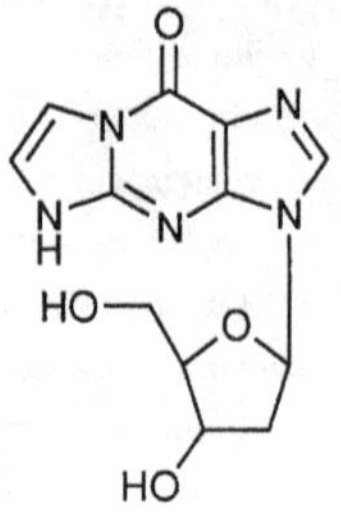 (199b)

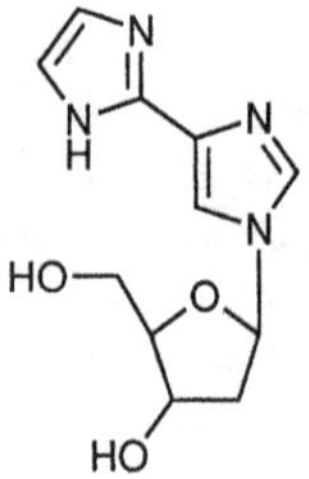

(200) 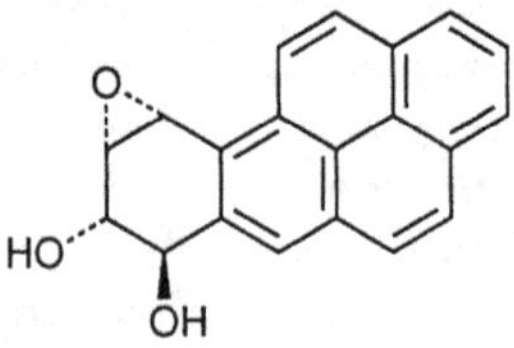 (+) (201)

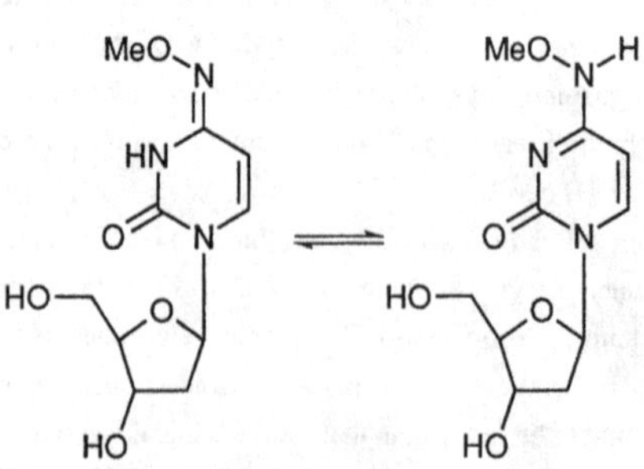

(202) 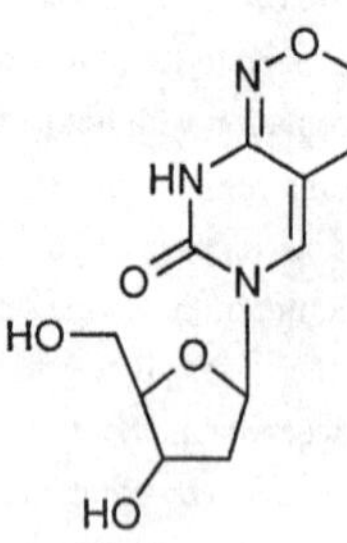 (203)

Dervan and Wemmer have studied recognition of DNA by netropsin and related analogues. A recognition code has been developed. The structural analysis of covalent peptide dimers *bis*(pyridine-2-carboxamidonetropsin)$(CH_2)_{3\text{-}6}$ complexed with 5'-TGACT-3' sites has been completed by 2-dimensional NMR.[192] Several other NMR studies of DNA-binding drugs or natural products have appeared this year.[193-196] The crystal structure of the anthracycline antibiotic idarubicin bound the d(TGATCA) has been solved at high resolution.[197]

The interactions of DNA with rhodium complexes have been studied by Barton and co-workers. The cleavage pattern of $tRNA^{(Phe)}$ with $Rh(phen)_2phi^{3+}$ (phen=phenanthroline, phi=9,10-phenanthrenequinonedimine) shows local differences to the DNA analogue $tDNA^{(Phe)}$ but that globally the two structures are similar.[198] ^{1}H-NMR studies of $Rh(NH_3)_4phi^{3+}$ indicate that the octahedral complex binds to the duplex in a manner consistent with classical intercalation.[199] Specificity arises from the non-intercalating functionalities of the complex. Photoinduced electron transfer has been demonstrated through a 15mer duplex with a $Ru(phen)_2dppz^{2+}$ (dppz, dipyridophenazine) donor and a $Rh(phi)_2(phen)^{3+}$ acceptor tethered to the 5'-termini of the oligonucleotides.[200]

7. Determination of Nucleic Acid Structures

There is great interest in the determination of the secondary structure of short synthetic oligodeoxynucleotides by both NMR and crystallography. The applications of X-ray analysis in the study of the conformation and the interactions of nucleic acids have been reviewed by Kennard and Salisbury.[201] Numerous crystal structures have appeared this year including that of an A-DNA octamer,[202] the structure of a DNA duplex containing 8-hydroxyguanine-adenine base pairs[203] and an oligonucleotide containing a guanine-guanine mismatch.[204] The high resolution crystal structure of a parallel stranded guanine tetraplex has been described.[205]

The study of oligodeoxynucleotides by NMR has revealed the structure of a DNA containing an extrahelical cytosine,[206] the arrangement of an inosine-adenosine mismatch in a DNA duplex,[207] an adjacent guanosine-adenosine mismatch [208] and a duplex containing a 3-base bulge.[209] NMR studies of $d(GAATTAAAATTCC)_2$ have revealed a wider minor groove in the TA region than that normally found in DNA and bending in the major groove in the same region.[210]

Alternating (C-T) sequences are involved in the formation of H-DNA structures associated with (GA)n.(CT) regions and are favoured by low pH. $d(CT)_4$, $d(TC)_4$ and $d(TC)_{15}$ have been studied by NMR.[211] It was found that their conformations are pH dependent; at pH7 an antiparallel duplex with C:T base pairs was found but at pH3, C+:T was observed.

Feigon and co-workers have studied the NMR spectra of oligodeoxynucleotides of the form dG_nT_n.[212-214] These DNA molecules form guanosine tetraplexes in solution and one $d(G_2T_2G_2TGTG_2T_2G_2)$ has been demonstrated to inhibit thrombin-catalysed fibrin clot formation *in vitro*.

The structure of an oligonucleotide designed to assemble a 4-way junction has been investigated by NMR.[215] Preliminary results in the structure determination of nucleic acid analogues containing an amide backbone hybridised to 2'-*O*-methyl containing RNA have been described.[216]

The crystal structures of only very few RNA molecules are currently known. A sparse-matrix for crystallisation of ribozymes and other small RNAs has been described[217] and crystals of a hammerhead ribozyme reported.[218]

The crystal and NMR structures of chimeric oligonucleotide duplexes containing both ribose and deoxyribose sugar residues have been reported.[218-220] The overall conformations were found to be A-type. The NMR structure of the self-complementary oligoribonucleotide $r(CGCGAAUUCGG)_2$ has been studied and compared to that of the DNA analogue.[221]

The selective isotopic enrichment of synthetic RNA has revolutionised the study of RNA structure by NMR. The HIV-1 TAR element,[222] a small RNA hairpin[223] and a base triple domain in RNA[224] have all been studied.

A number of other techniques have been applied to the study of nucleic acids. Triple helix formation has been visualised by electron-microscopy[225] and Raman spectroscopy has been applied to the study of Z-DNA.[226, 227]

There is growing interest in the primary structure determination and characterisation of small oligonucleotides by mass spectrometry. A review summarised the application of mass spectrometry to biochemical research.[228] A number of publications dealing with oligonucleotides including the observation of a duplex structure have appeared.[229-236]

References

1. P. Alexander and A. Holy, *Coll. Czech. Chem. Comm.*, 1994, **59**, 2127-2165.
2. C. McGuigan, D. Kinchington, M.F. Wang, S.R. Nicholls, C. Nickson, S. Galpin, D.J. Jeffries, and T.J. O'Connor, *FEBS Lett.*, 1993, **322**, 249-252.
3. C. McGuigan, D. Kinchington, S.R. Nicholls, S.R. Nickson, and T.J. O'Connor, *Bioorg. Med. Chem. Lett.*, 1993, **3**, 1207-1210.
4. C. McGuigan, P. Bellevergue, H. Sheeka, N. Mahmood, and A.J. Hay, *FEBS Lett.*, 1994, **351**, 11-14.
5. W. Thompson, D. Nicholls, W.J. Irwin, J.S. Al-Mushadania, S. Freeman, A. Karpas, J. Petrik, N. Mahmood, and A.J. Hay, *J. Chem. Soc. Perkin Trans 1*, 1993, 1239-1245.
6. C. McGuigan, R.N. Pathirana, J. Balzarini, and E. Declercq, *J. Med. Chem.*, 1993, **36**, 1048-1052.
7. C. McGuigan, R.N. Pathirana, M.P.H. Davies, J. Balzarini, and E. Declercq, *Bioorg. Med. Chem. Lett.*, 1994, **4**, 427-430.
8. C. McGuigan, M. Davies, R.N. Pathirana, N. Mahmood, and A.J. Hay, *Antiviral Res.*, 1994, **24**, 69-77.
9. J.H. Boal, R.P. Iyer, and W. Egan, *Nucleosides Nucleotides*, 1993, **12**, 1075-1084.
10. P. Franchetti , L. Cappellacci, M. Grifantini, L. Messini, G. Sheikha, A. Loi, E. Tramontano, A. Demontis, M. Spiga, and P. Lacolla, *J. Med. Chem.*, 1994, **37**, 3534-3541.
11. C. Perigaud, G. Gosselin, I. Lefebvre, J.L. Girardet, S. Benzaria, I. Barber, and J.L. Imbach, *Bioorg. Med. Chem. Lett.*, 1993, **3**, 2521-2526.

12 F. Puech, G. Gosselin, I. Lefebvre, A. Pompon, A. Aubertin, A. Kirn, and J. Imbach, *Antiviral Res.*, 1993, **22**, 2-3.

13 X. Pannecoucke, G. Parmentier, G. Schmitt, F. Dolle, and B. Luu, *Tetrahedron*, 1994, **50**, 1173-1178.

14 M.I. Balagopala, A.P. Ollapally, and H.J. Lee, *Nucleosides Nucleotides*, 1994, **13**, 1843-1853.

15 A. Namane, C. Gouyette, M.P. Fillion, G. Fillion, and T. Huynh-Dinh, *J. Med. Chem.*, 1992, **35**, 3039.

16 H. Schott, M.P. Haussler, and R.A. Schwendener, *Liebigs Ann. Chem.*, 1994, 277-282.

17 D.M. Coe, A. Garofalo, S.M. Roberts, R. Storer, and A.J. Thorpe, *J. Chem. Soc. Perkin 1*, 1994, 3061-3063.

18 C. Serra, G. Dewynter, J. Montero, and J. Imbach, *Tetrahedron*, 1994, **50**, 8427-8444.

19 K. Lee and D.F. Wiemer, *J. Org. Chem.*, 1993, **58**, 7808-7812.

20 L.L. Bennett, W.B. Parker, P.W. Allan, L.M. Rose, Y.F. Shealy, J.A. Secrist, J.A. Montgomery, G. Arnett, R.L. Kirkman, and W.M. Shannon, *Mol. Pharm.*, 1993, **44**, 1258-1266.

21 R.D. Elliott, G.A. Rener, J.M. Riordan, J.A. Secrist, L.L. Bennett, W.B. Parker, and J.A. Montgomery, *J. Med. Chem.*, 1994, **37**, 739-744.

22 J.E. Starrett, D.R. Tortolani, J. Russell , M.J.M. Hitchcock, V. Whiterock, J.C. Martin, and M.M. Mansuri, *J. Med. Chem.*, 1994, **37**, 1857-1864.

23 R.V. Srinivas, B.L. Robbins, M.C. Connelly, Y.-F. Gong , N. Bischofberger, and A. Fridland, *Antimicrob. Agents Chemother.*, 1993, **37**, 2247-2250.

24 P. Alexander, A. Holy, and M. Masojidkova, *Coll. Czech. Chem. Comm.*, 1994, **8**, 1853-1869.

25 K. Eger, E.M. Klunder, and M. Schmidt , *J. Med. Chem.*, 1994, **37**, 3057-3061.

26 J. Balzarini, A. Holy, J. Jindrich, L. Naesens, R. Snoeck, D. Schols, and E. Declercq, *Antimicrob. Agents Chemother.*, 1993, **37**, 332-338.

27 S.D. Chamberlain, K.K. Biron, R.E. Dornsife, D.R. Averett, L. Beauchamp, and G.W. Koszalka, *J. Med. Chem.*, 1994, **37**, 1371-1377.

28 M.R. Harnden and H.T. Serafinowska, *Nucleosides Nucleotides*, 1994, **13**, 903-913.

29 H. Dvorakova, A. Holy, and I. Rosenberg, *Coll. Czech. Chem. Comm.*, 1994, **9**, 2069-2094.

30 D.K. Kim, Y.W. Kim, and K.H. Kim, *Bioorg. Med. Chem. Lett.*, 1994, **4**, 2241-2244.

31 E.J. Reist, W.W. Bradford III, B.L. Ruhland-Fritsch, P.A. Sturm, and N.T. Zaveri, *Nucleosides Nucleotides*, 1993, **13**, 539.

32 J.L. Kelley, J.A. Linn, E.W. McLean, and J.V. Tuttle, *J. Med. Chem.*, 1993, **36**, 3455-3463.

33 J. Tomasz, B.R. Shaw, K. Porter, B.F. Spielvogel, and A. Sood, *Angew. Chem., Int. Ed. Engl.*, 1992, **31**, 1373-1375.

34 J. Matulic-Adamic and N. Usman, *Tetrahedron Lett.*, 1994, **35**, 3227.

35 G.M. Blackburn and M.J. Guo, *Tetrahedron Lett.*, 1993, **34**, 149-152.

36 X. Li , G.K. Scott, A.D. Baxter, R.J. Taylor, J.S. Vyle, and R. Cosstick, *J. Chem. Soc. Perkin 1*, 1994, 2123-2129.

37 M. Morr and V. Wray, *Angew. Chem., Int. Ed. Engl.*, 1994, **13**, 1394-1396.

38 M. Yoshikawa, T. Kato, and T. Takenishi, *Bull. Chem. Soc. (Japan)*, 1969, **42**, 3305-3308.

39 J. Ludwig, *Acta Biochim. Biophys. Acad. Sci. Hung.*, 1981, **16**, 131.

40 K.E. Ng and L.E. Orgel, *Nucl. Acids Res.*, 1987, **15**, 3573.

41 C. Battistini, S. Fustinoni, B.M. G, and D. Borghi, *Tetrahedron*, 1993, **49**, 1115-1132.

42 F. Seela, *Nucleosides Nucleotides*, 1991, **10**, 771.

43 X. Li and R. Cosstick, *J. Chem. Soc. Perkin 1*, 1993, 1091-1292.

44 M. Moor, L. Ernst, and L.R. Menge, *Liebigs Ann. Chem.*, 1982, 651.

45 F. Eckstein, *Ann. Rev. Biochem.*, 1985, **54**, 367-402.

46 L. Victorova, N. Dyatkina, D. Mozzherin, A. Atrazhev, A.A. Krayevsky, and M. Kukanova, *Nucl. Acids Res.*, 1982, **20**, 783.

47 A.A. Arzumanov and N.B. Dyatkina, *Nucleosides Nucleotides*, 1994, **13**, 1031-1037.

48 A.A. Krayevsky, L.S. Victorova, D.J. Mozzherin, and M.K. Kukhanova, *Nucleosides Nucleotides*, 1993, **12**, 83-93.

49 Shirokova Ea, Tarussova Nb, Shipitsin Av, Semizarov Dg, and Krayevsky Aa, *J. Med. Chem.*, 1994, **37**, 3739-3748.

50 Z.G. Chidgeavadge, R.S. Beabealashvilly, T.A. Rosovskaya, A.M. Atrazhev, N.B. Tarussova, S.K. Minassyan, N.B. Dyatkina, a.M.K. Kukhanov, A.V. Papchikhin, and A.A. Krayevsky, *Mol. Biol. Russian*, 1989, **23**, 1732-1742.

51 D.G. Semizarov, L.S. Victorova, A.A. Krayevsky, and M.K. Kukhanova, *FEBS Lett.*, 1993, **327**, 45-48.

52 D.G. Semizarov, L.S. Victorova, N.B. Dyatkina, M. Vonjantalipinski, and A.A. Krayevsky, *FEBS Lett.*, 1994, **354**, 187-190.

53 N. Ettner, U. Haak, M. Niederweis, and W. Hillen, *Nucleosides Nucleotides*, 1993, **12**, 757-771.

54 J. Ludwig and F. Eckstein, *J. Org. Chem.*, 1988, **54**, 631.

55 J. Wower, S.S. Hixson, L.A. Sylvers, X.Y. D, and R.A. Zimmermann, *Bioconjugate Chemistry*, 1994, **5**, 158-161.

56 M.K. Herrlein, R.E. Konrad, J.W. Engels, T. Holletz, and D. Cech, *Helv. Chim. Acta*, 1994, **77**, 586-596.

57 T. Kovacs and L. Otvos, *Tetrahedron Lett.*, 1988, **29**, 4525.

58 J. Hovinen, E. Azhayeva, A. Ázhayev, A. Guzaev, and H. Lönnberg, *J. Chem. Soc. Perkin Trans 1*, 1994, 211.

59 H. Aurup, D.M. Williams, and F. Eckstein, *Biochemistry*, 1982, **31**, 9636-9641.

60 B. Ortiz, A. Sillero, and M.A.G. Sillero, *Eur. J. Biochem.*, 1993, **212**, 263-270.

61 K. Fukuoka, F. Suda, R. Suzuki, H. Takaku, M. Ishikawa, and T. Hata, *Tetrahedron Lett.*, 1994, **35**, 1063-1066.

62 K. Fukuoka, S. Fuminori, R. Suzuki, M. Ishikawa, H. Takaku, and T. Hata, *Nucleosides Nucleotides*, 1994, **13**, 1557-1567.

63 K.Y. Hostetler, S. Parker, C.N. Sridhar, M.J. Martin, J.-L. Li, L.M. Stuhmiller, G.M.T. van Wijk, H. van den Bosch, M.F. Gardner, K.A. Aldern, and D.D. Richman, *Proc. Natl. Acad. Sci. USA*, 1993, **90**, 11835-11839.

64 S.L. Beaucage and P.I. Radhakrishnana, *Tetrahedron*, 1992, **48**, 2223-2311.

65 S.L. Beaucage and P.I. Radhakrishnana, *Tetrahedron*, 1993, **49**, 6123-6194.

66 D. Pendlebury, *Scientist*, 1993, **7**, 1.

67 R. Ramage and F.O. Wahl, *Tetrahedron Lett.*, 1993, **34**, 7133-7136.

68 M.M. Greenberg, *Tetrahedron Lett.*, 1993, **34**, 251-254.

69 W.K.-D. Brill, *Tetrahedron Lett.*, 1994, **35**, 3041-3044.

70 P. Wright, D. Lloyd, W. Rapp, and A. Andrus, *Tetrahedron Lett.*, 1993, **34**, 3373-76.

71 P. Kumar, N.N. Ghosh, K.L. Sadana, B.S. Garg, and K.C. Gupta, *Nucleosides Nucleotides*, 1993, **12**, 565-584.

72 F.X. Montserrat, A. Granda, and E. Pedroso, *Nucleosides Nucleotides*, 1993, **12**, 967-971.

73 L. Denapoli, A. Messere, D. Montesarchio, G. Piccialli, C. Santacroce, and G.M. Bonora, *Nucleosides Nucleotides*, 1993, **12**, 21-30.

74 C.L. Scremin and G.M. Bonoroa, *Tetrahedron Lett.*, 1993, **34**, 4663-4666.

75 T. Wada, K. Ishikawa, and T. Hata, *Tetrahedron*, 1993, **49**, 2043-2054.

76 J. Kaufman, M. Le, G. Ross, P. Hing, M. Budiansky, E. Yu, E. Campbell, V. Yoshimura, V. Fitzpatrick, K. Nadimi, and A. Andrus, *Biotechniques*, 1993, **14**, 834.

77 P.M. Hardy, D. Holland, S. Scott, A.J. Garman, C.R. Newton, and M.J. McLean, *Nucl. Acids Res.*, 1994, **22**, 2998-3004.

78 M. Sekine, T. Mori, and T. Wada, *Tetrahedron Lett.*, 1993, **34**, 8289-8291.

79 M. Sekine, *Nucleosides Nucleotides*, 1994, **13**, 6-7.

80 T. Wada and M. Sekine, *Tetrahedron Lett.*, 1994, **35**, 757-760.

81 J. Smrt, L. Arnold, J. Svoboda, R. Hak, and I. Rosenberg, *Coll. Czech. Chem. Comm.*, 1993, **58**, 1692-1698.

82 V.A. Efimov, A.L. Kalinkina, and O.G. Chakhmakhcheva, *Nucl. Acids Res.*, 1993, **21**, 5337-5344.

83 S.K. Awasthi, V. Khanna, G. Watal, and K. Misra, *Indian J. Chem. Section B Organic Chemistry Including Medicinal Chemistry*, 1993, **32**, 916-919.

84 M.V. Rao, C.B. Rees, V. Schehlmann, and P.S. Yu, *J. Chem. Soc. Perkin 1*, 1993, 43-55.

85 R. Vinayak, *METHODS: A Companion to Methods in Enzymology*, 1993, **5**, 7-18.

86 N.D. Sinha, P. Davis, N. Usman, J. Perez, R. Hodge, J. Kremsky, and R. Casale, *Biochimie*, 1993, **75**, 1-2.

87 F. Bergmann and W. Pfleiderer, *Helv. Chim. Acta*, 1994, **77**, 481-501.

88 Y. Palom, E. Alazzouzi, F. Gordillo, A. Grandas, and E. Pedroso, *Tetrahedron Lett.*, 1993, **34**, 2195-2198.

89 R.I. Hogrefe, A.P. McCaffrey, L.U. Borozdina, E.S. McCampbell, and M.M. Vaghefi, *Nucl. Acids Res.*, 1993, **21**, 4739-4741.

90 E. Westman and R. Stromberg, *Nucl. Acids Res.*, 1994, **22**, 2430-2431.

91 E. Rozners, E. Westman, and R. Stromberg, *Nucl. Acids Res.*, 1994, **22**, 94-99.

92 G. Loose, W. Naumann, A. Winkler, and G. Suptitz, *Journal Fur Praktische Chemie Chemiker Zeitung*, 1994, **336**, 233-236.

93 D.C. Capaldi and C.B. Reese, *Nucl. Acids Res.*, 1994, **22**, 2209-2216.

94 R. Kierzek, *Nucleosides Nucleotides*, 1994, **13**, 1757-1768.

95 L.M. Hoffman and M.G. Johnson, *Biotechniques*, 1994, **17**, 372-375.

96 Z.J. Lesnikowski, *Bioorganic Chemsitry*, 1993, **21**, 127-155.

97 A. Suska, A. Grajkowski, A. Wilk, B. Uznanski, J. Blaszczyk, M. Wieczorek, and W.J. Stec, *Pure And Applied Chemistry*, 1993, **65**, 707-714.

98 R.S. Varma, *Synlett*, 1993, 621-637.

99 W.J. Stec, A. Buznanski, A. Wilk, B. Hirschbein, and K.L. Fearon, *Tetrahedron Lett.*, 1993, **34**, 5317-5320.

100 Y. Hayakawa, M. Hirose, and R. Noyori, *Nucleosides Nucleotides*, 1994, **13**, 1337-1345.

101 T.K. Wyrzykiewicz and D.L. Cole, *Nucl. Acids Res.*, 1994, **22**, 2668-2669.

102 V.T. Ravikumar, T.K. Wyrzykiewicz, and D.L. Cole, *Nucleosides Nucleotides*, 1994, **13**, 1695-1706.

103 J.Y. Tang, J. Temsamani, and S. Agrawal, *Nucl. Acids Res.*, 1993, **21**, 2729-2735.

104 D.J. Earnshaw and R. Cosstick, *Abstracts of Papers of the American Chemical Society*, 1994, **207**, 53.

105 J.A. Piccirilli, J.S. Vyle, M.H. Caruthers, and T.R. Cech, *Nature*, 1993, **361**, 85-88.

106 J. Stawinski and M. Thelin, *J. Org. Chem.*, 1994, **59**, 130-136.

107 E.V. Vyazovkina, N.I. Komarova, and A.V. Lebedev, *Bioorg. Khim,* 1993, **19**, 86-94.

108 E.V. Vyazovkina, I.W. Engels, and A.V. Lebedev, *Bioorg. Khim.*, 1993, **19**, 197-210.

109 E.V. Vyazovkina, E.V. Savchenko, S.G. Lokhov, I.W. Engels, E. Wickstrom, and A.V. Lebedev, *Nucl. Acids Res.*, 1994, **22**, 2404-2409.

110 R.J. Jones, K.Y. Lin, J.F. Milligan, S. Wadwani, and M.D. Matteucci, *J. Org. Chem.*, 1993, **58**, 2983-2991.

111 J.M. Veal, X.L. Gao, and F.K. Brown, *J. Am. Chem. Soc.*, 1993, **115**, 7139-7145.

112 B. Meng, S.H. Kawai, D.G. Wang, G. Just, P.A. Giannaris, and M.J. Damha, *Angew. Chem., Int. Ed. Engl.*, 1993, **32**, 729-731.

113 X. Cao and M.D. Matteucci, *Tetrahedron Lett.*, 1994, **35**, 2325-2327.

114 A. Chur, B. Hols, O. Dahl, P. Valentinhansen, and E.B. Pedersen, *Nucl. Acids Res.*, 1993, **21**, 5179-5183.

115 R. Fathi, Q. Huang, J.L. Syi, W. Delaney, and A.F. Cook, *Bioconjugate Chemistry*, 1994, **5**, 47-57.

116 J.X. Huang, E.B. Mcelreoy, and T.S. Widlanski, *J. Org. Chem.*, 1994, **59**, 3520-3521.

117 W.D. Luo, E. Atrazheva, N. Fregeau, W.H. Gmeiner, and J.W. Lown, *Canadian Journal of Chemsitry*, 1994, **72**, 1548-1555.

118 Z.J. Lesnikowski, *Bioorganic Chemistry*, 1994, **22**, 128-139.

119 Z.J. Lesnikowski and R.F. Schinazi, *J. Org. Chem.*, 1993, **58**, 6531-6534.

120 F. Morvan, H. Porumb, G. Degols, I. Lefebvre, A. Pompon, B.S. Sproat, B. Rayner, C. Malvy, B. Lebleu, and J.L. Imbach, *J. Med. Chem.*, 1993, **36**, 280-287.

121 L. Chanteloup and N.T. Thuong, *Tetrahedron Lett.*, 1994, **35**, 877-880.

122 M.M. Vaghefi and R.I. Hogrefe, *Nucleosides Nucleotides*, 1993, **12**, 1007-1013.

123 M.L. Svendsen, J. Wengel, O. Dahl, F. Kirpekar, and P. Roepstorff, *Tetrahedron*, 1993, **49**, 11341-11352.

124 E.L. Hancox, B.A. Connolly, and R.T. Walker, *Nucl. Acids Res.*, 1993, **21**, 3485-3491.

125 M. Tarkoy, M. Bolli, B. Schweizer, and C. Leumann, *Helv. Chim. Acta*, 1993, **76**, 481-510.

126 K.-H. Altman, E. Kesseling, E. Francotte, and G. Rihs, *Tetrahedron Lett.*, 1994, **35**, 2331-2334.

127 F. Vandendriessche, K. Augustyns, A. Vanaerschot, R. Busson, J. Hoogmartens, and P. Herdewijn, *Tetrahedron*, 1993, **49**, 7223-7238.

128 T.F. Jenny and S.A. Benner, *Helv. Chim. Acta*, 1993, **76**, 826-841.

129 J. Hunziker, H.J. Roth, M. Bohringer, A. Giger, U. Diederichsen, M. Gobel, R. Krishnan, B. Jaun, C. Leumann, and A. Eschenmoser, *Helv. Chim. Acta*, 1993, **76**, 259-352.

130 S. Pitsch, S. Wendeborn, B. Jaun, and A. Eschenmoser, *Helv. Chim. Acta*, 1993, **76**, 2161-2183.

131 F. Seela and S. Lampe, *Helv. Chim. Acta*, 1994, **77**, 1003-1017.

132 T.S. Rao, A.F. Lewis, D.R. H, and G.R. Revankar, *Tetrahedron Lett.*, 1993, **34**, 6709-6712.

133 F. Seela and T. Wenzel, *Helv. Chim. Acta*, 1994, **77**, 1485-1499.

134 F. Seela and H. Winter, *Helv. Chim. Acta*, 1994, **77**, 597-607.

135 T.S. Rao, R.H. Durland, and G.R. Revankar, *J. Heterocyl. Chem.*, 1994, **31**, 935-940.

136 J.J. Voegel, M.M. Altorfer, and S.A. Benner, *Helv. Chim. Acta*, 1993, **76**, 2061-2069.

137 P.G. Kuimelis and K.P. Nambiar, *Tetrahedron Lett.*, 1993, **34**, 3813-3816.

138 T.J. Matray and M.M. Greenberg, *J. Am. Chem. Soc.*, 1994, **116**, 6931-6932.

139 Y.-Z. Xu and P.F. Swann, *Tetrahedron Lett.*, 1994, **35**, 303-306.

140 A. Ono, T. Okamoto, M. Inada, H. Nara, and A. Matsuda, *Chem. Pharm. Bull.,* 1994, **42**, 2231-2237.

141 T. Shida, H. Iwaori, M. Arakawa, and J. Sekiguchi, *Chem. Pharm. Bull.,* 1993, **41**, 961-964.

142 A. Guy, J. Dubet, and R. Téoule, *Tetrahedron Lett.*, 1993, **34**, 8101-8102.

143 S. Koizume, H. Kamiya, H. Inoue, and E. Ohtuska, *Nucleosides Nucleotides*, 1994, **13**, 1517-1534.

144 K.S. Ramasamy, M. Zounes, S.M. Gonzalez, S.M. Freier, S.M. Lesnik, L.L. Cummins, R.H. Griffney, B.P. Monica, and P.D. Cook, *Tetrahedron Lett.*, 1994, **35**, 215-218.

145 G.Y. Wang and D.E. Bergstrom, *Tetrahedron Lett.*, 1993, **34**, 6725-6728.

146 N.V. Heeb and S.A. Benner, *Tetrahedron Lett.*, 1994, **35**, 3045-48.

147 S. Gryaznov and R.G. Schultz, *Tetrahedron Lett.*, 1994, **35**, 2489-2492.

148 D.-J. Fu and L. McLaughlin, *Biochemistry*, 1992, **31**, 10941-10949.

149 D.-J. Fu and L. McLaughlin, *Proc. Natl. Acad. Sci. USA*, 1992, **89**, 3985-3989.

150 F. Seela, K. Mersmann, J.A. Grasby, and M.J. Gait, *Helvetica Chimica Acta*, 1993, **76**, 1809-1820.

151 F. Seela and K. Mersmann, *Helvetica Chimica Acta*, 1993, **76**, 1435-1449.

152 T. Tuschl, M.M.P. Ng, W. Pieken, F. Benseler, and F. Eckstein, *Biochemistry*, 1993, **32**, 11658-11668.

153 J. Grasby, C. Pritchard, K. Mersmann, F. Seela, and M. Gait, *Coll. Czech. Chem. Comm.*, 1993, **58**, 154-157.

154 J.A. Grasby, P.J.G. Butler, and M.J. Gait, *Nucleic Acids Research*, 1993, **21**, 4444-4450.

155 C.J. Adams, J.B. Murray, J.R. Arnold, and P.G. Stockley, *Tetrahedron Lett.*, 1994, **35**, 765-768.

156 C.J. Adams, J.B. Murray, J.R. Arnold, and P.G. Stockley, *Tetrahedron Lett.*, 1994, **35**, 1597-1600.

157 P. Limbach, P.F. Crain, and J.A. McCloskey, *Nucl. Acids Res.*, 1994, **22**, 2183-2196.

158 U. Pieles, B. Beijer, K. Bohmann, S. Weston, S. Oloughlin, V. Adam, and B.S. Sproat, *J. Chem. Soc. Perkin 1*, 1994, 3423-3429.

159 W.H. Gmeiner, P. Sahasrabudhe, and R.T. Pon, *J. Org. Chem.*, 1994, **59**, 5779-5783.

160 C. Saintomé, P. Clivio, J.-L. Fourrey, A. Woisard, and A. Favre, *Tetrahedron Lett.*, 1994, **35**, 873-876.

161 R. Kiezek, Y. Li, D.H. Turner, and P.C. Bevilacqua, *J. Am. Chem. Soc.*, 1993, **115**, 4985-4992.

162 K. Yamana, K. Nunota, H. Nakano, and O. Sangen, *Tetrahedron Lett.*, 1994, **35**, 2555-2558.

163 H. Urata and M. Akagi, *Tetrahedron Lett.*, 1993, **34**, 4015-4018.

164 Y. Nakamura, T. Akiyama, K. Bessho, and F. Yoneda, *Chem. Pharm. Bull.,* 1993, **41**, 1315-1317.

165 J. Hovinen, A. Guzaev, A. Azhayev, and H. Lönnberg, *Tetrahedrom Letters*, 1993, **34**, 8169-8172.

166 J. Hovinen, A. Guzaev, A. Azhayev, and H. Lönnberg, *Tetrahedron Lett.*, 1993, **34**, 5163-5166.

167 T.H. Keller and R. Haner, *Nucl. Acids Res.*, 1993, **21**, 4499-4505.

168 J. Grzybowski, D.W. Will, R.E. Randall, C.A. Smith, and T. Brown, *Nucl. Acids Res.*, 1993, **21**, 1705-1712.

169 Y. Ueno, R. Saito, and T. Hata, *Nucl. Acids Res.*, 1993, **21**, 4451-4457.

170 Z.P. Wei, C.H. Tung, T.M. Zhu, and S. Stein, *Bioconjugate Chemistry*, 1994, **5**, 468-474.

171 S.G. Zhang and M. Egli, *Origins Of Life And Evolution Of The Biosphere*, 1994, **24**, 495-505.

172 J.T. Goodwin and G.D. Glick, *Tetrahedron Lett.*, 1993, **34**, 5549-5552.

173 J. Milton, B.A. Connolly, T.T. Nikiforov, and R. Cosstick, *J. Chem. Soc., Chem. Comm.*, 1993, 779-780.

174 S.M. Gryaznov and R.L. Letsinger, *Nucl. Acids Res.*, 1993, **21**, 1403-1408.

175 N.G. Dolinnaya, M. Blumenfeld, I.N. Merenkova, T.S. Oretskaya, N.F. Krynetskaya, M.G. Ivanovskaya, M. Vasseur, and Z.A. Shabarova, *Nucl. Acids Res.*, 1993, **21**, 5403-5407.

176 J.T. Goodwin and G.D. Glick, *Tetrahedron Lett.*, 1994, **35**, 1647-1650.

177 J.T. Goodwin, S.E. Osbourne, and G.D. Glick, *Tetrahedron Lett.*, 1994, **35**, 4527-4530.

178 C.Y. Huang and P.S. Miller, *J. Am. Chem. Soc.*, 1993, **115**, 10456-10457.

179 C.Y. Huang, C.D. Cushman, and P.S. Miller, *J. Org. Chem.*, 1993, **58**, 5048-5049.

180 M.C. Jetter and F.W. Hobbs, *Biochemistry*, 1993, **32**, 3249-3254.

181 A.B. Staubli and P.B. Dervan, *Nucl. Acids Res.*, 1994, **22**, 2637-2642.

182 U. Asseline and N.T. Thuong, *Tetrahedron Lett.*, 1993, **34**, 4173-4176.

183 C.H. Tung, K.J. Breslauer, and S. Stein, *Nucl. Acids Res.*, 1993, **21**, 5489-5494.

184 U. Hansler and S.E. Rokita, *J. Am. Chem. Soc.*, 1993, **115**, 8554-8557.

185 A.K. Basu, M.L. Wood, L.J. Niedernhofer, L.A. Ramos, and J.M. Essigmann, *Biochemistry*, 1993, **32**, 12793-12801.

186 B. Mao, B. Li, S. Amin, M. Cosman, and N.E. Geacintov, *Biochemistry*, 1993, **32**, 11785-11793.

187 A.N.R. Nedderman, M.J. Stone, D.H. Williams, P.K.T. Lin, and D.M. Brown, *J. Mol. Biol.*, 1993, **230**, 1068-1076.

188 G.V. Fazakerley, Z. Gdaniec, and L.E. Sowers, *J. Mol. Biol.*, 1993, **230**, 6-10.

189 V. Brabec and M. Leng, *Proc. Natl. Acad. Sci. USA*, 1993, **90**, 5345-5349.

190 C.J. Vangarderen and L.P.A. Vanhoute, *Eur. J. Biochem.*, 1994, **225**, 1169-1179.

191 K. Vickery, A.M. Bonin, R.R. Fenton, S. Omara, P.J. Russell, L.K. Webster, and T.W. Hambley, *J. Med. Chem.*, 1993, **36**, 3663-3668.

192 T.J. Dwyer, B.H. Geierstanger, M. Mrksich, P.B. Dervan, and D.E. Wemmer, *J. Am. Chem. Soc.*, 1993, **115**, 9900-9906.

193 K.E. Barr, R.A. Russell, R.N. Warrener, and J.G. Collins, *FEBS Lett.*, 1993, **322**, 173-176.

194 M.A. Keniry, D.L. Banville, P.M. Simmonds, and R. Shafer, *J. Mol. Biol.*, 1993, **231**, 753-767.

195 W. Nerdal, O.M. Andersen, and E. Sletten, *Acta Chem. Scan.*, 1993, **47**, 658-662.

196 J.A. Parkinson, S.E. Ebrahimi, J.H. Mckie, and K.T. Douglas, *Biochemistry*, 1994, **33**, 8442-8452.

197 B. Gallois, B.L. Destaintot, T. Brown, and W.N. Hunter, *Acta. Cryst. Sec. D, Biol. Cryst.*, 1993, **49**, 311-317.

198 A.C. Lim and J.K. Barton, *Biochemistry*, 1993, **32**, 11029-11034.

199 J.G. Collins, T.P. Shields, and J.K. Barton, *J. Am. Chem. Soc.*, 1994, **116**, 9840-9846.

200 C.J. Murphy, M.R. Arkin, Y. Jenkins, N.D. Ghatlia, S.H. Bossmann, N.J. Turro, and J.K. Barton, *Science*, 1993, **262**, 1025-1029.

201 O. Kennard and S.A. Salisbury, *J. Biol. Chem.*, 1993, **268**, 10701-10704.

202 B.L. Destaintot, A. Dautant, C. Courseille, and G. Precigoux, *Eur. J. Biochem.*, 1993, **213**, 673-682.

203 K.E. McAuleyhecht, G.A. Leonard, N.J. Gibson, J.B. Thomson, and W.P. Watson, *Biochemistry*, 1994, **33**, 10266.

204 J.V. Skelly, K.J. Edwards, T.C. Jenkins, and S. Neidle, *Proc. Natl. Acad. Sci. USA*, 1993, **90**, 804-808.

205 G. Laughlin, A.I.H. Murchie, D.G. Norman, M.H. Moore, and P.C.E. Moody, *Science*, 1994, **265**, 520-524.

206 K.M. Morden and K. Maskos, *Biopolymers*, 1993, **33**, 27-36.

207 V.K. Rastogi, K.V.R. Chary, G. Govil, F.B. Howard, and H.T. Miles, *App. Mag. Res.*, 1994, **7**, 1-19.

208 K. Maskos, B.M. Gunn, D.A. Leblanc, and K.M. Morden, *Biochemistry*, 1993, **32**, 3583-3595.

209 F. Aboulela, A.I.H. Murchie, S.W. Homans, and D.M. Lilley, *J. Mol. Biol.*, 1993, **229**, 173-188.

210 V.P. Chuprina, E. Sletten, and O.Y. Fedoroff, *J. Biomol. Structure Dynamics*, 1993, **10**, 693-707.

211 T.N. Jaishree and A.H.J. Wang, *Nucl. Acids Res.*, 1993, **21**, 3839-3844.

212 F.W. Smith and J. Feigon, *Biochemistry*, 1993, **32**, 8682-8692.

213 F.W. Smith, F.W. Lau, and J. Feigon, *Proc. Natl. Acad. Sci. USA*, 1994, **91**, 10546-10550.

214 R.F. Macaya, P. Schultze, F.W. Smith, J.A. Roe, and J. Feigon, *Proc. Natl. Acad. Sci. USA*, 1993, **90**, 3745-3749.

215 J.A. Pikkemaat, H. Vandenelst, J.H. Vanboom, and C. Altona, *Biochemistry*, 1994, **33**, 14896-14907.

216 M.J.J. Blommers, U. Pieles, and A. Demesmaeker, *Nucl. Acids Res.*, 1994, **22**, 4187-4194.

217 J.A. Doudna, C. Grosshans, A. Gooding, and C.E. Kundrot, *Proc. Natl. Acad. Sci. USA*, 1993, **90**, 7829-7833.

218 H.W. Pley, D.S. Lindes, C. Delucaflaherty, and D.B. McKay, *J. Biol. Chem.*, 1993, **268**, 19656-19658.

219 M. Egli, N. Usman, and A. Rich, *Biochemistry*, 1993, **32**, 3221-3237.

220 T.N. Jaishree, G.A. Vandermarel, J.H. Vanboom, and A.H.J. Wang, *Biochemistry*, 1993, **32**, 4903-4911.

221 A.C. Wang, M.A. Kennedy, B.R. Reid, and G.P. Drobny , *J. Magnetic Resonance Series B*, 1994, **105**, 1-10.

222 M.J. Michnicka, J.W. Harper, and G.C. King, *Biochemistry*, 1993, **32**, 395-400.

223 P.W. Davis, W. Thurmes, and I. Tinoco, *Nucl. Acids Res.*, 1993, **21**, 537-545.

224 M. Chastain and I. Tinoco, *Biochemistry*, 1992, **31**, 12733-12741.

225 D.I. Cherny, V.A. Malkov, A.A. Volodin, and M.D. Frankkamenetskii, *J. Mol. Biol.*, 1993, **230**, 379-383.

226 J.M. Benevides, A.H.J. Wang, and G.J. Thomas, *Nucl. Acids Res.*, 1993, **21**, 1433-1438.

227 H. Sfihi, J. Liquier, L. Urpi, N. Verdaguer, J.A. Subirana, J. Igolen, and E. Taillandier, *Biopolymers*, 1993, **33**, 1715-1723.

228 G. Siuzdak, *Proc. Nat. Acad. Sci. USA*, 1994, **91**, 11290-11297.

229 J. Bai, Y.H. Liu, T.C. Cain, and D.M. Lubman, *Anal. Chem.*, 1994, **66**, 3423-3430.

230 M.C. Fitzgerald, G.R. Parr, and L.M. Smith, *Anal. Chem.*, 1993, **65**, 3204-3211.

231 B. Ganem, Y.T. Li, and J.D. Henion, *Tetrahedron Lett.*, 1993, **34**, 1445-1448.

232 K.J. Lightwahl, D.L. Springer, B.E. Winger, C.G. Edmonds, D.G. Camp, B.D. Thrall, and R.D. Smith, *J. Am. Chem. Soc.*, 1993, **115**, 803-804.

233 S.A. McLuckey and S. Habibigoudarzi, *J. Am. Chem. Soc.*, 1993, **115**, 12085-12095.

234 S.C. Pomerantz, J.A. Kowalak, and J.A. McCloskey, *Journal Of The American Society For Mass Spectrometry*, 1993, **4**, 204-209.

235 E.A. Stemmler, R.L. Hettich, G.B. Hurst, and M.V. Buchanan, *Rapid Comm. Mass Spec.*, 1993, **7**, 828-836.

236 K. Tang, S.L. Allman, and C.H. Chen, *Rapid Comm. Mass Spec.*, 1993, **7**, 943-948.

6
Ylides and Related Compounds

BY B. J. WALKER

1 Introduction

A new edition of Johnson's excellent book "Ylides and Imines of Phosphorus" has been published[1] almost thirty years after the original and is indispensable to anyone who studies or uses ylides, phosphorus-stabilised carbanions or imines.

The various phosphorus-based olefination methods continue to be used as widely as ever and further subtleties of stereo-control have been developed. Interest in the Wittig mechanism continues and evidence has been reported supporting the earlier suggestion that there is a contribution from single electron-transfer in certain cases. However, surprisingly few detailed investigations on the mechanisms of other phosphorus-based methods of alkene synthesis have been carried out. Interestingly, β-hydroxyalkylphosphines are reported to undergo *anti*-elimination on treatment with triethylamine and phosphorus trichloride to provide alkenes, probably *via* an *epi*-phosphonium salt intermediate.

2 Methylenephosphoranes

2.1 Preparation and Structure.- Equilibrium acidities (pK_{HA}) for a wide range of triphenylphosphonium cations and oxidation potentials of the corresponding ylides have been measured in DMSO solution.[2] It is estimated that the Ph_3P^+ group contributes about 29 kcalmol^{-1} to the acidity of the α-hydrogens, this is about 15 kcalmol^{-1} greater that the effect of Me_3N^+. It is proposed that the large contribution from Ph_3P^+ is due to a combination of field/inductive and polarisability effects. Contributions from phosphorus d-orbitals towards such effects are now definitely excluded, indeed the involvement of d-orbitals in ylide bonding or in phosphorus bonding in any way finally seems to have been laid to rest.[3]

Unstabilised α-iodoalkyl ylides (**1**) have been prepared and used in Wittig reactions to give 1-iodoalkenes with (*Z*)-stereoselectivity.[4] The P-fluoro-substituted ylides (**3**), containing a chiral alkoxy ligand, are formed as 1:1 mixtures of diastereoisomers from the corresponding alkoxydifluorophosphoranes (**2**).[5] However, (**3**) can be epimerised to predominately (>95% in one case) one diastereomer by treatment with lithium fluoride. The first carbodiphosphoranes (**4**) and (**6**) containing a P-H bond have been reported.[6] Compound (**4**) is thermally stable in the solid state but in solution at room temperature both (**4**) and (**6**) slowly isomerise to the ylides (**5**) and (**7**), respectively. The structures of all these products are supported by n.m.r. data.

Electrophilic substitution of the aryl ring in aryl-stabilised ylides (**8**) has not been observed previously, presumably because reactions with electrophiles generally take place preferentially on the ylidic carbon. It is now reported that the reaction of benzyltriphenylphosphonium salts with phosphorus trihalides leads to a mixture of ylides (**9**) and (**10**).[7] That the latter compound arises by

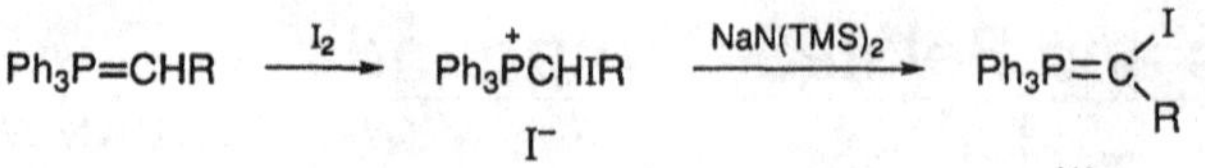

$Ph_3P{=}CHR \xrightarrow{I_2} Ph_3\overset{+}{P}CHIR\ I^- \xrightarrow{NaN(TMS)_2} Ph_3P{=}C(I)R$

(1)

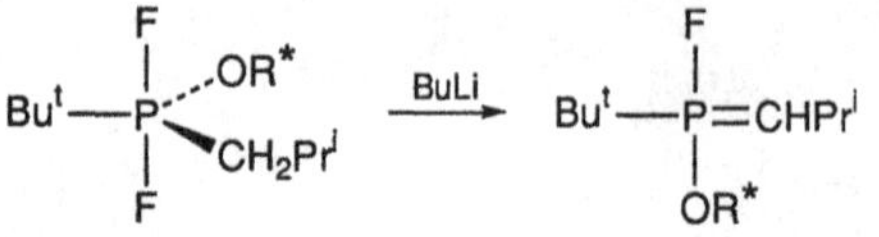

$Bu^tPF_2(OR^*)(CH_2Pr^i) \xrightarrow{BuLi} Bu^tP(F)(OR^*){=}CHPr^i$

(2) (3)

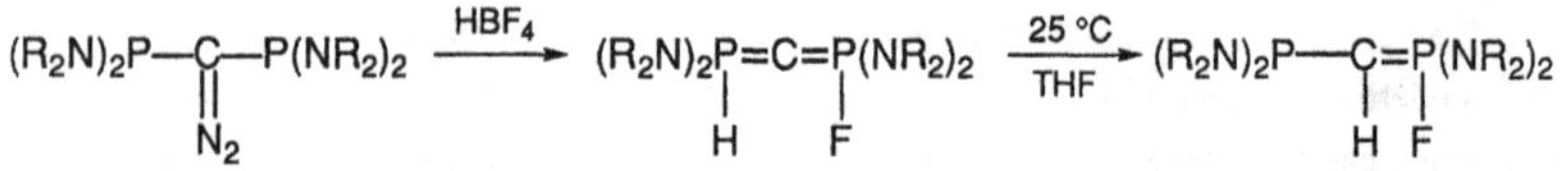

$(R_2N)_2P{-}C(=N_2){-}P(NR_2)_2 \xrightarrow{HBF_4} (R_2N)_2P(H){=}C{=}P(F)(NR_2)_2 \xrightarrow[THF]{25\ ^\circ C} (R_2N)_2P{-}C(H){=}P(F)(NR_2)_2$

(4) (5)

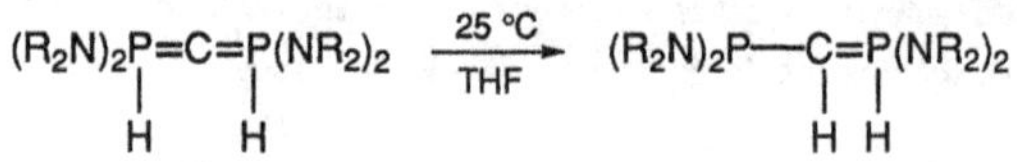

$(R_2N)_2P(H){=}C{=}P(H)(NR_2)_2 \xrightarrow[THF]{25\ ^\circ C} (R_2N)_2P{-}C(H){=}P(H)(NR_2)_2$

(6) (7)

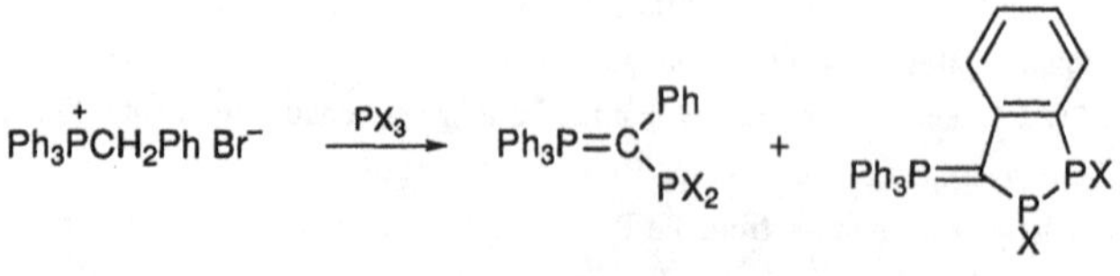

$Ph_3\overset{+}{P}CH_2Ph\ Br^- \xrightarrow{PX_3} Ph_3P{=}C(Ph)PX_2 + Ph_3P{=}$ (benzo-fused ring with PX and P–X)

(8) (9) (10)

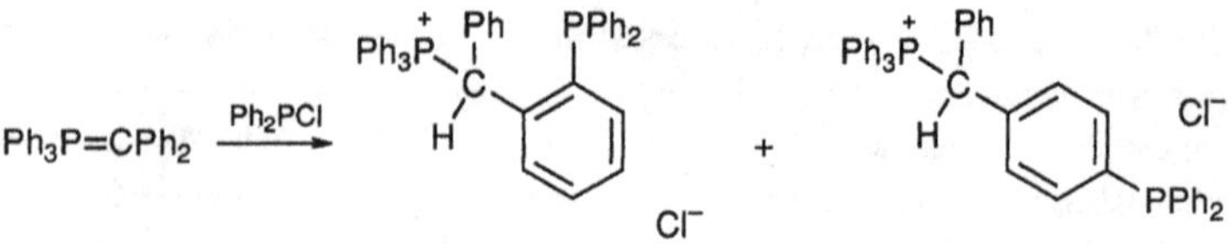

$Ph_3P{=}CPh_2 \xrightarrow{Ph_2PCl} Ph_3\overset{+}{P}{-}CH(Ph){-}C_6H_4{-}2{-}PPh_2\ Cl^- + Ph_3\overset{+}{P}{-}CH(Ph){-}C_6H_4{-}4{-}PPh_2\ Cl^-$

(11) (12)

$(R_2N)_2\ddot{P}{-}\ddot{C}{-}SiMe_3 \xrightarrow{Me_4Ga} (R_2N)_2P(Me){=}C(SiMe_3)(GaMe_3)$

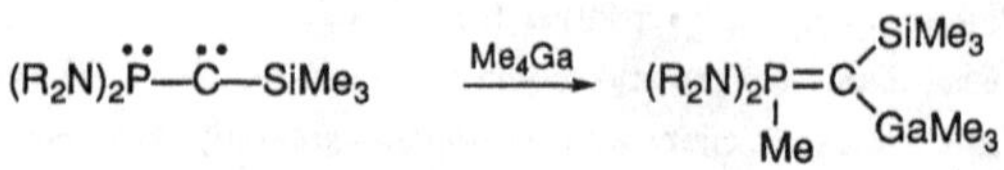

(13)

electrophilic substitution of the phenyl ring is supported by the formation of the aromatic substitution products (**11**) and (**12**) in similar reactions with (diphenylmethylene)triphenylphosphorane.

The first examples of gallium-substituted phosphonium ylides, e.g. (**13**), have been prepared by the reaction of phosphanyl carbenes with trimethyl gallium[8] and a further example (**14**) of a stable dithioxophosphorane has been reported.[9] Bestmann's group has synthesised the next members, C_3 (**15**) and C_4 (**16**), in his homologous carbon chain series and shown that for stability, as predicted, (**15**) requires two acceptor substituents while (**16**) requires a donor substituent at one end and an acceptor group at the other.[10]

The previously reported triphenylphosphine-catalysed isomerisation of ynoates (**17**) to dienoates has been adapted to provide a carbon-carbon bond-forming reaction which incorporates a reversal of reaction ("umpolung") at the γ-carbon of the yne structure to give, for example, (**18**).[11] The reaction is thought to proceed via ylide intermediates to give a vinylphosphonium salt which undergoes nucleophilic attack to provide the new C-C bond.

Phosphoranylideneaminoquinones, e.g. (**20**), have been prepared in good to excellent yields by the reaction of phosphines with [2,1]benzisoxazole-4,7-quinones, e.g. (**19**).[12] A safer alternative to the vinyl azide-based synthesis of (N-vinylimino)phosphoranes (**21**) has been reported (Scheme **1**).[13] N-(Alkylthiomethyl)iminophosphoranes (**23**) have been synthesised from the benzotriazole-substituted iminophosphorane (**22**) and used without isolation to provide routes to N-alkylthiomethyl-imines, -amides, and -ureas.[14] Tetraazolyl-substituted ylides (**25**) and (**26**) are the products of the reaction of vinylphosphonium salts (**24**) with excess sodium azide.[15] The novel 1,2-l^5-azaphosphete (**28**), which is potentially antiaromatic, has been prepared by the reaction of the azide (**27**) with dimethyl acetylenedicarboxylate, followed by heating the resulting adduct.[16] Compound (**28**) is stable in air and spectroscopic and X-ray structural data suggest that it is best represented by a non-antiaromatic zwitterionic structure.

An *ab initio* study of structure and bonding in diphosphinylcarbenes indicates that the non-symmetric ylide-type structure (**29**) is preferred.[17] The configuration and conformation of a number of polyfunctional 1-aminobuta-1,3-dienes (**30**), prepared by the reaction of dimethyl acetylenedicarboxylate with (*Z*)-β-enamino-l^5-phosphazenes, have been studied by n.m.r., X-ray, and theoretical methods.[18] The ^{31}P n.m.r. spectra of a number of protonated iminophosphorane-substituted proton sponges, e.g. (**31**), of known crystal structure have been studied in the solid state.[19]

2.2 Reactions of Methylenephosphoranes.-

2.2.1 Aldehydes.- Interest in mechanistic aspects of the Wittig reaction continues. Although it is not a new concept, substantial evidence has now been provided that the formation of oxaphosphetane intermediates in olefination reactions with non-stabilised, but not semi-stabilised, ylides with benzaldehyde and benzophenone involve contributions from an electron-transfer-radical coupling sequence (Scheme **2**) and is not exclusively via a polar addition mechanism.[20] A brief discussion of the implications of an SET mechanism for explanations of Wittig stereochemistry is also included. However, even the authors admit that this work offers only one of a number of possible answers to Wittig mechanistic questions. The effects of metal, particularly lithium, salts on Wittig reactions involving non-stabilised and semi-stabilised ylides have been extensively studied. Lithium salts are

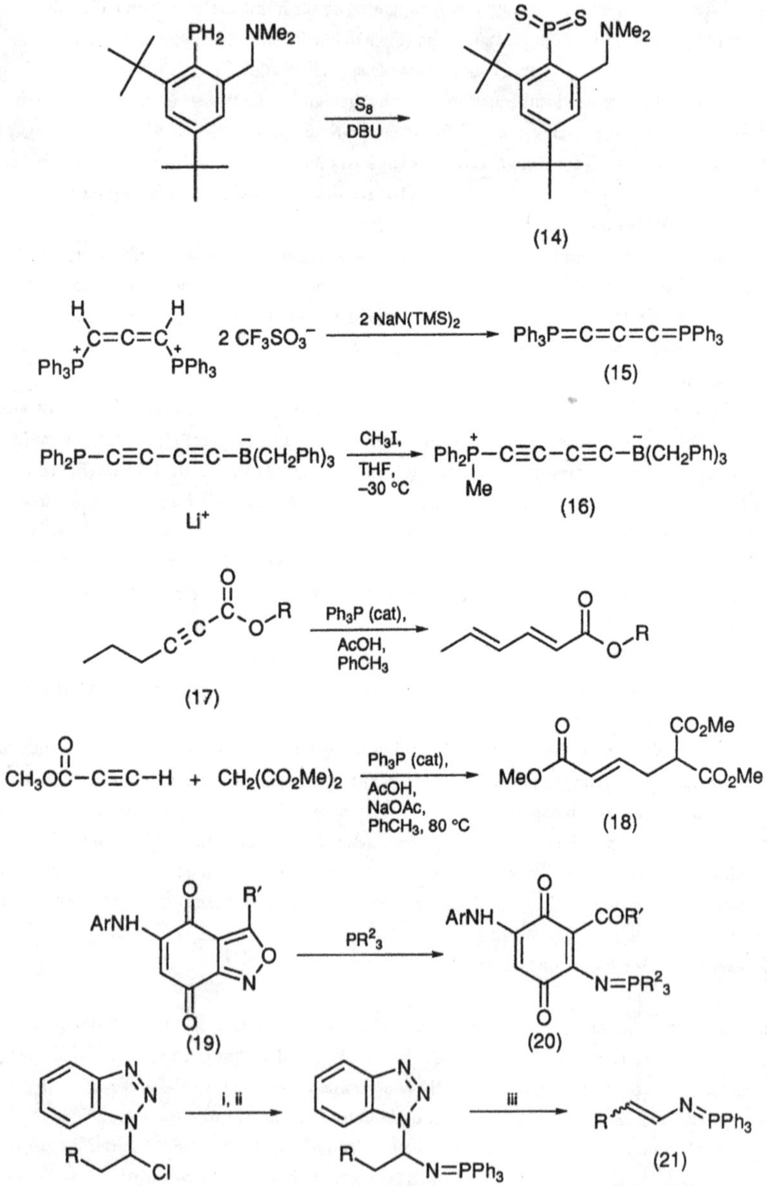

PH_2 NMe_2

S_8, DBU

S=P=S NMe_2

(14)

H H C=C=C $Ph_3\overset{+}{P}$ $\overset{+}{P}Ph_3$ 2 $CF_3SO_3^-$

2 NaN(TMS)$_2$

$Ph_3P{=}C{=}C{=}C{=}PPh_3$

(15)

$Ph_2P{-}C{\equiv}C{-}C{\equiv}C{-}\bar{B}(CH_2Ph)_3$ Li^+

CH_3I, THF, −30 °C

$Ph_2\overset{+}{P}(Me){-}C{\equiv}C{-}C{\equiv}C{-}\bar{B}(CH_2Ph)_3$

(16)

O C≡C–C–O–R

(17)

Ph_3P (cat), AcOH, $PhCH_3$

O O–R

$CH_3O\overset{O}{C}{-}C{\equiv}C{-}H$ + $CH_2(CO_2Me)_2$

Ph_3P (cat), AcOH, NaOAc, $PhCH_3$, 80 °C

O MeO CO_2Me CO_2Me

(18)

O R′ ArNH O N O

(19)

PR^2_3

O ArNH COR′ $N{=}PR^2_3$ O

(20)

N N N R Cl

i, ii

N N N R $N{=}PPh_3$

iii

R N PPh_3

(21)

Reagents: i, NaN_3, DMSO; ii, Ph_3P; iii, NaH, THF

Scheme 1

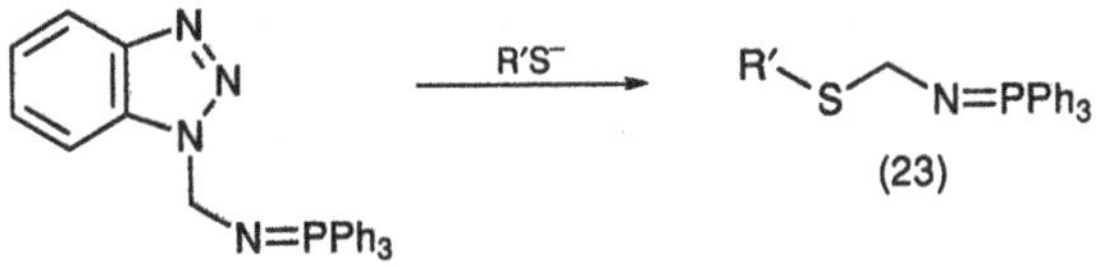

R'S⁻

(22)

(23)

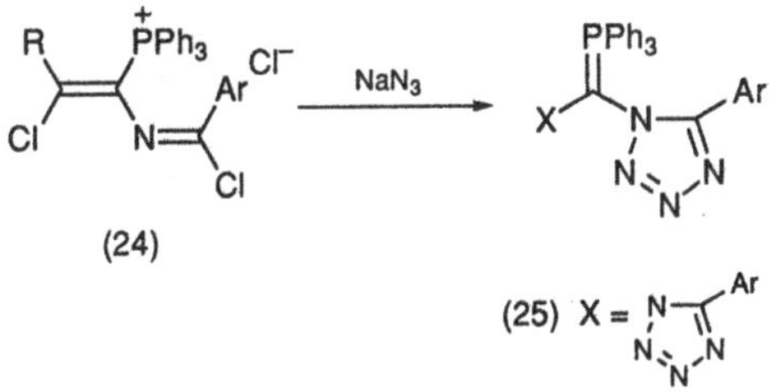

NaN_3

(24)

(25) X =

(26) X = CN

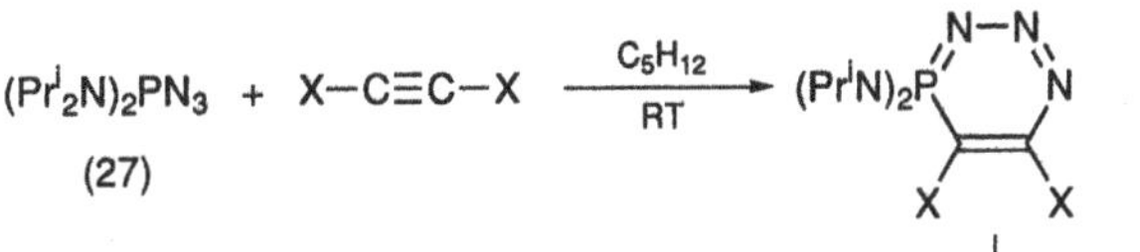

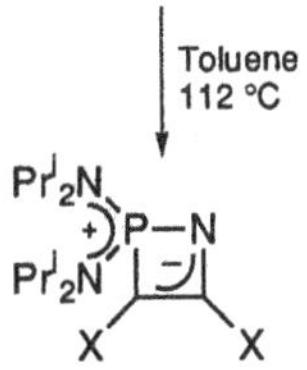

$(Pr^i_2N)_2PN_3$ + X−C≡C−X $\xrightarrow[RT]{C_5H_{12}}$

(27)

Toluene 112 °C

(29)

(28) X = CO_2Me

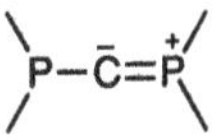

X−C≡C−X, CH_2Cl_2

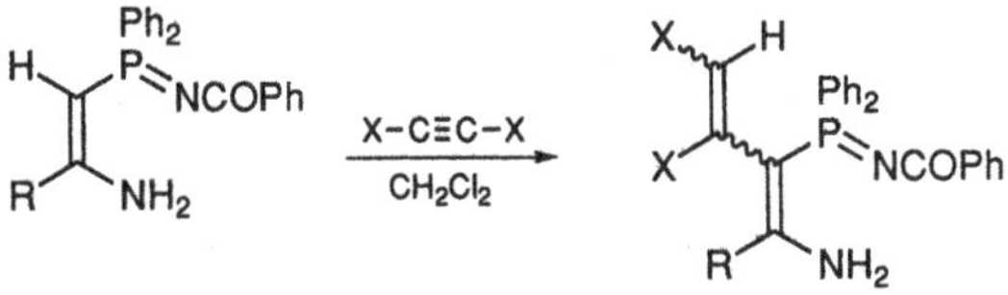

(30) X = CO_2Me

PF_6^-

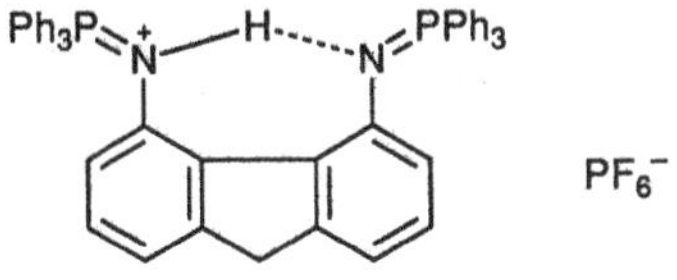

(31)

now shown to catalyse reactions of the stabilised ylide (**32**) with benzaldehyde and ketones, although in the latter case a large quantity of lithium cation is required.[21] Vedejs has now shown that, as with analogous reactions of aldehydes, high levels of (*E*)-stereoselectivity can be obtained from olefination reactions of ketones with ylides (**33**) and (**34**) in the absence of lithium salts.[22] Similar reactions with ethylidenetriphenylphosphorane show no simple stereochemical pattern, unlike those with aldehydes which have been previously reported to give predominately (*Z*)-alkenes. The pentacoordinate oxaphosphetane (**35**) has been synthesised and its structure determined by X-ray crystallography.[23] Thermolysis of (**35**) leads to the expected olefination reaction followed by isomerisation of the tetracoordinate 1,2-oxaphosphetane by-product (**36**) to give (**37**). Interestingly β-hydroxyalkyl-phosphines (**38**) are reported to undergo *anti*-elimination on treatment with triethylamine and phosphorus trichloride to provide alkenes.[24] A mechanism involving intermediate formation of the *epi*-phosphonium salt (**39**) is suggested to explain the results, although no phosphorus-containing products are isolated.

The 2H-pyran-substituted ylide (**40**) is presumably the species generated by the reaction of pyrylium tetrafluoroborate with methylene ylide followed by treatment with butyllithium.[25] The addition of aldehydes to (**40**) gave moderate to good yields of (*E,E,E*)-trienals (**41**) and thus offers a useful six-carbon homologation of aldehydes (Scheme **3**). Unsaturated aldehydes, e.g. (**42**), generated in acetone by the oxidation of furans with dimethyldioxirane, have been trapped by Wittig reactions with a variety of stabilised phosphorus ylides to provide routes to a range of dienes and trienes in moderate to excellent yield.[26] Allyl silanes (**43**) have been prepared with e.e. values of 90 to 98% and >90% (*E*)-stereochemistry by Wittig reactions of optically active α-silylaldehydes.[27] Both α-bromo-[28] and α-fluoro-[29] acrylic esters have been synthesised by Wittig methods. The former compounds (**45**) were obtained with good (*Z*)-selectivity by Wittig reactions of α-bromophosphonium ylides (**44**) while similar reactions of the corresponding α-fluorophosphonium ylide (**46**) gave equal amounts of (*E*)- and (*Z*)-isomers. However, reaction with the carbanion derived from the α-fluorophosphonate (**47**) gave only the (*E*)-isomer.

A Wittig reaction of the THF-swollen polystyrene resin (**48**) with the β-ketomethylene ylide (**49**) has been used to functionalise the resin for further transformations.[30] The aza-Wittig reaction of N-Boc-triphenyliminophosphorane (**50**) with 4-cyanobenzaldehyde followed by oxone oxidation provides the stable oxaziridine (**51**) (Scheme **4**) which acts as an electrophilic aminating agent towards amines and enolates.[31]

2.2.2 Ketones.- The reactions of alkoxycarbonylmethylenetriphenylphosphorane with 10-(methoxyimino)phenanthren-9-one[32] and 4-triphenylmethyl-1,2-benzoquinone[33] have been reported. The former reaction gives isomeric alkyl 10-(methoxyimino)phenanthren-9-ylidene acetates (**52**), the simple Wittig products, while the latter gives isolable zwitterions of structure (**53**). Heating (**53**) above 60°C leads to the elimination of triphenylphosphine and the formation of 2H-1-benzopyran-2-one and benzo[b]furan-2(3H)-one rings. Under phase transfer conditions benzylidene, and keto- and ester-stabilised ylides react with 7-(methoxyimino)-4-methyl-2H-chromene-2,8(7H)-dione (**54**) to give benzopyranooxazol-8-ones (**55**) or pyridones (**56**) depending on the ylide used.[34]

Allylidenetriphenylphosphoranes undergo γ-alkylation followed by intramolecular Wittig reaction with α-bromoketones to provide routes to cyclopentadienes,[35, 36] e.g. (**57**).[35] 2-Ethoxycyclo-

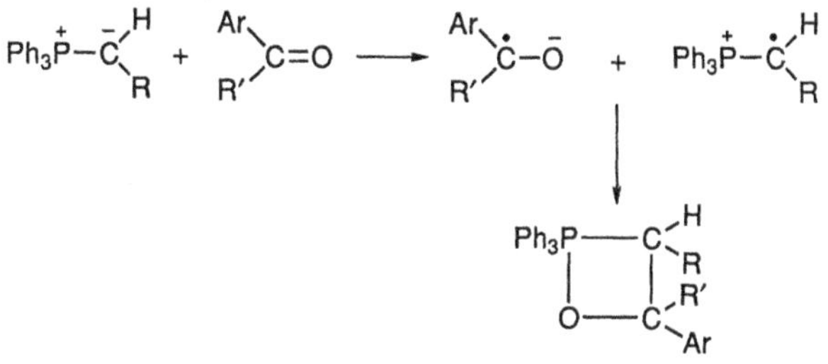

Scheme 2

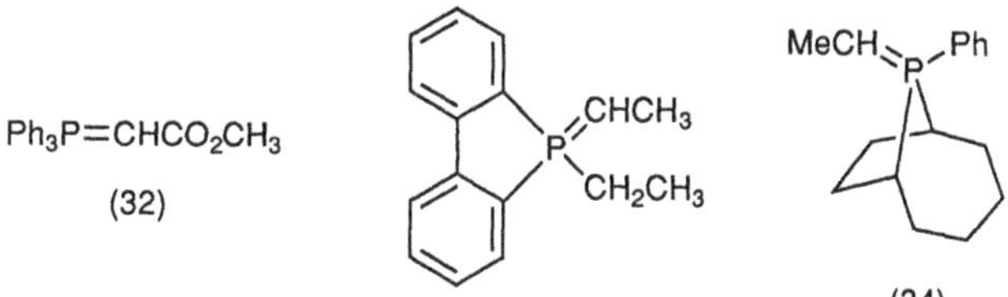

$Ph_3P{=}CHCO_2CH_3$

(32)

(33)

(34)

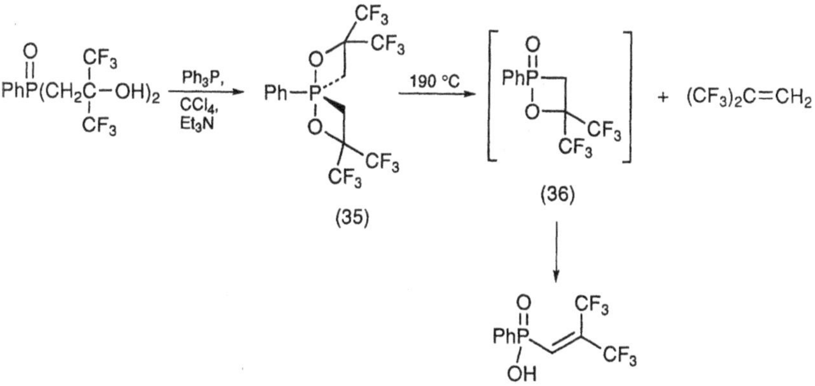

(35)

(36)

(37)

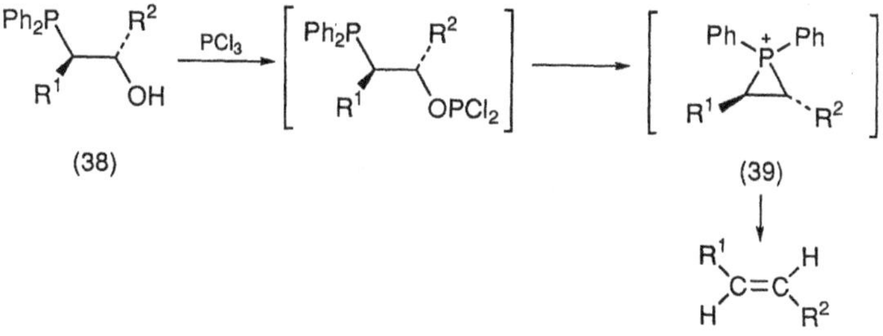

(38)

(39)

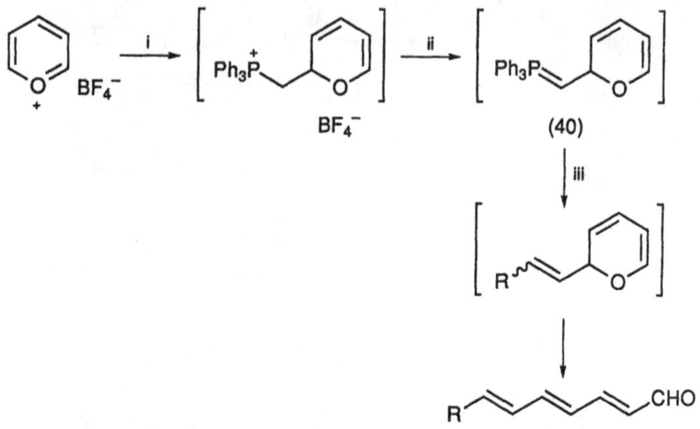

O^+ BF_4^- —i→ [$Ph_3\overset{+}{P}$ O] BF_4^- —ii→ [Ph_3P O] (40) —iii→ [R O] → R CHO (41)

Reagents: i, $Ph_3P{=}CH_2$, THF, –60 °C; ii, BuLi, TMEDA; iii, RCHO

Scheme 3

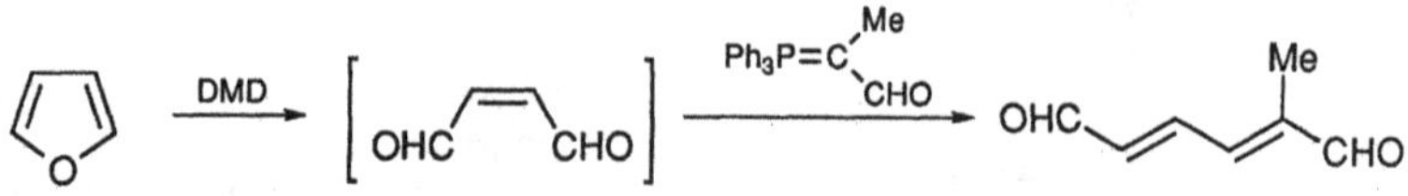

O —DMD→ [OHC CHO] (42) —$Ph_3P{=}C(Me)CHO$→ OHC Me CHO

R_3Si * CHO R^1 + $R^2CH{=}PPh_3$ → R_3Si * CH=CH R^2 R^1 (43)

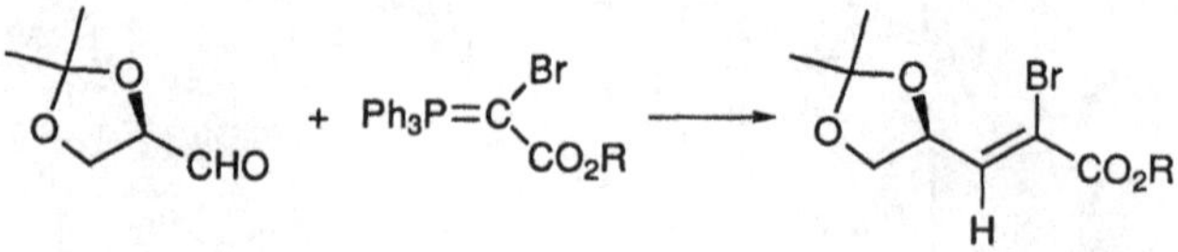

O O CHO + $Ph_3P{=}C(Br)CO_2R$ (44) → O O Br CO_2R H (45)

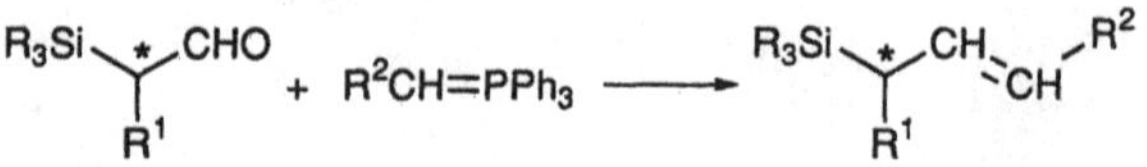

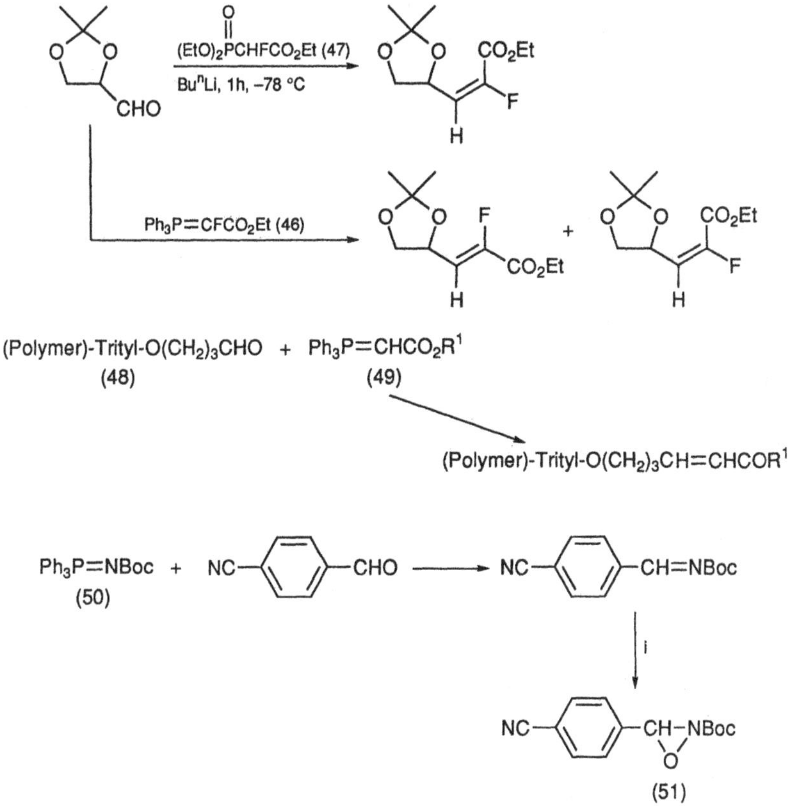

$(EtO)_2P(O)CHFCO_2Et$ (47)

Bu^nLi, 1h, –78 °C

CHO

CO_2Et

F

H

$Ph_3P{=}CFCO_2Et$ (46)

F

CO_2Et

H

+

CO_2Et

F

H

(Polymer)-Trityl-$O(CH_2)_3CHO$ + $Ph_3P{=}CHCO_2R^1$

(48) (49)

(Polymer)-Trityl-$O(CH_2)_3CH{=}CHCOR^1$

$Ph_3P{=}NBoc$ + NC–C_6H_4–CHO → NC–C_6H_4–CH=NBoc

(50)

i

NC–C_6H_4–CH–NBoc (oxaziridine, O)

(51)

Reagents: i, oxone, K_2CO_3, H_2O, $CHCl_3$, 0–4 °C

Scheme 4

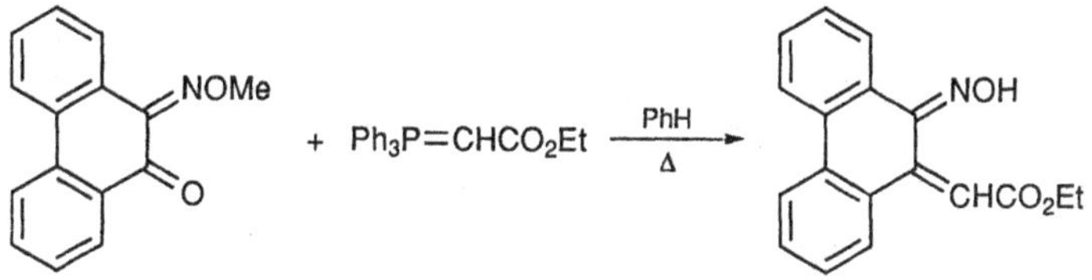

NOMe

O

+ $Ph_3P{=}CHCO_2Et$ —PhH, Δ→

NOH

$CHCO_2Et$

(52)

pentadienes (**59**) have been obtained by such a reaction of the 2-ethoxyallylic ylide (**58**) and, following acid treatment, converted into cyclopentenones (**60**) (Scheme **5**).[36] This sequence of reactions has been used in a synthesis of shogaol (**61**), the pungent principle of ginger. *Trans*-Hydroazulene derivatives (**63**) have been synthesised stereoselectively by a tandem Michael-intramolecular Wittig reaction of the cyclic ylide (**62**).[37]

Di-, tri-, and tetra-substituted 2-azadienes (**64**) carrying electron-withdrawing groups have been synthesised by the aza-Wittig reaction.[38] A one-pot synthesis of olefins from diorganyltelluride, diazo compounds, and carbonyl compounds in the presence of a Cu(I) catalyst is suggested to involve a Wittig-type reaction of an intermediate telluronium ylide.[39]

2.2.3 Miscellaneous Reactions.- New chemistry of ketenylidenetriphenylphosphorane (**65**) and its dimer and trimer has been reported. New structures, e.g. (**66**), thought to contain tin-carbon multiple bonds, have been prepared by the reaction of (**65**) with dialkyltin compounds.[40] The trimer (**68**) of (**65**) has been obtained by the reaction of the dimer (**67**) with a large excess of (**65**).[41] Compound (**68**) slowly undergoes a single Wittig reaction with reactive aldehydes and, on oxidation with oxaziridines, produces the quinone-type structure (**69**). Similar direct oxidation of the dimer (**67**) leads to the formation of various novel ylide structures, e.g. (**70**) and (**71**).[42] The reactions of stabilised ylides with carbon suboxide have been further investigated and shown to give either linear (**72**) or cyclic (**73**) adducts depending on the structure of the ylide used.[43] A variety of novel benzoannulated P-C four-membered ring ylides, e.g. (**76**), and other structures, e.g. (**75**), have been obtained from the halogenated ylides (**74**) (Scheme **6**).[44]

Enantiomerically pure (*R*)- and (*S*)-(methylenecyclopropyl)carbinol have been synthesised in low yield by a one-pot reaction of methylenetriphenylphosphorane with (*R*)- and (*S*)-epichlorohydrin, respectively, followed by the addition of paraformaldehyde (Scheme 7).[45] The yields were substantially improved in a two step reaction involving the isolation and further reaction of the (3,4-epoxybutyl)triphenylphosphonium salts (**77**).

Shen's group have continued their studies of the synthesis of perfluoroalkylated α,β-unsaturated esters and nitriles via the reaction of carbanions with perfluoroacylated ylides, e.g. (**78**). Further refinements of their previously published methods now allow the selective synthesis of (*Z*)- and (*E*)-β-alkyl-β-trifluoromethylacrylates from the ylide (**78**) (Scheme **8**)[46] and (*Z*)- and (*E*)-β-aryl-β-trifluoromethyl-acrylates and -acrylonitriles from the ylide (**79**) (Scheme **9**).[47] Wittig reactions of both stabilised and non-stabilised ylides with *S*-alkyl trifluorothioacetate (**80**) have been used to provide a general synthesis of trifluoromethyl vinyl sulfides (**81**) in moderate to excellent yield and predominantly as (*Z*)-isomers.[48] α-(Fluoroalkyl)enol ethers (**83**) have been prepared by (*Z*)-selective Wittig reactions of the corresponding perfluoroalkanoates (**82**).[49] Compounds (**83**) undergo Claisen rearrangement at 100 °C to give γ,δ-unsaturated perfluoroalkyl ketones. Fluorinated amides (**84**) initially undergo Wittig reactions with ethoxycarbonylmethylene ylide.[50] However, ultimately a variety of products are formed depending on the nature of (**84**).

There have been a number of further reports of flash vacuum pyrolysis (fvp) of ylides. This technique applied to β-oxoalkylidenetriphenylphosphorane (**85**) provides a route to alkynes which do not have electron-withdrawing substituents and which are not available by the simple pyrolysis of (**85**).[51] The procedure has been extended to 1,2,4-trioxo-3-triphenylphosphoranylidenebutanes (**86**) to

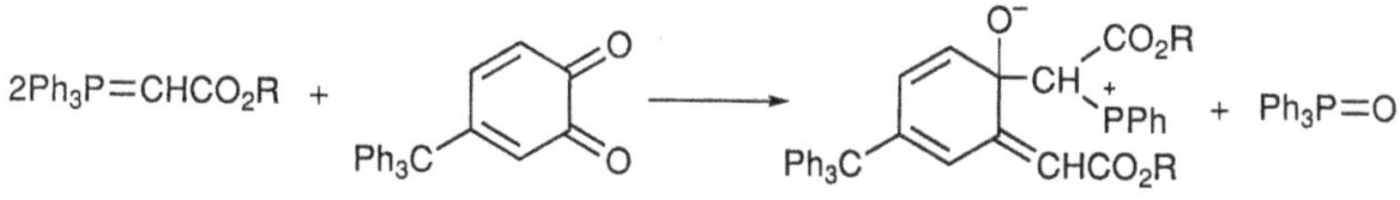

$2Ph_3P{=}CHCO_2R$ + (53) + $Ph_3P{=}O$

(53)

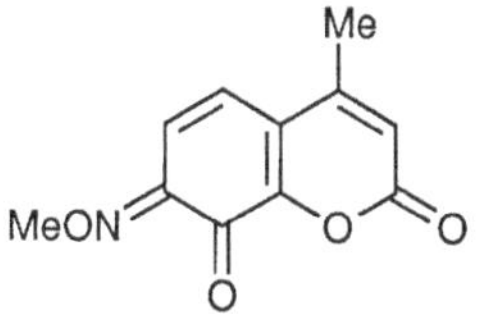

(54) 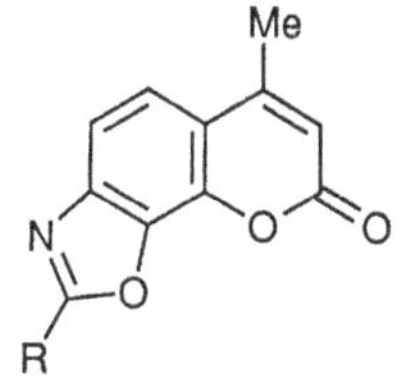 (55) 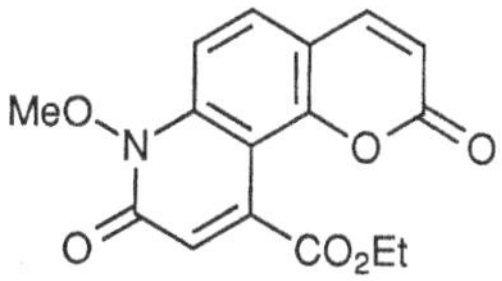 (56)

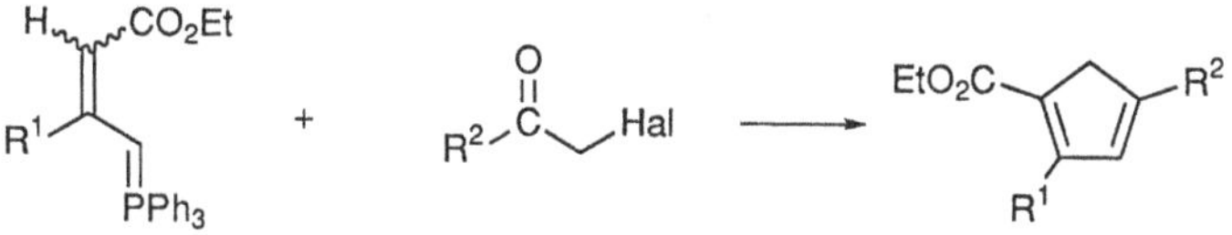

(57)

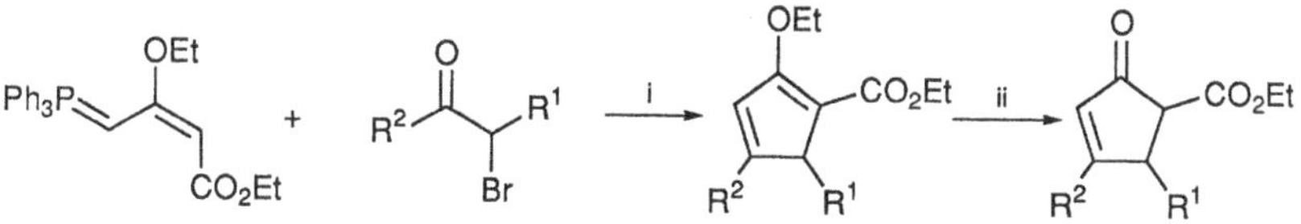

(58) (59) (60)

Reagents: i, Cs_2CO_3, CH_2Cl_2; ii, HCl, H_2O, $CHCl_3$

Scheme 5

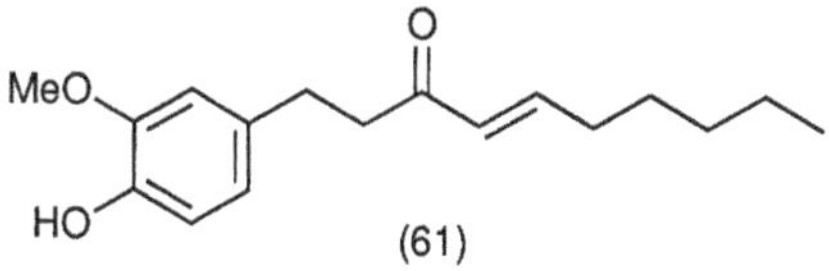

(61)

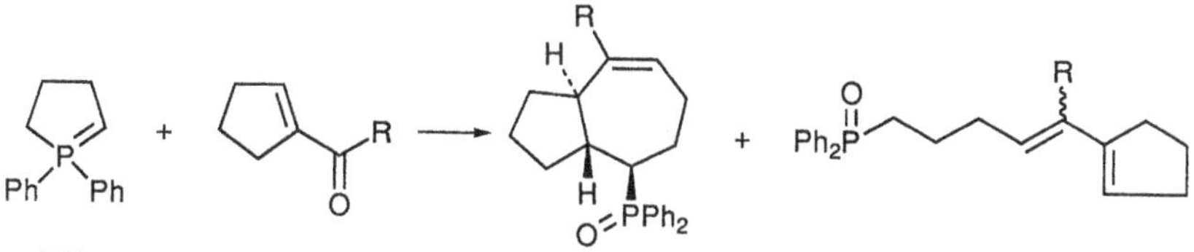

(62) (63)

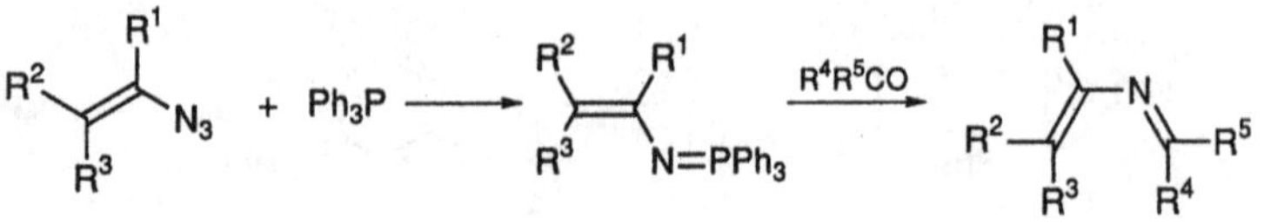

R^4R^5CO

(64)

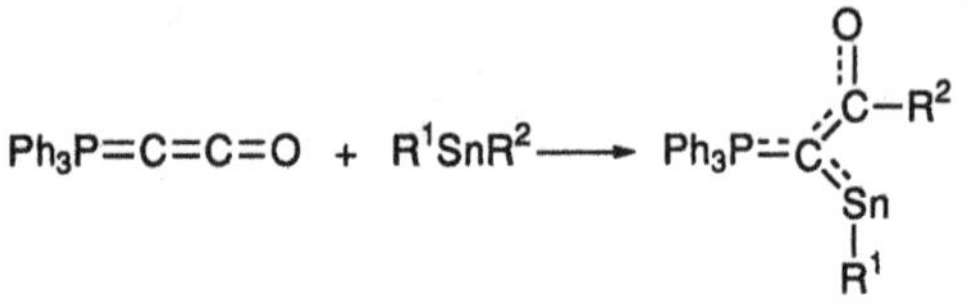

$Ph_3P{=}C{=}C{=}O$ + R^1SnR^2

(65)

(66)

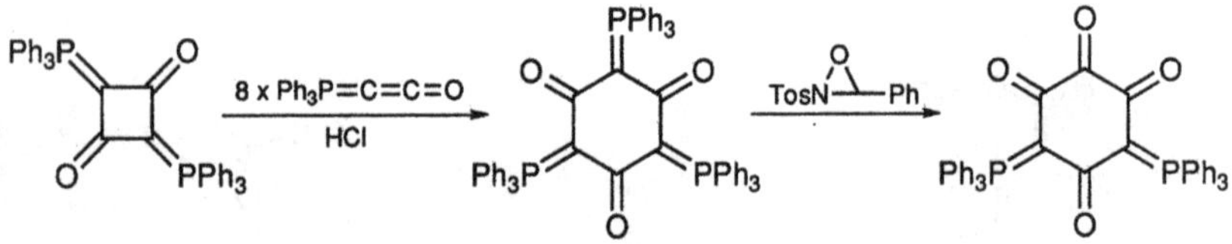

8 x $Ph_3P{=}C{=}C{=}O$

HCl

(67)

(68)

(69)

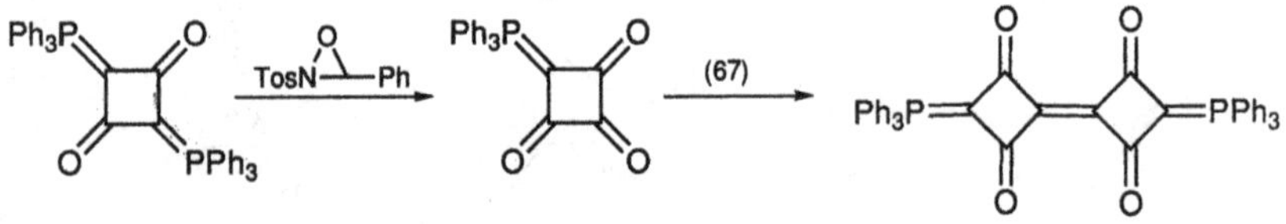

(67)

(70)

(71)

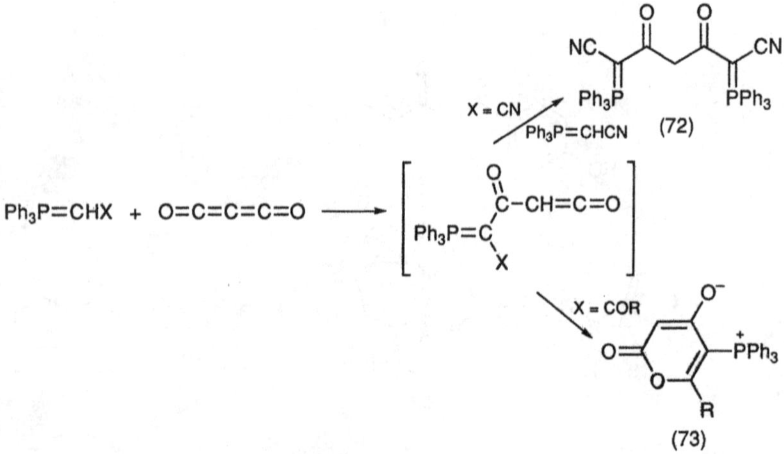

X = CN

$Ph_3P{=}CHCN$

(72)

$Ph_3P{=}CHX$ + $O{=}C{=}C{=}C{=}O$

X = COR

(73)

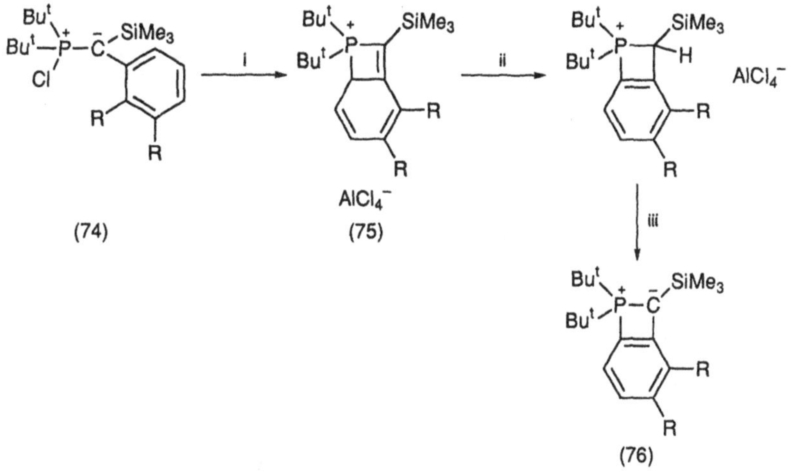

(74) (75) (76)

Reagents: i, $AlCl_3$; ii, pyridine; iii, $NaN(TMS)_2$

Scheme 6

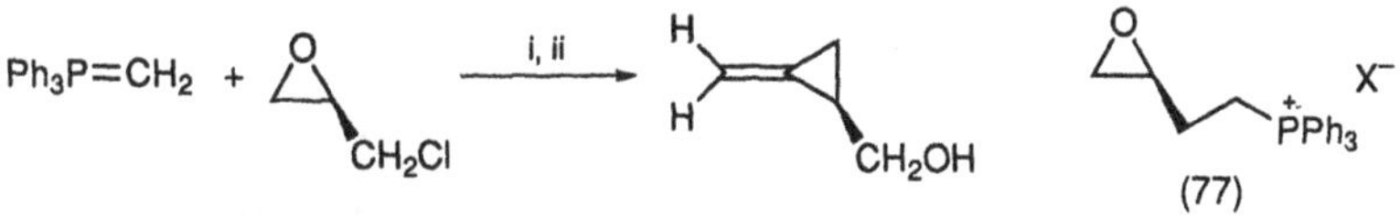

(77)

Reagents: i, Bu^nLi; ii, $(CH_2O)_n$

Scheme 7

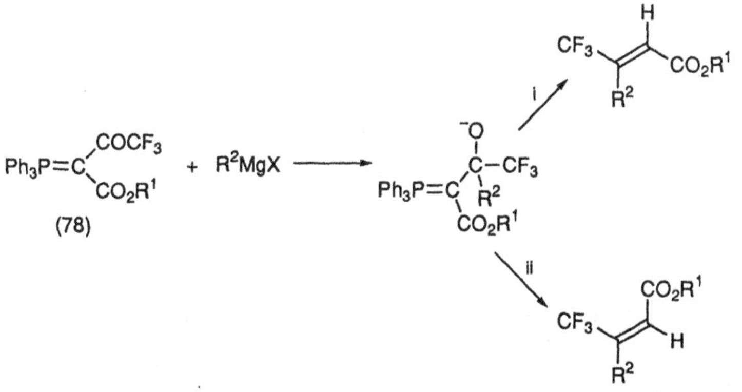

(78)

Reagents: i, 5% HCl, H_2O; ii, CH_3CO_2H

Scheme 8

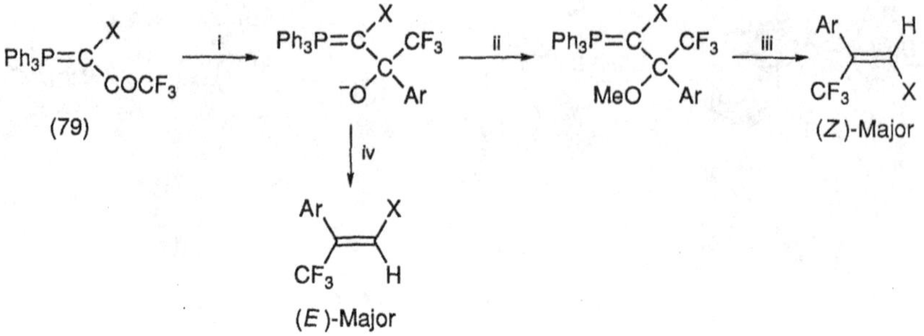

$Ph_3P{=}C(X)COCF_3$ (79) →(i) $Ph_3P{=}C(X)C(CF_3)(O^-)Ar$ →(ii) $Ph_3P{=}C(X)C(CF_3)(OMe)Ar$ →(iii) (*Z*)-Major

→(iv) (*E*)-Major

Reagents: i, ArLi; ii, MeI; iii, AcOH, PhH; iv, AcOH

Scheme 9

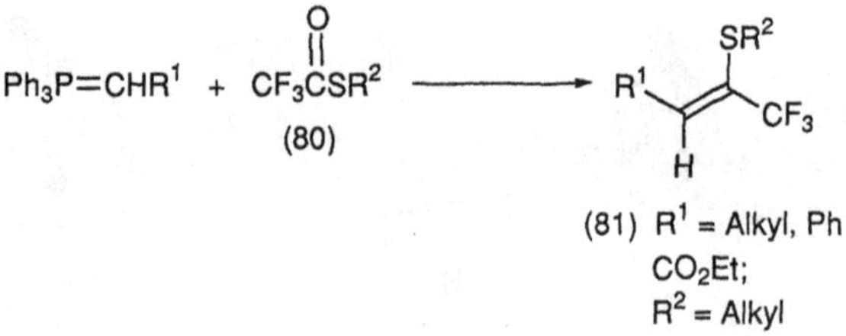

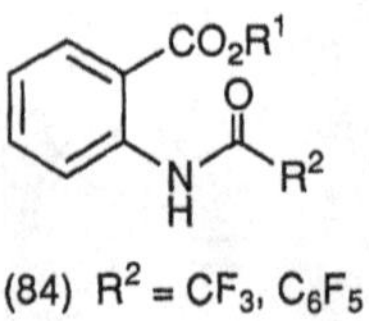

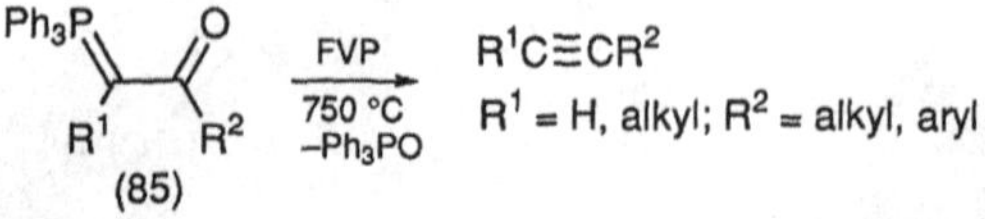

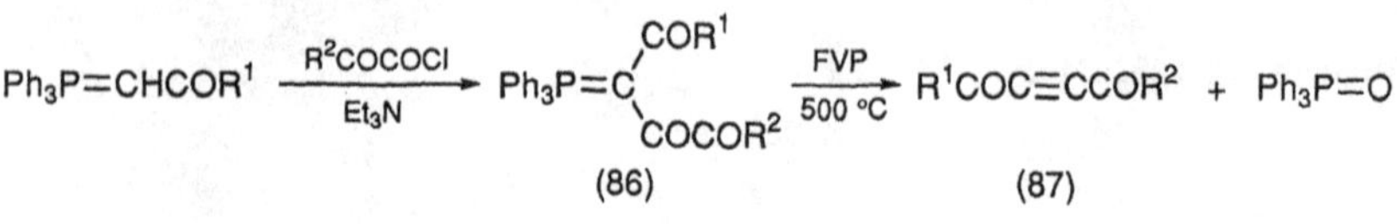

$Ph_3P{=}CHR^1$ + $CF_3C(O)SR^2$ (80) → (81) R^1 = Alkyl, Ph, CO_2Et; R^2 = Alkyl

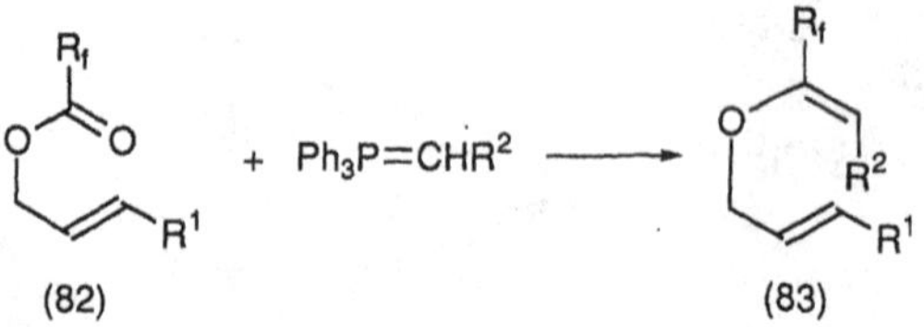

(82) + $Ph_3P{=}CHR^2$ → (83)

(84) $R^2 = CF_3, C_6F_5$

(85) →(FVP, 750 °C, $-Ph_3PO$) $R^1C{\equiv}CR^2$ R^1 = H, alkyl; R^2 = alkyl, aryl

$Ph_3P{=}CHCOR^1$ →($R^2COCOCl$, Et_3N) $Ph_3P{=}C(COR^1)COCOR^2$ (86) →(FVP, 500 °C) $R^1COC{\equiv}CCOR^2$ (87) + $Ph_3P{=}O$

provide a route to diacylalkynes (**87**) in variable yield.[52] Unlike the previous examples, fvp of (α-alkanesulfinyl)phosphonium ylides (**88**, R=alkyl) leads to the extrusion of triphenylphosphine, rather than triphenylphosphine oxide, to give thioesters (**89**, R=alkyl).[53] However, similar treatment of the corresponding (α-arylsulfinyl)phosphonium ylides (**88**, R=aryl) gives lower yields of the thioesters (**89**, R=aryl) together with a mixture of products derived from elimination of triphenylphosphine oxide. Similarly α-sulfinyl ylides (**90**) stabilised with alkoxycarbonyl groups generally give products derived from the elimination of triphenylphosphine oxide, and presumably the carbene (**91**), although again this depends to some extent on the nature of the sulfinyl substituents.[54]

Vinyltin derivatives (**92**) have been prepared by the reaction of acyltin compounds with phosphonium ylides or with phosphonate carbanions.[55] The tungsten complex (**93**) reacts with a wide range of aryl-substituted ylides (**94**), and the vinyl-substituted ylide (**95**), to give phosphorus to metal alkylidene transfer and hence provides a route to tungsten alkylidene complexes (**96**).[56]

The reaction of ester-, keto-, and nitrile-stabilised ylides with tetramethylthiuram disulfide (**97**) has been investigated and shown to give a variety of products including the ylides (**98**).[57] Ylide substituents have been used to stabilise the thioxophosphine (-P=S) function.[58] Such structures (**99**) are the first of these compounds to be stable at room temperature. In one case the structure has been confirmed by X-ray crystallography and a substantial contribution from the phosphalkene canonical form is indicated by ^{31}P n.m.r. spectra. Reaction of (**99**) with sulfur provides stable dithioxophosphoranes (**100**). 2,4-Di-*tert*-butyl-6-methoxyphenyldithioxophosphorane (**101**), in this case stabilised by steric effects, has been used as a probe for mechanistic studies of reactions involving Lawesson's reagent.[59] Evidence for the first generation of an iminophosphoranylidene carbenoid (**102**) has been obtained by trapping with phosphines to give novel ylide structures, e.g. (**103**) and (**104**) (Scheme **10**).[60]

3 The Structure and Reactions of Phosphine Oxide Anions

The carbanions of saturated and unsaturated β- and γ-(e.g. **105**)-hydroxyalkyldiphenyl-phosphine oxides give adducts (e.g. **106**) which, following protection of the primary alcohol function with trityl chloride, undergo base-induced decomposition to the expected alkene (Scheme **11**).[61] The diastereomeric adducts (**107**) and (**108**), produced from the cycloaddition of nitrile oxides to 1-alkylallylphosphine oxides, have been separated and reduced to the corresponding 4-amino-2-hydroxyalkylphosphine oxides which undergo base-induced decomposition to give the pure (*E*)-(**109**) and pure (*Z*)-(**110**)-3-amino alkenes.[62] The same method has been applied to the synthesis of (*E*)-enol ethers (**112**) of protected 4-amino aldehydes by isolation and reduction of the predominant *anti*-isomers of (**111**) derived from 1-alkoxyallylphosphine oxides (Scheme **12**).[63] An efficient route to 1-alkyl-3-phenyl-1,3-benzazaphospholine oxides (**115**) is available *via* the dilithiated species (**114**) generated from (**113**) by sequential treatment with methyllithium and *tertiary*butyllithium (Scheme **13**).[64] Attempts to cyclise (**113**) using one equivalent of base led to lower yields and mixtures of products.

4 The Structure and Reactions of Phosphonate Anions

The structure of the phosphorus-stabilised carbanions (**116**) and (**117**) has been investigated in detail at the HF/3-21G$^{(*)}$ level and a single-crystal X-ray structure of the carbanion (**118**) has been

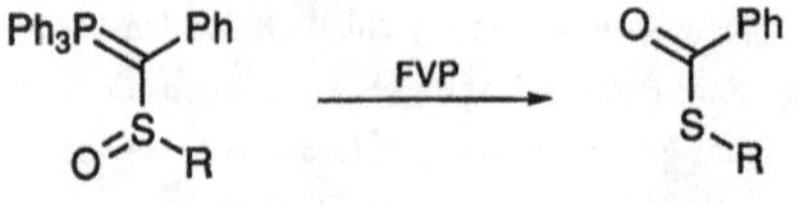

Ph_3P=C(Ph)S(=O)R (88) —FVP→ O=C(Ph)SR (89)

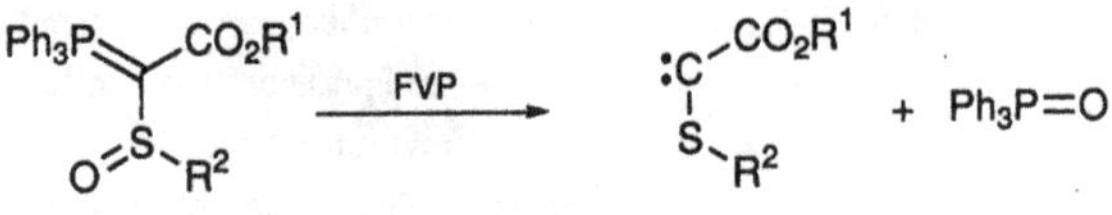

Ph_3P=C(CO_2R^1)S(=O)R^2 (90) —FVP→ :C(CO_2R^1)SR^2 (91) + Ph_3P=O

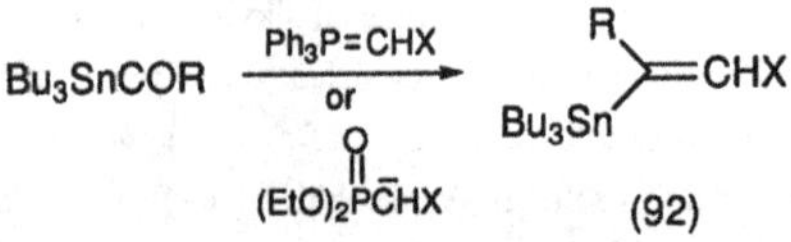

Bu_3SnCOR —Ph_3P=CHX or $(EtO)_2P(O)\bar{C}HX$→ Bu_3Sn(R)C=CHX (92)

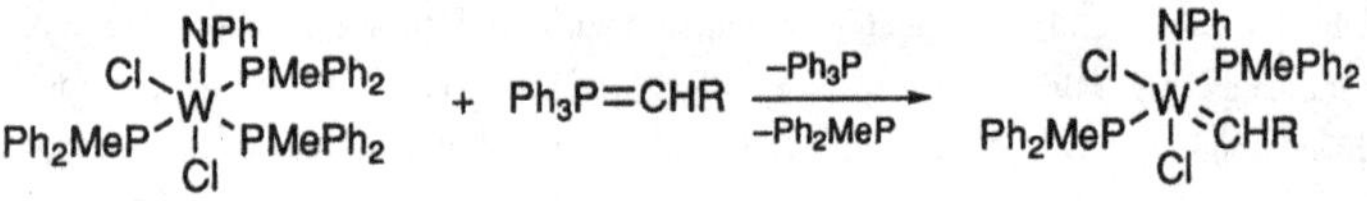

(93) + Ph_3P=CHR —$-Ph_3P$, $-Ph_2MeP$→ (96)

(94) R = Ar

(95) R = CH=CMe_2

Ph_3P=CHX + Me_2N–C(=S)–S–S–C(=S)–NMe_2 (97) —X = CO_2R→ Ph_3P=C(R)–S–C(=S)NMe_2 (98)

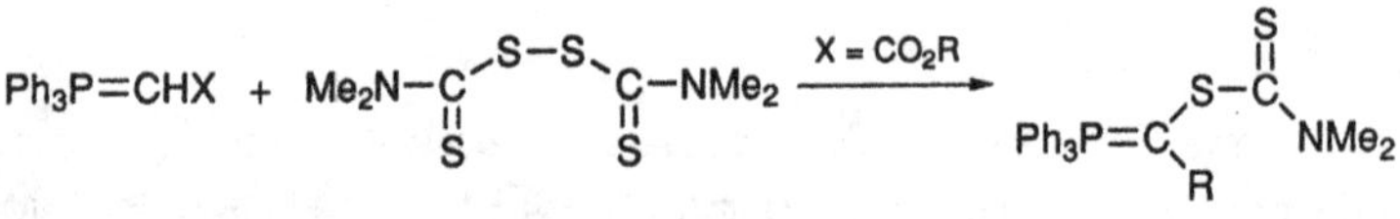

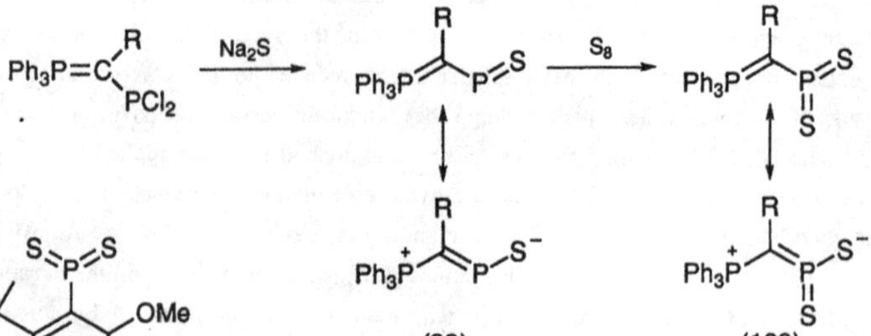

Ph_3P=C(R)PCl_2 —Na_2S→ (99) —S_8→ (100)

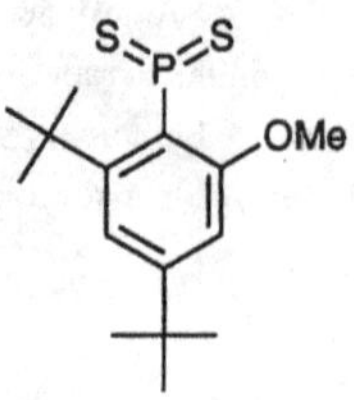

(101)

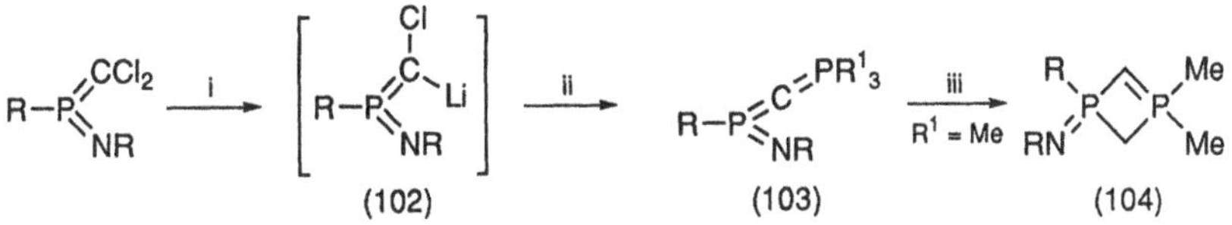

Reagents: i, Bu^nLi; ii, R^1_3P; iii, Heat

Scheme 10

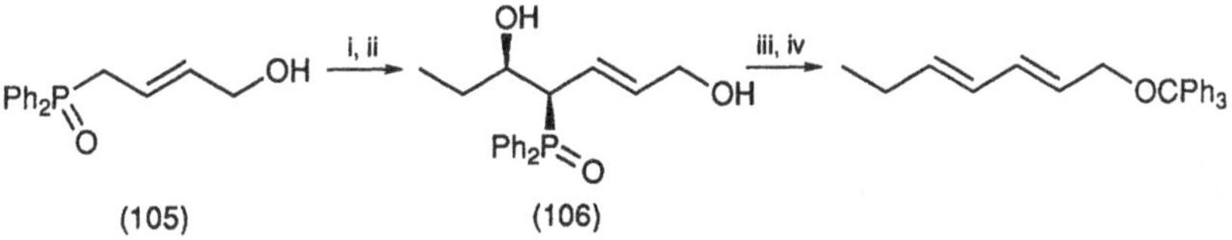

Reagents: i, BuLi; ii, RCHO; iii, Ph_3CCl; iv, NaH, DMF

Scheme 11

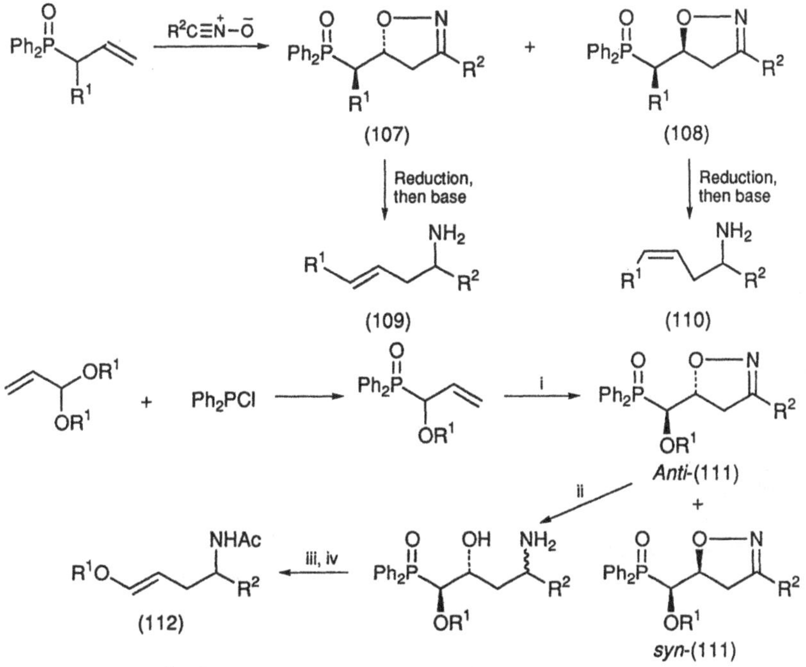

Reagents: i, $R^2C{\equiv}\overset{+}{N}{-}\overset{-}{O}$; ii, $NaBH_4$, $NiCl_2$; iii, Ac_2O; iv, NaH, DMF

Scheme 12

reported.[65] Further n.m.r. studies of thiophosphonamide-stabilised carbanions (**119**) have been carried out. The most recent study reveals that the structures vary dramatically depending on the α-substituents.[66] Thus while (**119**, R=Ph, ^{i}Pr, or adamantyl) have a planar (sp^2) carbanion centre, (**119**, R=H, Me or Et) have a tetrahedral (sp^3) one.

Continuing his studies of the use of chiral phosphonamide carbanions in asymmetric synthesis, Hanessian has reported the enantioselective construction of vicinally-substituted carbon centres through Michael addition reactions of (**120**) followed by γ-alkylation of the carbanion produced.[67] The chiral phosphonate (**121**) undergoes highly (*E*)-selective asymmetric mono-olefination reactions with *meso* dialdehydes (**122**) and (**123**) to give products with diastereoselectivities between 87% and 97%.[68] Olefinations with the chiral phosphonate (**121**) have also been used for the kinetic resolution of acrolein-Diels/Alder dimer (**124**).[69] The alkene stereochemistry and the diastereomeric ratios depend on the nature of the alkyl ester group of the phosphonate and on the temperature. Diastereomeric ratios >90% could be obtained. In spite of the increasing range of highly stereoselective phosphorus-based olefination methods, the stereoselective synthesis of trisubstituted alkenes still constitutes a problem. A solution to this using the chiral phosphonamidate (**125**) has now been reported.[70] Step-wise acylation-alkylation of (**125**) provides (**126**) in good yield and with excellent diastereoselectivity. Suitable choice of the reducing agent for the reduction of the β-keto group in (**126**) leads to either diastereoisomeric β-hydroxy compound, again with excellent selectivity. Finally the diastereomers (**127**) and (**128**) can be decomposed to provide the individual alkene isomers in excellent yield and generally >99% isomeric purity (Scheme **14**).

The reactions of dimethyl methylphosphonate and the corresponding carbanion in the gas phase have been investigated.[71] The carbanion displays a similar range of reactions to those encountered in solution, including olefination with carbonyl compounds. The effect on olefin stereochemistry of a variety of conditions in reactions of α-phosphono lactones (e.g. **129**) with ethanal and propanal has been studied and the results applied in syntheses of integerrinecic acid and senecic acid lactones.[72] Yet further minor modifications of the conditions for phosphonate-olefination reactions, involving the use of lithium hydroxide as the base, have been reported.[73]

Reaction of *N*-protected diethyl aspartate (**130**) or glutamate (**131**) with lithium trialkylphosphonoacetate in the presence of DIBALH leads to selective formation of the *N*-protected γ-amino-α,β-unsaturated dicarboxylates (**132**) as the major product.[74] 3-(Phosphonomethyl)cyclopentenones (**133**) and -hexenones (**134**) are the products of the reaction of dimethyl succinate and dimethyl glutarate, respectively, with excess dimethyl (lithiomethyl)phosphonate; presumably *via* two phosphonomethylations followed by an intramolecular olefination.[75] Examples of the many phosphonate olefination reactions carried out include the synthesis of (2*E*, 4*Z*)-4-aminoalkadienoates (**135**),[76] 2-aryloxy-3-phenylpropenoates (**136**),[77] and isoxazoles (**137**).[78]

The stereochemistry of alkylation of chiral phosphorus-stabilised carbanions (**138**) has been investigated and shown to be highly sensitive to the nature of the nitrogen substituent.[79] Phase-transfer catalysed alkylation of 2-(diethoxyphosphinyl)cyclohexanone (**139**) gives both *O*-(**140**) and *C*-(**141**) alkylated products.[80] The latter predominate when reactive, non-sterically demanding alkyl halides are used. α-Arylphosphonates (**142**) have been synthesised in good yields by the copper(1) salt-mediated arylation of phosphonate carbanions.[81] Under similar conditions *N*-(2-iodophenyl)-substituted phosphonates provide benzoxazole-(**143**) and oxindole-(**144**) substituted phosphonates.

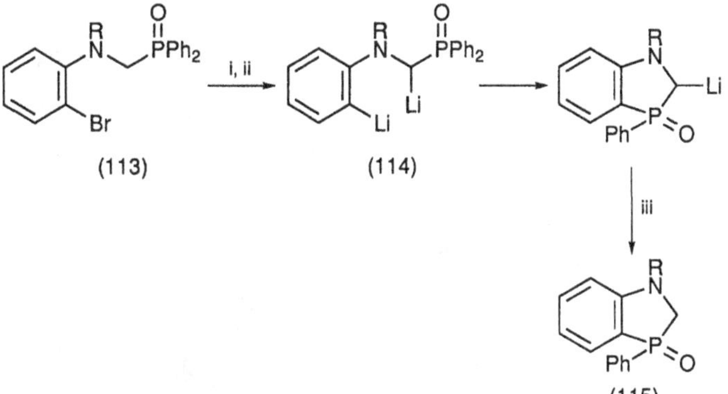

(113) (114) (115)

Reagents: i, LiMe, THF, –78 °C, 15 min; ii, Bu^tLi, 10 min; iii, NH_4Cl aq

Scheme 13

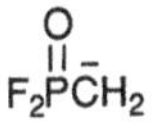

(116) 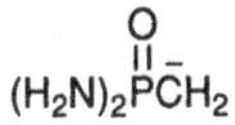 (117) 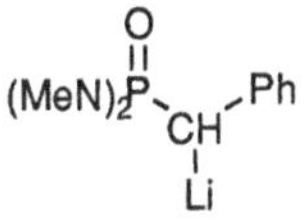 (118) 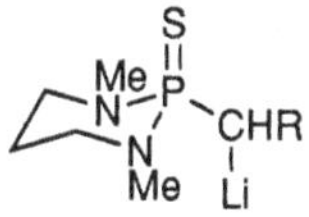 (119)

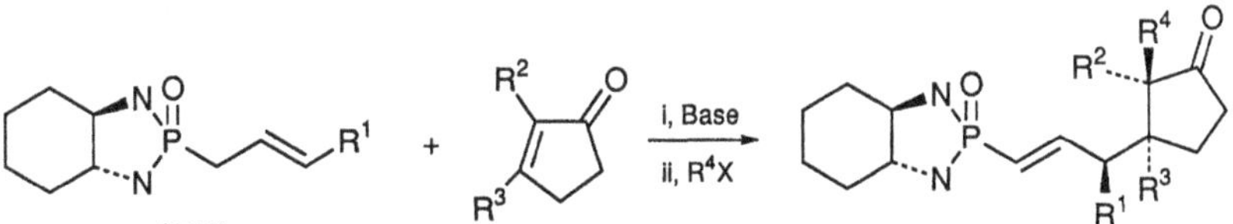

(120)

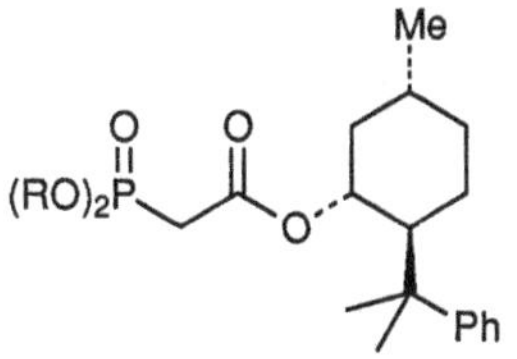

(121) 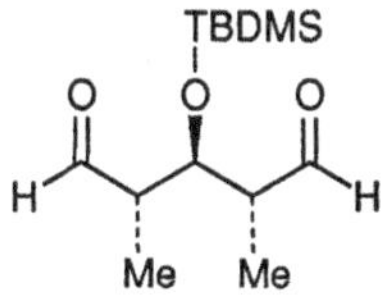 (122) 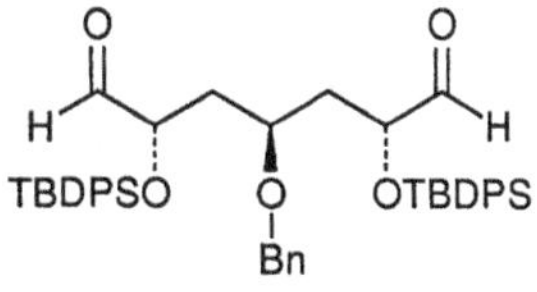 (123)

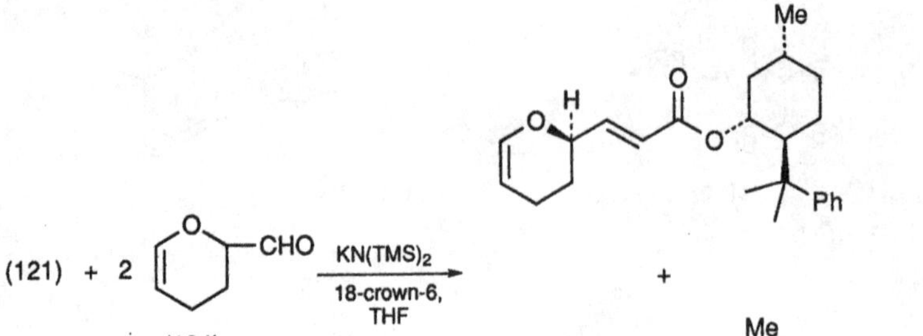

(121) + 2 (124) CHO; $KN(TMS)_2$, 18-crown-6, THF

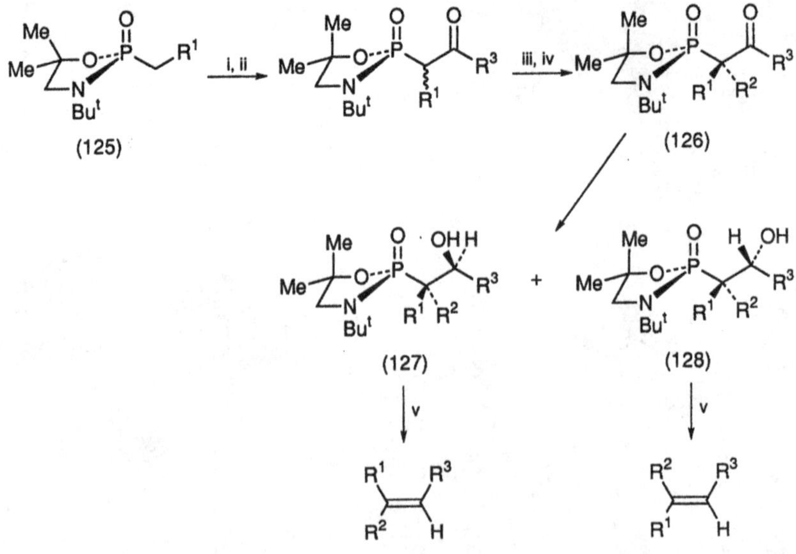

(125) → i, ii → iii, iv → (126) → (127) + (128); (127) → v; (128) → v

Reagent: i, Bu^nLi, THF; ii, R^3CO_2Me; iii, $KOBu^t$; iv, R^2I; v, THF, 105 °C

Scheme 14

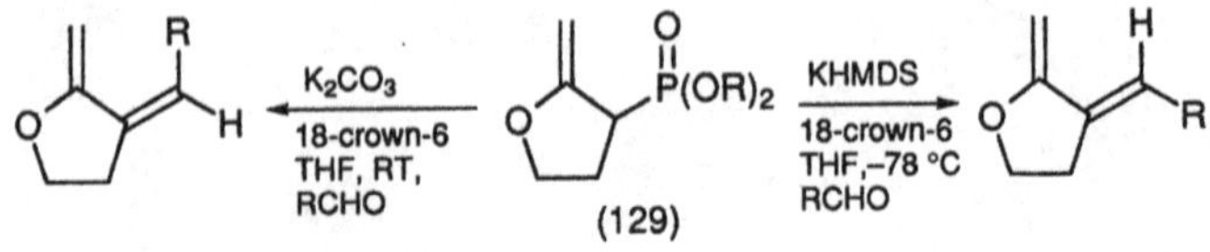

(129): K_2CO_3, 18-crown-6, THF, RT, RCHO; KHMDS, 18-crown-6, THF, –78 °C, RCHO

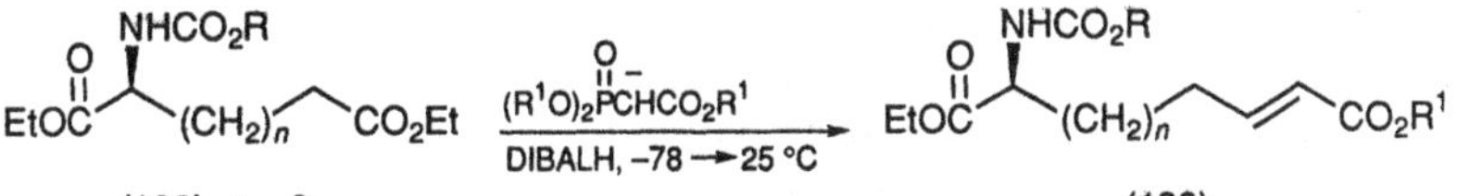

$NHCO_2R$, EtO_2C, $(CH_2)_n$, CO_2Et

$(R^1O)_2P(O)\bar{C}HCO_2R^1$

DIBALH, −78 → 25 °C

$NHCO_2R$, EtO_2C, $(CH_2)_n$, CO_2R^1

(130) $n = 0$
(131) $n = 1$

(132)

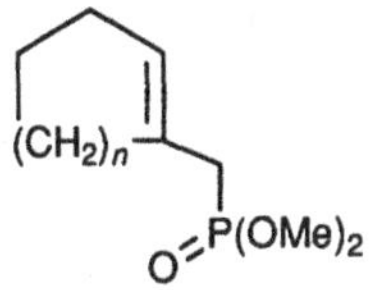

$(CH_2)_n$, $O{=}P(OMe)_2$

(133) $n = 1$
(134) $n = 2$

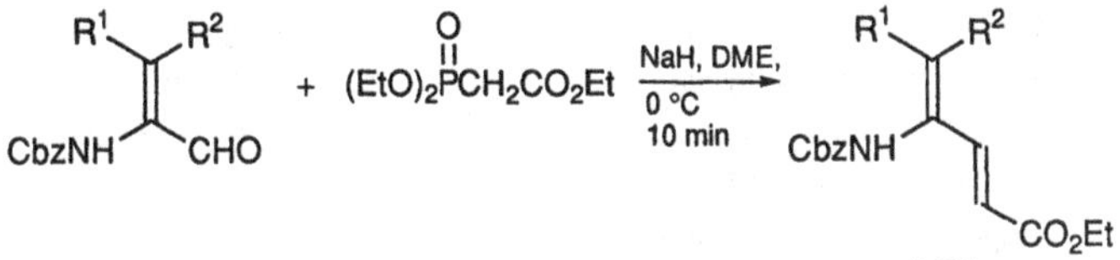

R^1, R^2, CbzNH, CHO + $(EtO)_2P(O)CH_2CO_2Et$

NaH, DME, 0 °C, 10 min

R^1, R^2, CbzNH, CO_2Et

(135)

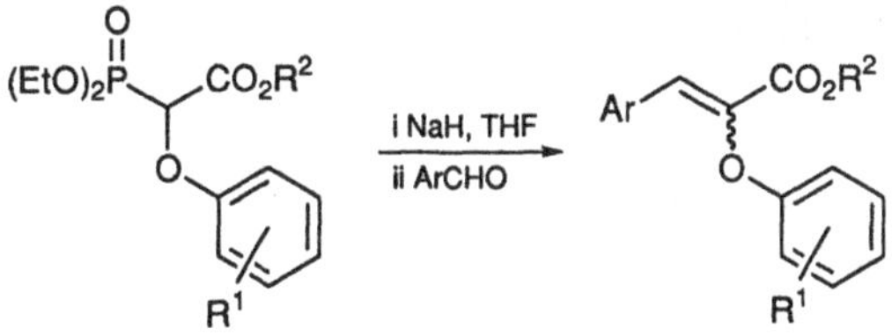

$(EtO)_2P(O)$, CO_2R^2, O, R^1

i NaH, THF
ii ArCHO

Ar, CO_2R^2, O, R^1

(136)

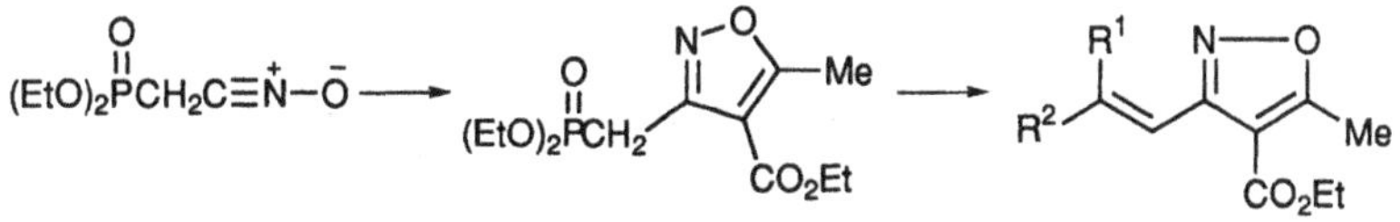

$(EtO)_2P(O)CH_2C{\equiv}\overset{+}{N}{-}\bar{O}$ → $(EtO)_2P(O)CH_2$, N–O, Me, CO_2Et → R^1, R^2, N–O, Me, CO_2Et

(137)

Me, HN, O, P, O, Li^+ ^-CH, Ph

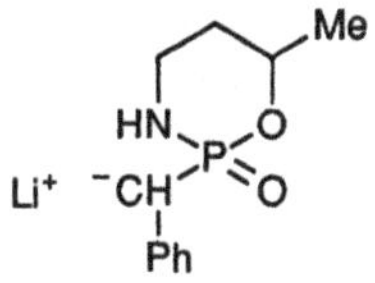

(138)

A range of α,α-dideuterated phosphonate esters, generally with >95% deuterium incorporation, have been prepared by treatment of α-(silylated)alkylphosphonate carbanions with deuterium oxide.[82] α-Fluoroalkylphosphonates (**146**) are increasingly important reagents and a new, flexible route to these compounds, in excellent yields from 1,1,1-dibromofluoromethylphosphonate, is provided by metal halogen exchange-silylation followed by protonation or alkylation of the carbanion (**145**) formed (Scheme **15**).[83] 1-Telluroalkylphosphonates (**147**), and hence vinyltellurides, have been prepared by the reaction of iodomethylphosphonates with lithium alkyltellurides or, conversely, by the reaction of phosphonate carbanions with benzenetellurenyl iodide.[84]

The addition of alk-3-en-1-ylphosphonate carbanions to α-nitroalkenes is highly diastereoselective to give exclusively the *erythro* isomers (**148**) which can be cyclised to (**149**) *via* intramolecular nitrile oxide-olefin cycloaddition (Scheme **16**).[85] Electrophilic amination of the appropriate phosphinate or phosphonate carbanion with di-*tert*-butyl azodicarboxylate has been used to prepare *t*-Boc-protected α-hydrazinoalkyl-phosphinate and -phosphonate esters (**150**) in moderate yield.[86] A similar reaction of α-cyano- and β-keto-alkylphosphonate carbanions with conjugated azoalkenes (**151**) has been used to provide routes to new 3-phosphonopyrrole derivatives (**152**).[87] The carbanion of the allylthiomethylphosphonate (**153**) undergoes a [3,2] sigmatropic shift to give the phosphonate (**154**) in excellent yield (Scheme **17**).[88] The base-induced rearrangement of the α-bromomethylphosphonamidate (**155**) to give (**158**) and (**159**) is suggested to take place *via* the intermediate azaphosphiridine (**157**) formed from the amide anion (**156**).[89]

5 Selected Applications in Synthesis

5.1 Carbohydrates.- (α,α-Difluoroalkyl)phosphonates (e.g. **160**) of a variety of monosaccharides have been synthesised in excellent yield by a displacement reaction of (lithiodifluoromethyl)-phosphonate with the corresponding triflate.[90] The thiazole-substituted carbonyl ylide (**161**) has been used in Wittig reactions to provide a masked α-ketoaldehyde function and applied to the synthesis of 3-deoxy-2-ulsonic acids[91] and α– and β–1-C-(2-thiazolacyl)-glycosides (**162**).[92] The reaction of arsonium ylides with pyranoses and furanoses gives mixtures of (*E*)-alkene derivatives (**163**) and the corresponding *C*-glycosyl (**164**).[93] The *C*-glycosyls (**164**) can be obtained as the exclusive product by a similar reaction of the corresponding arsonium salt in the presence of zinc bromide.

5.2 Carotenoids, Retenoids, Pheromones and Polyenes.- A convenient route to exclusively (*E*)-allylic phosphonates (e.g. **165**), compounds which are useful in polyene synthesis, is provided by the base-catalysed isomerisation of the corresponding vinylphosphonates.[94] The bis(trifluoromethyl) analogue (**167**) of a known, potent squalene synthase inhibitor has been synthesised using the Wittig reaction of (**166**) with hexafluoroacetone as a key step.[95]

Examples of the use of phosphonate-based and Wittig olefination in retinoid chemistry include the synthesis of (6*S*)-*cis*-(**168**)- and (6*S*)-*trans*-(**169**)-locked bicyclic retinals[96] and the preparation of (+)-(4*S*)- and (-)-(4*R*)-(11*Z*)-4-hydroxyretinals (**170**).[97]

Wittig reactions of deuterated salts (e.g. **171**) have been used as building blocks in the synthesis of a variety of deuterium-labelled Lepidoptera pheromones and analogues.[98] The main component (**175**) of the sex pheromone of the green stink bug *Nezara viridula* has been synthesised by phosphine oxide-based olefination of the ketone (**172**) (Scheme **18**).[99] The initial reaction gives

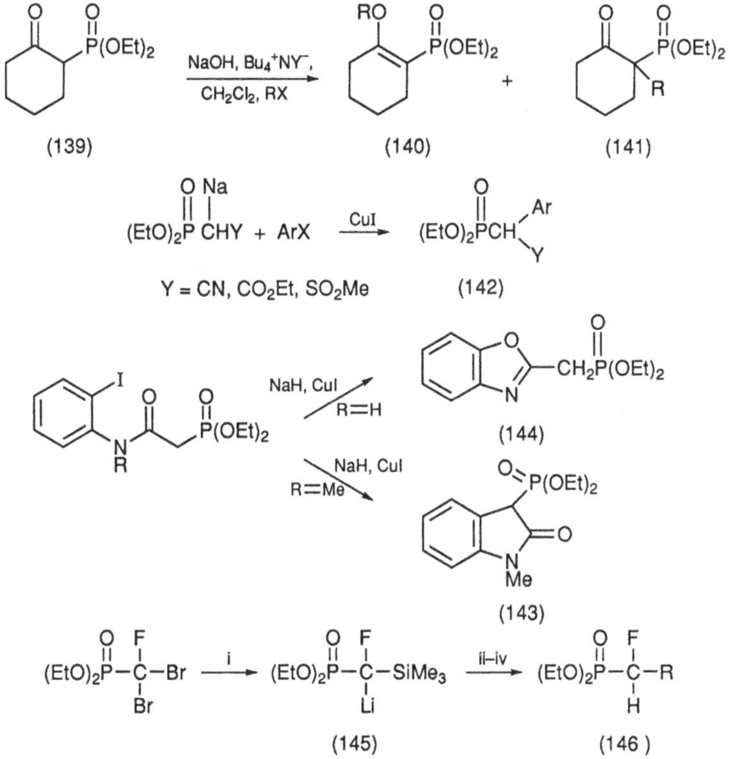

(139) → $NaOH, Bu_4{}^+NY^-, CH_2Cl_2, RX$ → (140) + (141)

$(EtO)_2P(O)CH(Na)Y$ + ArX —CuI→ $(EtO)_2P(O)CH(Ar)Y$

Y = CN, CO_2Et, SO_2Me (142)

NaH, CuI, R=H → (144)

NaH, CuI, R=Me → (143)

$(EtO)_2P(O)CFBr_2$ —i→ $(EtO)_2P(O)CF(Li)SiMe_3$ (145) —ii–iv→ $(EtO)_2P(O)CHFR$ (146)

Reagents: i, 2 x BuLi, Me_3SiCl, THF, –78 °C; ii, RI, –78 °C; iii, EtOH, EtOLi, 0 °C; iv, HCl aq, 0 °C

Scheme 15

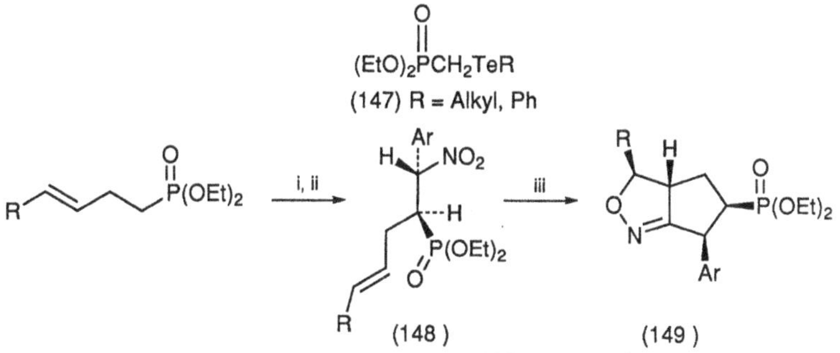

$(EtO)_2P(O)CH_2TeR$

(147) R = Alkyl, Ph

—i, ii→ (148) —iii→ (149)

Reagents: i, LDA, THF; ii, $ArCH=CHNO_2$; iii, PhN=C=O, Et_3N

Scheme 16

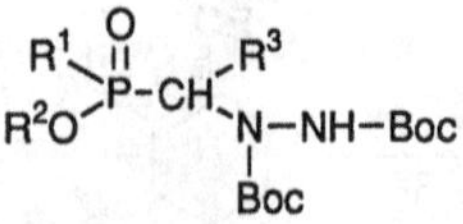

R^1 O R^3
R^2O P–CH N–NH–Boc
Boc

(150)

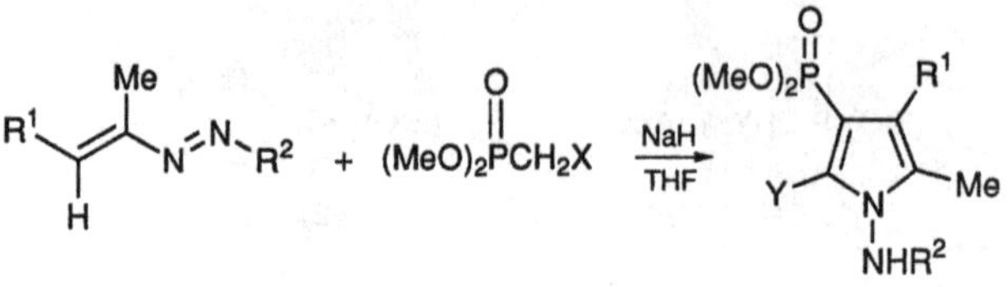

Me
R^1 N=N R^2
H
+ $(MeO)_2\overset{O}{P}CH_2X$ $\xrightarrow[THF]{NaH}$
$(MeO)_2\overset{O}{P}$ R^1
Y N Me
NHR^2

(151)

(152) Y = Me, NH_2

$(Pr^iO)_2\overset{O}{P}$ S $\xrightarrow{i, ii}$ $(Pr^iO)_2\overset{O}{P}$ SH

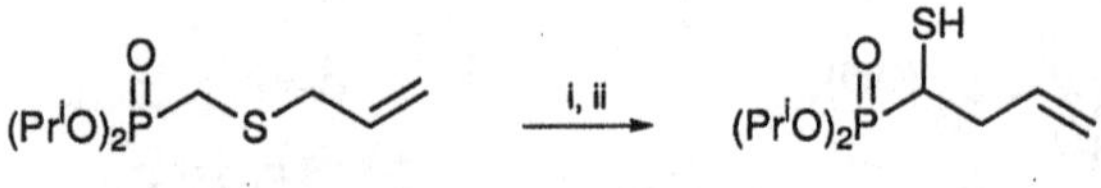

(153)

(154)

Reagents: i, Bu^nLi, THF, –15 °C; ii, H_3O^+

Scheme 17

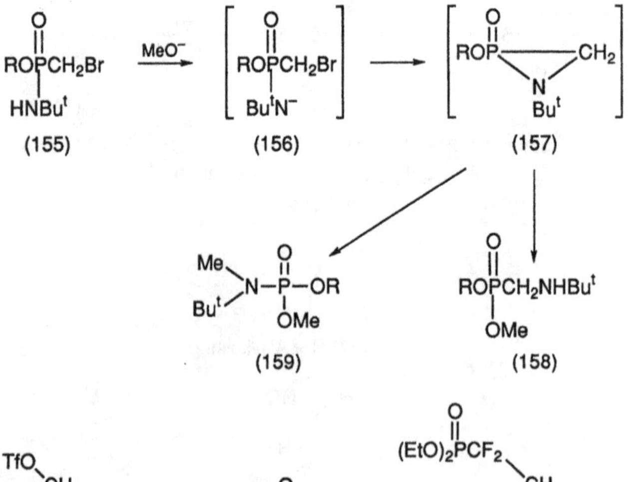

$RO\overset{O}{P}CH_2Br$
$HNBu^t$
$\xrightarrow{MeO^-}$
$[RO\overset{O}{P}CH_2Br$
$Bu^tN^-]$
⟶
$[RO\overset{O}{P}$ CH_2
N
$Bu^t]$

(155)

(156)

(157)

Me
N–P–OR
Bu^t O OMe

$RO\overset{O}{P}CH_2NHBu^t$
OMe

(159)

(158)

TfO
CH_2
O
BzO
BzO OBz
OBz
+ $(EtO)_2\overset{O}{P}\ CF_2Li$ ⟶
$(EtO)_2\overset{O}{P}CF_2$
CH_2
BzO O
BzO OBz
OBz

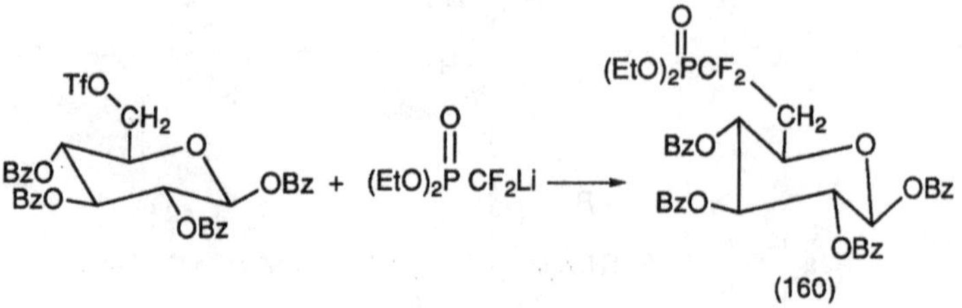

(160)

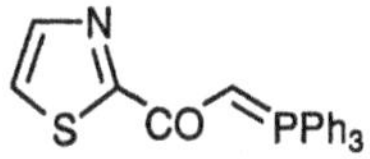

(161)

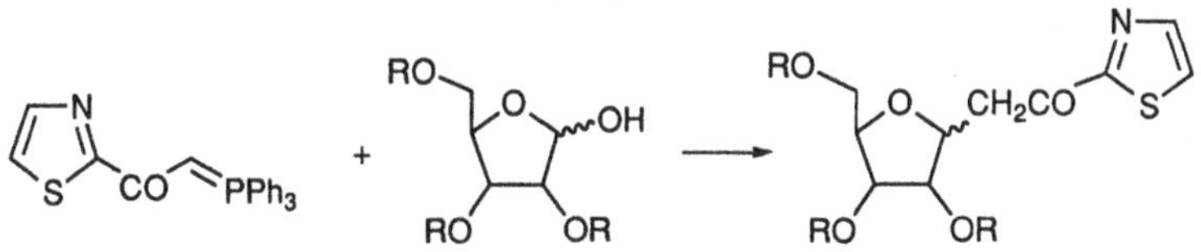

(161) (162)

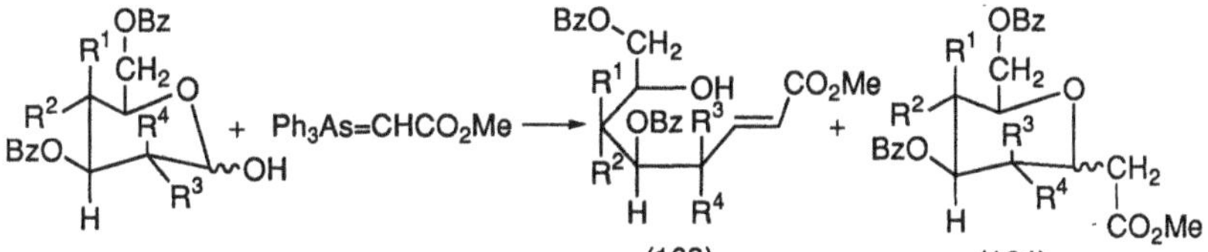

(163) (164)

(165)

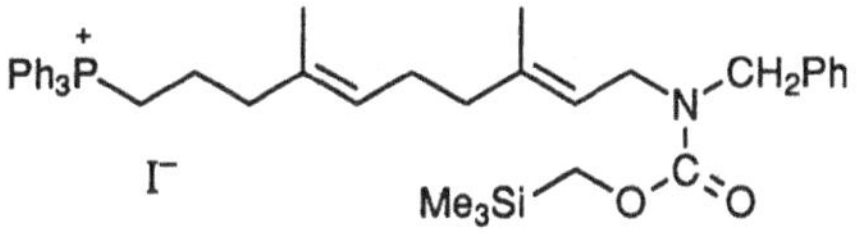

(166)

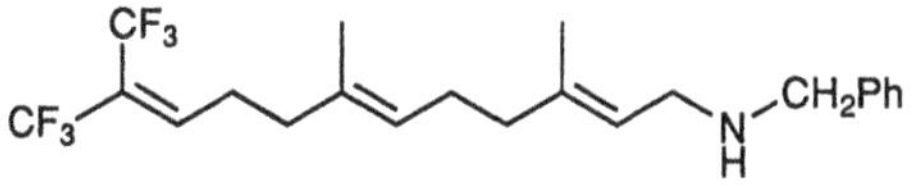

(167)

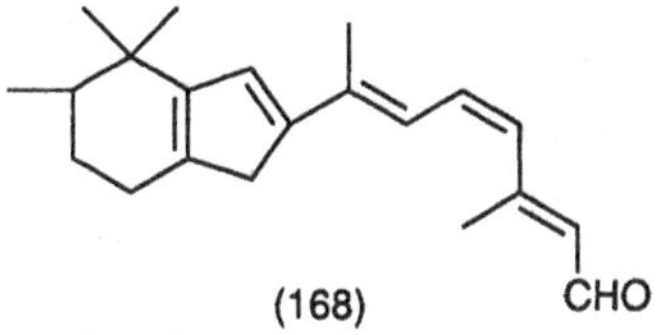

(168) 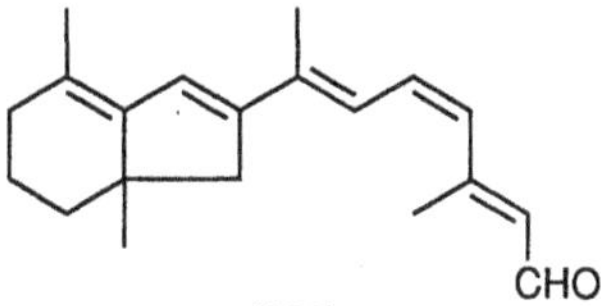 (169)

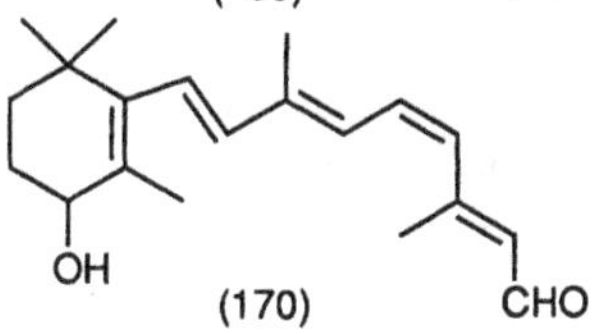

(170) (171)

the required *erythro*-β-hydroxyphosphine oxide (**173**) and the alkene (**174**) which is derived from the *threo*-β-hydroxyphosphine. However, the authors make no comment on the apparent much more facile decomposition of the *threo*- compared to the *erythro*-diastereomer. The *erythro*-isomer undergoes base-induced decomposition to give the required product (**175**). The β-hydroxyphosphine oxide (**176**) has been used to prepare all four stereoisomers of (2*E*,4*E*)-4,6,10,12-tetramethyl-2,4-tridecadien-7-one (**177**), the primary sex pheromone of the pine blast scale.[100] The oxide (**176**) was prepared (Scheme **19**) as a 6:1 mixture of the required *threo* and unwanted *erythro* forms. Base-induced decomposition of (**176**) provides a mixture which contains 85% of the required (*E*)-alkene. Two Wittig-based approaches to the two components (**178**) and (**179**) of the female sex pheromone of the Israeli pine blast scale have been reported.[101] Both approaches lead to a mixture of (**178**) and (**179**) which can be separated and reconstituted in the 75:25 natural ratio.

The linear polyenes (**181**), containing protected hydroquinone terminal groups, have been synthesised in moderate yield *via* olefination reactions of the bis(phosphonate) (**180**).[102]

5.3 Leukotrienes, Prostaglandins and Related Compounds.-Examples of the use of now standard Wittig methods in the synthesis of leukotrienes and related compounds include the synthesis of 12-ketoeicosatetraenoic acid (12-KETE) (**182**) and its 8,9-*trans*-isomer.[103] The stereospecific coupling of phosphonate (**183**) with the aldehyde (**184**) has been used to provide the essential carbon framework in a new, efficient total synthesis of leukotriene B_4 (LTB_4).[104] The arsonium ylide (**185**) has been used in a Wittig reaction *and* as a base to introduce the C_6-(*Z*)-alkene function in a synthesis of 3-hydroxyleukotriene B_4 (3-OH-LTB_4).[105] Use of the corresponding ylide gave much poorer yields and complex mixtures of products. The first reported synthesis of (-)-lipstatin (**186**), a potent inhibitor of pancreatic lipase, involves two (*Z*)-stereoselective Wittig reactions in the preparation of the key intermediate (**187**).[106] A cyclopropyl analogue (**188**) of arachidonic acid has been prepared using (*Z*)-selective Wittig reactions of the ylides (**189**) and (**190**) as key steps.[107] The first total synthesis of a barnacle hatching factor (**192**) has been achieved using a Wittig reaction of the ylide (**191**).[108]

5.4 Macrolides and Related Compounds.- Intramolecular phosphonate-based olefination continues to be widely used as a method of macrocyclisation. For example, a reported convergent synthesis of the polyene macrolide (-)-roxatin depends on ring closure via the complex phosphonate (**193**) to give the 30-membered ring[109] and an advanced intermediate (**195**) in a total synthesis of the diterpene lactone (+)-cleomeolide has been prepared from the phosphonate (**194**).[110] An investigation of the stereochemistry of macrocylic ring-closure using phosphonates has revealed that this is dependent on the base and conditions in a similar way to intermolecular examples.[111] Under most conditions the (*E*)-isomer predominates. However, the use of DBN/LiCl (Masamune/Roush conditions) provides the (*Z*)-alkene. An explanation for these results, based on differing rates of decompositon of the intermediate β-oxyanion phosphonates, is suggested. Although less used than phosphonates, phosphonium ylides have also been employed in macrocyclisation. For example, the macrocyclic diarylheptanoid garugamblin-2 (**197**) has been prepared by intramolecluar Wittig reaction of the phosphonium salt (**196**).[112]

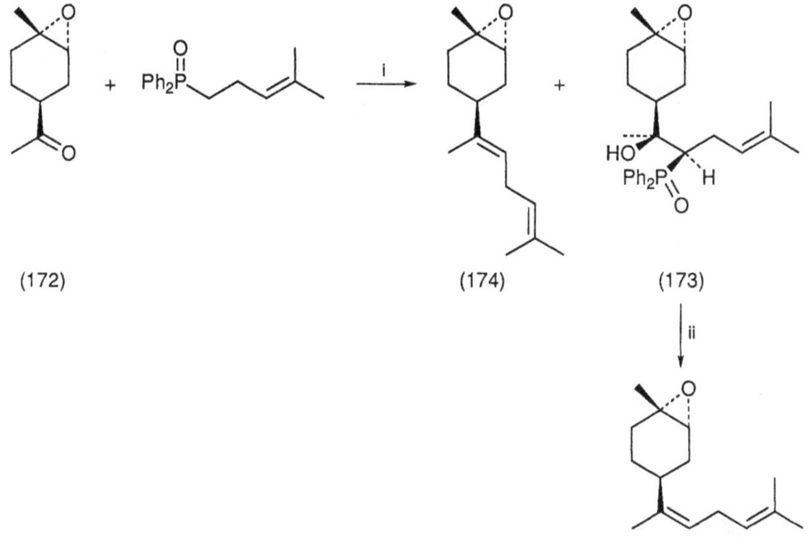

(172) (174) (173) (175)

Reagents: i, Bu^nLi, THF; ii, NaH, DMF

Scheme 18

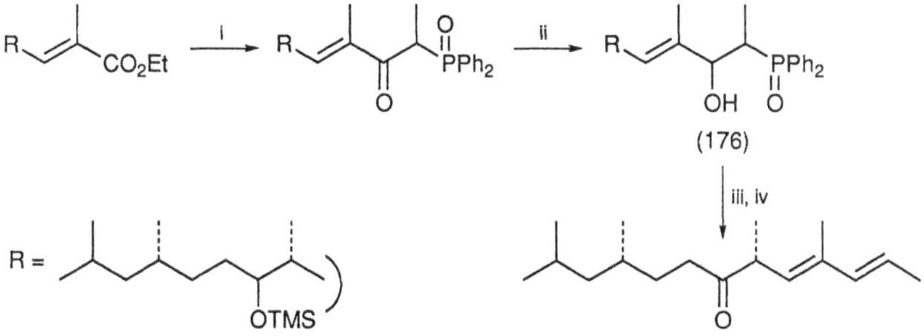

R = ... OTMS

(176) (177)

Reagents: i, $Ph_2P(O)CH_2CH_3$, Bu^nLi, THF, –78 °C; ii, DIBALH, Et_2O;
iii, NaH, DMF, 40 °C; iv, DMSO, $(COCl)_2$, CH_2Cl_2, Et_3N

Scheme 19

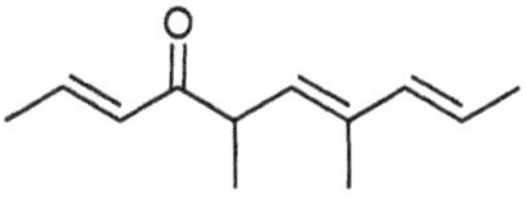

(178) 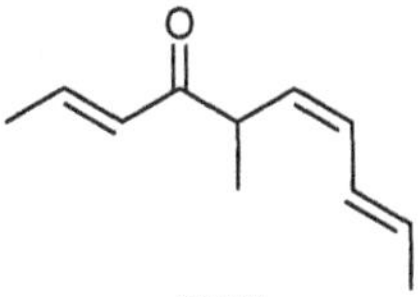 (179)

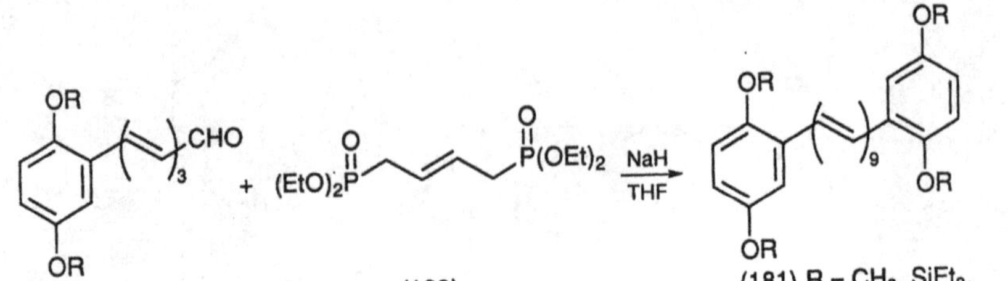

OR CHO OR + $(EtO)_2P$ O P(OEt)$_2$ O NaH THF OR OR OR OR

(180)

(181) R = CH_3, $SiEt_3$, $SiMe_2Bu^t$

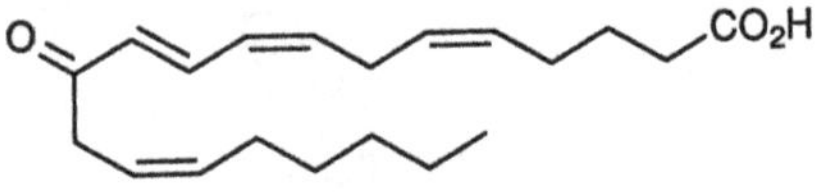

O CO_2H

(182)

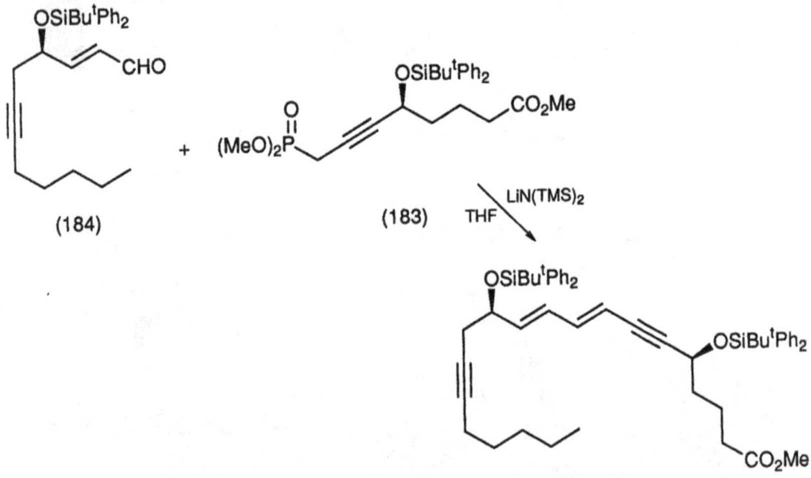

$OSiBu^tPh_2$ CHO + $(MeO)_2P$ O $OSiBu^tPh_2$ CO_2Me

(184) (183) THF $LiN(TMS)_2$

$OSiBu^tPh_2$ $OSiBu^tPh_2$ CO_2Me

$2Ph_3As$ TMS + O OH TSO → OH TMS

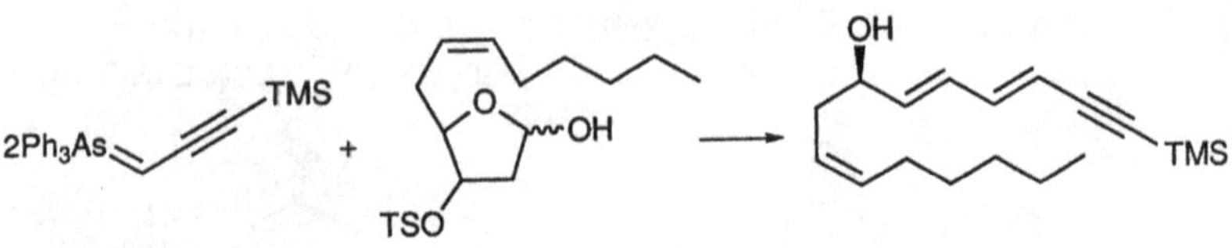

(185)

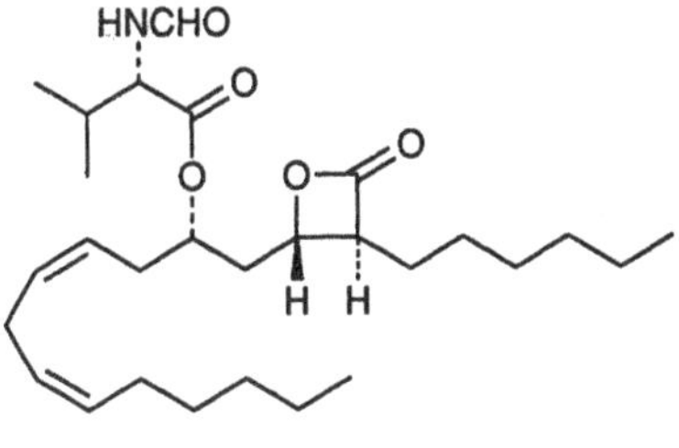

HNCHO O O O O H H

(186)

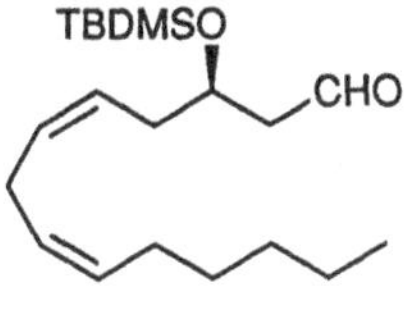

TBDMSO CHO

(187)

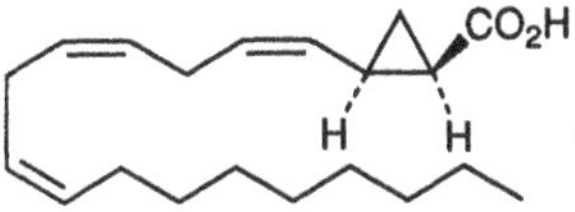

CO_2H H H

(188)

THPO PPh_3

(189)

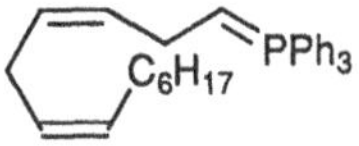

PPh_3 C_6H_{17}

(190)

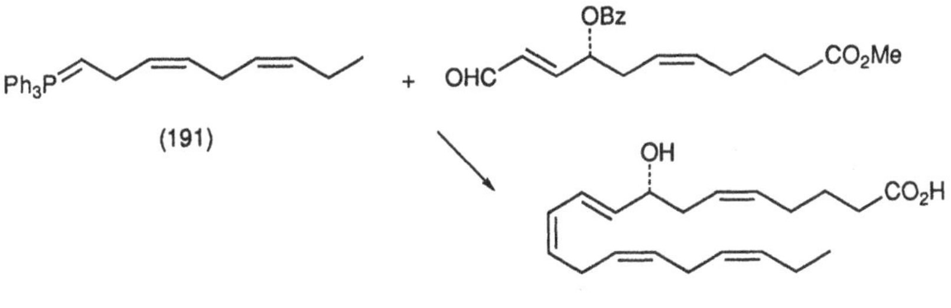

OBz

Ph_3P + OHC CO_2Me

(191)

OH CO_2H

(192)

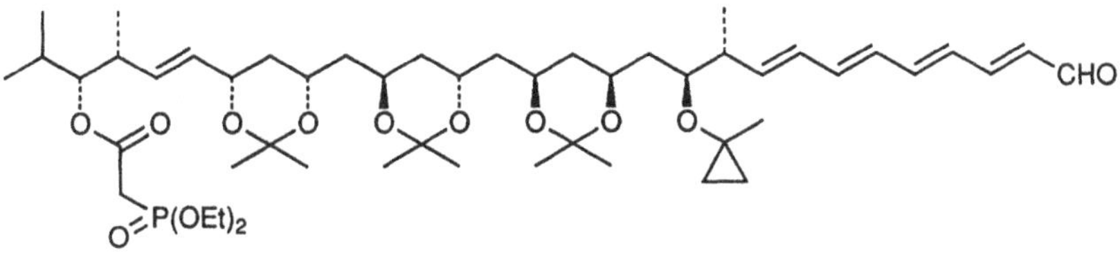

CHO O O O O O O O O O O $P(OEt)_2$

(193)

$(EtO)_2P{=}O$ CN Me Me O O

(194)

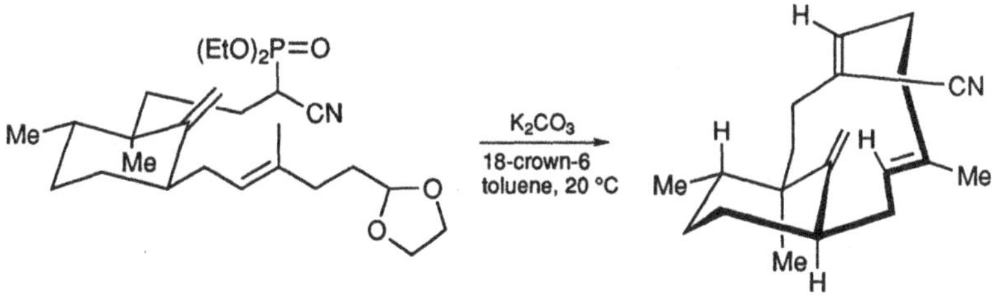

K_2CO_3
18-crown-6
toluene, 20 °C

H CN H H Me Me Me H

(195)

Alkylation of the lithioacetonylidene ylide (**198**) to give the β-keto ylide (**199**), followed by a Wittig reaction with a suitably protected aldehyde, has been used in studies directed towards the synthesis of the polyether macrolide halichondrin B.[113] Interestingly, attempts to use the ketophosphonate dianion (**200**) in place of (**198**) led to predominate attack on iodine, rather than carbon, during alkylation. The Wittig condensation of the phosphonium salt (**201**) with the 5-hydroxybutenolide (**202**), used in a reported synthesis of an analogue of β-milemycins, requires no less than four equivalents of base for optimum yield.[114] Barium hydroxide-induced phosphonate olefination has been used to prepare the C_{17}-C_{32} subunit (**203**) of the macrolide scytophycin C[115] and the Wadsworth-Emmons reaction has also been used in preparing the diene fragment as part of a synthesis of the right hand portion (**204**) of the type A streptogramin antibiotics.[116] A new synthesis of (-)-rapamycin uses a predominantly (*E*)-olefination reaction of the phosphine oxide (**205**) to construct the complex carbon chain.[117]

5.5 Nitrogen Heterocycles.- There continue to be a large number of reports of the use of the reactions of iminophosphoranes, primarily the aza-Wittig reaction, to synthesise a wide range of heterocyclic systems and the use of vinyliminophosphoranes in such reactions has further extended the range of heterocylic structures that can be synthesised by these methods.[118]

Examples of five-membered heterocyclic rings prepared by aza-Wittig and related methods include the alkaloid leucettamine B (**206**)[119] and pyrrolo[2,3-b]pyridines (**207**) (Scheme **20**).[120] The aza-Wittig reactions of the iminophosphorane (**208**) with methyl isocyanate, carbon dioxide, or carbon disulfide give heterocumulenes (**209**), (**210**), and (**211**), respectively, which react with amines or ammonia to provide a route to the framework of aplysinopsin alkaloids (**212**).[121]

Six-membered heterocyclic systems which have been synthesised by similar methods include 1,6-methano[10]annulenopyridines (**214**) from (**213**),[122] γ-carblines (**215**),[123] lavendamycin methyl ester,[124] a range of functionalised 2,3-dihydropyrido[3',2':4,5]thieno[3,2-d]pyrimidines (**216**),[125] and the quinazolino[3,4-a]perimidine derivatives (e.g. **218**) from 1,2-dihydro-properimidine azide (**217**).[126] The previously unreported 4-methylene-4H-3,1-benzozazine ring (**219**) has been prepared from *o*-azidoacetophenone[127] and the zwitterionic heteropolycyclic uracils (**220**) have been synthesised by a three-component reaction of iminophosphorane, isocyanate, and substituted pyridine.[128]

The synthesis of larger rings has also been accomplished including the fully unsaturated azolo-fused 1,3-diazepines (**222**) from bis(iminophosphoranes) (**221**)[129] and novel cyclic carbodiimides (e.g. **223**).[130] Iminophosphoranes are also intermediates in the synthesis of macrolactams (e.g. **225**) from the reaction of ω-azido acid anhydrides (e.g. **224**) with tributylphosphine.[131] The synthesis of iminophosphorane-containing cryptands and a spherand-type structure have been reported.[132]

A variety of 3-substituted indoles have been prepared in moderate yield by intramolecular olefination of the phosphine oxides (**226**) (Scheme **21**).[133] Reaction of the ylide (**227**) with succinaldehyde provides the double (**228**) and single (**229**) Wittig products.[134] Further Wittig reaction of the latter product with the ylide (**230**), derived from caprolactone, gave the unsymmetrical analogue (**231**) which was converted into a pentacyclic model (**232**) of ptilomycalin A (Scheme **22**).

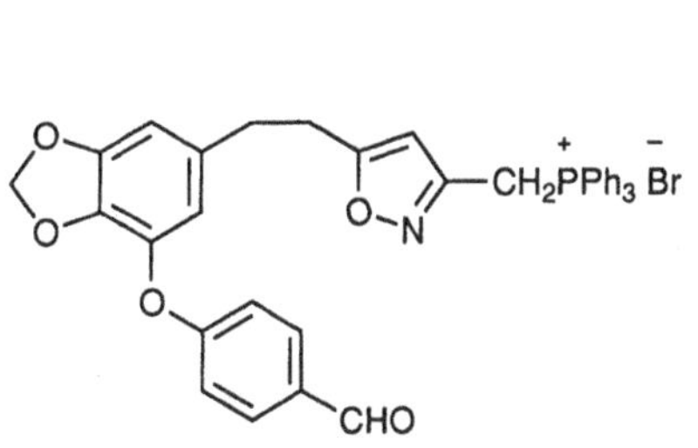

O O O CH$_2\overset{+}{P}$Ph$_3$ $\overset{-}{Br}$ O N O CHO

(196)

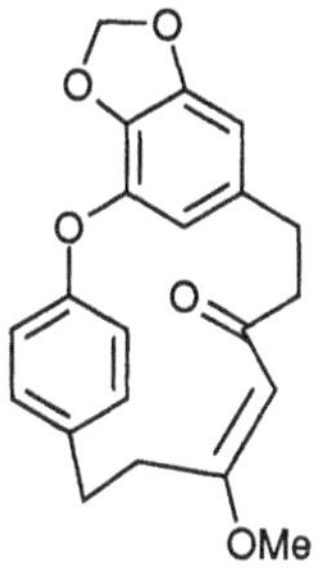

O O O O O OMe

(197)

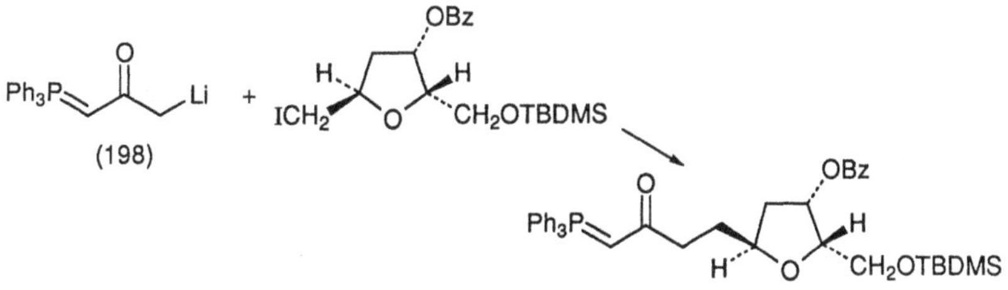

Ph_3P O Li + OBz H H ICH_2 O $CH_2OTBDMS$

(198)

Ph_3P O OBz H H O $CH_2OTBDMS$

(199)

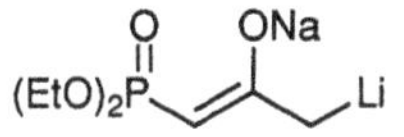

O ONa $(EtO)_2P$ Li

(200)

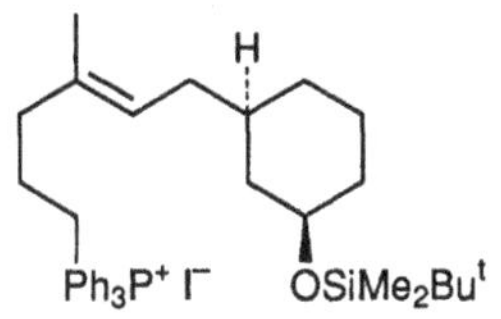

H Ph_3P^+ I^- $OSiMe_2Bu^t$

(201)

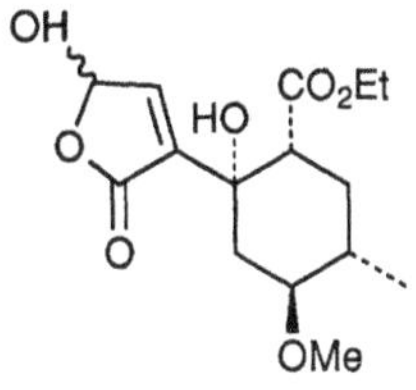

OH O O HO CO_2Et OMe

(202)

TIPSO OMe O O $P(OMe)_2$ + OHC OBz O Si O Bu^t Bu^t

TIPSO MeO O O Si O Bu^t Bu^t OMe CHO

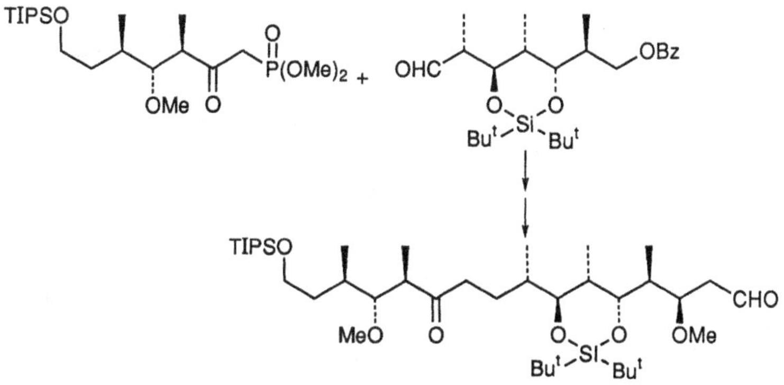

(203)

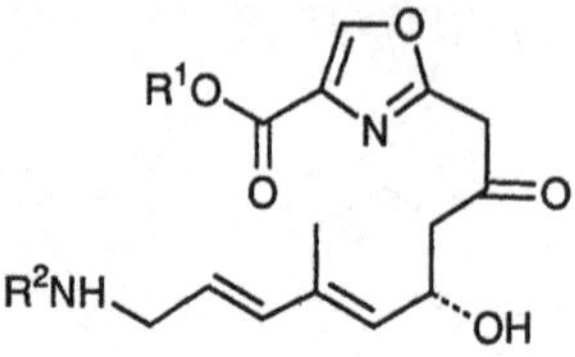

(204)

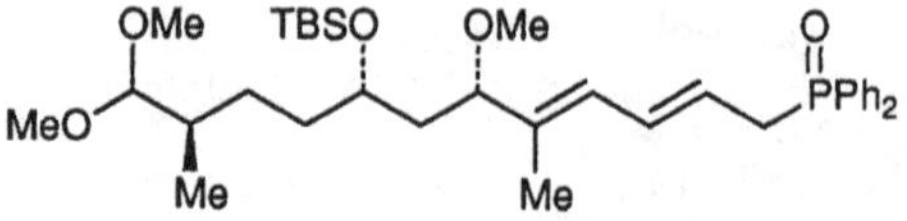

(205)

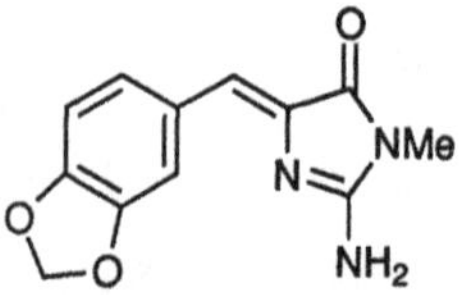

(206)

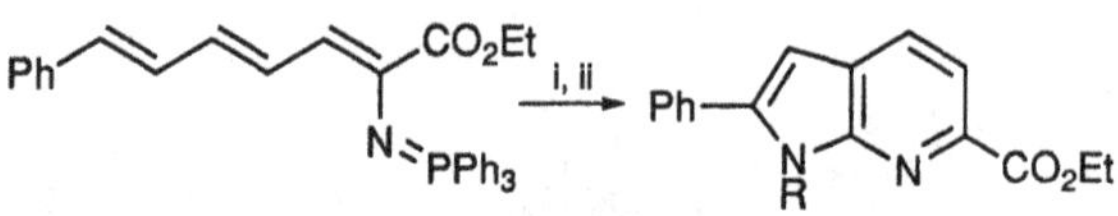

(207)

Reagents: i, RNCO; ii, $PhNO_2$, Δ

Scheme 20

(208)

(209) Z = NMe
(210) Z = O
(211) Z = S

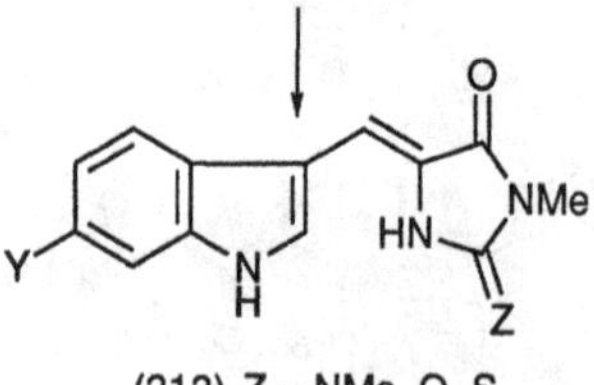

(212) Z = NMe, O, S

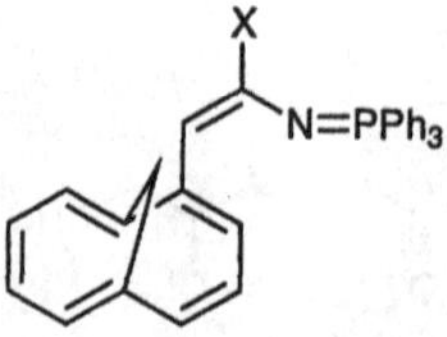

(213)

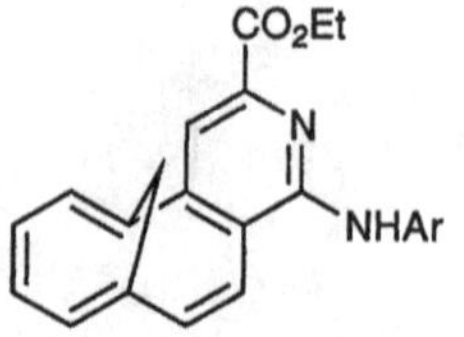

(214)

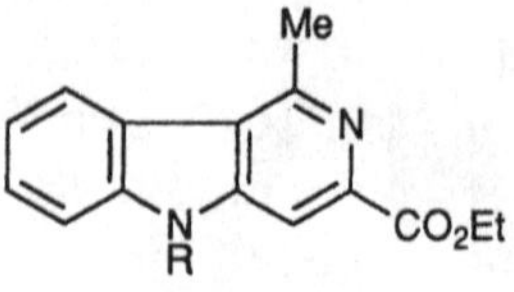

(215)

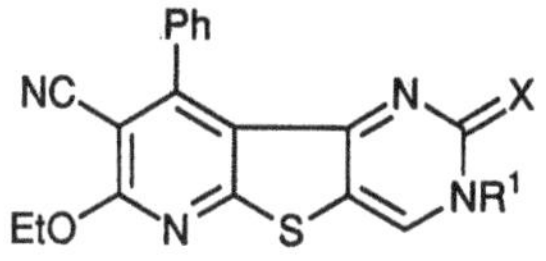

(216) X = NR, O, S

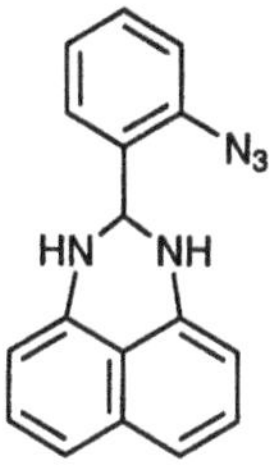

(217)

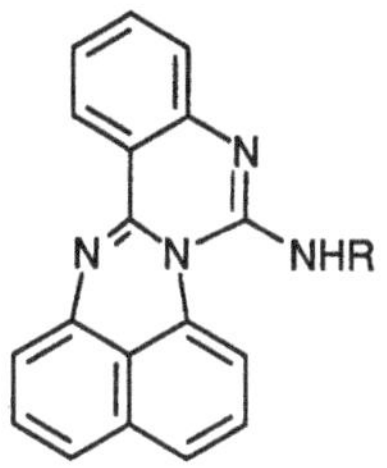

(218)

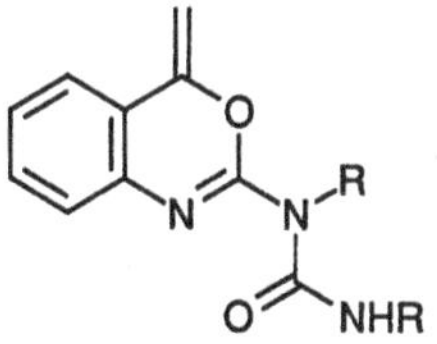

(219)

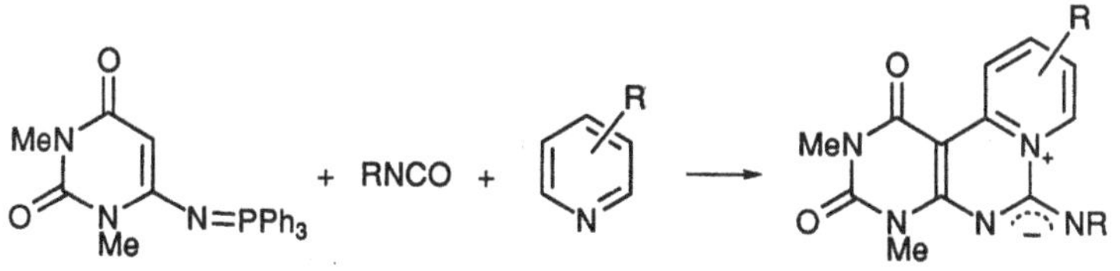

+ RNCO +

(220)

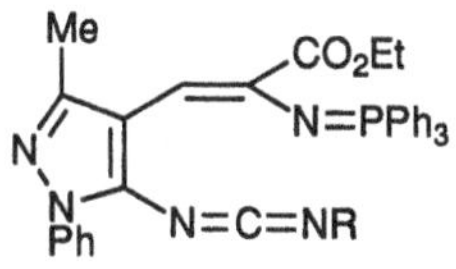

(221)

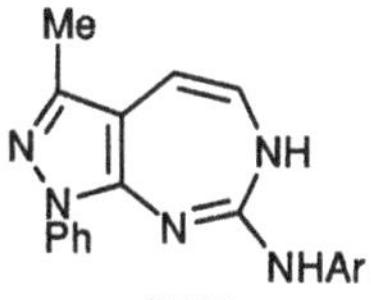

(222)

Boc_2O, CH_2Cl_2

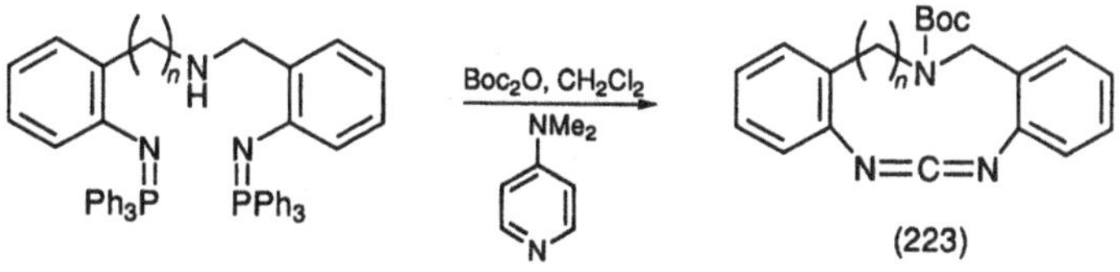

(223)

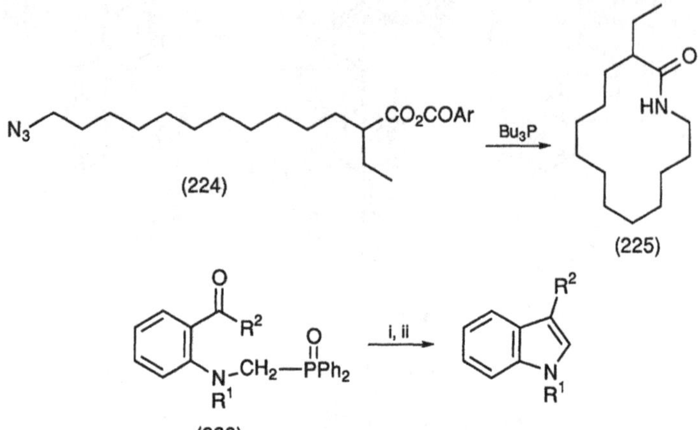

N_3 ... CO_2COAr (224) $\xrightarrow{Bu_3P}$ (225) HN, O

(226) $\xrightarrow{i, ii}$ indole (R^1, R^2)

Reagents: i, KHDMS, THF, –10 °C; ii, HCl, H_2O

Scheme 21

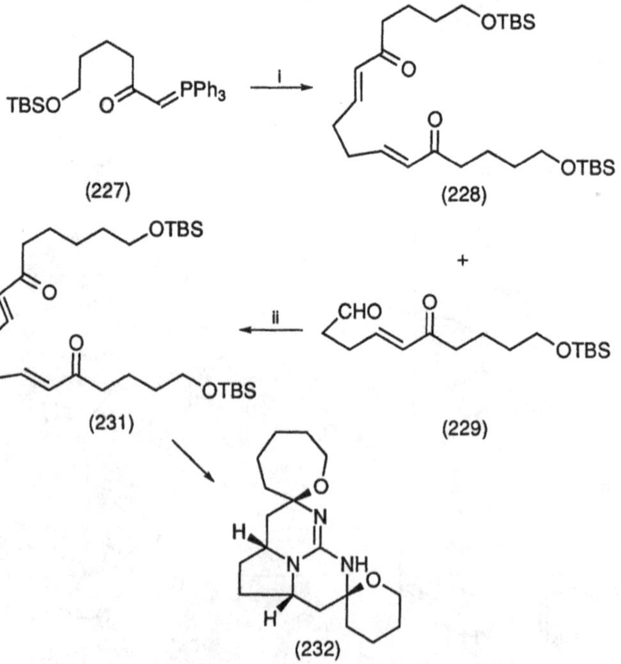

TBSO ... PPh_3 (227) $\xrightarrow{i}$ (228) OTBS, OTBS

(229) CHO, OTBS $\xrightarrow{ii}$ (231) OTBS, OTBS → (232)

Reagents: i, $OHC(CH_2)_2CHO$; ii, Ph_3P=...OTBS (230)

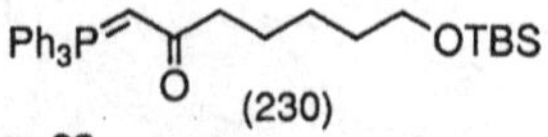

Scheme 22

5.6 Steroids and Related Compounds.- The previously reported (*E*)-17-[(diethylphosphono)-isocyanomethylene]-3-methoxyandrosta-3,5-diene (**233**) has been used in olefination reactions to provide a route to a series of polyfunctional unsaturated $D^{16,20}$-20-isocyanosteroids.[135] An intramolecular phosphonate-olefination of (**234**) to give (**235**) is a key step in a new approach to *trans-syn-cis*-perhydrophenanthrenic systems.[136] An attempted Wittig reaction of 4-pentynyltriphenyl-phosphonium bromide (**236**) with oestrone methyl ether gave the spiro-bicyclo[3.1.0]hexane (**237**) rather than the expected alkene.[137] The reaction appears to be general for ketones carrying α-quaternary carbon atoms and a mechanism involving cyclisation of the initially formed ylide (**238**) to give (**239**) is suggested.

5.7 Terpenes.- The first total synthesis of the biologically active marine natural product metachromin A (**241**) has been achieved using an (*E*)-stereoselective olefination with the phosphonate (**240**) as the key step.[138] The olefination reaction of the α-cyanophosphonate (**242**) with (*E,E*)-farnesal (**243**) has been studied under a wide variety of conditions with a view to optimising (*Z*)-stereoselectivity.[139] Excellent (*Z*)-stereoselectivity was achieved and the resulting procedure applied to a stereoselective synthesis of plaunotol (**244**). Epimerisation and β-elimination of base-sensitive substrates can cause problems in phosphonate- and phosphine oxide-based olefinations. Such reactions have been excluded in a recent report of enantiospecific syntheses of iridoid monoterpene lactones by intramolecular phosphonate-olefination through the use of mild bases such as DBU and di(*iso*propyl)ethyl amine/lithium chloride.[140]

5.8 Vitamin D Analogues and Related Structures.- What are now standard phosphine oxide-based olefination methods have been applied to the synthesis of 1-(1'-hydroxyethyl)-25-hydroxyvitamin D_3 analogues (**245**)[141] and to the preparation of the iodo analogue (**246**) and hence to other analogues with modified side chains.[142] A one-pot approach to the vitamin D_3 analogue (**252**) has been reported.[143] The method involves the reaction of the lithium ylide-anion (**247**) with epoxide (**248**) to give (**249**) followed by addition of the aldehyde (**250**) to give (**251**) and hence (**252**) (Scheme **23**).

5.9 Miscellaneous Reactions.- Phosphorus-based olefination continues to be frequently used in syntheses of tetrathiofulvalene (TTF) derivatives. For example, the phosphonium salts (**253**) and (**254**) have been prepared and used as Wittig reagents to provide routes to sulfur-rich and space-extended TTF derivatives[144] and both Wittig, involving (**255**), and phosphonate-based, involving (**256**), olefination, have been used in the synthesis of a range of new, extensively conjugated π-electron donors (e.g. **257**) combining both TTF and 2,2'-ethanediylidene-bis-(1,3-thiole) units.[145]

Examples of the use of the Wittig reaction in β-lactam synthesis include the construction of the five-membered ring in a preparation of 2-functionalised-methyl-1β-methyl-carbapenams (**258**).[146] Wittig reactions of the penam phosphonium salt (**259**), prepared from the corresponding iodide in a one-pot reaction, gave a series of (2-arylethenyl)penams with variable *E/Z* selectivity.[147]

Glyoxals (e.g. **260**) derived from *L*-amino acids and dipeptides have been trapped by Wittig reactions to provide (*E*)-unsaturated ester derivatives (**261**).[148] The reaction of (2*S*)-*N*-benzoyl-2-*tert*-butyl-4-methylene-1,3-oxazoline-5-one (**262**) with *iso*propylidenetriphenyl-phosphorane gives a

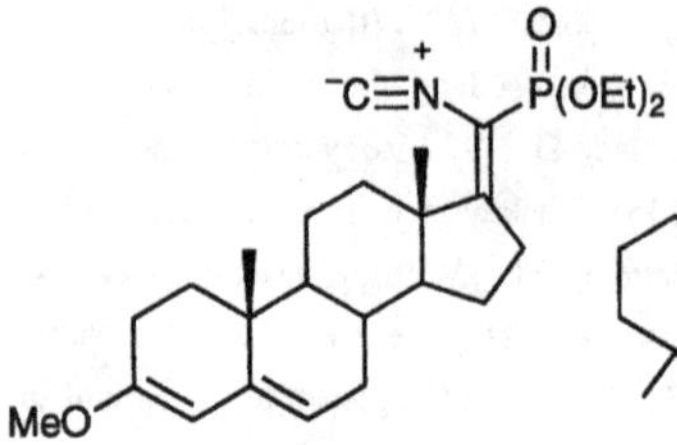

(233)

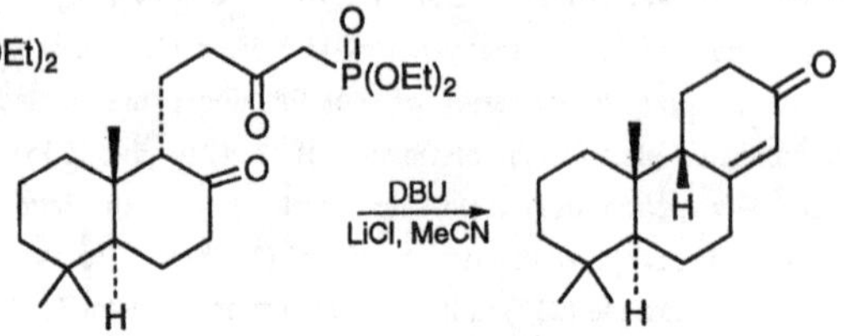
DBU
LiCl, MeCN

(234) (235)

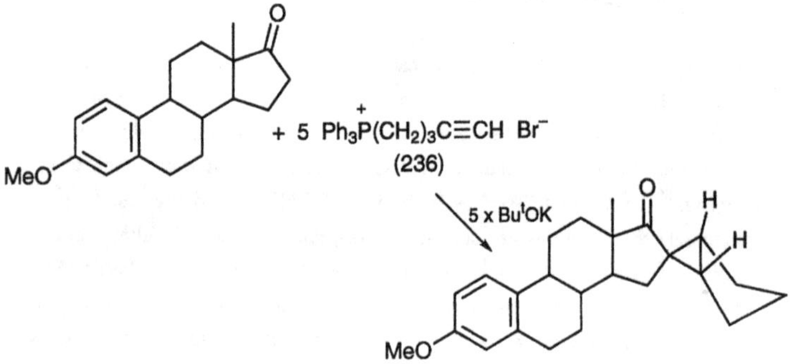
+ 5 $Ph_3\overset{+}{P}(CH_2)_3C{\equiv}CH$ Br^-
5 x Bu^{t}OK

(236)

(237)

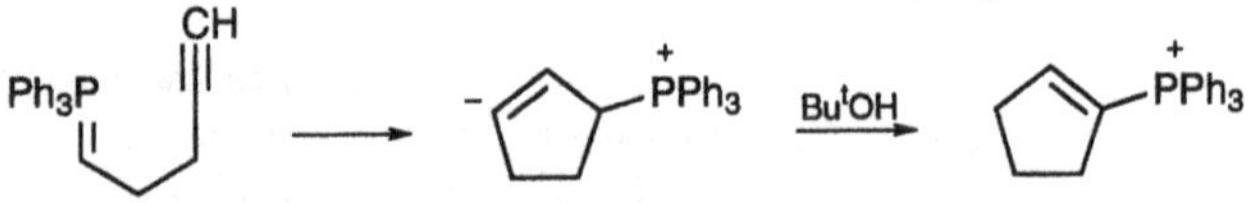
Bu^{t}OH

(238) (239)

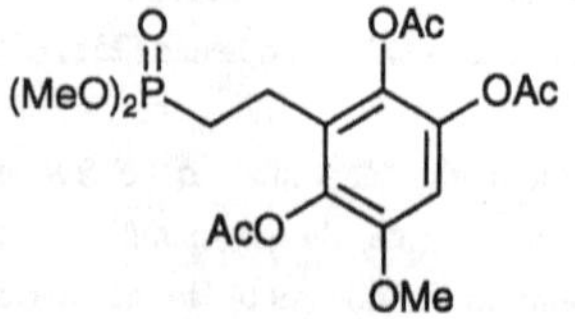

(240)

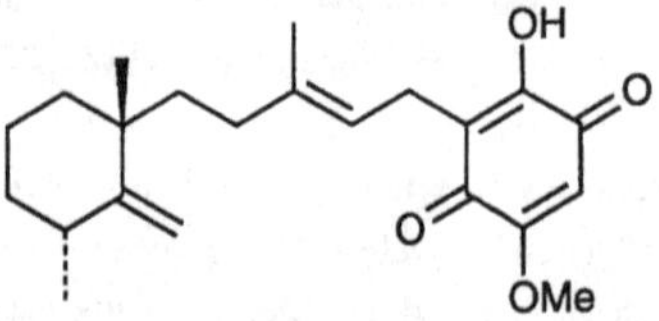

(241)

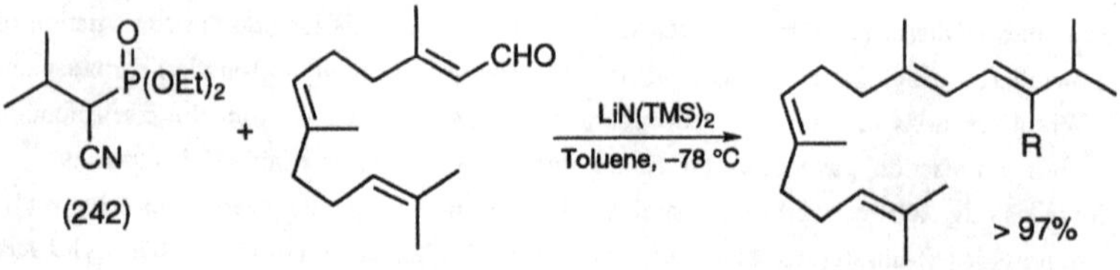
LiN(TMS)$_2$
Toluene, –78 °C
> 97%

(242) (243)

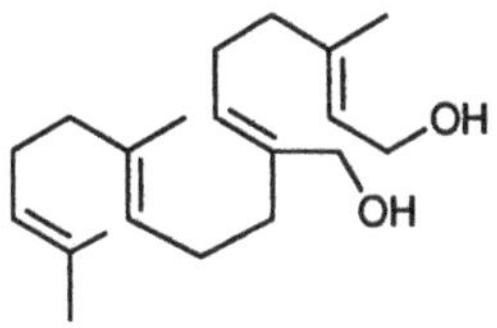

OH
OH
(244)

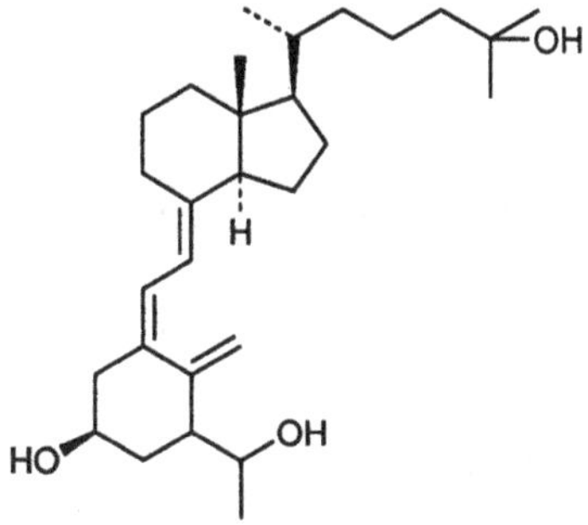

OH
H
HO
OH
(245)

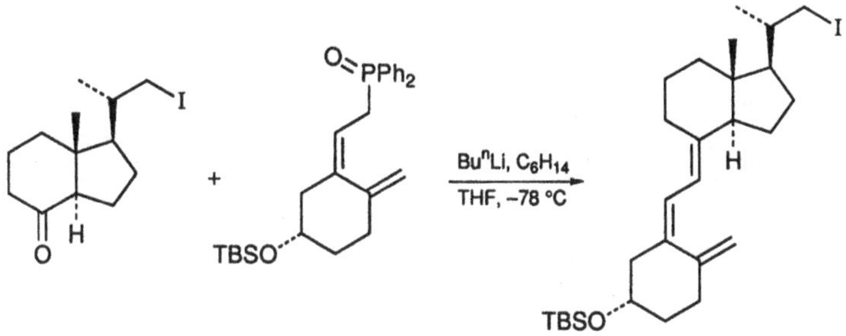

I
O
H
+
O
PPh_2
TBSO
Bu^nLi, C_6H_{14}
THF, –78 °C
I
H
TBSO
(246)

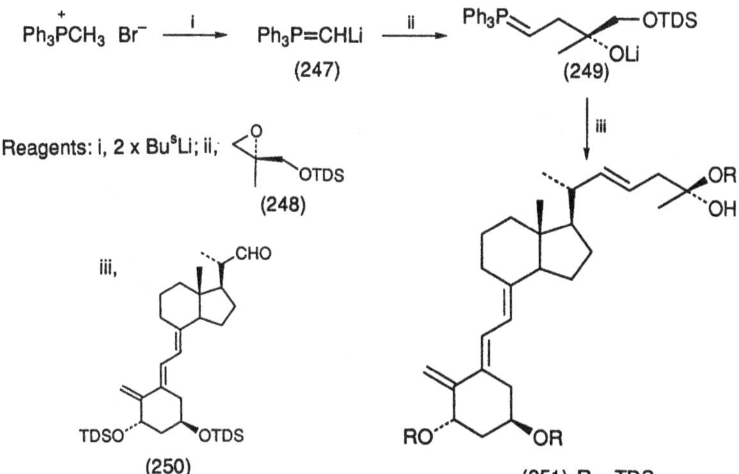

$Ph_3\overset{+}{P}CH_3$ Br^- —i→ $Ph_3P{=}CHLi$ (247) —ii→ Ph_3P ... OTDS, OLi (249)

iii

Reagents: i, 2 x Bu^sLi; ii, (248) O, OTDS

iii, (250) CHO, TDSO, OTDS

OR
OH
RO
OR
(251) R = TDS
(252) R = H

Scheme 23

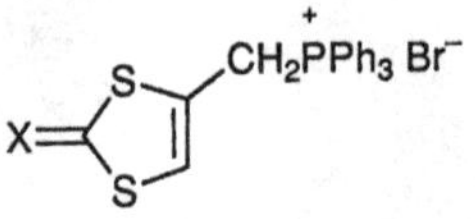

(253) X = O
(254) X = S

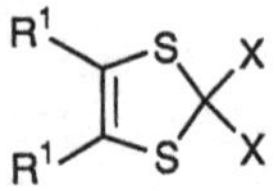

(255) X = Ph_3P^+
(256) X = $(MeO)_2P$(=O)

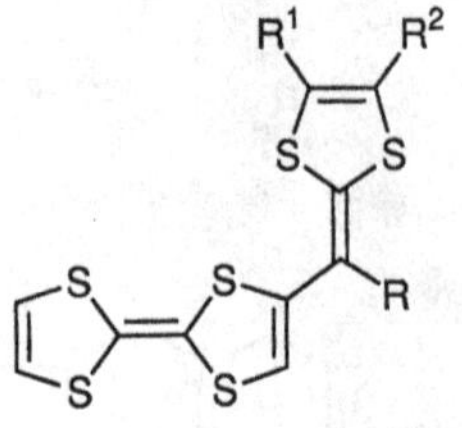

(257)

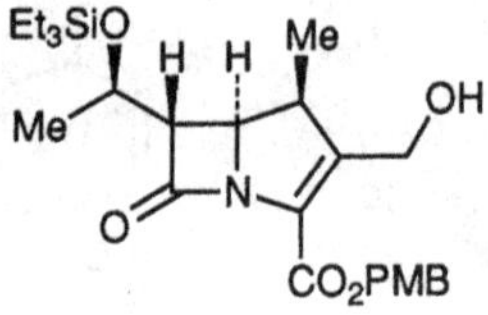

(258)

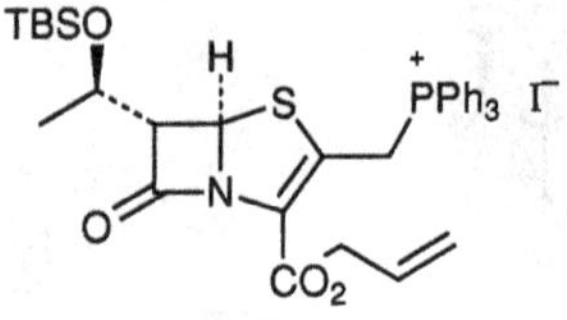

(259)

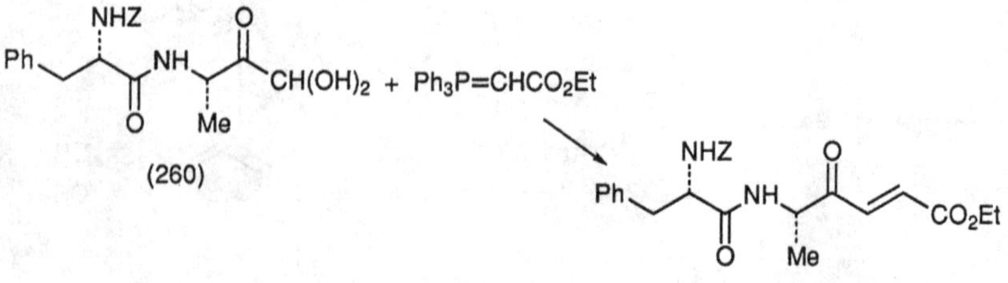

(260) + $Ph_3P{=}CHCO_2Et$ → (261)

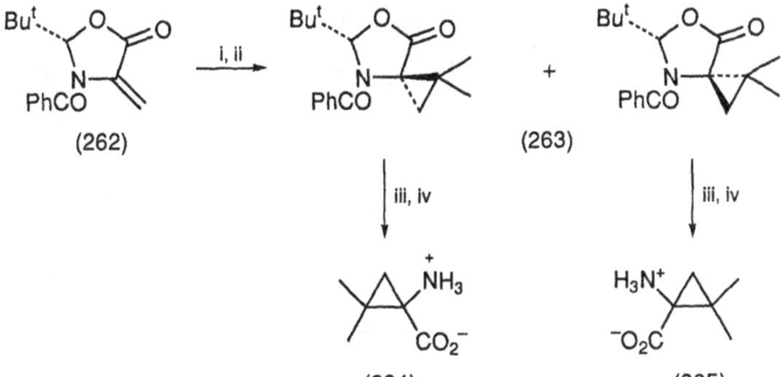

(262) (263) (264) (265)

Reagents: i, $Ph_3P{=}CMe_2$; ii, H_2O; iii, 2M HCl aq ; iv, propylene oxide

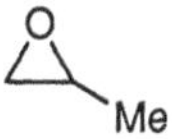

Scheme 24

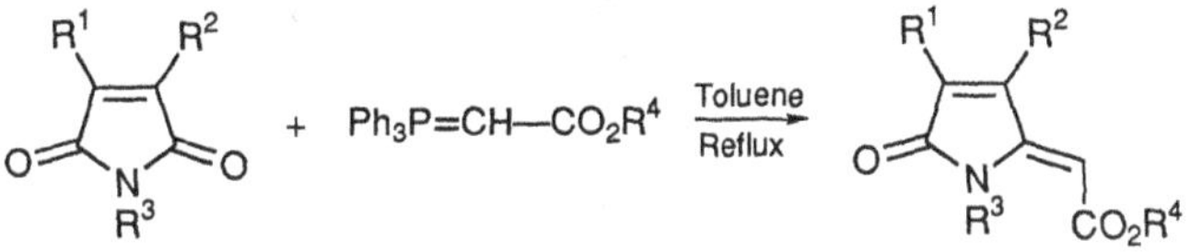

+ $Ph_3P{=}CH{-}CO_2R^4$ Toluene, Reflux → (266)

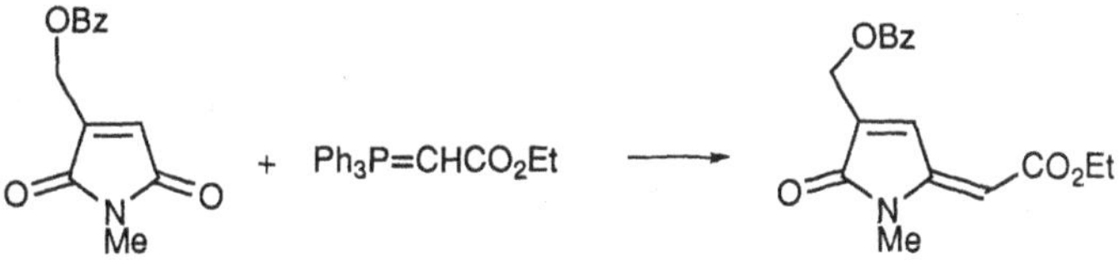

(267) + $Ph_3P{=}CHCO_2Et$ → (268)

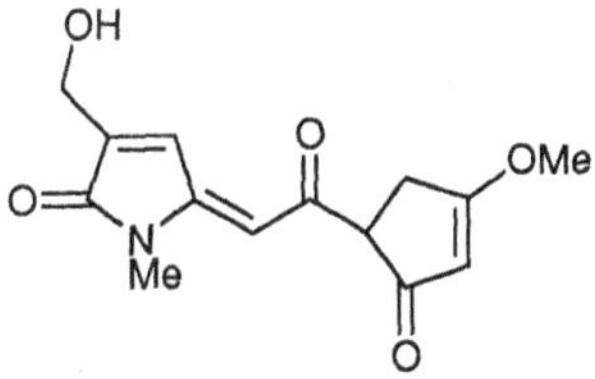

(269) 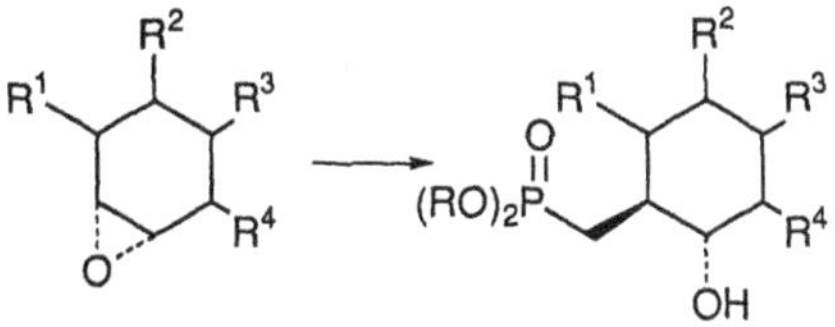 (270)

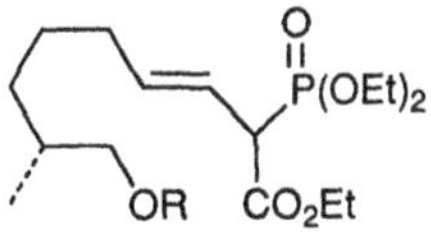

(271)

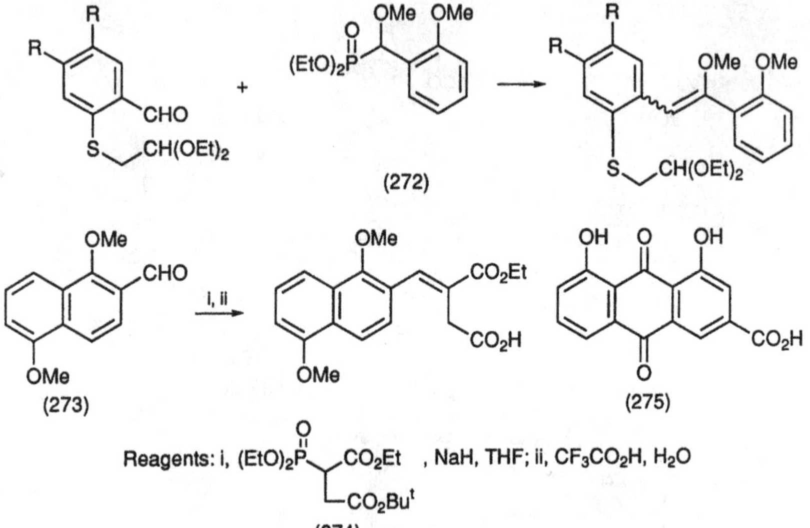

(272)

(273)

(275)

Reagents: i, $(EtO)_2P(O)CH(CO_2Et)CH_2CO_2Bu^t$ (274), NaH, THF; ii, CF_3CO_2H, H_2O

Scheme 25

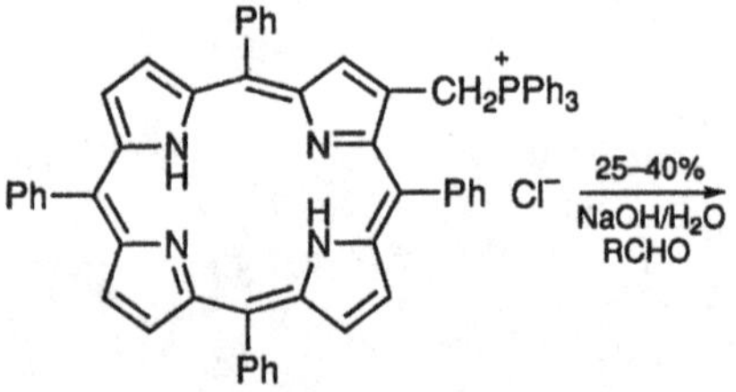

(276)

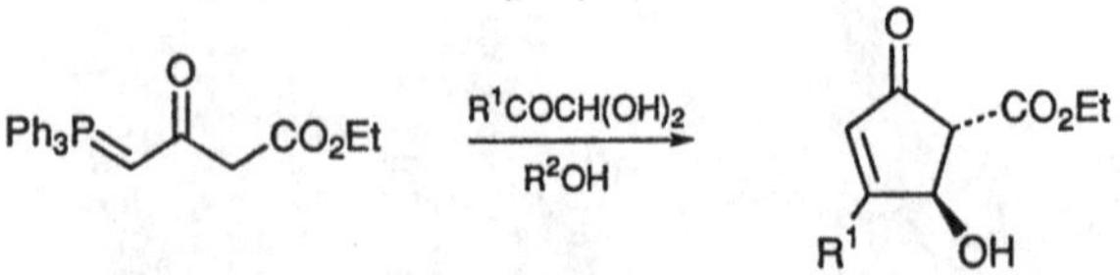

25–40% NaOH/H_2O RCHO

$R^1COCH(OH)_2$ R^2OH

(277)

(278)

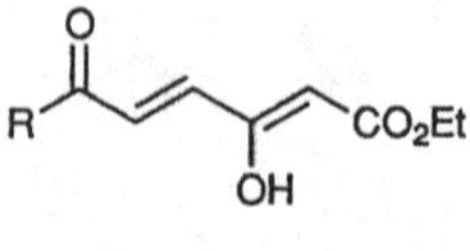

(279)

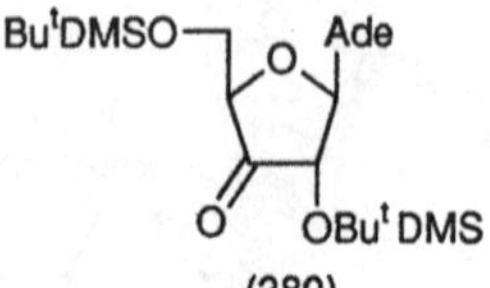

(280)

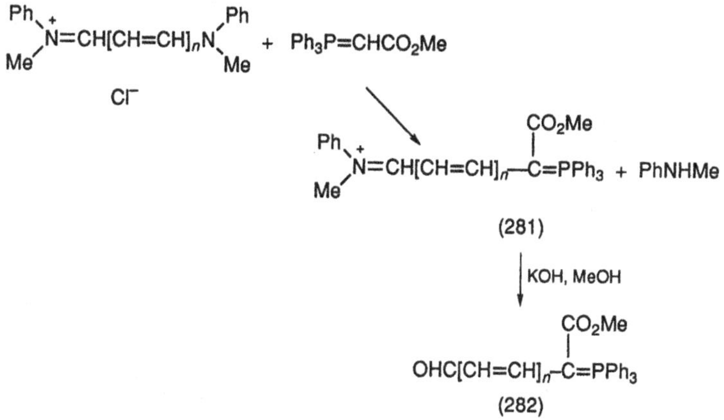

$$\mathrm{Ph(Me)\overset{+}{N}{=}CH[CH{=}CH]_{\mathit{n}}N(Ph)Me\ Cl^- + Ph_3P{=}CHCO_2Me}$$

$$\longrightarrow \mathrm{Ph(Me)\overset{+}{N}{=}CH[CH{=}CH]_{\mathit{n}}{-}C(CO_2Me){=}PPh_3 + PhNHMe}$$

(281)

$$\xrightarrow{\text{KOH, MeOH}} \mathrm{OHC[CH{=}CH]_{\mathit{n}}{-}C(CO_2Me){=}PPh_3}$$

(282)

mixture of diastereomeric cyclopropanes (**263**).[149] This mixture can be separated and the individual isomers converted into the enantiomerically pure (*R*)-(**264**) and (*S*)-(**265**)-2,3-methanovaline (Scheme **24**).

The synthesis of 5-ylidenepyrrol-2(5H)-ones (**266**) from the reaction of maleimides with stabilised phosphoranes under vigorous conditions has been reported.[150] Attempts to use similar reactions of unstabilised phosphoranes, phosphonate-, or phosphine oxide carbanions led to intractable products or recovered starting materials. The procedure developed in these studies has been applied to the synthesis of pukeleimide A (**269**) using the formation of the ylidenepyrrolone (**268**) from the maleimide (**267**) as a key step.[151]

The direct phosphonomethylation of cyclohexene oxides with lithiomethanephosphonate in the presence of boron trifluoride to give (**270**) has been compared with a step-wise approach to (**270**) involving initial reaction with (lithiomethyl)dimesitylborane followed by halogenation and phosphorylation.[152] The inconvenience of the step-wise approach must be weighed against its much higher degree of regioselectivity of epoxide ring-opening. Both ylide- and phosphonate-based olefination methods have been used in a synthetic approach to the tetronic acid ionophore antibiotic tetronasin.[153] An important step in the approach is the construction of a diene fragment using the novel phosphonate (**271**). Phosphonate-olefination involving (**272**) has been used to introduce synthons for all the structural components in a new approach to the synthesis of 5-thiorotenoids.[154] The stereospecific olefination of the aldehyde (**273**) with the differentially-protected phosphonate (**274**) is a key step in a novel synthesis of rhein (**275**) (Scheme**25**), the active principle of the anti-ostereoarthritic drug diacetyl rhein.[155]

The porphyrin-derived phosphonium salt (**276**) has been synthesised and used in Wittig reactions to prepare styryltetraporphyrins and butadiene porphyrin dimers.[156] 4-Hydroxycyclopentenones (**278**) have been obtained in moderate to good yield from the reaction of 3-ethoxycarbonyl-2-oxo-propylidenetriphenylphosphorane (**277**) with glyoxals.[157] The best yields of (**278**) were obtained from reactions in methanol or ethanol since the competing Wittig reaction to give (**279**) was suppressed in these solvents. A number of 3'-modified nucleosides have been prepared using initial Wittig or phosphonate-based olefination of the ketone (**280**) followed by hydrogenation.[158] Ylide derivatives (**281**) and (**282**) of poymethincyanines and polymethinmerocyanines, respectively, have been prepared.[159]

REFERENCES

1. A. W. Johnson, "Ylides and Imines of Phosphorus", Wiley Interscience, New York, 1993.
2. X-M. Zhang and F. G. Bordwell, *J. Am. Chem. Soc.*, 1994, **116**, 968.
3. D. G. Gilheany, *Chem. Rev.*, 1994, **94**, 1339.
4. J. Chen, T. Wang, and K. Zhao, *Tetrahedron Letters*, 1994, **35**, 2827.
5. O. I. Kolodiazhnyi, S. Ustenko, and O. Golovatyi, *Tetrahedron Letters*, 1994, **35**, 1755.
6. M. Soleilhavoup, A. Baceiredo, and G. Bertrand, *Angew. Chem. Int. Ed. Engl.*, 1993, **32**, 1167.
7. G. Jochem, A. Schmidpeter, M. Thomann, and H. Noth, *Angew. Chem. Int. Ed. Engl.*, 1994, **33**, 663.
8. A. H. Cowley, F. Gabbai, C. J. Carrano, L. M. Mokry, M. R. Bond, and G. Bertrand, *Angew. Chem. Int. Ed. Engl.*, 1994, **33**, 578.
9. M. Yoshifuji, K. Kamijo, and K.Toyota, *Tetrahedron Letters*, 1994, **35**, 3971.

10. H. J. Bestmann, D. Hadawi, H. Behl, M. Bremer, and F. Hampel, *Angew. Chem. Int. Ed. Engl.*, 1993, **32**, 1205.
11. B. M. Trost and C-J. Li, *J. Am. Chem. Soc.*, 1994, **116**, 3167.
12. S. Rodriguez-Morgade, T. Torres, and P. Vázquez, *Synthesis*, 1993, 1235.
13. A. R. Katritsky, R. Mazurkiewics, C. V. Stevens, and M. F. Gordeev, *J. Org. Chem.*, 1994, **59**, 2740.
14. A. R. Katritsky, J. Jiang, and J. V. Greenhill, *Synthesis*, 1994, 107.
15. L. V. Meervelt, R. N. Vydzhak, V. S. Bovarets, N. I. Mishchenko, and B. S. Drach, *Tetrahedron*, 1994, **50**, 1889.
16. J. Tejeda, R. Réau, F. Dahan, and G. Bertrand, *J. Am. Chem. Soc.*, 1993, **115**, 7880.
17. O. Treutler, R. Ahlrichs, and M. Soleilhavoup, *J. Am. Chem. Soc.*, 1993, **115**, 8788.
18. F. Lopez-Ortiz, E. Peláez-Arango, F. Palacios, J. Barluenga, S. Garcia-Granda, T. Baudilio, and A. Garcia-Fernádez, *J. Org. Chem.*, 1994, **59**, 1984.
19. A. L. Llamas-Saiz, C. Foces-Foces, J. Elguero, F. Aguilar-Parrilla, H-H. Limbach, P. Molina, M. Alajarin, A. Vidal, R. M. Claramont, and C. Lopez, *J. Chem. Soc., Perkin Trans. 2*, 1994, 209.
20. H. Yamataka, K. Nagareda, T. Takatsuka, K. Ando, T. Hanafusa, and S. Nagase, *J. Am. Chem. Soc.*, 1993, **115**, 8570.
21. D. L. Hooper, S. Garagan, and M. M. Kayser, *J. Org. Chem.*, 1994, **59**, 1126.
22. E. Vedejs, J. Cabaj, and M. J. Peterson, *J. Org. Chem.*, 1993, **58**, 6509.
23. T. Kawashima, H. Takami, and R. Okazaki, *J. Am. Chem. Soc.*, 1994, **116**, 4509.
24. N. J. Lawrence and F. Muhammad, *J. Chem. Soc., Chem. Commun.*, 1993, 1187.
25. K. Hemming and R. J. K. Taylor, *J. Chem. Soc., Chem. Commun.*, 1993, 1409.
26. B. J. Adger, C. Barrett, J. Brennan, P. McGuigan, M. A. McKervey, and B. Tarbit, *J. Chem. Soc., Chem. Commun.*, 1993, 1220.
27. V. Bhushan, B. B. Lohray, and D. Enders, *Tetrahedron Letters*, 1993, **34**, 5067.
28. H-D. Ambrosi, W. Duczek, M. Ramm, E. Gründemann, B. Schulz, and K. Jahnisch, *Liebigs Ann. Chem.*, 1994, 1013.
29. T. B. Patrick, M. V. Lanahan, C. Yang, J. K. Walker, C. L. Hutchinson, and B. E. Neal, *J. Org. Chem.*, 1994, **59**, 1210.
30. C. Chen, L. A. A. Randall, R. B. Miller, A. D. Jones, and M. J. Kurth, *J. Am. Chem. Soc.*, 1994, **116**, 2661.
31. J. Vidal, L. Guy, S. Sterin, and A. Collet, *J. Org. Chem.*, 1993, **58**, 4791.
32. D. N. Nicolaides, R. W. Awad, G. K. Papageorgiou, and J. Stephanidou-Stephanatou, *J. Org. Chem.*, 1994, **59**, 1083.
33. F. H. Osman, N. M. A. El-Rahmann, and F. A. El-Samahy, *Tetrahedron*, 1993, **49**, 8691.
34. C. Bezergiaunidou-Balouctsi, K. E. Litinas, E. Malamidou-Xenikaki, and D. N. Nicolaides, *Liebigs Ann. Chem.*, 1993, 1175.
35. M. Hatanaka, Y. Himeda, and I. Ueda, *J. Chem. Soc., Perkin Trans. 1*, 1993, 2269.
36. M. Hatanaka, Y. Himeda, R. Imashiro, Y. Tanaka, and I. Ueda, *J. Org. Chem.*, 1994, **59**, 111.
37. T. Fujimoto, Y-k. Uchiyama, Y-i. Kodama, K. Ohta, I. Yamamoto, and A. Kakehi, *J. Org. Chem.*, 1993, **58**, 7322.
38. F. Palacios, I. P. de Heredia, and G. Rubiales, *Tetrahedron Letters*, 1993, **34**, 4377.
39. Z-L. Zhou, Y-Z. Huang, and L-L. Shi, *Tetrahedron*, 1993, **49**, 6821.
40. H. Grützmacher, W. Deck, H. Pritzkow, and M. Sander, *Angew. Chem. Int. Ed. Engl.*, 1994, **33**, 456.

41. H.J. Bestmann, T-G. Fürst, and A. Schier, *Angew. Chem. Int. Ed. Engl.*, 1993, **32**, 1747.
42. H.J. Bestmann, T-G. Fürst, and A. Schier, *Angew. Chem. Int. Ed. Engl.*, 1993, **32**, 1746.
43. L. Pandolfo, G. Facchin, R. Bertani, P. Ganis, and G. Valle, *Angew. Chem. Int. Ed. Engl.*, 1994, **33**, 576.
44. U. Hein, H. Pritzkow, U. Fleischer, and H. Grützmacher, *Angew. Chem. Int. Ed. Engl.*, 1993, **32**, 1359.
45. K. Okuma, Y. Tanaka, K. Yoshihara, A. Ezaki, G. Koda, H. Ohta, K. Hara, and S. Kashimura, *J. Org. Chem.*, 1993, **58**, 5915.
46. T. Shen and S. Gao, *J. Org. Chem.*, 1993, **58**, 4564.
47. Y. Shen and S. Gao, *J. Chem. Soc., Perkin Trans. 1.*, 1994, 1473.
48. J-P. Bégué, D. Bonnet-Delpon, and A. M'Bida, *Tetrahedron Letters*, 1993, **34**, 7753.
49. J-P. Bégué, D. Bonnet-Delpon, S-W. Wu, A. M'Bida, T. Shintani, and T. Nakai, *Tetrahedron Letters*, 1994, **35**, 2907.
50. E. J. Latham, S. M. Murphy, and S. P. Stanforth, *Tetrahedron Letters*, 1994, **35**, 3395.
51. R. A. Aitken and J. I. Atherton, *J. Chem. Soc., Perkin Trans. 1.*, 1994, 1281.
52. R. A. Aitken, H. Hérion, A. Janosi, S. V. Raut, S. Seth, I. J. Shannon, and F. C. Smith, *Tetrahedron Letters*, 1993, **34**, 5621.
53. R. A. Aitken, M. J. Drysdale, and B. M. Ryan, *J. Chem. Soc., Chem. Commun.*, 1993, 1699.
54. R. A. Aitken, M. J. Drysdale, and B. M. Ryan, *J. Chem. Soc., Chem. Commun.*, 1994, 805.
55. J-B. Verlhac, H. Kwon, and M. Pereyre, *J. Chem. Soc., Perkin Trans. 1.*, 1993, 1367.
56. L. K. Johnson, M. Frey, T. A. Ulibarri, S. C. Virgil, R. H. Grubbs, and J. W. Ziller, *J. Am. Chem. Soc.*, 1993, **115**, 8167.
57. W. M. Abdou and E-S. M. A. Yakout, *Tetrahedron,* 1993, **49**, 6411.
58. G. Jochem, H. Nöth, and A. Schmidpeter, *Angew. Chem. Int. Ed. Engl.*, 1993, **32**, 1089.
59. M. Yoshifuji, D-L. An, K. Tryota, and M. Yasunami, *Tetrahedron Letters*, 1994, **35**, 4379.
60. W. Schilbach, V. von der Gönna, D. Gudat, M. Nieger, and E. Nieke, *Angew. Chem. Int. Ed. Engl.*, 1994, **33**, 982.
61. J. Clayden and S. Warren, *J. Chem. Soc., Perkin Trans. 1.*, 1994, 1529.
62. S. K. Armstrong, E. W. Collington, J. G. Knight, A. Naylor, and S. Warren, *J. Chem. Soc., Perkin Trans. 1.*, 1993, 1433.
63. S. K. Armstrong, E. W. Collington, and S. Warren, *J. Chem. Soc., Perkin Trans. 1.*, 1994, 515.
64. A. Couture, E. Deniau, and P. Grandclaudon, *J. Chem. Soc., Chem. Commun.*, 1994, 1329.
65. C. J. Cramer, S.E. Denmark, P. C. Miller, R. L. Dorrow, K.A. Swiss, and S.R. Wilson, *J. Am. Chem. Soc.*, 1994, **116**, 2437.
66. S.E. Denmark and K.A. Swiss, *J. Am. Chem. Soc.*, 1993, **115**, 12195.
67. S. Hanessian, A. Gimtsyan, A. Payne, Y. Hervé, and S. Beaudoin, *J. Org. Chem.*, 1993, **58**, 5032.
68. N. Kann and T. Rein, *J. Org. Chem.*, 1993, **58**, 3802.
69. T. Rein, N. Kann, R. Kreuder, B. Gangloff, and O. Reiser, *Angew. Chem. Int. Ed. Engl.*, 1994, **33**, 556.
70. S.E. Denmark and J. Amburgey, *J. Am. Chem. Soc.*, 1993, **115**, 10386.
71. R. C. Lunn and J. J. Grabowski, *J. Am. Chem. Soc.*, 1993, **115**, 7823.
72. K. Lee, J. A. Jackson, and D. F. Wiemer, *J. Org. Chem.*, 1993, **58**, 5967.
73. F. Bonadies, A. Cardilli, A. Lattanzi, L. R. Orelli, and A. Scettri, *Tetrahedron Letters*, 1994, **35**, 3383.
74. Z-Y. Wei and E. E. Knaus, *Tetrahedron Letters*, 1994, **35**, 2305.
75. E. Wenkert and M. K. Schorp, *J. Org. Chem.*, 1994, **59**, 1943.

76. Y. Nakamura and C-g. Shin, *Synthesis*, 1994, 552.

77. D. Haigh, *Tetrahedron*, 1994, **50**, 3177.

78. R. C. F. Jones, G. Bhalay, and P. A. Carter, *J. Chem. Soc., Perkin Trans. 1.*, 1993, 1715.

79. S. E. Denmark and C-T. Chen, *J. Org. Chem.*, 1994, **59**, 2922.

80. S. M. Ruder and V. R. Kulkarni, *Synthesis*, 1993, 945.

81. T. Minami, T. Isonaka, Y. Okada, and J. Ichikawa, *J. Org. Chem.*, 1993, **58**, 7009.

82. S. Berté-Verrando, F. Nief, C. Patois, and P. Savignac, *J. Chem. Soc., Perkin Trans. 1.*, 1994, 821.

83. C. Patois and P. Savignac, *J. Chem. Soc., Chem. Commun.*, 1993, 1711.

84. C-W. Lee, Y. J. Koh, and D. Y. Oh, *J. Chem. Soc., Perkin Trans. 1.*, 1994, 717.

85. C. Yuan and C. Li, *Tetrahedron Letters*, 1993, **34**, 5959.

86. D. Maffre, P. Dumy, J-P. Vidal, R. Escale, and J-P. Girard, *J. Chem. Research (S)*, 1994, 30.

87. O. A. Attanasi, P. Filippone, D. Giovagnoli, and A. Mei, *Synthesis*, 1994, 181.

88. H. Makomo, S. Masson, and M. Saquet, *Tetrahedron Letters*, 1993, **34**, 7257.

89. J. Fawcett, M. J. P. Harger, D. R. Russell, and R. Sreedharan-Menon, *J. Chem. Soc., Chem. Commun.*, 1993, 1826.

90. D. B. Berkowitz, M. Eggen, Q. Shen, and D. G. Sloss, *J. Org. Chem.*, 1993, **58**, 6174.

91. A. Dondoni, A. Marra, and P. Merino, *J. Am. Chem. Soc.*, 1994, **116**, 3324.

92. A. Dondoni and A. Marra, *Tetrahedron Letters*, 1993, **34**, 7327.

93. L. Dheilly, C. Lièvre, C. Fréchou, and G. Demailly, *Tetrahedron Letters*, 1993, **34**, 5895.

94. J. J. Kiddle and J. H. Babler, *J. Org. Chem.*, 1993, **58**, 3572.

95. C. F. Jewell, Jr., J. Brinkman, R. C. Petter, and J. R. Wareing, *Tetrahedron*, 1994, **450**, 3849.

96. Y. Katsuta, M. Sakai, and M. Ito, *J. Chem. Soc., Perkin Trans. 1.*, 1993, 2185.

97. Y. Katsuta, K. Yoshhara, K. Nakanishi, and M. Ito, *Tetrahedron Letters*, 1994, **35**, 905.

98. H. J. Bestmann, D. Fett, W. Garbe, N. Gunawardena, V. Martichonok, and O. Vostrowsky, *Leibigs Ann. Chem.*, 1994, 113.

99. L. H. B. Baptistella and A. M. Aleixo, *Leibigs Ann. Chem.*, 1994, 785.

100. L. Guo-qiang and X. Wei-chu, *Tetrahedron Letters*, 1993, **34**, 5931.

101. L. Zegelman, A. Hassner, Z. Mendel, and E. Dunkelblum, *Tetrahedron Letters*, 1993, **34**, 5641.

102. L. Duhamel, P. Duhamel, G. Ple, and Y. Ramondenc, *Tetrahedron Letters*, 1993, **34**, 7399.

103. S. S. Wang, J. Rokach, W. S. Powell, C. Dekle, and S. J. Feinmark, *Tetrahedron Letters*, 1994, **35**, 4051.

104. F. A. J. Kerdesky, S. P. Schmidt, and D. W. Brooks, *J. Org. Chem.*, 1993, **58**, 3516.

105. R. K. Bhatt, K. Chauhan, P. Wheelan, R. C. Murphy, and J. R. Falck, *J. Am. Chem. Soc.*, 1994, **116**, 5050.

106. J-M. Pons, A. Pommier, J. Lerpiniere, and P. Kocienski, *J. Chem. Soc., Perkin Trans. 1.*, 1993, 1549.

107. P. I. Butler, T. Clarke, C. Dell, and J. Mann, *J. Chem. Soc., Perkin Trans. 1.*, 1994, 1503.

108. T. K. M. Shing, K. H. Gibson, J. R. Wiley, and C. I. F. Watt, *Tetrahedron Letters*, 1994, **35**, 1067.

109. S. D. Rychnovsky and R. C. Hoye, *J. Am. Chem. Soc.*, 1994, **116**, 1753.

110. L. A. Paquette, T-Z. Wang, C. M. G. Philippo, and S. Wang, *J. Am. Chem. Soc.*, 1994, **116**, 3367.

111. M. L. Morin-Fox and M. A. Lipton, *Tetrahedron Letters*, 1993, **34**, 7899.

112. G. M. Keseru, M. Nógrádi, and M. Kajtár, *Leibigs Ann. Chem.*, 1994, 361.

113. A. J. Cooper, W. Pan, and R. G. Salomon, *Tetrahedron Letters*, 1993, **34**, 8193.

114. M. J. Hughes and E. J. Thomas, *J. Chem. Soc., Perkin Trans. 1.*, 1993, 1493.

115. I. Paterson and K. S. Yeung, *Tetrahedron Letters*, 1993, **34**, 5347.

116. M. Bergdahl, R. Hett, T. L. Friebe, A. R. Gangloff, J. Iqbal, Y. Wu, and P. Helquist, *Tetrahedron Letters*, 1993, **34**, 7371.

117. D. Romo, S. D. Meyer, D. D. Johnson, and S. L. Schreiber, *J. Am. Chem. Soc.*, 1993, **115**, 7906.

118. P. Molina, E. Aller, A. López-Lázaro, M. Alajarin, and A. Lorenzo, *Tetrahedron Letters*, 1994, **35**, 3817.

119. P. Molina, P. Almendros, and P. M. Fresneda, *Tetrahedron Letters*, 1994, **35**, 2235.

120. P. Molina, E. Aller, and M. A. Lorenzo, *Synthesis*, 1993, 1239.

121. P. Molina, P. Almendros, and P. M. Fresneda, *Tetrahedron*, 1994, **50**, 2241.

122. T. Bohn, W. Kramer, R. Neidlein, and H. Suschitzky, *J. Chem. Soc., Perkin Trans. 1.*, 1994, 947.

123. P. Molina, P. Almendros, and P. M. Fresneda, *Tetrahedron Letters*, 1993, **34**, 4701.

124. P. Molina, F. Murcia, and P. M. Fresneda, *Tetrahedron Letters*, 1994, **35**, 1453.

125. C. Peinador, M. J. Moreira, and J. M. Quintela, *Tetrahedron*, 1994, **50**, 6705.

126. P. Molina, A. Alias, A. Balado, and A. Arques, *Leibigs Ann. Chem.*, 1994, 745.

127. P. Molina, C. Conesa, A. Alias, A. Arques, M. D. Velasco, A. L. Llamas-Saiz, and C. Foces-Foces, *Tetrahedron*, 1993, **49**, 7599.

128. H. Wamhoff and A. Schmidt, *J. Org. Chem.*, 1993, **58**, 6976.

129. P. Molina, A. Arques, and A. Alias, *J. Org. Chem.*, 1993, **58**, 5264.

130. P. Molina, M. Alajarin, and P. Sanchez-Andrada, *Tetrahedron Letters*, 1993, **34**, 5155.

131. I. Bosch, P. Romea, F. Urpi, and J. Vilarrasa, *Tetrahedron Letters*, 1993, **34**, 4671.

132. J. Mitjaville, A-M. Caminade, R. Mathieu, and J-P. Majoral, *J. Am. Chem. Soc.*, 1994, **116**, 5007.

133. A. Couture, E. Deniau, Y. Gimbert, and P. Grandclaudon, *J. Chem. Soc., Perkin Trans. 1.*, 1993, 2463.

134. P. J. Murphy and H. L. Williams, *J. Chem. Soc., Chem. Commun.*, 1994, 819.

135. J. Stoelwinder and A. M. van Leusen, *J. Org. Chem.*, 1993, **58**, 3687.

136. J. M. Weibel and D. Heissler, *Tetrahedron Letters*, 1994, **35**, 473.

137. A. I. A. Broess, M. B. Groen, and H. Hamersma, *Tetrahedron Letters*, 1994, **35**, 335.

138. W. P. Almeida and C. R. D. Correia, *Tetrahedron Letters*, 1994, **35**, 1367.

139. H. Takayanagi, *Tetrahedron Letters*, 1994, **35**, 1581.

140. A. Nangia, G. Prasuna, and P. B. Rao, *Tetrahedron Letters*, 1994, **35**, 3755.

141. G. H. Posner, H. Dai, K. Afarinkia, N. N. Murthy, K. Z. Guyton, and T. W. Kensler, *J. Org. Chem.*, 1993, **58**, 7209.

142. J. Pérez-Sestelo, J. L. Mascareñas, L. Castedo, and A. Mouriño, *Tetrahedron Letters*, 1994, **35**, 275.

143. M. Okabe and R-C. Sun, *Tetrahedron Letters*, 1993, **34**, 6533.

144. T. Nozdryn, J. Cousseau, A. Gorgues, M. Jubault, J. Orduna, S. Uriel, and J. Garin, *J. Chem. Soc., Perkin Trans. 1.*, 1993, 1711.

145. M. Sallé, A. J. Moore, M. R. Bryce, and M. Jubault, *Tetrahedron Letters*, 1993, **34**, 7475.

146. S. Uyeo and H. Itani, *Tetrahedron Letters*, 1994, **35**, 4377.

147. M. Altamura and E. Perrotta, *Tetrahedron Letters*, 1994, **35**, 1417.

148. P. Darkins, N. McCarthy, M. A. McKervey, and T. Ye, *J. Chem. Soc., Chem. Commun.*, 1993, 1222.

149. R. Chinchilla, C. Nájera, S. Garcia-Granda, and Menéndez-Velázquez, *Tetrahedron Letters*, 1993, **34**, 5799.

150. G. B. Gill, G. D. James, K. V. Oates, and G. Pattenden, *J. Chem. Soc., Perkin Trans. 1.*, 1993, 2567.

151. G. D. James, S. D. Mills, and G. Pattenden, *J. Chem. Soc., Perkin Trans. 1.*, 1993, 2581.

152. J-L. Montchamp, M. E. Migaud, and J. W. Frost, *J. Org. Chem.*, 1993, **58**, 7679.

153. G. J. Boons, I. C. Lennon, S. V. Ley, E. S. E. Owen, J. Staunton, and D. J. Wadsworth, *Tetrahedron Letters*, 1994, **35**, 323.

154. L. Crombie and J. L. Josephs, *J. Chem. Soc., Perkin Trans. 1.*, 1993, 2599.

155. P. T. Gallagher, T. A. Hicks, A. P. Lightfoot, and W. M. Owton, *Tetrahedron Letters*, 1994, **35**, 289.

156. E. E. Bonfantini and D. L. Officer, *Tetrahedron Letters*, 1993, **34**, 8531.

157. M. Hatanaka, Y. Tanaka, Y. Himeda, and I. Ueda, *Tetrahedron Letters*, 1993, **34**, 4837.

158. K. Lee and D. F. Wiemer, *J. Org. Chem.*, 1993, **58**, 7808.

159. G. Märkl, A. Rehberger, and W. Schumann, *Tetrahedron Letters*, 1993, **34**, 7385.

7
Phosphazenes

BY C. W. ALLEN

1 Introduction

This chapter covers the literature of phosph(V)azenes with discussion of lower valent species restricted to molecules which can be transformed or related to a phosphorus(V) derivative. Both basic and applied studies have been perused leading to a significant volume of papers and patents. A global overview of phosphazene chemistry including acyclic, cyclic and polymeric derivatives with particular emphasis on recent advances has appeared. [1] The collected papers, including abstracts of posters, from the 12th International Conference of Phosphorus Chemistry (Toulouse, 1992) have been published.[2] Summaries of papers from a symposium on Inorganic and Organometallic Polymers at the 205th American Chemical Society Meeting (Denver, 1993) are available.[3] Highly focused reviews will be cited in the appropriate sections below.

2 Acyclic Phosphazenes

Various structural permutations of hypothetical XPN systems present appropriate goals for elucidation by ab initio molecular orbital calculations. In the cationic species RNP^+ (R=H,Me,Ph) location of the substituent on nitrogen is favored over the P-substituted isomer by 90.7 kcal (R=H) with the stability being sufficient to allow for gas phase existence. [4] The geometric parameters for the PNO/$HPNO^+$ and NPO/$HNPO^+$ pairs in a variety of forms (linear, cyclic, etc) have been obtained from SCF/3-21G* calculations.[5] Ab initio (G_1 and G levels) investigations of the series $[P,N,H_n]$ (n=0-2) and $[P,N,H_n]^+$ (n=0-3) have been carried out and related to possible mechanisms for the formation of PN. A systematic view of the PN bonding characteristics in the series was reported.[6] Larger species have also been investigated including MNDO-PM3 studies of $(CF_3)_2$ P=N=PPh_3[7] and ab initio studies on MeS(O)R_1=NPR_2=NPR_2=NMe (R_1 R_2=H,Me,Cℓ; R_1=F, R_2=Cℓ).[8] The latter investigation showed a clear distinction between single and double bonds and a slight variation in SNP and PNP fragments with change in substituent.[8] Spectroscopic studies of the phosphazene unit also continue to appear. The emission spectrum with rotational resolution of one of the main transitions in PN^+ has been reported.[9] The He(I)/He(II) photoelectron spectra of $(CF_3)_2$ P=N=PPh_3 do not fit the calculations suggesting a Koopman's theorem breakdown. The PNP unit has low ionization energies which are associated with the allenic type structure as a major contributor to ground state electronic structure.[7] E.I. and high resolution tandem mass spectrometry has shown the occurrence of the N-substituted phosphoazonium ions CH_3NP^+ and Mes*NP^+ [Mes*=(2,4,6-CMe_3)C_6H_2] in the fragmentation of PN(Me)CH_2CH_2N(Me)$^+GaC\ell_4^-$ and Mes*NPCℓ respectively.[10] Variable temperature ^{31}P NMR spectroscopy has shown dynamic behavior for $C\ell_3PN(PC\ell_2N)_nP(O)C\ell_2$ (n=2,3) arising from PN backbone rotation. Raman spectroscopy has demonstrated a planar cis-trans molecular conformation in the melt, solid state and in solution for these and related compounds.[11] Polarization transfer techniques lead to significant sensitivity enhancement in the ^{15}N NMR spectra of phosphorus-nitrogen compounds. ^{15}N data, from ^{31}P,^{15}N INEP experiments, shows a reduction in the bonding in the R^1NPNR^{11} fragment over the parent compound in the lithium complex $R^1NP(NR^1)(NR^{11})$ (R^1=CMe_3, R^{11}=Mes*).[12] The formation of inclusion compounds of **1** (host) and amines (guest) has been examined by ^{13}C CPMAS NMR spectroscopy. The host is in the anionic form and guests are either the protonated amine and one water molecule or the protonated amine and its neutral counterpart.[13] The strong Lewis base character of the nitrogen center in the phosphazene bond continues to be exploited. Proton sponge like materials **2** (R=H, NH_2, NMe_2, $N=PPh_3$) have been examined with respect to protonation behavior. The separately reported crystal structures were corrected using semiempirical (AM1 or PM3) calculations to separate crystal packing from intrinsic configurations. Solution properties were probed using both NMR ($^1H/^{13}C$) spectroscopy and pKa determination. The **2** series are stronger bases than classical proton sponges but decompose on removal of the acidic proton.[14] ^{31}P MAS NMR of **2** and related materials show that in the solid state proton transfer does not occur, even in strongly hydrogen bonded systems.[15] Linear phosphazene bases such as **3** have continued to attract attention and are of sufficient value to be offered commercially (Aldrich, Fluka). The basicities (in acetonitrile) of several uncharged hindered systems have been determined. Very high values, $(R_2N)_3$ $P=NP(N=R^1)[N=P(NR_2)_3]N=P(NR_2)_3$ or **3** (R=Me, $R^1=CMe_3$, CH_2CMe_3; $R_2=(CH_2)_4$; $R^1=CMe_3$) 42.6-44 and 45-47 respectively, have been reported.

The Staudinger reaction is still the premier synthetic method for synthesis of new acyclic phosphazenes. In selected cases the intermediate phosphazides can be isolated as in the case of **4** ($R=P(NEt_2)_3$). Thermal decomposition of **4** ($R=(NEt_2)_3P=NN=NPh$) gives the expected phosphoranimine.[17] Sterically crowded azides RN_3 ($R=CMe_3$, 1-Ad, Ph_3C) react with R'_3P($R'=NMe_2$, C-pr) to give $RN=NN=PR'_3$. The corresponding reaction with CMe_3PF_2 goes to $CMe_3PF_2(=NR)$.[18] Selective oxidation of one phosphorus (III) center in Ph_2PRPPh_2 [$R=o-C_6H_4$, cis-CH=CH, R'N (R'=alkyl)] with AN_3 ($A=Me_3Si$, $P(O)(OPh_2)_2$, $p-CNC_6F_4$) can be achieved.[19,20] The course of the Staudinger reaction of Ph_3P with ortho substituted arylazides depends on the ortho substituent. When R in $1,2-C_6H_4(N_3)R$ is $CH=CBr_2$ or $C(O)CH_3$ the expected phosphoranimine is formed, however when R= CH=CHX ($X=C(COEt)_2$, CH_2NO_2) a cyclic phosphazide, $Ph_3P=N\overline{NN=C_6H_4=C}CH_2X$ is obtained.[21] Numerous N-silylphosphoranimines with various perturbations of substituents have been prepared by the Staudinger reaction for investigation as poly(phosphazene) precursors (see Section 6). Variation at the silicon center of

$(F_3CH_2O)_3$ $P{=}NSiR_3$ (R_3=$PhMe_2$, Et_3, Ph_3, $Me_2(CMe_3)$, Me_2H).[22] While variation at the phosphorus center to provide $(RO)_{3-n}(R'O)_nP{=}NSiMe_3$ ($R{=}MeOCH_2CH_2O$, $R'{=}CF_3CH_2O$, n=1,2; $R{=}MeOCH_2CH_2OCH_2CH_2O$, $R'{=}CF_3CH_2O$, n=1,2)[23] as well as other mixed alkoxy-trifluoroalkoxy[24] and aryloxy/alkoxy[25] has been reported. The reaction of Me_3SiN_3 with R_2PF where R is a bulky substituent gives the Staudinger product $R_2P{=}NSiMe_3$ while other R groups give more complex products due to azide exchange. [26,27] $Me_3CPh_2SiN_3$ is less reactive and more typically provides phosphoranimines.[27] The azide exchange can give rise to $R_3P{=}NR'$, $R_{3-n}P(N_3)_n{=}NR'$ (n=1,2) and subsequently $R_{3-n}P(N{=}PR_3)_n{=}NR'$ when the chlorophosphines are employed as starting materials.[26] The tricyclic phosphine, $P(MeNCH_2CH_2)_3N$ can be oxidized by organoazides to the very strong non-ionic bases, $RN{=}P(MeNCH_2CH_2)_3N$ (R=Me,Ph). The relative basicity of these and selected other phosphoranimines was established by ^{31}P NMR equilibration studies.[28] The β-chloroethylamine species $MeNC_6H_4C(O)NMePR$ [$R{=}N(CH_2CH_2C\ell)_2$, $NHCH_2CH_2C\ell$] can be converted to $MeNC_6H_4C(O)NMeP(R){=}NC(O)C_6H_4NO_2$ upon reaction with $N_3C(O)C_6H_4NO_2$.[29] Spin-labeled nitroxides with exocyclic carbonyl azides react with aziridino phosphines to gives antineoplastic agents with low toxicity and activity against leukemia cell lines[30]

Other routes to acyclic phosphazenes are based on phosphous (III) starting materials. Diazo derivatives $R_1R_2CN_2$ (R_1/R_2=H/CO_2R, H/H, CO_2Et/CO_2Et) react with the bicyclic phosphite $EtC(CH_2O)_3P$ to give the phosphazides, $EtC(CH_2O)_3P{=}N{-}N{=}CR_1R_2$. A ^{31}P NMR study of the kinetics of the reaction show that it is a two step mechanism with a second-order term followed by a first-order one.[31] The reaction of the NH_2 center in **5** ($X{=}H_2$) with $Ph_3P/C_2C\ell_6/NEt_3$ gives **5** ($X{=}PPh_3$).[32] The reaction of $(Me_3Si)_2$ $NPMe_2$ with propargyl chloride gives $Me_3SiN{=}PMe_2CH_2C{\equiv}CH$ and with $[(Me_3Si)_2N](Me_3SiNH)PC{\equiv}CCH_2C\ell$ gives $[(Me_3Si)_2N](Me_3SiNH)PC{\equiv}CCH_2PMe_2{=}NSiMe_3$.[33] The addition of aldehydes RCHO (R=Ph,CMe_3) to the heterocyclic phosphine $(SiMe_3)_2NPNHCH_2CH_2NH$ produces $Me_3SiOCRP({=}NSiMe_3)NHCH_2CH_2NH$.[34] The photolysis of R_2PN_3 ($R{=}ipr_2N$, $(C_6H_{11})_2N$) in the presence of boranes proceeds through transient phosphanylnitrene borane adduct to give phosphoranimines Thus in the presence of Mes_2BF ,

$R_2P(F)=NBMes_2$ is obtained and with R'_3B (R'=Et, C_6H_{11}), $R_2P(H)NBR_2'$ is the product.[35] The reaction of the coordinated PMe_3 in $Pd(PMe_3)_3I^+$ with the anilide anion, $NHPh^-$, gives $Me_3P=NPh$ which crystallizes as the dimer of the aniline adduct, $[Me_3P=NPh \cdot PhNH_2]_2^-$.[36]

Phosphorus(V) reagents may also be transformed to phosphoranimines. The reaction of $C\ell PNPC\ell_3^+PC\ell_6^-$ or $PC\ell_5$ with $(Me_3Si)_2NH$ provides the ionic oligomers $C\ell(PC\ell_2=N)_nPC\ell_3^+PC\ell_6^-$ in addition to cyclic and polymeric species in a process which can be followed by ^{31}P NMR.[37] The transformation of $P(O)C\ell_3$ to $C\ell_2P(=NH)OSiR_3$ may be accomplished using the hexaorganodisilazanes. The phosphoranimine when allowed to react with chlorophosphoranes $R_nPC\ell_{5-n}$ gives $R_nC\ell_{3-n}P=NP(O)C\ell_2$ (n=1-3) in high yield.[38] Desilation of $Ph_2P(Se)N(SiMe_3)_2$ with $KOCMe_3$ followed by coupling with $Ph_2P(S)C\ell$ provides $Ph_2P(S)SePPh_2=NSiMe_3$.[39] The double Kirsanov reaction of $(C\ell_3PNMe)_2$ with primary aromatic amines gives the bis phosphinimine hydrochloride $(RNH)_2P(=NR)NP(NHR)_2NHMe^+C\ell^-$. The structure observed (Section 7) has been rationalized on the basis of hyperconjugation of the NMe lone pair and the adjacent PN bond. Ab initio calculations on $(H_2N)_3P=NH$ and its protonated form give added support for this proposal. Methanolic KOH deprotonation leads to the highly basic $(RNH)_2P(=NR)NMeP(=NR)(NHR)_2$ derivatives. Primary aromatic amines do not react with the dimeric phosphinimines $[PhN=P(NMe_2)NPh]_2$ which are obtained via Staudinger reaction of $(Me_2NPNPh)_2$.[40]

A significant increase in the study of the relationship between phospha(V)zenes and phospha(III)zenes has been noted for this time period. In addition to the expected products observed in the photolysis of $MesP(O)CMe_3N_3$ in the presence of Me_2S, products derived from the capture of the intermediate phosphonamidate, $MesP(O)=NCMe_3$, are observed.[41] The trihalogenophosphine imides $Mes^*N=PX_3$ ($X=C\ell$, Br) are obtained nicely by oxidation of the phosphorus(III) precursor by $C\ell_2$, Br_2 or $PhIC\ell$. Mixed halogen species are also available, including $Mes^*N=PC\ell_2I$, from the disproportion reaction of the chlorophosphorus(III) derivative with I_2. The structures (Section 7) show large bond angles and short phosphorus-nitrogen distances.[42] Oxidation of Mes*N=PN(Mes*)$SiMe_3$ with $C\ell_2$ or Br_2 provides X_2P(=NMes*)N(Mes*)$SiMe_3$, which upon heating, eliminates Me_3SiX to give XP(=NMes*)$_2$. Exchange with Me_3SiI leads to IP(=NMes*)$_2$.[43] The addition of the carbonyl group to R_2NP=NR (R=$SiMe_3$) gives **6** (mixture of isomers). Substrates and products include: CF_3C(O)ICHF_2 (**6**, R'=CHF_2); CF_3(=CH_2)C(O)OMe (**6**, R'=$COMe_2$) and (MeO)C=C(CF_3)OP(=NR)(NR_2)O; o-C_6H_4(OH)C(O)CF_3, o-C_6H_4OP(RNH)(=NR)C(OR)(CF_3), which adds $(CF_3)_2$C=O to give phosophoranes.[44] The reaction of $C\ell_3$CLi, to Mes*P=NMes* gives Mes*P(=NMes*)=$CC\ell_2$ which can be lithiated to give the intermediate carbenoid type species Mes*P(=NMes*)=CCℓLi which can be trapped with various nucleophiles e.g. THF gives Mes*P(=NMes*)=CHTHF, $Me_3SiC\ell$ gives Mes*P(=NMes*)=CC$\ell$$SiMe_3$ and Me_3P leads to Mes*P(=NMes*) CH_2PMe_2CH via Mes*P(=NMes*)=CHPMe_2CH_2.[45] The Staudinger reaction on **7** (X=PMes) gives the intermediate **7** (X=P(=NR)Mes) which after rearrangement to Me_3CC=C(CMe_3)P(Mes)C(=NR) can add RN_3 to provide Me_3CC=C(CMe_3)P(Mes)=NRC(=NR). A byproduct of the formation of **7** is the bicyclic derivative CMe_3C=C(CMe_3)CP(=NR)C(CMe_3)=C(CMe_3)PMe_3.[46] Ambient temperature thermolysis of RPNMes*PMes* (R=Me,n-Bu,CMe_3) leads to RP(=NMes*)=PMes*.[47] A two plus one thermal cyclodimerization of RP=NR' (where R and R' are sterically demanding alkyl, anyl) gives **8**. Mixtures of two different substituted phospha(III)zenes also have been investigated.[48] Addition of RP=NR' with PhC≡CC≡CPh provides RP(=NR')C(Ph)=CC≡CPh and the spiro derivative RP(=NR')C(Ph)=CC=C(Ph)PR(=NR') subsequent addition of RP=NR' to the monocyclic entity gives the four-membered ring RN=P(R')C(Ph)=C(C≡CPh)P(=NR)R' Addition of another mole of the monomer toeither thefour-membered ring or the spirocycle gives **9** which undergoes hydrolysis to (ArN=)P(Me)C(Ph)=CHC[=CPhP(O)MeNHAr]P(=NAr)Me. Mixed substituent four-membered rings can also be obtained from two different monomers. When the R' group is a chlorine atom further derivatization with alkoxide ions gives either MePC(Ph)=C(Ph)P(=NAr)P(Cℓ)NAr or the bicyclic (ArN=)P(Et)C=C(OMe)OP(=NAr)EtC=C(OMe)O.[49] A novel elimination of HNR_2 from the trisaminophosphines $(R_2N)_2PNHMes*$ gives the intermediate $(R_2N)_2P(H)=NMes*$ on the way to the phosphorus(III) species $R_2NP=NMes*$.[50] Deprotonation of RNHP=NMes* by butyllithium gives $(RNPNMes*Li)_2$ which upon addition of $Ph_2PC\ell$ gives $Ph_2PP(=NR)=NRr$ or $P(=NAr)NRPPh_2$ depending on the solvent. Azides add to the lithiated dimer to provide the anionic allylic analogs $Mes*N=P(NR)NR'^-$ which upon addition of $CpFe(CO)_2C\ell$ gives $CpFe(CO)_2P(=NMes*)=NR$. Similar organometallic chemistry using $\eta^3C_3H_5Fe(CO)_3C\ell$ on $[Mes*N=P=PMes*]^-$ gives $\eta^3C_3H_5Fe(CO)_3P(=NMes*)=PMes*$ which upon heating leads to the sigma allyl $CH_2=CHCH_2P(=NMes*)=PMes$ and finally by cyclilzation to phosphinoazaphospholene.[51] The reaction of $(Et_3P)_2Ni(C\ell P=NMes*)$ with aluminum chloride gives the coordinated iminophosphenium ion $[(Et_3P)_2NiP{\equiv}NMes*]^+AlC\ell_4^-$.[52] Neutral species $Ph_3P(X)Rh(P{\equiv}NMes^+)$ are formed by oxidative addition of $RhC\ell(PPh_3)_3$ to $XP{\equiv}NMes*$ (X=Cℓ,Br,I).[53]

Reaction chemistry based on the phospha(V)zene unit continues to be an important area of investigation. (Vinylimino)phosphoranes have attracted particular attention in this time period. The basic synthetic strategy often involves enamine based carbon-carbon bond formation followed by aza-Wittig chemistry to provide nitrogen heterocycles. This synthetic method has been thoroughly reviewed.[54] Benzotriazole protecting group removal from (triphenylphosphoranylalkyl) benzotriazoles by the action of sodium hydride presents a new, safe route to N-vinyliminophosphoranes $RCH=CHN=PPh_3$ which may be trapped with chalcone to give 2,4-diphenylpyridines.[55] Addition of dimethylacetylene dicarboxylate (DMAD) to N-acyl-(Z)-β-enaminophosphazenes gives $MeOC(O)CH=C(C(O)OMe)C(PPh_2=NCOPh)=C(R)NR$. The sterochemistry and conformation was established by NOE experiments which were confirmed by calculations and x-ray (Section 7) studies. The subsequent chemistry of the phosphazene was not reported.[56]

Annulation reactions of **5** ($X=PPh_3$) with isocyanates, carbon disulfide or carbon dioxide give rise to pyridino[3',2':4,5]-thieno[3,2-d]pyrimidine derivatives.[32] Interesting selectivity is noted in the reactions of **10** with RNCX(X=O,S) in that when R=alkyl cyclization involving both phosphoranimines occurs whereas when R=aryl cyclilzation involves the phosphoranimine directly attached to the arene and the =CHAr center leaving one phosphoranimine intact. Hydrolysis of the remaining center results in formation of the primary amine.[57] Vinyliminophosphoranes derived from 1,6-methanol[10]-annulenes react with isocyanates to give 1,6-methanol[10]-annulenopyridines.[58] Bis(vinyliminophosphoranes) combine with isocyanates to provide azolofused 1,3-diazepines. Differences on the periphery of the fused ring systems occur based on isocyanato (alkyl vs aryl) substituents.[59] The thermal reaction of [(2-azulenyl)imino] phosphoranes on [(1-azaazulen-z-)imino]-phosphoranes with 2-bromotropone allow for the preparation of 6-aza- and 6,7-diazaazuleno[1,2-a]azulenes.[60] [(1-Acenaphthyl)imino]tributylphosphoranes, [(2-azulenyl)imino]-tributylphosphorane or [(3-indenyl)imino]tributylphosphoranes interact with substrates such as α,β-unsaturated ketones or $BrCH_2C(O)Ph$ to give acenaphthopyridine, azulenopyridine or pyrrole derivatives.[61] Aza-Wittig reactions of ortho(vinyl) substituted aryliminotriphenyl phosphoranes provide routes to arylquinolines or benzoxazines.[21]

The use of phosphazene bases in organic synthesis is slowly attacting more attention. Deprotonation of pyrido[1,2-a]-azepinone by $Me_3CN=P[N=P(NMe_2)_3]_2N=P(NMe_2)_2N=P(NMe_3)_3$ occurs α to the carbonyl to provide the enolate anion.[62] A similar four phosphorus base was used in combination with ethyl acetate as a metal-free initiator for the anionic polymerization of methyl methacrylate.[63] The addition of Mes_2BMe to $R_2P(F)=NBMes_2$(R=ipr) gives $R_2PN(Mes)B(Me)Mes$ while $R_2P(H)NBR'_2$ ($R'=Et,C_6H_{11}$) undergoes tautomerization to the aminophosphine, R_2PNHBR_2'.[35] Sequential treatment of $C\ell_2P(O)N=PC\ell_3$ with sec-butanol and methanol saturated with NH_3 gives a mixture of $(NH_2)P(O)NHP(O)(OMe)NH_2$,

$(NH_2)_2P(O)NHP(O)(OMe)_2$ and $HN[P(O)(OMe)_2]_2$. If liquid ammonia is used in place of the methanol/NH_3 reagent, $HN[P(O)(NH_2)_2]_2$ is obtained. These p-amido-imidodiphosphoric acid esters are effective flame retardants for cotton.[64] Hydrolysis reactions of phosphoranimines can lead to PN bond retention as in the case of the transformation of $CMe_3PF_2(=NR)$ (where R is a sterically demanding substituent) to $CMe_3PF(O)NHR$[18]. The more typical route is PN cleavage as noted in the removal of phosphorus from $Ph_3P=NC(O)CHRCMe_2OH$ to give $NH_2C(O)CHRCMe_2OH$. The phosphazene precursor is available from $RCH_2C(O)Cl$ in high yield by sequential treatment with NaN_3/Ph_3P and n-butylLi/Me_2CO.[65] Protonation also occurs at the basic nitrogen center of phosphoranimines. Treatment of $RN=P(MeNCH_2CH_2)_3N$ (R=Me,Ph) with CF_3CO_2H gives $RNHP(MeNCH_2CH_2)_3N^+CF_3CO_2^-$. The analagous reaction occurs with $PhN=P(NMe_2)_3$.[28] Amine hydrochlorides are available from $RN=P(OEt)_3$ by reaction with 20% HCl or paratoluene sulfuric acids.[66]

An alternative focus of reaction chemistry is that where the phosphorus-nitrogen bond is unchanged in synthetic transformations. Certain of these reactions have been mentioned above. Extension of the length of short chain phosphazenes can be accomplished by reactions of $Cl_2P(O)NPCl_3NHSiMe_3$ which is available from $Cl_2P(O)NPCl_3$ and $(Me_3Si)_2NH$. This reaction with $Cl_3PNPCl_3^+PCl_6^-$ or $Cl_3PNPCl_2NPCl_3^+PCl_6^-$ gives $Cl_2P(O)N(PCl_2N)_nPCl_3$ (n=2,3).[11] New hindered highly basic phosphazenes such as **3** can be prepared by coupling $HN=P(NMe_2)_2N=P(NMe_2)_3$ with $Cl_3P=NCMe_3$ or $Cl_3P=NCMe_3$ and $HN=P(NMe_2)_2$ sequentially. The resulting cationic materials are deprotonated with KNH_2.[16] Photolysis of the $Cl_3P=NP(O)Cl_2/Me_3SnCH=CH_2$ mixture gives high yields of $(CH_2=CH)_nCl_{3-n}P=NP(O)Cl_2$ where n=1 or 2 depending on the mole ration of the mixture.[67,68] Analogous introduction of the allyl function can be accomplished.[67,69] The reactions of $Ph_3P=NLi$ with halogenated substrates provide for a rich diversity of new compounds. The use of α-bromoesters gives $Ph_3PN(CH_2CO_2Me)_2^+Br^-$ when in excess, otherwise $Ph_3P=NCHRCO_2Me$ is formed and gives $Ph_3PNHCHRCO_2Me^+Cl^-$ upon addition of HCl. Similar additions of RCl

($R=PH_2P(O)$, tosylate, $SO_2C\ell$) give $PH_3P{=}NR$, which in the case of $R=SO_2C\ell$, reacts with HE (E=OEt, NEt_2) to give $Ph_3P{=}NSO_2E$. Direct bromination gives $Ph_3P{=}NBr$ which reacts with Ar_3P to give the PNP^+ analogs $Ph_3PNPAr_3{}^+Br^-$.[70] The reaction of $Me_3SiN{=}PMe_2CH_2C{\equiv}CH$ with $Me_3SiC\ell$ gives $Me_3SiN{=}PMe_2CH_2C{\equiv}CSiMe_3$ while $HC\ell$ causes rearrangement to $Me_3SiN{=}PMe_2CH{=}C{=}CH_2$ and $Me_3SIN{=}PMe_2C{\equiv}CMe$.[33] The $Ph_3P{=}N$ moiety may be transferred from $Ph_3P{=}NSiMe_3$ to various nitrogen heterocycles by reaction with a chloro or nitro substitutent on the heterocycle.[71] Nucleophilic substitution of $P_3PN{=}PC\ell$ with X^- (X=OEt,OPh, SEt,NEt_2, F) gives Ph_3PNPX_2 which in turn can yield $Ph_3P{=}NP(O)HX$ by hydrolysis, $Ph_3P{=}NP(O)X_2$ by the action of NO_2 or $Ph_3P{=}NP(S)X_2$ from S_8.[26] The reaction of aryloxides, OC_6H_4-p-X(X=F,Cℓ,Br) with $C\ell_2P(O)NPC\ell_3$ gives $(ArO)_2P(O)NP(OAr)_3$ and with $C\ell_2P(O)NPC\ell_2NPC\ell_3$, $(ArO)_2P(O)NP(OAr)_2NP(OAr)_3$ is obtained. The structures (Section 7) of these short chain species are good for modeling the backbone conformation, but not the packing, of the high polymers.[72]

One particular area of focus of reactions wherein the phosphazene bond is left formally undisturbed is the formation of acyclic metallophosphazenes The reaction of $TiC\ell_4$ with $Me_3SiNPPh_3$ gives $TiC\ell_3N{=}PPh_3$ which is formulated as having significant TiN double bond character.[73] The interaction of Cp^*TiF_3 with $Ph_3PNSiMe_3$ and $C_2H_2(Ph_2PNSiMe_3)_2$ leads to $Cp^*TiF_2NPPh_3$ and $(Cp^*TiF_2NPPh_2)_2C_2H_2$ resectively.[74] The formation of $Sb(NPPh_3)_4{}^+SbX_6{}^-$ (X=F,Cℓ) from SbX_5 and $Me_3SiNPPh_3$ has been reported. The combination of a short antimony-nitrogen distance and low Mössbauer isomer shift leads to the suggestion of significant SbN π bonding.[75] If $SbC\ell_3$ is used in place of SbX_5, $SbC\ell_2(NPR_3)$ (R=Me,Ph) is obtained which upon treatment with $SbC\ell_5$ gives $Sb_2C\ell_5(NPMe_3)_2{}^+SbC\ell_6{}^-\cdot CH_3CN$ or $[SbC\ell(NPPh_3)]_2(SbC\ell_6)_2\cdot 6CH_3CN$. The mixed valence Sb(III)/Sb(V) configuration was confirmed by Mössbauer spectroscopy.[76] The distorted tetrahedral species $MC\ell_2(Me_3SiNPMe_3)_2$ (M=Zn,Co) are obtained from the appropriate dichlorides. If the $CoC\ell_2$ reaction is carried out in acetonitrile in the presence of NaF, $CoC\ell(HNPMe_3)_2$ is obtained.[77] Simple Cu(I) and Cu(II) chlorides give rise to octahedral clusters (Section 7) upon treatment with $Me_3SiNPMe_3$. The formulations are $[Cu_6Cl_6(NPMe_3)_4][Cu(Me_3SiNPMe_3)_2]Cl_2$ and $[Cu_6Cl_6NPMe_3)_4]Cl\cdot Me_3SiNPMe_3\cdot CH_2Cl_2$ respectively. Each species has been shown by cyclic voltametry to undergo a reversible oxidation.[78] The reaction of $RPCl_4$ with $LiN(SiMe_3)_2$ gives the potentially useful species $RPCl{=}NSiMe_3$ (R=Cl,Ph). The trichlorospecies provides cis-$WCl_4(N{=}PCl_3)_2$ or $WCl_5N{=}PCl_3$ starting with WCl_6. The latter derivative undergoes chloride abstraction upon treatment with $GaCl_3$ to give $WCl_4(N{=}PCl_3)^+GaCl_4^-$. In a similar fashion cis-$WCl_4(N{=}P(Ph)Cl_2)_2$ may be obtained.[79] Building blocks for radiopharmaceuticals are obtained from the reaction of MO_4^- (M=Re, ^{188}Re, ^{99m}Tc) and $Ph_3P{=}NSiMe_3$ which gives $Ph_3PNH_2^+MO_4^-$. The products convert to $Ph_3P{=}NMO_3$ on heating.[80,81] Similar chemistry of the $Bu_3PNSiMe_3$ species occurs and the stability of biomacomolecules labeled with these substrates was reported.[82] The bifunction ligands **11** coordinate to $CuCl_2$. In the case of **11** (R=H), the $LCuCl_2$ adduct (L=**11**,R=H) is obtained with strong pyridine and weak phosphazene nitrogen coordination. In solution $CuLCl_2$ and CuL_2Cl_2 are detected by esr spectroscopy and the phosphazene coordination is believed to remain intact.[83] When L=**11** (R=Me) a dimer, $(LCuCl_2)_2$, is isolated with a similar pattern of nitrogen donor behavior.[84] More elaborated phosphoranimines have also been explored as donors to metals. In addition to acyclic species, metallocyclic derivatives are often obtained in these systems (Section 5). The reaction of NH_4VO_3 with $HN[P(NMe_2)_2NSiMe_3]_2$ is accompanied by migration of trimethylsilyl to the oxygen of the vanadyl ion giving $N[P(NMe_2)_2NH_2]_2^+VO_4(SiMe_3)_2^-$.[85] Only a single center is bonded to titanium in complexes derived from $(Me_3Si)_2NPPh_2NPPh_2NSiMe_3$ with $TiCl_4$ $[TiCl_3NPPh_2NPPh_2N(SiMe_3)_2]$ and $Ph_2P(=NSiMe_3)NHSiMe_3$ with $RTiX_3$ ($RTiX_2NPPh_2NHSiMe_3$; R=Cp*, X=Cl,F; R=Cp, X=Cl).[86] Similar chemistry occurs in the $R_2PN(R')PR_2{=}NSiMe_3/CpTiCl_3$ system with the formation of $R_2PN(R')PR_2{=}NTiCpCl_2$.[19] Ligands derived from phosphorus oxyacids having an imidophosphorus center are of interest in solvent extraction work. The fundamental formation constants for the N,N,N^1,N^1 - octabutylimidodiphosphotetramide/

nitric acid in a two/phase solvent system (water/toluene) has been determined using potentiometric titrations. Extraction equilibria, in a two phase system, of the same ligand coordinated to trivalent rare earth ions have also been examined.[88] Polyalkylphosphonitrilic acid has been used in extraction of arsenic and antimony from sulfate solutions.[89] Linear phosphazenes such as $C\ell_3PN(PC\ell_2N)_nPC\ell_3^+SbC\ell_6^-$ (n=1,2) have been used as catalysts for the preparation of dispersions of high-viscosity polysiloxanes in cyclic siloxanes.[90]

3. Cyclophosphazenes

Reviews of cyclophosphazene chemistry include an overview of cyclophosphazene chemistry with reference to synthesis, structure, bonding and reactivity[1], a summary of fire retardant additives and materials based on cyclophosphazenes [91], polymerization of main group vinyl derivatives including cyclophoshazenes with an olefinic moiety as part of the exocyclic group [92] and a survey which compares ^{31}P NMR vs x-ray data which suggests that the former is a better predictor for the propensity of cyclophosphazenes to be precursors to new polymers.[93]

Theoretical and experimental approaches to understanding structure and properties of cyclophosphazenes have appeared. The most significant of these is a report of the observed (x-ray diffraction) and calculated (ab initio) deformation electron densities in the benzene solvate of hexa(aziridinyl)-cyclotriphosphazene. The electron densities in the phosphazene clearly show the classic island model pattern with nodes in the π system at nitrogen. The dual π/π^1 system is also clearly demonstrable.[94] Ab initio calculations on $(NPX_2)_3$ (X=F,$C\ell$) have been reported and compared to previous studies. No qualitative differences exist between the two with the out of plane π system being roughly twice the estimated π charge density of the in plane system.[95] Molecular mechanics methods have been carried out on cyclophophazenes with glyme type exocyclic groups and their

Group I cation (Li^+-Rb^+) ion pairs in order to understand the phase transfer ability of these materials.[96] The structural features of bicyclic materials derived from cyclotetraphosphazenes, $P_4N_4(NRR')_6NR''$, have been compared to bicyclic phosphazanes with negative hyperconjugation, $N_{lp}\rightarrow\sigma^*$, being significant only in the former.[97] The role of different calculational and analytical methods in establishing reactivity ratios for copolymerization reactions involving cyclophosphazenes with olefinic moieties as part of the exocyclic group has been explored. Several previously reported systems have been reexamined.[98] The electrical properties of the resin produced by vinyl polymerization of $N_3P_3(OCH_2CF_3)_3(OCH_2CH_2OC(O)CH{=}CH_2)_3$ show that it goes from an insulator to a semiconductor with increase in temperature. Dielectric constant and dielectric loss measurements are also reported. The observed behavior has been associated with a charge transfer complex.[99] Resins from hydroxyethylmethacrylato cyclophosphazenes have been examined with respect to transverse and tensile strength which show these materials to be suitable for denture base resins. Ultramark 1621, $N_3P_3[OCH_2(CF_2CF_2)_xH]_6$ (x=1,2 or 3), is a useful calibrant for negative and position ion fast-atom bombardment high resolution mass spectrometry.[101] Diatomaceous earths with deposited $N_3P_3Cl_6$ show increased water absorption.[102] Spectral (IR and UV-VIS) and thermal analysis (DTA) studies of $N_3P_3Cl_{6-n}(C_6H_5)_n$ (n=2,4,6)[103] and presumed hexaaryl derivatives[104] have been interpreted in terms of structural, symmetry and phase changes. The effects of protonation were also investigated. Novel properties of phosphazene clathrate tunnel systems continue to attract attention. Vinyl and acrylic monomers can act as guests in a tris(o-phenylenedioxy)cyclo-triphosphazene host. Polymerization of these monomers is initiated by ^{60}Co γ rays. Notable observations include enhanced degrees of stereoregularity and absence of radiation cross-linked materials.[105] Similar studies on diene monomers show sterospecific formation of the 1,4-trans addition products.[105] Two types of clathrates are formed by tris(naphthalene-2,3-dioxy) cyclotriphosphazene, benzene is included in a channel-type area while p-xylene forms a cage type. Given the choice of benzene and para-xylene, the substituted arene will be absorbed suggesting that the lattice stability of the cage is greater than the channel. In mixtures of disubstituted benzenes, the para isomer is selectively included and hence may be separated in this fashion.[107] The (naphthalene-dioxyo) phosphazene may be deposited as thin films which absorb organic molecules in clathrates by a vapor-solid reaction.[108]

Reports of the synthesis of cyclophosphazenes from acyclic precursors sporadically occur. The classic ammonolysis method is used in the reaction of 9-trichloro-9-phosphafluorene with ammonium chloride to give **12** which is isolated as the chloroform clathrate.[109] The reaction of $C\ell_3PNPC\ell_3^+C\ell^-$ with $(Me_3Si)_2NH$ gives rise to cyclo- and polyphosphazenes.[37] Fluorophosphines R_2PF (R_2=Ph_2, $C_5H_{10}N$, Et_2N) and RPF_2(R=PhO, Et_2N) react with Me_3SiN_3 to give $R_{3-n}P(N_3)_n$ which go on to form oligo- and polyphosphazenes, $(RR'PN)_n$.[27] The trimethylsilylazide reaction with $\overline{C_6H_4N(MePC\ell NRC}(O)$ surprisingly produces the cyclophosphazene **13**.[29]

Substitution reactions of cyclophosphazene still attract interest. Solvent free $N_3P_3(NH_2)_6$ can be obtained from the semiammoniate by treatment with KNH_2 in an ammonia atmosphere in an autoclave. The reactions of $(NPC\ell_2)_3$ with amino acid ester hydrochlorides and triethylamine provide $N_3P_3[NHXCO_2R]_6$ (X=$(CH_2)_3$, R=CH_2Ph, ipr, n-Bu, Me; X=$(CH_2)_5$, R=ipr; X=nil, R=ipr) and $N_3P_3[N(CH_2CO_2CHMe_2)_2]_6$ which can be hydrolyzed to the sodium salts of the acid by treatment with NaOH.[111] Reactions of diamines continue to be reported. The preparation of the mono-spiro derivative, $N_3P_3C\ell_4[NH(CH_2)_2S(CH_2)_6S(CH_2)_2NH]$, from $N_3P_3C\ell_6$ and 3,10-dithiadodecane-1,12-diamine is available.[112] The synthesis on a solid support (KOH, Alumina or Na_2CO_3, talc) leads to high yields of spirocyclic $N_3P_3C\ell_4[NH(CH_2)_3O(CH_2)_nO(CH_2)_3NH]$ (n=3,6). The corresponding solution phase reaction gives mixtures of spiro, transannular bridging and coupled rings.[113]
Octachlorocyclotetraphosphaze and 1,4-butanediamine give $N_4P_4C\ell_7NH(CH_2)_4NHC\ell_4P_4C\ell_7$ which can be hydrolyzed to the tetrametaphosphinate $(NH)_4P_7(O)_4(OH)_3NH(CH_2)_4NH(OH)_3(O)_4P_4(NH)_4$. Some

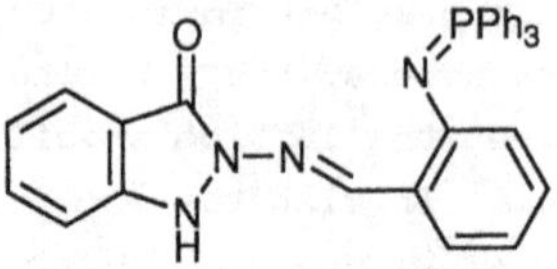

(1) 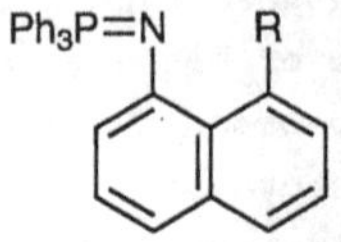 (2)

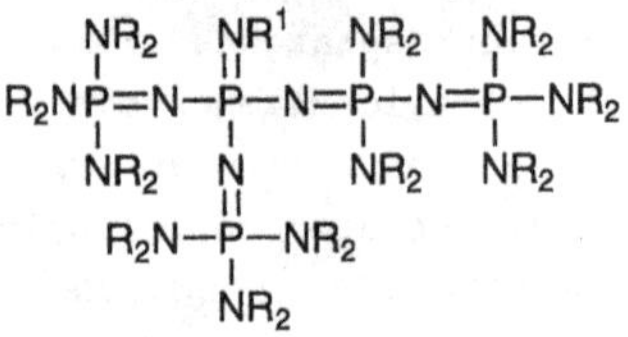

(3) 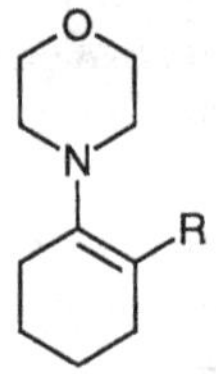 (4)

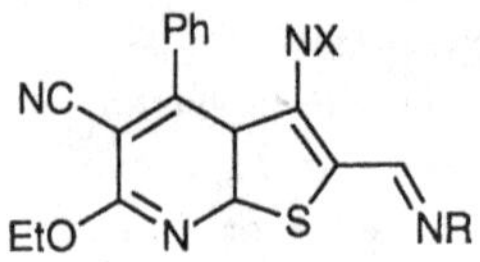

(5) 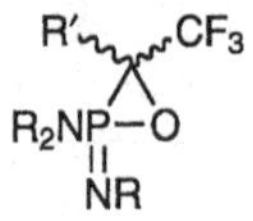 (6) 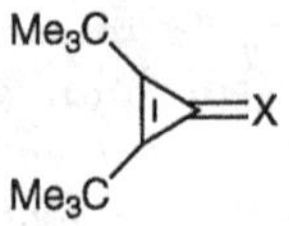 (7)

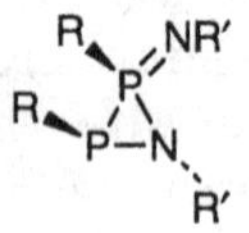

(8) 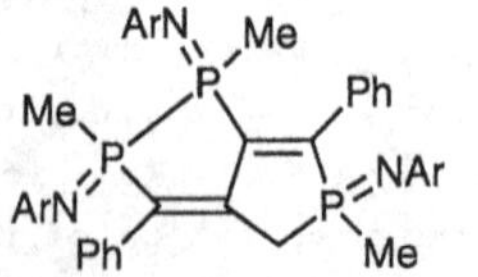 (9) 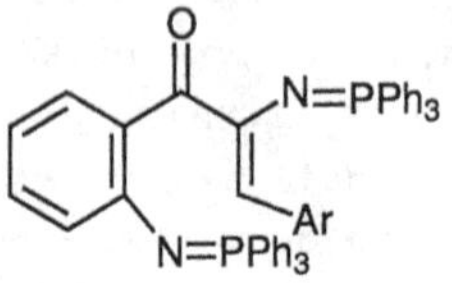 (10)

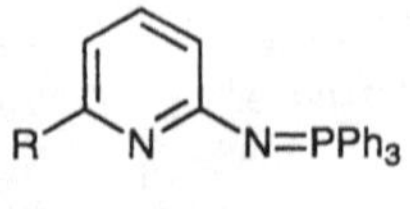

(11) 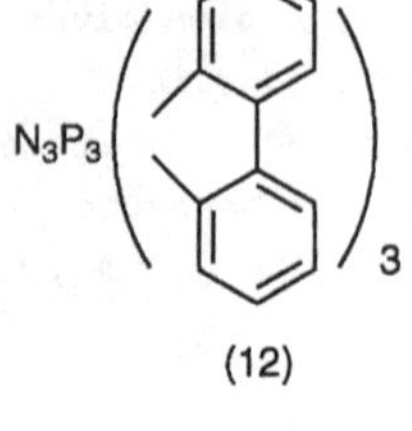 (12) 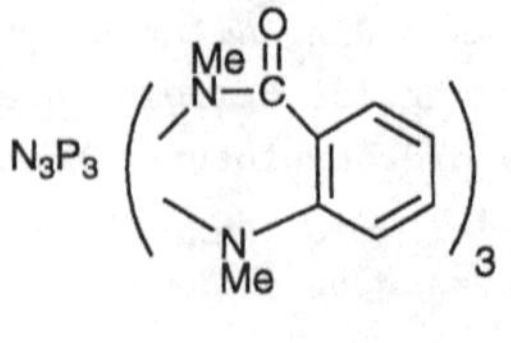 (13)

exocyclic phosphosphorus-nitrogen cleavage occurs.[114] As in recent years, oxygen based anions have been the most widely explored nucleophiles. In spite of numerous studies, new results on the hydrolysis of $N_3P_3Cl_6$ can be obtained. In THF, several products are indicated with the major one arising from a reaction involving the solvent, $(NPCl_2)_2NHP(O)O(CH_2)_4X$ where X is either Cl or OR.[115] The reaction of $N_3P_3Cl_5OAr$ with ArO^-M^+ (Ar=nitroaryenes, M^+=Na, Bu_4N, Et_3BzN) in a two solvent system under phase transfer catalysis conditions shows an increase in cis stereoselectivity over the reaction run in bulk.[116] The series $N_3P_3(OC_6H_5\text{-}p\text{-}X)_6$ (X=F,Cl,Br,I) has been prepared and structurally characterized (Section 7). It has been suggested that considerable PO π bonding occurs.[72] The reaction of $N_3P_3Cl_6$ with sodium 2,6-(ditertbutyl)phenoxide NaOR gives, at least two products, $N_3P_3Cl_5OR$ and the phosphorus-phosphorus bridged dimer **14**. Chalcone, $PhC(O)CH{=}CH\text{-}C_6H_4(R)$, derivatives of $N_3P_3Cl_6$ have been reported. The reaction of ROH/NEt_3 in dioxane gives $N_3P_3Cl_5OR$ which can be converted to $N_3P_3(OR')_5OR$ ($R'=CH_2CF_3$, Ph). Use of the sodium salt, NaOR, allows for preparation of $N_3P_3(OR)_6$. Photolysis of $N_3P_3Cl_5OR$ gives the two plus two cycloaddition product $PhC(O)CHCH(C_6H_4ON_3P_3Cl_5)CH(C_6H_4ON_3PCl_5)CHC(O)Ph$.[118] The reaction of $N_3P_3Cl_6$ with 2,4-dihydroxybenzophenone in the presence of Et_3N gives $N_3P_3Cl_5OR$ and $N_3P_3(OR)_6$ [R=$PhC(O)C_6H_3(OH)$]. The UV stabilizing (in a polystyrene film) ability of the new phosphazene is comparable to the parent ROH but low thermal loss of the phosphazene derivatives was noted.[119] The reaction of hydroquinone with $N_3P_3Cl_6$ in the cyclohexane/pyridine system gives a complex prepolymer $[P(OC_6H_4OH)_2N]_3$ of unknown structure.[120] The biphenyl mesogenic groups in $N_3P_3(OC_6H_4C_6H_4R)_6$ (R=CN, $C_6H_{11}O$), prepared from $N_3P_3Cl_6$/$NaOC_6H_4C_6H_4R$, show smetic C and monotropic nematic C liquid crystal behavior respectively.[121] The synthesis and liquid crystal behavior of $[NP(OC_6H_4CO_2(CH_2)_{10}CO_2R)_2]_{3,4}$ (R=cholesterol) have been reported. The trimer forms are isotropic liquids at the melting point but the tetramer gives a cholesteric mesophase.[122] The phosphate substituted derivative, $N_3P_3Cl_5OP(O)Ph_2$ has been prepared and considered as a dispersant in ceramic processing.[123] A new and promising synthetic method for the synthesis of fluoroalkoxy fluorocyclotriphosphazenes involves liberation of the anion from the trimethylsilyl derivative by the action of CsF in THF. Thus the reactions of $Me_3SiOCH_2(CF_2)_n$-CH_2OSiMe_3 (n=2,3) and $N_3P_3F_6$ give both spirocycles and bridged dimers, i.e. $N_3P_3F_4OCH_2(CF_2)_nCH_2O$ or $N_3P_3F_5OCH_2(CF_2)_nCH_2ON_3P_3F_5$. Lower catalyst concentration and temperature favors the bridged species. Some derivatives undergo a transformation from the bridged to the spiro form on sublimation. The reaction of $CF_3CH_2OSiMe_3$ gives a quantitative yield of $N_3P_3(OCH_2CF_3)_6$ but no fluoroalkoxide is formed when $N_3P_3C\ell_6$ is the substrate. The products were examined by ^{31}P NMR with broadband fluorine decoupling.[124] The spirocycle formed from $N_3P_3C\ell_6$ and 1,3 butanediol and its dimethylamino derivative, $N_3P_3X_4[OCHMe(CH_2)_2O]$ (X=Cℓ, NMe_2) show an AMX ^{31}P NMR spectrum due to the unsymmetrical nature of the spirocyclic entity.[125] The reaction of LiC≡CR (R=Me,n-Bu,Ph) with $N_3P_3F_6$ have been examined. The alkylacetylenic derivative have all isomers with the non-geminal pathway being favored. Mixed phenylacetylene, alkylacetylene derivatives have also been prepared and demonstrate substituent control of the reaction pathway.[126] The Pd(O) catalalyzed coupling of Me_4Sn and $N_3P_3C\ell_6$ gives 2,2-$N_3P_3C\ell_4Me_2$ in 90% yield.[67]

Much of the new chemistry of cyclophosphazenes comes from exploiting the synthetic potential of organic substituents on the cyclophosphazenes. A subset of this approach is the use of cyclophosphazenes as ligands in coordination or organometallic chemistry. The reactions of $N_3P_3(OPh)_5(OC_6H_4R)$ or $N_3P_3(OC_6H_4R)_6$ (R=Li) with $CpFe(CO)_2I$ give the organometallic phosphazene of the same general formula where R=$CpFe(CO)_2$.[127] The addition of $PdC\ell_2$ to the bis(3,5-dimethylpyrazolyl)phosphazenes 2,2-$N_3P_3(MeNCH_2CH_2O)_2(dmp)_2$ and **15** gives $PdC\ell_2$•L (where L is the pyrazolylphosphazene). The ligands act as bidentate donors involving the 2-nitrogen centers of the pyrazolyl. An additional weak intermolecular interaction with an endocyclic nitrogen also occurs in **15**•$PdC\ell_2$. The ^{31}P NMR spectra for $N_3P_3(MeNCH_2CH_2O)_2(dmp)_2$•$PdC\ell_2$ indicate a fluxional system giving rise, in the slow exchange limit, to inequivalent $P(MeNCH_2CH_2O)$ centers.[128] The reactions of the related phosphazenes 2,2-$P_3N_3Ph_4(dmp)_2$ and **16** with $M(CO)_6$ gives the complexes $M(CO)_3L$ (M=Mo,W). The phosphazene acts as a tridentate donor involving two pyrazolyl nitrogens in the two position and an endocyclic phosphazene nitrogen center. The donor character of the endocyclic nitrogen is altered by substitution on the remote phosphorus atoms.[129] The persubstituted 3,5-dimethylpyrazolyl phosphazene $N_3P_3(dmp)_6(L)$ reacts with Cu(II)perchlorate to yield $[CuL(C\ell O_4)(H_2O)_2]C\ell O_4$ which upon addition of bipyridyl gives $[CuL(bipy)(C\ell O_4)]C\ell O_4$. Optical EPR and IR spectroscopy were used to characterize these derivatives. The phosphazene appears to act as a tridentate ligand as was noted above. Attempts to obtain $(CuC\ell_2)_2L$ resulted in ligand hydrolysis to yield the $(CuC\ell_2)_2N_3P_3O(dmp)_5^-$ complex. The coordination environment about Cu(II) is a distorted trigonal bipyramid with coordination from a nitrogen center of the pyrazoyl group on each of two adjacent phosphorus centers, an endocyclic nitrogen center and two chloride ions (Section 7).[130] The ethylenediamine and propylenediamine derivatives (L), $N_3P_3(NMe_2)_4[NH(CH_2)_nNH]$ (n=2,3), combine with $M(CO)_6$ to form $M(CO)_4L$ (n=2,3;M=Mo,W). Spectroscopic comparisons to previously structurally characterized complexes show that the phosphazene ligand acts as a bidentate donor using one endocyclic nitrogen center and one of the nitrogen centers in the diamine spirocycle.[131] The reaction of the amino acid derivatives $N_3P_3[NH(X)CO_2^-Na^+]_6$ (vide supra) with Gd^{3+} forms insoluble oligiomers $N_3P_3[NH(X)CO_2^-]_6 \cdot 2Gd^{3+}$ with stability constants greater than 10^{28}.[111] The interactions of tetrametaphosphimates derived from hydrolysis of 1,4-butanediamine bridged cyclotetraphosphazenes (vide supra) with $Co(NH_3)_6C\ell_3$, $NiC\ell_2 \cdot 6H_2O$, $CoC\ell_2 \cdot 6H_2O$ and $LnC\ell_3$(Ln=Pr,Y_6) have been examined.[114] A brief mention of solid cyclolinear polymers of cyclophosphazenes and metal ions is available.[132] Instead of the expected metal complex, the reaction of $Cu(C\ell O_4)_2$ with hexamorpholino-cyclotriphosphazene in THF/MeOH produces a THF clathrate of the protonated phosphazene with a perchlorate counterion.[133] Metal ion coordination does occur in the product of $Zn(BH_4)_2$ and $N_3P_3(NHPh)_6$. The exocyclic nitrogen centers and BH_4 moiety are coordinated to the Zn center in $N_3P_3(NHPh)_6 \cdot Zn(BH_4)_2$.[134]

Organic and polymer chemistry at the exocyclic position is also a profitable route to new phosphazene derivatives. The reactions of $(NPCl_2)_3$ with NaNCO in the presence of aliphatic alcohols gives the monourethanes $N_3P_3Cl_5NHC(O)OR$. The disubstituted derivatives have a predominately geminal configuration. The reaction is more complex when AgNCO is used in acetonitrile. In this case, **17** is formed presumably by attack of a phosphonium ion on the solvent. Other products include $N_3P_3Cl_4[NHC(O)OR]_2$ and $N_3P_3Cl_4[NHC(O)OR](OR)$. In a repeat of older work, the reaction of 2,2-$N_3P_3Cl_4(NH_2)_2$ with phosgene in the presence of NEt_3 was actually shown to produce $N_3P_3Cl_4(NH_2)NCO$ which could be trapped with alcohols or amines to give urethanes or amides.[135] The reaction of excess paraformaldehyde with $N_3P_3(NMeNH_2)_6$ gives $N_3P_3(NMeN{=}CH_2)_6$.[136] A wide range of phosphazenes can be sulfopropylated by reaction with 1,3-propane sultone, $O_2\overline{SOCH_2C}H_2$. If no amino groups are available, in favorable cases endocyclic nitrogen centers can be derivatized. Amino groups directly bound to the phosphazene or as part of a side chain can undergo reaction.[137] The sequential reaction of $N_3P_3Cl_6$ with $NaOC_6H_4C(O)H$, $NaBH_4$ and HBr/H_2SO_4 gives $N_3P_3(OC_6H_4CH_2Br)_6$ which initiates ring opening polymerization of 2-methyl-2-oxazoline to give the first phosphazene based star-branched polymer, $N_3P_3(OC_6H_4CH_2[N(COMe)CH_2CH_2]_nOH)_6$. A linear analog is formed by $N_3P_3(OPh)_5OC_6H_4CH_2Br$. The N-acetyl groups can be removed to give N-protonated poly(ethylenimine) branches. The surfactant behavior of these systems has also been explored.[138] Cinnamate derivatives, e.g. $N_3P_3(OCH_2CF_3)_5O(CH_2)_2O(CH_2)O(CO)CH{=}CHPh$, have been obtained from the reaction of $N_3P_3Cl_6$ with $NaO(CH_2)_2O(CH_2)_2OTHF$ (THF = tetrahydropyranyl) and $NaOCH_2CF_3$ followed by deprotection and coupling of the terminal alcohol with $ClC(O)CH{=}CHPh$. The olefin center undergoes a photochemically allowed two plus two cycloaddition to the cyclobutane.[139] The hydroxy function in

$N_3P_3(OC_6H_4OH)_6$ can be coupled with $BrCH_2CH{=}CH_2$ or $CO_2H(CH_2)_nCH{=}CH_2$ (n=1,2,8) to give olefins anchored to the arene. Oxidation with m-chloroperbenzoic acid gives the expoxides $N_3P_3(OC_6H_4OCH_2\overline{CHCH_2O})_6$ and $N_3P_3(OC_6H_4OC(O)(CH_2)_nCH{=}CH_2)_6$.[140] The $N_3P_3(OCH_2CF_3)_5O^-$ ion is available from $N_3P_3(OCH_2CF_3)_6$ by hydrolysis and in turn can couple with acrylchlorides to give $N_3P_3(OCH_2CF_3)_5OC(O)CR{=}CH_2$ (R=H, Me) which in turn can undergo homo- and copolymerization by a radical olefin addition process.[98] The sol-gel process has been applied to $(EtO)_4Si$ reactions with $N_3P_3(OPh)_5OC_6H_4OH$ and $N_3P_3(OC_6H_4OH)_6$ to give new phosphazene containing silica materials.[141] Hydrogenation of alkynylcyclophosphazenes presents a route to high yielding syntheses of alkylcyclophosphazenes.[126] The 4-aminophenoxy substituent is a useful center for generating phosphazene containing polymers. The synthesis of $2,2\text{-}N_3P_3Ph_2(OC_6H_4NH_2)_4$ from the dinitrophenoxy precursor and its subsequent reaction with maleic anhydrides has been examined. The in situ cyclodehydration and thermal polymerization of these precursors give phosphazene based polyimide matrix polymers which have heat and flame resistant properties.[142,143] Similar polymerization chemistry has been explored on the two isomeric forms of $N_3P_3(OPh)_3(OC_6H_4NH_2)_3$. Properties such as decomposition onset temperature and Tg vary with the isomeric identity of the phosphazene.[143] Displacement of nitro groups from tris[(3 or 4-nitrophthalimido)phenoxy]triphenoxycyclotriphosphazene with the dianion of bisphenol A gives cyclomatrix cyclophosphazenes containing poly(ether imides). Thermal properties of these materials have been reported.[145] Linear polymers are obtained from the reaction of $N_3P_3(OPh)_4(OC_6H_4NH_2)_2$ with diacid chlorides.[146] Phosphazenes with reactive functionalities can be used to cure epoxy resins. Systems which have been explored include $2,2\text{-}N_3P_3R_4(NH_2)_2$ (R=OR′,NHR′,NR'_2,SR′)[147,148] and reactive oligomers obtained from the interaction of diamines with mono-spiro-o-carboranylenecyclotriphosphazene.[149] Selected electrical, mechanical and chemical properties of the cured systems in the former case have been reported.[147] Exocyclic olefins are the other major reactive center which has been extensive explored as a route to new derivatives and polymers. Hydrosilation of cis-$N_3P_3Cl_4[O\text{-}2,4\text{-}C_6H_3(OMe)CH_2CH{=}CH_2]_2$ with $HSiMe_2OSiMe_3$ gives rise to the siloxane substituted system. The monosubstituted precursor, $N_3P_3Cl_5OC_6H_3(OMe)CH_2CH{=}CH_2$ reacts with $(HSiMeO)_4$ to give a novel mixed ring system, $[N_3P_3Cl_4OC_6H_3(OMe)(CH_2)_3Si(Me)O]_4$, and with $PhMe_2SiH$ or $MeCl_2SiH$ to give $N_3P_3Cl_4OC_6H_3(OMe)(CH_2)_3SiRR'R''$.[150] Copolymerization of $2,4\text{-}N_3P_3F_4(C{\equiv}CPh)C_6H_4CMe{=}CH_2$ and styrene followed by $Co_2(CO)_8$ addition to the acetylene in the pendant phosphazene gives a polymer which undergoes a reversible one electron reduction to a radical anion.[126] Suspension polymerization of the now well-known hexaHEMA derivative, $N_3P_3(OCH_2CH_2OC(O)CMe{=}CH_2)_6$ gives hard, heat resistant particles.[151] Mixed derivatives of $P_4N_4Cl_8$ with p-bromophenol and HEMA such as $P_4N_4(OC_6H_4Br)_3(OCH_2CH_2OC(O)CMe{=}CH_2)_5$ were subjected to bulk polymerization to produce visible light-cured radiopaque resins.[152] UV-irradiation of $N_3P_3(OCH_2CF_3)_3(OCH_2CH_2C(O)CMe{=}CH_2)_3$ also produces resin systems.[99]

In addition to certain of the articles cited above, numerous new applications of cyclophosphazenes have been described, almost exclusively in the patent literature. The ever popular flame retardancy property of phosphazenes is again an important area. The use of amidophosphazenes $[(NH_2)_2PN]_n$ ($n \geq 3$) for cellulose fibers is noted,[153,154] including an approach which allows the use of the amidophosphazene without previous removal of the difficult to separate NH_4Cl.[154] Alkoxy (Et,Pr,Bu) phosphazenes combined with siloxanes have also been used in cellulosic fibers.[155] Triallyl-phenoxytriphenoxycyclotriphosphazene is an additive in the production of flame retardant poly(siloxanes).[156] Cyclophosphazene based lubricating fluids are also of interest. Alkoxy and aryloxy trimers and tetramers are proposed as lubricating oils for metal working [157,158] and fluoroalkoxy derivatives for refrigerators.[159] Antioxidant additives for cyclophosphazene fluids used under higher temperature conditions have been reported.[160,161] A large body of applications are derived from the production of matrix materials from cyclophosphazenes with polymerizable side chains. The most commonly employed exocyclic unit is the hydroxyethylmethacrylate (HEMA) function.[162] Applications of these materials include the following: intermediate layers in electrophotographic photoreceptors,[163-165] wear proof carrier for an electrophotographic developer,[166] binder for holographic recording material,[167] heat resistant layer for thermal-transfer ink sheet,[168] protective layer for a photocell,[169] abrasion resistant coatings for hard surfaces,[170] anti-static layers [171], durable anti-fogging surfaces[172,173], low-shrinkage, weather-resistant coatings [174], thermosetting decorative boards for counter tops, etc.[175], transfer resin substrates for CVD plasma deposited silica [176] and emulsions to produce curable hard layers on steel.[177] Cyclomatrix materials from step polymerization involved oligoethylene glycol/$N_3P_3Cl_6$ for polymeric solid electrolytes [178] and bis(4-aminophenyl) ether combined with $N_3P_3R_4Cl_2$ (R=morpholino) for monofiltration membranes.[179] Fluoroalkoxyphosphazenes have been proposed as water-repellent, skin-protecting cosmetics.[180] A tert-butoxycarbonyl protected phosphazene is a component of a photoacid chemical application photoresist material.[181]

4 Cyclophosha(thia)zenes and Related Systems

This section includes ring systems with both phosphazene and thiazene (selenazene) components. A review of thiazenes, including phosphathiazenes, covering aromatic, antiaromatic and radical systems focused on inter/intramolecular bonding in relation to the design and synthesis of molecular conductors. Phosphathiazenes are best considered as exhibiting isolated PNP^+ and NSN^- units.[182] A readable overview of selenium-nitrogen chemistry, including cyclophospha(selena)zenes, is available.[183] The reaction of **18** (E=S) with RLi R=Me,CMe_3Ph produces $Li[Ph_4P_2N_4S_2R]$ in which the cation binds two rings together by coordination of two adjacent nitrogen atoms on one ring and to one nitrogen center on the other ring. Variable temperature ^7Li and ^{31}P NMR spectroscopy show that this is a fluxional system.[184] The coordination chemistry of **19** (E=S, R=Ph; E=Se, R=Me, Et, Ph) with trans - $Pt(PEt_3)_2Cl_2$ involves complexes where a nitrogen atom replaces one of the phosphines in the platinum coordination sphere. A 1:1 complex is formed in the sulfur system while a 2:1 (metal/ligand) complex involving distant (2,6) nitrogen centers is obtained for the selenium and sulfur systems. The six-membered ring $Ph_2\overline{PNPNSN}$ forms a 1:1 complex utilizing the nitrogen atom in the PNS fragment.[185] Both static and dynamic ^{31}P and ^{77}Se NMR have been used to investigate the reactions of **18** (E=Se) with platinum and palladium complexes. Insertion into the Se-Se bond occurs with $M(PPh_3)_2L_2$ (M=Pt, $L=C_2H_4$; M=Pd, $L=Ph_3P$). The product, $M(PPh_3)_2\cdot1,5\text{-}Ph_4P_2N_4Se_2$ undergoes a reversible thermal 1,3-metallotropic shift to give $[M(PPh_3)\cdot1,5\text{-}Ph_4P_2N_4Se_2]_2$ in which the metal ions bridge the two phospha(selena)zene rings. The structure of the sulfur analog has previously been reported. Addition of $MCl_2(PEt_3)_2$ (M=Pt,Pd) gives the 1:1 and 2:1 (metal: ligand) adducts $[MCl_2(PEt_3)]_n \cdot$ **18** (E=Se), n=1,2). The 2:1 adduct occurs in the platinum system and involves 2,6-nitrogen centers. The reaction of **18** (E=Se) with PhLi followed by $MCl_2(PEt_3)_2$ gives $MCl(PEt_3)_2Ph_4P_2N_4Se_2Ph$ (M=Pt,Pd). Coordination in this case is to the anionic selenium center. Slow rotation about the Se-Pt bond has been detected.[186] The coordination chemistry of the anion obtained from **18** (E=S) and CMe_3Li reaction is of interest. Addition of Cp_2ZrCl_2 gives $Cp_2ZrClPh_4P_2N_4S_2CMe_3$ which exhibits a three-membered metallacycle, $\overline{ZrNS}$, as the point of attachment for the metal. Reaction of this material with YX (Y=Me, X=I; $Y=X=Br_2$) gives $X\overline{SNP(Ph)_2NS(CMe_3)NP(Ph)_2N}$ while HCl(g) gives $Me_3C\overline{SNP(Ph)_2N(H)SClN(H)P(Ph)_2N^+}Cl$ which can be dehydrohalogenated to the neutral species.[187] Sulfur-nitrogen rings with exocyclic phosphazene units are also known. The reaction of S_4N_4 and diphenyl(2-pyridyl)phosphine gives $(NC_5H_4)Ph_2P{=}NS_3N_3$.[188] The corresponding reaction with $Ph_2(R)P$ (R=2-pyridylamino, NC_5H_4NH) at ambient temperature gives $Ph_2(R)P{=}NS_3N_3$ which slowly rearranges in solution to **18** (E=S) and $Ph_2\overline{PNSNSN}$. The rearranged products are obtained directly in refluxing acetonitrile. When PhRR'P ($R=NC_5H_4NH$, $R'=(C_6H_{11})_2N$) is employed, $Ph(R')\overline{PNSNSN}$ is obtained. This material forms a 1:1 cycloaddition product with nonbornadiene.[189] The reaction of dilithioferrocene with cis-$NPF_2[NS(O)Ph]_2$ gives a ferrocene with a cyclophosphathiazene on each cyclopentadienyl ring $(NPF[NS(O)Ph]_2C_5H_4)_2Fe$. The mass spectrometric fragmentation path and cyclic voltammetry have been reported. The oxidation occurs at a high potential (+1.19v vs SCE).[190]

5 Miscellaneous Phosphazene Ring Systems including Metallocycles

This section covers remaining systems containing at least one phospha(V)zene moiety in the ring. A few selected metallocyclic systems can be found in previous sections but the majority of these are covered below. A review of some work on metallocyclic derivatives and a comparison with metallosiloxanes has appeared.[191] MNDO calculations and ^{35}Cl NQR data (frequencies and asymmetry parameters) for diazaphosphorines, **20** (A=N, B=C, C=D=Cl, F=H; A=N, B=C, C=H, E=F=Cl; A=C, B=N, C=E=CCl_3, D=CN). The phosphorus-chlorine distances are predicted to be invariant in the series.[192] The reaction of Me_3CPCl_2 and $F_3CC_6H_4N(SiMe_3)_2$=$NSiMe_3$ gives **21**. The PR′ bridge in **22** is cleaved by SO_2Cl_2 to RC=NP(R′)Cl=NCR=NP(R′)Cl=N. Thermal analytical data (TGA/DSC) of the three ring systems are reported.[193] The combination of $(MeO)_2PNCO$ and RC=NNC_6H_4R' gives the monophosphazene, $(MeO)_2P$=$NCH_2N(C_6H_4R')N$=CR, which decomposes to $(MeO)_2P(O)CR$=$NNHC_6H_4R'$.[194] The 1,4-dipolar cycloaddition of R_2PN_3 (R=iPr) and dimethylacetylene dicarboxylate gives R_2P=NN=$NC(CO_2Me)$=$C(CO_2Me)$, which thermally eliminates nitrogen to give R_2P=$NC(CO_2Me)$=$C(CO_2Me)$, which is a non-aromatic dipole stabilized species.[195] Cycloaddition reactions of the low coordinate phosphazyne $P{\equiv}NAr^+AlCl_4^-$ $(Ar=(Me_3C)_3C_6H_2)$ give species which, from structural data and ab initio calculations, are best formulated as intramolecular donor/acceptor complexes. Substrates/products include: $R_2'NP$=$NCMe_3(R'$=iPr,$Me_3Si)$/ $PNCMe_3P(NR_2')NAr^+AlCl_4^-$ and $RN_3(CME_3,CEt_3)$/$PNArNNNR^+AlCl_4^-$.[196] Nucleophilic substitution reactions of pentachloro cyclocarbophosphazene, $ClC{=}NPCl_2{=}NPCl_2{=}N$, have been explored with $(RO)C{=}NP(OR)_2{=}NP(OR)_2{=}N$ ($R{=}Me, CH_2CF_3, Ph$) and $ClC{=}NPCl_2{=}NP(Cl)FeCp(CO)_2{=}N$.[197] Analogous reactions with amines give $(R_2N)C{=}NP(NR_2){=}P(NR_2){=}N$ ($R_2{=}Me_2, Et_2$, piperidine). Primary amines give the hydrolytically unstable $N_3P_2C(NHR)_5{\cdot}HCl$. Interesting site selectivity has also been noted. If $HNPh_2$ and $NaOCH_2CF_3$ are allowed to react sequentially with $N_3P_2CCl_5$, the carbon center is the initial site of attack providing $[NP(OCH_2CF_3)_2]_2NCNPh_2$ while the sequential addition of RONa ($R{=}2,6{-}Ph_2C_6H_3$) and $HNMe_2$ gives $NP(NMe_2)_2NP(NMe_2)(OR)NCNMe_2$ i.e. the phosphorus center of initial attack.[198]

The number of both main group and transition element metallocyclic phosphazenes reported has increased. The reaction of $Ph_2P(NHSiMe_3){=}NSiMe_3$ with alkali metal sources (BuLi, NaH, KH, Rb, Cs) leads to the chelated species, $Ph_2PN(SiMe_3)M(THF)_nN(SiMe_3)$, which have varying degrees of solid state structural complexity.[199] The reactions of $HN[P(NMe_2)_2NSiMe_3]_2$ with MH ($M{=}Na, K$) gives the dimer, $[NaN(P(NMe_2)_2NSiMe_3)_2]_2$, and polymer, $KN[P(NMe_2)_2NSiMe_3]_2$, both based on chelating phosphazenes using the central and terminal nitrogen atoms. Addition of $Ca[N(SiMe_3)_2]_2{\cdot}2THF$ to the neutral ligand gives $Ca(N[P(NMe_2)_2NSiMe_3]_2)_2$ with a tridentate phosphazene ligand.[200] The analogous reaction of the barium precursor with the ligand noted above and $H_2NP(NMe_2)_2{=}NP(NMe_2)_2{=}NP(NMe_2)_2{=}NSiMe_3$ gives $Ba(N[P(NMe_2)_2NSiMe_3]_2)_2$ in which the phosphazene acts as a tridentate chelate and $[BaNP(NMe_2)_2NP(NMe_2)_2NSiMe_3]_4$ which has a cubane-type structure. These barium compounds can undergo sol-gel chemistry to give hydrated barium(II) hydroxide gels.[201] The reactions of $HN(PR_2NSiMe_3)_2$ ($R{=}Ph, NMe_2$) with MMe_3 ($M{=}Al, Ga, In$) gives $Me_2MNSiMe_3PR_2NPR_2NSiMe_3$ which exhibits unusual arrangements of the cyclohexane like ring atoms.[202] The aforementioned $LiPh_2P(NSiMe_3)_2$ species combines with group IV metal dihalides to give $[Ph_2P(NSiMe_3)_2]_2M$ ($M{=}Sn, Pb$) which exhibit bidentate chelation and sterochemically active lone electron pairs on the metal.[203] The acyclic ligand $Me_3SiNP(NMe_2)_2NP(NMe_2)_2NH_2$ was prepared from $N[P(NMe_2)_2NH_2]_2^+Cl^-$ by sequential reactions with NaH and Me_3SiNMe_2.

Reaction of this ligand with antimony(III) acetate gives $(AcO)_2SbNHP(NMe_2)_2NP(NMe_2)_2NH$. The structure shows an extended array of rings linked by hydrogen bonding and having cis/trans isomerism in the relations of the acetate groups about the Sb(III) center.[205] The chemistry of $Me_3SiNC(4\text{-}RC_6H_4)NPPh_2NSiMe_3$ (R=H,CF_3) has been explored. With $TeC\ell_4$, 4-$CF_3C_6H_4CNPPh_2NTe(pyridine)C\ell N$ is obtained while two six-membered rings connected via four-membered Re-N-Re-N adducts are obtained from the reaction with Re_2O_7. An acyclic rhenium (VII) derivative, $Me_3SiNC(4\text{-}CF_3C_6H_4)NPPh_2NReO_2OSiMe_3$, is obtained from $Me_3SiOReO_3$. Zinc can also be part of a six-membered phosphazene ring , $Me_3SiNP(NMe_2)_2NP(NMe_2)_2N(SiMe_3)ZnR$, as shown in the products of the $NH[P(NMe_2)_2NSiMe_3]_2/ZnR_2(R{=}N(SiMe_3)_2,Me,Et)$ interaction.[206] Oxidative scrambling of the $LiN(PPh_2)_2$ ligand occurs in reactions with selected transition metal halides. When $CoC\ell_2$ is employed, $Co(PPh_2NPPh_2NPPh_2)_2$ is obtained, which involves coordination to the two terminal phosphorus atoms of each ligand and a nitrogen center of one of the phosphazene chelates.[207] With $MC\ell_2$ (M=Ni,Pd), $M(PPh_2NPPh_2NPPh_2)_2$ is formed.[208] A thorough examination of the synthesis, reactivity and spectroscopy of the bis(iminophosphoranyl) methanide, $CH(PPh_2{=}NC_6H_4\text{-}4\text{-}R)_2^-$, complexes of rhodium (I) and iridium (I) has been reported. The $M[CH_2(PPh{=}NC_6H_4R)_2L_2$ complexes which are formed by two different routes, exhibit σ coordination to each of a N and C center. The complexes are fluxional in solution via a process involving intramolecular attack of the free iminophosphorane unit. Reactivity studies range from simple ligand (L) exchange to more complex processes. The aza-Wittig reaction of CO_2 occurs only at the free iminophosphorane center giving $M[CH(PPh_2{=}NC_6H_4CH_3)PPh_2{=}O]L_2$. Protonation with CF_3CO_2H occurs at the nitrogen of the free iminophosphorane. The analogous reaction with $HC\ell$ follows both the above pathway and by protonation of the carbon center to give **23.** This reaction goes via an oxidative addition process in which the iridium hydride has been observed.[209] The reaction of $RN{=}PPh_2CH_2PPh_2$ with $[Rh(CO)_2C\ell]_2$ gives $C\ell(CO)RhN(R){=}PPh_2CH_2PPh_2$[20]. The above ligand

(R=$SiMe_3$) in combination with Re_2O_7 yields (Ph_2PCH_2N)Re(O)$OSiMe_3$ which contains a Ph_2PCH_2P=NRe metallocycle.[19] The structure of $C\ell_2Pt(PPh_2N(Ph)P(S)Ph_2 \cdot H_2O$ which has a $PtPPh_2N(Ph)P(S)Ph_2$ chelate is not formally a phosphazene but has a short PN (167.4) bond.[210]

6. Poly(phosphazenes)

This section covers polymers derived from open-chain phosphazenes and related cross-linked materials. Cyclolinear and cyclomatrix materials as well as carbon-chain polymers with cyclophosphazene substituents are covered in Section 3. A wide range of review material has appeared including the following: an overview of poly(phosphazenes) in the context of phosphazene chemistry[1]; a two part comprehensive examination of photochemistry and photophysics of polyorganophosphazenes devoted to monomolecular processes in solution and solid state[211] along with bimolecular events and practical applications[212]; phosphazene polymers as advanced materials with emphasis on tailored design to fit specified properties [213-216]; recent developments of main-group inorganic polymers [217]; the derivatization chemistry of dimethyl- and methylphenyl phosphazene polymers[218]; a valuable survey of dilute solution characterization such as viscosity, scaling laws and unperturbed dimensions [219]; non-metal nitride solid state chemistry with a strong emphasis on phosphorus-nitrogen species[220,221]; the synthesis of structurally well defined poly(phosphazenes) using the phosphoranimine route [222]; recent advances in poly(thionylphosphazenes)[223]; polymer electrolyte systems [224] and a brief discussion of inorganic polymer nomenclature [225]. Review materials in less accessible sources and languages include Japanese summaries focused on preparative chemistry [226-228] and fiber technologies in Russian.[229]

Information on the formation of poly(phosphazenes) from small molecule precursors continues to become available. New results in ring-opening polymerization are limited in mixed ring systems. The polymerization of the cabophosphazene, $C\ell CNPC\ell_2NPC\ell_2N$, occurs at relatively mild conditions and can be

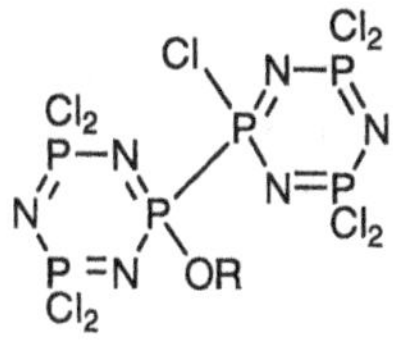

(14)

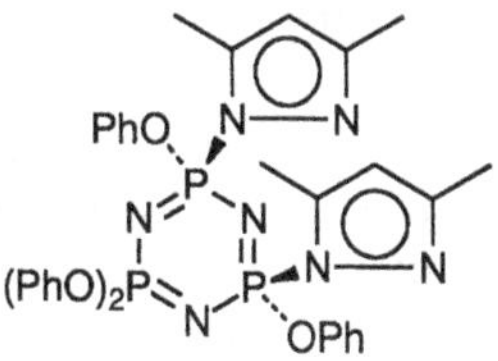

(15)

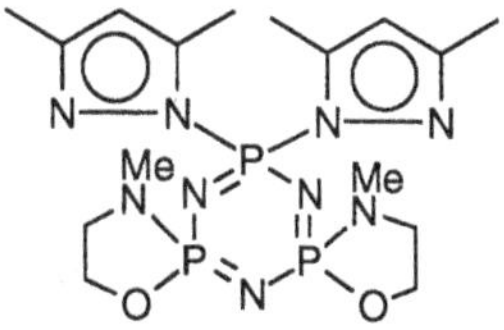

(16)

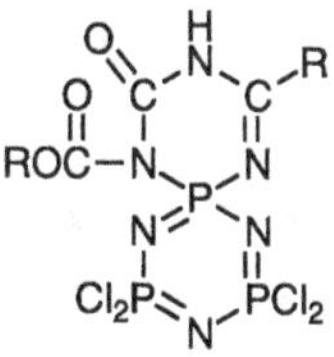

(17)

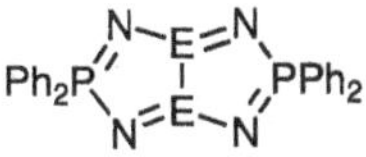

(18)

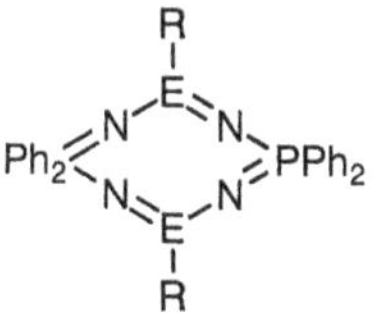

(19)

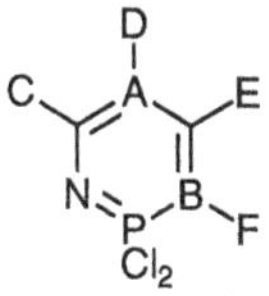

(20)

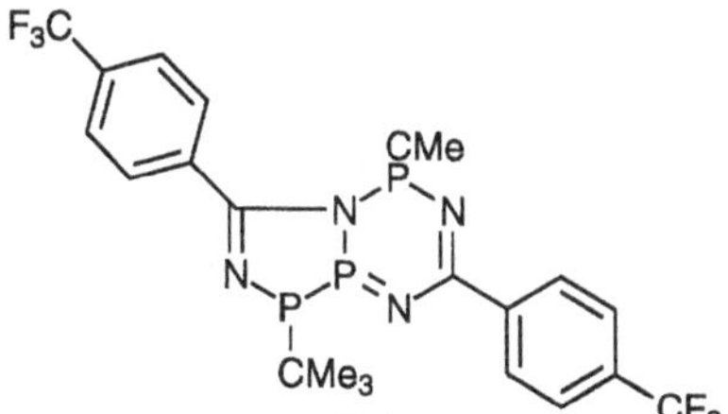

(21)

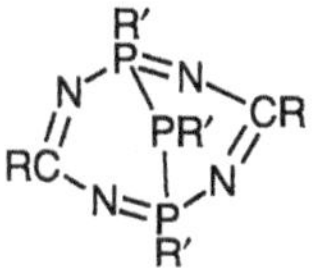

(22)

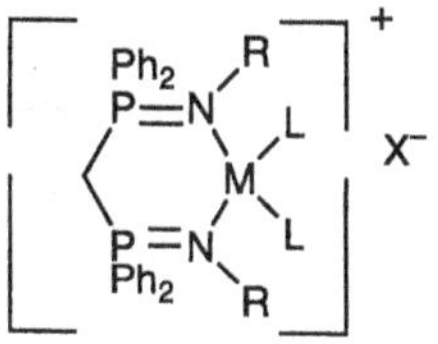

(23)

followed by ^{31}P NMR spectroscopy. Significant ring strain due to the presence of the carbon atom in the ring plays a significant role in the polymerizabilitiy of this material.[197,198,230] Ring-opening processes also occur for the phospha(thia)zenes, $(NPX_2)_2NS(O)X$(X=Cℓ,F) to give rise to the mixed phosphorus/ nitrogen sulfur/nitrogen polymers. [231,232] More extensive activity has been focused on step polymerization methods. Details of the polycondensation reaction of $Cℓ_3P{=}NP(O)Cℓ_2$ continue to appear. No cross-linking occurs in the process and conditions for molecular weight control [233] as well as the catalytic role of additives [234] have been explored. Alkyl or aryl substituted polymers are available from $RPCℓ_2{=}NP(O)Cℓ_2$. The remaining chlorine in the polymer atoms can be derivatized by reaction with $4MeOC_6H_4ONa$.[235] Poly(dichlorophosphazene) and cyclic oligomers arise in the $Cℓ_3PNPCℓ_3^+Cℓ^-/PCℓ_5$ reaction.[37] The use of $R_3P{=}NSiMe_3$ phosphoranimine precursors has received the most attention. The effect of catalysts (F^-, OPh^-, HMPA) in the preparation of poly(alkyl/arylphosphazenes) is to reduce polymerization time and broaden the molecular weight (MW) distribution. Preparation of $[(CF_3CH_2O)_2PN]_n$ from $(CF_3CH_2O)_3PNSiR_3$ with fluoride, N-methylimidazole [237] on $(R_2N)_3S^+X^-$ [238] catalysts has been reported. The reaction is first order in monomer, sensitive to the substituents on the silicon center with a decrease in MW with increased size of the substituent being note.[22] If a very bulky nitrogen substituent, e.g. R=CMe_3CPh_3, adamantyl, in $(CF_3CH_2O)_3P{=}NR$ is present, the phosphoranimine acts as a chain terminator in the $(CF_3CH_2O)_3P{=}NSiMe_3$ system.[239] Mixed substituent alkoxyphosphazene polymers can be obtained from the mixed substituent phosphoranimine precursors.[23,25] Kinetic studies of these systems have also been conducted.[23] Copolymers may also be obtained by simultaneous reaction of two different phosphoranimine precursors. The kinetic data for this reaction shows simultaneous loss of each monomer and hence formation of random copolymers. Thermal properties in this system vary in a regular fashion with composition.[240,241] Sequential addition of phosphoranimines gives block copolymers indicating a living polymer system. Thermal properties vary in a regular fashion and are different from the random copolymers.[241,242] An alternative reaction of Me_3SiN_3 with phosphorus(III) derivatives is azide exchange. The reaction of $Ph(R)POCH_2CF_3$ with Me_3SiN_3 gives $Ph(R)PN_3$ (R=Me-4-C_6H_4) which is converted to $(Ph(R)PN)_n$ by thermolysis.[243,244] Various fluorophosphines $R_{3-n}PF_n$ (n=1,2; R=Ph, PhO, $C_5H_{10}N$, Et_2N) undergo a similar process but the resulting polymers have not been extensively characterized.[27] Microporous[245] and glassy[246] solid state materials based on tetrahedral phosphorus centers with nitrogen atoms at some of the corners have been prepared.

Displacement of halogens from halophosphazene polymers is a classic synthesis route which is still utilized extensively. Irradiation of $(NPC\ell_2)_n/CH_2=CHCH_2SnMe_3$ mixtures gives rise to partial chlorine atom replacement. The remaining reactive centers can be removed by reaction with NaOPh to give $[NP(OPh)_{2-x}(CH_2CH=CH_2)_x]_n$ where the maximum degree of substitution (x) is 0.75.[67-69] Aminolysis reactions of $(NPC\ell_2)_n$ gives rise to mixed substituent allylamino/ethylanilino derivatives which can be cross-linked upon exposure to γ-radiation. The resulting materials were evaluated as resists.[247] The synthesis (from $(NPC\ell_2)_n$ with the appropriate oxyanions) and characterization (MW, thermal properties, viscous flow temperature) of $[NP(OC_6H_4C_6H_5)_2]_n$[248] and $[NP(OC_6H_4C_6H_5)_{1-x}(OEt)_x]_n$[249] have been reported. Films for electroluminescent devices have been prepared from the reaction of $(NPC\ell_2)_n$ and $Ph_2NC_6H_4ONa$ which provides $[NP(OC_6H_4NPh_2)_2]_n$.[250] Heterophosphazene polymers also exhibit halogen displacement chemistry. The chlorine atoms in poly(carbophazene), $[NPC\ell_2NPC\ell NCC\ell]_n$, can all be displaced by primary H_2NR (R=iPr, nPr, Me) and secondary amines HNR_2 (R=Me,Et). Sequential reactions involving $HNPh_2$ followed by $N_2OCH_2CF_3$ give $[NP(OCH_2CF_3)_2NP(OCH_2CF_3)NCNPh_2]_n$ while the $HNPh_2/HNMe_2$ system gives $[NP(NMe_2)_2NP(NMe_2)(NPh_2)NCNPh_2]_n$. Thermal analysis data are also reported.[198] Various aryloxides react with $[(NPC\ell_2)_2NS(O)X]_n$ (x=F,Cℓ) to give $[(NP(OAr)_2NS(O)OAr]_n$. The chloro derivative (X=Cℓ) was also converted to mixed alkoxy/aryloxy derivatives.[232]

The side groups in phosphazene polymers can be modified by further synthetic transformations. Sulfopropylation of $[NP(OCH_2CH_2OCH_2CH_2NH_2)_x(OCH_2CF_3)_y]_n$ with 1,3-propane sulfone gives polymers with up to half of the amino substituents converted to $OCH_2CH_2OCH_2CH_2NH(CH_2)_3SO_3H$ groups which impart water solubility. The glass transition temperatures, Tg, are similar to the non-sulfonated precursors.[137] Organometallic polymers of the type $[NP(OC_6H_4FeCp(CO)_2)_x(OC_6H_4Br)_{2-x}]_n$ were prepared from treatment of the para-bromophenoxy derivative with butyllithium and $CpFe(CO)_2I$. The insoluble materials show bridging carbonyls and are easily oxidized to paramagnetic materials. Analogous chemistry can be performed on cross-linked films of $[NP(OC_6H_4Br)_2]_n$. Various surface analytical techniques show that modification occurs and extends to a depth of 30 um.[127] Radiation cross-linking of aryloxyphosphazenes using both γ and electron beam irradiation has been thoroughly examined. The incorporation of allylic groups increased the cross-linking efficiency by an order of magnitude. The grafted films can be extracted from the precursors by extraction of the former with THF.[251] Photochemistry of phosphazene substituents was the subject of several investigations. Chalcone ($R=C_6H_5CH=CHC(O)Ph$) systems $[NP(OR)(OR')]_n$ ($R'=$ Ph, CH_2CF_3) were prepared by the synthetic route used for the cyclic trimer models (Section 3). The polymers can be cross-linked photochemically presumably by a two plus two cycloaddition process.[118] Cinnimate derivatives ($R = (CH_2)_2OC(O)CH=CHPh$) have also been obtained (Section 3) and also undergo photocross-linking.[139] Photografting of $CH_2=CRR'$ and $[NP(OC_6H_4Me)_2]_n$ in the presence of 10% benzophenone results in the formation of $OC_6H_4CH_2(CH_2CRR')_n$ groups. Various reactive groups in the R,R' set have been exploited for further chemical modification. Proteins can be immobilized by ring opening of an expoxide substituent by amines and a 4-vinylpyridine graft allows for metal binding.[252] The photooxidation of $[NP(OC_6H_4CH_2Ph)_2]_n$ films at short and long wave lengths occurs at the benzylic methylene groups giving rise to benzophenone, benzaldehyde and benzoic acid via hydroperoxide intermediates.[253] Photochemical grafting of

N,N-dimethylacrylamide into $[NPR_2]_n$ (R=OC_6H_4-iPr, OC_6H_4-sec-Bu, $OC_6H_4CH_2Ph$, OCh_2CF_3) introduces cross-linking and increased hydrophobicity into the phosphazene.[254] Grafting of acrylic monomers onto alkoxyphosphazenes gives bound polyacrylates.[255] The coordination of M^+(M=Li,Ag) to $(NPMeR)_n$ (R=Me,Ph) occurs at backbone sites. The metal ions are weakly coordinated and exhibit mobility. At low temperatures both free and coordinated Li^+ can be detected.[256,257] The reaction of $(NPMe_2)_n$ with BuLi gives reactive carbanions which have been allowed to combine with α-haloethers, Me_2CO followed by α-haloethers, $R_2(RO)SiC\ell$ and tert-butylmethacrylate. The reaction gives oxygen containing substituents in general and grafted poly(tert-butylmethacrylate) in the last case.[258] Analogous reactions with acid chlorides give keto derivatives.[259] The availability of a hydroxyl group on an organic side chain allows for a wide range of chemical modification of organophosphazenes. Coupling with allyl containing halides or acid chlorides gives materials which can be converted to epoxides by m-chloroperbenzoic acid oxidation.[140] Alkoxyhydroxy side chains have been prepared and reacted with diisocyanate or aryloxyphosphazenes with succinic anhydride groups.[260] Phosphazene containing silica materials are obtained by the reaction of $(EtO)_4Si$ with $[NP(OC_6H_4OH)_2]_n$.[141] Alumina particles can be coated with amino or alkoxyphosphazenes in order to produce preceramic disperants. The interaction is accompanied by loss of side chain or depolymerization.[261] Poly(phosphazenes) with 2-hydroxylethyl- methacrylate side chains can be combined with powdered marble and photocross-linked to give building materials.[262] An area of particular concern in organofunction phosphazene polymers is the production of polymers related to biological systems. Important aspects of this work have been reviewed.[263] Sodium salts of the esters of glycolic and lactic acids (NaOCHRC(O)OR'; R=H,Me, R'=Et,Benzyl) react with $(NPC\ell_2)_n$ to give the appropriate substituted phosphazenes which readily undergo hydrolysis.[264] The mixed imadazoyl/methylphenoxy or ethyl glycinatomethylphenoxy systems have been examined for skeletal tissue regeneration materials.[265] Amino acid esters

($NH_2CHR'C(O)OR$; R'=H,Me; R=Me,Ph) have been attached to poly(dichlorophosphazene) and their hydrolysis rates and decomposition pathways have been explored.[266,267] Several of these materials are microcrystalline and have high Tg values. Similar materials with a depsipeptide
co-substituent have been explored. The degration process for these polymers leads to the amino acid, ethanol and polymer backbone cleavage.[268,269] The bis(glycine ethylester) phosphazene hydrolytic degration reaction has been examined in detail on bulk samples. The results are compatible with applications as a support for short-term controlled drug release.[270] The hydrogels obtained from the cation including gelation of $[NP(OC_6H_4CO_2H)_2]$ have generated considerable excitement as biomaterials. Further details on the encapsulation of cells and enzymes by this system are available.[271] The controlled release of drugs [272] and proteins [273] by these hydrogels has been established. Molecular engineering of the polymer by inducing hydrolytic susceptibility via reactive side groups has also been investigated.[273,274] The specific details of the degradation path of amino acid co-substituent systems have been examined.[274] Studies of cell adhesion and spread on bioerodible phosphazenes show that useful osteoblast-polymer composite materials suitable for combination with bone are possible. [275] Various cell lines have been found to adhere and spread on the $[NP(OC_6H_5)_{2-x}(NHBu)_x]_n$ polymer surface. [276] The alkoxide ions obtained from methoxypolyethylene glycol can replace some of the groups on the surface of $[NP(OCH_2CF_3)_2]_n$. The exchange reaction products are biocompatible and have been investigated by a range of surface techniques.[271] A few miscellaneous, ionic modifications of polyphosphazene substituents include ionomers from the combination of $[NP(OCH_2CF_3)_x(OC_6H_5CO_2H)_y]_n$ and $M(OH)_2$ (M=Ba,Sr) [278] and the use of $HOP(OR)=N(P(OR)_2=N)_nP(OR)_2(O)$ for extraction of scandium from sulfuric acid solutions.[279]

Physiochemical studies on poly(phosphazenes)are also an area of active investigation. Certain of these have been mentioned above. MNDO modeling for $[NP(OPh)_2]_n$ has been carried out by extrapolating results from small linear, EL[NP(OPh)$_2$]$_2$ER [EL, ER=H,NO,SNPCℓ_2, etc.], H[NP(OPh)$_2$]$_n$H (n=1-6), and cyclic (NPR$_2$)$_n$ (R=OH,n=1-8; R=OPh, n=1-3) systems. Most structural features were transferrable to the polymer. The polymer backbone is not planar but rather has a helical configuration with a long pitch for each helix.[280] MNDO calculations of the vibrational spectra (IR,Raman) of (NPX$_2$)$_n$ (X=Cℓ, OCH$_2$CF$_3$, OPh) agree with experimental values which were also obtained.[281] The photochemistry and photophysics of [NP(Me)Ph]$_n$ and the same system with grafted poly(styrene) chains shows a red shifted emission spectrum which can be quenched with CCℓ_4. The photochemistry leading to chain cleavage originates in a singlet state.[282] Solid state structure and its relation to phase changes is a major theme in physiochemical investigations. Two mesophase transitions have been observed (by DSC) for (NPPh$_2$)$_n$; the transition temperatures are a function of molecular weight. Both the high and low temperature forms undergo transitions to other forms on continued cycling.[283] Complex mesophase behavior has also been noted and evaluated on the basis of structure for [NP(OC$_6$H$_4$-p-Me)$_2$]$_n$.[284] Single crystals in films of [NP(OC$_6$H$_4$-p-Et)$_2$]$_n$ have been examined by wide-angle x-ray diffraction, electron diffraction and electron micrographs. The chains are in a trans-cis conformation.[285] The [NP(OC$_6$H$_3$-3,4-Me$_2$)$_2$]$_n$ structure in cast films has been modeled. Thermotropic transitions to a mesomorphic state have been detected.[286] Various phase states including a conformational disordered (condis) mesophase have been identified for [NP(OCH$_2$CF$_2$CF$_3$)$_2$]$_n$. Modifications of the elementary cell have been observed in both high and low temperature ranges.[287] The mesophase and condis crystalline phases have been identified separately for [NP(OCH$_2$CF$_2$CF$_2$CF)$_2$]$_n$.[288] A new crystalline form for [NP(OCH$_2$CF$_3$)$_2$]$_n$ has been detected leading to a scheme for annealing including two types of mesomorphic structures being proposed.[289] The phase diagram for the [NP(OCH$_2$CF$_3$)$_2$]$_n$/solvent (EtOAc or DMSO) has been constructed.[290] The thermodynamics of the trifluoroethoxy polymer-solvent (acetone, EtOAc) system have been evaluated and related to viscometric characterization.[291] Mixed random aryloxy copolymers, $[NP(OAr)_x(OAr')_{2-x}]_n$ have been shown to be semicrystalline. Films have been cross-linked by γ radiation and thermomechanical properties evaluated. Above the melting point, stress induced crystallization has been observed.[292] Members of the series $[NP(OPh)(OC_6H_4\text{-}4\text{-}Et)_{2-x}]_n$ are amorphous and films can be cross-linked by γ radiation. Stress-strain and thermoelastic studies show a decrease in unperturbed dimensions with increased temperature. Excellent network reinforcement can be obtained by silica generated in situ by the hydrolysis of absorbed $(EtO)_4Si$.[293] The synthesis and ferroelectric liquid crystal behavior of $[NP(O[CH_2]_{10}OC_6H_4C_6H_4CO_2CH_2CH(Me)Et)_2]_n$ has been reported and a wide range of the liquid crystal phase was noted.[294] The introduction of a 4'-nitrostilbene unit onto the phosphazene side chain gives NLO properties. The solubility is low but can be increased in the mixed substituent systems, $[NP(OC_6H_4CH{=}CHC_6H_4NO_2)_x(OR)_{2-x}]_n$ (R=Ph, $OCH_2(CF_2)_4H$, OCH_2CF_3).[295] The electron donor properties of substituents in films of poly(phosphazenes) have been correlated to the measured non-linear optical properties.[296] Dielectric properties of $[NP(OR)_2]_n$ ($R{=}OCH_2CF_3$, $C_6H_4\text{-}3\text{-}Me$) have been measured and correlated to glass-rubber transition and local motion of short segments, the latter not being detected in the aryloxy case.[297] A careful and detailed study of the electrochemistry of a series of ferrocenylpolyphosphazenes, $[(NP(Me)Ph)_x(NPPhCH_2CR(OH)C_5H_4FeCp)_y]_n$, has shown that in films deposited on electrodes the oxidation potential is independent of degree of substitution. Diffusion coefficient data suggest that both physical diffusion and electron hopping occur.[298] Ionic conductivity in phosphazene based polymeric electrolytes is an ongoing area of interest. Films of lithium triflate salt of $[NP(OCH_2CH_2OCH_2CH_2OCH_3)_2]_n$ (MEEP) can be photochemically cross-linked. This inhibits slow loss of electrolyte from the cell.[299] Similar results are obtained from $[NP[(OCH_2CH_2O)_7R]_2]_n$ (R=Me, $CH_2CH{=}CH_2$).[300] The polymer electrolyte, MEEP/poly(propylene oxide), combined with several lithium salts has been employed as an electrolyte and conductivities have been reported.[301] Electric relaxation in the mixed polymer system noted above (as its NaSCN salt) shows both slow and fast ion migration processes depending on temperature.[302] Characterization of the NaSCN system by conductivity, IR and DSC measurements has also been reported.[303] MEEP or its composites with poly(propylene oxide) or poly(ethylene glycol diacrylate) provide $Li/SOC\ell_2$ batteries which alleviate problems of voltage delay.[304] The oligo-(ethyleneglycol) system $[NP(OCH_2CH_2O)_yCH_2CH{=}CH_2)_x[(O(CH_2CH_2O)_zCH_3]_{2-x}]_n$ undergoes reactions with lithium bisulfide which converts the allyl residues to $CH_2CH_2CH_2SO_3Li$ units. The dc conductivities of these and related systems were compared to $MEEP{\bullet}LiSO_2CF_3$ (which has higher values).[305,306] Doping of anilinophosphazenes $[NP(NHC_6H_3XY)_2]_n$ and the N-alkylated analogs with Bu_4NBr give a 10^6 increased electrical conductivity.[307] The advantages of various phosphazene containing batteries have been noted in patents.[308,309] Blends and composites containing poly(phosphazenes) are also attracting interest. The compatibility of poly(vinylphenol) and several poly(phosphazenes) has been probed and miscibility was noted with MEEP and other proton acceptors but not with phenoxide or trifluoroethoxide substituted polymers.[310] Up to 50% of a blend of $[NP(OC_6H_5\text{-}4\text{-}C\ell)_2]_n$ can be tolerated by poly(styrene). Both thermal stability and flame retardancy were improved by blending with the phosphazene.[311] Blends of $[NP(Me)Ph]_n$ with poly(methylmethacrylate, carbonate or acrylonitrile) have been studied by DSC.[312] The polyethylene oxide/PNP-2000 system (a commercial fluoalkoxy copolymer) was examined in terms crystallization kinetics, spherulite growth, glass transition temperature and Flory-Huggins interaction parameter. The system is not 100% compability over the composition range.[313] Addition of poly(epichlorohydrin) to the blend leads to compability which is proportional to the additive concentration.[314] Sol-gel polymerization of $(EtO)_4Si$, $Ti(OiPr)_4$, $Zr(O\text{-}nBu)_4{\bullet}BuOH$ and $Al(O\text{-}SecBu)_3$ in MEEP gave clear composites with improved mechanical properties as determined by dynamic mechanical and stress-strain measurements.[315] The final major area of interest for poly(phosphazenes) is in membranes. The structural range available from side group selectivity in poly(organo)-phosphazenes has been examined in terms of gas permeation and selectivity. Mixed aryloxy, fluoralkoxy/aryloxy and organometallic/fluoralkoxy systems have been studied and high diffusion coefficients were noted.[316] Helium from natural gas wells may be separated from methane by permeation through poly(phosphazene) membranes. The role of chemical and structural changes on the separation performance was explored.[317] The gas sorption behavior through mesophase $[NP(OC_6H_4\text{-}4\text{-}CMe_3)_2]_n$ shows it to be one of the most selective poly(phosphazenes).[318] Poly(organo)phosphazenes have been doped with a cobalt-porphyrin complex. The cobalt complex reversibly binds oxygen. The combination of a high phosphazene diffusion constant and large equilibrium constant for binding of oxygen to the cobalt complex gives a system with high oxygen permeability.[319] The oxygen gas permeability in water has been correlated to mechanical properties (Young's modulus, tensile strength) for the $[NP(NHBu)(NHHex)]_n$ membrane.[320] Microporous materials coated with poly(phosphazenes) such as $[NP(NHBu)_2]_n$ make effective filtration membranes for separation of gases and liquids.[321] Poly(phosphazenes) with hydrophilic groups embedded in a zirconia matrix provide good monofiltration membranes.[322] Spraying of nebulized drops containing $[NP(OPh)_2]_n$ on to a substrate layer allows for forming polymer membrane coatings.[323] A few miscellaneous items include the solution characterization of $[NP(OC_6H_4CO_2Na)_2]_n$. The polymer behaves as a flexible random coil and can effect strong binding to a cationic chromophore.[324] Kieselguhr supports with deposited $[NP(OCH_2CF_3)_2]_n$ show increased water vapor absorption capacity.[102]

7. Crystal Structures of Phosphazenes

The following compounds have been examined by diffraction methods. All distances are in picometers and angles in degrees.

Compound	Coments	Ref.
1•$Me_2CHNH_3^+$•H_2O	PN 155.4(4) ∠PNC 134.5(4) Inclusion Compound	13
3•HPF_6 (R_2=Me_2; R'=CMe_2CMe_3)	∠PNP 138-158 Preliminary	16
Ph_3CN=N-N=P$(NMe_2)_3$	P=N 162.04(16) PN 163.99-164.77(17)	18
$Ph_2PCH_2PPh_2$=NC_6F_4-p-CN	PN 156.7(4) ∠PNC 132.9(3)	20
$Ph_2PCH_2PPh_2$=$NC_6H_2F(NO_2)_2$	PN 158.9(3) ∠PNC 128.8(4)	20
[K(Ph_2P(Se)$NSiMe_3$)•THF$]_2$	KNPSe rings fused via K and Se coordination PN 155.6(13), 159.4(12) ∠NPSe 114.3(9) ∠PNSi 137.7(9), 138.8(8)	39
$(C_6H_{11})_2$NP(F)=$NBMes_2$	P=N 152.0(3) PNR_2 165.0(3), 163.8(4) ∠BNP 150.2(3)	35
$[Me_3P$=NPh•$PhNH_2]_2$	Tetramer of two phosphazenes linked by two anilines via H-bonding PN 157.2(3) ∠PNC 128.4(2)	36
$(MeC_6H_4NH)_3$P(=NC_6H_4Me)-NMeP$(NHC_6H_4Me)_2NHMe^+Cl^-$	P=N 154.3(9) PNHAr 159.8-163.9(6) PNHMe 159.4(6)	40
Mes*$NPCl_3$	PN 146.7(4) ∠CNP 160.9(3)	42
Mes*$NPBr_3$	PN 146(1) ∠CNP 160.1(9)	42

$IP(=NMes^*)_2$	PN 152.4(1) ∠NPN 121.9(2) endo/exo NPN config.	43
$Mes^*P(=NMes^*)=CC\ell SiMe_3$	PN 152.3(3) ∠CNP 122.6(1)	45
$CMe_3P(=NMes^*)=PMes^*$	PN 155.0(5) ∠NPC 107.7(3)	47
8 ($R,R'=Et_3C$)	P=N 154.3(2) PN 164.5-179.6(2)	48
8 ($R=Et_3C$, $R'=Mes^*$)	P=N 154.6(3) PN 164.5(2), 166.0(3)	48
9	P=N 152.9(7), 152.8(5), 150.6(5)	49
$(Et_2N)_2P(C\ell)=NMes^*$	P=N 149.2(2) ∠PNC 161.5(2)	50
$RCH=CRC(PPh_2=NC(O)Ph)=C(R')NH_2$ $R=MeOC(O)$, $R'=4\text{-}MeC_6H_4$	P=N 160.3(5), 160.7(3) ∠PNC 125.1(4), 121.1(2)	56
$(FC_6H_4O)_2P(O)NP(OC_6H_4F)_3$	PN 152.8(1), 160.0(2) ∠PNP 135.39(9)	72
$(C\ell C_6H_4O)_2P(O)NP(OC_6H_4C\ell)_3$	PN 151.5(2), 158.1(2) ∠PNP 149.5(2)	72
$(BrC_6H_4O)_2P(O)NP(OC_6H_4Br)_3$	PN 152.6(7), 157.4(6) ∠PNP 131.5(5)	72
$TiC\ell_3N=PPh_3$	PN 161.4(4) ∠TiNP 180	73
$CP^*TiF_2N=PPh_3$	PN 156.7(6) ∠TiNP 152.7(4)	74
$(Cp^*TiF_2NPPh_2)_2C_2H_2$	PN 156.2(5) ∠TiNP 150.4(3)	74
$Sb(NPPh_3)_4^+SbF_6^-$	PN 157.5(7) ∠SbNP 132.2(4)	75
$Sb_2C\ell_5(NPMe_3)_2SbC\ell_6\cdot CH_3CN$	$C\ell_2Sb(III)/C\ell_4Sb(V)$ bridged by $NPMe_3$ PN 163.6(5), 163.7(5) ∠SbNP 125.4(3)-131.9(3)	76

$[SbCl(NPPh_3)(CH_3CN)_2]_2$-$(SbCl_6)\cdot 2CH_3CN$	$SbCl(CH_3CN)_2$ bridged by $NPPh_3$ PN 162.01(3) ∠SbNP 123.6(2), 132.0(2)	76
$ZnCl_2(Me_3SiNPMe_3)_2$	PN 160.6(3), 160.3(3) ∠ZnNP 115.3(1), 115.6(1)	77
$CoCl_2(Me_3SiNPMe_3)_2$	PN 159.4(6), 159.8(6) ∠CoNP 116.5(4), 115.6(3)	
$CoCl_2(HNPMe_3)_2$	PN 160.2(7), 159.7(9) ∠CoNP 128.3(5), 127.7(5)	77
$Cu_6Cl_6(NPMe_3)_4\cdot$ $Cu(Me_3SiNPMe_3)_2Cl_2$	Cu_6 Oh Cluster with 4 trigonal faces capped with $NPPh_3$(PN 156-163(1)). linear $Cu(Me_3SiNPMe_3)_2$ counter ion (PN 162(1), 165(2))	78
$Cu_6Cl_6(NPMe_3)_4Cl\cdot Me_3SiNPMe_3\cdot$ Ch_2Cl_2	Oh cluster as above PN 162.0(9)-162.6(8) with $Me_3SiNPMe_3$ trapped in lattice PN 163(1)	78
$WCl_5N{=}PCl_3$	PN 157.6(18) ∠WNP 176.2(12)	79
Cis-$WCl_4(N{=}PCl_2Ph)_2$	PN 157.7(5), 156.5(5) ∠WNP 158.5(3), 162.0(3)	79
$WCl_4(N{=}PCl_3)^+GaCl_4^-$	Weak W-Cl---Ga bridge PN 156.2(11) ∠WPN 161.6(14)	79
$Ph_3PNH_2^+ReO_4^-$	Strong hydrogen bonding PN 163.6(7)	81
$LCuCl_2$ (L=**11**, R=H)	Strongly distorted T_d at Cu PN 160.0(2), 159.8(3) ∠CNP 124.1, 122.2(2)	83
$(LCuCl_2)_2$ (L=**11**, R=Me)	Chloro bridged dimer of tetragonal Cu atoms PN 166(2), 165(1) ∠CNP 124(2), 128(1)	84

$HN[P(NMe_2)_2NH_2]^+VO_4(SiMe_3)_2^-$	PN 158.4(4), 157.4(4) PN (terminal) 161.7(4)-164.7(5) ∠PNP (internal) 129.4(3)	85
$Cl_3TiNPPh_2N(SiMe_3)_2$	P=N 162.3(3), 157.7(4), 158.2(4) PN 164.4(3) ∠TiNP 177.5(2) ∠ NPN 116.2(3), 109.4(3)	86
$Cp^*TiCl_2NPPh_2NHSiMe_3$	P=N 158.1(3) PN 164.2(3) ∠ TiNP 166.1(1) ∠ NPN 117.3(2)	86
$N_3P_3(NC_2H_4)_6 \cdot C_6H_6$	P=N 158.7(1), 160.1(1) PN(exocyclic) 167.5(1), 168.0(1) Deformation densitites determined	94
$N_3P_3(O_2C_{10}H_6)_3 \cdot p\text{-}C_6H_4Me_2$	Cage type clathrate	107
$N_3P_3(O_2C_{10}H_6)_3 \cdot S$ $S = C_6H_6, C_6F_6, p\text{-}C_6H_4Me_2$, Cp_2Fe	Unit cell and space groups determined	108
12$\cdot CHCl_3$	PN 157.4(8)-160.2(8) Slight boat relatively short intermolecular π aromatic stacking	109
$N_3P_3(NH_2)_6$	PN(endo) 159.3-160.9 PN(exo) 163.6-165.9 ∠NPN 115.8, ∠ PNP 122.9 boat conformation strong H-bond network	110
$N_3P_3Cl_4[NH(CH_2)_2S(CH_2)_6S(CH_2)_2NH]$	PN(endo) 154.7(5)-161.9(9) PN(exo) 161.7(5), 162.7(5) ∠NPN(exo) 104.7(3) Planar N_3P_3 15-membered macrocycle	112
$N_3P_3(OC_6H_4\text{-}p\text{-}F)_6$	Av. PN 158.1(5) ∠PNP 121.3-122.5(3) ∠NPN 117.2-117.7(1) Planar	72

$N_3P_3(OC_6H_4\text{-p-}C\ell)_6$	Av. PN 157.2(12) Mean $\angle$PNP 122.3(5) Mean $\angle$NPP 117.6(5) slightly puckered	72
$N_3P_3(OC_6H_4\text{-p-Br})_6$	Av.PN 157.1(19) $\angle$PNP 121.7(8) $\angle$PNP 116.6(7) essentially planar	72
$N_3P_3(OC_6H_4\text{-p-I})$	Av. PN 157.0(5) $\angle$PNP 121.0(2) $\angle$PNP 118.0(2) Planar	72
14	Av. PN 157.6(5) PP 2.193(2) long PCℓ at P-P Center	117
$N_3P_3F_4OCH_2(CF_2)_2CH_2O$	PN 156.5(4)-158.1(4) $\angle$NPN 118.2-119.9(2) distorted at spiro etner	124
$[N_3P_3(OC_6H_4F)_5]_2OCH_2(CF_2)_2CH_2O$	PN (Av) 157.2(6) $\angle$NPN 118.9(4) 115.5(4)	124
15	PN(endo) 157.7(2)-158.6(2) PN(exo) 167.8(7)-169.1(2) $\angle$NPN 116.9(1)-117.9(2) nearly planar	128
15$\cdot PdC\ell_2$	PN(endo) 157.4(3)-158.4(3) PN(exo) 170.2(3),170.5(3) $\angle$NPN 166.9(2)-118.7(2) N(endo)-Pd 286.0 P_3N_3 planar	128
16	PN(endo) 156.9(3)-159.2(3) PN(exo) 168.8(3),169.9(3) $\angle$NPN 115.4(2)-119.4(2) slight distortion from planar	129
$N_3P_3Ph_4(dmp)_2Mo(CO)_3$	PN(endo) 155.7(2)-163.6(2) PN (exo) 170.1(2),170.9(3) $\angle$NPN 114.4(2),115.9(2),122.3(1) non-planar	129
16 $\cdot W(CO)_3$	PN(endo) 154.5(7)-162.8(7) PN(exo) 169.6(7),170.2(7) $\angle$NPN 113.6(4),115.4(4),121.6(9) non-planar	129

Compound	Data	Ref.
$(H_2dmp)^+[(CuCl_2)_2N_3P_3O(dmp)_5]^- \cdot MeCN$ dmp=3,5 dimethylpyrazoyl group	PN(endo) 152.2(12)-163.0(12) PN(exo) 167.2-170.8(13) PO 150.5(11)	130
$N_3P_3(NC_4H_8O)_6H^+ClO_4^- \cdot THF$ (morpholino substituent)	THF clathrate PN(endo) 158(Av), 168-169	132
17 (R=Me)	PN(endo) 154.1(3)-158.8(3) PN(exo) 160.5(4) 175.1(3) ∠NP(R_2)N 113.8(8) N_3P_3 almost planar	135
$[LiPh_4P_2N_4S_2Ph \cdot THF]_2$	PN 160.8(9)-164.9(9) Dimer of two $Ph_4P_2N_4S_2Ph^-$ rings with fused Li_2N_2 ring.	184
$[trans\text{-}PtCl_2(PEt_3)]_2 \cdot$**19** (E=S, R=Ph)	disordered due to different ring conformations PN (mean) 161	185
$Cp_2ZrClPh_4P_2N_4S_2CMe_3$	PN 160.0(7)-164.8(5) Zr coordinated to adjacent N and S Preliminary	187
$Me_3CSNP(Ph)_2N(H)SNP(Ph)_2N$	PN 160.3(6)-166.1(6) ∠NPN 114.2(3),112.1(3)	187
$(NC_5H_4)Ph_2P{=}NS_3N_3$	PN 162,1(3) ∠PNS 122.5(2)	188
$Ph(R_2N)PNSNSN$ ($R=C_6H_{11}$)	PN(endo) mean 161.8(3) skew boat conformation	189
21	structural diagram only	193
$R_2P{=}NC(CO_2Me){=}C(CO_2Me)$ R=iPr	PN(endo) 170.2(3) PN(exo) 162.6(3), 162.9(3) ∠NPC 81.2(2)	195
$PNCMe_3P(NR')_2NAr^+AlCl_4^-$ R'=iPr, $Ar=(Me_3C)_3C_6H_2$	PN(endo) 163.6(3), 163.2(3) PN(exo) 162.9(3) ∠NPN(endo) 86.6(1),77.2(1)	196
$PNArNNNr^+AlCl^-$ $R=Et_3C, Ar=(Me_3C)_3C_6H_2$	PN 165.3(3). 164.4(3) ∠NPN 86.7(2)	196

5	PN 157.6(3)-161.4(3) ∠NPN 115.2(2), 114.8(1) ∠PNP 118.2(3)	197
	PN(Av.) 159 ∠NPN 117.4 ∠PNP 119.8 ring strain	197
$e_3)_2$Li(THF)$_2$	PN 157.9(4), 158.0(4) ∠NPN 109.4(2)	199
$Ie_3)_2]_2$$Na^-$$Na(THF)_6^+$	PN 156.7(5)-157.6(6) ∠NPN 112.8(3), 111.2(3)	199
$e_3)_2$K(THF)$_4$	PN 156.7(6), 157.1(5) ∠NPN 114.1(3)	199
$Ie_3)_2$Rb(THF)$]_2$	PN 158.5(3), 156.7(4) ∠NPN 113.7(2)	199
$Ie_3)_2Cs]_n$	PN 157.5(1), 157.6(1) ∠NPN 113.7 polymer by weak association	199
$e_2)_2NSiMe_3]_2)_2$	PN(in PNP) 156.4-158.4(2) PN (in $NSiMe_3$) 168.3-170.2(2) ∠PNP 158.16(14) 4 coordinate Na^+	200
$_2NSiMe_3]_2$	PN(in PNP) 155.1-160.0(2) PN (in $NSiMe_3$) 168.4-170.0(2) ∠PNP 134.45(11) polymer, 7 coordinate K^+	200
$e_2)_2NSiMe_3]_2)_2$	PN(in PNP) 157.0-158.1(3) PN(in $NSiMe_3$) 165.2-167.7(3) ∠PNP 162.0(2), 161.4(2) 6 coordinate Ca^{2+}	200
$e_2)_2NSiMe_3]_2)_2$	PN(Av) 158,155 ∠NPN 108 ∠PNP 158	201
$_2)_2NP(NMe_3)_2NSiMe_3]_4$	PN 153.7(7)-161.0(7) ∠NPN 117.8(3)-122.8(3) two ligands tetradentate; two ligands tri- and pentadentate	201

$Me_2AlNSiMe_3PPh_2NPPh_2NSiMe_3$	PN 158.5(2) 161.7(2) ∠ NPN 113.47(8), 113.44(8)	202
$Me_2InNSiMe_3PPh_2NPPh_2NSiMe_3$	PN 157.8(2)-160.4(2) ∠ NPN 116.89(10), 116.10(11)	202
$[Ph_2P(NSiMe_3)_2]_2Sn$	PN 158.9(6)-161.9(6) ∠ NPN 106.7(3), 105.8(3)	203
$[Ph_2P(NSiMe_3)_2]_2Pb$	PN 158.2(4)-161.5(4) ∠ NPN 108.3(2), 107.9(2)	203
$(AcO)_2SbNHP(NMe_2)_2NP(NMe_2)_2NH$	P=N 157.9(3), 156.9(3) PNH 161.6(3), 162.1(3) $PNMe_2$ 163.2-164.9(3)	204
$CF_3C_6H_4CNPPh_2NTe(Py)C\ell N \cdot$ ½toluene py=pyridine	PN 161.6(7),161.3(7) ∠ NPN 117.5(3)	205
$[4\text{-}CF_3C_6H_4CNPPh_2NReO_2N]_2 \cdot CH_2C\ell_2$	PN 161.1(12), 165.6(15) ∠NPN 114.6(8) dimer	205
$Me_3SiNP(NMe_2)_2NP(NMe_2)_2$- $N(SiMe_3)ZnN(SiMe_3)_2$	PN(endo) 157.78(13) PN(exo) 160.7(2),165.7(2)	206
$Co(PPh_2NPPh_2NPPh_2)_2$	PN 158.1-158.9 163.6-165.9 bi- and tridentate ligands	207
$Ni(PPh_2NPPh_2NPPh_2)_2$	PN 158.8(7)-163.7(2) ∠ NPN 119.5(4), 119.1(4) Ni-P coordination	208
$Pd(PPh_2NPPh_2NPPh_2)_2$	PN 158.8(6)-164.2(6) ∠ NPN 119.8(3), 120.3(3) Pd-P coordination	208
$IrCH(PPh_2{=}NC_6H_4\text{-}4\text{-}Me)_2COD$	PN 155(1), 162(1) 4 membered IrNPC	209
$C\ell(CO)RhN(C_6F_4\text{-}4\text{-}CN){=}PPh$- CH_2PPh_2	PN 161.6(2) ∠PNRh 113.2(1)	20
$C\ell_2Pt(PPh_2N(Ph)P(S)Ph_2] \cdot H_2O$	PN 167.4(7)	210

$(NPPh_2)_n$	x-ray and electron diffraction cell dimensions for several forms	283
$[NP(OC_6H_4Me)_2]_n$	cell dimensions for several forms	284
$[NP(OC_6H_4Et)_2]_n$	four chains per unit cell trans-cis arrangement	285
$[NP(OC_6H_3\text{-}3,4\text{-}Me_2)_2]_n$	$(trans_3cis)_2$ array other cells detected on heating	286
$[NP(OCH_2CF_2CF_3)_2]_n$	cell dimensions for several forms	287
$[NP(OCH_2(CF_2)_2CF_3)_2]_n$	cell dimensions for several forms	288
$[NP(OCH_2CF_3)_2]_n$	new crystalline form detected	289
$Cl_3PNP(O)Cl_2$	measured at 100°K PN 158.3(3)-159.3(3) 151.7(3)-153.0(3) ∠PNP 137.8(2)-144.1(2) cisoid configuration	325
$Ph_3PNH_2^+N_3^-$	PN 160.3(3) ∠CNP 106.9, 109.5,113.2(2) hydrogen bonding to azide	326
$1,2\text{-}C_6H_4(N{=}PPh_3)NMe_2$	PN 156.4(2) ∠PNC 127.2(2) Ph_3PN twisted 8.3° rel to benzene ring	327
$N_3P_3(OR)_6$ OR=2,3-naphthylenedioxy	PN 156.7(6)-160.0(7) boat conformation; distorted from that observed for clathrated analogs	328
$NP(OR)_6$ $1,4\text{-}C_6H_4Me_2$ OR=2,3-naphthylenedioxy	PN 156.7(2)-157.4(2) p-xylene in cage-like cavity	329
$PhCNPPh_2NPPh_2N$ (cyclic)	PN 160.1(2), 162.6(2) ∠PNP 116.4(2) ∠NPN 116.4(1) nearly planar	330

$Na_3AlP_3O_9N$	$N(PO_3)_3$ around 6 coord. Al^{3+} PN 170.66, ∠ PNP 115.4 N slightly pyramidal	331

References

1. C.W. Allen, Coord. Chem. Rev., **1994**, 130, 137.
2. Phosphorus, Sulfur Silicon Relat. Elem., **1993**, 75-77.
3. Polym. Prepr. (Am. Chem. Soc., Div. Polym. Chem.), **1993**, 34(1).
4. R. Glaser, C.J. Horan, G.S. -C. Choy and B.L. Harris, Phosphorus, Sulfur Silicon Relat. Elem., **1993**, 77, 73.
5. G.M. Chaban, N.M. Klimenko and O.P. Charkin, Bull. Russ. Acad. Sci. (Engl. Transl.), **1992**, 41, 99.
6. M. Esseffar, A. Luna, O.Mo and M. Yanez, J. Phys. Chem., **1993**, 97, 6607.
7. H.G. Ang, W.L. Kwik and I. Novak, J. Chem. Res. Synop., **1993**, 422.
8. J.B. Lagowski, R. Jaeger, I. Manners and G.J. Vansco, Polym. Prepr. (Am. Chem. Soc., Div. Polym. Chem.), **1993**, 34(1), 326.
9. I.K. Ahmad and P.A. Hamilton, J. Mol. Spectr., **1994**, 163, 214.
10. J.M. Curtis, N. Burford and T.M. Parks, Org. Mass Spectrom., **1994**, 29, 414.
11. D. Bougeard, C. Bremand, R. DeJaeger and Y. Lemmouchi, Phosphorus, Sulfur Silicon Relat. Elem., **1993**, 79, 147.
12. D. Gudat, Magn. Reson. Chem., **1993**, 31, 925.
13. C. Foces-Foces, A.L. Llamas-Saitz, R.M. Charamunt, C. Lopez, J. Elguero, P. Molina, A. Argues and R. Obon, J. Inclus. Phenom. Mol. Recogn. Chem., **1993**, 16, 155.
14. J. Laynez, M. Menendez, J.L. Saiz Velasco, A.L. Llamas-Saiz, C. Foces-Foces, J. Elguero, P. Molina, M. Alajarin and A. Vidal, J. Chem. Soc., Perkin Trans. 2, **1993**, 709.
15. A.L. Llamas-Saiz, C. Foces-Foces, J. Elguero, M. Alajarin, A. Vidal, R.M. Charamunt and C. Lopez, J. Chem. Soc., Perkin Trans. 2, **1994**, 209.
16. R. Schweisinger, C. Hasenfratz, H. Schlemper, L. Walz, E. Peters, K. Peters and H.G. von Schnering, Angew. Chem., Int. Ed. Engl., **1993**, 32, 1361.
17. A. A. Tolmachev, A.N. Kostyuk, E.S. Kozlov, A.P. Polishchu and A.N. Chernega, Russ. J. Gen. Chem. (Engl. Transl.), **1992**, 62, 2207.
18. J.R. Goerlich, M. Farkens, A. Fisher, P.G. Jones and R. Schmutzler, Z. Anorg. Allgem. Chem., **1994**, 620, 707.
19. R. G. Cavell, R.W. Reed, K.V. Katti, M.S. Balakrishna, P.W. Collins, V. Mozol and I. Bartz, Phosphorus, Sulfur Silicon Relat. Elem., **1993**, 76, 9.
20. K. V. Katti, B.D. Santarsieno, A.A. Pinkerton and R.G. Cavell, Inorg. Chem. **1993**, 32, 5919.

21. P. Molina, C. Conesa, A. Alias, A. Argues, M.D. Velasco, A.L. Llamas-Saiz and C. Foces-Foces, Tetrahedron, **1993**, 49, 7599.
22. F.S. Burhus II, R.A. Montague and K. Matyjasewski, Polym. Prepr. (Am. Chem. Soc., Div. Polym. Chem.), **1994**, 35(1), 460.
23. K. Matyzasjewski, M.S. Lindenberg, M.K. Moore, and M. L. White, J. Poly. Sci., Part A: Polym. Chem., **1994**, 32, 465.
24. H. Suzuki, S. Katayama and F. Okada, Jpn. Kokai Tokkyo Koho, JP 05186482 (Chem. Abstr. **1994**, 120, 9131).
25. K. Matyjaszewski, M.S. Lindenberg and M.L. White, Polym. Prepr. (Am. Chem. Soc., Div. Polym. Chem.), **1992**, 33(1), 1096.
26. L. Riesel, R. Friebe and D. Strum, Phosphorus, Sulfur Silicon Relat. Elem., **1993**, 76, 207.
27. L. Riesel, R. Friebe and D. Strum, Z. Anorg. Allgem. Chem., **1993**, 619, 1685.
28. J. Tang, J. Dopke and J.G. Verkade, J. Am. Chem. Soc., **1993**, 115, 5015.
29. I. Neda, T. Kauhorat and R. Schmutzler, Z. Naturforsch., B: Chem. Sci., **1994**, 49, 171.
30. N.A. Buina, R.R. Sadreyeva, I.A. Nuretdinov, N.P. Konovalova, R.F. Dyachovskaya and L.N. Volkova, Khim-Farm. Zh., **1992**, 26, 66 (Chem. Abst., **1993**, 119, 27977).
31. C. Pei and X. Xu, Phosphorus, Sulfur Silicon Relat. Elem., **1993**, 84, 143.
32. C. Peinador, M.J. Moreira and J. M. Quinteda, Tetrahedron, **1994**, 50, 6705.
33. C.M. Angelov and R.H. Neilson, Inorg. Chem., **1993**, 32, 2279.
34. V. Sun and T.P. Kee, J. Chem. Soc., Perkin Trans. 1, **1993**, 1369.
35. C. Alcaraz, A. Baceiredo, F. Dahan and G. Bertrand, J. Am. Chem. Soc., **1994**, 116, 1225.
36. K.A. Abboud, L.A. Villanueva and J.M. Boncella, Acta Crystallogr., Sect. C: Cryst. Struct. Commun. **1993**, C49, 1848.
37. P.Y. Hammoutou, J. Heubel and R.DeJager, Phosphorus, Sulfur Silicon, Relat. Elem., **1993**, 79, 97.
38. G. Bertrand, J.P. Majoral, H. Rolland and P. Potin, Fr. Demande, FR 268237 (Chem. Abst., **1993**, 119, 181021).
39. T. Chivers, M. Parvez and M.A. Seay, Inorg. Chem., **1994**, 33, 2147.
40. R. Murugavel, S.S. Kumaravel, S.S. Krishnamurthy, M. Nethaji and J. Chandrasekhar, J. Chem. Soc., Dalton Trans., **1994**, 847.
41. M.J.P. Harger, J. Chem. Res., Synop., **1993**, 334.
42. N. Burford, J.A. Clyburne, D.P. Derek, M.J. Schriver and J.F. Richardson, J. Chem. Soc., Dalton Trans., **1994**, 997.
43. A. Ruban, M. Nieger and E. Nieke, Agnew. Chem., Int. Edn. Engl., **1993**, 32, 1419.
44. F.U. Seifert and G.V. Röschenthaler, Z. Naturforsch, B: Chem. Sci., **1993**, **48**, 1089.
45. W. Schilbach, V. von der Gönna, D. Gudat, M. Nieger and E. Niecke, Angew. Chem., Int. Ed. Engl., **1994**, 33, 982.

46. W. Eisfeld, M. Slany, U. Bergsträsser and M. Regitz, Tetrahedron Lett., **1994**, 35, 1527.
47. E. Niecke, B. Kramer, M. Nieger and H. Severin, Tetrahedron Lett., **1993**, 34, 4627.
48. B. Barion, G. Gärtner-Winkhans, M. Link, M. Nieger and E. Niecke, Chem. Ber., **1993**, 126, 2187.
49. M. Link, E. Niecke and M. Nieger, Chem. Ber., **1994**, 127, 313.
50. N. Burford, J.A.C. Clyburne, S. Mason and J.F. Richardson, Inorg. Chem., **1993**, 32, 4988.
51. E. Niecke, G. David, R. Detsch, B. Kramer, M. Nieger and P. Wenderoth, Phosphorus, Sulfur Silicon Relat. Elem., **1993**, 76, 25.
52. E. Niecke, J.F. Nixon, P. Wenderoth, B.F. Trigo Passos and M. Nieger, J. Chem. Soc., Chem. Commun., **1993**, 846.
53. V.D. Romancnko, A.V. Ruban, A.B. Roshenko, M.I. Povolotskii, T.V. Sarina, D. Gudat and E. Niecke, Mendeleev Commun., **1993**, 7 (Chem. Abst., **1993**, 119, 39398).
54. M. Nitta, Rev. Heteroat. Chem., **1993**, 9, 87 (Chem. Abst., **1994**, 120, 269955).
55. A.R. Katritzky, R. Mazurkiewicz, C.V. Stevens and M.F. Gordeev, J. Org. Chem., **1994**, 59, 2740.
56. F. Lopez-Ortiz, E. Pelaez-Arango, F. Palacios, J. Barluenga, S. Garcia-Granda, B. Tejerina and A. Garcia-Fernandez, J. Org. Chem., **1994**, 59, 1984.
57. P. Molina, E. Aller, A. Lopez-Lazaro, M. Alajarin and A. Lorenzo, Tetrahedron Lett., **1994**, 35, 1994.
58. T. Bohn, W. Kramer, R. Neidlein and H. Suschitzky, J. Chem. Soc., Perkin Trans. 1, **1994**, 947.
59. P. Molina, A. Argues and A. Alias, J. Org. Chem., **1993**, 58, 5264.
60. M. Nitta, Y. Iino, T. Sugiyama and A. Akaogi, Tetrahedron Lett., **1993**, 34, 831.
61. M. Nitta and Y. Iino, Rikogaku Kenkyusho Hokoku, Waseda Diagaku, **1993**, 140, 24 (Chem. Abst., **1994**, 120, 8451).
62. W. Maier, M. Keller and W. Eberbach, Heterocycles, **1993**, 35, 817.
63. T. Pietzonka and D. Seebach, Agnew. Chem., Int. Ed. Engl., **1993**, 32, 716.
64. A. Manenc, R. DeJaeger and P. Potin, Fr. Demande, FR 2682385 (Chem. Abst., **1993**, 119, 181022).
65. P. Froyen, Phosphorus, Sulfur Silicon Relat. Elem., **1993**, 78, 161.
66. A. Zwierzak, Phosphorus, Sulfur Silicon Relat. Elem., **1993**, 75, 51.
67. H. Rolland, P. Potin, J.P. Majoral and G. Bertrand, Phosphorus, Sulfur Silicon Relat. Elem., **1993**, 76, 211.
68. H. Rolland, P. Potin, J.P. Majoral and G. Bertrand, Inorg. Chem., **1993**, 32, 4679.
69. V. Adrover, G. Bertrand, J.P. Majoral, H. Rolland and P. Potin, Fr. Demande, FR 2682386 (Chem. Abst., **1994**, 120, 31595).
70. H.J. Cristan, A. Perrand, E. Manginot and E. Torreilles, Phosphorus, Sulfur Silicon, Relat. Elem., **1993**, 75, 7.

71. O.A. Rakitin, N.V. Obnuchinikova and L.I. Khmeinitski, Phosphorus Sulfur Silicon Relat. Elem., **1993**, 78, 309.
72. H.R. Allcock, D.C. Ngo and K.B. Visscher, Inorg. Chem., **1994**, 33, 2090.
73. T. Rubenstahe, D.W. vonGudenberg, F. Weller, K. Dehnicke and H. Goesmann, Z. Naturforsch., B: Chem. Sci., **1994**, 49, 15.
74. M. Sotoodeh, I. Leichweis, H.W. Roesky, M. Noltemeyer and H.G. Schmidt, Chem. Ber., **1993**, 126, 913.
75. D. Nusshair, R. Garbe, F. Weller, J. Pebler and K. Dehnicke, Z. Anorg. Allgem. Chem., **1994**, 620, 67.
76. R. Garbe, J. Pebler, K. Dehnicke, D. Fenske, H. Goesmann and G. Baum, Z. Anorg. Allgem. Chem., **1994**, 620, 592.
77. R. M. zu Kocher, G. Frenzen, B. Neumüller, K. Dehnicke and J. Magull, Z. Anorg. Allgem. Chem., **1994**, 620, 431.
78. R.M. zu Kocher, A. Behrendt and K. Dehnicke, Z. Naturforsch., B: Chem Sci., **1994**, 49, 301.
79. C. H. Honeyman, A.J. Lough and I. Manners, Inorg. Chem., **1994**, 33, 2988.
80. K.V. Katti, P.R. Singh, K.K. Katti, C.L. Barnes, K. Kopicka, A. R. Ketring and W.A. Volkert, Phosphorus, Sulfur Silicon Relat. Elem., **1993**, 75, 55.
81. K.V. Katti, P.R. Singh, C.L. Barnes, K.K. Katti, K. Kopicha, A.R. Ketring and W.A. Volkert, Z. Naturfosch., B: Chem. Sci., **1993**, 1381.
82. K.V. Katti, W.A. Volkert, A.R. Ketring and P.R. Sing, PCT Int. Appl., WO9308839 (Chem. Abst., **1993**, 119, 134570).
83. D.V. Tolkachev, R.U. Amanov, M.Yu. Antipin, A.A. Khodak, S.P. Solodovnikov, N.N. Bubnov, Yu. T. Struchkov and M.I. Kabachnick, Bull. Russ. Acad. Sci. (Engl. Transl.), **1992**, 41, 93.
84. R.U. Amanov, D.V. Tolkachev, M.Yu. Antipin, A.A. Khodak, Kh. T. Sharipov, Yu. T. Struchkov and M.I. Kabachnik, Bull. Russ. Acad. Sci. (Engl. Transl.), **1993**, 42, 295.
85. S.K. Pandey, H.W. Roesky, D. Stalke, A. Steiner, H.G. Schmidt and M. Noltemeyer, Phosphorus, Sulfur Silicon Relat. Elem. **1993**, 84, 231.
86. R. Hasselbring, I. Leichtweis, M. Noltemeyer, H.W. Roesky, H.G. Schmidt and A. Herzog, Z. Anorg. Allgem. Chem., **1993**, **619**, 1543.
87. N.V. Jarvis, L. Krueger and J.G.H. du Preez, Solvent. Extr. Ion Exch., **1993**, 11, 811.
88. N.V. Jarvis, L. Kruger and J.G.H. du Preez, Solvent. Extr. Ion Exch., **1994**, 12, 423.
89. V.F. Travkin, A.N. Kravehenko and G.P. Miroevskii, Tsvetn. Met., **1993**, 14, (Chem. Abst., **1994**, 120, 12155).
90. J. dela Croi Habimana and S. Westall, U.S. Pat. US5210129 (Chem. Abst., **1993**, 119, 118750).
91. C.W. Allen, J. Fire Sci., **1993**, 11, 320.
92. L. McNally and C.W. Allen, Heteroatom. Chem., **1993**, 4, 159.
93. M. C. Labarre and J.F. Labarre, J. Mol. Struct., **1993**, 300, 593.
94. T.S. Cameron, B.Borecka and W. Kwiatkowski, J. Am. Chem. Soc., **1994**, 116, 1211.

95. M. Breza and S. Biskupic, J. Mol. Struct. (Theochem), **1994**, 309, 305.
96. A. A. Varnek, A. Maia, D. Landini, A. Gamba, G. Morosi and G. Podda, J. Phys. Org. Chem., **1993**, 6, 113.
97. R. Murugavel, S.S. Krishnamurthy and M. Nethaji, J. Chem. Soc., Dalton Trans., **1993**, 2569.
98. C.W. Allen, D.E. Brown, R. Hayes, R. Tooze and G.L. Poyser, Polym. Prepr. (Am. Chem. Soc., Div. Polym. Chem.), **1993**, 34(1), 339.
99. T. Kimura and M. Kajiwara, J. Inorg. Organomet. Polym., **1992**, 2, 431.
100. K. Ishigami, R. Maeda, M. Maeda, H. Hamada, K. Shoh, A. Shimada, H. Morii, K. Fukushi, T.Takeda and K. Ohki, J. Nihon Univ. Sch. Dent., **1993**, 35, 36 (Chem. Abst., **1993**, 119, 103262).
101. L. Jiang and M. Moini, J. Am. Soc. Mass Spectrom., **1992**, 3, 842.
102. E. Gocheva, P. Vassileva, L.Lakov and Q. Peshev, J. Mater. Sci., **1993**, 28, 5251.
103. A.H. Al-Kubaisi and M.B. Sayed, Thermochim. Acta, **1993**, 222, 241.
104. M.B. Sayed, Thermochim. Acta, **1994**, 232, 249.
105. H.R. Allcock, E.N. Silverberg and G.K. Dudley, Macromolecules, **1994**, 27, 1033.
106. H.R. Allcock, G.K. Dudley and E.N. Silverberg, Macromolecules, **1994**, 27, 1039.
107. K. Kubono, N. Asaka, T.Taga, S. Isoda and T. Kobayashi, J. Mater. Chem., **1994**, 4, 291.
108. K. Kubono, H. Kurata, S. Isoda and T. Kobayashi, J. Mater. Chem., **1993**, 3, 615.
109. C. Combes, H.J. Cristau, W. S. Li, M. McPartlin, F. Plenat and I.J. Scowen, J. Chem. Res. (S), **1994**, 134.
110. F. Golinski and H. Jacobs, Z. Anorg. Allgem. Chem., **1994**, 620, 965.
111. C. Valerio, M.C. Labarre and J.F. Labarre, J. Mol. Struct., **1993**, 299, 171.
112. J. J. and B. Raymond, S. Scheidecker and J.F. Labarre, J. Mol. Struct., **1992**, 271, 289.
113. S. Scheidecker, D. Semenzin, G. Etemad-Moghadau, F. Sourines, M. Koening and J.F. Labarre, Phosphorus, Sulfur Silicon Relat. Elem., **1993**, 80, 85.
114. I.A. Rozanov, D. Murashov, V.I. Moiseev and L. Ya. Medvedeva, Russ. J. Inorg. Chem. (Engl. Transl.), **1993**, 38, 707.
115. G. Hagele, M. Murray, and C. PaPadopoulos, Phosphorus, Sulfur Silicon Relat. Elem., **1993**, 77, 89.
116. N.G. Razumova, A.E. Shumeiko, A.A. Afonikin and A.F. Popov, Mendeleev Commun. (Engl. Transl.), **1994**, 64.
117. T. Hokelck, Z. Kilic and A. Kilic, Acta Crystallogr. Sect. C: Cryst. Struct. Commun., **1994**, C50, 453.
118. H.R. Allcock and C.G. Cameron, Macromolecules, **1994**, 27, 3131.
119. K. Inoue, H. Takahata and T.Tanigaki, J. Appl. Polym. Sci., **1993**, 50, 1857.

120. D.F. Femec and R.R. McCaffrey, J. Appl. Polym. Sci., **1994**, 52, 501.
121. K. Moriya, S. Miyata, S. Yano and M. Kajiwara, J. Inorg. Organometal. Polym., **1992**, 2, 443.
122. Y.S. Freidzon, M.V. D'yachenko, D.R. Tur and V.P. Shibaev, Polym. Prepr. (Am. Chem. Soc., Div. Polym. Chem., **1993**, 34(1), 146.
123. S.D. Burton, W.D. Samuels, G.J. Exarhos and J. M. Pleva, Mater. Res. Soc. Symp. Proc., **1992**, 249, 267.
124. A.J. Elias, R.L. Kirchmeier and J.M. Shreeve, Inorg. Chem., **1994**, 33, 2727.
125. M.G. Muralidhara, N. Grover and V. Chandrasekhar, Polyhedron, **1993**, 12, 1509.
126. C.W. Allen and M. Bahadur, Phosphorus, Sulfur Silicon Relat. Elem., **1993**, 76, 203.
127. H.R. Allcock, E.N. Silverberg, C.J. Nelson and W.D. Coggio, Chem. Mater., **1993**, 5, 1307.
128. A. Chandrasekaran, S.S. Krishnamurthy and M. Nethaji, Inorg. Chem., **1993**, 32, 6102.
129. A. Chandrasekaran, S.S. Krishnamurthy and M. Nethaji, J. Chem. Soc., Dalton Trans., **1994**, 63.
130. J.K.R. Thomas, V.Chandrasekhar, S.R. Scott, R. Hallford and W.A. Cordes, J. Chem. Soc., Dalton Trans., **1993**, 2589.
131. A. Chandrasekaran and S.S. Krishnamurthy, Indian J. Chem., Sect. A, **1994**, **33**, 391.
132. G. Davies and N. El-Kady, US Pat., US5286469A (Chem. Abst., **1994**, 20, 259968).
133. J.Jand, S. Scheidecker, M. Graffeuil and J.F. Labarre, J. Mol. Struct., **1993**, 295, 133.
134. N.S. Kedrova, B.V. Levin, N.N. Mal'tseva and B.I. Saidov, Koord. Khim., **1993**, 19, 278 (Chem. Abst., **1993**, 119, 216238).
135. A.A. van der Huizen, P.L. Buwalda, T. Wilting, H.Pol, A.P. Jekel, A. Meetsma and J.C. van de Grampel, J. Chem. Soc., Dalton Trans., **1994**, 577.
136. C. Galliot, A.M. Caminade, F. Dahan and J.P. Majoral, Angew. Chem., Int. Ed. Engl., **1993**, 32, 1476.
137. H.R. Allcock, E.H. Klingenberg and M. F. Welker, Macromolecules, **1993**, 26, 5512.
138. J.Y. Chang, H.J. Ji, M.J. Han, S.B. Rhee, S. Cheong and M. Yoon, Macromolecules, **1994**, 27, 1376.
139. H.R. Allcock and C.G. Cameron, Macromolecules, **1994**, 27, 3125.
140. G. Fantin, A. Medici, M. Fogagnolo, P.Pedrini, M. Gleria, R. Bertani and G. Facchin, Eur. Polym. J., **1993**, 29, 1571.
141. G. Facchin, G. Fantin, M. Gleria, M. Guglielmi and F. Spizzo, Polym. Prepr. (Am. Chem. Soc., Div. Polym. Chem.), **1993**, 34(1), 322.
142. D. Kumar, A.D. Gupta and M. Khullar, Polym. Prepr. (Am. Chem. Soc., Div. Polym. Chem.), **1993**, 34(1), 312.
143. D. Kumar, A.D. Gupta and M. Khullar, J. Polym. Sci., Part A: Polym. Chem., **1993**, 31, 2739.
144. D. Kumar, A.D. Gupta and M. Khullar, J. Inorg. Organomet. Polymers, **1993**, 3, 259.

145. D. Kumar, M. Khullar and A.D. Gupta, Polymer, **1993**, 34, 3025.
146. Y. W. Chen-Yang and Y.H. Chuang, Phosphorus, Sulfur Silicon Relat. Elem., **1993**, 76, 261.
147. M. Kajiwara, J. Mater. Sci. Lett., **1993**, 12, 1803.
148. S. Kubota, O. Ito, I. Maeda and Y. Horiuchi, Jpn. Kokai Tokkyo Koho, JP04279625 (Chem. Abst., **1993**, 119, 73850).
149. M. P. Prigozhina, L.G. Komarova and A.L. Rusanov, Plast. Massy, **1992**, 38 (Chem. Abst., **1994**, 120, 219244).
150. J.G. van de Grampel, A.P. Jekel, R. Puyenbroek, T.J. Arling, M. C. Faber, W. Fransen, A. Meetsma and J.H. Wubbels, Phosphorus, Sulfur Silicon Relat. Elem., **1993**, 76, 215.
151. S. Mori, Jpn. Kokai Tokkyo Koho, JP 05186538 (Chem. Abst., **1994**, 120, 10068).
152. M. Anzai, M. Kobori, K. Yoshihashi, H. Kikuchi, H. Hirose and M. Nishiyama, J. Nihon Univ. Sch. Dent., **1992**, 34, 196 (Chem. Abst., **1993**, 119, 146517).
153. K. Morimoto, Jpn. Kokai Tokkyo Koho, JP 0533268, (Chem. Abst., **1993**, 119, 29943).
154. K. Morimoto and T. Nunoo, Jpn. Kokai Tokkyo Koho, JP 0533269 (Chem. Abst., **1993**, 119, 29944).
155. J. Fukuoka and F. Tanimoto, Jpn. Kokai Tokkyo Koho, JP 05005278 (Chem. Abst., **1993**, 119, 10427).
156. L.M. Babcock, J.K. Band and R. T. Leibfried, Eur. Pat. Appl., EP 556844 (Chem. Abst., **1994**, 120, 136160).
157. K. Ookaito, Jpn. Kokai Tokkyo Koho, JP 05059381 (Chem. Abst., **1993**, 119, 76155).
158. K. Ookaito, Jpn. Kokai Tokkyo Koho, JP 05059382 (Chem. Abst., **1993**,119, 76156).
159. T. Nakanaga and J. Tada, Jpn. Kokai Tokkyo Koho, JP 05070783 (Chem. Abst., **1993**, 119, 121075).
160. B.S. Nader, US Pat., US 5194652 (Chem. Abst., **1993**, 119, 121066).
161. B.S. Nader and M.N. Inbasekaran, US Pat., US 5219477 (Chem. Abst., **1993**, 119, 184567).
162. M. Namiki and S. Mori, Jpn. Kokai Tokkyo Koho, JP 05306342 (Chem. Abst., **1994**, 120, 194194).
163. Y. Haskimoto, T. Koyama, A. Maryyama, S. Nagahara, and S. Mayama, Jpn. Kokai Tokkyo Koho, JP 04353860 (Chem. Abst., **1993**, 119, 82844).
164. A. Kurahashi, PCT Int., WO 9307540 (Chem. Abst., **1994**, 120, 65829).
165. Y. Hashimoto, T. Koyama, S. Mayama, A. Maruyama and S. Nagahara, Jpn. Kokai Tokkyo Koho, JP 04353859 (Chem. Abst., **1993**, 119, 82843).
166. H. Kawada, M. Funato, N. Kono and I. Kimura, Jpn. Kokai Tokkyo Koho, JP 05134465 (Chem. Abst., **1994**, 120, 284909).
167. M. Tani, Jpn. Kokai Tokkyo Koho, JP 05165389 (Chem. Abst., **1994**, 120, 311768).
168. M. Kitayama, Jpn. Kokai Tokkyo Koho, JP 05309957 (Chem. Abst., **1994**, 120, 285109).
169. Y. Hashimoto and H. Tanaka, Eur. Pat. Appl., EP562517 (Chem. Abst., **1994**, 120, 111696).

170. H. Tokaji, M. Matsuda, M. Kamatsu and T. Koyanagi, Jpn. Kokai Tokkyo Koho, JP 05008350 (Chem. Abst., **1993**, 119, 119046).
171. M. Kitayama, Jpn. Kokai Tokkyo Koho, JP 05230442 (Chem. Abst., **1994**, 120, 136251).
172. H. Hosono, S. Kurasaki and T.Taniguchi, Jpn. Kokai Tokkyo Koho, JP 05051471 (Chem. Abst., **1993**, 119, 97605).
173. H. Hosono, S. Kurasaki and T. Taniguchi Jpn. Kokai Tokkyo Koho, JP 05051472 (Chem. Abst., **1993**, 119, 97606).
174. H. Ando, Eur. Pat. Appl., EP 557943 (Chem. Abst., **1994**, 120, 194166).
175. H. Nakagawa, Jpn. Kokai Tokkyo Koho, JP 05208468 (Chem. Abst., **1994**, 120, 193574).
176. M. Shinoda, R. Nishide and T. Nagashima, Jpn. Kokai Tokkyo Koho, JP 05222224 (Chem. Abst., **1994**, 120, 56855).
177. H. Ando, Jpn. Kokai Tokkyo Koho, JP 05222142 (Chem. Abst., **1994**, 120, 56854).
178. M. Takeuchi and H. Yashima, Jpn. Kokai Tokkyo Koho, JP 05074467 (Chem. Abst., **1993**, 119, 142958).
179. A. Boye, A. Grangeon and C. Guizard, Eur. Pat. Appl., EP 5119900 (Chem. Abst., **1993**, 119, 10080).
180. A. Kuroda and A. Mizuguchi, Jpn. Kokai Tokkyo Koho, JP 05124951 (Chem. Abst., **1993**, 119, 167488).
181. S.K. Yang, C.E. Park, K.D. Ahn, Pollimo, **1993**, 17, 335 (Chem. Abst., **1994**, 120, 90544).
182. R.T. Oakley, Can. J. Chem., **1993**, 71, 1775.
183. T.Chivers, Main Group Chem. News, **1993**, 1(3), 6.
184. T. Chivers, M. Edwards, R.W. Hilts, M. Parvez and R. Vollmerhaus, J. Chem.Soc., Chem. Commun., **1993**, 1483.
185. T. Chivers, D.D. Doxsee, R.W. Hilts and M. Parvez, Can. J. Chem., **1993**, 71, 1821.
186. T. Chivers, D.D. Doxsee and R.W. Hilts and M. Parvez, Inorg. Chem., **1993**, 32, 3244.
187. T. Chivers, R.W. Hilts and M. Parvez, Inorg. Chem., **1994**, 33, 997.
188. C.J. Thomas, K.K. Bhandary, L.M. Thomas, S.E. Senadhi and S. Vijay-Kumar, Bull. Chem. Soc. Jpn., **1993**, 66, 1830.
189. T. Mohan, C.J. Thomas, M.N. Sudheendra Rao, G. Aravamudan, A. Meetsma and J.C. van de Grampel, Heteroatom. Chem., **1994**, 5, 19.
190. P. Zanello, G. Opromolla, G. Giorgi, J.C. van de Grampel and H. F.M. Schoo, Polyhedron, **1993**, 12, 1329.
191. H.W. Roesky, P.Olms, R. Hasselbring, W. Reinhard, F.Q. Liu and M. Noltemeyer, Phosphorus, Sulfur Silicon Relat. Elem., **1993**, 76, 255.
192. G.V. Dolgushin, P.A. Nikitin, Yu. E. Sapozhikov, M.Yu. Dmitrichenko, V.G. Rozino and M.G. Voronkov, Z. Naturforsch. A: Phys. Sci., **1994**, 49, 171.
193. M.R. Estra and C. Vazquez, Phosphorus, Sulfur Silicon Relat. Elem., **1993**, 76, 265.
194. T.S. Dolgushina, K.A. Potekhin, Yu. T. Struchkov, V.A. Galishev and A.A. Petrov, Russ. J. Chem. (Engl. Transl), **1993**, 63, 111.

195. J. Tejeda, R. Reau. F. Dahan and G. Bertrand, J. Am. Chem. Soc., **1993**, 115, 7880.
196. G. David, E.Niecke, M. Nieger, V. von der Gonna and W.W. Schueller, Chem Ber., **1993**, 126, 1513.
197. H.R. Allcock, S.M. Coley, I. Manners, K.B. Visscher, M. Parvez, O. Nuyken and G. Renner, Inorg. Chem., **1993**, 32, 5088.
198. H.R. Allcock, S.M. Coley and C. T. Morrissey, Macromolecules, **1994**, 27, 2904.
199. A. Steiner and D. Stalke, Inorg. Chem., **1993**, 32, 1977.
200. R. Hasselbring, S.K. Pandley, H.W. Roesky, D. Stalke and A. Steiner, J. Chem. Soc., Dalton Trans., **1993**, 3447.
201. S.K Pandey, A. Steiner, H.W. Roesky and D. Stalke, Angew. Chem., Int. Ed. Engl., **1992**, 32, 596.
202. R. Hasselbring, H.W. Roesky, A. Heine, D. Stake and G.M. Sheldrick, Z. Naturforsch., B: Chem. Sci., **1994**, 49, 43.
203. U. Kilimann, M. Noltemeyer and F.T. Edelmann, J. Organomet. Chem., **1993**, 443, 35.
204. S. K. Pandey, R. Hasselbring, A. Steiner, D. Stalke and H.W. Roesky, Polyhedron, **1993**, 12, 2941.
205. I. Leichtweis, R. Hasselbring, H.W. Roesky, M. Noltemeyer and A. Herzog, Z. Naturforsch., B: Chem. Sci., **1993**, 48, 1234.
206. S.K. Pandey, A. Steiner, H.W. Roesky and D. Stalke, Inorg. Chem., **1993**, 32, 5444.
207. J. Ellerman, J. Sutter, F.A. Knoch and M. Moll, Angew. Chem., Int. Ed. Engl., **1993**, 32, 700.
208. J. Ellermann, J. Sutter, C. Schelle, F.A. Knoch and M. Moll, Z. Anorg. Allgem. Chem., **1993**, 619, 2006.
209. P. Imhoff, R. van Asselt, J.M. Ernsting, K. Vrieze, C.J. Elsevier, W.J.J. Smeets, A.L. Spek and A.P.M. Kentgens, Organometallics, **1993**, 12, 1523.
210. M.S. Balakrishna, R. Klein, S. Uhlenbrock, A.A. Pinkerton and R.G. Cavell, Inorg. Chem., **1993**, 32, 5676.
211. P. Bortolus and M. Gleria, J. Inorg. Organomet. Polym., **1994**, 4, 1.
212. P. Bortolus and M. Gleria, J. Inorg. Organomet. Polym., **1994**, 4, 95.
213. H.R. Allcock, Adv. Mater., **1994**, 6, 106.
214. H.R. Allcock, Prog. Pac. Polym. Sci. 2, Proc. Pac. Polym. Conf., 2nd, **1992**, 89 (Chem. Abst., **1994**, 120, 271199).
215. H.R. Allcock, Phosphorus, Sulfur Silicon Relat. Elem., **1993**, 76, 199.
216. H.R. Allcock, Polym. Prepr. (Am. Chem. Soc., Div. Polym. Chem.), **1993**, 34(1), 261.
217. I. Manners, Annu. Rep. Prog. Chem., Sect. A: Inorg. Chem., **1993**, 88, 77.
218. P. Wisian-Neilson, Contemp. Top. Polym. Sci., **1992**, 7, 333 (Chem. Abst., **1994**, 120, 324220).
219. M.P. Tanazona, Polymer, **1994**, 35, 819.
220. W. Schnick, Phosphorus, Sulfur Silicon Relat. Elem., **1993**, 76, 183.
221. W. Schnick, Angew. Chem., Int. Ed. Engl., **1993**, 32, 806.

222. K. Matyjaszewski, Polym. Prepr. (Am. Chem. Soc., Div. Polym. Chem.), **1992**, 33(1), 178.
223. R. Ziembinski, C. Honeyman, O. Mourad, D. Foucher, R. Rulkens, M. Liang, Y. Ni and I. Manners. Phosphorus, Sulfur Silicon Relat. Elem., **1993**, 76, 219.
224. K.M. Abraham and M. Alamgir, Proc. Int. Power Sources Symp., **1992**, 35th, 264 (Chem. Abst., **1993**, 119, 229943).
225. K. Matyjaszewski and W.V. Metanomski, Polym. Prepr. (Am. Chem. Soc., Div. Polym. Chem., **1993**, 34(1), 6.
226. M. Kajiwara, Kubunshi, **1993**, 42, 568 (Chem. Abst., **1994**, 120, 218566).
227. M. Kajiwara, Shinsozai, **1992**, 3, 23 (Chem. Abstr., **1993**, 119, 139807).
228. M. Kajwara and K. Murahkami, Eds., Inorganic Polymers , Sangyo (Tokyo), 1992, 3090 (Chem. Abst., **1993**, 119, 73278).
229. V.S. Matveev, Khim. Volokna, **1992**, 3 (Chem. Abstr., **1993**, 119, 119353).
230. S.M. Coley, H.R. Allcock, I. Manners, K. Visscher, O. Nuyken and G. Renner, Polym. Prepr. (Am. Chem. Soc., Div. Polym. Chem.), **1992**, 33(2), 166.
231. M. Liang, D. Foucher and I. Manners, Polym. Prepr. (Am. Chem. Soc., Div. Polym. Chem), **1992**, 33(1), 1134.
232. M. Edwards, Y. Ni, M. Liang, A. Stammer, J. Massey, J.G. Vansco and I. Manners, Polym. Prepr. (Am. Chem. Soc., Div. Polym. Chem.), **1993**, 34(1), 324.
233. P. Potin and R. DeJaeger, Phosphorus, Sulfur Silicon Relat. Elem., **1993**, 76, 227.
234. R. DeJaeger and P. Potin, Phosphorus, Sulfur Silicon Relat. Elem., **1993**, 76, 221.
235. G. Bertrand, J.P. Marjoral, P.Potin, R. Herue and M. Robba, Fr. Demande, FR 2682391 (Chem. Abst., **1994**, 120, 77971).
236. C.E. Wood, R. Samuel, W.R. Kucera, C.M. Angelov and R.H. Neilson, Polym. Prepr. (Am. Chem. Soc., Div. Polym. Chem.), **1993**, 34(1), 263.
237. K. Matyjaszewski, M. S. Lindenberg, J.L. Spearman and M.L. White, Polym. Prepr. (Am. Chem. Soc., Div. Polym. Chem.), **1992**, 33(2), 176.
238. S. Katayama, H. Suzuki and F. Okada, Jpn. Kokai Tokkyo Koho, JP 04337328 (Chem. Abst., **1993**, 119, 96503).
239. R.A. Montague, F. Burkus II and K. Matyiaszewski, Polym. Prepr. (Am. Chem. Soc., Div. Polym. Chem., **1993**, 34(1), 316.
240. K. Matyjaszewski, M.S. Lindenberg, M. K. Moore, M.L. White and M. Kojima, J. Inorg. Organomet. Polym., **1993**, 3, 317.
241. K. Matyjaszewski, M.S. Lindenberg, M. K. Moore and M.L. White, Polym. Prepr. (Am. Chem. Soc., Div. Polym. Chem.), **1993**, 34(1), 274.
242. K. Matyjaszewski, M.K. Moore and M.L. White, Macromolecules, **1993**, 26, 6741.
243. U. Franz, O. Nuyken and K. Matyjaszewski, Macromol. Rapid Commun. **1994**, 15, 169.
244. U. Franz, O. Nuyken and K. Matyjasewski, Macromolecules, **1993**, 26, 3723.
245. W. Schnick and J. Luecke, Ber. Offen., DE 4201484 (Chem. Abst., **1993**, 119, 163488).

246. R. Marchand, R. Connanec, E. Gueguen and Y. Laurent, Phosphorus, Sulfur Silicon Relat. Elem., **1993**, 76, 277.
247. M. F. Welker, H.R. Allcock, G.L. Grune, R.T. Chern and V.T. Stannett, Polym. Mater. Sci. Eng., **1992**, 66, 259(Chem. Abst., **1993**, 119, 227037).
248. F. Zheng, Z. Lin, C.Yie and X. Tang, Gaodeng Xuexiao Huaxue Xuebao, **1992**, 13, 1469 (Chem. Abst., **1993**, 119, 117948).
249. Z. Lin, F. Zheng, F. Liu and X. Tang, Jilin Daxue Ziran Kexue Xuebao, **1992**, 93 (Chem. Abst., **1993**, 119, 50025).
250. H. Sato, M. Sugihara and J. Ito, Jpn. Kokai Tokkyo Koho, JP 05310949 (Chem. Abst., **1994**, 120, 246110).
251. G.L. Grune, V.T. Stannett, R.T. Chern and J. Harada, Polym. Adv. Technol., **1993**, 4, 34 (Chem. Abst., **1994**, 120, 193798).
252. H.R. Allcock, C.J. Nelson and W.D. Coggio, Chem. Mater., **1994**, 6, 516.
253. M. Scoponi, F. Pradella, V. Carassiti, M. Gleria and F. Minto, Makromol. Chem., **1993**, 194, 3047.
254. F. Minto, M. Scopoini, M. Gleria, F. Pradella and P. Bortolus, Eur. Polym. J., **1994**, 30, 375.
255. M. Gleria, F. Minto, P. Burtolus, G. Fachin, R. Bertani, M. Scoponi and F. Pradella, Polym. Prepr. (Am. Chem. Soc., Div. Polym. Chem.), **1993**, 34(1), 270.
256. F.J. Garcia-Alonso and P. Wisian-Neilson, Polym. Prepr. (Am. Chem. Soc. Div. Polym. Chem.), **1993**, 34(1), 264.
257. P. Wisian-Neilson and F.J. Garcia-Alonso, Macromolecules, **1993**, 26, 7156.
258. L. Bailey, M. Bahadur and P. Wisian-Neilson, Polym. Prepr. (Am. Chem. Soc., Div. Polym. Chem.), **1993**, 34(1), 318.
259. M.Q. Islam, J. Bangladesh Chem. Soc., **1993**, 6, 109 (Chem. Abst., **1994**, 120, 299437).
260. C. Francart-Delprato, R. DeJager and D. Houalla, Polym. Prepr. (Am. Chem. Soc. Div. Polym. Chem.), **1993**, 34(1), 320.
261. W.D. Samuels, G.J. Exarhos, S.D. Burton, B.J. Tarasevich and D.J. Stasko, Mater. Res. Soc. Symp. Proc., **1992**, 249, 371 (Chem. Abst., **1993**, 119, 277149).
262. S. Mori, M. Kitayama and H. Ando, Jpn. Kokai Tokkyo Koho, JP 05202209 (Chem. Abst., **1994**, 120, 32578).
263. S. Cohen, M.C. Bano, L.G. Cima, H.R. Allcock, J. P. Vacanti, C.A. Vacanti and R. Langer, Clin. Mater., **1993**, 13, 3.
264. H.R. Allcock, S.R. Pucher and A.G. Scopelianos, Macromolecules, **1994**, 27, 1.
265. C.T. Laurencin, M.E. Norman, H.M. Elgendy, S.F. El-Amin, H.R. Allcock, S.R. Pucher and A.A. Ambrosio, J. Biomed. Mater. Res., **1993**, 27, 963.
266. S.R. Pucher and H.R. Allcock, Polym. Prepr. (Am. Chem. Soc., Div. Polym. Chem.), **1992**, 33(2), 108.
267. H.R. Allcock, S.R. Pucher and A.G. Scopelianos, Macromolecules, **1994**, 27, 1071.
268. J. Crommen, J. Vandorpe, S. Vansteenkiste and E. Schacht, ACS Symp. Ser., **1993**, 520, 297.
269. J. Crommen, J. Vandorpe and E. Schacht, J. Controlled Release, **1993**, 24, 167.
270. E. M. Ruiz, C.A. Ramirez, M. A. Aponte and G.V. Barbosa-Canovas, Biomaterials, **1993**, 14, 491.

271. S. Cohen, H.R. Allcock and R. Langer, Recent. Adv. Pharm. Ind. Biotechol., Minutes Int. Pharm. Technol. Symp., 8th, Eds., A. Hincal, K. Atilla, and H.S. Kas, Sante (Paris), **1993**, 36 (Chem. Abst., **1993**, 119, 158276).
272. A.K. Andrianov, S. Cohen, K.B. Visscher, L.G. Payne, H.R. Allcock and R. Langer, J. Controlled Release, **1993**, 27, 69.
273. A.K. Andrianov, R. Langer, L.G. Payne, B.E. Robats, S.A. Jenkins and H.R. Allcock, Proc. Int. Symp. Controlled Release Bioact. Mater., 20th, Eds., T.J. Rossman, N.A. Pappas and H.L. Gabelnich, Controlled Release Soc. (Deerfield, IL), **1993**, 26 (Chem. Abst., **1993**, 119, 233952).
274. A.K. Andrianov, L.G. Payne, K.B. Visscher, H.R. Allcock and R. Langer, Polym. Prepr. (Am. Chem. Soc., Div. Polym. Chem.), **1993**, 34(2), 233.
275. C.T. Laurencin, C.D. Morris, H. Pierre-Jacques, E.R. Schwartz, A.R. Keaton and L. Zou, Polym. Adv. Technol., **1992**, 3, 359.
276. M. Kajiwara, Phosphorus, Sulfur Silicon Relat. Elem., **1993**, 76, 163.
277. S. Lora, G. Palma, R. Bozio, P. Caliceti and G. Pezzin, Biomaterials, **1993**, 14, 430.
278. J. Jin II, Y.H. Lee, K. K. Yoon and Y.S. Shon, Pollimo, **1993**, 17, 498 (Chem. Abst., **1993**, 119, 271891).
279. V.A. Petrova, A.A. Palant and V.A. Reznichenko, Russ. J. Inorg. Chem. (Engl. Transl.), **1993**, 38, 1278.
280. R.C. Boehm, J. Phys. Chem., **1993**, 97, 13877.
281. D. Bougeard, C. Bremand, R. DeJaeger and Y. Lemmouchi, Macromol. Chem. Phys., **1994**, 195, 105.
282. C.E. Hoyle, D. Creed, P. Subramanian, P. Chatterton, I.B. Rufus and P. Wisian-Neilson, Polym. Prepr. (Am. Chem. Soc., Div. Polym. Chem.), **1993**, 34(1), 276.
283. M. Kojima, J.H. Magill, M. Cypryk, M.L. White, U. Franz and K. Matyjaszewski, Macromol. Chem. Phys., **1994**, 195, 1823.
284. T. Miyata, T. Masuko,, M. Kojima and J.H. Magill, Macromol. Chem. Phys., **1994**, 195, 253.
285. T. Anzai, H. Nakamura, T. Miyata, K. Yonetake, T. Masuko, J. Mater. Sci. Lett., **1993**, 12, 1803.
286. T. Miyata, K. Yonetake and T. Masuko, J. Mater. Sci., **1994**, 29, 2467.
287. A.N. Zadorin, E.M. Antipov, V.G. Kulichikin and D.R. Tur, Vysokomol. Soedin., Ser. A., **1993**, 35, 675 (Chem. Abst., **1994**, 120, 31666).
288. E. M. Antipov, A.N. Zadorin, S.A. Kuptsov, N.P. Tsukanova, D.R. Tur and V. Kulichikhin, Vysokomol. Soedin., Ser. B, **1993**, 35, 1540 (Chem. Abst., **1994**, 120, 108218).
289. S.A. Kuptsov, L.K. Golova, A.N. Zadorin, N.P. Kruchinin, N.V. Vasil'eva, G.Ya. Rudinskaya, D.R. Tur and E.M. Antipov, Vysokomol. Soedin., Ser. A, **1993**, 35, 541 (Chem. Abst., **1993**, 119, 161192).
290. S.A. Kuptsov, L.K. Golova, G.Ya. Rudinskaya, N.V. Vasil'eva, D.R. Tur and M.E. Antipov, Vysokomol. Soedin., Ser. A, **1992**, 34, 62 (Chem. Abst., **1993**, 119, 9449).

291. M.M. Iovleva, L.Ya Konovalova, L.K. Golova, V.N. Smirnova and D. R. Tur, Vysokomol. Soedin., Ser. B., **1992**, 34, 64 (Chem. Abst., **1993**, 119, 118315).
292. G.B. Schoni and J.E. Mark, J. Inorg. Organomet. Polym., **1993**, 3, 331.
293. J.E. Mark and G.B. Schoni, J. Inorg. Organomet. Polym., **1993**, 3, 347.
294. Y. Nakada, S. Maruoka, I. Tsuchida and T. Saito, Jpn. Kokai Tokkyo Koho, JP 04257594 (Chem. Abst., **1993**, 119, 106360).
295. A.J. Jaglowski and R.E. Singler, Polym. Prepr. (Am. Chem. soc., Div. Polym. Chem.), **1993**, 34, 314.
296. G.J. Exarhos, W.D. Samuels and S.D. Burton, Mater. Res. Soc. Symp. Proc., **1992**, 244, 269 (Chem. Abst., **1993**, 119, 36928).
297. K. Moriya, Y. Nishibe and S. Yano, Macromol. Chem. Phys., **1994**, 195, 713.
298. A.L. Crumbliss, D. Cooke, J. Castillo and P. Wisian-Neilson, Inorg. Chem., **1993**, 32, 6088.
299. H.R. Allcock, C.J. Nelson and W.D. Coggio, Polym. Mater. Sci. Eng., **1993**, 68, 76.
300. Y. Tada, M. Sato, N. Takeno, T. Kameshima, V. Nakacho and K. Shigehara, Macromol. Chem. Phys., **1994**, 195, 571.
301. M. K. Abraham and M. Alamgir, US Pat., US 5166009 (Chem. Abst., **1993**, 119, 99857).
302. J.A. Subramony and A.R. Kulkarni, Solid State Ionics, **1994**, 67, 235.
303. J.A. Subramony and A.R. Kulkarni, Mater. Sci. Eng., B, **1994**, B22, 206.
304. K.M. Abraham. D.M. Pasquariello, M. Alamgir and W.P. Kilroy, J. Power Sources, **1993**, 44, 385.
305. Y. Tada, M. Sato, N. Takeno, Y. Nakacho and K. Shigehara, Makromol. Chem., **1993**, 194, 2163.
306. Y. Tada, M. Sato, N. Takeno, Y. Nakacho and K. Shigehara, Chem. Mater., **1994**, 6, 27.
307. H. Sato, Jpn. Kokai Tokkyo Koho, JP 05148424 (Chem. Abst., **1994**, 120, 78593).
308. K. Higashimoto, K. Nakai, K. Hironaka, T. Hayakawa, A. Komaki, T. Nakanaga and M. Taniguchi, Jpn. Kokai Tokkyo Koho, JP 05205777 (Chem. Abst., **1993**, 119, 253675).
309. N. Kajiwara, T. Ogino, T. Myazuki and T. Kawagoe, Jpn. Kokai Tokkyo Koho, JP 06013108 (Chem. Abst., **1994**, 120, 303349).
310. C.J. T. Landry, W.T. Ferrar, D.M. Teeganden and B.K. Coltrain, Polym. Prepr. (Am. Chem. Soc., Div. Polym. Chem.), **1992**, 33(2), 557.
311. Y.M. Chen-Yang and T.T. Wu, Polym. Prepr. (Am. Chem. Soc., Div. Polym. Chem.), **1993**, 34(1), 272.
312. L. Braswell, Annu. Tech. Conf.-Soc. Plast. Eng., **1992**, 50th (Vol. 1), 407 (Chem. Abst., **1994**, 120, 271760).
313. C.R. Herrero and J.L. Acosta, Polym. Int., **1993**, 32, 349.
314. C.R. Herrero and J.L. Acosta, Polym. J., **1994**, 26, 786.
315. B.K. Coltrain, W.T. Ferrar, C.J.T. Landry, T.R. Molaire, D.E. Schildrant and V.K. Smith, Polym. Prepr. (Am. Chem. Soc., Div. Polym. Chem.), **1993**, 34(1), 266.

316. C.J. Nelson, W.D. Coggio, I. Manners, H.R. Allcock, D. Walker, L. Pessan and W.J. Koros, Polym. Prepr. (Am. Chem. Soc., Div. Polym. Chem.), **1992**, 33(2), 319.
317. E.S. Peterson and M.L. Stone, J. Membr. Sci., **1994**, 86, 57.
318. Yu. P. Yampol'skir, S.M. Shishatskii, V.I. Bondar, G. Golemme, E.M. Antipov, V.A. Gubanov, G. Golemme, E. Drioli and N.A. Plate, Vysokomol. Soedin., Ser. A., **1993**, 35, 1486 (Chem. Abst., **1994**, 120, 108320).
319. H. Nishide, H. Kawakami, S. Nishihara, K. Tsuda, T. Suzuki and E. Tsuchida, Polym. Adv. Technol., **1993**, 4, 17.
320. M. Kajiwara, Polym. Prepr. (Am. Chem. Soc., Div. Polym. Chem.), **1993**, 34(1), 268.
321. R. Soria, C. Defalgue and J. Gillot, Eur. Pat. Appl., EP524592 (Chem. Abst., **1993**, 119, 97459.
322. C. Guizard, A. Boye, A. Larbot, L. Cot and A. Grangeon, Recents Prog. Genie Procedes, **1992**, 6, 27 (Chem. Abst., **1993**, 119, 140638).
323. K.M. McHugh, L.D. Watson, R.E. McAtee and S.A. Ploger, U.S. Pat., US5252212 (Chem. Abst., **1994**, 120,, 56166).
324. G. Masci, S. Contadini, V. Crescenzi and M. Dentini, Polym. Prepr. (Am. Chem. Soc., Div. Polym. Chem.), **1993**, 34(1), 1067.
325. F. Belaj, Acta Crystallogr., Sect. B: Struct. Sci., **1993**, B49, 254.
326. W. Clegg and C. Bleasdale, Acta Crystallogr., Sect. C: Cryst. Struct. Commun., **1994**, C50, 740.
327. A.L. Llamas-Saiz and C. Foces-Foces, Acta Crystallogr., Sect. C: Cryst. Struct. Commun., **1994**, C50, 255.
328. K. Kubono, N. Asaka, S. Isoda, T. Kobayashi and T.Taga, Acta Crystllogr., Sect. C: Cryst, Struct. Commun., **1994**, C50, 324.
329. K. Kubono, N. Asaka, S. Isoda and T. Kobayashi, Acta Crystallogr., Sect. C: Cryst. Struct. Commun., **1993**, C49, 404.
330. V. Chandrasekhar, T. Chivers and M. Parvez, Acta Crystallogr., Sect. C: Cryst. Struct. Commun., **1993**, C49, 393.
331. R. Conanec, P.L. H'Aridon, W. Feldmann, R. Marchand and Y. Laurent, Eur. J. Solid State Inorg. Chem., **1994**, 31, 13.

Author Index

In this index the number in parenthesis is the Chapter number of the citation and this is followed by the reference number or numbers of the relevant citations within that Chapter.

Abad, M.M. (1) 215
Abboud, K.A. (7) 36
Abdou, W.M. (6) 57
Abdreimova, R.R. (4) 4, 8
Abe, M. (1) 346, 347
Aboulela, F. (5) 209
About-Jaudet, E. (4) 160
Abraham, K.M. (7) 224, 301, 304
Abushanab, E. (4) 165
Achiwa, K. (1) 108, 111
Acosta, J.L. (7) 313, 314
Adam, V. (5) 158
Adam, W. (2) 16
Adamov, V.M. (1) 256
Adams, C.J. (5) 155, 156
Adger, B.J. (6) 26
Adrover, V. (7) 69
Afarinkia, K. (6) 141
Afonikin, A.A. (7) 116
Agbossou, F. (1) 229
Aghabozorg, H. (1) 122
Agrawal, S. (5) 103
Aguilar-Parrilla, F. (6) 19
Ahlrichs, R. (1) 387; (6) 17
Ahmad, I.K. (7) 9
Ahn, K.D. (7) 181
Aibasov, E.Zh. (1) 249
Aitken, R.A. (6) 51-54
Akagi, M. (5) 163
Akaogi, A. (7) 60
Akiba, K. (2) 20
Akinc, M. (1) 76
Akiyama, T. (5) 164
Aksinenko, A.Yu. (1) 261
Aladzheva, I.M. (1) 309; (4) 222
Alajarin, M. (6) 19, 118, 130; (7) 14, 15, 57
Alam, M. (4) 246
Alamgir, M. (7) 224, 301, 304
Alazzouzi, E. (5) 88
Albanov, A.I. (1) 248
Albericio, F. (3) 81, 115
Albinati, A. (1) 106
Albouy, D. (1) 84
Alcaraz, G. (1) 175, 380; (7) 35
Alcock, N.W. (1) 481, 497
Aldern, K.A. (5) 63
Aleixo, A.M. (6) 99
Aleksanyan, M.S. (4) 282
Alekseiko, L.N. (1) 278, 391
Aleshkova, M.M. (4) 8
Alexakis, A. (3) 118
Alexander, P. (4) 245; (5) 1, 24
Alfer'ev, I.S. (4) 281
Al'fonsov, V.A. (4) 46-49, 229, 230
Al-Hafidh, J. (4) 22
Alias, A. (1) 171; (6) 126, 127, 129; (7) 21, 59
Alisi, M.A. (3) 108
Al-Juboori, M.A.H.A. (2) 10
Al-Kubaisi, A.H. (7) 103
Allan, P.W. (5) 20
Allcock, H.R. (7) 72, 105, 106, 118, 127, 137, 139, 197, 198, 213-216, 230, 247, 252, 263-267, 271-274, 299, 316
Allen, C.W. (7) 1, 91, 92, 98, 126
Allen, J.V. (1) 56
Allen, T.L. (1) 434
Aller, E. (6) 118, 120; (7) 57
Allman, S.L. (5) 236
Allured, V.S. (3) 38
Almeida, W.P. (6) 138
Almendros, P. (6) 119, 121, 123
Al'metkina, L.A. (4) 49, 229
Al-Mushadania, J.S. (5) 5
Alonso, R.A. (1) 26
Al-Resayes, S.I. (1) 414-416
Altamura, M. (6) 147
Altmann, K.-H. (3) 92; (5) 126
Altona, C. (5) 215
Altorfer, M.M. (5) 136
Alvarez, M. (1) 33
Alyea, E.C. (1) 481
Amanov, R.U. (7) 83, 84
Ambrosi, H.-D. (6) 28
Ambrosio, A.A. (7) 265
Amburgey, J. (6) 70
Amin, S. (5) 186
An, D.L. (1) 349, 451; (3) 138; (4) 226; (6) 59
Andersen, O.M. (5) 195
Anderson, L.W. (3) 82
Andersson, C. (1) 19
Ando, H. (7) 174, 177, 262
Ando, K. (6) 20
Andreae, S. (3) 6
Andrews, P.C. (3) 88
Andriamiadanarivo, R. (4) 133
Andrianov, A.K. (7) 272-274
Andrieu, J. (1) 6
Andrjewski, G. (4) 205
Andrus, A. (3) 98; (5) 70, 76
Ang, H.G. (7) 7
Angelici, R.J. (1) 402, 403, 421
Angelov, C.M. (7) 33, 236
Angermaier, K. (1) 135
Antipin, M.Yu. (1) 309, 389; (3) 31, 56; (7) 83, 84
Antipov, E.M. (7) 287-290, 318
Antipova, V.V. (4) 251
Anzai, M. (7) 152
Anzai, T. (7) 285
Aoki, S. (3) 145
Aponte, M.A. (7) 270
Arakawa, M. (5) 141

Aravamudan, G. (7) 189
Araya-Maturana, R. (1) 308
Arbasov, E.Zh. (4) 5, 6
Arbuzova, S.N. (1) 63, 248
Argues, A. (7) 13, 21, 59
Arif, A.M. (2) 27; (3) 50
Arkin, M.R. (5) 200
Arkipov, V.P. (4) 150
Arling, T.J. (7) 150
Armour, M.A. (1) 251; (4) 169
Armstrong, S.K. (1) 269, 272; (6) 62, 63
Arnett, G. (5) 20
Arnold, J.R. (5) 155, 156
Arnold, L. (5) 81
Arques, A. (1) 171; (6) 126, 127, 129
Arredondo, Y. (1) 158
Arsanious, M.H.N. (4) 159
Arzumanov, A.A. (5) 47
Asaka, N. (7) 107, 328, 329
Asam, A. (1) 34, 35
Ascher, A.K. (3) 57
Ashby, E.C. (1) 23, 24
Ashe, A.J. (1) 471
Asseline, U. (3) 114; (5) 182
Assmann, B. (1) 135
Atamas', L.I. (4) 124
Atherton, J.I. (6) 51
Athey, P.S. (4) 68
Atmyan, L.O. (4) 52
Atrazhev, A.M. (5) 46, 50
Atrazheva, E. (3) 75; (5) 117
Atrees, S.S. (4) 228
Attanasi, O.A. (4) 130, 131; (6) 87
Atwood, D.A. (1) 72
Atwood, V.O. (1) 72
Aubertin, A. (5) 12
Aubuchon, S.R. (1) 74
Augustyns, K. (5) 127
Aurup, H. (5) 59
Averett, D.R. (5) 27
Averin, A.D. (1) 426
Avino, A.M. (3) 81, 115
Awad, R.W. (6) 32
Awasthi, S.K. (5) 83
Ayllon, J.A. (1) 304
Azhayev, A. (5) 58, 165, 166
Azhayeva, E. (5) 58

Baasov, T. (4) 60
Baban, J.A. (4) 192, 211
Babcock, L.M. (7) 156
Babler, J.H. (4) 121; (6) 94
Baceiredo, A. (1) 175, 380, 461, 485; (6) 6; (7) 35
Bachrach, S.M. (1) 191, 193, 474, 475
Badanyan, Sh.O. (4) 282
Bader, A. (1) 16
Baer, F. (1) 506
Bätcher, M. (1) 225
Baeten, W. (1) 342
Bagautdinova, D.B. (4) 150
Bahadur, M. (7) 126, 258
Bai, J. (5) 229
Bailey, L. (7) 258
Bakshi, P.K. (1) 433, 444; (3) 132
Balado, A. (6) 126
Balagopala, M.I. (5) 14
Balakrishna, M.S. (1) 240; (7) 19, 210
Balasheva, T.M. (4) 275
Balasubramanian, K.K. (1) 180
Bałczewski, P. (4) 261
Balegroune, F. (1) 5
Balitzky, Y.V. (3) 135
Ball, R.G. (1) 176
Balzarini, J. (5) 6, 7, 26
Bampos, N. (1) 87
Band, J.K. (7) 156
Bankman, M. (1) 365
Bano, M.C. (7) 263
Bansal, R.K. (3) 5
Banville, D.L. (5) 194
Baptistella, L.H.B. (6) 99
Baraniak, J. (4) 1
Baranov, G.M. (4) 189
Baranov, V.G. (3) 56
Baranova, L.I. (1) 252
Barber, I. (5) 11
Barbosa-Canovas, G.V. (7) 270
Barelay, F. (3) 111
Barion, B. (7) 48
Barion, D. (1) 395, 425; (3) 128
Barloo, M. (4) 197, 220
Barluenga, J. (6) 18; (7) 56
Barnes, C.L. (3) 36; (7) 80, 81
Barnum, B.A. (1) 20
Barr, K.E. (5) 193
Barrans, J. (1) 489; (3) 60
Barrett, C. (6) 26
Barron, A.R. (1) 294
Barth, A. (1) 35
Bartik, B. (1) 110
Bartik, T. (1) 110
Barton, J.K. (5) 198-200
Bartsch, R. (1) 492
Bartz, I. (7) 19
Bashilov, V.V. (4) 100, 101
Basu, A.K. (5) 185
Batra, B.S. (4) 225
Bats, J.W. (3) 70
Battistini, C. (5) 41
Batyeva, É.S. (4) 46-49, 229, 230
Baudilio, T. (6) 18
Baudler, M. (1) 242
Baudry, D. (1) 493
Bauermeister, S. (4) 51, 82
Baum, G. (7) 76
Baumstark, A.L. (1) 163
Bausch, J.W. (1) 20
Baxter, A.D. (5) 36
Beabealashvilly, R.S. (5) 50
Beachley, O.T. (1) 79
Beaucage, S.L. (3) 3, 4, 96; (5) 64, 65
Beauchamp, L. (5) 27
Beaudoin, S. (6) 67
Becker, G. (1) 27, 404, 413
Beckers, H. (3) 32
Beckmann, H. (4) 227
Beder, S.M. (1) 324
Beggs, J.D. (3) 79
Bégué, J.-P. (6) 48, 49
Behl, H. (1) 17; (6) 10
Behrendt, A. (1) 505; (7) 78
Behrens, S. (3) 109; (4) 34
Beijer, B. (5) 158
Bekker, A.R. (1) 230, 231, 233; (3) 31, 63, 65, 66; (4) 73
Belaj, F. (7) 325
Belaug, M.-P. (4) 163
Belaya, S.L. (1) 340
Beletskaya, I.P. (1) 426; (4) 122
Bellan, J. (1) 489
Bellevergue, P. (5) 4
Bel'skii, V.K. (4) 58
Ben Alloum, A. (4) 105
Bender, H.R.G. (3) 37
Benevides, J.M. (5) 226
Benner, S.A. (5) 128, 136, 146
Bennett, L.L. (5) 20, 21
Bennett, N.S. (1) 207
Benseler, F. (5) 152
Bentolila, A. (4) 255
Bentrude, W.G. (2) 27; (3) 50, 143, 144
Benyei, A. (1) 146, 310
Benzaria, S. (5) 11
Berbruggen, C. (4) 197
Berclaz, T. (1) 377
Bergdahl, M. (6) 116
Berger, D.J. (1) 438
Bergmann, F. (5) 87
Bergot, B.J. (3) 24
Bergsträsser, U. (1) 126, 201, 373, 408; (7) 46
Bergstrom, D.E. (3) 76, 88; (5) 145
Berkowitz, D.B. (4) 137; (6) 90
Bermak, J.C. (1) 190; (4) 254
Bernardinelli, G. (1) 396
Bernhardt, H. (1) 22

Bertani, R. (6) 43; (7) 140, 255
Berté-Verrando, S. (4) 95, 117; (6) 8, 82
Bertrand, G. (1) 175, 200, 202, 380, 461, 485; (3) 142; (6) 6, 16; (7) 35, 38, 67-69, 195, 235
Besser, S. (1) 238
Bessho, K. (5) 164
Bestmann, H.J. (1) 17; (6) 10, 41, 42, 98
Bévierre, M.-O. (3) 93
Bevilaqua, P.C. (5) 161
Bezergiaunidou-Balouctsi, C. (6) 34
Bhagwat, S.S. (1) 185
Bhalay, G. (4) 293; (6) 78
Bhandary, K.K. (7) 188
Bhat, S.N. (1) 377
Bhatia, D. (3) 78
Bhatt, R.K. (6) 105
Bhushan, V. (6) 27
Bickelhaupt, F. (1) 374
Bieger, K. (1) 507, 508
Bieler, S. (4) 285
Bienest, R. (4) 15
Bilger, E. (1) 506
Bilow, U. (1) 303
Binder, H. (1) 64
Binger, P. (1) 417
Birkel, M. (1) 194
Biron, K.K. (5) 27
Bischofberger, N. (5) 23
Biskupic, S. (7) 95
Bissinger, P. (1) 118, 121, 159, 161
Bittman, R. (4) 238
Blachnik, R. (1) 244
Blackburn, G.M. (5) 35
Blagborough, T.C. (1) 288
Blaschette, A. (1) 277
Błaszczyk, J. (4) 45, 237; (5) 97
Blazis, V.J. (4) 153
Bleasdale, C. (7) 326
Blommers, M.J.J. (5) 216
Blumenfeld, M. (5) 175
Blumenthal, A. (1) 161
Boal, J.H. (5) 9
Bock, H. (1) 365
Boduszek, B. (4) 203
Böge, O. (3) 34
Boehm, R.C. (7) 280
Börner, A. (1) 32, 112, 113
Boese, R. (1) 392, 409; (4) 104
Bohmann, K. (5) 158
Bohn, T. (6) 122; (7) 58
Bohringer, M. (1) 413; (5) 129
Boisdon, M.-T. (1) 488; (3) 130
Bollaert, W. (3) 19; (4) 197
Bolli, M. (3) 91; (5) 125
Bonadies, F. (6) 73
Boncella, J.M. (7) 36
Bond, M.R. (1) 202; (6) 8
Bondar, V.I. (7) 318
Bonfantini, E.E. (6) 156
Bonin, A.M. (5) 191
Bonnet-Delpon, D. (6) 48, 49
Bonora, G.M. (5) 73, 74
Bonrath, W. (1) 39
Bookham, J.L. (1) 89, 239
Boons, G.J. (6) 153
Borangazieva, A.K. (1) 133; (4) 5
Bordwell, F.G. (1) 327; (6) 2
Borecka, B. (7) 94
Borghi, D. (5) 41
Borloo, M. (3) 19, 20
Borozdina, L.U. (5) 89
Bortolus, P. (7) 211, 212, 254, 255
Bosch, I. (6) 131
Bossmann, S.H. (5) 200
Bott, R.C. (4) 235
Bott, S.G. (1) 213
Botta, M. (1) 186
Botteghi, C. (1) 54
Bottle, S.E. (4) 235
Bougeard, D. (7) 11, 281
Boulos, L.S. (4) 159
Bounja, Z. (2) 22
Boutonnet, F. (1) 125, 445
Bovarets, V.S. (6) 15
Bovin, A.N. (1) 260
Bowmaker, G.A. (1) 45
Bowsher, D.K. (1) 7
Boyd, E.A. (3) 8; (4) 102
Boye, A. (7) 179, 322
Bozio, R. (7) 277
Brabec, V. (5) 189
Brack, A. (3) 114
Bradford, W.W., III (5) 31
Brandi, A. (1) 265
Brankovan, V. (4) 246
Braswell, L. (7) 312
Brauer, D.J. (1) 397
Braunstein, P. (1) 6
Breen, T.L. (1) 441, 442
Breit, B. (1) 371, 409
Brel', A.K. (4) 103
Bremand, C. (7) 11, 281
Bremer, M. (1) 17; (6) 10
Brenchley, G. (1) 8; (3) 124
Brennan, J. (6) 26
Breslauer, K.J. (5) 183
Breuer, E. (4) 278
Breutel, C. (1) 130
Breza, M. (7) 95
Bricklebank, N. (1) 150; (2) 9, 10
Brill, W.K.-D. (3) 97; (5) 69
Brinkman, J. (6) 95
Brisset, H. (1) 160
Broess, A.I.A. (6) 137
Bronson, J.J. (4) 246
Brooks, D.W. (6) 104
Broschk, B. (1) 410
Brost, R.D. (1) 210
Brougham, D.F. (1) 388
Brovarets, V.S. (1) 315-319
Brown, D.E. (7) 98
Brown, D.M. (3) 89; (5) 187
Brown, F.K. (5) 111
Brown, T. (3) 79; (5) 168, 197
Bruce, G.C. (1) 210
Brufani, M. (3) 108
Bruget, D.N. (1) 405
Brunel, J.-M. (3) 117; (4) 89
Brunet, J.-J. (1) 466, 467
Brunet, L. (1) 495
Brunette, J.-P. (1) 5
Brunner, H. (1) 9, 10, 138
Brusdeilins, N. (1) 170
Bryce, M.R. (6) 145
Bryson, C. (4) 239
Bubnov, N.N. (4) 100, 101; (7) 83
Buchanan, M.V. (5) 235
Buchwald, S. (1) 353
Budiansky, M. (5) 76
Budnikova, Yu.G. (1) 226; (4) 7
Buina, N.A. (7) 30
Buisman, G.J.H. (3) 122
Buono, G. (3) 117; (4) 89, 119
Burford, N. (1) 424, 433, 444; (3) 129, 132; (7) 10, 42, 50
Burgos-Lepley, C.E. (4) 177
Burhus, F.S., II (7) 22
Burk, M.J. (1) 31
Burkus, F., II (7) 239
Burnaeva, L.A. (4) 32
Burns, B. (4) 234
Burns, J.A. (1) 75, 78
Burrows, A.D. (1) 6
Burton, D.J. (1) 299, 326; (4) 136, 142
Burton, J.M. (2) 3, 33; (4) 16
Burton, S.D. (7) 123, 261, 296
Busson, R. (5) 127
Butcher, R.J. (1) 69
Butenschoen, H. (1) 39
Butin, B.M. (4) 148
Butler, I.R. (1) 14
Butler, P.I. (6) 107
Butler, P.J.G. (5) 154
Buwalda, P.L. (7) 135
Buyniski, J.P. (4) 239
Buznanski, A. (5) 99
Buzykin, B.I. (4) 265, 266, 294
Byers-Hill, J. (1) 75, 78

Bykhovskaya, O.V. (1) 309
Byrne, L.T. (4) 14
Byun, H.-S. (4) 238

Cabaj, J. (6) 22
Cabioch, J.-L. (1) 367
Cabras, M.A. (1) 54
Cabrera, A. (1) 284
Cagnet, R. (4) 54
Cai, B.Z. (4) 178
Cain, T.C. (5) 229
Calas, N.J. (1) 276
Caliceti, P. (7) 277
Cameron, C.G. (7) 118, 139
Cameron, D.G. (4) 202
Cameron, T.S. (1) 433, 444; (3) 132; (7) 94
Caminade, A.M. (4) 59, 236; (6) 132; (7) 136
Cammers-Goodwin, A. (1) 208
Camp, D.G. (5) 232
Campagne, J.-M. (4) 253
Campbell, D.A. (1) 190; (4) 254
Campbell, E. (5) 76
Cananagh, R.L. (4) 239
Cantrill, A.A. (4) 301
Canty, A.J. (1) 214
Cao, R. (2) 26
Cao, X. (5) 113
Capaldi, D.C. (5) 93
Cappellacci, L. (5) 10
Carassiti, V. (7) 253
Cardilli, A. (6) 73
Carmichael, D. (1) 414, 494, 500
Carrano, C.J. (1) 72, 202; (6) 8
Carruthers, N.I. (4) 114
Carter, P.A. (4) 293; (6) 78
Carty, A.J. (1) 214
Caruthers, M.H. (5) 105
Casabo, J. (1) 215
Casale, R.A. (3) 102; (5) 86
Castaneda, F. (1) 308
Castedo, L. (6) 142
Castillo, J. (7) 298
Castro, J.L. (1) 176
Cateni, F. (4) 134
Cavell, R.G. (1) 205, 240; (7) 19, 20, 210
Cech, D. (5) 56
Cech, T.R. (5) 105
Cellai, L. (3) 108
Cen, W. (4) 146
Cesta, M.C. (3) 108
Cevasco, G. (4) 256, 257
Chaban, G.M. (7) 5
Chabanenko, K.Y. (3) 26
Chae, M.Y. (4) 76
Chakhmakhcheva, O.G. (5) 82
Challet, S. (1) 241
Chamberlain, S.D. (5) 27
Champness, N.R. (1) 52
Chan, C.S. (1) 297
Chan, T.M. (4) 114
Chandler, A.J. (4) 64
Chandrasekaran, A. (7) 128, 129, 131
Chandrasekhar, J. (3) 61; (7) 40
Chandrasekhar, V. (7) 125, 130, 330
Chang, J.Y. (7) 138
Chantegrel, B. (4) 133
Chanteloup, L. (5) 121
Chao, S.-H.L. (1) 79
Charamunt, R.M. (7) 13, 15
Charette, A.B. (1) 187
Charkin, O.P. (7) 5
Charrier, C. (1) 462
Chary, K.V.R. (5) 207
Chastain, M. (5) 224
Chatterton, P. (7) 282
Chattopadhyaya, J. (4) 86
Chau, F.T. (1) 423
Chauhan, K. (6) 105
Che, C.-M. (1) 138
Chelucci, G. (1) 54
Chen, B. (3) 82
Chen, C. (6) 30
Chen, C.H. (5) 236
Chen, C.-T. (4) 263; (6) 79
Chen, G. (1) 250
Chen, J. (4) 217, 258; (6) 4
Chen, K. (4) 28
Chen, M.-R. (4) 248
Chen, R.Y. (4) 178, 194, 248
Chen, S. (4) 140, 214
Chen, S.Q. (4) 87
Chen, T. (1) 66
Chen, W. (4) 85
Chen, X.-R. (4) 194
Chen, Y. (1) 163
Chen, Z. (1) 141; (3) 45
Cheng, W.C. (1) 138
Chen-Yang, Y.W. (7) 146, 311
Cheong, S. (7) 138
Cherezova, E.N. (1) 128
Cherkasov, R.A. (4) 279
Chern, R.T. (7) 247, 251
Chernega, A.N. (1) 153, 389, 390, 412, 429; (7) 17
Chernov, P.P. (4) 32
Cherny, D.I. (5) 225
Chertanova, L.F. (4) 265, 266
Chiba, M. (1) 108
Chida, S. (4) 41
Chidgeavadge, Z.G. (5) 50
Chiesi-Villa, A. (1) 203
Chinchilla, R. (6) 149
Chivers, T. (7) 39, 183-187, 330
Cho, D.Y. (1) 147
Cho, I.H. (1) 147
Cho, N.S. (1) 164
Chooi, S.Y.M. (1) 37
Choy, G.S.-C. (7) 4
Christau, H.J. (7) 109
Christe, K.O. (2) 8
Chrystal, E. (3) 111
Chuang, Y.H. (7) 146
Chudakova, T.I. (4) 127
Chuit, C. (1) 1; (2) 35
Chung, K.Y. (1) 321
Chuprina, V.P. (5) 210
Chur, A. (5) 114
Churchill, M.R. (1) 79
Chvertkin, B.Ya. (4) 44, 262
Chvertkina, L.V. (4) 44
Cicchi, S. (1) 265
Ciccu, A. (1) 397
Cima, L.G. (7) 263
Claereboudt, J. (1) 342
Claeson, G. (4) 192, 211
Claeys, M. (1) 342
Claramont, R.M. (6) 19
Clarke, T. (6) 107
Clayden, J. (1) 270, 271, 273; (6) 61
Clegg, W. (1) 89; (7) 326
Clement, J.-C. (1) 209
Clivio, P. (5) 160
Cloke, F.G.N. (1) 419
Clyburne, J.A.C. (1) 424, 433; (3) 129; (7) 42, 50
Coe, D.M. (5) 17
Coggio, W.D. (7) 127, 252, 299, 316
Cohen, J.S. (3) 82
Cohen, S. (7) 263, 271, 272
Cole, D.L. (3) 71; (5) 101, 102
Coley, S.M. (7) 197, 198, 230
Collet, A. (6) 31
Collignon, N. (4) 160
Collington, E.W. (1) 269, 270, 272; (6) 62, 63
Collins, J.G. (5) 193, 199
Collins, P.W. (7) 19
Coltrain, B.K. (7) 310, 315
Combellas, C. (1) 58
Combes, C. (7) 109
Conanec, R. (7) 331
Conesa, C. (1) 171; (6) 127; (7) 21
Conn, L.M. (1) 207
Connanec, R. (7) 246
Connelly, M.C. (5) 23
Connolly, B.A. (5) 124, 173

Contadini, S. (7) 324
Conti, F. (1) 239
Contreras, R. (2) 2
Cook, A.F. (5) 115
Cook, P.D. (3) 83; (5) 144
Cook, P.F. (4) 249
Cooke, D. (7) 298
Cooper, A.J. (6) 113
Coote, S.J. (1) 55-57
Cordes, W.A. (7) 130
Corelli, F. (1) 186
Corey, E.J. (1) 141; (3) 45
Correia, C.R.D. (6) 138
Corriu, R.J.P. (1) 1; (2) 35
Cosman, M. (5) 186
Cosseau, J. (1) 296
Cosstick, R. (5) 36, 43, 104, 173
Coste, J. (4) 253
Costisella, B. (3) 18; (4) 15, 111, 132, 149, 210
Cot, L. (7) 322
Côté, B. (1) 187
Coumbe, T. (1) 100
Courseille, C. (5) 202
Cousseau, J. (6) 144
Couthon, H. (4) 204
Couture, A. (1) 259; (6) 64, 133
Cowley, A.H. (1) 72, 202; (6) 8
Crain, P.F. (5) 157
Cramer, C.J. (6) 65
Cramer, F. (1) 219; (4) 12
Creed, D. (7) 282
Crescenzi, V. (7) 324
Cristau, H.J. (1) 311; (7) 70
Crittell, C.M. (1) 194
Crochet, P. (1) 50
Crombie, L. (6) 154
Crommen, J. (7) 268, 269
Crumbliss, A.L. (7) 298
Cummins, L.L. (3) 83; (5) 144
Curtis, J.M. (7) 10
Cushman, C.D. (5) 179
Cypryk, M. (7) 283

Dabkowski, W. (1) 219; (3) 67, 68; (4) 12
Daffner, R. (1) 379
Dahan, F. (1) 125, 461; (3) 142; (6) 16; (7) 35, 136, 195
Dahl, O. (5) 114, 123
Dai, D. (1) 166
Dai, H. (6) 141
Dai, X. (4) 146
d'Alarcao, M. (4) 19
Dalby, K.N. (4) 63
Dalcanale, E. (4) 17
Dam, M.A. (1) 374
Damha, M.J. (3) 95; (5) 112
Dangyan, Yu.M. (4) 282
Darai, M.M. (4) 134
Daran, J.-C. (1) 399
Darensbourg, D.J. (1) 146, 310
Dargatz, M. (1) 22
Darkins, P. (6) 148
Darrow, J.W. (4) 173
Datema, R. (4) 246
Dautant, A. (5) 202
David, G. (1) 382, 432; (3) 133; (7) 51, 196
Davies, G. (7) 132
Davies, M.P.H. (5) 7, 8
Davis, P. (5) 86
Davis, P.W. (5) 223
Davis, R. (1) 288
Davy, W. (1) 295
Dawson, G.J. (1) 55-57
Day, R.O. (1) 457; (2) 4, 18, 19; (4) 298
Deadman, J. (4) 192, 211
Dechamps, M. (3) 87
Deck, W. (6) 40
Declercq, E. (5) 6, 7, 26
Declercq, J.-P. (1) 1; (2) 35
Deeming, A.J. (1) 212
Defalgue, C. (7) 321
Degols, G. (5) 120
De Groot, A. (3) 19; (4) 197
de Heredia, I.P. (6) 38
Dehnicke, K. (1) 30, 486; (7) 73, 75-78
DeJaeger, R. (7) 11, 37, 64, 233, 234, 260, 281
de Jesus, R. (4) 247
de Jongh, C.P. (4) 164, 276
Dekle, C. (6) 103
dela Croi Habimana, J. (7) 90
Delaney, W. (5) 115
de la Torre, B.G. (3) 81, 115
del Bano, M.J. (1) 172
Dell, C. (6) 107
Delucaflaherty, C. (5) 218
De Lucchi, O. (1) 4, 103, 473
Demailly, G. (6) 93
Demarcq, M.C. (1) 167
Demerseman, B. (1) 50
De Mesmaeker, A. (3) 93; (5) 216
Demik, N.N. (4) 122
de Mont, W.W. (1) 211
Demontis, A. (5) 10
Denapoli, L. (5) 73
De Nardo, M.M. (4) 134
Deng, R.M.K. (1) 237
Deniau, E. (1) 259; (6) 64, 133
Denis, J.-M. (1) 366-368
Denmark, S.E. (4) 263; (6) 65, 66, 70, 79
Denney, D.B. (2) 13
Denney, D.Z. (2) 13
Dentini, M. (7) 324
Derek, D.P. (7) 42
Dervan, P.B. (5) 181, 192
des Abbayes, H. (1) 209
Deschamps, B. (1) 463, 465
Deshayesand, C. (4) 133
Desper, J.M. (1) 279
Destaintot, B.L. (5) 197, 202
Detsch, R. (7) 51
Deubelly, B. (1) 120
Deutsch, E.A. (1) 18, 289
Devitt, P.G. (3) 16
de Vries, N.K. (3) 119
Dewynter, G. (5) 18
Dheilly, L. (6) 93
Diederichsen, U. (5) 129
Diemert, K. (1) 38
Dillingham, M.D.B. (1) 75, 78
Dillon, K.B. (1) 237, 359
Diver, S.T. (1) 207
Dixneuf, P.H. (1) 50
Dixon, D.A. (2) 8
Dmitrichenko, M.Yu. (7) 192
Dmitriev, V.I. (1) 63, 248
Dockery, K.P. (3) 143
Dodd, D.S. (1) 182
Dodonov, V.A. (2) 14
Dogadina, A.V. (4) 96
Dolgushin, G.V. (4) 97; (7) 192, 194
Dolinnaya, N.G. (5) 175
Dolle, F. (5) 13
Dommisse, R. (3) 19; (4) 197
Dondoni, A. (6) 91, 92
Dong, W. (1) 406
Dopke, J. (7) 28
Dore, A. (1) 4, 473
Dorfman, Ya.A. (1) 133, 249; (4) 3-6, 8
Dornsife, R.E. (5) 27
Doroshkevich, D.M. (4) 3, 8
Dorrow, R.L. (6) 65
Dou, D. (1) 66, 67
Doudna, J.A. (5) 217
Dougherty, S.W. (1) 124
Douglas, K.T. (5) 196
Doutheau, A. (4) 133
Doxsee, D.D. (7) 185, 186
Doyle, R.J. (1) 47
Drabowitz, J. (4) 205
Drach, B.S. (1) 314-319; (6) 15
Drake, R.J. (1) 83
Drapailo, A.B. (4) 307
Dreschel, K. (3) 74
Driess, M. (1) 28, 68, 435

Drioli, E. (7) 318
Drobny, G.P. (5) 221
Drueckhammer, D.G. (4) 173
Drysdale, M.J. (6) 53, 54
Dubet, J. (5) 142
Dubourg, A. (1) 1; (2) 35
Duczek, W. (6) 28
Dudley, G.K. (7) 105, 106
Duesler, E.N. (1) 66, 67
Duff, J.-L. (1) 24
Duhamel, L. (6) 102
Duhamel, P. (6) 102
du Mont, W.-W. (1) 225, 352
Dumy, P. (4) 212; (6) 86
Dunbar, K.R. (1) 140
Dunkelblum, E. (6) 101
Dunn, B.B. (1) 110
du Preez, J.G.H. (7) 87, 88
Durland, R.H. (5) 132, 135
Dvořáková, H. (4) 244; (5) 29
Dwyer, T.J. (5) 192
D'yachenko, M.V. (7) 122
Dyachenko, V.I. (3) 55
Dyachovskaya, R.F. (7) 30
Dyatkina, N.B. (5) 46, 47, 50, 52
Dzhiembaev, B.Zh. (4) 148

Earnshaw, D.J. (5) 104
Ebata, S. (3) 123
Ebel, S. (3) 79
Eberbach, W. (7) 62
Ebrahimi, S.E. (5) 196
Eckstein, F. (5) 45, 54, 59, 152
Edelmann, F.T. (7) 203
Edmonds, C.G. (5) 232
Edmundson, R.S. (4) 18
Edwards, A.J. (1) 81
Edwards, K.J. (5) 204
Edwards, M. (7) 184, 232
Edwards, M.L. (1) 177
Edwards, P.G. (1) 82
Efimov, V.A. (5) 82
Egan, W. (5) 9
Eger, K. (5) 25
Eggen, M. (4) 137; (6) 90
Egli, M. (5) 171, 219
Eguchi, K. (1) 206
Eisenhardt, S. (3) 70
Eisfeld, W. (1) 373; (7) 46
Elachqar, A. (4) 292
El-Amin, S.F. (7) 265
Elgendy, H.M. (7) 265
Elgendy, S. (4) 192, 211
Elguero, J. (6) 19; (7) 13-15
El Hallaoui, A. (4) 292
Elias, A.J. (7) 124
El-Kady, N. (7) 132
Ellerman, J. (7) 207, 208
Elliott, J. (1) 270
Elliott, R.D. (5) 21
El Malouli Bibout, M. (4) 128
El-Rahmann, N.M.A. (6) 33
El-Samahy, F.A. (6) 33
Elschenbroich, C. (1) 505, 506
Elsevier, C.J. (7) 209
El-Subbagh, H. (4) 165
Emga, T.J. (3) 126
Emmerich, C. (1) 60
Enders, D. (6) 27
Endo, T. (1) 322
Engel, R. (1) 297
Engels, J.W. (3) 70; (5) 56, 108, 109
Englert, U. (1) 411
Ephritikhine, M. (1) 493
Eritja, R. (3) 81, 115
Erkelens, C. (3) 110
Erker, G. (1) 417
Ernst, L. (5) 44
Ernsting, J.M. (7) 209
Escale, R. (4) 212; (6) 86
Escarceller, M. (3) 81
Eschbach, B. (1) 27
Eschenmoser, A. (5) 129, 130
Eshhar, Z. (4) 255
Espinosa, A. (1) 172
Essassi, E.M. (1) 489; (3) 60
Esseffar, M. (7) 6
Esser, L. (1) 216
Essigmann, J.M. (5) 185
Estra, M.R. (7) 193
Etemad-Moghadam, G. (1) 84, 383; (4) 191; (7) 113
Ettner, N. (5) 53
Evans, S.A., Jr. (2) 12; (4) 31, 154, 181
Exarhos, G.J. (7) 123, 261, 296
Ezaki, A. (6) 45

Fabbri, D. (1) 4, 103, 473
Faber, M.C. (7) 150
Facchin, G. (6) 43; (7) 140, 141, 255
Facklam, T. (1) 126
Failla, S. (4) 195
Faizova, F.Kh. (4) 5
Falck, J.R. (6) 105
Falsone, G. (4) 134
Fan, M. (1) 67
Fantin, G. (7) 140, 141
Farkens, M. (1) 173; (3) 22, 57-59; (7) 18
Farooqui, F. (3) 99
Fathi, R. (5) 115
Fattakhov, S.G. (1) 36
Fauchère, J.L. (4) 125
Fauq, A.H. (4) 21
Favre, A. (5) 160
Fawcett, J. (1) 134; (4) 302; (6) 89
Fawzi, R. (1) 483
Fayet, J.-P. (1) 488; (3) 130
Fazakerley, G.V. (5) 188
Fearon, K.L. (3) 24; (5) 99
Feaster, J.E. (1) 31
Fedoroff, O.Y. (5) 210
Fedouloff, M. (1) 8; (3) 124
Feigon, J. (5) 212-214
Feinmark, S.J. (6) 103
Feldmann, W. (7) 331
Femec, D.F. (7) 120
Fen, H. (4) 62
Feng, H. (4) 217
Fenske, D. (1) 105; (7) 76
Fenton, R.R. (5) 191
Ferao, A.E. (1) 463
Feringa, B.L. (3) 119
Fernandez-Galan, R. (2) 34
Ferrar, W.T. (7) 310, 315
Feshchenko, N.G. (1) 131, 221, 223
Fett, D. (6) 98
Field, J.S. (1) 48
Field, L.D. (1) 87, 88
Fields, L.B. (1) 114
Filippone, P. (4) 130, 131; (6) 87
Fillion, G. (5) 15
Fillion, M.P. (5) 15
Filocamo, L. (3) 108
Finet, J.P. (4) 200
Fink, W.H. (1) 434
Finocchiaro, P. (4) 195
Fischer, A. (1) 15, 173, 228, 286, 287, 492; (3) 22, 34, 51, 54
Fischer, J. (1) 397, 497
Fisher, A. (7) 18
Fister, T. (4) 68
Fitzgerald, M.C. (5) 230
Fitzpatrick, N.J. (1) 388
Fitzpatrick, V. (5) 76
Fleischer, U. (1) 381; (6) 44
Florent, J.C. (4) 54
Floriani, C. (1) 203
Flower, K.R. (1) 419
Fluck, E. (1) 64, 507, 508
Foces-Foces, C. (1) 171; (6) 19, 127; (7) 13-15, 21, 327
Fogagnolo, M. (7) 140
Foley, M.A. (4) 239
Fominskii, D.G. (2) 14
Ford, K.L. (1) 323
Forner, K. (3) 18
Foucher, D. (7) 223, 231

Fourrey, J.-L. (5) 160
Francart-Delprato, C. (7) 260
Franchetti, P. (5) 10
Francois, P. (3) 87
Francotte, E. (3) 92; (5) 126
Frankhauser, P. (1) 68
Frankkamenetskii, M.D. (5) 225
Fransen, W. (7) 150
Franz, U. (7) 243, 244, 283
Fréchou, C. (6) 93
Frederick, J.H. (1) 497
Fredericks, E.J. (1) 199
Freeman, R.N. (1) 247
Freeman, S. (5) 5
Fregeau, N. (3) 75; (5) 117
Freibe, R. (7) 26
Freidzon, Y.S. (7) 122
Freier, S.M. (3) 83, 93; (5) 144
Frejaville, C. (4) 200
Frenking, G. (1) 499
Frenzen, G. (7) 77
Fresneda, P.M. (6) 119, 121, 123, 124
Frey, M. (6) 56
Freyer, A.J. (1) 69
Fridland, A. (5) 23
Friebe, R. (1) 227; (7) 27
Friebe, T.L. (6) 116
Frische, K. (4) 168
Fritz, G. (1) 49
Fritz, M. (1) 61
Frost, C.G. (1) 55-57
Frost, J.W. (4) 166; (6) 152
Frost, M. (1) 370
Froyen, P. (7) 65
Früh, Th. (4) 38
Frutos, J.C. (3) 118
Fu, D.-J. (5) 148, 149
Fu, T.Y. (1) 263
Fuchs, E. (1) 372
Fürst, J. (1) 9, 10, 138
Fürst, T.-G. (6) 41, 42
Fujii, M. (1) 197
Fujimoto, T. (1) 268; (6) 37
Fujita, M. (1) 97
Fukase, K. (3) 105
Fukoyo, E. (1) 12
Fukuoka, J. (7) 155
Fukuoka, K. (5) 61, 62
Fukushi, K. (7) 100
Fuminori, S. (5) 62
Funahashi, M. (3) 123
Funato, M. (7) 166
Funk, R.L. (4) 71
Furin, G.G. (4) 141
Furlong, P.J. (1) 53
Furukawa, N. (2) 15
Furuta, T. (4) 30, 67
Fustinoni, S. (5) 41

Gabbai, F.P. (1) 202; (6) 8
Gabbitas, N. (1) 46
Gadek, T.R. (1) 189
Gärtner-Winkhaus, C. (1) 425; (3) 128; (7) 48
Gait, M.J. (5) 150, 153, 154
Gal, Y.-S. (1) 328
Galenko, T.G. (4) 52
Galishev, V.A. (7) 194
Gallagher, M.J. (4) 65
Gallagher, P.T. (6) 155
Gallegos, A. (4) 167
Galliot, C. (7) 136
Gallois, B. (5) 197
Galloway, C.P. (1) 169
Galpin, S. (5) 2
Gamba, A. (7) 96
Gambarotta, S. (1) 122
Ganapathy, S. (3) 143
Gancarz, R. (4) 199
Ganem, B. (5) 231
Gangloff, A.R. (6) 116
Gangloff, B. (6) 69
Gani, D. (3) 111
Ganis, P. (6) 43
Gao, J.-X. (1) 138
Gao, S. (6) 46, 47
Gao, X.L. (5) 111
Garagan, S. (6) 21
Garbe, R. (7) 75, 76
Garbe, W. (6) 98
Garcia-Alonso, F.J. (7) 255, 257
Garcia-Fernández, A. (6) 18; (7) 56
Garcia-Granda, S. (6) 18, 149; (7) 56
Gardner, M.F. (5) 63
Garg, B.S. (3) 78; (5) 71
Garifzyanova, G.G. (4) 49
Garin, J. (1) 296; (6) 144
Garman, A.J. (3) 73; (5) 77
Garofalo, A. (5) 17
Garrigou-Lagrange, C. (1) 344
Garrigues, B. (4) 59, 191
Garton, D. (3) 40
Gaspar, P.P. (1) 438
Gates, P.N. (2) 10
Gaumont, A.-C. (1) 366
Gavrilova, E.L. (1) 305
Gazikasheva, A.A. (4) 266
Gdaniec, Z. (5) 188
Geactinov, N.E. (5) 186
Gefflaut, T. (4) 29
Geierstanger, B.H. (5) 192
Geise, H. (1) 342
Gellman, S.H. (1) 279-282
Geoffroy, M. (1) 377, 396
Gerber, J.P. (4) 260
Gerber, K.P. (4) 162
Gerbling, K.-P. (4) 206
Ghatlia, N.D. (5) 200
Ghodsi, T. (1) 216
Ghosh, N.N. (5) 71
Giannaris, P.A. (3) 95; (5) 112
Gibson, K.H. (6) 108
Gibson, N.J. (5) 203
Gielen, M. (1) 292
Giger, A. (5) 129
Gilbertson, S.R. (1) 250
Gilheany, D.G. (1) 53; (6) 3
Gill, G.B. (6) 150
Gillot, J. (7) 321
Gimbert, Y. (6) 133
Gimtsyan, A. (6) 67
Gindling, M.J. (1) 199
Gingras, M. (1) 207
Giorgi, G. (7) 190
Giovagnoli, D. (4) 130, 131; (6) 87
Girard, J.-P. (4) 212; (6) 86
Girardet, J.L. (5) 11
Givens, R.S. (4) 68
Gladiali, S. (1) 4, 103, 473
Glaser, R. (7) 4
Glebova, Z.I. (4) 264
Gleiter, R. (1) 413
Gleria, M. (7) 140, 141, 211, 212, 253-255
Glick, G.D. (5) 172, 176, 177
Gloede, J. (2) 17; (4) 15
Glowacki, Z. (4) 158
Gmeiner, W.H. (3) 75; (5) 117, 159
Gobel, M. (5) 129
Gobran, H.R. (1) 72
Gocheva, E. (7) 102
Goddard, R. (1) 417
Godfrey, S.M. (1) 150, 285; (2) 9, 10
Goede, S.J. (1) 374
Goerlich, J.R. (1) 51, 102, 129, 165, 173; (3) 22; (7) 18
Goesmann, H. (1) 105; (7) 73, 76
Goggins, G.D. (4) 239
Goldman, A.S. (3) 126
Golemme, G. (7) 318
Golinski, F. (7) 110
Gololobov, Yu.G. (3) 11, 135; (4) 80
Golova, L.K. (7) 289-291
Golovatyi, O. (6) 5
Gomez, M. (1) 466, 467
Gonbeau, D. (1) 367, 383, 406
Gong, Y.-F. (5) 23

Gonzalez, C. (3) 83
Gonzalez, S.M. (5) 144
Gooding, A. (5) 217
Goodwin, H.P. (1) 359
Goodwin, J.T. (5) 172, 176, 177
Gordeev, M.F. (6) 13; (7) 55
Gordillo, F. (5) 88
Gordon, N.J. (4) 31, 154, 181
Gorgues, A. (1) 296; (6) 144
Gorla, F. (1) 106
Goryunov, E.I. (4) 69
Gosselin, G. (5) 11, 12
Gostoli, G. (3) 108
Goti, A. (1) 265
Gourdel, Y. (1) 90, 160
Gouyette, C. (5) 15
Gouygou, M. (1) 383, 399
Govil, G. (5) 207
Grabowski, J.J. (6) 71
Grabowski, M.J. (4) 179, 180
Grachev, M.K. (1) 230, 231, 233; (3) 62-66
Graczyk, P.P. (1) 274, 343; (4) 261
Gradoz, P. (1) 493
Graffeuil, M. (7) 133
Grajkowski, A. (5) 97
Granda, A. (5) 72
Grandas, A. (3) 116; (5) 88
Grandclaudon, P. (1) 259; (6) 64, 133
Grandjean, D. (1) 5
Grangeon, A. (7) 179, 322
Grasby, J.A. (5) 150, 153, 154
Grauvogl, G. (1) 118, 120, 121
Gray, G.M. (1) 7, 440
Green, B.S. (4) 255
Green, D. (4) 192, 211
Green, M. (1) 418
Greenberg, M.M. (5) 68, 138
Greene, N. (3) 17
Greenhill, J.V. (6) 14
Grelet, D. (1) 311
Grenlich, P. (4) 221
Grev, R.S. (1) 438
Grifantini, M. (5) 10
Griffney, R.H. (3) 83; (5) 144
Gripper, K.D. (1) 75, 78
Groarke, P.J. (1) 388
Grobe, J. (1) 375, 376, 410, 413
Groen, M.B. (6) 137
Groshens, T.J. (1) 69
Gross, H. (3) 18; (4) 111, 132, 149
Gross, R.S. (4) 144
Grosshans, C. (5) 217
Grossmann, G. (1) 238; (4) 104, 227
Grosspietsch, T. (1) 375
Grover, N. (7) 125
Grubbs, R.H. (6) 56
Gruber, M. (1) 232; (3) 41-44
Gruendemann, E. (6) 28
Grützmacher, H. (1) 381; (6) 40, 44
Grundy, S.L. (1) 210
Grune, G.L. (7) 247, 251
Gryaznov, S. (3) 84; (5) 147, 174
Gryaznova, T.V. (3) 136, 137
Grzybowski, J. (5) 168
Guastini, C. (1) 203
Gubanov, V.A. (7) 318
Gudat, D. (1) 456; (6) 60; (7) 12, 45, 53
Gude, C. (1) 185
Gudima, A.O. (1) 429
Gueguen, E. (7) 246
Guenot, P. (1) 367, 368
Gugliemi, M. (7) 141
Guilbert, B. (1) 50
Guillemin, J.-C. (1) 366-368
Guizard, C. (7) 179, 322
Gulea-Purcarescu, M. (4) 160
Gulliver, D.J. (1) 52
Gunawardena, N. (6) 98
Guneratne, R.D. (4) 142
Gunger, A.A. (4) 103
Gunn, B.M. (5) 208
Guo, M.J. (5) 35
Guo-qiang, L. (6) 100
Gupta, A.D. (7) 142-145
Gupta, K.C. (3) 78; (5) 71
Gurevich, I.E. (4) 96
Gurumurthy, R. (1) 23
Gusarova, N.K. (1) 63, 248
Guseinov, F.I. (4) 184, 267
Guy, A. (5) 142
Guy, L. (6) 31
Guyard, C. (4) 125
Guyton, K.Z. (6) 141
Guzaev, A. (5) 58, 165, 166

Haak, U. (5) 53
Haase, C. (1) 483, 484
Habibigoudarzi, S. (5) 233
Hachemi, M. (4) 107
Hadawi, D. (1) 17; (6) 10
Haddad, M. (3) 60
Haefner, S.C. (1) 140
Haegele, G. (4) 104, 195
Haemers, A. (3) 19, 20; (4) 197, 220
Hafner, M. (1) 360, 361
Hagele, G. (1) 255; (7) 115
Hahn, E. (1) 216
Hahn, T. (1) 38
Haigh, D. (6) 77
Haines, R.J. (1) 48
Hajèk, M. (4) 289
Hajouji, H. (1) 466, 467
Hak, R. (5) 81
Halcomb, R.L. (3) 145
Hallford, R. (7) 130
Haltiwanger, R.C. (3) 39
Hamada, H. (7) 100
Hamann, T. (1) 277
Hambley, T.W. (5) 191
Hamersma, H. (6) 137
Hamilton, P.A. (7) 9
Hamilton, R. (4) 185
Hammerschmidt, F. (4) 272-274
Hammond, G.B. (4) 143
Hammond, P.J. (2) 13
Hammoutou, P.Y. (7) 37
Hampel, F. (1) 17; (6) 10
Han, I.-S. (1) 306
Han, M.J. (7) 138
Han, Y. (1) 76
Hanafusa, T. (6) 20
Hanaya, T. (1) 251; (4) 169-171
Hancox, E.L. (5) 124
Haner, R. (5) 167
Hanessian, S. (6) 67
Hanna, M.T. (1) 324
Hanna, N.B. (3) 99
Hannioui, A. (4) 128
Hansert, B. (1) 357
Hansler, U. (5) 184
Hanson, B.E. (1) 110
Hao, S. (1) 122
Hara, K. (6) 45
Harada, J. (7) 251
Harde, C. (4) 206
Hardy, P.M. (3) 73; (5) 77
Harger, M.J.P. (4) 233, 300, 302; (6) 89; (7) 41
H'Aridon, P.L. (7) 331
Haristos, D.A. (1) 234
Harlow, R.L. (1) 31
Harnden, M.R. (5) 28
Harper, J.W. (5) 222
Harrer, H.M. (2) 16
Harris, B.L. (7) 4
Harrowfield, J.M. (4) 14
Haruyama, T. (1) 369
Hasek, J. (1) 290
Hasenfratz, C. (7) 16
Hashimoto, Y. (7) 163, 165, 169
Hasselbring, R. (7) 86, 191, 200, 202, 204, 205
Hassner, A. (6) 101
Hata, T. (5) 61, 62, 75, 169
Hatanaka, M. (6) 35, 36, 157
Hattori, T. (1) 258
Haueisen, R.H. (1) 331

Haug, W. (1) 384
Haupt, E.T.K. (4) 183
Haussler, M.P. (5) 16
Hawthorne, M.F. (3) 74
Hay, A.J. (5) 4, 5, 8
Hayakawa, K. (4) 38
Hayakawa, T. (7) 308
Hayakawa, Y. (5) 100
Hayashi, T. (1) 12, 94, 96
Hayashizaka, N. (1) 258
Hayes, R. (7) 98
Haynes, R.K. (1) 247
He, B.L. (4) 201
Heckmann, G. (1) 64, 507, 508
Heeb, N.V. (3) 106; (5) 146
Heesche-Wagner, K. (1) 86
Hegemann, M. (1) 413
Heim, B. (1) 399
Heim, U. (1) 381
Hein, U. (6) 44
Heine, A. (7) 202
Heissler, D. (6) 136
Helinski, J. (3) 67, 68
Hellberg, L.H. (1) 152
Heller, D. (1) 112
Helquist, P. (6) 116
Hemming, K. (6) 25
Henderson, W. (1) 204
Hendrickson, J.B. (1) 333
Henion, J.D. (5) 231
Henschel, D. (1) 277
Herbowski, A. (1) 18, 289
Herbst-Irmer, R. (1) 238, 460
Herd, O. (1) 62
Herdewijn, P. (5) 127
Hérion, H. (6) 52
Herrero, C.R. (7) 313, 314
Herrlein, M.K. (5) 56
Herrmann, A. (1) 126
Herrmann, E. (1) 22, 238
Herrmann, W.A. (1) 109
Herue, R. (7) 235
Hervé, M.J. (1) 383
Hervé, Y. (6) 67
Herzog, A. (7) 86, 205
Hessler, G. (1) 397
Hett, R. (6) 116
Hettich, R.L. (5) 235
Heubel, J. (7) 37
Heuer, L. (1) 224
Heydt, H. (1) 126, 201, 362, 372
Hey-Hawkins, E. (1) 358
Hibino, H. (4) 70
Hicks, T.A. (6) 155
Higa, K.T. (1) 69
Higashimoto, K. (7) 308
Higashizima, T. (1) 95; (3) 121
Hill, T.G. (3) 39
Hillen, W. (5) 53
Hilts, R.W. (7) 184-187
Himeda, Y. (6) 35, 36, 157
Hing, P. (5) 76
Hirano, M. (1) 454; (4) 231
Hirata, M. (4) 25
Hironaka, K. (7) 308
Hirose, H. (7) 152
Hirose, K. (4) 171
Hirose, M. (5) 100
Hirotsu, K. (1) 347, 378, 400; (4) 232
Hirschbein, B.L. (3) 24; (5) 99
Hitchcock, M.J.M. (4) 246; (5) 22
Hitchcock, P.B. (1) 395, 418-420, 490
Hixson, S.S. (5) 55
Ho, J. (1) 83, 441
Hobbs, F.W. (5) 180
Hobson, L.J. (1) 14
Hocking, M.B. (1) 479
Hockless, D.C.R. (4) 14
Hodge, R. (5) 86
Hodosi, G. (1) 157
Hoeschele, J.D. (1) 168
Hoffman, L.M. (5) 95
Hoffmann, M. (4) 158
Hogg, A.M. (1) 251; (4) 169
Hogrefe, R.I. (5) 89, 122
Hokelck, T. (7) 117
Holland, D. (3) 73; (5) 77
Holletz, T. (5) 56
Hollfelder, F. (4) 63, 64
Holmes, R.R. (2) 4, 5, 18, 19
Holmgren, S.K. (1) 279, 280, 282
Hols, B. (5) 114
Holý, A. (4) 244, 245; (5) 1, 24, 26, 29
Holz, J. (1) 32
Homans, S.W. (5) 209
Honeyman, C.H. (7) 79, 223
Hoogmartens, J. (5) 127
Hooper, D.L. (6) 21
Horan, C.J. (7) 4
Horiuchi, T. (1) 99
Horiuchi, Y. (7) 148
Hoshino, M. (1) 115, 116
Hosono, H. (7) 172, 173
Hostetler, K.Y. (5) 63
Hosztafi, S. (1) 179
Hotei, Y. (1) 268
Hou, Z. (1) 83, 442
Houalla, D. (2) 22-25; (7) 260
Hough, D. (4) 290
Houlton, A. (1) 496
Hovinen, J. (5) 58, 165, 166
Howard, F.B. (5) 207
Hoye, P.A.T. (1) 134
Hoye, R.C. (6) 109
Hoyle, C.E. (7) 282
Hu, W. (4) 62
Huaang, W. (4) 135
Huang, B. (4) 135
Huang, C.Y. (5) 178, 179
Huang, G.Y. (4) 224
Huang, J.X. (5) 116
Huang, L. (1) 166
Huang, Q. (5) 115
Huang, T.B. (3) 134
Huang, W.Q. (4) 201
Huang, W.S. (4) 196
Huang, Y. (2) 27; (3) 50; (4) 28
Huang, Y.-Z. (6) 39
Hubert, C. (4) 191
Hubler, K. (1) 404
Huck, N.P.M. (3) 110
Hudson, H.R. (3) 9; (4) 202
Hünsch, S. (4) 86
Hughes, M.J. (6) 114
Hulst, R. (3) 119
Hung, J.-T. (1) 440
Hunter, R. (1) 331
Hunter, W.N. (5) 197
Hunziker, J. (5) 129
Hurst, G.B. (5) 235
Hursthouse, M.B. (1) 82
Hussoin, M.S. (1) 333
Hutchinson, C.L. (6) 29
Huttner, G. (1) 28, 34, 35, 60, 61
Huy, N.H.T. (1) 439, 464
Huyhn-Dinh, T. (5) 15

Iacobucci, S. (4) 19
Iannelli, M.A. (3) 108
Ichikawa, J. (6) 81
Ichikawa, Y. (3) 145
Igau, A. (1) 125, 445
Ignat'ev, Yu.A. (4) 7
Igolen, J. (5) 227
Iida, A. (1) 266
Iino, Y. (7) 60, 61
Ikeda, T. (1) 322
Imamoto, T. (1) 107
Imashiro, R. (6) 36
Imbach, J.L. (5) 11, 12, 18, 120
Imhof, W. (1) 28, 91, 450
Imhoff, P. (7) 209
Imrie, C. (1) 334-337
Inada, M. (5) 140
Inbasekaran, M.N. (7) 161
Indzhikyan, M.G. (1) 329
Inoue, H. (5) 143
Inoue, K. (7) 119
Ioannou, P.V. (3) 7
Ionin, B.I. (4) 96

Ionkin, A.S. (1) 457; (4) 298
Iorish, V.Yu. (1) 233; (3) 63
Iovleva, M.M. (7) 291
Iqbal, J. (6) 116
Irving, A. (1) 331
Irwin, W.J. (5) 5
Isab, A.A. (1) 168
Ishigami, K. (7) 100
Ishiguro, K. (1) 162
Ishii, A. (1) 115, 116
Ishii, M. (1) 472
Ishikawa, K. (5) 75
Ishikawa, M. (5) 61, 62
Ishikawa, Y. (1) 378; (4) 232
Ishmaeva, E.A. (1) 431
Islam, M.Q. (7) 259
Ismagilov, R.K. (4) 150
Ismailov, V.M. (4) 184
Isoda, S. (7) 107, 108, 328, 329
Isonaka, T. (6) 81
Itani, H. (6) 146
Ito, J. (7) 250
Ito, M. (6) 96, 97
Ito, O. (7) 148
Ito, Y. (1) 98, 137
Itoh, Y. (4) 81
Ivanova, T.A. (4) 52
Ivanovskaya, M.G. (5) 175
Ivison, P. (1) 288
Ivomin, S.P. (1) 132
Iwamura, H. (1) 96
Iwamura, M. (4) 30, 67
Iwanaga, H. (1) 246
Iwaori, H. (5) 141
Iwasaki, G. (4) 38
Iyer, R.P. (3) 3, 4; (5) 9

Jack, K.S. (1) 276
Jackson, J.A. (4) 35, 277; (6) 72
Jackson, R.D. (1) 2
Jacobs, H. (7) 110
Jacobsen, E.N. (1) 114
Jacquier, R. (4) 208, 216
Jaeger, R. (7) 8
Jaglowski, A.J. (7) 295
Jahnisch, K. (6) 28
Jaishree, T.N. (5) 211, 220
James, G.D. (6) 150, 151
James, K. (3) 8
James, S. (1) 2
Janati, T. (1) 368
Jand, J. (7) 133
Jankowski, S. (1) 437; (3) 127
Janosi, A. (6) 52
Jansen, M. (1) 303
Janssen, B.C. (1) 34
Jarvis, N.V. (7) 87, 88
Jasinski, J.P. (1) 74
Jaud, J. (1) 445; (4) 59
Jaun, B. (5) 129, 130
Jayaraman, K. (3) 77
Jeannin, Y. (1) 399
Jedlicka, B. (1) 13
Jeffries, D.J. (5) 2
Jekel, A.P. (7) 135, 150
Jenkins, I.D. (4) 235
Jenkins, S.A. (7) 273
Jenkins, T.C. (5) 204
Jenkins, Y. (5) 200
Jennings, L.J. (4) 243
Jenny, T.F. (5) 128
Jeong, I.N. (1) 326
Jeske, J. (1) 225, 460
Jetter, M.C. (5) 180
Jewell, C.F., Jr. (6) 95
Ji, H.J. (7) 138
Jiang, J. (6) 14
Jiang, L. (7) 101
Jiang, M. (3) 40
Jiao, X.-Y. (3) 19, 20; (4) 197, 220
Jin, J., II (7) 278
Jin, S. (4) 106
Jindrich, J. (4) 244; (5) 26
Jochem, G. (1) 21, 436; (3) 125; (6) 7, 58
Johnson, A.E. (6) 1
Johnson, D.D. (6) 117
Johnson, D.K. (1) 214
Johnson, L.K. (6) 56
Johnson, M.D. (1) 330
Johnson, M.G. (5) 95
Johnson, R.A. (4) 177
Jones, A.D. (6) 30
Jones, C. (1) 415, 420
Jones, L.J. (1) 77
Jones, P.G. (1) 15, 173, 217, 225, 228, 232, 277, 286, 287, 460, 492; (3) 22, 34, 42, 51, 54, 57; (7) 18
Jones, R.A. (1) 72
Jones, R.C.F. (4) 293; (6) 78
Jones, R.J. (5) 110
Jones, V. (1) 499
Jonin, P. (4) 253
Joo, F. (1) 146, 147, 310
Josephs, J.L. (6) 154
Jouati, A. (1) 377, 396
Jubault, M. (1) 296; (6) 144, 145
Jun, H. (1) 402, 403, 421
Just, G. (5) 112
Jutzi, P. (1) 170

Kabachnik, M.I. (1) 309; (4) 69, 122, 222; (7) 83, 84
Kabat, M.M. (4) 187
Kachkovskaya, L.S. (1) 430
Kadyrov, R. (1) 112
Kadyrova, V.Kh. (1) 128
Kählig, H. (1) 43
Käshammer, D. (1) 27
Kagan, H.B. (1) 32, 113
Kajiwara, M. (7) 99, 121, 147, 226-228, 276, 320
Kajiwara, N. (7) 309
Kajiyama, K. (2) 20
Kajtár, M. (6) 112
Kakehi, A. (4) 172; (6) 37
Kakkar, V.V. (4) 192, 211
Kal'chenko, V.I. (4) 124
Kalchhauser, H. (1) 43
Kalinina, N.V. (4) 268, 269
Kalinkina, A.L. (5) 82
Kam, C.-M. (4) 203
Kamalov, R.M. (4) 304
Kamatsu, M. (7) 170
Kamer, P.C.J. (3) 122
Kameshima, T. (7) 300
Kamijo, K. (1) 348, 452; (6) 9
Kamiya, H. (5) 143
Kanazawa, A. (1) 322
Kane, R.R. (3) 74
Kang, Y.B. (1) 41
Kann, N. (6) 68, 69
Karaghiosoff, K. (3) 5
Kargin, Yu.M. (1) 226; (4) 7
Karpas, A. (5) 5
Karsch, H.H. (1) 118-121
Kashemirov, B.A. (4) 262, 270
Kashimura, S. (6) 45
Kaslina, N.A. (4) 275
Kataeva, O.N. (2) 30
Kataky, J.C.S. (4) 57
Katalenic, D. (2) 31
Katayama, S. (7) 24, 238
Katho, A. (1) 146, 310
Kato, T. (5) 38
Katritsky, A.R. (6) 13, 14; (7) 55
Katsuta, Y. (6) 96, 97
Katti, K.K. (7) 80, 81
Katti, K.V. (1) 205; (3) 35, 36; (7) 19, 20, 80-82
Katz, S.A. (3) 38, 39
Kaufman, J. (5) 76
Kaufmann, B. (1) 66
Kaukorat, T. (2) 29; (3) 52, 53; (7) 29
Kawada, H. (7) 166
Kawagoe, T. (7) 309
Kawai, M. (1) 196
Kawai, S.H. (5) 112
Kawakami, H. (7) 319
Kawano, C. (1) 197

Kawashima, T. (1) 246; (2) 21; (6) 23
Kawecki, M. (1) 121
Kayser, M.M. (6) 21
Kazankova, M.A. (1) 117, 426
Keaton, A.R. (7) 275
Kedrova, N.S. (7) 134
Kee, T.P. (3) 14-17; (4) 156, 157; (7) 34
Keglevich, G. (1) 458, 459, 478, 480, 503; (4) 94, 287, 288
Keitel, I. (3) 18; (4) 111, 132, 149, 210
Keller, H. (1) 126; (4) 215
Keller, M. (7) 62
Keller, T.H. (5) 167
Keller, W. (1) 20
Kelley, J.I. (4) 123
Kelley, J.L. (5) 32
Kellner, K. (4) 285
Kelly, D.G. (1) 285
Kemmitt, R.D.W. (1) 134
Kemmler, M. (1) 139, 483
Kemp, R.A. (2) 1
Kendall, N.I. (1) 243
Keniry, M.A. (5) 194
Kennard, C.H.L. (1) 276; (4) 235
Kennard, O. (5) 201
Kennedy, M.A. (5) 221
Kensler, T.W. (6) 141
Kentgens, A.P.M. (7) 209
Kerdesky, F.A.J. (6) 104
Keseru, G.M. (6) 112
Kesseling, E. (5) 126
Kesselring, R. (3) 92
Kessenikh, A.V. (4) 275
Kessler, H. (3) 109; (4) 34
Ketring, A.R. (7) 80-82
Kettenbach, R.T. (1) 39
Khachatryan, R.A. (1) 329
Khalikova, S.F. (4) 148
Khalil, F.Y. (1) 324
Khaliullin, R.R. (4) 295
Khanna, V. (5) 83
Kharchenko, A.V. (1) 132
Kharchenko, V.I. (1) 391
Kharitonov, A.V. (1) 260
Kharstan, M.A. (4) 50
Khasnis, D.V. (2) 3, 33; (4) 16
Khlopushina, G.A. (4) 32
Khmeinitski, L.I. (7) 71
Khodak, A.A. (7) 83, 84
Khodorkovskii, B.O. (1) 300
Khokhlov, P.S. (4) 44, 262, 268-270
Khotinen, A.V. (4) 129
Khullar, M. (7) 142-145
Kibardin, A.M. (2) 6; (3) 136, 137
Kiddle, J.J. (4) 121; (6) 94
Kierzek, R. (5) 94, 161
Kihara, S. (1) 293
Kikuchi, H. (7) 152
Kikugawa, Y. (1) 156
Kilburn, J.D. (3) 27; (4) 115
Kilic, A. (7) 117
Kilic, Z. (7) 117
Kilimann, U. (7) 203
Kilroy, W.P. (7) 304
Kim, D.K. (5) 30
Kim, D.Y. (4) 188
Kim, I.W. (1) 147
Kim, J. (1) 145
Kim, J.C. (4) 56
Kim, J.H. (4) 65
Kim, J.S. (1) 306
Kim, K.D. (1) 164
Kim, K.H. (5) 30
Kim, K.S. (1) 147
Kim, M. (1) 147
Kim, S.H. (4) 56
Kim, Y.W. (5) 30
Kimpton, B.R. (1) 44
Kimura, I. (7) 166
Kimura, T. (7) 99
Kimura, Y. (4) 25
Kinchington, D. (5) 2, 3
King, G.C. (5) 222
Kirby, A.J. (4) 63, 64
Kirchmeier, R.L. (7) 124
Kirchner, K. (1) 320
Kirchoff, R. (1) 355, 392, 398
Kirillov, G.A. (4) 99
Kirkman, R.L. (5) 20
Kirn, A. (5) 12
Kirpekar, F. (3) 94; (5) 123
Kishikawa, K. (1) 107
Kitayama, K. (1) 137
Kitayama, M. (7) 168, 171, 262
Kivekas, R. (1) 304
Klaic, B. (2) 31
Klein, R. (1) 240; (7) 210
Klein, Th. (1) 61
Klimenko, N.M. (7) 5
Klimentova, G.Yu. (4) 267
Klingenberg, E.H. (7) 137
Klintschar, G. (1) 42
Klunder, E.M. (5) 25
Knaus, E.E. (6) 74
Knight, J.G. (1) 269; (6) 62
Knobel', Yu.K. (4) 250
Knoch, F.A. (7) 207, 208
Kobayashi, T. (7) 107, 108, 328, 329
Kobori, M. (7) 152
Kobrina, L.S. (1) 410
Koch, P. (1) 242
Kocienski, P. (6) 106
Koda, G. (6) 45
Kodama, Y.-i. (6) 37
Koeller, K.J. (4) 153
Koenig, M. (1) 84, 383; (4) 191
Koening, M. (7) 113
Koga, Y. (4) 25
Koguchi, K. (1) 266
Koh, Y.J. (4) 11; (6) 84
Kohlpainter, C.W. (1) 109
Koidan, G.N. (1) 412
Koide, T. (1) 206
Koizumi, S. (4) 38; (5) 143
Kojima, M. (7) 240, 283, 284
Kojima, S. (2) 20
Kokpanbaeva, A.O. (4) 5
Kolbina, V.E. (4) 97
Kolesova, V.A. (4) 33
Kolodyazhnyi, O.I. (1) 153, 154, 252; (6) 5
Kolomeitsev, A.A. (3) 26
Kolomiets, A.F. (3) 55
Kolomnikova, G.D. (3) 11; (4) 80
Kolyamshin, O.A. (4) 286
Komaki, A. (7) 308
Komarova, L.G. (7) 149
Komarova, N.I. (5) 107
Komatsu, H. (3) 123
Kondo, H. (3) 145
Kondo, M. (1) 162
Kono, N. (7) 166
Konovalova, I.V. (2) 11; (4) 32
Konovalova, L.Ya. (7) 291
Konovalova, N.P. (7) 30
Konrad, R.E. (5) 56
Koo, Y.J. (3) 29
Kopicka, K. (7) 80, 81
Koprowski, M. (1) 245, 265
Kopylova, L.Yu. (4) 150
Korkin, A.A. (1) 412
Kormachev, V.V. (4) 286
Kornath, A. (1) 218
Koros, W.J. (7) 316
Koroteev, A.M. (3) 31
Koroteev, M.P. (3) 31; (4) 72, 73
Kortus, K. (1) 113
Koser, G.F. (4) 28
Kosikowski, A.P. (4) 21
Kossev, K. (4) 175
Kost, D. (4) 271
Koster, R. (1) 235
Kostitsyn, A.B. (1) 362
Kostka, K. (4) 179, 180
Kostyuk, A.N. (7) 17
Koszalka, G.W. (5) 27
Kotani, S. (3) 105
Kovács, A. (1) 503; (4) 287, 288
Kovacs, I. (1) 49

Kovacs, T. (5) 57
Kowalak, J.A. (5) 234
Koyama, T. (7) 163, 165
Koyanagi, T. (7) 170
Koziara, A. (4) 61
Kozikowski, A.P. (1) 182; (4) 167
Kozioł, A.E. (4) 77
Kozlov, E.S. (1) 132, 222; (7) 17
Kramer, B. (1) 351; (3) 140; (7) 47, 51
Kramer, W. (4) 221; (6) 122; (7) 58
Krasil'nikova, E.A. (1) 305
Kratky, C. (1) 13
Kravehenko, A.N. (7) 89
Krawczyk, E. (3) 23; (4) 308
Krayevsky, A.A. (5) 46, 48-52
Krebs, B. (1) 375
Krebs, F. (1) 372
Kreidler, K. (4) 104
Kreitmeier, P. (1) 379
Kremsky, J. (5) 86
Kreuder, R. (6) 69
Krill, J. (1) 15
Krinitskaya, L.V. (4) 275
Krishnamurthy, S.S. (3) 61; (7) 40, 97, 128, 129, 131
Krishnan, R. (5) 129
Kroker, J. (1) 505
Kroll, L.C. (1) 199
Krolovets, A.A. (4) 99, 251
Kroner, J. (1) 148
Kropp, H.W. (1) 255
Kruchinin, N.P. (7) 289
Krüger, C. (1) 417
Krueger, L. (7) 87, 88
Kruijtzer, J.A.W. (3) 110
Krutikov, V.I. (4) 92, 198
Krylova, T.O. (3) 11; (4) 80
Krynetskaya, N.F. (5) 175
Krzyzanowska, B. (4) 45, 237
Ksander, G.M. (4) 247
Kubono, K. (7) 107, 108, 328, 329
Kubota, S. (7) 148
Kucera, W.R. (7) 236
Kuchen, W. (1) 38
Kudelska, W. (4) 77
Kudzin, Z.H. (4) 205
Kueper, L.W. (4) 68
Kuhnigk, J. (1) 417
Kuijpers, W.H.A. (3) 80, 100
Kuimelis, R.G. (3) 86; (5) 137
Kukhanova, M.K. (5) 46, 48, 50, 51
Kuk'har, V.P. (4) 190
Kukhareva, T.S. (3) 55
Kulichikhin, V.G. (7) 287, 288
Kulkarni, A.R. (7) 302, 303
Kulkarni, V.R. (4) 259; (6) 80
Kumar, D. (7) 142-145
Kumar, P. (3) 78; (5) 71
Kumaravel, S.S. (7) 40
Kumar Das, V.G. (1) 339
Kunath, A. (4) 132
Kundrot, C.E. (5) 217
Kuptsov, S.A. (7) 288-290
Kurahashi, A. (7) 164
Kuramochi, T. (1) 97
Kurasaki, S. (7) 172, 173
Kurata, H. (7) 108
Kurdyukov, A.I. (4) 295
Kurita, J. (1) 3, 472
Kurochkina, G.I. (1) 230, 231; (3) 64-66
Kuroda, A. (7) 180
Kurth, M.J. (6) 30
Kurtikyan, T.S. (4) 282
Kurz, S. (1) 358
Kusumoto, S. (3) 105, 112
Kuszmann, J. (1) 157
Kuwano, R. (1) 98
Kuyl-Yeheskiely, E. (3) 100, 101
Kuznetsov, V.A. (4) 46-48, 230
Kuznikowski, M. (1) 245
Kwiatkowski, W. (7) 94
Kwik, W.L. (7) 7
Kwon, H. (6) 55
Kyrimis, V. (1) 444; (3) 132

Laali, K.K. (1) 194
Labarre, J.F. (7) 93, 111-113, 133
Labarre, M.C. (7) 93, 111
Laber, B. (4) 206
Lachmann, J. (1) 70, 80
Lacolla, P. (5) 10
Lacombe, S. (1) 367, 406
Läge, M. (1) 375
Lagowski, J.B. (7) 8
Lake, C.H. (1) 79
Lakov, L. (7) 102
Lammertsma, K. (1) 440
Lampe, S. (5) 131
Lanahan, M.V. (6) 29
Lance, M. (1) 493
Landert, H. (1) 130
Landini, D. (7) 96
Landry, C.J.T. (7) 310, 315
Landuyt, L. (1) 385, 386
Lanfranchi, M. (2) 34
Lang, H. (1) 91, 236, 446, 448-450
Lange, G. (1) 376
Langer, R. (7) 263, 271-274
Langhans, K.P. (1) 62
Laporte, F. (1) 468, 476
Lappa, S. (3) 108
Larbot, A. (7) 322
Larelina, A.P. (4) 69
Larpent, C. (1) 144
Latham, E.J. (6) 50
Lattanzi, A. (6) 73
Lattman, M. (2) 3, 33; (4) 16
Latypov, S.K. (3) 136
Lau, F.W. (5) 213
Laughlin, G. (5) 205
Laurencin, C.T. (7) 265, 275
Laurent, Y. (7) 246, 331
Lauricella, R. (4) 200
Lauritsen, K. (1) 298, 325, 338; (3) 10
Lavrent'ev, A.N. (4) 92, 198
Law, D.J. (1) 134
Lawrence, N.J. (1) 100, 104; (6) 24
Lay, U. (1) 91, 236
Laynez, J. (7) 14
Le, M. (5) 76
Lebedev, A.V. (5) 107-109
Leblanc, D.A. (5) 208
Leblanc, J.C. (1) 241
Lebleu, B. (5) 120
Le Corre, M. (1) 90, 160
Lee, C.-H. (4) 88
Lee, C.W. (1) 164; (6) 84
Lee, H.J. (5) 14
Lee, J.H. (4) 88
Lee, J.S. (1) 164
Lee, J.W. (4) 188
Lee, K. (4) 174, 242, 277; (5) 19; (6) 72, 158
Lee, K.R. (1) 147
Lee, P.H. (1) 306
Lee, S.G. (1) 321
Lee, S.M. (1) 164
Lee, Y.H. (4) 88; (7) 278
Lefebvre, I. (5) 11, 12, 120
Le Floch, P. (1) 498, 500-502
Le Goffic, F. (3) 103; (4) 10
Legon, A.C. (1) 341
Leibfried, R.T. (7) 156
Leichtweis, I. (7) 74, 86, 205
Leise, M. (1) 236, 446, 448-450
Le Lagadec, R. (1) 50
Lemmouchi, Y. (7) 11, 281
Le Moigne, F. (4) 200
Leng, M. (5) 189
Lennhoff, D. (1) 218
Lennon, I.C. (6) 153
Leonard, G.A. (5) 203
Leont'eva, I.V. (1) 309
Lequan, M. (1) 344
Lequan, R.M. (1) 344
Lermontov, S.A. (4) 13
le Roux, C. (4) 82
Lerpiniere, J. (6) 106

Lesnik, E.A. (3) 83; (5) 144
Lesnikowski, Z.J. (5) 96, 118, 119
Letsinger, R.L. (5) 174
Leumann, C. (3) 90, 91; (5) 125, 129
Leung, P.-H. (1) 37
Le Van, D. (1) 375, 376, 410, 413
Levason, W. (1) 52
Levin, B.V. (7) 134
Levina, L.V. (1) 133, 249; (4) 6, 8
Lewis, A.F. (5) 132
Lewis, L. (3) 77
Ley, S.V. (6) 153
Lhassani, M. (4) 208
Li, B. (5) 186
Li, C. (4) 108, 116, 283, 284; (6) 85
Li, C.-J. (1) 143; (6) 11
Li, D. (1) 85
Li, G. (2) 26
Li, H.Y. (4) 178
Li, J.-L. (5) 63
Li, S. (4) 62, 193, 209
Li, W.S. (7) 109
Li, X. (5) 36, 43
Li, Y. (5) 161
Li, Y.-F. (4) 274
Li, Y.T. (5) 231
Li, Z. (3) 141; (4) 79, 165, 224
Liang, M. (7) 223, 231, 232
Li Dan, (4) 165
Liebeskind, L.S. (4) 252
Lièvre, C. (6) 93
Lightfoot, A.P. (6) 155
Lightwahl, K.J. (5) 232
Lilley, D.M. (5) 209
Lim, A.C. (5) 198
Limbach, H.-H. (6) 19
Limbach, P. (5) 157
Lin, G.Y. (4) 224
Lin, J. (4) 35
Lin, K.Y. (5) 110
Lin, P.K.T. (5) 187
Lin, S. (1) 207
Lin, Z. (7) 248, 249
Lindenberg, M.S. (7) 23, 25, 237, 240, 241
Lindes, D.S. (5) 218
Lindner, E. (1) 139, 483, 484
Ling, L. (4) 25
Linitas, K.E. (6) 34
Link, M. (1) 425, 427; (3) 128; (7) 48, 49
Linn, J.A. (4) 123; (5) 32
Lippmann, T. (4) 17
Lipton, M.A. (6) 111
Liquier, J. (5) 227
Lis, T. (1) 289; (4) 77
Liskamp, R.M.J. (3) 109, 110; (4) 34, 207
Litvinov, I.V. (2) 30
Liu, C. (4) 20, 26
Liu, F. (7) 249
Liu, F.Q. (7) 191
Liu, L. (2) 26
Liu, L.Z. (2) 13
Liu, Y.H. (5) 229
Liu, Z. (1) 263; (4) 85
Livantsov, M.V. (3) 21
Livinghouse, T. (3) 40
Llamas-Saiz, A.L. (1) 171; (6) 19, 127; (7) 13-15, 21, 327
Lloyd, D. (5) 70
Loakes, D. (3) 89
Locke, J. (1) 168
Loeckritz, A. (4) 297
Lönnberg, H. (5) 58, 165, 166
Lönnecke, P. (1) 244
Lohray, B.B. (6) 27
Loi, A. (5) 10
Lokhov, S.G. (5) 109
Loose, G. (5) 92
Lopez, C. (6) 19; (7) 13, 15
Lopez, L. (1) 488, 489; (3) 60, 130
Lopez, M. (1) 215
López-Lázaro, A. (6) 118; (7) 57
Lopez-Ortiz, F. (6) 18; (7) 56
Lora, S. (7) 277
Lorenzo, A. (6) 118, 120; (7) 57
Lorthioir, O. (3) 114
Losier, P. (1) 444; (3) 132
Lough, A.J. (7) 79
Lown, J.W. (3) 75; (5) 117
Lu, R. (4) 53
Lu, S. (1) 12
Lu, Y. (3) 103
Lubman, D.M. (5) 229
Luck, I.J. (1) 88
Ludwig, J. (5) 39, 54
Luecke, J. (7) 245
Lugan, N. (1) 33
Lukashev, N.V. (1) 117, 426
Lukes, I. (1) 457; (4) 298
Lukyanenko, S.N. (3) 33
Luk'yanov, N.V. (4) 52
Luk'yanov, O.A. (4) 83, 303
Luna, A. (7) 6
Lunn, R.C. (6) 71
Luo, W.D. (3) 75; (5) 117
Luu, B. (3) 104; (5) 13
Luzikov, Yu.N. (1) 117
Lynch, D.E. (1) 276
Lysenko, S.A. (4) 50

Ma, F.P. (4) 79
Ma, Y. (4) 193
Maah, M.J. (1) 415, 418
Macan, S.M.E. (1) 14
McAtee, R.E. (7) 323
McAuleyhecht, K.E. (5) 203
McAuliffe, C.A. (1) 150, 285; (2) 9, 10
Macaya, R.F. (5) 214
McCaffrey, A.P. (5) 89
McCaffrey, R.R. (7) 120
McCallum, J.S. (4) 252
McCampbell, E.S. (5) 89
McCarthy, J.R. (1) 177; (4) 144
McCarthy, N. (6) 148
McCloskey, J.A. (5) 157, 234
McCollum, C. (3) 98
Macdonald, C. (1) 444; (3) 132
Macdonald, J.E. (4) 239
McEldoon, W.L. (4) 174
McElroy, A.B. (1) 270
McElroy, E.B. (5) 116
McFarlane, H.C.E. (1) 239
McFarlane, W. (1) 44, 89, 239
McGuigan, C. (5) 2-4, 6-8
McGuigan, P. (6) 26
McHugh, K.M. (7) 323
McIntosh, C.A. (1) 330
McKay, D.B. (5) 218
McKervey, M.A. (4) 185; (6) 26, 148
Mackie, A.G. (2) 10
McKie, J.H. (5) 196
McKinstry, L. (3) 40
McLaughlin, L. (5) 148, 149
McLean, E.W. (4) 123; (5) 32
McLean, M.J. (3) 73; (5) 77
McLoughlin, M. (1) 250
McLuckey, S.A. (5) 233
McNally, L. (7) 92
McNaughton, D. (1) 405
McNeil, J.D. (2) 3, 33; (4) 16
McPartlin, M. (3) 9; (7) 109
McPhail, A.T. (1) 71, 74, 77
Maeda, H. (1) 206
Maeda, I. (7) 148
Maeda, M. (7) 100
Maeda, R. (7) 100
Märkl, G. (1) 379; (6) 159
Maeshima, T. (3) 13
Maffei, M. (4) 119
Maffre, D. (4) 212; (6) 86
Magill, J.H. (7) 283, 284
Magomedova, N.S. (4) 58
Magull, J. (7) 77
Mahajna, M. (4) 278
Mahmood, N. (5) 4, 5, 8
Mahon, M.F. (4) 234
Mahrwald, D. (1) 506

Maia, A. (7) 96
Maier, L. (4) 214, 217
Maier, W. (7) 62
Majewski, P. (1) 155
Majima, T. (1) 262
Majoral, J.-P. (1) 125, 445; (4) 59, 236; (6) 132; (7) 38, 67-69, 136, 235
Maki, T. (1) 206
Makleit, S. (1) 179
Makomo, H. (6) 88
Malamidou-Xenikaki, E. (6) 34
Malavaud, C. (1) 488; (3) 130
Malenko, D.M. (3) 33; (4) 176
Malenkovskaya, M.A. (4) 36
Malik, K.M.A. (1) 82
Malkov, V.A. (5) 225
Malkova, G.Sh. (4) 279
Mal'tseva, N.N. (7) 134
Malvy, C. (5) 120
Maly, K. (1) 290
Malysheva, S.F. (1) 63, 248
Manassero, M. (1) 4
Manenc, A. (7) 64
Mang, J.Y. (4) 188
Mangeney, P. (3) 118
Manginot, E. (7) 70
Manhart, S. (1) 59, 127
Mann, G. (4) 17
Mann, J. (6) 107
Manners, I. (7) 8, 79, 197, 217, 223, 230-232, 316
Mano, S. (3) 120
Mansuri, M.M. (5) 22
Manzano, B.R. (2) 34
Manzo, P.G. (1) 26
Mao, B. (5) 186
Marchand, R. (7) 246, 331
Marchenko, A.P. (1) 412
Marchetti, M. (1) 54
Marinetti, A. (1) 40, 92, 401
Mark, J.E. (7) 292, 293
Markovskii, L.N. (1) 431; (4) 124, 307
Marra, A. (6) 91, 92
Martens, R. (1) 211
Martichonok, V. (6) 98
Martin, J.C. (4) 246; (5) 22
Martin, M.J. (5) 63
Martin, S.F. (3) 107; (4) 37
Martin, S.J. (1) 270
Martynov, B.I. (4) 99
Martynov, I.V. (4) 13, 251
Maruoka, S. (7) 294
Maruyama, A. (7) 163, 165
Marzouk, H. (1) 58
Masauji, H. (1) 196
Mascareñas, J.L. (6) 142
Masci, G. (7) 324
Maskos, K. (5) 206, 208
Maskva, V.V. (4) 295
Masojidkova, M. (5) 24
Mason, S. (1) 424; (3) 129; (7) 50
Massa, W. (1) 505
Massey, J. (7) 232
Masson, S. (4) 126; (6) 88
Mastryukova, T.A. (1) 309; (4) 33, 222
Masuko, T. (7) 284-286
Matassa, V.G. (1) 176
Mathey, F. (1) 40, 401, 438, 439, 462-465, 468, 476, 477, 494, 495, 498, 500-502
Mathieu, R. (1) 33; (4) 236; (6) 132
Mathieu-Pelta, I. (2) 12
Matos, R.M. (1) 395, 490
Matray, T.J. (5) 138
Matrosov, E.I. (1) 278
Matsuda, A. (5) 140
Matsuda, M. (7) 170
Matsuda, Y. (1) 504
Matsui, M. (1) 293
Matsumoto, H. (4) 182
Matsumoto, Y. (1) 12, 96
Matsuo, M. (1) 107
Matsuyama, H. (1) 312
Matt, D. (1) 5
Matteucci, M.D. (5) 110, 113
Matthews, R.W. (3) 9
Matulic-Adamic, J. (5) 34
Matuschek, B. (4) 186
Matuszowski, B. (4) 68
Matveev, V.S. (7) 229
Matyjaszewski, K. (7) 22, 23, 25, 222, 225, 237, 239-244, 283
Mauthner, K. (1) 320
Mayama, S. (7) 163, 165
Mayer, F. (1) 411
Mayer, H.A. (1) 139, 483
Mazières, M.R. (1) 429; (2) 23
Mazurkiewicz, R. (6) 13; (7) 55
M'Bida, A. (6) 48, 49
Medici, A. (7) 140
Medvedeva, L.Ya. (7) 114
Meervelt, L.V. (6) 15
Meetsma, A. (7) 135, 150, 189
Meeuwenoord, N.J. (3) 101
Mehdi, S. (4) 144
Mei, A. (4) 130, 131; (6) 87
Meidine, M.F. (1) 395
Meignan, G. (1) 144
Melino, G. (3) 108
Men, A.J. (4) 201
Mendel, Z. (6) 101
Menedez-Velazquez, J. (6) 149
Menendez, M. (7) 14
Meng, B. (5) 112
Menge, L.R. (5) 44
Menneret, C. (4) 54
Mercier, B. (4) 238
Mercier, F. (1) 468, 476, 477, 495
Mercier, H.P.A. (2) 8
Mereiter, K. (1) 320
Merenkova, I.N. (5) 175
Merifield, E. (1) 8; (3) 124; (4) 234
Merino, P. (6) 91
Mersmann, K. (5) 150, 151, 153
Messere, A. (5) 73
Messerle, B.A. (1) 87
Messini, L. (5) 10
Metanomski, W.V. (7) 225
Metz, B. (1) 506
Meunier-Piret, J. (1) 291
Meyer, H.U. (1) 352
Meyer, S.D. (6) 117
Meyer, T.G. (1) 217; (3) 57
Michalik, M. (1) 105, 112
Michalska, M. (4) 77
Michalski, J. (1) 219; (3) 23, 67, 68; (4) 12, 308
Michaud, D.P. (3) 102
Michnicka, M.J. (5) 222
Mierke, D.F. (3) 109; (4) 34
Migaud, M.E. (4) 166; (6) 152
Mikhalin, N.V. (4) 281
Mikityuk, A.D. (4) 268, 269
Mikołajczyk, M. (1) 274, 275, 343, 430; (4) 261, 307
Mikulcik, P. (1) 118
Miles, H.T. (5) 207
Miller, N.E. (1) 123, 124
Miller, P.C. (6) 65
Miller, P.S. (5) 178, 179
Miller, R.B. (6) 30
Milligan, J.F. (5) 110
Mills, S.D. (6) 151
Mills, S.J. (4) 22, 24
Milton, J. (5) 173
Minami, T. (6) 81
Minassyan, S.K. (5) 50
Minck, K.-O. (4) 240
Minkwitz, R. (1) 218
Minto, F. (7) 253-255
Miranda, R. (1) 284
Miroevskii, G.P. (7) 89
Mironov, V.F. (2) 11; (4) 32
Miroshnichenko, V.V. (1) 221
Misharev, A.D. (1) 256
Mishchenko, N.I. (1) 319; (6) 15
Misiak, H. (1) 353
Misra, K. (5) 83
Misra, R. (3) 85

Mitchell, C.R. (1) 247
Mitchell, T.N. (1) 86
Mitjaville, J. (4) 236; (6) 132
Mitrasov, Yu.N. (4) 286
Mitsiura, K. (3) 69
Mitsui, K. (1) 156
Miyahara, I. (1) 347, 400
Miyamoto, Y. (1) 266
Miyano, S. (1) 258
Miyata, S. (7) 121
Miyata, T. (7) 284-286
Miyazawa, M. (3) 123
Miyoshi, K. (1) 447
Mizsak, S.A. (4) 177
Mizuguchi, A. (7) 180
Mizuno, H. (1) 266
Mkrtchyan, G.A. (1) 329
Mo, O. (7) 6
Modak, A.S. (4) 142
Modranka, R. (4) 179, 180
Modro, A.M. (4) 276
Modro, T.A. (1) 334-337; (4) 51, 82, 161-164, 260, 276
Mohan, T. (7) 189
Moini, M. (7) 101
Moise, C. (1) 241
Moiseev, V.I. (7) 114
Mok, K.F. (1) 37
Mokry, L.M. (1) 202; (6) 8
Molaire, T.R. (7) 315
Molchanova, G.N. (4) 69
Molina, P. (1) 171, 172; (6) 19, 118-121, 123, 124, 126, 127, 129, 130; (7) 13, 14, 21, 57, 59
Moll, M. (7) 207, 208
Molloy, K.C. (4) 234
Momose, S. (3) 123
Monakov, Yu.B. (1) 220
Monforte, P. (1) 1; (2) 35
Mongustova, G.V. (1) 307
Monia, B.P. (3) 83
Monica, B.P. (5) 144
Monje, M.-C. (2) 22, 23
Montague, R.A. (7) 22, 239
Montchamp, J.-L. (4) 166; (6) 152
Monteagudo, E. (1) 186
Montero, J. (5) 18
Montesarchio, D. (5) 73
Montgomery, C.D. (2) 32
Montgomery, J.A. (5) 20, 21
Montserrat, F.X. (5) 72
Moody, P.C.E. (5) 205
Moon, S.H. (4) 56
Moor, M. (5) 44
Moore, A.J. (6) 145
Moore, M.H. (5) 205
Moore, M.K. (7) 23, 240-242
Morden, K.M. (5) 206, 208
Moree, W.J. (4) 207
Moreira, M.J. (6) 125; (7) 32
Moreno-Manas, M. (1) 158
Mori, I. (4) 38
Mori, S. (7) 151, 162, 262
Mori, T. (5) 78
Morii, H. (7) 100
Morimoto, K. (7) 153, 154
Morimoto, T. (1) 108, 111
Morin-Fox, M.L. (6) 111
Morise, X. (1) 367
Morishita, T. (1) 266
Moriya, K. (7) 121, 297
Morosi, G. (7) 96
Morozik, Yu.I. (4) 250
Morr, M. (5) 37
Morris, C.D. (7) 275
Morrissey, C.T. (7) 198
Mortreux, A. (1) 229
Morvan, F. (5) 120
Moskva, V.V. (4) 150, 184, 267
Moss, R.A. (4) 66
Motoki, S. (1) 482; (3) 131
Mourad, O. (7) 223
Moureau, L. (2) 24, 25
Mouriño, A. (6) 142
Mozol, V. (7) 19
Mozzherin, D.J. (5) 46, 48
Mrksich, M. (5) 192
Mueller, C. (1) 492; (3) 57
Müller, G. (1) 70, 80, 118, 120, 235
Münchenberg, J. (3) 34
Muhammad, F. (1) 100, 104; (6) 24
Muir, A.S. (1) 44; (2) 10
Mukmeneva, N.A. (1) 128
Mullah, K.B. (3) 144
Muller, C. (1) 129
Muller, E.L. (4) 161
Mundt, C. (3) 46-49
Mundt, O. (1) 27
Murahkami, K. (7) 228
Muralidhara, M.G. (7) 125
Murashov, D. (7) 114
Murata, T. (4) 41
Murchie, A.I.H. (5) 205, 209
Murcia, F. (6) 124
Murphy, C.J. (5) 200
Murphy, P.J. (6) 134
Murphy, R.C. (6) 105
Murphy, S.M. (6) 50
Murray, J.B. (5) 155, 156
Murray, M. (7) 115
Murthy, N.N. (6) 141
Murugavel, R. (3) 61; (7) 40, 97
Musin, R.Z. (3) 136, 137
Mustafin, A.Kh. (4) 110
Muth, A. (1) 35, 60, 61
Mutti, S. (3) 118
Muzzin, P. (3) 87
Myazuki, T. (7) 309

Nader, B.S. (7) 160, 161
Nadimi, K. (5) 76
Naesens, L. (5) 26
Nagahara, S. (7) 163, 165
Nagar, P.N. (4) 91
Nagareda, K. (6) 20
Nagase, S. (6) 20
Nagashima, T. (7) 176
Nahorski, S.R. (4) 21, 24
Naiini, A.A. (1) 76
Nair, H.K. (4) 142
Najera, C. (6) 149
Nakacho, Y. (7) 300, 305, 306
Nakada, Y. (7) 294
Nakagawa, H. (7) 175
Nakai, K. (7) 308
Nakai, T. (6) 49
Nakamoto, C. (3) 146
Nakamura, H. (7) 285
Nakamura, M. (4) 299
Nakamura, S. (1) 93
Nakamura, Y. (5) 164; (6) 76
Nakanaga, T. (7) 159, 308
Nakanishi, K. (6) 97
Nakano, H. (5) 162
Nakao, R. (1) 268
Nakayama, J. (1) 115, 116
Nakazawa, H. (1) 447
Namane, A. (5) 15
Nambiar, K.P. (3) 86, 106; (5) 137
Namiki, M. (7) 162
Nangia, A. (6) 140
Nara, H. (5) 140
Naumann, W. (5) 92
Naumov, V.A. (2) 30
Naylor, A. (1) 269; (6) 62
Nazareno, M.A. (1) 25
Neal, B.E. (6) 29
Neda, I. (1) 228; (2) 29; (3) 51-54, 57-59; (7) 29
Nedderman, A.N.R. (5) 187
Neef, G. (1) 188
Nees, H.-J. (1) 201
Nefedov, O.M. (1) 362
Neff, K.-H. (4) 206
Neibecker, D. (1) 466, 467
Neidle, S. (5) 204
Neidlein, R. (4) 186, 215, 218, 219, 221; (6) 122; (7) 58
Neilson, R.H. (7) 33, 236
Nelson, C.J. (7) 127, 252, 299, 316
Nelson, J.H. (1) 481, 497

Nerdal, W. (5) 195
Ness, H.J. (1) 126
Nesterova, L.I. (3) 33
Nethaji, M. (3) 61; (7) 40, 97, 128, 129
Neumann, B. (1) 170, 353, 354, 393, 422
Neumüller, B. (1) 30, 73, 486, 507, 508; (7) 77
Newkome, G.R. (1) 195
Newton, C.R. (3) 73; (5) 77
Ng, K.E. (5) 40
Ng, M.M.P. (5) 152
Ng, S.W. (1) 339
Ngo, D.C. (7) 72
Nguyen, M.T. (1) 192, 385, 386, 388
Ni, Y. (7) 223, 232
Nicholls, D. (5) 5
Nicholls, S.R. (5) 2, 3
Nichols, R. (3) 88
Nickson, C. (5) 2, 3
Nicolaides, D.N. (6) 32, 34
Niecke, E. (1) 351, 364, 370, 382, 395, 425, 427, 432, 455, 456, 470, 485; (3) 37, 128, 133, 140; (6) 60; (7) 43, 45, 47-49, 51-53, 196
Niedernhofer, L.J. (5) 185
Niederweis, M. (5) 53
Niediek, K. (1) 73
Nief, F. (1) 469; (4) 117; (6) 82
Nieger, M. (1) 175, 351, 364, 370, 382, 425, 427, 432, 455, 456, 470, 485; (3) 37, 128, 133, 140; (6) 60; (7) 43, 45, 47-49, 51, 52, 196
Nielsen, P. (3) 94
Nierlich, M. (1) 493
Nifant'ev, E.E. (1) 230, 231, 233; (3) 31, 55, 56, 62-66; (4) 36, 50, 58, 72, 73
Nikiforov, T.T. (5) 173
Nikitidis, A. (1) 19
Nikitin, E.V. (4) 7
Nikitin, P.A. (7) 192
Nishibe, Y. (7) 297
Nishide, H. (7) 319
Nishide, R. (7) 176
Nishihara, S. (7) 319
Nishimura, Y. (1) 197
Nishiyama, M. (7) 152
Nissan, R. (1) 69
Nitta, M. (7) 54, 60, 61
Nixon, J.F. (1) 395, 414-416, 418-420, 490, 491; (7) 52
Nizamov, I.S. (4) 46-49, 229, 230
No, B.I. (1) 220
Nöth, H. (1) 65-67, 148, 302, 436; (3) 125; (6) 7, 58
Nógrádi, M. (6) 112
Noltemeyer, M. (7) 85, 86, 191, 203, 205
Nomura, N. (4) 70
Nonomura, T. (1) 107
Nordhoff, E. (4) 206
Norman, A.D. (3) 38, 39
Norman, D.G. (5) 205
Norman, M.E. (7) 265
Norman, N.C. (1) 345
Norton, S.J. (4) 249
Novak, I. (7) 7
Novikova, V.G. (4) 39, 40, 305
Novikova, Z.S. (4) 122
Novoselova, T.R. (1) 226
Nowalinska, M. (4) 61
Nowotny, M. (1) 505, 506
Noyori, R. (5) 100
Nozaki, K. (1) 95; (3) 120, 121
Nozdryn, T. (1) 296; (6) 144
Nugent, R.A. (4) 177
Nugent, W.A. (1) 31
Nunoo, T. (7) 154
Nunota, K. (5) 162
Nuretdinov, I.A. (7) 30
Nuretdinova, O.N. (4) 39, 40, 305
Nusshair, D. (7) 75
Nuyken, O. (7) 197, 230, 243, 244

Oae, S. (1) 196
Oakley, R.T. (7) 182
Oates, K.V. (6) 150
Obnuchinikova, N.V. (7) 71
Obon, R. (7) 13
O'Carroll, F. (4) 64
Ochoa de Retana, A.M. (4) 291
O'Connor, T.J. (5) 2, 3
Odinets, I.L. (4) 222
Officer, D.L. (6) 156
Ogawa, S. (2) 15; (4) 41
Ogino, T. (7) 309
Ogiwara, A. (1) 94
Ognyanov, V.I. (4) 21
Ogura, K. (1) 97, 293
Oh, D.Y. (3) 29; (4) 11, 188; (6) 84
Oh, J.S. (1) 164
Ohki, K. (7) 100
Ohmori, H. (1) 206
Ohms, G. (1) 238; (4) 227
Ohmura, H. (3) 123
Ohno, A. (1) 12, 197, 264; (3) 13
Ohta, H. (1) 3, 312, 313; (6) 45
Ohta, K. (1) 268; (6) 37
Ohta, T. (1) 99
Ohtsuka, E. (5) 143
Ojea, V. (4) 118
Okabe, M. (6) 143
Okada, F. (7) 24, 238
Okada, K. (1) 378; (4) 232
Okada, Y. (6) 81
Okamoto, R. (1) 251; (4) 169
Okamoto, T. (5) 140
Okamoto, Y. (4) 93, 299
Okazaki, R. (1) 246; (2) 21; (6) 23
Okuda, M. (1) 181
Okuma, K. (1) 312, 313; (6) 45
Oldroyd, R.D. (1) 52
Oleksyszyn, J. (4) 203
Oliver, P.A. (1) 207
Ollapally, A.P. (5) 14
Ollig, J. (1) 255
Olms, P. (7) 191
Oloughlin, S. (5) 158
Olszak, T.A. (4) 180
Omara, S. (5) 191
Omelanczuk, J. (1) 267, 430; (4) 306, 307
Ono, A. (5) 140
Ono, M. (1) 313
Onys'ko, P.P. (4) 127
Ookaito, K. (7) 157, 158
Oparin, D.A. (1) 300
Opromolla, G. (7) 190
Orduna, J. (1) 296; (6) 144
Orelli, L.R. (6) 73
Oretskaya, T.S. (5) 175
Orgel, L.E. (5) 40
Orpen, A.G. (1) 2
Ortiz, B. (5) 60
Orzhekovskaya, E.I. (3) 56
Osawa, T. (4) 30, 67
Osborn, H.M.I. (4) 301
Osbourne, S.E. (5) 177
Oshikawa, T. (1) 266; (4) 172
Osipov, V.N. (4) 262
Osman, F.H. (4) 228; (6) 33
Osowska-Pacewicka, K. (4) 84
Otani, K. (1) 369
Otero, A. (2) 34
Otoguro, A. (1) 453
Otvos, L. (5) 57
Oussaid, B. (4) 59, 191
Ovchinnikov, V.V. (4) 280
Owen, E.S.E. (6) 153
Owton, W.M. (6) 155
Ozaki, S. (3) 146; (4) 25
Ozarowski, A. (1) 441
Ozawa, F. (1) 96
Ozegowski, S. (3) 18; (4) 111, 149, 210
Ozerov, A.A. (4) 103
Ozon, V. (1) 311

Pabel, M. (1) 16
Pack, H.D. (4) 56
Paetzold, P. (1) 411
Paine, R.T. (1) 65-67
Palacios, F. (4) 291; (6) 18, 38; (7) 56
Palacios, S.M. (1) 25, 26
Paladino, J. (4) 125
Palant, A.A. (7) 279
Palma, G. (7) 277
Palom, Y. (5) 88
Pan, H. (1) 292
Pan, W. (6) 113
Panchishin, S.Ya. (1) 314
Pandey, S.K. (7) 85, 200, 201, 204, 206
Pandolfo, L. (6) 43
Pang, J. (1) 166
Pannecoucke, X. (3) 104; (5) 13
Pannell, K.H. (4) 128
Panosyan, G.A. (4) 282
Papadopoulos, C. (7) 115
Papageorgiou, G.K. (6) 32
Papchikhin, A.V. (5) 50
Papel, M. (1) 41
Paquette, L.A. (6) 110
Parg, R.P. (3) 27; (4) 115
Park, C.E. (7) 181
Parker, S. (5) 63
Parker, W.B. (5) 20, 21
Parkinson, J.A. (5) 196
Parks, T.M. (7) 10
Parmentier, G. (5) 13
Parr, G.R. (5) 230
Parratt, M.J. (4) 243
Parry, C.J. (1) 48
Parry, J.S. (1) 82
Parvez, M. (7) 39, 184-187, 197, 330
Pasquariello, D.M. (7) 304
Patel, G. (4) 192, 211
Patel, P.G. (1) 44
Paterson, I. (6) 115
Pathirana, R.N. (5) 6-8
Patick, A. (4) 246
Patil, R. (4) 74, 75
Patois, C. (4) 95, 112, 117, 138, 139; (6) 82, 83
Patrick, T.B. (6) 29
Patsanovskii, I.I. (1) 431
Pattenden, G. (6) 150, 151
Paver, M.A. (1) 81
Pavlov, V.A. (4) 295
Payne, A. (6) 67
Payne, L.G. (7) 272-274
Peachey, B.J. (4) 14
Pebler, J. (7) 75, 76
Pedersen, E.B. (5) 114
Pedrini, P. (7) 140
Pedroso, E. (3) 116; (5) 72, 88
Pei, C. (7) 31
Peiffer, G. (4) 128
Peinador, C. (6) 125; (7) 32
Peláez-Arango, E. (6) 18; (7) 56
Pellinghelli, M.A. (2) 34
Pellon, P. (1) 90, 160
Penco, S. (4) 256
Pendlebury, D. (5) 66
Pen'kovskii, V.V. (1) 390, 391
Pennington, W.T. (1) 75, 78
Perekalin, V.V. (4) 189
Perera, S.D. (1) 11
Peresypkina, L.P. (4) 98
Pereyre, M. (6) 55
Perez, J. (5) 86
Pérez-Sestelo, J. (6) 142
Périe, J. (4) 29
Perigaud, C. (5) 11
Perlikowska, W. (4) 261, 307
Perrand, A. (7) 70
Perriott, L.M. (1) 193, 475
Perrotta, E. (6) 147
Peshev, Q. (7) 102
Pessan, L. (7) 316
Petach, H.H. (1) 204
Peters, E. (7) 16
Peters, K. (7) 16
Peters, R. (4) 104
Peterson, E.S. (7) 317
Peterson, M.J. (1) 207; (6) 22
Petit, F. (1) 229
Petrik, J. (5) 5
Petrosyan, V.S. (3) 21
Petrov, A.A. (4) 96; (7) 194
Petrova, J. (4) 183
Petrova, T.V. (4) 3, 8
Petrova, V.A. (7) 279
Petrovskii, P.V. (1) 309; (4) 33, 69, 222, 251
Petruneva, R.M. (1) 220
Petrus, C. (4) 208, 216
Petrus, F. (4) 208
Petter, R.C. (6) 95
Pezzin, G. (7) 277
Pfeifer, K.-H. (1) 413
Pfister-Guillouzo, G. (1) 367, 383, 406
Pfitzner, A. (1) 64
Pfleiderer, W. (5) 87
Philippo, C.M.G. (6) 110
Pianka, M. (4) 202
Piccialli, G. (5) 73
Piccirilli, J.A. (5) 105
Pickersgill, I.F. (3) 16
Pieken, W. (5) 152
Pieles, U. (5) 158, 216
Pieper, U. (1) 350; (3) 6, 139
Pierlikowska, W. (1) 430
Pierre-Jacques, H. (7) 275
Pietrusiewicz, K.M. (1) 125, 245, 265, 283, 445; (4) 23
Pietzonka, T. (7) 63
Pike, A.J. (3) 12
Pikkemaat, J.A. (5) 215
Pinchuk, A.M. (1) 222; (4) 52
Pinkerton, A.A. (1) 205, 240; (7) 20, 210
Pipko, S.E. (3) 135
Pirozhenko, V.V. (4) 124
Pisarnitskii, D.A. (3) 21
Pitsch, S. (5) 130
Piulats, J. (3) 115
Plate, N.A. (7) 318
Ple, G. (6) 102
Pleixats, R. (1) 158
Plemenkov, V.V. (4) 295
Plenat, F. (1) 311; (7) 109
Pleva, J.M. (7) 123
Pley, H.W. (5) 218
Ploger, S.A. (7) 323
Plotnikova, O.M. (4) 58
Pluskowski, J. (4) 77
Podanyi, B. (1) 157
Podda, G. (7) 96
Podlahova, J. (1) 290
Pohl, S. (1) 225
Pohlenz, H.-D. (4) 206
Poindexter, G.S. (4) 239
Pol, H. (7) 135
Polborn, K. (1) 148; (4) 304
Polimbetova, G.S. (4) 5, 8
Polishchu, A.P. (7) 17
Polozov, A.M. (4) 110, 129
Polubentsev, A.V. (1) 248
Polushin, N.N. (3) 82
Polyakova, I.A. (4) 275
Pombo-Villar, E. (4) 118
Pomerantz, S.C. (5) 234
Pommier, A. (6) 106
Pompon, A. (5) 12, 120
Pon, R.T. (5) 159
Pons, J.-M. (6) 106
Popinski, J. (1) 7
Popov, A.F. (7) 116
Popov, A.V. (4) 13
Porter, K. (5) 33
Porumb, H. (5) 120
Posner, G.H. (6) 141
Postula, J.F. (4) 76
Potekhin, K.A. (7) 194
Potin, P. (7) 38, 64, 67-69, 233-235
Potter, B.V.L. (4) 20, 22, 24, 26
Povolotskii, M.I. (7) 53

Powell, W.S. (6) 103
Power, M.B. (1) 294
Power, P.P. (1) 443
Powers, J.C. (4) 203
Powis, G. (4) 167
Poyser, G.L. (7) 98
Pradella, F. (7) 253-255
Prakasha, T.K. (2) 4, 5, 18, 19
Prashed, M. (4) 241
Prasuna, G. (6) 140
Precigoux, G. (5) 202
Predvoditelev, D.A. (4) 36
Prescher, D. (4) 135
Preut, H. (1) 218
Prigozhina, M.P. (7) 149
Prikhod'ko, Y.V. (1) 251; (4) 169
Pringle, P.G. (1) 2
Prishchenko, A.A. (3) 21
Pritchard, C. (5) 153
Pritchard, R.G. (1) 150; (2) 9, 10
Pritzkow, H. (1) 68, 381, 435; (6) 40, 44
Pryce, M.A. (3) 9
Pucher, S.R. (7) 264-267
Pudovik, A.N. (2) 6, 30; (3) 136, 137; (4) 46-49, 229, 280
Pudovik, M.A. (2) 30
Puech, F. (5) 12
Pugashova, N.M. (4) 73
Pujol, B. (4) 133
Purnanend, (4) 225
Puschner, B. (1) 22
Pushin, A.N. (1) 261; (4) 13
Puyenbroek, R. (7) 150

Qamar, R. (4) 249
Qi, M. (4) 145, 147
Qiu, W. (1) 299; (4) 136
Qu, J. (4) 258
Quin, G.S. (1) 458, 459
Quin, L.D. (1) 437, 457-459; (3) 127; (4) 298
Quinteda, J.M. (6) 125; (7) 32

Racha, S. (4) 165
Rachon, J. (4) 158
Raddatz, P. (4) 240
Radhakrishnana, P.I. (5) 64, 65
Radseck, J. (1) 382
Raevskii, O.A. (4) 52
Rafferty, M.D. (4) 185
Rahn, J.A. (1) 497
Raithby, P.R. (1) 81
Rakhmatulina, T.N. (1) 63
Rakhov, I.A. (4) 33
Rakitin, O.A. (7) 71
Ramage, R. (5) 67
Ramasamy, K.S. (3) 83; (5) 144
Ramirez, C.A. (7) 270
Ramm, M. (4) 15; (6) 28
Ramondenc, Y. (6) 102
Ramos, L.A. (5) 185
Randall, L.A.A. (6) 30
Randall, R.E. (5) 168
Randina, L.V. (3) 33
Ranger, G.C. (1) 285
Rao, A.R.R. (1) 183
Rao, G.K. (1) 198
Rao, M.V. (5) 84
Rao, P.B. (6) 140
Rao, T.S. (5) 132, 135
Raper, E.S. (1) 89
Rapisardi, R. (4) 195
Rapp, W. (5) 70
Rastogi, V.K. (5) 207
Rath, N.P. (4) 152
Ratovskii, G.V. (1) 340
Rauchfuss, T.B. (1) 169
Raushel, F.M. (4) 76
Raut, S.V. (6) 52
Ravikumar, V.T. (3) 25, 71; (5) 102
Raymond, B. (7) 112
Raymond, J.J. (7) 112
Rayner, B. (5) 120
Raza, S.K. (3) 85
Razumova, N.G. (7) 116
Réau, R. (3) 142; (6) 16; (7) 195
Reck, G. (1) 338
Reddy, M.P. (3) 99
Reddy, V.S. (3) 35, 36
Reed, R. (1) 380; (7) 19
Reese, C.B. (5) 84, 93
Reetz, M.T. (1) 136
Regan, A.C. (3) 8; (4) 102
Regitz, M. (1) 126, 194, 201, 360-362, 371-373, 401, 407-409; (7) 46
Rehberger, A. (6) 159
Reibenspies, J.H. (1) 146
Reid, B.R. (5) 221
Rein, T. (6) 68, 69
Reinhard, W. (7) 191
Reinsborough, V.C. (1) 330
Reisch, J. (1) 183
Reiser, O. (6) 69
Reist, E.J. (5) 31
Reitel, G.V. (1) 430
Rener, G.A. (5) 21
Rengen, K. (1) 297
Renner, G. (7) 197, 230
Renouard, C. (1) 50
Renz, M. (4) 180
Revankar, G.R. (5) 132, 135
Reyé, C. (1) 1; (2) 35
Reznichenko, V.A. (7) 279
Reznik, V.S. (1) 36
Rhee, S.B. (7) 138
Ricard, L. (1) 40, 92, 401, 462, 465, 468, 469, 476, 494, 495, 500-502
Rich, A. (5) 219
Richardson, J.-F. (1) 424; (3) 129; (7) 42, 50
Richie, K.A. (1) 7
Richman, D.D. (5) 63
Richmond, M.G. (1) 213
Richter, F. (1) 257
Richter, L.S. (1) 189
Richter, R. (1) 119
Richter, W. (4) 86
Ridlehuber, R.W. (1) 23
Riegel, B. (1) 64
Riesel, L. (1) 227, 298, 325, 338; (3) 10, 46-49; (7) 26, 27
Rihs, G. (3) 92; (5) 126
Riordan, J.M. (5) 21
Rippmann, F. (4) 240
Ritzen, H. (3) 145
Rivero, I.A. (1) 152
Rizopoulos, A.L. (1) 356
Robats, B.E. (7) 273
Robba, M. (7) 235
Robbins, B.L. (5) 23
Roberts, R.M.G. (1) 496
Roberts, S.M. (5) 17
Robinson, G.H. (1) 75, 78
Robles, J. (3) 116
Rodi, Y.K. (1) 488, 489; (3) 60, 130
Rodriguez-Morgade, S. (1) 174; (6) 12
Roe, J.A. (5) 214
Roepstorff, P. (5) 123
Röschenthaler, G.-V. (1) 428; (2) 28; (3) 26; (7) 44
Roesky, H.W. (1) 350; (3) 139; (7) 74, 85, 86, 191, 200-202, 204-206
Rohwer, E.R. (1) 334
Rokach, J. (6) 103
Rokita, S.E. (5) 184
Rolland, H. (7) 38, 67-69
Romakhin, A.S. (4) 7
Romanenko, V.D. (1) 390, 429-431; (4) 307; (7) 53
Romanova, I.P. (1) 36
Romea, P. (6) 131
Romo, D. (6) 117
Roos, H.M. (4) 162
Rose, L.M. (5) 20
Rosenberg, I. (5) 29, 81

Roshenko, A.B. (7) 53
Roskamp, E.J. (1) 323
Rosmanitz, P. (3) 70
Rosovskaya, T.A. (5) 50
Ross, G. (5) 76
Rossi, R.A. (1) 25
Roth, H.J. (5) 129
Roucoux, A. (1) 229
Roumestant, M.L. (4) 292
Roundhill, D.M. (4) 175
Rousseau, R. (1) 83
Roy, S.K. (3) 102
Royo, M. (3) 81
Rozanov, I.A. (7) 114
Rozhenko, A.B. (1) 142, 429
Rozinov, V.G. (4) 97; (7) 192
Rozners, E. (5) 91
Ruban, A.V. (1) 455; (4) 307; (7) 43, 53
Rubenstahe, T. (7) 73
Rubiales, G. (6) 38
Ruchkina, N.G. (4) 58
Ruder, S.M. (4) 259; (6) 80
Rudinskaya, G.Ya. (7) 289, 290
Rudolph, J. (1) 136
Rudzevich, V.L. (1) 429
Rudzinski, J. (1) 437; (3) 127
Rühlicke, A. (1) 393, 394
Rufus, I.B. (7) 282
Ruhland-Fritsch, B.L. (5) 31
Ruiz, E.M. (7) 270
Ruiz, M. (4) 118
Rukachaisirikul, T. (1) 214
Rulkens, R. (7) 223
Rusanov, A.L. (7) 149
Russell, C.A. (1) 81
Russell, D.R. (1) 134; (4) 302; (6) 89
Russell, J. (5) 22
Russell, P.J. (5) 191
Russell, R.A. (5) 193
Ryan, B.M. (6) 53, 54
Ryan, T.G. (3) 27; (4) 115
Rybakov, V.B. (1) 149
Rychnovsky, S.D. (1) 145; (6) 109

Saak, W. (1) 225
Sabirov, Z.M. (1) 220
Sadana, K.L. (5) 71
Sadanani, N.D. (1) 457; (4) 298
Sadreyeva, R.R. (7) 30
Sadybakasov, B.K. (3) 31, 56
Saegusa, T. (1) 369
Safina, Yu.G. (4) 279
Safrany, S.T. (4) 24
Sagandykova, R.R. (4) 3
Sagar, A.D. (4) 9
Sahasrabudhe, P. (5) 159
Saidov, B.I. (7) 134
Saika, H. (4) 38
Saint-Clair, J.-F. (4) 126
Saintomé, C. (5) 160
Saito, R. (5) 169
Saito, T. (1) 482; (3) 131; (7) 294
Saiz Velasco, J.L. (7) 14
Sakai, M. (6) 96
Sakai, N. (1) 95; (3) 120, 121
Sakai, T. (1) 482; (3) 131
Sakamoto, J. (1) 258
Sakoda, R. (4) 182
Salamończyk, G.M. (4) 23
Salari, H. (4) 238
Salas, I. (1) 284
Salceda, C. (1) 284
Salem, G. (1) 46, 47
Salisbury, S.A. (5) 201
Sallé, M. (6) 145
Salmon, M. (1) 284
Salomon, R.G. (6) 113
Salunkhe, M.M. (4) 9
Samat, G. (4) 128
Samuel, R. (7) 236
Samuels, W.D. (7) 123, 261, 296
Sanchez, M. (1) 429; (2) 23
Sanchez-Andrada, P. (6) 130
Sander, M. (6) 40
Sanders, J.C.P. (2) 8
Sanders, T.C. (4) 143
Sandmeyer, F. (1) 417
Sangen, O. (5) 162
Sangu, S. (1) 454; (4) 231
Santacroce, C. (5) 73
Santarsiero, B.D. (1) 205; (7) 20
Santhosh, K.C. (1) 180
Santini, C.J. (2) 33; (4) 16
Sapozhikov, Yu.E. (7) 192
Saquet, M. (4) 126; (6) 88
Sarfo, K. (1) 204
Sarina, T.V. (7) 53
Sarmah, C.S. (4) 57
Sarmento, J.G. (4) 239
Saruta, K. (3) 112
Sasaki, M. (4) 81
Sasmor, H. (3) 71
Sata, N. (1) 266
Sato, H. (7) 250, 307
Sato, M. (7) 300, 305, 306
Sato, R. (4) 41
Sauvaget, F. (4) 289
Savage, P.B. (1) 279-282
Savchenko, E.V. (5) 109
Savenkov, N.F. (4) 262
Savignac, P. (1) 367, 368; (4) 95, 112, 117, 138, 139; (6) 82, 83
Savla, P.M. (1) 471
Sawaki, Y. (1) 162
Sawamura, M. (1) 98, 137
Sawasaki, K. (4) 299
Sayed, M.B. (7) 103, 104
Scettri, A. (6) 73
Schacht, E. (7) 268, 269
Schädler, D. (1) 238
Scheffer, J.R. (1) 263
Schehlmann, V. (5) 84
Scheidecker, S. (7) 112, 113, 133
Schelle, C. (7) 208
Schier, A. (1) 59, 119, 127, 159; (6) 41, 42
Schilbach, W. (1) 456; (6) 60; (7) 45
Schildrant, D.E. (7) 315
Schinazi, R.F. (5) 119
Schlemper, H. (7) 16
Schlenker, T. (1) 484
Schlewer, G. (4) 109
Schlosser, M. (1) 301
Schmid, R. (1) 320; (3) 44
Schmidbaur, H. (1) 59, 127, 135, 159, 161
Schmidpeter, A. (1) 21, 436, 487; (3) 5, 125; (4) 304; (6) 7, 58
Schmidt, A. (6) 128
Schmidt, H.G. (7) 85, 86
Schmidt, M. (5) 25
Schmidt, R.R. (4) 168
Schmidt, S.P. (6) 104
Schmidt, T. (1) 422
Schmitges, C.-J. (4) 240
Schmitt, G. (3) 104; (5) 13
Schmitt, L. (4) 109
Schmutzler, R. (1) 15, 51, 102, 129, 165, 173, 217, 224, 228, 232, 286, 287; (2) 29; (3) 22, 34, 41-43, 51-54, 57-59; (7) 18, 29
Schnabel, W. (1) 262
Schnell, M. (4) 297
Schnick, W. (7) 220, 221, 245
Schnyder, A. (1) 130
Schoeller, W.W. (1) 384, 432; (3) 133
Schols, D. (5) 26
Schomburg, D. (1) 224; (3) 43
Schoni, G.B. (7) 292, 293
Schoo, H.F.M. (7) 190
Schorp, M.K. (6) 75
Schott, H. (5) 16
Schreiber, S.L. (6) 117
Schriver, M.J. (7) 42
Schrobilgen, G.J. (2) 8
Schueller, W.W. (7) 196
Schulteis, P. (1) 168
Schultz, R.G. (3) 84; (5) 147

Schultze, P. (5) 214
Schulz, B. (6) 28
Schumann, H. (1) 216
Schumann, I. (1) 354, 422
Schumann, W. (6) 159
Schwabacher, A.W. (1) 295
Schwartz, E.R. (7) 275
Schwarz, W. (1) 29
Schweisinger, R. (7) 16
Schweizer, B. (5) 125
Schwendener, R.A. (5) 16
Scopelianos, A.G. (7) 264, 267
Scoponi, M. (7) 253-255
Scott, G.K. (5) 36
Scott, S. (3) 73; (5) 77
Scott, S.R. (7) 130
Scowen, I.J. (7) 109
Scremin, C.L. (3) 96; (5) 74
Seay, M.A. (7) 39
Sebald, A. (1) 225
Secrist, J.A. (5) 20, 21
Seebach, D. (7) 63
Seeger, A. (1) 188
Seela, F. (5) 42, 131, 133, 134, 150, 151, 153
Seidel, G. (1) 235
Seifert, F.U. (1) 428; (2) 28; (7) 44
Seitz, Th. (1) 61
Seiukar, R.S. (4) 9
Sekiguchi, J. (5) 141
Sekine, M. (3) 72; (5) 78-80
Self, M.F. (1) 71, 74, 77
Selve, C. (3) 30
Semenova, M.G. (1) 131, 142, 223
Semezin, D. (1) 84; (7) 113
Semizarov, D.G. (5) 49, 51, 52
Senadhi, S.E. (7) 188
Sentemov, V.V. (1) 305
Seo, K. (4) 55
Serafinowska, H.T. (5) 28
Sernau, V. (1) 34
Serra, C. (5) 18
Serves, S.V. (3) 7
Seth, S. (6) 52
Seto, H. (1) 332
Seto, K. (4) 182
Severengiz, T. (1) 352
Severin, H. (1) 351; (3) 140; (7) 47
Sfihi, H. (5) 227
Shabana, R. (4) 228
Shabarova, Z.A. (5) 175
Shafer, R. (5) 194
Shaikhudinova, S.I. (1) 63, 248
Shannon, I.J. (6) 52
Shannon, W.M. (5) 20
Shapiro, G. (4) 118
Sharashkina, M.V. (4) 83, 303
Sharifullin, A.S. (1) 128
Sharipov, Kh.T. (7) 84
Sharma, M.V. (4) 74, 75
Shatzmiller, S. (4) 218, 219
Shaw, B.L. (1) 11
Shaw, B.R. (5) 33
Shaw, G.F. (1) 168
Shcherbina, T.M. (4) 69
Shealy, Y.F. (5) 20
Sheehan, S.K. (3) 40
Sheeka, H. (5) 4
Sheffer-Dec-Noor, S. (4) 60
Sheikha, G. (5) 10
Sheldrick, G.M. (1) 350; (3) 139; (7) 202
Sheldrick, W.S. (1) 62
Shen, Q. (4) 137; (6) 90
Shen, T. (6) 46
Shen, Y. (4) 120, 145-147; (6) 47
Shevchenko, I.V. (1) 15, 228
Shi, L.-L. (6) 39
Shibaev, V.P. (7) 122
Shibuya, S. (4) 151, 155, 213
Shida, T. (5) 141
Shields, T.P. (5) 199
Shigehara, K. (7) 300, 305, 306
Shimada, A. (7) 100
Shin, C.-G. (6) 76
Shin, W. (1) 321
Shinde, C.P. (4) 74, 75
Shing, T.K.M. (6) 108
Shinoda, I. (1) 482; (3) 131
Shinoda, M. (7) 176
Shintani, T. (6) 49
Shioji, K. (1) 197, 264; (3) 13
Shiotsuka, M. (1) 504
Shipitsin, A.V. (5) 49
Shirabe, K. (1) 378; (4) 232
Shiratori, S. (1) 3
Shirokova, E.A. (5) 49
Shishatskii, S.M. (7) 318
Sho, K. (1) 369
Shode, O.O. (3) 9
Shoh, K. (7) 100
Shon, Y.S. (7) 278
Shreeve, J.M. (7) 124
Shulezhko, V.A. (1) 222
Shumeiko, A.E. (7) 116
Shurubura, G.V. (1) 252
Siah, S.-Y. (1) 37
Siany, M. (1) 401
Sieler, J. (4) 227
Sierbert, W. (1) 68
Sierra, M.L. (1) 462
Sigalas, M.P. (1) 356
Sillanpaa, R. (1) 304
Sillero, A. (5) 60
Sillero, M.A.G. (5) 60
Sillett, G.J.D. (1) 491
Silver, J. (1) 496
Silverberg, E.N. (7) 105, 106, 127
Simger, M. (1) 333
Simmonds, P.M. (5) 194
Simon, C. (1) 179
Simurova, V.V. (4) 176
Singh, P. (3) 77; (7) 80-82
Singler, R.E. (7) 295
Sinha, N.D. (3) 102; (5) 86
Sinitsa, A.D. (3) 33, 135; (4) 127, 176
Siuzdak, G. (5) 228
Skaric, V. (2) 31
Skelly, J.V. (5) 204
Skelton, B.W. (4) 14
Skouta, S. (2) 22, 24, 25
Skowronska, A. (3) 23; (4) 308
Slabzhennikov, S.N. (1) 278
Slany, M. (1) 373; (7) 46
Sletten, E. (5) 195, 210
Sloss, D.G. (4) 137; (6) 90
Smeal, T.M. (1) 72
Smeets, W.J.J. (7) 209
Smernik, R.J. (1) 87
Smirnova, S.B. (4) 44
Smirnova, V.N. (7) 291
Smith, C.A. (5) 168
Smith, F.C. (6) 52
Smith, F.W. (5) 212-214
Smith, G. (1) 276; (4) 235
Smith, J.M. (1) 213
Smith, L.M. (5) 230
Smith, M.B. (1) 212
Smith, R.D. (5) 232
Smith, V.K. (7) 315
Smoli, O.B. (1) 314
Smrt, J. (5) 81
Sneddon, L.G. (1) 20
Snoeck, R. (5) 26
Sobanov, A.A. (4) 280
Sobolev, A.N. (4) 58
Sobti, A. (1) 184
Sokolov, M.P. (4) 265, 266, 294
Sokolov, V.B. (1) 261
Sokolov, V.I. (4) 101
Soleilhavoup, M. (1) 387, 461; (6) 6, 17
Solodenko, V.A. (4) 190
Solodovnikov, S.P. (4) 100, 101; (7) 83
Soloshonok, V.A. (4) 190
Solotnov, A.F. (4) 52
Somanathan, R. (1) 152
Sommese, A.G. (1) 437; (3) 127
Song, H. (1) 85
Song, J.-I. (1) 122
Song, S. (1) 481
Song, X. (1) 423

Sonnenburg, R. (3) 54, 57
Sonnenschein, H. (4) 132
Sonveaux, E. (3) 87
Sood, A. (5) 33
Sookraj, S.H. (1) 48
Sopchik, A.E. (2) 27; (3) 143
Soria, R. (7) 321
Sosul'nikov, M.L. (1) 256
Sotiropoulos, D.N. (3) 7
Sotoodeh, M. (7) 74
Sourines, F. (7) 113
Sowers, L.E. (5) 188
Spearman, J.L. (7) 237
Spek, A.L. (7) 209
Sperbeck, D.M. (4) 247
Spielvogel, B.F. (5) 33
Spiess, B. (4) 109
Spiga, M. (5) 10
Spilling, C.D. (4) 152, 153
Spindler, F. (1) 130
Spizzo, F. (7) 141
Sprecher, M. (4) 271
Springer, D.L. (5) 232
Sproat, B.S. (5) 120, 158
Sreedharan-Menon, R. (4) 302; (6) 89
Sridhar, C.N. (5) 63
Srinivas, R. (1) 198; (5) 23
Srinivasachar, K. (3) 96
Srivastava, S.C. (3) 85
Stafford, J.N.W. (1) 146, 310
Stalke, D. (7) 85, 199-202, 204, 206
Stallman, J.B. (4) 71
Stammer, A. (7) 232
Stammler, H.-G. (1) 170, 353, 354, 393, 422
Stanforth, S.P. (6) 50
Stang, P.J. (1) 194
Stannek, J. (1) 408
Stannett, V.T. (7) 247, 251
Stanton, J.L. (4) 247
Starich, M. (1) 168
Starrett, J.E. (5) 22
Stasko, D.J. (7) 261
Staubli, A.B. (5) 181
Staude, S. (1) 65
Staunton, J. (6) 153
Stawinski, J. (3) 28; (5) 106
Stec, W.J. (3) 24, 69; (4) 1, 2; (5) 97, 99
Steglich, F. (4) 104
Steimann, M. (1) 483
Stein, S. (5) 170, 183
Steiner, A. (7) 85, 199-201, 204, 206
Steinmueller, F. (1) 487
Stelzer, O. (1) 62, 397
Stemerick, D.M. (1) 177
Stemmler, E.A. (5) 235
Stenzel, V. (1) 225
Stepanova, Yu.Z. (1) 431
Stephan, D.W. (1) 83, 441, 442
Stephan, M. (1) 99
Stephanidou-Stephanatou, J. (6) 32
Stepień, A. (4) 179, 180
Sterin, S. (6) 31
Sterns, M. (1) 46
Stevens, C.V. (6) 13; (7) 55
Stieglitz, G. (1) 30, 486
Stipa, P. (4) 200
Stobart, S.R. (1) 210
Stockley, P.G. (5) 155, 156
Stoelwinder, J. (6) 135
Stolpmann, H. (1) 148, 302
Stone, M.J. (5) 187
Stone, M.L. (7) 317
Storer, J.W. (2) 7
Storer, R. (5) 17
Storhoff, B.N. (1) 199
Stowell, J.K. (4) 27
Strelenko, Yu.A. (4) 83, 303
Streubel, R. (1) 370, 460
Stromberg, R. (4) 64; (5) 90, 91
Strubel, R. (1) 439
Struchkov, Yu.T. (1) 309, 389; (3) 31, 56; (4) 282; (7) 83, 84, 194
Strum, D. (7) 26, 27
Strutwolf, J. (1) 384
Stützer, A. (1) 159
Stuhmiller, L.M. (5) 63
Sturm, D. (1) 227
Sturm, P.A. (5) 31
Sturtz, G. (4) 204
Suarez, A.I. (4) 35
Suba, C. (1) 58
Subirana, J.A. (5) 227
Subramanian, P. (7) 282
Subramony, J.A. (7) 302, 303
Suda, F. (5) 61
Sudheendra Rao, M.N. (7) 189
Sugihara, M. (7) 250
Sugiyama, T. (7) 60
Sukhojenko, I.I. (4) 13
Sulikowski, G.A. (1) 184
Sum, V. (3) 14, 15; (4) 156, 157
Summa, V. (1) 186
Summers, C.A. (4) 28
Sun, R.-C. (6) 143
Sun, V. (7) 34
Sun, X. (1) 24
Sun, Y. (1) 214
Suptitz, G. (5) 92
Suschitzky, H. (6) 122; (7) 58
Suska, A. (5) 97
Sutter, J. (7) 207, 208
Sutter-Beydoun, N. (1) 5
Suvalova, E.A. (4) 127
Suzuki, H. (7) 24
Suzuki, M. (1) 369
Suzuki, N. (1) 94
Suzuki, R. (5) 61, 62
Suzuki, S. (7) 238
Suzuki, T. (7) 319
Svendsen, M.L. (5) 123
Sviridon, A.I. (1) 223
Svoboda, J. (5) 81
Swann, P.F. (5) 139
Sweeney, J.B. (4) 301
Swiss, K.A. (6) 65, 66
Syi, J.L. (5) 115
Sylvers, L.A. (5) 55

Tacke, M. (1) 148
Tada, J. (7) 159
Tada, Y. (7) 300, 305, 306
Taga, T. (7) 107, 328
Taillandier, E. (5) 227
Taira, K. (2) 7
Tajiri, Y. (2) 15
Takahashi, C. (4) 55
Takahata, H. (7) 119
Takaki, I. (1) 115
Takaku, H. (5) 61, 62
Takami, H. (2) 21; (6) 23
Takamuku, S. (4) 93, 299
Takatsuka, T. (6) 20
Takaya, H. (1) 95, 99; (3) 120, 121
Takayanagi, H. (6) 139
Takeda, T. (7) 100
Takenishi, T. (5) 38
Takeno, N. (7) 300, 305, 306
Takeuchi, M. (7) 178
Talzi, V.P. (1) 307
Tanabe, K. (2) 7
Tanaka, H. (7) 169
Tanaka, Y. (1) 312; (6) 36, 45, 157
Tanazona, M.P. (7) 219
Tang, C.C. (4) 42, 43, 78, 79, 223, 224
Tang, J. (7) 28
Tang, J.S. (1) 458
Tang, J.Y. (5) 103
Tang, K. (5) 236
Tang, X. (7) 248, 249
Tang, Y.W. (1) 423
Tani, K. (1) 93
Tani, M. (7) 167
Tanigaki, T. (7) 119
Taniguchi, M. (7) 308
Taniguchi, T. (7) 172, 173
Tanimoto, F. (7) 155
Tanoury, G.J. (1) 141; (3) 45

Tarasenko, E.A. (1) 117
Tarasevich, B.J. (7) 261
Tarbit, B. (6) 26
Tarköy, M. (3) 90, 91; (5) 125
Tarraga, A. (1) 172
Tarrason, G. (3) 115
Tarussova, N.B. (5) 49, 50
Tashiro, K. (1) 363
Tattershall, B.W. (1) 243
Tawfik, D.S. (4) 255
Taylor, N.J. (1) 214
Taylor, R.J. (5) 36; (6) 25
Tebby, J.C. (4) 183
Teeganden, D.M. (7) 310
Tegge, W. (3) 113
Teigelkamp, S. (3) 79
Teixidor, F. (1) 215, 304
Tejeda, J. (3) 142; (6) 16; (7) 195
Tejerina, B. (7) 56
Teloniati, A.G. (3) 7
Temsamani, J. (5) 103
Téoule, R. (5) 142
Terent'eva, S.A. (2) 30
Ternansky, R.J. (3) 12
Teterin, Yu.A. (1) 256
Thea, S. (4) 256, 257
Theisen, P. (3) 98
Thelin, M. (3) 28; (5) 106
Thibault-Starzyk, F. (4) 107, 296
Thiébault, A. (1) 58
Thiebaut, S. (3) 30
Thomann, M. (1) 65, 148, 302; (6) 7
Thomas, C.J. (5) 226; (7) 188, 189
Thomas, E.J. (6) 114
Thomas, J.K.R. (7) 130
Thomas, L.M. (7) 188
Thomas, N.F. (4) 20
Thompson, C.M. (4) 35
Thompson, M.L. (3) 39
Thompson, W. (5) 5
Thomson, J.B. (5) 203
Thorat, M.T. (4) 9
Thorn, J.C. (1) 341
Thornton-Pett, M. (1) 11, 239; (3) 15
Thorpe, A.J. (5) 17
Thrall, B.D. (5) 232
Thuong, N.T. (3) 114; (5) 121, 182
Thurieau, C. (4) 125
Thurmes, W. (5) 223
Tiekink, E.R.T. (1) 292
Tijani, A. (1) 130
Tillack, A. (1) 105
Tinoco, I. (5) 223, 224
Tirakyan, M.R. (4) 282
Tkachev, V.V. (4) 52
Töke, L. (1) 503; (4) 287, 288
Togni, A. (1) 130
Toia, R.F. (4) 65
Tokaji, H. (7) 170
Tolkachev, D.V. (7) 83, 84
Tolmachev, A.A. (1) 131, 132, 142, 222, 223; (7) 17
Toma, J.M.D.R. (3) 76
Tomasz, J. (5) 33
Tomilov, A.P. (4) 7
Tomioka, K. (1) 181
Tommes, P. (1) 38
Tooze, R. (7) 98
Tordo, P. (4) 200
Torgomyan, A.M. (1) 329
Torigai, H. (4) 30, 67
Torreilles, E. (7) 70
Torres, T. (1) 174; (6) 12
Tortolani, D.R. (5) 22
Tóth, G. (1) 503; (4) 287
Toulhoat, C. (1) 253, 254
Toupet, L. (1) 90
Toyota, K. (1) 346-349, 363, 378, 400, 451-454; (3) 138; (4) 226, 231, 232; (6) 9
Tramontano, E. (5) 10
Trapassi, G. (1) 186
Travkin, V.F. (7) 89
Treiber, A. (2) 16
Treutler, O. (1) 387; (6) 17
Trigo Passos, B.F. (1) 395; (7) 52
Troev, K. (4) 90, 175
Trofimov, B.A. (1) 63, 248
Troitskaya, L.B. (4) 39
Trojanov, S.I. (1) 149
Trost, B.M. (1) 143; (6) 11
Trotter, J. (1) 263
Truffert, J.-C. (3) 114
Tryota, K. (6) 59
Tsuchida, E. (7) 319
Tsuchida, I. (7) 294
Tsuchiya, T. (1) 3, 472
Tsuda, K. (7) 319
Tsuji, S. (1) 162
Tsukanova, N.P. (7) 288
Tsuzuki, S. (2) 7
Tsvetkov, E.N. (1) 260
Tüchmantel, W. (4) 167
Tukanova, S.K. (4) 148
Tumanski, B.L. (4) 100, 101
Tung, C.H. (5) 170, 183
Tur, D.R. (7) 122, 287-291
Turner, D.H. (5) 161
Turro, N.J. (5) 200
Tuschl, T. (5) 152
Tuttle, J.V. (4) 123; (5) 32

Uchibori, Y. (1) 332
Uchida, T. (1) 482; (3) 131
Uchida, Y. (1) 196
Uchimaru, T. (2) 7
Uchiyama, Y.-K. (6) 37
Ueda, I. (6) 35, 36, 157
Ueno, Y. (5) 169
Uesugi, T. (1) 400
Ugi, I. (4) 86
Uhlenbrock, S. (1) 240; (7) 210
Uhlig, F. (1) 22, 238
Uhlig, W. (1) 22
Újszaszy, K. (1) 459, 503; (4) 287, 288
Ulibarri, T.A. (6) 56
Umeno, M. (1) 332
Umetani, S. (1) 293
Uozumi, Y. (1) 94, 96
Urata, H. (5) 163
Urazbaev, V.N. (1) 220
Uriel, S. (1) 296; (6) 144
Urpi, F. (6) 131
Urpi, L. (5) 227
Uryupin, A.B. (4) 33
Usifoh, C.O. (1) 183
Usman, N. (5) 34, 86, 219
Ustenko, S.N. (1) 154; (6) 5
Uyeo, S. (6) 146
Uznanski, B. (3) 24; (5) 97

Vacanti, C.A. (7) 263
Vacanti, J.P. (7) 263
Vaghefi, M.M. (5) 89, 122
Vahrenkamp, H. (1) 357
Valentinhansen, P. (5) 114
Valerio, C. (7) 111
Valle, G. (6) 43
Van Aerschot, A. (5) 127
van Asselt, R. (7) 209
van Boeckel, C.A.A. (3) 80, 100
van Boom, J.H. (3) 100, 101, 109, 110; (4) 34; (5) 215, 220
van de Grampel, J.C. (7) 135, 150, 189, 190
van den Bosch, H. (5) 63
Van den Driessche, F. (5) 127
Van den Elst, H. (5) 215
van der Huizen, A.A. (7) 135
van der Marel, G.A. (3) 101; (4) 207; (5) 220
van der Voort Maarschalk, F.W. (1) 479
Vandorpe, J. (7) 268, 269
Van Garderen, C.J. (5) 190
Van Houte, L.P.A. (5) 190
van Leeuwen, P.W.N.M. (3) 122
van Leusen, A.M. (6) 135
Van Meervelt, L. (1) 319

van Oijen, A.H. (3) 109, 110; (4) 34
Van Praet, E. (1) 192
Vanquickenborne, L.G. (1) 192, 385, 386
Van Rooyen, P.H. (1) 337
Vansco, G.J. (7) 8, 232
Vansteenkiste, S. (7) 268
van Wijk, G.M.T. (5) 63
Varbanov, S. (1) 255
Varma, R.S. (5) 98
Varnek, A.A. (7) 96
Vasil'eva, N.V. (7) 289, 290
Vasquez, P.C. (1) 163
Vasseur, M. (5) 175
Vassileva, P. (7) 102
Vasyanina, L.K. (1) 230, 231; (3) 56, 62, 64-66
Vau'kin, G.I. (4) 52
Vazquez, C. (7) 193
Vázquez, P. (1) 174; (6) 12
Veal, J.M. (5) 111
Vedejs, E. (1) 207; (6) 22
Vega, F.R. (1) 209
Velasco, L. (1) 284
Velasco, M.D. (1) 171; (6) 127; (7) 21
Venanzi, L.M. (1) 106
Verbruggen, C. (3) 19, 20; (4) 220
Verdaguer, N. (5) 227
Verkade, J.G. (1) 76; (7) 28
Verlhac, J.-B. (6) 55
Veya, P. (1) 203
Viallefont, P. (4) 292
Vickery, K. (5) 191
Victorova, L.S. (5) 46, 48, 51, 52
Vidal, A. (6) 19; (7) 14, 15
Vidal, J.-P. (4) 212; (6) 31, 86
Vidal, M. (1) 253, 254
Vierhufe, H. (1) 188
Vigner, J. (1) 493
Vijay-Kumar, S. (7) 188
Vilarrasa, J. (6) 131
Vilitkevich, A.G. (4) 113
Villanueva, L.A. (7) 36
Villemin, D. (4) 105, 107, 289, 296
Vinas, C. (1) 215, 304
Vinayak, R. (5) 85
Vincens, M. (1) 253, 254
Vinogradova, T.K. (1) 316
Virgil, S.C. (6) 56
Visscher, K.B. (7) 72, 197, 230, 272, 274
Viswanadha, R.V. (1) 198
Voegel, J.J. (5) 136
Voelker, H. (1) 350; (3) 139
Vogt, H. (1) 149, 298, 325, 338; (3) 10
Vojtisek, P. (1) 290
Volkert, W.A. (7) 80-82
Volkova, L.N. (7) 30
Vollmerhaus, R. (7) 184
Volodin, A.A. (5) 225
von der Gönna, V. (1) 432, 456; (3) 133; (6) 60; (7) 45, 196
von Gudenberg, D.W. (7) 73
Vonjantalipinski, M. (5) 52
von Loewis, M. (1) 338
von Schnering, H.G. (7) 16
Vonwiller, S.C. (1) 247
Vorob'ev-Desyatovskii, N.V. (1) 256
Voronkov, M.G. (7) 192
Vostrowsky, O. (6) 98
Vrieze, K. (7) 209
Vu, H. (3) 77
Vyazovkina, E.V. (5) 107-109
Vydzhak, R.N. (1) 315-319; (6) 15
Vyle, J.S. (5) 36, 105
Vysotskii, V.I. (4) 113

Wada, T. (3) 72; (5) 75, 78, 80
Wadgaonkar, P.P. (4) 9
Wadsworth, D.J. (6) 153
Wadwani, S. (5) 110
Wagener, C.C.P. (1) 334-336
Wagman, A.S. (3) 107; (4) 37
Wahl, F.O. (5) 67
Wakamiya, T. (3) 112
Walker, B.J. (4) 185
Walker, D. (7) 316
Walker, J.K. (6) 29
Walker, M.A. (1) 178
Walker, R.T. (5) 124
Walsh, J.G. (1) 53
Walter, O. (1) 60, 61, 446
Walz, L. (7) 16
Wamhoff, H. (6) 128
Wang, A.C. (5) 221
Wang, A.H.J. (5) 211, 220, 226
Wang, D.G. (5) 112
Wang, G. (4) 193, 209, 217
Wang, G.Y. (5) 145
Wang, K. (3) 126; (4) 135
Wang, M.F. (5) 2
Wang, S. (6) 110
Wang, S.S. (6) 103
Wang, T. (6) 4
Wang, T.-Z. (6) 110
Wang, X.-H. (1) 301
Wang, Y.P. (2) 13
Ward, J. (1) 32, 113
Wareing, J.R. (6) 95
Warren, S. (1) 269-273; (6) 61-63
Warrener, R.N. (5) 193
Wasiak, J. (1) 219; (4) 12
Watal, G. (5) 83
Watanabe, Y. (3) 146; (4) 25
Waterson, D. (1) 270
Watson, L.D. (7) 323
Watson, W.P. (5) 203
Watt, C.I.F. (6) 108
Webb, K.M. (1) 123
Weber, H.-P. (4) 118
Weber, L. (1) 353-355, 392-394, 398, 422
Webster, L.K. (5) 191
Weferling, N. (1) 62
Wegmann, T. (1) 360, 361
Wegner, P. (1) 139
Wei, Z.P. (5) 170
Wei, Z.-Y. (6) 74
Weibel, J.M. (6) 136
Weichmann, H. (1) 257, 291
Wei-chu, X. (6) 100
Weik, C. (4) 218, 219
Weissensteiner, W. (1) 13
Welker, M.F. (7) 137, 247
Well, M. (1) 286, 287
Weller, F. (7) 73, 75
Wells, A. (1) 151; (4) 235
Wells, R.L. (1) 71, 74, 77
Welsch, J. (1) 126
Wemmer, D.E. (5) 192
Wendeborn, S. (5) 130
Wenderoth, P. (1) 364, 470; (7) 51, 52
Wengel, J. (3) 94; (5) 123
Wenkert, E. (6) 75
Wentrup, C. (1) 200
Wenzel, T. (5) 133
Wessels, P.L. (4) 163
Westall, S. (7) 90
Westerhausen, M. (1) 29
Westman, E. (5) 90, 91
Weston, S. (5) 158
Westuick, J. (4) 22
Wettling, T. (1) 101
Wheelan, P. (6) 105
White, A.H. (4) 14
White, M.L. (7) 23, 25, 237, 240-242, 283
Whiterock, V. (5) 22
Wiaterek, C. (1) 242
Wickstrom, E. (5) 109
Widhalm, M. (1) 13, 42, 43
Widlanski, T.S. (4) 27; (5) 116
Wieczorek, M.W. (1) 274, 283; (4) 45, 237; (5) 97
Wiemer, D.F. (4) 174, 242, 277; (5) 19; (6) 72, 158
Wilcox, R.A. (4) 21, 24

Wild, S.B. (1) 16, 41
Wiley, J.R. (6) 108
Wilk, A. (3) 24; (4) 2; (5) 97, 99
Will, D.W. (3) 79; (5) 168
Williams, D.H. (5) 187
Williams, D.J. (1) 14
Williams, D.M. (5) 59
Williams, H.L. (6) 134
Williams, J.M.J. (1) 55-57
Williams, J.P. (1) 45
Willis, A.C. (1) 41, 46, 47, 276
Wills, M. (1) 8; (3) 124; (4) 234
Wilson, S.R. (6) 65
Wilson, W.L. (1) 481, 497
Wilson, W.W. (2) 8
Wilting, T. (7) 135
Winger, B.E. (5) 232
Winkler, A. (5) 92
Winkler, U. (1) 28
Winter, H. (5) 134
Winter, M. (1) 446
Wisian-Neilson, P. (7) 218, 255, 257, 258, 282, 298
Wittaker, A.K. (1) 276
Wocadlo, S. (1) 505
Woisard, A. (5) 160
Wolf, R. (2) 24, 25; (3) 93
Wong, C.-H. (3) 145
Wong, S.C. (4) 114
Wong, W.-K. (1) 138
Wong, W.-T. (1) 138
Wood, C.E. (7) 236
Wood, M.L. (5) 185
Wos, J.A. (4) 71
Woudenberg, R.C. (1) 74
Wower, J. (5) 55
Wray, V. (5) 37
Wright, D.S. (1) 81
Wright, P. (5) 70
Wu, G. (1) 166
Wu, G.P. (4) 42, 43, 223, 224
Wu, H. (4) 209
Wu, S.-W. (6) 49
Wu, T.T. (7) 311
Wu, X.-P. (1) 437; (3) 127; (4) 298
Wu, Y. (6) 116
Wu, Z.-P. (1) 457
Wubbels, J.H. (7) 150
Wyrzykiewicz, T.K. (3) 25; (5) 101, 102

Xia, C. (1) 85
Xiao, J. (1) 168
Xu, J.C. (4) 87
Xu, X. (7) 31
Xu, Y.-Z. (5) 139

Yabui, A. (1) 266; (4) 172
Yabuta, M. (1) 93
Yagupolskii, Y.L. (3) 26
Yakout, E.-S.M.A. (6) 57
Yamada, Y. (4) 81
Yamagata, T. (1) 93
Yamagishi, T. (4) 151, 155, 213
Yamaguchi, Y. (1) 447
Yamamoto, H. (1) 251; (4) 70, 169-171
Yamamoto, I. (1) 268; (6) 37
Yamamoto, K. (3) 123
Yamamoto, N. (4) 81
Yamamoto, T. (3) 146
Yamana, K. (5) 162
Yamashita, M. (1) 266; (4) 55, 172
Yamataka, H. (6) 20
Yamazaki, A. (1) 111
Yampol'skir, Yu.P. (7) 318
Yan, Z.B. (4) 196
Yanagawa, M. (1) 107
Yanagi, K. (1) 12
Yanagisawa, A. (4) 70
Yanez, M. (7) 6
Yang, B. (4) 120
Yang, C. (4) 249; (6) 29
Yang, H. (4) 53, 246
Yang, K. (1) 213
Yang, S.K. (7) 181
Yang, S.-W. (1) 440
Yang Lu, (4) 10
Yano, S. (7) 121, 297
Yarkevich, A.N. (1) 260
Yartsev, V.M. (1) 344
Yashima, H. (7) 178
Yasuda, K. (4) 170
Yasui, S. (1) 197, 264; (3) 13
Yasuike, S. (1) 3, 472
Yasunami, M. (1) 349, 451; (3) 138; (4) 226; (6) 59
Yasuoka, J. (3) 112
Yazaki, J. (1) 97
Ye, T. (6) 148
Yeung, K.S. (6) 115
Yie, C. (7) 248
Yokomatsu, T. (4) 151, 155, 213
Yoneda, F. (5) 164
Yonetake, K. (7) 285, 286
Yoon, K.H. (7) 278
Yoon, M. (7) 138
Yoon, T.S. (1) 321
Yoshifuji, M. (1) 346-349, 363, 378, 400, 451-454; (3) 138; (4) 226, 231, 232; (6) 9, 59
Yoshihara, K. (6) 45, 97
Yoshihara, M. (1) 264; (3) 13
Yoshihashi, K. (7) 152
Yoshikawa, M. (5) 38
Yoshimura, T. (3) 105
Yoshimura, V. (5) 76
Yoshioka, H. (1) 332
Yoshioka, R. (1) 116
Young, V.G. (1) 403
Yu, E. (5) 76
Yu, K.-L. (4) 246
Yu, P.S. (5) 84
Yuan, C. (4) 62, 106, 108, 116, 140, 193, 209, 214, 217, 283, 284; (6) 85
Yuan, C.Y. (4) 196
Yurchenko, A.A. (1) 131, 142, 222, 223
Yurchenko, R.I. (1) 221; (4) 52, 98
Yurchenko, V.G. (4) 52

Zablocka, M. (1) 125, 445
Zabotina, E.Ya. (1) 487
Zaburdyaeva, S.N. (2) 14
Zadorin, A.N. (7) 287-289
Zakharov, L.S. (4) 69
Zakrzewski, J. (1) 496
Zanello, P. (7) 190
Zaripov, I.M. (4) 7
Zavalishina, A.I. (3) 56
Zaveri, N.T. (5) 31
Zefirov, N.S. (4) 13
Zegelman, L. (6) 101
Zemlianoy, V.N. (1) 252
Zendegui, J.G. (3) 77
Zhan, C.G. (3) 134
Zhang, D. (4) 78
Zhang, H. (2) 3, 33; (4) 16, 66
Zhang, J.L. (3) 134
Zhang, M. (4) 78, 79
Zhang, P. (3) 88
Zhang, R. (4) 258
Zhang, S. (1) 295; (2) 26
Zhang, S.G. (5) 171
Zhang, X.-M. (1) 327; (6) 2
Zhang, Y.H. (4) 201, 248
Zhang, Y.J. (3) 134
Zhang, Z. (2) 26
Zhao, K. (6) 4
Zhao, Y. (3) 141
Zhdanov, B.V. (4) 275
Zhdanov, Yu.A. (4) 264
Zheng, F. (7) 248, 249
Zhichkin, P.E. (1) 117
Zhou, D.B. (4) 116
Zhou, H. (4) 214
Zhou, L. (3) 96
Zhou, Z.-L. (6) 39
Zhu, C. (3) 141
Zhu, G. (1) 166
Zhu, Q. (1) 323

Zhu, T.M. (5) 170
Ziegler, J. (1) 9
Ziembinski, R. (7) 223
Ziller, J.W. (1) 294; (6) 56
Zimmermann, R.A. (5) 55
Zinchenko, S.V. (1) 63, 340
Zlotin, S.G. (4) 83, 303
Zobel, B. (1) 22
Zolravkova, Z. (4) 183
Zolutukhin, M.M. (4) 92
Zotov, Yu.L. (1) 220
Zou, L. (7) 275
Zounes, M. (3) 83; (5) 144
Zowaszki, S. (4) 84
Zsolnai, L. (1) 28, 34, 35, 60, 61, 236, 446, 448, 450
Zue, J. (4) 68
zu Kocher, R.M. (7) 77, 78
Zvistunova, N.Yu. (4) 190
Zwierzak, A. (4) 51, 61, 84; (7) 66
Zyablikova, T.A. (3) 136; (4) 265
Zyuz, K.V. (1) 317